A COMPREHENSIVE SURVEY OF INTERNATIONAL SOYBEAN RESEARCH - GENETICS, PHYSIOLOGY, AGRONOMY AND NITROGEN RELATIONSHIPS

Edited by **James E. Board**

A Comprehensive Survey of International Soybean Research - Genetics, Physiology, Agronomy and Nitrogen Relationships
http://dx.doi.org/10.5772/45867
Edited by James E. Board

Contributors

Minobu Kasai, Denis M. Sytnikov, Huynh Viet Khai, Zhanyuan Zhang, Gustavo Souza, Suzana Bertolli, Tiago Catuchi, Rogerio Soratto, Luciano Fietto, Murilo Alves, Cristiane Fortes Gris, Alexana Baldoni, Motoki Kubo, Pedro Reis, Elizabeth Fontes, Takeo Yamakawa, Celia R. Carlini, Rafael Real-Guerra, Fernanda Stanisçuaski, Brett Ferguson, Takuji Ohyama, Laura C. Hudson, Kevin C. Lambirth, Kenneth L. Bost, Kenneth J. Piller, Ana Maria Heuminski De Avila, Srinivasan Ramachandran, Tzi-Bun Ng, Jack Ho Wong, Arvind M. Kayastha, Alka Dwevedi, Marco Arruda, Herbert Barbosa, Lidiane Mataveli, Silvana Ruella Oliveira, Sandra Arruda, Ricardo Azevedo, Priscila Gratão, Eduardo Antonio Gavioli, Akira Kanazawa, Hilton Silveira Pinto, Lidia Skuza, Ewa Filip, Izabela Szućko, Donald Smith, Sowmya Subramanian, Isao Kubo, Kuniyoshi Shimizu, Man-Wah Li, Yee Shan Ku, Yuk Lin Yung, Chao Qing Wen, Hon-Ming Lam, Xueyi Liu, Wan-Kin Au-Yeung, Jeandson Silva Viana, Edilma Pereira Gonçalves, Abraão Cícero Da Silva, Valderez Matos

Published by InTech

Janeza Trdine 9, 51000 Rijeka, Croatia

Edition 2014

Publishing Process Manager Ana Pantar
Technical Editor InTech DTP team
Cover InTech Design team

Additional hard copies can be obtained from orders@intechopen.com

A Comprehensive Survey of International Soybean Research - Genetics, Physiology, Agronomy and Nitrogen Relationships, Edited by James E. Board
p. cm.
ISBN 978-953-51-0876-4

INTECH

open science | open minds

Contents

Preface

Soybean is the most important oilseed and livestock feed crop in the world, accounting for 58% of total world oilseed production and 69% of protein meal consumption by livestock. These dual uses are attributed to the crop's high protein content (nearly 40% of seed weight) and oil content (approximately 20%); characteristics that are not rivaled by any other agronomic crop. Besides its use as a high-protein livestock and poultry feed, and oilseed crop (used in margarines, cooking oils, and baked and fried food products), soybean has various other industrial uses such as biodiesel, fatty acids, plastics, coatings, lubricants, and hydraulic fluids. In Asian countries such as China, Japan and Indonesia, the whole seed is directly consumed as human food; or it is incorporated into human food items such as tofu, tempeh, soy milk, soy cheese, or other products. Soybean consumption as human food is increasing outside of Asia. Recently, health benefits for soybean have been recognized for heart disease, cancer, osteoporosis, and menopause. The American Heart Association recommends daily human consumption of 25 mg of soybean to help prevent heart and circulatory diseases.

In 2010, 258.4 million metric tons of soybean were produced in the world, having a value of $111 billion. Over 80% of the world's soybeans are produced in three countries: the USA, Brazil, and Argentina. These three countries are also the main exporters of soybean to the world market. Major importing countries are China, Japan, the European Union, and Mexico. A testimony to the increasing importance of soybean on the world agricultural stage is in the stunning growth of production shown by Argentina and Brazil over the last 25 years. Between 1986 and 2010, the production has risen from 17.3 to 70 million metric tons in Brazil (a four-fold increase) and from 7 to 49.5 million metric tons in Argentina (a seven-fold increase). Both countries have demonstrated to the world how an organized effort of research, education and extension can create an entire industry around production and use of an agricultural commodity.

Against the backdrop of soybean's striking ascendancy is the increased research interest in the crop throughout the world. The objective of this book is to provide readers with a view of the high quality of soybean research being conducted in so many different parts of the world. With all the dissension and rancor in the world (wars, terrorism, financial panic, etc.) it is truly heartening to see the efforts being made to create a greater understanding of soybean in so many diverse parts of the world. Such efforts will go a long way to meeting increased demand for soybeans; a demand driven by increased world population and rising living standards. Because expansion of agricultural land to meet this demand is limited, the only way to meet increased world demand for soybean is by greater production per area of currently available land. This is why research, such as that contained in this book, is so vital for future soybean production.

It is in this light that I would like to acknowledge all the authors for their outstanding efforts in composing these chapters. The information presents a comprehensive view of research efforts in genetics, plant physiology, agronomy, agricultural economics, and nitrogen relationships that will benefit soybean stakeholders and scientists throughout the world. We hope you enjoy the book.

James E. Board
Professor of Agronomy
School of Plant, Environmental, and Soil Sciences
Louisiana State University Agricultural Center
Baton Rouge, Louisiana, USA

Section 1

Soybean Nitrogen Relationships

Chapter 1

A Proteomics Approach to Study Soybean and Its Symbiont *Bradyrhizobium japonicum* –A Review

Sowmyalakshmi Subramanian and Donald L. Smith

Additional information is available at the end of the chapter

http://dx.doi.org/10.5772/53728

1. Introduction

Soil is a dynamic environment due to fluctuations in climatic conditions that affect pH, temperature, water and nutrient availability. These factors, along with agricultural management practices, affect the soil micro-flora health and the capacity for effective plant-microbe interactions. Despite these constant changes, soil constitutes one of the most productive of earth's ecospheres and is a hub for evolutionary and other adaptive activities.

1.1. Biological nitrogen fixation

Biological nitrogen fixation (BNF) is one of the most important phenomena occurring in nature, only exceeded by photosynthesis [1,2]. One of the most common limiting factors in plant growth is the availability of nitrogen [3]. Although 4/5ths of earth's atmosphere is comprised of nitrogen, the ability to utilize atmospheric nitrogen is restricted to a few groups of prokaryotes that are able to covert atmospheric nitrogen to ammonia and, in the case of the legume symbiosis, make some of this available to plants. Predominantly, members of the plant family Leguminosae have evolved with nitrogen fixing bacteria from the family Rhizobiaceae. In summary, the plants excrete specific chemical signals to attract the nitrogen fixing bacteria towards their roots. They also give the bacteria access to their roots, allowing them to colonize and reside in the root nodules, where the modified bacteria (bacteroids) can perform nitrogen fixation [1,4,5]. This process is of great interest to scientists in general, and agriculture specifically, since this highly complex recognition and elicitation is co-ordinated through gene expression and cellular differentiation, followed by plant growth and development; it has the potential to minimize the use of artificial nitrogen fertilizers and pesticides in crop management. This biological nitrogen fixation process is complex, but has been best examined in some detail in the context of soybean-*Bradyrhizobium* plant-microbe interactions.

1.2. Soybean – The plant

Soybean (*Glycine max* (L.) Merrill) is a globally important commercial crop, grown mainly for its protein, oil and nutraceutical contents. The seeds of this legume are 40% protein and 20% oil. Each year soybean provides more protein and vegetable oil than any other cultivated crop in the world.

Soybean originated in China, where it has been under cultivation for more than 5000 years [6]. The annual wild soybean (*G. soja*) and the current cultivated soybean (*G. max*) can be found growing in China, Japan, Korea and the far east of Russia, with the richest diversity and broadest distribution in China, where extensive germplasms are available. The National Gene Bank at the Institute of Crop Germplasm Resources, part of Chinese Academy of Agriculture Sciences (ICGR-CAAS), Beijing, contains close to 24,000 soybean accessions, including wild soybean types. Soybean was introduced into North America during the 18th century, but intense cultivation started in the 1940s – 1950s and now North America is the world's largest producer of soybean [7,8]. Although grown worldwide for its protein and oil, high value added products such as plant functional nutraceuticals, including phospholipids, saponins, isoflavones, oligosaccharides and edible fibre, have gained importance in the last decade. Interestingly, while genistein and diadzein are signal molecules involved in the root nodulation process, the same compounds can attenuate osteoporosis in post-menopausal women. The other isoflavones have anti-cancer, anti-oxidant, positive cardiovascular and cerebrovascular effects [9]. More recently soybean oil has also been used as an oil source for biodiesel [10-14].

Table 1 provides the latest statistics on soybean cultivation and production as available at FAOSTAT [15]

	World	Africa	Americas	Asia	Europe	Oceania	Canada
Area harvested (Ha)	102,386,923	1,090,708	78,811,779	19,713,738	2,739,398	31,300	1,476,800
Yield (Hg/Ha)	25,548	13,309	28,864	14,100	17,491	19,042	29,424
Production (Tonnes)	261,578,498	1,451,646	227,480,272	27,795,578	4,791,402	59,600	4,345,300
Seeds (Tonnes)	6,983,352	43,283	4,838,633	1,906,313	193,870	1,252	154,300
Soybean oil (Tonnes)	39,761,852	390,660	24,028,558	12,442,496	2,890,760	9,377	241,300

Table 1. Soybean production statistics (FAOSTAT 2010)

Soybean is a well-known nitrogen fixer and has been a model plant for the study of BNF. Its importance in BNF led to the genome sequencing of soybean; details of the soybean genome are available at soybase.org (*G. max* and *G. soja* sequences are available at NCBI as well). Although considerable work has been conducted on other legumes with respect to biological nitrogen fixation, we focus only on soybean for this review.

The efficiency of BNF depends on climatic factors such as temperature and photoperiod [16]; the effectiveness of a given soybean cultivar in fixing atmospheric nitrogen depends on the interaction between the cultivar's genome and conditions such as soil moisture and soil nutrient availability [17,18]; and the competitiveness of the bacterial strains available, relative to indigenous and less effective strains, plus the amount and type of inoculants applied, and interactions with other, possibly antagonistic, agrochemicals that are used in crop protection [19]. The most important criteria, however, is the selection of an appropriate strain of *B. japonicum* since specific strains can be very specific to soybean cultivar, and subject to influence by specific edaphic factors [20,21,22]. Under most conditions, soybean meets 50-60% of its nitrogen demand through BNF, but it can provide 100% from this source [23].

1.3. *Bradyrhizobium japonicum*

B. japonicum, is a gram negative, rod shaped nitrogen fixing member of the rhizobia and is an N_2-fixing symbiont of soybean. *B. japonicum* strain USDA110, was originally isolated from soybean nodules in Florida, USA, in 1957 and has been widely used for the purpose of molecular genetics, physiology, and ecology, owing to its superior symbiotic nitrogen fixation activity with soybean, relative to other evaluated strains. The genome sequence of this strain has been determined; the bacterial genome is circular, 9.11 Million bp long and contains approximately 8373 predicted genes, with an average GC content of 64.1% [24,25].

Initially attached to the root-hair tips of soybean plants, rhizobia colonize within the roots and are eventually localized within symbiosomes, surrounded by plant membrane. This symbiotic relationship provides a safe niche and a constant carbon source for the bacteria while the plant derives the benefits of bacterial nitrogen fixation, which allows for the use of readily available nitrogen for plant growth. Inoculation of soybean with *B. japonicum* often increases seed yield [eg. 26].

B. japonicum synthesize a wide array of carbohydrates, such as lipopolysaccharides, capsular polysaccharides, exopolysaccharides (EPS), nodule polysaccharides, lipo-chitin oligosaccharides, and cyclic glucans, all of which play a role in the BNF symbiosis. Bacteria produce polysaccharide degrading enzymes, such as polygalacturonase and carboxymethylcellulase, cleave glycosidic bonds of the host cell wall at areas where bacteria are concentrated, creating erosion pits in the epidermal layer of the roots, allowing the bacteria gain entry to the roots [27]. The energy source for *B. japonicum* is the sugar trehalose, which is taken up readily and converted to CO_2 [28,29,30,31]. On the other hand UDP-glucose is taken up in large quantities but metabolized slowly, like sucrose and glucose. Promotion of plant growth causes more O_2 to be released and more CO_2 to be taken up [24,27].

1.4. Lipo-chitooligosaccharide (LCO) from *Bradyrhizobium japonicum*

As mentioned earlier in this review, the process of nodulation in legumes begins with a complex signal exchange between host plants and rhizobia. The first step in rhizobial establishment in plant roots is production of isoflavonoids as plant-to-bacterial signals; the most common in the soybean-*B. japonicum* symbiosis being genestin and diadzein [32], which trigger the *nod* genes in the bacteria which, in turn, produce LCOs, or Nod factors, that act as return signals to the plants and start the process of root hair curling, leading to nodule formation. Some recent literature has also shown that jasmonates can also cause *nod* gene activation in *B. japonicum* although the strain specificities are very different from those of isoflavonoids such as genistein [33-36]. LCOs are oligosaccharides of β-1,4-linked N-acetyl-D-glucosamine coded for by a series of *nod* genes and are rhizobia specific [37,38]. The nodDABCIJ genes, conserved in all nodulating rhizobia [37,39,40] are organized as a transcriptional unit and regulated by plant-to-rhizobia signals such isoflavanoids [41-43].

Nodulation and subsequent nitrogen fixation are affected by environmental factors. It has been observed that, under sub-optimal root zone temperatures (for soybean 15-17 ºC), pH stress and in the presence of nitrogen, isoflavanoid signal levels are reduced; while high temperature (39 ºC) increases non-specific isoflavanoid production and reduces *nod* gene activation, thereby affecting nodulation [44]). Our laboratory has isolated and identified the major LCO molecule produced by *B. japonicum* 532C as Nod Bj V (C18:1;MeFuc) [45]. This Nod factor contains a methyl-fucose group at the reducing end that is encoded by the host-specific *nodZ* gene [46], which is an essential component for soybean-rhizobia interactions.

LCOs also positively and directly affect plant growth and development in legumes and non-legumes. The potential role of LCOs in plant growth regulation was first reported by Denarie and Cullimore [47]). Nod genes A and B from *R. meliloti*, when introduced into tobacco, altered the phenotype by producing bifurcated leaves and stems, suggesting a role for *nod* genes in plant morphogenesis [48]. The development of somatic embryos of Norway spruce is enhanced by treatment with purified Nod factor from *Rhizobium* sp. NGR234. It has been suggested that these Nod factors can substitute for auxin and cytokinin like activities in promoting embryo development, and that the chitin core of the nod factor is an essential component for regulation of plant development [49,50]. Some of the LCO induced enod genes in non-legumes seem to encode for defence related responses, such as chitinase and PR proteins [42,43], peroxidase [51] and enzymes of phenylpropanoid pathway, such as L-phenylalanine ammonia-lyase (PAL) [52]. Seed gemination and seedling establishment is enhanced in soybean, common bean, maize, rice, canola, apple and grapes, accompanied by increased photosynthetic rates [53]. Hydroponically grown maize showed an increase in root growth when LCO was applied to the hydroponic solution [54,55] and foliar application to greenhouse grown maize resulted in increases in photosynthetic rate, leaf area and dry matter [56]. Foliar application to tomato, during early and late flowering stages, increased flowering and fruiting and also fruit yield [57]. An increase in mycorrhizal colonization (*Gigaspora margarita*) was observed in *Pinus abies* treated with LCO [50,58]. Recent research in our laboratory, on soybean leaves treated with LCOs under sub-optimal growth conditions, revealed the up-regulation of over 600 genes, many of which are defense and stress response related, or transcription factors; microarray results show that the transcriptome of the leaves is highly responsive to LCO treatment at 48 h post treatment [59]. These results suggest the need to investigate more carefully the mechanisms by which microbe-to-plant signals help plants accommodate abiotic and biotic stress conditions.

Since the protein quality of soybean plays an important role in overall agricultural and in nutraceuticals production, it is imperative that we study the proteomics of soybean and its symbiont *B. japonicum*, not only for better understanding of the crop, but also for the betterment of agriculture practices and production of better high value added food products for human consumption.

1.5. Proteomics as a part of integrative systems biology

The "omics" approach to knowledge gain in biology has advanced considerably in the recent years. The triangulation approach of integrating transcriptomics, proteomics and metabolomics is being used currently to study interconnectivity of molecular level responses of crop plants to various conditions of stress tolerance and adaptation of plants, thus improving systems level understanding of plant biology [60, 61].

While transcriptomics is an important tool for studying gene expression, proteomics actually portrays the functionality of the genes expressed. Several techniques are available for studying differential expression of protein profiles, and can be broadly classified as gel-based and MS (mass spectrometry)-based quantification methods. The gel based approach uses conventional, two-dimensional (2-D) gel electrophoresis, and 2-D fluorescence difference gel electrophoresis (2D-DIGE), both based on separation of proteins according to isoelectric point, followed by separation by molecular mass. The separated protein spots are then isolated and subjected to MS analysis for identification. Major drawbacks of these techniques are laborious sample preparation and inability to identify low abundance, hydrophobic and basic proteins.

The MS based approach can be a label-based quantitation, where the plants or cells are grown in media containing ^{15}N metabolite label or using ^{15}N as the nitrogen source. Label-free quantitation, however, is easier and allows analysis of multiple and unlimited samples. This technique, also referred to as MudPIT (multidimensional protein identification technology), is a method used to study proteins from whole-cell lysate and/or a purified complex of proteins [62,63]. The total set of proteins or proteins from designated target sites are isolated and subjected to standard protease digestions (eg. such as tryptic digestion). In brief, flash frozen leaf samples are ground in liquid nitrogen and polyphenols; tannins and other interfering substances such as chlorophyll are removed. The processed tissue is resuspended in a chaotropic reagent to extract proteins in the upper phase, and the plant debris is discarded [64-70]. The total protein set, in the resulting solution, is further quantified using the Lowry method [71]. The protein samples (2 μg of total protein each), once digested with trypsin, can then be loaded onto a microcapillary column packed with reverse phase and strong cation exchange resins. The peptides get separated in the column, based on their charge and hydrophobicity. The columns are connected to a quarternary high-performance liquid chromatography pump and coupled with an ion trap mass spectrometer, to ionize the samples within the column and spray them directly into a tandem mass spectrometer. This allows for a very effective and high level of peptide separation within the mixture, and detects the eluting peptides to produce a mass spectrum. The detected peptide ions, at measured mass-to-charge (m/z) ratios with sufficient intensity, are selected for collision-induced dissociation (CID). This procedure allows for the fragmenting of the peptides to produce a product ion spectrum, the MS/MS spectrum. In addition, the fragmentation occurs preferentially at the amide bonds, to generate N-terminal fragments (b ions) and C-terminal fragments (y ions) at specific m/z ratios, providing structural information about the amino acid sequence and sites of modification. The b ion and y ion patterns are matched to a peptide sequence in a translated genomic database to help identify the proteins present in the sample [72-75]. A variety of database searching and compiling algorithms are used to interpret the data obtained for structure and function of the identified proteins.

2. Analyses of soybean proteomics

2.1. Physiological and biological changes in the soybean proteome

2.1.1. Whole plant organs

The various tissues of soybean have specific groups of associated proteins at each developmental stage. While leaves at various developmental stages showed 26 differentially expressed proteins, the first trifoliate stage manifested the greatest increase in protein types of the outer/inner envelope of choloroplast membrane and also of the protein transport machineries. Young leaves showed abundant chaperonin-60, while HSP 70 and TP-synthase b were present in all the tissues analyzed. Age dependent correlation was observed in net photosynthesis rate, chlorophyll content and carbon assimilation. During the flowering stage, flower tissue expressed 29 proteins that were exclusively involved in protein transport and assembly of mitochondria, secondary metabolism and pollen tube growth (Ahsan and Komatsu., 2009 [76]. Soybean peroxisomal adenine nucleotide carrier (GmPNC1) is associated with the peroxisomal membrane and facilitates ATP and ADP importing activities. The proteins At PNC1 and At PNC2 are arabidopsis orthologs of Gm PNC1. Under constant darkness, Gm PNC1 increased in cotyledons up to 5 days post germination and the levels were rapidly reduced when the seedlings were exposed to light. RNA interference studies on arabidopsis At PNC1 and At PNC2 suggests that PNC1 assists with transport of ATP/ADP in the peroxisomal fatty acid-b oxidation pathway post germination (Arai et al., 2008 [77]. This probably helps the seedling establish vigour for future growth.

In order to establish if xylem proteins and the apoplast conduit are involved in long distance signalling in autoregulation of nodulation (AON) in the soybean-*B. japonicum* symbiosis, xylem and apoplast fluids were collected from hypocotyl, epicotyl and stem tissues. In addition, proteins from imbibing seeds were evaluated to determine possible relationships of these proteins with the xylem and apoplast proteins, especially during the seed to seedling stage transition. The proteins secreted from imbibing seeds were different from the set of xylem-related proteins. Hypocotyl, epicotyl and stem xylem proteins were generally similar. Comparison of wild type and nts1007 plants showed no difference in xylem protein profiles, suggesting that xylem proteins were not involved in AON. However, a lipid transfer protein and Kunitz trypsin inhibitor, both known to have roles in plant signalling, were identified within the xylem proteins [78].

Proteomic studies on chasmogamous (CH) CH cv. Toyosuzu and cleistogamous (CL) CL cv. Karafuto-1 flowerbuds using 2D gel revealed differential protein levels of β-galactosidase and protein disulfide isomerase. Cleistogamy occurs in plants under diverse stress conditions, such as drought and cold, and can also vary with temperature and light [79]. Soybean cv Maverick was used to study proteomics during seed filling stages, at 2, 3, 4, 5 and 6 weeks after flowering, using 2D and MALDI-TOF-MS. Storage proteins, proteins involved in metabolism and metabolite transport and defense related proteins were the most abundant, along with cysteine and methionine biosynthesis proteins, lipoxygenases and 14-3-3-like proteins [80,81].

Based on these findings, it is clear that the plant partitions its proteomics based on ontogeny and this specificity probably plays a crucial role in organ maturation and transition from one stage to another in the plants life cycle. Understanding this is of fundamental importance in agriculture, global food production, biofuel production and issues such as plant responses to climate change.

2.1.2. Seeds

Both 2D gel and peptide mass fingerprinting techniques (MALDI-TOF-MS) were used to study the proteins of mature and dry soybean (cv. Jefferson) seeds. Sucrose binding proteins, alcohol dehydrogenase and seed maturation proteins were some of the key proteins identified (Mooney and Thelen 2004 [82]. A comparison of four methods for protein isolation and purification from soybean seed was one of the first reports on soybean proteomics; thiourea/urea and TCA protocols were found to be the best. Proteins extracted with these two methods and further characterized by MALDI-TOF-MS and LC-MS helped identify proteins such as β-conglycinin, glycinin, Kunitz trypsin inhibitor, alcohol dehydrogenase, Gm Bd 28K allergen and sugar binding proteins in seeds [83]. The two major soybean storage proteins are α-conglycinin and glycinin. While the α-conglycinin subunits separated well in the pH range 3.0-10.0, glycinin polypeptides could be separated in pH ranges 4.0-7.0 and 6.0-11.0. Apart from these major storage proteins, this combined proteomic approach (2D-PAGE and immobilized pH gradient strips) also identified 44 storage proteins in wild soybean (*G. soja*) and 34 additional storage proteins in its cultivated counterpart (*G. max*) [84]. A comparative proteome analysis of soybean seed and seedling tissue suggested that there were dramatic changes in the protein profiles during seed germination and during seedling growth. The seed storage proteins β-conglycinin and glycinin were seen to degrade rapidly and their degradation products were either accumulated or degraded further as the seeds germinated. This degradation of the storage proteins indicates that the proteolysis process provides amino acids and energy for the growing seedlings, and gives access to new detail regarding these processes [85].

Synthesis of soybean glycinin and conglycinin, was suppressed by RNA interference. The storage protein knockdown (SP2) seeds were very similar to the wild type during development and at maturity. Proteomic analysis of the SP2 soybean genotypes and next-generation transcript sequencing (RNA-Seq) suggested that the seeds could rebalance their transcriptome and metabolome in the face of at least some alterations. GFP quantification for glycinin allele mimics further revealed that glycinin was not involved in proteome rebalance and that seeds are capable of compensating through increases in other storage proteins, to maintain normal protein content, even if the major storage proteins were not available [86].

Transgenic soybean seeds have higher amounts of malondialdehyde, ascorbate peroxidase, glutathione reductase, and catalase (29.8, 30.6, 71.4, and 35.3%, respectively) than non-transgenic seeds. Precursors of glycinin, allergen Gly m Bd 28k, actin and sucrose binding proteins were the other proteins identified [87,88]. High protein accessions of soybean (with 45 % or more protein in seeds) were compared with soybean cultivar Williams 82. 2-DE-MALDI-TOF-MS followed by Delta2D image analysis showed huge differences in 11S storage globulins amongst the accessions. In addition, the trait for high protein from PI407788A was moved to experimental line LG99-469 and was stable upon transformation [89,90].

2.1.3. *Roots, root hairs and nodules*

Since the root apical meristem (RAM) is responsible for the growth of the plant root system and root architecture plays and important role in determining the performance of crop plants, a proteome reference map of the soybean root apex and the differentiated root zone was established. The root apex samples comprised of 1 mm of the root apex, encasing the RAM, the quiescent center and the root cap. The predominant proteins in the root belonged to those of stress response, glycolysis, redox homeostasis and protein processing machinery. The root apex contained key proteins, such as those involved in redox homeostasis and flavonoid biosynthesis, but was underrepresented in glycolysis, stress response and TCA cycle related proteins [91]. Analysis of the proteome of isolated soybean root hair cells using 2-D gel and shotgun proteomics approaches identified proteins involved in basic cell metabolism, those whose functions are specific to root hair cell activities, including water and nutrient uptake, vesicle trafficking, and hormone and secondary metabolism [92, 93]. Proteomic studies of soybean roots and root hairs after *B. japonicum* inoculation explains the importance of initial plant-bacteria symbiotic interaction. A 2-D, MALDI-TOF, MS based approach shows that enzymes such as chitinase and phosphoenolpyruvate carboxylase are differentially expressed in root hairs. As well as peroxidase and phenylalanine-ammonia lyase, found to be expressed during rhizhobial inoculation, other novel proteins such as phospholipase D and phosphoglucomutase were found to be expressed [94]. Nodule cytosol proteins from soybean cv. Williams 82 were found to be 28% related to carbon metabolism, 12% related to nitrogen metabolism, 12% related to reactive oxygen metabolism and 11% related to vesicular trafficking proteins. The vesicular trafficking proteins could be involved in the exchange of micro- and macro-molecules during the process of nodulation, while carbon, nitrogen and reactive oxygen species are related to physiological functions during nitrogen fixation [95]. The peribacteroid membrane (PBM) of the soybean symbiosome contains chaperonins such as HSP60, BiP (HSP70) and PDI, and serine and thiol protease, all of which are involved in protein translocation, folding, maturation and degradation of proteins related to the symbiosomes. Nodulin proteins 53b and 26B, associated with the PBM, were also present, although their function is not clear [96].

2.2. Soybean proteomics under stress conditions

Like all plants, soybean also encounters various stressors during its life cycle. Work related to flooding, drought, salt, heat, biotic stressors, metal toxicity, ozone, phosphorous deficiency and seed protein allergens are reviewed here.

2.2.1. Flooding stress

Plasma membrane proteins from the root and hypocotyl of soybean seedlings were purified and subjected to 2-D gel electrophoresis, followed by MS and protein sequencing, and also using nanoliquid chromatography followed by nano-LC-MS/MS based proteomics. The two techniques were used to compare the proteins present, and this indicated that during flooding stress proteins typically found in the cell wall were up-regulated in the plasma membrane. Also, the anti-oxidative proteins were up-regulated to protect the cells from oxidative damage, heat shock proteins to protect protein degradation and signaling proteins to regulate ion homeostasis [97]. MS based proteomics applied to root tips of two-day-old seedlings flooded for 1 day showed increased levels of proteins involved in energy production. Proteins involved in cell structure maintenance and protein folding were negatively affected, as was their phosphorylation status [98].

Two-day-old germinated soybean seeds were subjected to water logging for 12 h and total RNA and proteins were analyzed from the root and hypocotyl. At the transcriptional level, the expression of genes for alcohol fermentation, ethylene biosynthesis, pathogen defense, and cell wall loosening were all significantly up-regulated, while scavengers and chaperons of reactive oxygen species were seen to change only at the translational level. Transcriptional and translational level changes were observed for hemoglobin, acid phosphatase, and Kunitz trypsin protease inhibitors. This adaptive strategy might be for both hypoxia and more direct damage of cells by excessive water [99]). Proteins from 2-day-old soybean seedlings flooded for 12 h were analyzed using 2-D gel MS, 2-D fluorescence difference gel electrophoresis, and nanoliquid chromatography. Early responses to flooding involved proteins related to glycolysis and fermentation, and inducers of heat shock proteins. Glucose degradation and sucrose accumulation increased due to activation of glycolysis and down-regulation of sucrose degrading enzymes, in addition the methylglyoxal pathway, a detoxification system linked to glycolysis, was up-regulated. 2-D gel based phosphoproteomic analysis showed that proteins involved in protein synthesis and folding were dephosphorylated under flooding conditions [100]. Water logging stress imposed on very early soybean seedlings (V2 stage) resulted in a gradual increase of lipid peroxidation and *in vivo* H_2O_2 production. Proteomic studies of the roots using 2-D gel, MALDI-TOF-MS or electrospray ionization tandem mass spectrometry (ESI-MS/MS) analysis, identified 14 up-regulated and 5 down-regulated proteins. Five newly discovered proteins were associated with water logging, a known anaerobic stress. The proteins included those associated with signal transduction, programmed cell death, RNA processing, redox homeostasis and energy metabolism.

Increases in glycolysis and fermentation pathway associated proteins were indicative of adaptation of the plant to this alternate energy provision pathway. Other novel proteins, such as a translation initiation factor, apyrase, auxin-amidohydrolase and coproporphyrinogen oxidase, were also identified [101]. Mitochondrial proteomics from 2-day-flooded 4-day-old soybean seedlings identified increases in the levels of proteins and metabolites associated with TCA cycle and the γ-amino butyrate shunt. Increases in NADH and NAD and a decrease in ATP during the stress suggest that the electron transport chain is disrupted, although NADH production increases through TCA cycle activity [102].

Soybean seeds germinated for 48 h were subjected to water logging stress for 6-48 h. In addition to general stress responses due to increases in reactive oxygen species scavengers, several glycolytic enzymes were up-regulated, suggesting changes in energy generation [103].

2.2.2. Water stress – Drought

Soybean root activities are affected during water stress. The root can be partitioned into zones 1 (apical 4 mm zone) and 2 (4-8 mm zone), based on maximum elongation during well watered conditions. Soluble proteins from these regions, studied under both well-watered and water deficit stress conditions, revealed region-specific regulation of the phenylpropanoid pathway. Zone 1 of roots manifested increases in isoflavanoid biosynthesis related enzymes and proteins that contribute to growth and maintenance of the roots under water stress conditions. However, zone 2 of water stressed roots manifested up-regulation of caffeoyl-CoA *O*-methyltransferase (a protein involved in lignin biosynthesis), protective proteins related to oxidative damage, ferritin proteins that sequester iron, and 20S proteasome α-subunit A. Increases in lignin accumulation and ferritin proteins preventing availability of free iron in this zone were suggested to be the factors affecting root growth during water stress [104]. An investigation of the soybean plasma membrane proteome, under osmotic stress, was conducted using 2-day-old seedlings subjected to 10% PEG for 2 days; both gel- and nano-LC MS/MS-based proteomics methods were utilized to analyze the samples. Out of the 86 proteins identified by nano-LC MS/MS approach, 11 were up-regulated and 75 proteins down-regulated under PEG mediated stress. Three homologues of plasma membrane transporter proteins H1-ATPase and calnexin were prominent [105]. Similarly, 3-day-old soybean seedlings were subjected to 10% PEG treatment or water withdrawal and samples collected from roots, hypocotyl and leaves, 4-days after treatment, for proteome analysis. The root was the most responsive and affected organ for both drought stress induction methods. The leaves showed increases in metabolism-related proteins, while the energy production and protein synthesis machineries were negatively affected. HSP70, actin isoform B and ascorbate peroxidase were up-regulated in all the tissues analyzed. Importantly, methionine synthase, a drought response protein, decreased, suggesting negative effects of drought stress on these seedlings [106].

2.2.3. High temperature stress

Tissue specific proteomics under high temperature stress revealed 54, 35 and 61 differentially expressed proteins in the leaves, stems and roots, respectively. Heat shock proteins and those involved in antioxidant defense were up-regulated while proteins for photosynthesis, amino acid and protein synthesis and secondary metabolism were down- regulated. HSP70 and other low molecular weight HSPs were seen in all the tissues analyzed. ChsHSP and CPN-60 were tissue specific and the sHSPs were found only in tissues under heat stress, and were not induced by other stresses such as cold or hydrogen peroxide exposure [107].

2.2.4. Salt stress

Salt stress is also an important abiotic stressor that affects crop growth and productivity. Of the 20% of agricultural land available globally, 50% of the cropland is estimated by the United Nations Environment Program (The UNEP) to be salt-stressed [108]. As the plant grows under salt stresses conditions, depending on the severity of the stress, the plants can experience reduced photosynthesis, protein and energy production, and changes in lipid metabolism [109,110]. As soil salinity increase, the effects on seed germination and germinating seedlings are profound. Responses to salinity and drought stress are similar; they affect the osmotic activity of the root system, thereby affecting the movement of water and nutrients into the plants. In Canadian soils, salinity varies between spring and fall and the most saline conditions are seen at the soil surface just after spring thaw. In the Canadian prairies, the dominant salts of saline seeps include calcium (Ca), magnesium (Mg) and sodium (Na) cations, and sulphate (SO_4^-) anions [111]. Soybean is very sensitive to Cl^-, but not greatly affected by Na^+, because of its ability to restrict movement of Na^+ to leaves [112].

This first report regarding soybean seedling proteomic responses to salt stress evaluated length and fresh weight of the hypocotyl and roots of soybean exposed to a series of NaCl concentrations. At 200 mM NaCl, the length and fresh weight of hypocotyl and roots were greatly reduced, with a simultaneous increase in proline content, suggesting activation of mechanisms for coping with salt stress. In addition, hypocotyl and root samples from 100 mM NaCl treated seedlings up-regulated seven key proteins, such as late embryogenesis-abundant protein, b-conglycinin, elicitor peptide three precursor, and basic/helix-loop-helix protein. The same treatment caused down-regulation of protease inhibitor, lectin, and stem 31-kDa glycoprotein precursor. This combination of up- and down-regulated proteins indicates a metabolic shift and could represent a strategy used by soybean seedlings to enhance tolerance of, or adapt to, salt stress [113].

Sobhanian et al. [110,114] found that treatment of soybean seedlings with 80 mM NaCl arrests the growth and development of both hypocotyl and roots. This study assessed effects on leaf, hypocotyl and root proteomics of salt treated soybean seedlings and found that reduction of glyceraldehyde-3-phospahte dehydrogenase was indicative of reduction in ATP production, and down-regulation of calreticulin was associated with disruption in the calcium signalling pathway, both of which are associated with decreased plant growth. The levels of other proteins, such as kinesin motor protein, trypsin inhibitor, alcohol dehydrogenase and annexin, were also found to change, suggesting that these proteins might play different roles in soybean salt tolerance and adaptation [110,114].

Soybean cultivars Lee68 and N2899 are salt-tolerant and salt-sensitive, respectively. The percentage germination was not affected when exposed to 100 mmol L^{-1} NaCl, however, the mean germination time for Lee68 (0.3 days) and N2899 (1.0 day) was delayed, compared with control plants. Hormonal responses to salt stress differed between these cultivars. Both cultivars, increased abscisic acid levels and decreased giberrelic acid (GA 1, 3) and isopentyladenosine concentrations; auxin (IAA) increased in Lee68, but remained unchanged in N2899. 2-D gel electrophoresis, followed by MALDI-TOF-MS analysis, of the proteins from germinated seeds suggested increases in ferritin and the 20S proteasome subunit β-6 in both the cultivars. Glyceraldehyde 3-phosphate dehydrogenase, glutathione *S*-transferase (GST) 9, GST 10, and seed maturation protein PM36 were down-regulated in Lee68, but these proteins were naturally present in low concentrations in N2899 and were seen to up-regulate following exposure to salt stress [115].

2.2.5. Biotic stress

The soybean-*Phytophthora soje* plant-oomycete interaction is of agriculture and economic importance, as this oomycete causes soybean root and stem rot, translating to an annual global loss of $1-2 billion US. Twenty-six proteins were significantly affected in a resistant soybean cultivar (Yudou25) and 20 in a sensitive one (NG6255), as determined by 2-D gel analysis, followed by MALDI-TOF-MS. The distribution pattern of the affected proteins were - 26% energy regulation, 15% protein destination and storage, 11% defense against disease, 11% metabolism, 9% protein synthesis, 4% secondary metabolism, and 24% unknown/hypothetical proteins [116].

Soybean mosaic virus (SBMV) causes one of the most serious viral infections of soybean; leaves of infected plants were studied at a series of time points using 2-D gel electrophoresis, followed by MALDI-TOF-MS and tandem TOF/TOF-MS. Proteins expressed in the inoculated leaves were identified and were seen to be involved in protein degradation, defense signalling, coping with changes in the levels of reactive oxygen species, cell wall reinforcement, and energy and metabolism regulation. Quantitative real time PCR was used to focus on gene expression related to some of these proteins. Photosynthesis and metabolism related genes were down-regulated at all the time points, while most of the energy related genes (respiration in this case) were up-regulated for at least five of the six time points studied [117]. At the time of this writing, this report is the only one addressing the proteomic approach to molecular understanding of soybean-SBMV interaction.

2.2.6. Other miscellaneous stress related reports

Aluminium toxicity is often observed in acidic soils and Baxi 10 (BX10) is an Al-resistant cultivar. One-week-old soybean seedlings treated with 50 mM $AlCl_3$ for 24, 48 and 72 h were studied for characterization of root proteins in response to Al; and 2-D gel electrophoresis followed by MS revealed 39 proteins expressed differentially following Al treatment. Of these 21 were up-regulated (such as heat shock proteins, glutathione S-transferase, chalcone related synthetase, GTP-binding protein, ABC transporters and ATP binding proteins). Five proteins were also down-regulated and 15 newly induced proteins were present following AL treatment [118].

The process of nitrogen fixation demands large amounts of phosphorus [119]. When soybean plants are starved of phosphorus, 44 phosphate starvation proteins are expressed in soybean nodules [120].

Label free proteomics, coupled with multiple reaction monitoring (MRM) with synthetic isotope labelled peptides, was used to study 10 allergens from 20 non-genetically modified commercial varieties of soybean. The concentration of these allergens varied between 0.5-5.7 $\mu g\ mg^{-1}$ of soybean protein. At the time of this writing, this is the only proteomic report on soybean allergens [121].

The responses of soybean plants exposed to 116 ppb O_3 involved significant changes to carbon metabolism, photosynthesis, amino acid, flavanoid and isoprenoid biosynthesis, signaling, homeostasis, anti-oxidant and redox pathways [122], as indicated by shifts in expression of the relevant proteins.

More information regarding soybean functional genomics and proteomics is available at the publicly accessible Soybean Knowledgebase (SoyKB) http://soykb.org/ [123].

3. *Bradyrhizobium japonicum* and its proteomics/exoproteomics

Culturing bacteria *in vitro* can cause changes in the bacterial physiology and genetics. In order to discriminate between types of these differences, *B. japonicum* cultivated in HM media and those isolated from root nodules were studied for their protein profile using 2-D PAGE and MALDI-TOF. The cultured cells showed greater levels of proteins related to fatty acid, nucleic acid and cell surface synthesis. While carbon metabolism proteins related to global protein synthesis, maturation and degradation and membrane transporters seemed to be similar in both cultured and nodule isolated bacteria, nitrogen metabolism was more pronounced in the bacteroids. Despite the quantitative differences in some proteins in the cultured and nodule isolated bacteria, it was observed that the various proteins in common between them performed similar functions [124]. A high resolution 2-D gel electrophoresis analysis of these bacteroids revealed a number of proteins, of which about 180 spots could be identified using the *B. japonicum* database (http://www.kazusa.or.jp/index.html) [125]. The bacteroids showed a lack of defined fatty acid and nuclei acid metabolic pathways, but were rich in proteins related to protein synthesis, scaffolding and degradation. Other proteins with high expression levels were associated with cellular detoxification, stress regulation and signalling, all of which clearly establishes that differentiation into bacteroids results in a clear shift on metabolism and expression of metabolic pathways required by the bacteroids for their specialized activities [126].

Since competitiveness plays an important role in this symbiotic relationship, 2-D gel electrophoresis, image and data analysis, and in-gel digestion proteomic studies, were conducted on *B. japonicum* 4534, a strain with high competitiveness, and *B. japonicum* 4222, with low competitiveness, for nodulation. When treated with diadzein, both the strains showed upregulation of proteins: 24 in *B. japonicum* 4534 and 10 in *B. japonicum* 4222. Upon treatment with diadzein and other extracellular materials such as extracellular enzymes and polysaccharides involved in nodulation of the strains tested, the numbers increased to 78 (43 up-regulated and 35 down-regulated) and 47 (25 up-regulated and 22 down-regulated) in these two strains. Proteins not related to nodulation were also present, and the higher number of proteins expressed by *B. japonicum* 4534 may be the reason for increased competitiveness during symbiosis [127]. Comparative studies on whole cell extracts of genistein induced and non-induced cultures of a strain used in commercial inoculants in Brazil, *B. japonicum* CPAC 15 (=SEMIA 5079), and of two genetically related strains grown *in vitro* were conducted using 2-D gel electrophoresis followed by mass spectrometry. Some of the noteworthy proteins belonged to the cytoplasmic flagellar component FliG, periplasmic ABC transporters, proteins related to the biosynthesis of exopolysaccharides (ExoN), proteins that maintain redox state and the regulon PhyR$^{-\sigma EcfG}$, which is known to increase the competitiveness of *B. japonicum* and also help the bacteria under stress conditions, and several other hypothetical proteins [128].

B. japonicum utilizes the bacterial Type III secretion system (TTSS). In order for TTSS to be effective it requires a flavonoid inducer. The *tts* gene cluster of *B. japonicum* is regulated by the isoflavone genistein. In its presence NodD1 and NodW activate the *ttsI*, which is a two-component response regulator, necessary for expression of other genes in the *tts* cluster. In addition, the operons governing the TtsI regulon have a conserved motif in the *tts* box promotor region, which underscores the importance of regulation of TTSS in *B. japonicum*. Flagellin is a bulk protein synthesized by *B. japonicum* that plays an important role in TTSS. Mutant *B. japonicum* cells created by deleting the flagellin genes *bll6865* and *bll6866* were studied for their exoprotein profiles, in comparison with the non-mutated strains. Upon induction using genistein, it was observed that amongst the identifiable proteins, Blr1752 similar to NopP of *Rhizobium* sp. strain NGR234, Blr1656 (GunA2) having endoglucanase activity and three other proteins having similarity to proteins of the flagellar apparatus were detected. However, none of these proteins were detected in the mutant exoproteome, suggesting that these proteins are the products of a highly conserved *tts* box motif containing genes that encode these secreted proteins [129 and references therein].

A study of 2-D gel electrophoresis combined with MALDI-TOF MS for the identification of *B. japonicum* strains 110, BJDΔ283 and BJD567 exoproteomes revealed a high frequency of substrate-binding proteins of the ABC transporter family. Addition of genistein to the cultures altered the exoproteome; three flagellar proteins and a nodulation outer protein, Pgl, were identified. Further shotgun mass spectrometry of the genistein induced exoproteome revealed the presence of nodulation outer proteins, NopB, NopH, NopT and type III-secreted protein GunA2. Addition of diadzein or coumerstrol, instead of genistein, to the cell culture showed a reduction in the type III-secreted protein GunA2 [130]. *B. japonicum* cell lines derived from strain SEMIA 566 are adapted to stressful environmental conditions in Brazil. They also vary in their capacity for symbiotic nitrogen fixation. A representational difference analysis study was conducted on the strains S 370 and S 516, derived from SEMIA 566. Strain S 370 produces the nodulation outer protein P gene, which is strongly associated with the TTSS, and is also the major determinant of effective nodulation [131].

B. japonicum strain CPAC 15 (5SEMIA 5079) is a strain used in commercial inoculants; it belongs to the same serogroup as strain USDA 123 and is used in Brazil on soybean. Both of these strains are known to be highly competitive and saprophytic. Apart from *B. japonicum* strain USDA 110, which has been sequenced [24,25], CPAC 15 is the only stain that has been partially sequenced in any significant measure [132]. CPAC 15 and two related strains, S 370 and S 516, were studied using whole-cell 2-D protein gel electrophoresis and spot profiles of selected proteins using MS. Cytoplasmic and periplasmic proteins found to occur in diverse metabolic pathways related to the saprophytic properties of CPAC 15; 26 hypothetical proteins were identified [133].

B. japonicum strain USDA 110 from soybean plants cultivated in growth chambers were harvested at 21 days of symbiosis and subjected to transcriptomics studies and proteomics using gelLC-MS/MS. Through this integrated approach 27.8% of the theoretical proteome and 43% of the predicted genes and proteins were detected. Analysis of the biological and functional pathways highlighted proteins involved in carbon and nitrogen metabolism: several enzymes of the TCA cycle, gluconeogenesis and pentose phosphate pathway. Experiments with bacteroids obtained from soybean plants grown under field conditions showed identical results [134 and references therein].

4. Other dimensions to soybean-rhizobacteria interactions

Apart from *B. japonicum*, which produces LCOs, other rhizobacteria, such as *Bacillus thuringiensis* NEB17 reside in the rhizosphere of higher plants [135], forming a phyto-microbiome, much like the human microbiome, now realized to be so important in human health [136]. *Bacillus thuringiensis* NEB17 is symbiotic with *B. japonicum*, produce bacteriocins. *Bacillus* species were first reported to produce bacteriocins in 1976. The low-molecular-weight bacteriocins of gram-positive bacteria have bactericidal activity, mainly against certain other gram-positive bacteria [137]. Bacteriocins are ribosomally produced peptides which affect the growth of related bacterial species. The most studied bacteriocin is colicin, produced by members of the Enterobacteriaceae [138]. Due to their commercial importance as natural preservatives and as therapeutic agents against pathogenic bacteria, these antimicrobial peptides have been a major area of scientific research [137,139].

Bacteriocins are grouped into four distinct classes based on the peptide characteristics such as post translational modifications, side chains, heat stability, N-terminal sequence homology and molecular weight [140]. *Bacillus thuringiensis* NEB17 was isolated from soybean root nodules as putative endophytic bacteria in 1998 in our laboratory. When co-inoculated with *B. japonicum* under nitrogen free conditions this bacterium promoted soybean growth, nodulation and grain yield [141, 142]. Subsequently, the causative agent of plant growth promotion, a bacteriocin, was isolated from *B. thuringiensis* NEB17, and is now referred to as thuricin 17 [143]. Initially, its partial sequence was determined [144], and its full sequence has been more recently reported [145]. Thuricin 17 is a low molecular weight peptide of 3162 Da, stable across a pH range of 1.0–9.25, highly heat resistant and is inactivated by treatment with proteolytic enzymes. Based on its N-terminal sequence homology of thuricin 17 and that of the also newly isolated bacthuricin F4, a new class of bacteriocins, class IId was proposed [143]. The bacteriocins produced by *B. thuringiensis* strain NEB17 (Th17) and *B. thuringiensis* subsp. kurstaki BUPM4 (bacthuricin F4 - 3160.05 Da) have been reported to show functional similarities and anti-microbial activities [146]. In addition, thuricin 17, applied as leaf spray and root drench, has positive effects on soybean and corn growth, which was first reported from our laboratory [145]; this constituted the first report of plant growth stimulation by a bacteriocin.

Proteomic profiling of both these bacteria are underway in our laboratory and we hope to acquire some indications of plant proteomic shifts related to biological nitrogen fixation through these experiments over the next few months.

5. Conclusions and future perspectives

Soybean is an important protein and oil seed crop and BNF is an important source of nitrogen for the crop. Considerable work has been conducted regarding soybean proteomics, facilitated by recent advancements in technology, but a more systematic approach to this method is required in order to understand the intricacies of plant growth and development in the face of interactions with various symbionts. There is wide variation in the ability of *B. japonicum* strains to fix atmospheric nitrogen and screening of the various strains known to us, in the light of specific agro-climatic conditions, will help improve effective BNF at a very low cost. In this regard, the proteomic profile can be of immense help in highlighting the protein-protein interactions that are involved during the process of nodule initiation, formation and sustenance. This in-depth knowledge of the role of proteins in nodulation and plant growth promotion processes will assist in further improvement of soybean cultivars and their associated *B. japonicum* strains, for a better and more sustainable agriculture.

Author details

Sowmyalakshmi Subramanian and Donald L. Smith*

*Address all correspondence to: donald.smith@mcgill.ca

Department of Plant Sciences, Macdonald Campus, McGill University, Quebec, Canada

References

[1] Vance CP. Legume symbiotic nitrogen fixation: agronomic aspects 1998. In The Rhizobiaceae: Molecular biology of model Plant-Associated bacteria, eds., Spaink HP,

Kondorosi A, Hooykaas PJJ, Dordrecht, The Netherlands: Kluwer Academic Publishers, pp. 509-530.

[2] Graham PH, Vance CP. Nitrogen fixation in perspective: an overview of research and extension needs. Field Crops Research 2000;65: 93-106.

[3] Newbould P. The use of nitrogen fertilizers in agriculture. Where do we go practically and ecologically? Plant Soil 1989;115: 297-311.

[4] Sadowsky MJ, Graham PH. Soil biology of the Rhizobiaceae 1998. In: (Spaink HP, Kondorosi A, Hooykaas PJJ. ed. The Rhizobiaceae, 155-172. (Kluwer: Dordrecht, The Netherlands).

[5] Graham PH, Vance CP. Legumes: Importance and constraints to greater use. Plant Physiology 2003;131: 872-877.

[6] Cui Z, Carter TE, Gai J, Qui J, Nelson RL. Origin, description, and pedigree of Chinese soybean cultivars released from 1923 to 1995. U.S. Department of Agriculture, Agricultural Research Service, Tech. Bull 1999; No. 1871.

[7] Hymowitz T, Harlan JR. Introduction of soybean to North America by Samuel Bowen in 1765. Economic Botany 1983;37:371-379.

[8] Qui L-J, Chang R-Z: The origin and history of soybean 2010. In The soybean : botany, production and uses / edited by Guriqbal Singh. Pp- 1-23, CAB International.

[9] Lui K. Soybeans as a powerhouse of nutrients and phytochemicals. In Soybeans as functional foods and ingredients; Lui K., Ed.;AOCS Press: Champaign, IL 2004;p1-53.

[10] Mandal KG, Sahab KP, Ghosha PH, Hatia KM and Bandyopadhyaya KK. Bioenergy and economic analysis of soybean-based crop production systems in central India. Biomass and Bioenergy 2002; 23:337-345.

[11] Du W, Xu Y, Liu D. Lipase-catalysed transesterification of soya bean oil for biodiesel production during continuous batch operation. Biotechnology and Applied Biochemistry 2003;38(Pt 2):103-6.

[12] Mushrush GW, Wynne JH, Willauer HD, Lloyd CL. Soybean-derived biofuels and home heating fuels. Journal of Environmental Science and Health. Part A-Toxic/Hazardous Substances and Environmental Engineering 2006;41(11):2495-502.

[13] Huo H, Wang M, Bloyd C, Putsche V. Life-cycle assessment of energy use and greenhouse gas emissions of soybean-derived biodiesel and renewable fuels. Environmental Science and Technology 2009;43(3):750-6.

[14] Pestana-Calsa MC, Pacheco CM, de Castro RC, de Almeida RR, de Lira NP, Junior TC. Cell wall, lignin and fatty acid-related transcriptome in soybean: Achieving gene expression patterns for bioenergy legume. Genetics and Molecular Biology 2012;35(1 (suppl)):322-330.

[15] FAO (2009) *FAOSTAT*. Food and Agriculture Organization of the United Nations, Rome, Italy. Available at: http://faostat.fao.org (last accessed 8 July 2012).

[16] Shiraiwa T, Sakashita M, Yagi Y, Horie T. Nitrogen fixation and seed yield in soybean under moderate high-temperature stress. Plant Production Science 2006;9: 165–167.

[17] Sridhara S, Thimmegowda S, Prasad TG. Effect of water regimes and moisture stress at different growth stages on nodule dynamics, nitrogenase activity and nitrogen fixation in soybean [*Glycine max* (L.) Merrill]. Journal of Agronomy and Crop Science 1995;174: 111-115.

[18] Jung G, Matsunami T, Oki Y, Kokubun M. Effects of water logging on nitrogen fixation and photosynthesis in supernodulating soybean cultivar Kanto 100. Plant Production Science 2008;11: 291-297.

[19] Campo RJ, Hungria M. Sources of nitrogen to reach high soybean yields 2004: In: *Proceedings of VII World Soybean Research Conference, IV International Soybean Processing and Utilization Conference, III Congresso Brasileiro de Soja Brazilian Soybean Congress*, Foz do Iguassu, PR, Brazil, 29 February–5 March 2004, pp. 1275-1280.

[20] Hughes RM, Herridge DF. Effect of tillage on yield, nodulation and nitrogen fixation of soybean in far north-coastal New South Wales. Australian Journal of Experimental Agriculture 1989;29: 671-677.

[21] Alves BJR, Boddey RM, Urquiaga S. The success of BNF in soybean in Brazil. Plant and Soil 2003;252: 1-9.

[22] Abaidoo RC, Keyser HH, Singleton PW, Dashiell KE, Sanginga N. Population size, distribution, and symbiotic characteristics of indigenous *Bradyrhizobium* spp. that nodulate TGx soybean genotypes in Africa. Applied Soil Ecology 2007;35: 57-67.

[23] Salvagiotti F, Cassman KG, Specht JE, Walters DT, Weiss A, Dobermann A. Nitrogen uptake, fixation and response to fertilizer N in soybeans: A review. Field Crops Research 2008;108: 1-13.

[24] Kaneko T, Nakamura Y, Sato S, Minamisawa K, Uchiumi T, Sasamoto S, Watanabe A, Idesawa K, Iriguchi M, Kawashima K, Kohara M, Matsumoto M, Shimpo S, Tsuruoka H, Wada T, Yamada M, Tabata S. Complete genomic sequence of nitrogen-fixing symbiotic bacterium *Bradyrhizobium japonicum* USDA110. DNA Research 2002a; 9:189-197.

[25] Kaneko T, Nakamura Y, Sato S, Minamisawa K, Uchiumi T, Sasamoto S, Watanabe A, Idesawa K, Iriguchi M, Kawashima K, Kohara M, Matsumoto M, Shimpo S, Tsuruoka H, Wada T, Yamada M, Tabata S. Complete genomic sequence of nitrogen-fixing symbiotic bacterium *Bradyrhizobium japonicum* USDA110 (supplement). DNA Research 2002b;9: 225-256.

[26] Ndakidemi PA, Dakora FD, Nkonya EM, Ringo D, Mansoor H. Yield and economic benefits of common bean (*Phaseolus vulgaris*) and soybean (*Glycine max*) inoculation

in northern Tanzania. Australian Journal of Experimental Agriculture 2006;46: 571-577.

[27] Mateos PF, Baker DL, Petersen M, Velázquez E, Jiménez-Zurdo JI, Martínez-Molina E, Squartini A, Orgambide G, Hubbell DH, Dazzo FB. Erosion of root epidermal cell walls by rhizobium polysaccharide-degrading enzymes as related to primary host infection in the rhizobium–legume symbiosis. Canadian Journal of Microbiology 2001;47: 475-487.

[28] Salminen SO, Streeter JG. Uptake and metabolism of carbohydrates by *Bradyrhizobium japonicum* bacteroids. Plant Physiology 1986; 83: 535-540.

[29] Müller J, Boller T, Wiemken A. Trehalose becomes the most abundant non-structural carbohydrate during senescence of soybean nodules. Journal of Experimental Botany 2001;52(358):943-7.

[30] Streeter JG, Gomez ML. Three enzymes for trehalose synthesis in *Bradyrhizobium* cultured bacteria and in bacteroids from soybean nodules. Applied and Environmental Microbiology 2006;72(6):4250-5.

[31] Sugawara M, Cytryn EJ, Sadowsky MJ. Functional role of *Bradyrhizobium japonicum* trehalose biosynthesis and metabolism genes during physiological stress and nodulation. Applied and Environmental Microbiology 2010;76(4):1071-81.

[32] Rao JR, Cooper JE () Rhizobia catabolize gene-inducing flavanoid via C-ring fission mechanisms. Journal of Bacteriology 1994;176: 5409-5413.

[33] Mabood F, Souleimanov A, Khan W and Smith DL. Jasmonates induce Nod factor production by *Bradyrhizobium japonicum.* Plant Physiology and Biochemistry 2006a; 44:759–765.

[34] Mabood F, Zhou X and Smith DL. Pre-incubation of *Bradyrhizobium japonicum* cells with methyl jasmonate (MeJA) increases soybean nodulation and nitrogen fixation under short season field conditions. Agronomy Journal 2006b;98:289–294.

[35] Mabood F, Zhou X, Lee KD, Smith DL. Methyl jasmonate, alone or in combination with genistein, and *Bradyrhizobium japonicum* increases soybean (*Glycine max* L.) plant dry matter production and grain yield under short season conditions. Field Crops Research 2006c;95:412-419.

[36] Mabood F and Smith DL. Pre-incubation of *Bradyrhizobium japonicum* with jasmonates accelerates nodulation and nitrogen fixation in soybean (*Glycine max*) at optimal and suboptimal root zone temperatures. Physiologia Plantarum 2005;125:311-325.

[37] Spaink H, Wijfjes A, Lugtenberg B. Rhizobium NodI and NodJ proteins play a role in the efficiency of secretion of lipochitin oligosaccharides. Journal of Bacteriology 1995;177: 6276-6281.

[38] Perret X, Staehelin C, Broughton WJ. Molecular basis of symbiotic promiscuity. Microbiology and Molecular Biology Reviews 2000;64: 180-201.

[39] Kamst E, Spaink HP, Kafetzopoulos D (1998) Biosynthesis and secretion of rhizobial lipochitin-oligosaccharide signal molecules. Pages 29-71 in: Subcellular Biochemistry 29: Plant-Microbe Interactions. B. B. Biswas and H. K. Das, eds. Plenum Press, New York

[40] Vazquez M, Santana O, Quinto C. The NodI and NodJ proteins from *Rhizobium* and *Bradyrhizobium* strains are similar to capsular polysaccharide secretion proteins from gram-negative bacteria. Molecular Microbiology 1993;8: 369-377.

[41] Carlson R, Price N, Stacey G. The biosynthesis of rhizobial lipo-oligosaccharide nodulation signal molecules. Molecular Plant Microbe Interactions 1994;7: 684-95.

[42] Schultze M, Kondorosi Á. The role of lipochitooligosaccharides in root nodule organogenesis and plant cell growth. Current Opinion in Genetics and Development 1996;6: 631-638.

[43] Schultze M, Kondorosi Á. Regulation of symbiotic root nodule development. Annual Reviews in Genetics 1998;32: 33-57.

[44] Bai Y, D'Aoust F, Smith DL, Driscoll BT. Isolation of plant-growth-promoting *Bacillus strains* from soybean root nodules. Canadian Journal of Microbiology 2002a;48: 230-238.

[45] Prithiviraj B, Souleimanov A, Zhou X, Smith DL. Differential response of soybean (*Glycine max* (L.) Merr.) genotypes to lipo-chito-ligosaccharide Nod Bj-V ($C_{18:1}$ Me-Fuc). Journal of Experimental Botany 2000;51: 2045-2051.

[46] López-Lara I, Drift K, Brussel A, Haverkamp J, Lugtenberg B, Thomas-Oates J, Spaink H. Induction of nodule primordia on *Phaseolus* and *Acacia* by lipo-chitin oligosaccharide nodulation signals from broad-host-range rhizobium strain GRH2. Plant Molecular Biology 1995;29: 465-477.

[47] Denarie J, Cullimore J. Lipo-oligosaccharide nodulation factors: a minireview new class of signaling molecules mediating recognition and morphogenesis. Cell 1993;74: 951-954.

[48] Schmidt J, Rohrig H, John M, Wieneke U, Stacey G, Koncz C, Schell J. Alteration of plant growth and development by *Rhizobium* nodA and nodB genes involved in the synthesis of oligosaccharide signal molecules. Plant Journal 1993;4: 651-658.

[49] Dyachok J, Tobin A, Price N, von Arnold S. Rhizobial Nod factors stimulate somatic embryo development in *Picea abies*. Plant Cell Reports 2000;19: 290-297.

[50] Dyachok J, Wiweger M, Kenne L, von Arnold S. Endogenous nod-factor-like signal molecules promote early somatic embryo development in Norway spruce. Plant Physiology 2002;128: 523-533.

[51] Cook D, Dreyer D, Bonnet D, Howell M, Nony E, VandenBosch K. Transient induction of a peroxidase gene in *Medicago truncatula* precedes infection by *Rhizobium meliloti*. The Plant Cell 1995;7: 43-55.

[52] Inui H, Yamaguchi Y, Hirano S. Elicitor actions of N-acetylchitooligosaccharides and laminarioligosaccharides for chitinase and L-phenylalanine ammonia-lyase induction in rice suspension culture. Bioscience, Biotechnology and Biochemistry 1997;61: 975-978.

[53] Zhang F, Smith DL. Interorganismal signaling in suboptimum environments: the legume–rhizobia symbiosis. Advances in Agronomy 2001;76: 125–161.

[54] Souleimanov A, Prithiviraj B, Smith DL. The major Nod factor of *Bradyrhizobium japonicum* promotes early growth of soybean and corn. Journal of Experimental Botany 2002a;53: 1929-1934.

[55] Souleimanov A, Prithiviraj B, Carlson RW, Jeyaretnam B, Smith DL. Isolation and characterization of the major nod factor of *Bradyrhizobium japonicum* strain 532C. Microbiological Research 2002b;157: 25-28.

[56] Khan W (2003) Signal compounds involved with plant perception and response to microbes alter plant physiological activities and growth of crop plants. PhD Thesis, McGill University.

[57] Chen C, McIver J, Yang Y, Bai Y, Schultz B, McIver A. Foliar application of lipo-chitooligosaccharides (Nod factors) to tomato (*Lycopersicon esculentum*) enhances flowering and fruit production. Canadian Journal of Plant Science 2007;87: 365-372.

[58] Oláh B, Brière C, Bécard G, Dénarié J, Gough C. Nod factors and a diffusible factor from arbuscular mycorrhizal fungi stimulate lateral root formation in *Medicago truncatula* via the DMI1/DMI2 signalling pathway. Plant Journal 2005;44: 195-207.

[59] Wang N, Khan W, Smith DL. Changes in soybean global gene expression after application of Lipo-chitooligosaccharide from *Bradyrhizobium japonicum* under sub-optimal temperature. PLoS ONE 2012;7(2): e31571.

[60] Sha Valli Khan PS, Hoffmann L, Renaut J, Hausman JF Current initiatives in proteomics for the analysis of plant salt tolerance, Current Science 2007;93: 6-11.

[61] Nanjo Y, Nouri M-Z, Komatsu S. Quantitative proteomic analyses of crop seedlings subjected to stress conditions; a commentary. Phytochemistry 2011;72: 1263-1272.

[62] Paoletti AC, Zybailov B, Washburn MP. Principles and applications of multidimensional protein identification technology. Expert Review of Proteomics 2004;1: 275-282.

[63] Delahunty C, Yates JR. Protein identification using 2D-LC-MS/MS. Methods 2005;35(3): 248–255.

[64] Herbert BR, Molloy MP, Gooley AA, Walsh BJ, Bryson WG, Williams KL. Improved protein solubility in two-dimensional electrophoresis using tributyl phosphine as reducing agent. Electrophoresis 1998;19(5):845-51.

[65] Molloy MP, Herbert BR, Walsh BJ, Tyler MI, Traini M, Sanchez JC, Hochstrasser DF, Williams KL, Gooley AA. Extraction of membrane proteins by differential solubilization for separation using two-dimensional gel electrophoresis. Electrophoresis 1998;19(5):837-44.

[66] Ferro M, Seigneurin-Berny D, Rolland N, Chapel A, Salvi D, Garin J, Joyard J. Organic solvent extraction as a versatile procedure to identify hydrophobic chloroplast membrane proteins. Electrophoresis 2000;21: 3517-3526.

[67] Cilia M, Fish T, Yang X, McLaughlin M, Thannhauser TW, Gray S. A comparison of protein extraction methods suitable for gel-based proteomic studies of aphid proteins. Journal of Biomolecular Techniques 2009;20(4): 201-215.

[68] Amalraj RS, Selvaraj N, Veluswamy GK, Ramanujan RP, Muthurajan R, Agrawal GK, Rakwal R, Viswanathan R. Sugarcane proteomics: establishment of a protein extraction method for 2-DE in stalk tissues and initiation of sugarcane proteome reference map. Electrophoresis 2010;31(12): 1959-1974.

[69] Dawe AL, Mu R, Rivera G, Salamon JA. Molecular methods for studying the Cryphonectria parasitica - hypovirus experimental system. Methods in Molecular Biology 2011;722, 225-236.

[70] Koay SY, Gam LH. Method development for analysis of proteins extracted from the leaves of *Orthosiphon aristatus*. Journal of Chromatography B. Analytical Technologies in The Biomedical and Life Sciences 2011;879: 2179-2183.

[71] Lowry OH, Rosebrough NJ, Farr AL, Randall RJ. Protein measurement with Folin phenol reagent. Journal of Biological Chemistry 1951;193: 265-275.

[72] Washburn MP, Ulaszek R, Deciu C, Schieltz DM, Yates JR. III. Analysis of quantitative proteomic data generated via multidimensional protein identification technology. Analytical Chemistry 2002;74: 1650-1657.

[73] Aebersold R, Mann M. Mass spectrometry-based proteomics. Nature 2003;422(6928): 198-207.

[74] Lill J. Proteomic tools for quantitation by mass spectrometry. Mass Spectrometry Reviews 2003;22(3): 182-194.

[75] Liu H, Sadygov RG, Yates JR. A model for random sampling and estimation of relative protein abundance in shotgun proteomics. Analytical Chemistry 2004;76: 4193-4201.

[76] Ahsan N, Komatsu S. Comparative analyses of the proteomes of leaves and flowers at various stages of development reveal organ specific functional differentiation of proteins in soybean. Proteomics 2009;9: 4889-4907.

[77] Arai Y, Hayashi M, Nishimura M. Proteomic identification and characterization of a novel peroxisomal adenine nucleotide transporter supplying ATP for fatty acid b-oxidation in soybean and arabidopsis. The Plant Cell 2008;20: 3227-3240.

[78] Djordjevic MA, Oakes M, Li DX, Hwang CH, Hocart CH, and Gresshoff PM. The glycine max xylem sap and apoplast proteome. Journal of Proteome Research 2007;6: 3771-3779.

[79] Khan NA, Takahashi R, Abe J, Komatsu S. Identification of cleistogamy-associated proteins in flower buds of near-isogenic lines of soybean by differential proteomic analysis. Peptides 2009;30: 2095-2102.

[80] Hajduch M, Ganapathy A, Stein JW, Thelen JJ. A systematic proteomic study of seed filling in soybean. Establishment of high-resolution two-dimensional reference maps, expression profiles, and an interactive proteome database. Plant Physiology 2005;137: 1397-1419.

[81] http://www.oilseedproteomics.missouri.edu/soybean.php

[82] Mooney BP, Thelen JJ. High-throughput peptide mass fingerprinting of soybean seed proteins: automated workflow and utility of UniGene expressed sequence tag databases for protein identification. Phytochemistry 2004;65: 1733-1744.

[83] Natarajan SS, Xu C, Caperna TJ, Garrett WM. Comparison of protein solubilization methods suitable for proteomic analysis of soybean seed proteins. Analytical Biochemistry 2005;342: 214-220.

[84] Natarajan SS, Xu C, Bae H, Caperna TJ, Garrett WM. Characterization of storage proteins in wild (*Glycine soja*) and cultivated (*Glycine max*) soybean seeds using proteomic analysis. Journal of Agriculture and Food Chemistry 2006;54: 3114-3120.

[85] Kim HT, Choi U-K, Ryu HS, Lee SJ, Kwon O-S. Mobilization of storage proteins in soybean seed (*Glycine max* L.) during germination and seedling growth. Biochimica et Biophysica Acta 2011;1814: 1178-1187.

[86] Schmidt MA, Barbazuk WB, Sandford M, May G, Song Z, Zhou W, Nikolau BJ, Herman EM. Silencing of soybean seed storage proteins results in a rebalanced protein composition preserving seed protein content without major collateral changes in the metabolome and transcriptome. Plant Physiology 2011; 156: 330-345.

[87] Brandao AR, Barbosa HS, Arruda MAZ. Image analysis of two-dimensional gel electrophoresis for comparative proteomics of transgenic and non-transgenic soybean seeds. Journal of proteomics 2010;73: 1433-1440.

[88] Barbosa HS, Arruda SCC, Azevedo RA, Arruda MAZ. New insights on proteomics of transgenic soybean seeds: evaluation of differential expressions of enzymes and proteins. Analytical and Bioanalytical Chemistry 2012;402: 299-314.

[89] Krishnan HB. Evidence for accumulation of the β-subunit of β-conglycinin in soybean [*Glycine max* (L) Merr.] embryonic axes. Plant Cell report 2002;20: 869-875.

[90] Krishnan HB, Nelson RL. Proteomic analysis of high protein soybean (*Glycine max*) accessions demonstrates the contribution of novel glycinin subunits. Journal of Agriculture and Food Chemistry 2011;59: 2432-2439.

[91] Mathesius U, Djordjevic MA, Oakes M, Goffard N, Haerizadeh F, Weiller GF, Singh MB, Bhalla PL. Comparative proteomic profiles of the soybean (*Glycine max*) root apex and differentiated root zone. Proteomics 2011;11: 1707-1719.

[92] Brechenmacher L, Lee J, Sachdev S, Song Z, Nguyen THN, Joshi T, Oehrle N, Libault M, Mooney B, Xu D, Cooper B, Stacey G. Establishment of a protein reference map for soybean root hair cells. Plant Physiology 2009;149: 670-682.

[93] Toorchi M, Yukawa K, Nouri M-Z, Komatsu S. Proteomics approach for identifying osmotic-stress-related proteins in soybean roots. Peptides 2009;30: 2108-2117.

[94] Wan J, Torres M, Ganapathy A, Thelen J, DaGue BB, Mooney B, Xu D, Gary Stacey. Proteomic analysis of soybean root hairs after infection by *Bradyrhizobium japonicum*. Molecular Plant Microbe Interactions 2005;18: 458–467.

[95] Oerhle NW, Sarma AD, Waters JK, Emerich DW. Proteomic analysis of soybean nodule cytosol. Phytochemistry 2008;69: 2426-2438.

[96] Panter S, Thomson R, de Bruxelles G, Laver D, Trevaskis B, Udvardi M. Identification with proteomics of novel proteins associated with the peribacteroid membrane of soybean root nodules. Molecular Plant Microbe Interactions 2000;13: 325-333.

[97] Komatsu S, Wada T, Abale´a Y, Nouri M-Z, Nanjo Y, Nakayama N, Shimamura S, Yamamoto R, Nakamura T, Furukawa K. Analysis of plasma membrane proteome in soybean and application to flooding stress response. Journal of Proteome Research 2009a;8: 4487-4499.

[98] Nanjo Y, Skultety L, Uva´c◎kova´ L, Klubicova´ K, Hajduch M, Komatsu S. Mass spectrometry-based analysis of proteomic changes in the root tips of flooded soybean seedlings. Journal of Proteome Research 2012;11: 372-385.

[99] Komatsu S, Yamamoto R, Nanjo Y, Mikami Y, Yunokawa H, Sakata K. A comprehensive analysis of the soybean genes and proteins expressed under flooding stress using transcriptome and proteome techniques. Journal of Proteome Research 2009b;8: 4766-4778.

[100] Nanjo Y, Skultety L, Ashraf Y, Komatsu S. Comparative proteomic analysis of early-stage soybean seedlings responses to flooding by using gel and gel-free techniques. Journal of Proteome Research 2010;9: 3989-4002.

[101] Alam I, Lee D-G, Kim K-H, Park C-H, Sharmin S A, Lee H, Oh K-W, Yun B-W, Lee B-H. Proteome analysis of soybean roots under water logging stress at an early vegetative stage. Journal of Biosciences 2010;35: 49-62.

[102] Komatsu S, Yamamoto A, Nakamura T, Nouri M-Z, Nanjo Y, Nishizawa K, Furukawa K. Comprehensive analysis of mitochondria in roots and hypocotyls of soybean

under flooding stress using proteomics and metabolomics techniques. Journal of Proteome Research 2011;10: 3993-4004.

[103] Hashiguchi A, Sakata K, Komatsu S. Proteome analysis of early-stage soybean seedlings under flooding stress. Journal of Proteome Research 2009;8: 2058-2069.

[104] Yamaguchi M, Valliyodan B, Zhang J, Lenoble ME, Yu O, Rogers EE, Nguyen HT, Sharp RE. Regulation of growth response to water stress in the soybean primary root. I. Proteomic analysis reveals region-specific regulation of phenylpropanoid metabolism and control of free iron in the elongation zone. Plant, Cell and Environment 2010;33: 223-243.

[105] Nouri M-Z, Komatsu S. Comparative analysis of soybean plasma membrane proteins under osmotic stress using gel-based and LC MS/MS-based proteomics approaches. Proteomics 2010;10: 1930-1945.

[106] Mohammadi PP, Moieni A, Hiraga S, Komatsu S. Organ specific proteomic analysis of drought-stressed soybean seedlings. Journal of proteomics 2012;75: 1906-1923.

[107] Ahsan N, Donnart T, Nouri M-Z, Komatsu S. Tissue-specific defense and thermo-adaptive mechanisms of soybean seedlings under heat stress revealed by proteomic approach. Journal of Proteome Research 2010;9: 4189-4204.

[108] Yan L. Effect of salt stress on seed germination and seedling growth of three salinity plants. Pakistan Journal of Biological Sciences 2008;11: 1268-1272.

[109] Parida AK, Das AB. Salt tolerance and salinity effects on plants: a review. Exotoxicology and Environmental Safety 2005;60:324-49.

[110] Sobhanian H, Aghaei K, Komatsu S. Changes in the plant proteome resulting from salt stress: Towards the creation of salt tolerant crops? Journal of proteomics 2011;74: 1323-1337.

[111] Agri-Facts: Salt tolerance of Plants: http://www1.agric.gov.ab.ca Nov 2001(last accessed 12th July 2012).

[112] Dabuxilatu MI, Ikeda M. Distribution of K, Na and Cl in root and leaf cells of soybean and cucumber plants grown under salinity conditions. Soil Science Plant Nutrition 2005;51:1053-7.

[113] Aghaei K, Ehsanpour AA, Shah AH, Komatsu S. Proteome analysis of soybean hypocotyl and root under salt stress. Amino Acids 2009;36: 91-98.

[114] Sobhanian H, Razavizadeh R, nanjo Y, Ehsanpour AA, Rastgar Jazii F, Motamed N, Komatsu S. Proteome analysis of soybean leaves, hypocotyls and roots under salt stress. Proteome Science 2010; 8: 1-15.

[115] Xu X-Y, Fan R, Zheng R, Li C-M, Yu D-Y. Proteomic analysis of seed germination under salt stress in soybeans. Journal Zhejiang Univ-Sci B (Biomedicine and Biotechnology) 2011;12: 507-517.

[116] Zhang YM, Zhao JM, Xiang Y, Bian XC, Zuo QM, Shen Q, Gai JY, Xing H. Proteomics study of changes in soybean lines resistant and sensitive to *Phytophthora sojae*. Proteome Science 2011;9: 52 – doi:10.1186/1477-5956-9-52

[117] Yang H, Huang Y, Zhi H, Yu D. Proteomics-based analysis of novel genes involved in response toward soybean mosaic virus infection. Molecular Biology Reports 2011;38:511-521.

[118] Zhena Y, Qia J-L, Wanga S-S, Sua J, Xua G-H, Zhanga M-S, Miaoa L, Pengb X-X, Tiana D and Yanga Y-H. Comparative proteome analysis of differentially expressed proteins induced by Al toxicity in soybean. Physiologia Plantarum 2007;131: 542-554.

[119] Vance CP. Symbiotic nitrogen fixation and phosphorus acquisition: plant nutrition in a world of declining renewable resources. Plant Physiology 2001;127: 390-397.

[120] Chen Z, Cui Q, Liang C, Sun L, Tian J, Liao H. Identification of differentially expressed proteins in soybean nodules under phosphorus deficiency through proteomic analysis. Proteomics 2011;11: 4648-4659.

[121] Houston NL, Lee D-G, Stevenson SE, Ladics GS, Bannon GA, McClain S, Privalle L, Stagg N, Herouet-Guicheney C, MacIntosh SC, Thelen JJ. Quantitation of soybean allergens using tandem mass spectrometry. Journal of Proteome Research 2011;10: 763-773.

[122] Galant A, Koester RP, Ainsworth EA, Hicks LM, Jez JM. From climate change to molecular response: redox proteomics of ozone-induced responses in soybean. New Phytologist 2012;194: 220-229.

[123] Joshi T, Patil K, Fitzpatrick MR, Franklin LD, Yao Q, Jeffrey R Cook JR, Wang Z, Libault M, Brechenmacher L, Valliyodan B, Wu X, Cheng J, Stacey G, Nguyen HT, Xu D. Soybean Knowledge Base (SoyKB): a web resource for soybean translational genomics. BMC Genomics 2012; 13 (Suppl 1):S15.

[124] Sarma AD, Emerich DW. A comparative proteomic evaluation of culture grown vs nodule isolated *Bradyrhizobium japonicum*. Proteomics 2006;6(10):3008-28.

[125] http://www.kazusa.or.jp/index.html (last accessed 12th July 2012).

[126] Sarma AD, Emerich DW. Global protein expression pattern of *Bradyrhizobium japonicum* bacteroids: a prelude to functional proteomics. Proteomics 2005;5(16):4170-84.

[127] Jun L, Wen-Li X, Ming-chao MA, Da-wei G, Xin J, Feng-ming C, Delong S, Hui-jun C, Li L. Proteomic study on two *Bradyrhizobium japonicum* strains with different competitivenesses for nodulation. Journal of Integrative Agriculture 2011;10(7):1072-1079.

[128] da Silva Batista JS, Hungria M. Proteomics reveals differential expression of proteins related to a variety of metabolic pathways by genistein-induced *Bradyrhizobium japonicum* strains. Journal of Proteomics. 2012;75(4):1211-9. Epub 2011 Nov 16.

[129] Süss C, Hempel J, Zehner S, Krause A, Patschkowski T, Göttfert M. Identification of genistein-inducible and type III-secreted proteins of *Bradyrhizobium japonicum*. Journal of Biotechnology 2006;126(1):69-77.

[130] Hempel J, Zehner S, Göttfert M, Patschkowski T. Analysis of the secretome of the soybean symbiont *Bradyrhizobium japonicum*. Journal of Biotechnology 2009;10:140(1-2):51-8.

[131] Barcellos FG, Batista JS, Menna P, Hungria M. Genetic differences between *Bradyrhizobium japonicum* variant strains contrasting in N(2)-fixation efficiency revealed by representational difference analysis. Archives in Microbiology 2009;191(2):113-22.

[132] Godoy, L. P., Vasconcelos, A. T. R., Chueire, L. M. O., Souza, R. C. *et al.*, Genomic panorama of *Bradyrhizobium japonicum* CPAC 15, a commercial inoculant strain largely established in Brazilian soils and belonging to the same serogroup as USDA 123. *Soil Biol. Biochem.* 2008; 40: 2743–2753.

[133] Batista JS, Torres AR, Hungria M. Towards a two-dimensional proteomic reference map of *Bradyrhizobium japonicum* CPAC 15: spotlighting "hypothetical proteins". Proteomics 2010;10(17):3176-89.

[134] Delmotte N, Ahrens CH, Knief C, Qeli E, Koch M, Fischer HM, Vorholt JA, Hennecke H, Pessi G. An integrated proteomics and transcriptomics reference data set provides new insights into the *Bradyrhizobium japonicum* bacteroid metabolism in soybean root nodules. Proteomics 2010;10(7):1391-400.

[135] Gray EJ, Smith DL. Intracellular and extracellular PGPR: commonalities and distinctions in the plant-bacterium signaling processes. Soil Biology and Biochemistry 2005;37: 395-412.

[136] Kinross JM, von Roon AC, Holmes E, Darzi A and Nicholson JK 2008. The human gut microbiome: Implications for future health care. Current Gastroenterology Reports 10:396-403.

[137] Tagg J, Daiani A, Wannamaker L. Bacteriocins of gram-positive bacteria. Bacteriological Reviews 1976;40: 722-756.

[138] Pugsley AP. The ins and outs of colicins. II. Lethal action, immunity and ecological implications. Microbiological Sciences 1984;1: 203-205.

[139] Jack RW, Tagg J, Ray B. Bacteriocins of gram positive bacteria. Microbiological Reviews 1995;59: 171-200.

[140] Klaenhammer T. Genetics of bacteriocins produced by lactic acid bacteria. FEMS Microbiology Letters 1993;12: 39-86.

[141] Bai Y, Souleimanov A, Smith DL. An inducible activator produced by a *Serratia proteamaculans* strain and its soybean growth-promoting activity under greenhouse conditions. Journal of Experimental Botany 2002b;373:1495-1502.

[142] Bai Y, Zhou X, Smith DL Enhanced soybean plant growth resulting from co-inoculation of *Bacillus* strains with *Bradyrhizobium japonicum*. Crop Science 2003;43: 1774-1781.

[143] Gray EJ, Lee K, Di Falco M, Souleimanov A, Zhou X, Smith DL. A novel bacteriocin, thuricin 17, produced by PGPR strain *Bacillus thuringiensis* NEB17: isolation and classification. Journal of Applied Microbiology 2006b;100: 545-554.

[144] Gray EJ, Di Falco M, Souleimanov A, Smith DL. Proteomic analysis of the bacteriocin, thuricin-17 produced by *Bacillus thuringiensis* NEB17. FEMS Microbiology Letters 2006a;255: 27-32.

[145] Lee K, Gray EJ, Mabood F, Jung W, Charles T, Clark SRD, Ly A, Souleimanov A, Zhou X, Smith DL. The class IId bacteriocin thuricin-17 increases plant growth. Planta 2009;229: 747-755.

[146] Jung W, Mabood F, Souleimanov A, Zhou X, Jaoua S, Kamoun F, Smith DL. Stability and antibacterial activity of bacteriocins produced by *Bacillus thuringiensis* and *Bacillus thuringiensis* ssp. *Kurstaki*. Journal of Microbiology and Biotechnology 2008;18: 1836-1840.

Chapter 2

The Development and Regulation of Soybean Nodules

Brett James Ferguson

Additional information is available at the end of the chapter

http://dx.doi.org/10. 5772/52573

1. Introduction

1.1. Legumes: Environmentally and agronomically important plant species

Legumes represent the third largest family of angiosperms, with > 18,000 species worldwide (*Leguminosae* or *Fabaceae*). Due to their high nutritional value, they have been cultivated by many cultures for use in agriculture. Indeed, legumes currently represent some of the most important food, feed and fuel crops grown around the world, second only to cereals. They have a global-production value of > $200 billion per year and are cultivated on 12–15% of the world's available, arable land (Graham and Vance 2003; Peoples et al., 2009; Jensen et al., 2012). Collectively, ~247 million tons of legumes are produced each year, and represent > 25% of the world's primary crop production (Graham and Vance 2003; European Association for Grain Legume Research 2009). Some key legume crop species include: soybean, bean, pea, chickpea, cowpea, pigionpea and alfalfa.

Legumes play a significant role in both agriculture and in natural ecosystems by having a unique influence on the nitrogen cycle. They do so via a highly-specialized symbiotic relationship with soil bacteria commonly called rhizobia. Via a sophisticated signaling exchange, the bacteria infect the plant root and induce the formation of novel organs, termed nodules (Ferguson et al., 2010). The nodule is colonized by the bacteria and provides them with an ideal habitat to convert or 'fix' atmospheric di-nitrogen into other nitrogen-based compounds, such as ammonia, which can be used by the plant. This benefits the plant directly, as nitrogen is essential for growth. It also benefits the rhizosphere as the soil nitrogen content is replenished once the plant dies and subsequently decomposes (referred to as 'green manure'). As a result, legumes are often used in crop rotations in agriculture. In addition, some legumes are being exploited as a source of sustainable biofuel. Species such as soybean, and the tree legume *Pongamia pinnata*, offer the most promise due to their high seed oil content (Scott et al., 2008).

2. Nitrogen fertilizer: An unsustainable reliance in modern agriculture

Current agriculture practices are heavily reliant on nitrogen-based fertilizers to achieve high yields (Peoples *et al.*, 2009; Jensen et al., 2012). Indeed, it is estimated that half of the world's population is fed as a direct result of nitrogen fertilizer use (Erisman et al., 2008). Although this approach works well to maximize crop production, it is a very risky dependence that is not sustainable, and is inefficient, expensive, and often pollutes. Hence, there are a number of very sound reasons to lessen these nitrogen inputs and alleviate the current reliance.

The production of industrial nitrogen fertilizers requires a great deal of fossil fuel, involving what is known as the Haber-Bosch process. Once manufactured, additional fossil fuel is needed to transport the fertilizer to its place of sale, and then even more fossil fuel is needed to apply it in the field. Collectively, this accounts for a staggering 50% of fossil fuel use in modern agriculture and 5% of the world's annual consumption of natural gas (Crutzen et al., 2007; Canfield et al., 2010). As a direct result, the purchase and application of nitrogen fertilizer can be quite expensive, preventing many farmers in developing regions of the world from being able to use it. Moreover, as the cost of fossil fuel continues to rise, so too does the cost of nitrogen fertilizer. This can considerably cut into farmer profits and often leads to costs being passed on to the consumer.

In addition to the monetary costs associated with nitrogen fertilizer use, there are often considerable environmental costs. It is reported that industrial manipulation of the nitrogen cycle has already exceeded safe boundaries for global environmental change (Rockstrøm et al., 2009; Canfield et al., 2010; Charpentier and Oldroyd 2010; Beatty and Good 2011). Due to the combustion of fossil fuel, the production and application of nitrogen fertilizer results in large quantities of carbon dioxide (CO_2) being released into the atmosphere. This contributes to the greenhouse effect, as does the release of nitrous oxide (N_2O), which is produced during the decomposition of nitrogen fertilizer in the soil and is estimated to be 292 times more active as a greenhouse gas than CO_2 (Crutzen et al., 2007).

Each year >100 million tonnes of nitrogen are applied to crops worldwide (Glass 2003). This excessive use of nitrogen reduces the levels of trace nutrients in the soil and increases soil acidity. Moreover, the application of nitrogen-based fertilizers is a largely inefficient process, with as much as 30 - 50% lost to leaching. This nitrogen run-off can cause significant environmental damage, including the eutrophication of waterways resulting from associated algal blooms, *etc.* (Vance 2001). Recent reports indicate that the cost to fix this environmental damage will significantly outweigh the economic benefit gained from nitrogen fertilizer use (Sutton et al., 2011).

The excess runoff from nitrogen fertilizer can also contaminate drinking water. High levels of nitrogen in drinking water can lead to methemoglobinaemia, commonly referred to as

Blue-baby syndrome because of the blue-grey skin colour displayed by affected infants (Murphy 1991; Knobeloch et al., 2000). Blue-baby syndrome is potentially fatal and occurs when the hemoglobin of an infant's red blood cells is oxidized to methemoglobin, which is unable to properly transport oxygen.

3. Biological nitrogen fixation: A sound alternative to nitrogen fertilizer

Attempts to reduce the use of nitrogen fertilizers need to incorporate responsible, cost effective and environmentally sound options; an improved use of legumes and an increased understanding of legume nodulation represent such options. Indeed, this symbiotic relationship between legumes and rhizobia represents the most important nitrogen-fixation association in the world, with an annual production of approximately 200 million tons of nitrogen (Peoples et al., 2009; Jensen *et al.,* 2012). Optimising this symbiosis can increase crop yields and enhance soil fertility, whilst reducing the negative monetary costs and environmental impacts associated with nitrogen fertilizer use (Hirel et al., 2007; Peoples et al., 2009; Canfield et al., 2010). Hence, increasing our use of legume crops and identifying critical factors required to control nodulation are seen as pivotal steps towards reducing our reliance on nitrogen fertilizers and improving agricultural sustainability (*e. g.* , Giller and Cadisch 1995; Vance 2001; Peoples et al., 2009; Rockstrøm *et al.,* 2009; Canfield et al., 2010; Jensen et al., 2012).

Soybean alone is estimated to produce up to 200 kg N ha^{-1} in aboveground biomass in a single growing season. Of the soybean nitrogen content, 58-68% is estimated to be derived from symbiotic nitrogen fixation (Salvagiotti et al., 2008; Peoples et al., 2009; Jensen et al., 2012). Following harvesting, the remaining portions of the plant, including roots and nodules which represent 30-60% of the nitrogen content, are left to replenish the nitrogen content of the surrounding soil (Mahieu et al., 2007; McNeill and Fillery 2008).

4. Legume nodules: The perfect environment for rhizobia nitrogen fixation

Nodulation is a complex process orchestrated by a multitude of bacteria and plant signals (reviewed in Ferguson and Mathesius 2003; Ferguson et al., 2010). The process is initiated by plant roots secreting flavonoid molecules into the soil. This attracts compatible rhizobia and concomitantly stimulates them to synthesize a highly specific signal molecule called Nod factor. The rhizobia strain that is compatible with soybean is *Bradyrhizobium japonicum*. The plant perceives Nod factor via LysM receptors on the root. In soybean, these receptors are called GmNFR1 and GmNFR5 (Indrasumunar et al., 2010, 2011). Nod factor perception triggers a subsequent signaling cascade that is required for proper nodule establishment. The known legume and rhizobia genes/signals that are involved in this signaling cascade have recently been thoroughly reviewed in Ferguson (2012).

The presence of the rhizobia together with their Nod factor signal molecule initiates the nodulation infection process. Root hair penetration is the most common form of rhizobia invasion. The bacteria attach to emerging root hairs, which begin to deform and eventually encapsulate some of the bacteria, which are continuously dividing (Callaham and Torrey 1981; Turgeon and Bauer 1985). This process happens in as little as 6 – 8 h post-inoculation (Yao and Vincent 1969; Bhuvaneswari et al., 1981; Bhuvaneswari and Solheim 1985; Turgeon and Bauer 1982, 1985). Specialized structures, called infection threads, begin to form and provide a passage way for the bacteria to enter the root (reviewed by Gage 2004). These infection threads are predominately comprised of plant cell wall components and they permit the bacteria to continue proliferating within the host plant.

As the process of rhizobia infection occurs, cortical cells in the root begin to divide and eventually give rise to the nodule primordium (Calvert et al., 1984; Mathews et al., 1989). The position of the nodule primordium is typically adjacent to the radial cells of the xylem, and away from the phloem. This positioning is thought to be largely dependent on plant hormone levels, namely gradients of the gaseous hormone, ethylene (Heidstra et al., 1997; Gresshoff et al., 2009; Lohar *et al.,* 2009). Additional tissues, including vascular tissues and central nodule tissues that are composed of both invaded and non-invaded cells, also develop to form the nodule structure (Newcomb et al., 1979; Calvert et al., 1984; Ferguson and Reid 2005).

Infection threads initiating in the root hair eventually grow and extend towards the dividing nodule primordium located in the root cortex. Once there, rhizobia located at the tip of the infection threads are released into an infection droplet that separates and is released into the cytoplasm of the host cell. Within the cytoplasm, the rhizobia are encapsulated by a specialized plant-derived membrane, known as the peribacteroid membrane, making what is commonly referred to as the symbiosome (Udvardi and Day 1997).

Ultimately, the dividing bacteria differentiate into what are known as bacteroids, which are highly specialized and whose main purpose is to fix atmospheric di-nitrogen gas. Inside the mature nodule, the bacteroids use a nitrogenase enzyme complex to fix the di-nitrogen into forms of nitrogen that the plant can use, such as ammonia. The ammonia, which is toxic to the plant, is then quickly converted into compounds such as glutamate or ureides that are non-toxic and are safely transported throughout the plant. Legume nodules provide the ideal setting for this process as they establish a peripheral oxygen barrier, via physical and metabolic barriers, to create a low-oxygen environment that is essential for nitrogenase activity to occur.

The nodules formed on the roots of soybean plants are referred to as 'determinate' nodules. They are spherical and lack a persistent meristem, unlike indeterminate nodule structures that can form on other legume species, particularly those from temperate growing regions (Ferguson et al., 2010). The life-span of a soybean nodule is typically a few weeks, after which they senesce and are replaced by new nodule structures developing on the growing root system. Following nodule senescence, the bacteroids can re-differentiate and become new inoculum for the soil (Gresshoff and Rolfe 1978).

http://dx.doi.org/10.5772/52573

5. Autoregulation of nodulation: Too much of a good thing can be bad

A number of genes that are required for proper nodule formation have been elucidated (reviewed in Caetano-Anollés and Gresshoff 1991; Ferguson et al., 2010; Ferguson, 2012). The loss of any of these genes typically results in a reduced, or a complete lack of, nodule development. In addition to these positive regulators of nodule formation, there are also a number of external and internal factors that act as negative regulators of nodulation. Mutants unable to synthesise or perceive these factors exhibit increased nodule numbers, often referred to as a hyper- or super-nodulation phenotype (Figure 1). Many of these factors function in the Autoregulation of Nodulation (AON) pathway, which is a mechanism used by the host plant to regulate its nodule numbers (reviewed in Reid et al., 2011a). Indeed, less than 10% of all rhizobia infection events result in the establishment of a fully functional nodule, largely due to AON. By controlling nodule development in this way, the host plant can balance its need to acquire nitrogen against its ability to expend energy establishing and maintaining nodules. Supernodulating plants lacking AON are typically developmentally-stunted (when inoculated with a compatible rhizobia strain) and yield poorly as a result of this balance being disrupted (Figure 2).

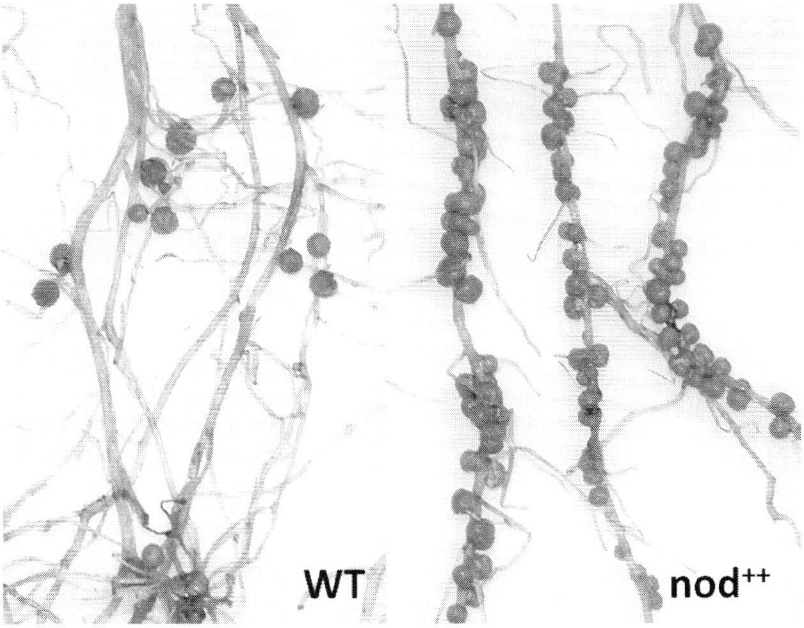

Figure 1. Roots of wild-type (WT) and supernodulating mutant (nod^{++}) soybean plants exhibiting mature nodule structures as a result of a symbiotic interaction with *Bradyrhizobium japonicum*.

Figure 2. Soybean plants growing in a field in Toowoomba, Queensland, Australia. Mutants unable to form nodules (nod^-) are stunted and pale compared with wild-type (WT) plants due to their inability to establish a symbiotic interaction with nitrogen-fixing *Bradyrhizobium japonicum*. Supernodulating mutants (nod^{++}) are significantly stunted in stature as a result of investing too much energy into forming nodule structures.

The AON pathway involves long-distance root-shoot signaling initiated during nodule development by the synthesis of a root-derived signal (Gresshoff and Delves, 1986; Delves et al, 1986; Reid et al., 2011a). Recent work has indicated that this signal is likely a CLV3/ESR-related (CLE) peptide(s) hormone (Okamoto et al., 2009; Mortier et al., 2010; Reid et al., 2011b; Lim et al., 2011). In soybean, these CLE peptides are called Rhizobia Induced CLE1 (RIC1) and RIC2 (Reid et al 2011b; Lim et al., 2011). Grafting and over-expression experiments have shown that these signals travel to the shoot (Delves et al., 1986; Reid et al., 2011b), likely via the xylem, where they, or a product of their action, are perceived by a LRR receptor kinase, called the Nodulation Autoregulation Receptor Kinase (NARK) in soybean (*e. g.* , Searle et al., 2003). NARK may act in a complex with other receptors, such as CLAVATA2 and KLAVIER (Miyazawa et al., 2010; Krusell et al., 2011). This perception results in the production of a novel Shoot-Derived Inhibitor (SDI). The SDI signal subsequently travels from the shoot back down to the roots, likely via the phloem, where it acts to inhibit further nodulation events (Reid et al 2011a). It has recently been established in soybean that SDI is small (<1 kDa), Nod factor- and NARK-dependent, heat stable, and is unlikely to be an RNA or a protein (Lin et al., 2010, 2011a). Recent work using soybean has also identified a number of novel components that may interact with NARK directly, or that may function downstream of NARK to regulate the AON process. These include genes identified using site-directed mutagenesis that encode Kinase-Associated Protein Phosphatases, called *GmKAPP1* and *GmKAPP2* (Miyahara et al., 2008), and genes identified using complete transcriptome analyses (RNAseq), such as the putative Ubiquitin Fusion Degradation protein, *GmUFD1a* (Reid et al., 2012).

Additional genes and factors also regulate nodule numbers. Root-specific genes in pea (*PsNOD3*; Postma et al., 1988) and *L. Japonicas* (*LjRDH1*, Ishikawa et al., 2008; *LjTML*, Magori et al., 2009) may have a role in the biosynthesis or translocation of RIC1 and RIC2, or in the perception of SDI. However, the identity of these genes is not yet known. Loss of function of *LjASTRAY*, which encodes a bZIP transcription factor (Nishimura et al., 2002), or *MtEFD*, which encodes an ERF transcription factor (Vernié et al., 2008), also results in increased nodule numbers. Whether these genes function in the AON pathway remains to be determined.

Ethylene and nitrate are also known to inhibit nodule development (Carroll et al., 1985; Lorteau et al., 2001; Ferguson et al., 2005 a,b; Ferguson et al., 2011). Mutations to ethylene sensitivity or response genes, such as *LjETR1* and *LjEIN2/MtEIN2*, result in increased nodule formation (Penmetsa et al., 2008; Gresshoff et al., 2009; Lohar et al., 2009). Interestingly, an additional CLE peptide identified in soybean that negatively regulates nodule development, called Nitrogen-Induced CLE1 (NIC1) is highly similar to RIC1 and RIC2, but is induced by nitrate, not rhizobia (Reid et al., 2011b). Both the RICs and NIC1 appear to be perceived by GmNARK, only the nitrate-induced CLE exhibits little-to-no mobility and is perceived in the root, whereas the rhizobia-induced CLE undergoes long distance transport and is perceived in the shoot.

6. Soybean: A model species for legume research

Soybean has been the subject of a great deal of research in an effort to identify unique traits and to isolate superior cultivars offering improved growth and yields (Gresshoff 2012). It also represents an excellent model species for legumes in general (Ferguson and Gresshoff 2009), with outcomes frequently extrapolated to the other important food and feed legume crops, such as bean, pea, chickpea, faba bean, lentil, peanut, clover and lucerne (*e. g.* , Rispail et al., 2010).

Soybean represents one of the best characterized legumes species, both physiologically and biochemically. It grows quickly, is high yielding, and has a size and stature that are well suited for most field and laboratory studies (Figure 3). Its relatively large size enables the harvest of large quantities of tissues and it is ideal for studies involving grafting (*e. g.* , Delves et al., 1986; Reid et al., 2011b), xylem sap analyses (*e. g.* , Djordjevic et al., 2007), *Agrobacterium rhizogenes*-mediated transformation for gene over-expression and RNA interference (*e. g.* , Kereszt et al., 2007; Reid et al., 2011b; Lin et al., 2011b) and Virus-Induced Gene

Silencing (VIGS) for functional genomics approaches (*e. g.* , Zhang and Ghabrial, 2006). Furthermore, soybean has a large germplasm, including vast mutant (Figure 2; *e. g.* , Carroll et al., 1985; Bolon et al., 2011) and TILLING populations (*e. g.* , Cooper et al., 2008; Batley et al., 2012).

Figure 3. Glasshouse grown soybean plants 1 and 3 weeks after germination. The fast, uniform growth of soybean, together with the availability of its genome sequence and its amenability to most physiological, molecular and biochemical analyses makes it an ideal model species for legume research.

http://dx.doi.org/10. 5772/52573

Recently, the soybean genome was sequenced (Schmutz et al., 2010) and complete transcriptome analyses have been performed, including the generation of transcriptome atlases (Libault et al., 2010a,b; Severin et al., 2010; Reid et al., 2012; Hayashi et al., 2012). Together, these resources provide an efficient non-targeted tool to identify new genes and patterns of gene expression. Analyses between the genome of soybean and those available for other legumes species, including *M. truncatula*, *L. japonicus*, pigionpea and bean, also provide an excellent opportunity for comparative legume genomics (Cannon et al., 2009).

Understanding the genes and genomes of legumes will help to establish elite cultivars that benefit sustainable farming practices. Integrating central regulators of nodulation is essential for such targeted legume crop improvement. Indeed, outcomes derived via soybean research could help to underpin future advances in managing the legume-rhizobia symbiosis (*e. g.* , Rispail et al., 2010). This could lead to improved nitrogen use efficiency and reduced nitrogen-fertilizer inputs, thus helping to reduce the monetary and environmental costs associated with nitrogen-fertilizer use.

7. Conclusions and future perspectives

The processes of nodulation and nitrogen fixation have long been recognized for their agricultural benefits and ability to improve soil-health. Recent work identifying the genes and signals involved in nodule formation and AON have significantly enhanced our understanding of the processes and could provide targets for plant breeding and engineering programs aiming to develop legume, and perhaps even non-legume, varieties with an improved ability to acquire nitrogen.

Abbreviations

AON (Autoregulation Of Nodulation);

CLE (CLV3/ESR-related peptide);

NARK (Nodulation Autoregulation Receptor Kinase);

NIC (Nitrogen-Induced CLE peptide);

Nod Factor (Nodulation Factor);

RIC (Rhizobia-Induced CLE peptide);

RNAseq (RNA sequencing);

SDI (Shoot Derived Inhibitor)

Acknowledgements

Thanks to Prof. Peter M. Gresshoff for his helpful discussions and kind advice. The Australian Research Council, The University of Queensland and the Centre of Excellence for Integrative Legume Research (CEO348212) are thanked for financial assistance.

Author details

Brett James Ferguson

Centre of Excellence for Integrative Legume Research, School of Agriculture & Food Sciences, The University of Queensland, St Lucia, Brisbane, Australia

References

[1] Batley J, Liu S, Mirzaei S, El-Mellouki T, Ferguson BJ, Gresshoff PM, Meksem K (2012) TILLING in soybean reveals contrasting mutant phenotypes of LRR-receptor kinase genes *GmNARK* and *GmCLAVATA1A*. BMC Genomics, submitted.

[2] Beatty PH, Good AG (2011) Future prospects for cereals that fix nitrogen. Science 333: 416-417.

[3] Bolon Y-T, Haun W, Xu W, David Grant, Stacey MG, Nelson RT, Gerhardt DJ, Jeddeloh JA, Stacey G, Muehlbauer GJ, Orf JH, Naeve S, Stupar RM, Vance CP (2011) Phenotypic and genomic analyses of a fast neutron mutant population resource in soybean. Plant Physiology 156: 240–253.

[4] Bhuvaneswari T V, Bhagwat AA, Bauer WD (1981) Transient susceptibility of root cells in four common legumes to nodulation by rhizobia. Plant Physiology 68: 1144-1149.

[5] Bhuvaneswari TV, Solheim B (1985) Root hair deformation in the white clover/*Rhizobium trifolii* symbiosis. PhysiologiaPlantarum 63: 25-34.

[6] Caetano-AnollésG, Gresshoff PM (1991) Plant genetic control of nodulation. Annual Review of Microbiology45: 345–382.

[7] Cannon SB, May GD, Jackson SA (2009) Three sequenced legume genomes and many crop species: rich opportunities for translational genomics. Plant Physiology 151: 970-977.

[8] Callaham DA, Torrey JG (1981) The structural basis for infection of root hairs of *Trifolium repens* by *Rhizobium*. Canadian Journal of Botany 59: 1647– 1664.

[9] Calvert, H.E., Pence, M.K., Pierce, M., Malik, N.S.A. and Bauer, W.D. (1984) Anatomical analysis of the development and distribution of *Rhizobium* infections in soybean roots. Canadian Journal of Botany 62: 2375-2384.

[10] Canfield DE, Glazer AN, Falkowski PG (2010) The evolution and future of earth's nitrogen cycle. Science 330: 192-196.

[11] Charpentier M, Oldroyd G (2010) How close are we to nitrogen-fixing cereals? Current Opinion in Plant Biology 13: 556-564.

[12] Carroll BJ, McNeil DL, Gresshoff PM (1985) Isolation and properties of soybean [*Glycine max* (L.) Merr.] mutants that nodulate in the presence of high nitrate concentrations. Proceeding of the National Academy of Science USA 82: 4162–4166.

[13] Cooper JL, Till BJ, Laport RG, Darlow MC, Kleffner JM, Jamai A, El-Mellouki T, Liu S, Ritchie R, Nielsen N, Bilyeu KD, Meksem K, Comai L, Henikoff S(2008) TILLING to detect induced mutations in soybean. BMC Plant Biology 8: 9.

[14] Crutzen P, Mosier AR, Smith KA, Winiwarter W (2007) N2O release from agro-fuel production negates global warming reduction by replacing fossil fuels. Atmos Chem Phys Discussions 7: 11191-11205.

[15] Delves, A.C., Mathews, A., Day, D.A., Carter, A.S., Carroll, B.J. and Gresshoff, P.M. (1986) Regulation of the soybean-*Rhizobium* nodule symbiosis by shoot and root factors. Plant Physiology 82: 588-590.

[16] Djordjevic MA, Oakes M, Li D, Hwang CH, Hocart CH, Gresshoff PM (2007) The *Glycine max* xylem sap and apoplast proteome. Journal of Proteome Research 6: 3771-3779.

[17] European Association for Grain Legume Research (www.grainlegumes.com), accessed January 2009.

[18] Erisman JW, Sutton MA, Galloway J, Klimont Z, Winiwarter W (2008) How a century of ammonia synthesis changed the world. Nature Geosciences 1: 636-639.

[19] Ferguson, B.J. (2012) Rhizobia and legume nodulation genes. In: *Brenner's Online Encyclopedia of Genetics,* 2nd edn (Maloy, S. and Hughes, K., eds), Oxford, UK: Elsevier/ Academic Press, In press.

[20] Ferguson, B.J., Foo, E., Ross, J.J. and Reid, J.B. (2011) Relationship between gibberellin, ethylene and nodulation in *Pisum sativum*. New Phytologist 189: 829-842.

[21] Ferguson, B.J. and Gresshoff, P.M. (2009) Soybean as a model legume. Grain Legumes: 53: 7.

[22] Ferguson, B.J., Indrasumunar, A., Hayashi, S., Lin, M-H., Lin, Y-H., Reid, D.E. and Gresshoff, P.M. (2010) Molecular analysis of legume nodule development and autoregulation. Journal of Integrative Plant Biology 52: 61-76.

[23] Ferguson, B.J. and Mathesius, U. (2003) Signaling interactions during nodule development. Journal of Plant Growth Regulation 22: 47-72.

[24] Ferguson BJ, Reid JB (2005) *cochleata*: getting to the root of legume nodules. Plant Cell Physiology46: 1583–1589.

[25] Ferguson, B.J., Ross, J.J. and Reid, J.B. (2005) Nodulation phenotypes of gibberellin and brassinosteroid mutants of pea. Plant Physiology 138: 2396-2405.

[26] Ferguson BJ, Wiebe EM, Emery RJN, Guinel FC (2005b) Cytokinin accumulation and an altered ethylene response mediate the pleiotropic phenotype of the pea nodulation mutant R50 (*sym16*). Canadian Journal of Botany83: 989–1000.

[27] Food and Agriculture Organization of the United Nations (www.faostat.fao.org), accessed January 2010.

[28] Gage DJ (2004) Infection and invasion of roots by symbiotic, nitrogenfixing rhizobia during nodulation of temperate legumes.Microbiol.Mol. Biol.Rev. 68: 280–300.

[29] Giller KE, Cadisch G (1995) Future benefits from biological nitrogen fixation: An ecological approach to agriculture. Plant and Soil 174: 255-277.

[30] Glass A (2003) Nitrogen use efficiency of crop plants: physiological constraints upon nitrogen absorption. Critical Reviews in Plant Sciences 22: 453–470.

[31] Graham PH, Vance CP (2003) Legumes: Importance and constraints to greater use. Plant Physiology131: 872–877.

[32] Gresshoff, P.M. (2012) *Glycine max* (soybean). In: *Brenner's Online Encyclopedia of Genetics,* 2nd edn (Maloy, S. and Hughes, K., eds), Oxford, UK: Elsevier/Academic Press, In press.

[33] Gresshoff PM, Delves AC (1986) Plant genetic approaches to symbiotic nodulation and nitrogen fixation in legumes. In: Blonstein AD, King PJ, eds. Plant Gene Research III. A Genetical Approach to Plant Biochemistry. Springer Verlag, Wien. pp. 159–206.

[34] Gresshoff PM, Lohar D, Chan PK, Biswas B, Jiang Q, Reid D, Ferguson B, Stacey G (2009) Genetic analysis of ethylene regulation of legume nodulation. Plant Signalling and Behaviour 4: 818–823.

[35] Gresshoff PM, Rolfe, BG (1978) Viability of *Rhizobium* bacteroids isolated from soybean nodule protoplasts. Planta 142: 329-333.

[36] Hayashi S, Reid DE, Lorenc M, Stiller J, Edwards D, Gresshoff PM, Ferguson BJ (2012) Transient Nod-factor dependent gene expression in the nodulation competent zone of soybean (*Glycine max* L. Merr.) roots. Plant Biotechnology Journal, doi: 10.1111/j.1467-7652.2012.00729.x

[37] Heidstra R, Yang WC, Yalcin Y, Peck S, Emons AM, Van Kammen A, Bisseling T (1997) Ethylene provides positional information on cortical cell division but is not in-

volved in Nod factor induced root hair tip growth in *Rhizobium*-legume interaction. Development 124: 1781-1787.

[38] Hirel B, Le Gouis J, Ney B, Gallais A (2007) The challenge of improving nitrogen use efficiency in crop plants: towards a more central role for genetic variability and quantitative genetics within integrated approaches. Journal of Experimental Botany 58: 2369-2387.

[39] Indrasumunar, A., Kereszt, A., Searle, I., Miyagi, M., Li, D., Nguyen, C.D.T., Men, A., Carroll, B.J. and Gresshoff, P.M. (2010) Inactivation of duplicated *Nod factor receptor 5* (*NFR5*) genes in recessive loss-of-function non-nodulation mutants of allotetraploid soybean (*Glycine max* L. Merr.). Plant Cell Physiology 51: 201-214.

[40] Indrasumunar, A., Searle, I., Lin, M-H., Kereszt, A., Men, A., Carroll, B.J. and Gresshoff, P.M. (2011) *Nodulation factor receptor kinase 1a* controls nodule organ number in soybean (*Glycine max* L. Merr.) The Plant Journal 65: 39-50.

[41] Jensen ES, Peoples MB, Boddey RM, Gresshoff PM, Hauggaard-Nielsen H, Alves BJR, Morrison MJ (2012) Legumes for mitigation of climate change and feedstock in a bio-based economy – A review. *Agronomy for Sustainable Development 32: 329-364.*

[42] Ishikawa K, Yokota K, Li YY, Wang Y, Liu CT, Suzuki S, Aono T, Oyaizu H (2008) Isolation of a novel root-determined hypernodulation mutant rdh1 of *Lotus japonicus*. Soil Science and Plant Nutrition54: 259–263.

[43] Kereszt, A., Li, D., Indrasumunar, A., Nguyen, C.D.T., Nontachaiyapoom, S., Kinkema, M., and Gresshoff, P.M. (2007).*Agrobacterium rhizogenes*-mediated transformation of soybean to study root biology. Nature Protocols 2: 948–952.

[44] Krusell L, Sato N, Fukuhara I, *et al.* 2011. The *Clavata2* genes of pea and *Lotus japonicus* affect autoregulation of nodulation. The Plant Journal 65: 861–871.

[45] Lohar D, Stiller J, Kam J, Stacey G, Gresshoff PM (2009) Ethylene insensitivity conferred by a mutated *Arabidopsis* ethylene receptor gene alters nodulation in transgenic *Lotus japonicus*. Annals of Botany104: 277–285.

[46] Lorteau MA, Ferguson BJ, Guinel FC (2001) Effects of cytokinin on ethylene production and nodulation in pea *(Pisumsativum)* cv. Sparkle. Physiologia Plantarum 112: 421-428.

[47] Knobeloch L, Salna B, Hogan A, Postle J, Anderson H (2000)Blue babies and nitrate-contaminated well water. Environmental Health Perspectives 108: 675-678.

[48] Libault. M., Farmer, A., Brechenmacher, L., Drnevich, J., Langley, R.J., Bilgin, D.D., Radwan, O., Neece, D.J., Clough, S.J., May, G.D. and Stacey, G.(2010a) Complete transcriptome of the soybean root hair cell, a single-cell model, and its alteration in response to *Bradyrhizobium japonicum* infection.Plant Physiology 152: 541-552.

[49] Libault, M., Farmer, A., Joshi, T., Takahashi, K., Langley, R.J., Franklin, L.D., He, J., Xu, D., May, G. and Stacey, G. (2010b) An integrated transcriptome atlas of the crop

model *Glycine max*, and its use in comparative analyses in plants. The Plant Journal 63: 86-99.

[50] Lim, C.W., Lee, Y.W. and Hwang, C.H. (2011) Soybean nodule-enhanced CLE peptides in roots act as signals in GmNARK-mediated nodulation suppression. Plant Cell Physiol. 52: 1613-1627.

[51] Lin, M-H., Gresshoff, P.M., Indrasumunar, A., and Ferguson, B.J. (2011b) pHairyRed: a novel binary vector containing the *DsRed2* reporter gene for visual selection of transgenic hairy roots. Molecular Plant 4: 537-545.

[52] Lin, Y-H., Ferguson, B.J., Kereszt, A. and Gresshoff, P.M. (2010) Suppression of hypernodulation in soybean by a leaf-extracted, NARK- and Nod factor-dependent small molecular fraction. New Phytologist 185: 1074-1086.

[53] Lin, Y-H., Lin, M-H., Gresshoff, P.M. and Ferguson, B.J. (2011a) An efficient petiole-feeding bioassay for introducing aqueous solutions into dicotyledonous plants. Nature Protocols 6: 36-45.

[54] Magori S, Oka-Kira E, Shibata S, Umehara Y, Kouchi H, Hase Y, Tanaka A, Sato S, Tabata S, Kawaguchi M (2009) *TOO MUCH LOVE*, a root regulator associated with the long-distance control of nodulation in *Lotus japonicus*. Molecular Plant-Microbe Interactions22: 259–268.

[55] Mahieu S, Fustec J, Faure M, Corree-Hellou G, Grozat Y (2007) Comparison of two ^{15}N labeling methods for assessing nitrogen rhizodeposition of pea. Plant and Soil 295: 193-205.

[56] Mathews A, Carroll BJ, Gresshoff PM (1989) Development of *Bradyrhizobium* infection in supernodulating and non-nodulating mutants of soybean (*Glycine max* [L.] Merrill). Protoplasma150: 40–47.

[57] McNeill AM,Fillery IRP (2008) Field measurement of lupin belowground nitrogen accumulation and recovery in the subsequent cereal-soil system in a semi-arid Mediterranean-type climate. Plant and Soil 302: 297-316.

[58] Miyahara A, Hirani TA, Oakes M, Kereszt A, Kobe B, Djordjevic MA, Gresshoff PM (2008) Soybean nodule autoregulation receptor kinase phosphorylates two kinase-associated protein phosphatases *in vitro*. Journal of Biological Chemistry283: 25381–25391.

[59] Miyazawa, H, Oka-Kira E, Sato N, Takahashi H, Wu GJ, Sato S, Hayashi M, Betsuyaku S, Nakazono M, Tabata S, Harada K, Sawa S, Fukuda H, Kawaguchi M (2010). The receptor-like kinase KLAVIER mediates systemic regulation of nodulation and non-symbiotic shoot development in *Lotus japonicus*. Development 137: 4317–4325.

[60] Mortier V, Den Herder G, Whitford R, Van de Velde W, Rombauts S, D'haeseleer K, Holsters M, Goormachtig S (2010) CLE peptides control *Medicago truncatula* nodulation locally and systemically. Plant Physiology 153: 222-237.

[61] Murphy AP (1991) Chemical removal of nitrate from water. Nature 350: 223-225.

[62] Newcombe W, Sippel D, Peterson RL (1979) The early morphogenesis of *Glycine max* and *Pisumsativum*root nodules. Canadian Journal of Botany57: 2603–2616.

[63] Nishimura R, Ohmori M, Fujita H, Kawaguchi M (2002b) A Lotus basic leucine zipper protein with a RING-finger motif negatively regulates the developmental program of nodulation. Proceeding of the National Academy of Science USA 99: 15206–15210.

[64] Okamoto S, Ohnishi E, Sato S, Takahashi H, Nakazono M, Tabata S, Kawaguchi M (2009) Nod factor/nitrate-induced CLE genes that drive HAR1-mediated systemic regulation of nodulation. Plant and Cell Physiology 50: 67-77.

[65] Peoples MB, Brockwell J, Herridge DF, Rochester IJ, Alves BJR, Urquiaga S, Boddey RM, Dakora FD, Bhattarai S, Maskey SL, Sampet C, Rerkasem B, Khans DF, Hauggaard-Nielsen H, Jensen BS (2009) The contributions of nitrogen-fixing crop legumes to the productivity of agricultural systems. Symbiosis 48: 1-17.

[66] Penmetsa RV, Uribe P, Anderson J, Lichtenzveig J, Gish JC, Nam YW, Engstrom E, Xu K, Sckisel G, Pereira M, Baek JM, Lopez- Meyer M, Long SR, Harrison MJ, Singh KB, Kiss GB, Cook DR (2008) The *Medicago truncatula* of the *Arabidopsis EIN2* gene, sickle, is a negative regulator of symbiotic and pathogenic microbial interactions. The Plant Journal 55: 580–595.

[67] Postma JG, Jacobsen E, Feenstra WJ (1988) Three pea mutants with an altered nodulation studied by genetic-analysis and grafting. Journal of Plant Physiology132: 424–430.

[68] Reid, D.E., Ferguson, B.J., Hayashi, S., Lin, Y-H., Gresshoff, P.M. (2011a) Molecular mechanisms controlling legume autoregulation of nodulation. Annals of Botany 108: 789-795.

[69] Reid, D.E., Ferguson, B.J., Gresshoff, P.M. (2011b) NARK-dependent nodule regulation is activated in the shoot and root in response to inoculation and nitrate induced CLE peptides of soybean. Molecular Plant-Microbe Interactions 24: 606-618.

[70] Reid D.E., Hayashi S., Lorenc M., Stiller J., Edwards D., Gresshoff P.M., Ferguson B.J. (2012) Identification of systemic responses in soybean nodulation by xylem sap feeding and complete transcriptome sequencing reveal a novel component of the autoregulation pathway. Plant Biotechnology Journal 10: 680-689.

[71] Rispail N, Kalo P, Kiss GB, Ellis THN, Gallardo K, Thompson RD, Prats E, Larrainzar E, Ladrera R, González EM, Arrese-Igor C, Ferguson BJ, Gresshoff PM, Rubiales D (2010) Model legumes contribute to faba bean breeding. Field Crops Research 115: 253-269.

[72] Rockstrøm J, Steffen W, Noone K, Persson A, Chapin III FS, Lambin EF, Lenton TM, Scheffer M, Folke C, Schellnhuber HJ, Nykvist B, de Wit CA, Hughes T, van der

Leeuw S, Rodhe H, Sörlin S, Snyder PK, Costanza R, Svedin U, Falkenmark M, Karlberg L, Corell RW, Fabry VJ, Hansen J, Walker B, Liverman D, Richardson K, Crutzen P, Foley JA (2009) A safe operating space for humanity. Nature 461: 472-475.

[73] Salvagiotti F, Cassman KG, Specht JE, Walters DT, Weiss A, Dobermann A (2008) Nitrogen uptake, fixation and response to fertilizer N in soybeans: A review. Field Crops Research 108: 1-13.

[74] Schmutz, J., Cannon, S.B., Schlueter, J., Ma, J., Mitros, T., Nelson, W., Hyten, D.L., Song, Q., Thelen, J.J., Cheng, J., Xu, D., Hellsten, U., May, G.D., Yu, Y., Sakurai, T., Umezawa, T., Bhattacharyya, M.K., Sandhu, D., Valliyodan, B., Lindquist, E., Peto, M., Grant, D., Shu, S., Goodstein, D., Barry, K., Futrell-Griggs, M., Abernathy, B., Du, J., Tian, Z., Zhu, L., Gill, N., Joshi, T., Libault, M., Sethuraman, A., Zhang, X.C., Shinozaki, K., Nguyen, H.T., Wing, R.A., Cregan, P., Specht, J., Grimwood, J., Rokhsar, D., Stacey, G., Shoemaker, R.C., Jackson, S.A. (2010) Genome sequence of the palaeopolyploid soybean. Nature 463: 178-183.

[75] Scott P, Pregelj L, Chen N, Hadler J, Djordjevic M, Gresshoff P (2008). *Pongamia pinnata:* An untapped resource for the biofuels industry of the future. Bio Energy Research 1: 2-11

[76] Searle, I.R., Men, A.E., Laniya, T.S., Buzas, D.M., Iturbe-Ormaetxe, I., Carroll, B.J. and Gresshoff, P.M. (2003) Long-distance signaling in nodulation directed by a CLAVATA1-like receptor kinase. Science 299: 109-112.

[77] Severin, A.J.,Woody, J.L., Bolon, Y.T., Joseph, B., Diers, B.W., Farmer, A.D., Muehlbauer, G.J., Nelson, R.T., Grant, D., Specht, J.E., Graham, M.A., Cannon, S.B., May, G.D., Vance, and Shoemaker, R.C. (2010) RNA-seq atlas of *Glycine max*: A guide to the soybean transcriptome. BMC Plant Biology 10: 160.

[78] Sutton MA, Oenema O, Erisman JW, Leip A, van Grinsven H, *et al.* (2011) Too much of a good thing. Nature 472: 159-162.

[79] Turgeon BG, Bauer WD (1982) Early events in the infection of soybean by *Rhizobium japonicum*. Time course and cytology of the initial infection process. Canadian Journal of Botany60: 152–161.

[80] Turgeon BG, Bauer WD (1985) Ultrastructure of infection thread development during infection of soybean by *Rhizobium japonicum*. Planta163: 328–349.

[81] Udvardi M, Day D (1997) Metabolite transport across symbiotic membranes of legume nodules. Annual Review of Plant Physiology and Plant Molecular Biology 48: 493–523.

[82] Vance CP (2001) Symbiotic nitrogen fixation and phosphorus acquisition.Plant nutrition in a world of declining renewable resources. Plant Physiology 127: 390-397.

[83] Vernie T, Moreau S, de Billy F, Plet J, Combier JP, Rogers C, Oldroyd G, Frugier F, Niebel A, Gamas P (2008) EFD Is an ERF transcription factor involved in the control

of nodule number and differentiation in *Medicago truncatula*. The Plant Cell20: 2696–2713.

[84] Yao PJ, Vincent JM (1969) Host specificity in the root "curling factor" of *rhizobium* spp. Australian Journal of Biological Science 22: 413-423.

[85] Zhang C, Ghabrial SA (2006) Development of Bean pod mottle virus-based vectors for stable protein expression and sequence-specific virus-induced gene silencing in soybean. Virology 344: 401-411.

Chapter 3

Soybean as a Nitrogen Supplier

Matsumiya Yoshiki, Horii Sachie,
Matsuno Toshihide and Kubo Motoki

Additional information is available at the end of the chapter

http://dx.doi.org/10.5772/51017

1. Introduction

Soybean has been cultivated all over the world since ancient times for its high protein and lipid content. It is one of the most important agricultural products in the world and its global production is more than 220 million tons per year [1]. Vegetable oil production from soybean is the highest among plant oils (30%) [2].

Soybean is used directly as food in Japan and several Asian countries. Recently, soybean protein was recognized as both healthy and tasty and is used in food such as Tofu and soy sauce. Soybean-meal, which remains after extraction of the vegetable oil, contains about 50% protein with well balanced amino acids. Therefore, soybean-meal is often re-utilized as animal foodstuff.

Soybean waste was utilized as an organic fertilizer prior to the 1940s [3-6]. However, a chemical fertilizer took the place of the organic fertilizer because it produced faster results. Organic fertilizers are now gradually being used again for increased food production safety and the protection of the environment.

Soybean cultivation is well known for improving soil fertility [3, 7, 8]. Root-nodules are formed by the soybean plant, and atmospheric N_2 is fixed by the nitrogen fixing bacteria in the root-nodule [9]. N_2 is converted to NH_4^+ by nitrogenase from these nitrogen fixing bacteria, and this NH_4^+ is supplied to the soil environment.

Recently, investigations into the utilization of proteins from soybean waste have been carried out for the development of high quality foods. Protein fractions, such as soy protein isolates and whey protein are industrially produced, and these fractions are used as additives for the improvement of food nutrition [10]. Moreover, several soybean proteins and peptides have been purified and utilized as medicines for hypotension, rheumatism, and cholesterol control [11-13]. The bioactive peptides of soybean protein have also been investigated [5, 6].

This chapter explains how soybean cultivation and soybean protein are nitrogen suppliers and describes the production of novel bioactive peptides from soybean and legumes.

2. Nitrogen supply by soybean cultivation

2.1. Nitrogen fixing bacteria

N_2 is fixed by nitrogen fixing bacteria in the soil environment [14-17]. These bacteria convert N_2 to NH_4^+. The biological reduction of atmospheric N_2 to NH_4^+ (nitrogen fixation) provides about 65% of the biosphere's available nitrogen [18].

As long ago as 1890, a nitrogen fixing bacteria was isolated from a root nodule and identified as *Rhizobium leguminosarum* [19, 20]. Shortly after this, *Clostridium pasteurianum* and *Azotobacter* sp. were also isolated as nitrogen fixing bacteria in the soil environment [21-23]. Now, more than 100 genera have been isolated and identified as nitrogen fixing bacteria. Among them, genera *Rhizobium, Bradyrhizobium, Azorhizobium,* and *Frankia* lead to the formation of root-nodules in legumes [16].

Nitrogenase (EC 1.18.6.1) from nitrogen fixing bacteria catalyzes N_2 to NH_4^+ ($N_2 + 8H_2 + 8e^- + 16ATP + 16H_2O \rightarrow 2NH_3 + H_2 + 16ADP + 16Pi$). NH_4^+ is further converted to NO_2^- and NO_3^- by ammonia oxidizing bacteria (AOB) and nitrite oxidizing bacteria (NOB).

Figure 1. Soybean root nodule

2.2. Relationship between nitrogen fixing bacteria and soybean cultivation

The roots of soybean secrete flavonoids and enhance the growth of nitrogen fixing bacteria around the root [24]. The nitrogen fixing bacteria infect the soybean root, and the root-nodule is formed. Bacteroids in the root-nodule fix and provide nitrogen from the air [25]. *Bradyrhizobium japonicum, B. elkanii, B. lianigense,* and *Sinorhizobium fredii* have been identified as the root-nodule forming bacteria in soybean cultivation [16, 26, 27].

The change in soil microbial diversity after soybean cultivation has been analyzed by PCR-DGGE. Root-nodules were shown to be formed and specific bacteria were increased during cultivation (Figures 1 and 2) but not the total number of bacteria in the soil. Soybean cultivation caused nitrogen accumulation in the soil environment.

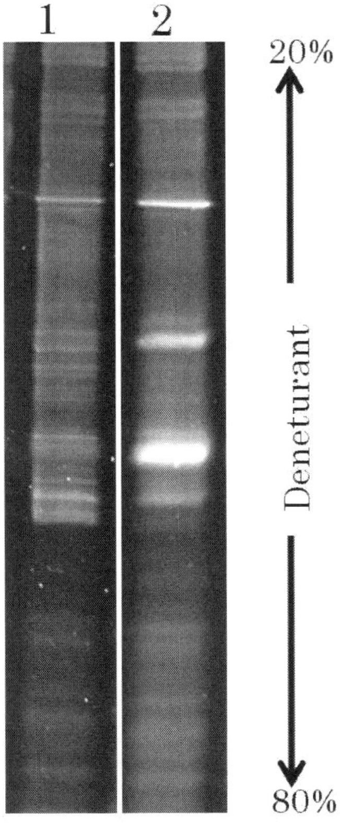

Figure 2. PCR-DGGE profiles of soybean cultivated soil, 1: Before cultivation, 2: after cultivation.

3. Enhancement of nitrogen circulation by soybean cultivation and soybean protein

3.1. Evaluation of nitrogen circulation in soil environment

The nitrogen cycle is illustrated in Figure 3. Organic forms of nitrogen such as protein are degraded to peptides and amino acids by soil microorganisms, and these peptides and amino acids are then converted to NH_4^+. Subsequently, NH_4^+ is further converted to NO_2^- and NO_3^- (nitrification). NO_2^- is denitrified to N_2 by denitrifying bacteria and this N_2 is converted to NH_4^+ by the nitrogen fixing bacteria, and NH_4^+ is accumulated in the soil environment again.

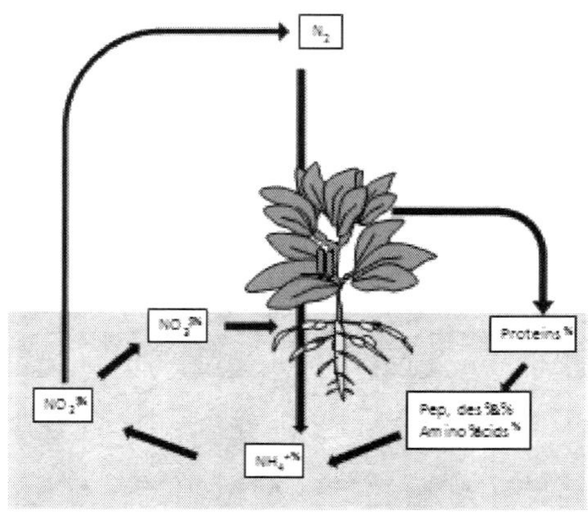

Figure 3. The soil nitrogen cycle

The nitrification process is the rate limiting step in the nitrogen cycle [28]. To further investigate the soil nitrogen cycle, a new method for the evaluation of nitrogen circulation activity was constructed based on bacterial number, ammonium oxidizing activity (AOA), and nitrite oxidizing activity (NOA) (Figure 4) [29]. These three indices were used to construct a radar chart of nitrogen circulation in the soil. The area of the radar chart was calculated, and then the value was treated as a nitrogen circulation activity (0–100 points).

3.2. Enhancement of nitrogen circulation

A database of nitrogen circulation activity was constructed using 155 agricultural soils (Figure 5). The nitrogen circulation activity of agricultural soil ranges from 0 to 99.6 points with an average of 26 points.

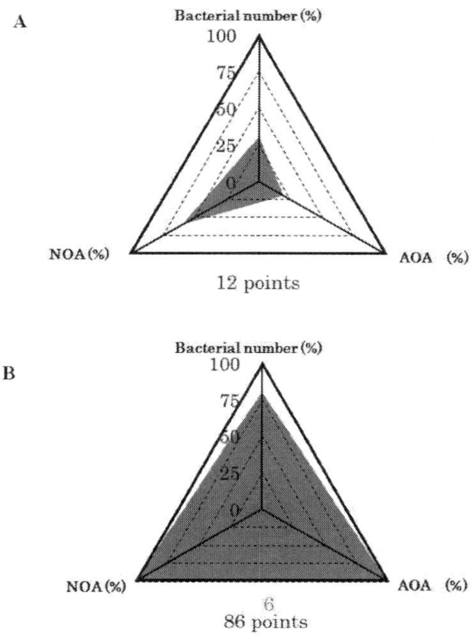

A: Un-fertile soil, B: fertile soil.

Figure 4. Values of nitrogen circulation activity in soil environments

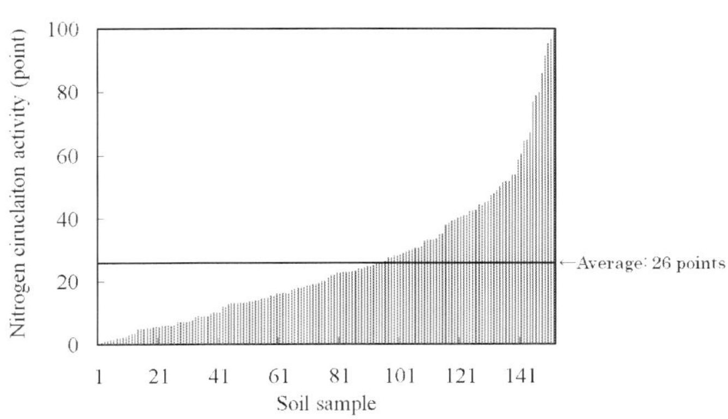

Figure 5. Database of nitrogen circulation activity in 155 agricultural soils

Soybean cultivation leads to nitrogen accumulation in the soil environment, and therefore nitrogen circulation activity should be enhanced by soybean cultivation. This enhancement was further analyzed (Figure 6) and activity was shown to be enhanced 26 to 95 points after soybean cultivation.

Soybean waste is also rich in nitrogen (Table 1), and is often used as an organic fertilizer. Soil nitrogen is increased by using soybean waste as fertilizer, and consequently nitrogen circulation is increased. Soybean waste is also rich in carbon (C/N values; 5.1), and therefore soil bacteria and bacterial activity may also be increased by the addition of soybean waste.

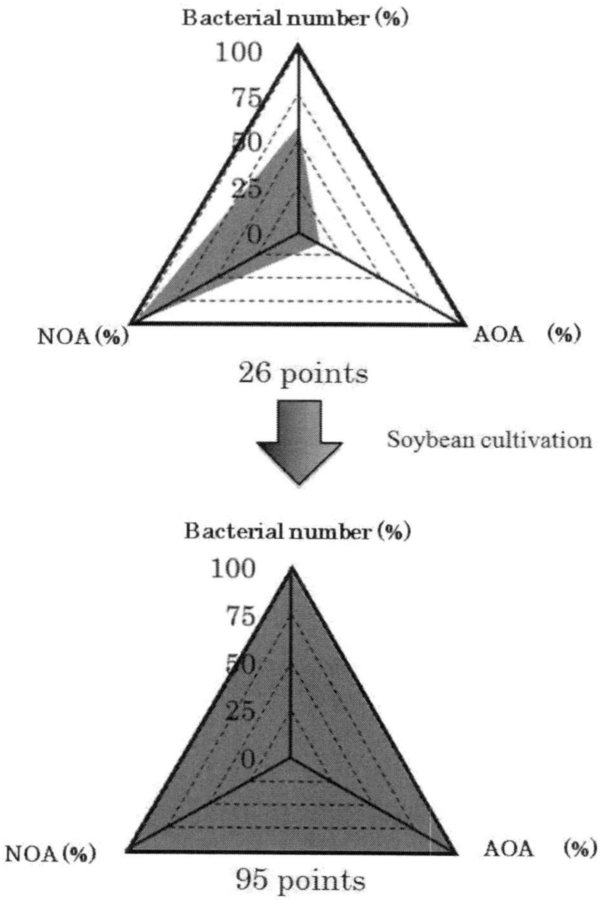

Figure 6. Effect of soybean cultivation on nitrogen circulation activity in soil

http://dx.doi.org/10.5772/51017

Component	Value
Total carbon	450,000 mg/kg
Total nitrogen	87,500 mg/kg
Total phosphorous	6,100 mg/kg
Total potassium	18,900 mg/kg
C/N ratio	5.1

Table 1. Components of soybean meal

4. Bioactive peptides from soybean protein

4.1. Plant growth promoting peptides from soybean waste

For efficient use of soybean waste, it is treated with an alkaline protease from *Bacillus circulans* HA12 (degraded soybean meal products; DSP) [4, 6]. Plant growth promotion by DSP has been investigated using various plant species [30]. The fresh weight of *Brassica rapa* was shown to be increased by 25% through the addition of DSP (12 mg-peptides/kg-soil) (Figure 7). The growth of *Solanum tuberosum* L., *Solanum lycopersicum*, and *Brassica juncea* were also promoted by addition of DSP. Moreover, DSP also produced thicker roots than a chemical fertilizer, indicating that DSP contains bioactive peptides for plant growth.

Figure 7. Plant growth-promoting effect of DSP, A: Chemical fertilizer, B; DSP.

4.2. Root hair promoting peptide in DSP

The number of root hairs in *B. rapa* was increased and each was elongated when DSP (30 μg/ml) was added (Figure 8) to the soil. In order to analyze the root hair promoting effect by DSP, the structure of the root hair promoting peptide (RHPP) in DSP was investigated [6]. Degraded products of Kunitz trypsin inhibitor (KTI) in soybean protein showed higher root hair promoting activity, and the RHPP was purified by several chromatographic steps from degraded products of KTI.

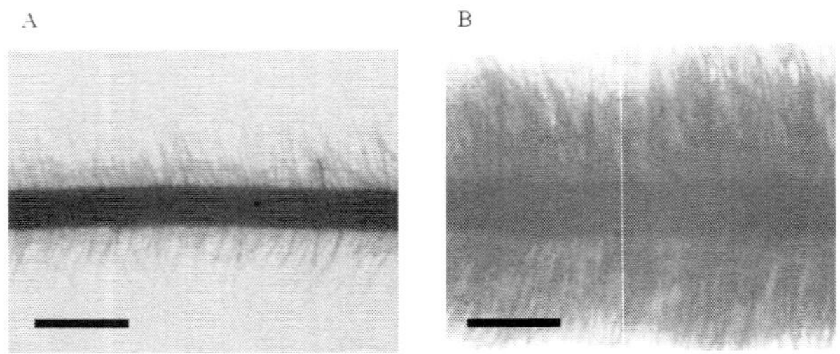

Figure 8. Root hair promoting effect of DSP, A: Root of *Brassica rapa* grown in plant growth medium, B: root of *B. rapa* grown with DSP in plant growth medium. Bar denotes 1 mm.

The molecular mass of RHPP was analyzed by matrix-assisted laser desorption/ionization time-of-flight mass spectrometry (MALDI-TOF MS) [6]. The molecular weight of the bioactive peptide was 1,198.2 Da (Figure 9), and the molecular weight of the amino acid sequence in KTI was searched. Positions 27–38 in KTI (Gly-Gly-Ile-Arg-Ala-Ala-Pro-Thr-Gly-Asn-Glu-Arg) were identical to this molecular weight, and this peptide was thus designated as the RHPP (Figure 9). The RHPP that was chemically synthesized was also shown to have root hair promoting activity (data not shown).

^{1}DFVLDNEGNPLENGGTYYILSDITAFGGIRAAPTGNERCP^{40}

^{41}LTVVQSRNELDKGIGTIISPSYRIRFIAEGHPLSLKFDSF^{80}

^{81}AVIMLCVGIPTEWSVVEDLPEGPAVKIGENKDAMDGWFRL^{120}

^{121}ERVSDDEFNNYKLVFCPQQREDDKCGDIGISIDDDGHTRR^{160}

^{161}LVVSKNKPLVVQFQKLDKESL^{181}

Figure 9. Amino acid sequence of RHPP in Kunitz trypsin inhibitor, the RHPP amino acid sequence is shown by gray box.

5. Novel plant bioactive peptides from other legume

Many other legumes form root-nodules with nitrogen fixing bacteria. The nitrogen fixing bacteria related to legume cultivation are classified into 13 genera (*Rhizobium, Ensifer, Mesorhizobium, Bradyrhizobium, Methylobacterium, Azorhizobium, Devosia, Burkholderia, Phyllobacterium, Microvirga, Ochrobactrum, Cupriavidus,* and *Shinella*) and 98 species [31].

Legumes such as *Astragalus sinicus, Trifolium repens,* and *Arachis hypogaea* are cultivated as green manure for the improvement of soil fertility. The host specificity of the nitrogen fixing bacteria, *M. huakuii, R. trifolii,* and *Bradyrhizobium* sp., are very high, infecting *A. sinicus, T.*

repens, and *A. hypogaea*, respectively [32]. These legumes are rich in proteins and form root-nodules via the same mechanisms as soybean.

In order to find novel bioactive peptides, attempts to degrade protein biomass from *A. hypogaea* by various proteases (thermolysin, subtilisin, proteinaseK, and trypsin) were made. Bioactivities of root hair and lateral root formation were found by degradation with proteinaseK (Figure 10). Degraded products of *A. hypogaea* by proteinase K (30 μg/ml) showed strong root hair promoting activity at the same level as DSP. Moreover, degraded products of *A. hypogaea* promoted lateral root growth in *B. rapa*, suggesting that degradation of legume proteins has a possibility to produce new bioactive peptides.

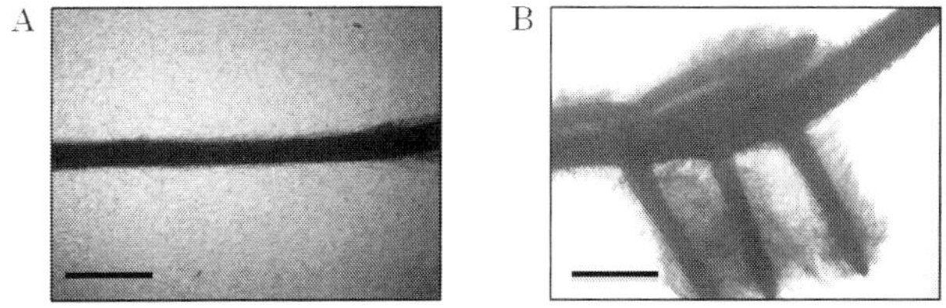

Figure 10. Bioactive effect of degraded products of *A. hypogaea* on root of *B. rapa*, A: Root of *Brassica rapa* grown in plant growth medium, B: root of *B. rapa* grown with degraded products of *A. hypogaea*. Bar denotes 1 mm.

6. Conclusion

Soybean supplies nitrogen into the soil environment by forming root nodules and accumulating protein in its seed. Soybean cultivation has been shown to enhance nitrogen circulation by about 3.6 times accompanied with increases in nitrogen fixing bacteria.

DSP has been shown to increase the fresh weight of plants, and a peptide from DSP promoted root hair formation in *B. rapa*. Moreover, other bioactive peptides were found by degradation of proteins from A. hypogaea with proteinaseK treatment. The proteins of legumes will also become nitrogen sources for plant growth and the soil environment.

Author details

Matsumiya Yoshiki, Horii Sachie, Matsuno Toshihide and Kubo Motoki

Department of Biotechnology, College of Lifesciences, Ritsumeikan University Nojihigashi 1-1-1, Kusatsu, Shiga 525-8577, Japan

References

[1] Uchida, M. (2007). Prospects for World Trade in the Soybean Sector. *Journal of Nagoya Bunri University*, 7, 97-102.

[2] USDA. (2007). *Oilseeds: World Markets and Trade.*

[3] Okuda, A. (2011). *Dojou hiryou sousetu*, Tokyo: Yokendou, Co. LTD.

[4] Kubo, M., Okajima, J., & Hasumi, F. (1994). Isolation and Characterization of Soybean Waste-Degrading Microorganisms and Analysis of Fertilizer Effects of the Degraded Products. *Applied and Environmental Microbiology*, 60(1), 243-247.

[5] Matsumiya, Y., & Kubo, M. (2011). Soybean Peptide: Novel Plant Growth Promoting Peptide from Soybean. *Soybean and Nutrition*, Rijeka, Intech.

[6] Matsumiya, Y., & Kubo, M. (2008). Utilization of Biomass Based on Biorefinery: Development of Novel Bioactive Peptides from Soybean Waste. *Current Topics in Biotechnology*, 4.

[7] Cass, A., Gusli, S., & Mac, Leod. D. (1994). Sustainability of Soil Structure Quality in Rice Paddy-Soya-Bean Cropping Systems in South Sulawesi Indonesia. *Soil and Tillage Research*, 31(4), 339-52.

[8] Tago, K., Ishii, S., Nishizawa, T., Otsuka, S., & Senoo, K. (2011). Phylogenetic and Functional Diversity of Denitrifying Bacteria Isolated from Various Rice Paddy and Rice-Soybean Rotation Fields. *Microbes and Environments*, 26(1), 30-5.

[9] Nagatani, H., Shimizu, M., & Valentine, R. (1971). The Mechanism of Ammonia Assimilation in Nitrogen Fixing bacteria. *Archives of Microbiology*, 79(2), 164-75.

[10] Malhotra, A., & Coupland, J. N. (2004). The Effect of Surfactants on the Solubility, Zeta Potential, and Viscosity of Soy Protein Isolates. *Food Hydrocolloids*, 18(1), 101-8.

[11] Yonekura, M., & Yamamoto, A. (2004). Isolation and Application of Physiologically Active Peptides from Soybean Whey and Okara Proteins. *Soy Protein Research*, 7, 79-84.

[12] Yonekura, M., & Tanaka, A. (2003). Isolation and Application of Physiologically Active Peptides from Soybean Whey and Okara Proteins. *Soy Protein Research*, 6, 88-93.

[13] Farzamirad, V., & Aluko, R. E. (2008). Angiotensin-Converting Enzyme Inhibition and Free-Radical Scavenging Properties of Cationic Peptides Derived from Soybean Protein Hydrolysates. *International Journal of Food Sciences and Nutrition*, 59(5), 428-37.

[14] Steenhoudt, O., & Vanderleyden, J. (2000). *Azospirillum*, a Free-Living Nitrogen-Fixing Bacterium Closely Associated with Grasses: Genetic, Biochemical and Ecological Aspects. *FEMS Microbiology Reviews*, 24(4), 487-506.

[15] Orr, C. H., James, A., Leifert, C., Cooper, J. M., & Cummings, S.P. (2011). Diversity and Activity of Free-Living Nitrogen-Fixing Bacteria and Total Bacteria. *Organic and Conventionally Managed Soils. Applied and Environmental Microbiology*, 77(3), 911-9.

[16] Fred, E. B., Baldwin, I. L., & Mc Coy, E. (2002). Root Nodule Bacteria and Leguminous Plants. *UW-Madison Libraries Parallel Press*.

[17] Berman-Frank, I., Lundgren, P., & Falkowski, P. (2003). Nitrogen Fixation and Photosynthetic Oxygen Evolution in Cyanobacteria. Research. *Microbiology*, 154(3), 157-64.

[18] Lodwig, E. M., Hosie, A. H., Bourdes, A., Findlay, K., Allaway, D., Karunakaran, R., Downie, J. A., & Poole, P. S. (2003). Amino-Acid Cycling Drives Nitrogen Fixation in the Legume-Rhizobium Symbiosis. *Nature*, 422(6933), 722-6.

[19] Nutman, P.S. (1998). Biological Nitrogen Fixation. *Foundations of Modern Biochemistry*, 4, 199-216.

[20] Frank, B. (1890). Ueber die Pilzymbiose der Leguminosen. *Paul Parey Publishing*.

[21] Winogradski, S. (1895). Recherches Sur L'assimilation de L'azote Libre de L'atmosphere Par les Microbes. *Archives des Sciences Biologiques*, 3, 297-352.

[22] Beijerinck, M. (1901). Ueber Oligonitrophile Mikroben" (in German). Zentralblatt für Bakteriologie, Parasitenkunde, Infektionskrankheiten und Hygiene. *Abteilung II.*, 7, 561-582.

[23] Coyne, M. (1996). A Cartoon History of Soil Microbiology. *Journal of Natural Resources and Life Sciences Education*, 25(1), 30-36.

[24] Van de Sande, K., & Bisseling, T. (1997). Signalling . *Symbiotic Root Nodule Formation, Essays in Biochemistry*, 32, 127-42.

[25] Nakayama, T., & Botany, W. E. B. (2005). http://www.biol.tsukuba.ac.jp/~algae/BotanyWEB/.

[26] Kaneko, T., Nakamura, Y., Sato, S., Minamisawa, K., Uchiumi, T., Sasamoto, S., Watanabe, A., Idesawa, K., Iriguchi, M., & Kawashima, K. (2002). Complete Genomic Sequence of Nitrogen-Fixing Symbiotic Bacterium *Bradyrhizobium japonicum* USDA110. *DNA Research*, 9(6), 189-197.

[27] Scholla, M.H., & Elkan, G.H. (1984). *Rhizobium fredii* sp. nov., a Fast-Growing Species that Effectively Nodulates Soybeans. *International Journal of Systematic Bacteriology*, 34(4), 484-6.

[28] Højberg, O., Binnerup, S. J., & Sørensen, J. (1996). Potential Rates of Ammonium Oxidation, Nitrite Oxidation, Nitrate Reduction and Denitrification in the Young Barley Rhizosphere. *Soil Biology and Biochemistry*, 28(1), 47-54.

[29] Tsuda, H., Matsuno, T., Kubota, K., Matsumiya, Y., & Kubo, M. (2010). A New Method for an Evaluation of a Nitrogen Circulation Activity in Soil. *Memoirs of the Institute of Science & Engineering*, 69, 39-46.

[30] Hasegawa, N., Fukumoto, Y., Minoda, M., Plikomol, A., & Kubo, M. (2002). Promotion of Plant and Root Growth by Soybean Meal Degradation Products. *Biotechnology Letters*, 24(18), 1483-6.

[31] Weir, B. (2008). *The Current Taxonomy of Rhizobia*, http://www.rhizobia.co.nz/taxonomy/rhizobia.

[32] Nuswantara, S., Fujie, M., Yamada, T., Malek, W., Inaba, M., Kaneko, Y., & Murooka, Y. (1999). Phylogenetic Position of Mesorhizobium huakuii subsp. rengei, a Symbiont of Astragalus sinicus cv. Japan. *Journal of Bioscience and Bioengineering*, 87(1), 49-55.

Chapter 4

How to Increase the Productivity of the Soybean-Rhizobial Symbiosis

Denis M. Sytnikov

Additional information is available at the end of the chapter

http://dx.doi.org/10.5772/51563

1. Introduction

Fixation of molecular nitrogen is one of the processes determining the biological productivity of our planet, which is why its study is one of the primary tasks of modern biology. The nitrogen cycle in nature is one of the key components of biogeochemical cycles of the Earth. The atmosphere consists of almost 80% (by volume) of the chemical element and is its main source. Nitrogen makes up a part of proteins and other molecules that form the basis of the structural organization of all levels of life. Humans and animals need it in the form of animal and vegetable protein. Plants need it in the form of salts of nitric acid and ammonium ions. [1-3].

The economic and environmental crisis, the decline in the quality of crop production, and the deterioration of the natural fertility of soil - are all reasons for increased attention to biological farming, which focuses on using the potential of natural ecosystems, in particular, nitrogen-fixing microorganisms.

Microorganisms which assimilate molecular nitrogen in the atmosphere are diazotrophs and they have a similar biochemical mechanism of nitrogen fixation. There are two main groups of nitrogen-fixing microorganisms, those that enter into a symbiosis with plants and those living freely in the soil. The second group includes associative nitrogen fixers and microorganisms, more adapted to a free existence in the soil. The division of nitrogen fixers into freeliving and associative is conditional, since the ability to live freely in the soil is typical to all nitrogen-fixing bacteria, whilst only symbiotic nitrogen fixers are able to assimilate molecular nitrogen very closely with plants. Links of cyanobacteria with other organisms can be quite diverse: there are phycobionts in lichens which live in the air-cells of mosses, in the leaves of water ferns, and in other places. It should be noted that the potential of symbiotic nitrogen fixers is significantly higher than of those that are free living [4-7].

The symbiotic and associative systems of plants and diazotrophs are an example of the evolution of the interaction of living organisms. Their study is particularly relevant with the implementation of highly productive and environmentally friendly farming. The biological fixation of molecular nitrogen from the air is a process of fixation and assimilation of nitrogen by microorganisms. It is of great practical importance, since the industrial production of chemical nitrogen fertilizer requires significant use of costly energy resources, which by themselves can be harmful for the environment. A comprehensive study of this problem is necessary due to the need to create new and effective biological preparations. The creation and use of biological agents on the basis of nitrogen-fixing microorganisms is the most justifiable method of increasing the productivity of plants and the quality of their harvest, which allows maintenance of the natural fertility of soils and the ecological balance of the environment. Their use makes it possible to regulate the number and activity of beneficial microflora in the rhizosphere of crops, and to provide plants with nitrogen fixed from the atmosphere. For example, in addressing the shortage of high-grade protein the key role belongs to the soybean. However, the soils on which the crop is grown for the first time [8] usually do not have nodule bacteria compatible with soybean or bacteria number is small (up to 20 per gram of soil).

2. Legume Lectins as a Factor in an Effective Symbiosis

The full establishment and functioning of the legume-rhizobial symbiosis depends significantly on a number of abiotic, biotic and anthropogenic factors. In particular, the effectiveness of symbiotic nitrogen-fixing is affected by temperature, aeration, pH level and moisture content of the substrate, the presence of pesticides, the content of nitrogen and other macro- and micronutrients. Of further importance is the virulence and activity of root nodule bacteria. The study of the influence and the establishment of the degree of importance of different factors to the effectiveness of symbiosis, will determine the conditions, and help to develop methods of optimizing the functioning of the symbiotic systems.

Wide dissemination of lectins in a variety of plants and the presence of these proteins in virtually all organs and tissues of plant organisms demonstrate the importance of their role in life processes. Initially it was assumed that the presence of lectins was a distinctive feature of seeds from the legume family. However, the number of organisms in which lectins have been found is increasing every day and there is reason to believe that all the above mentioned plants as well as algae, lichens and fungi contain lectins [9-11]. Nevertheless, in the present day researchers focus much of their attention on lectins of leguminous plants, the increased study of which leads us to understand the activity of these proteins in relation to the structure of the latter.

Legume lectins are a large group of carbohydrate binding proteins derived mainly from seeds. Lectins contained in the seeds of legumes, are localized in the proteins and make up approximately 10% of soluble protein extract. In recent years it has been found that these proteins are present in other parts of plants, including stems, leaves, bark, roots, and root nodules. Lectins accumulate in the vacuoles of cells and can come to the surface of the plants. They can also be associated with membranes and cell walls [12-15].

The basis of the biological activity of lectins is a reversible reaction with carbohydrates, which defines the different types of biological reactions [16] and their physiological significance. Lectins do not have uniform structural characteristics. In legumes, these proteins are generally composed of two or four subunits with a molecular mass of 25 000 - 30 000 and each having one carbohydrate binding site [15, 17]. Lectins can interact with both mono-and oligosaccharides, as well as the remnants of carbohydrates present in the complex organic substances - glycoproteins, polysaccharides, and glycosides. In the early 1980's there was established a carbohydrate specificity for many legume seed lectins. However indisputable physiological carbohydrate ligands for any of these proteins have not been identified [14].

Over the past half-century many hypotheses were developed regarding the role of lectins in the life of legumes, but so far none of them have been fully confirmed. These proteins were detected due to their ability to bind carbohydrates, and most hypotheses about their functions are based solely on this wonderful property. Since it is known that the plant may contain a significant number of lectins, which are localized in different tissues and may form a gene duplication, while carrying out a variety of functions, it is believed that any attempt to dedicate a specific role to these proteins is doomed to failure [14].

An important feature of most of the representatives of the family *Fabaceae* (Legumes) is the ability to enter into symbiotic relationships with rhizobia. They belong to the family of *Rhizobiaceae* (bacteria genera *Rhizobium, Bradyrhizobium, Mezorhizobium, Sinorhizobium, Azorhizobium*), and form special structures on the roots called nodules. Lectins are considered as a component of the molecular and chemical interactions underlying the formation of symbiotic structures [6, 9, 11]. The symbiosis between rhizobia and legumes is based on a complex sequence of morphophysiological changes in the cells of both partners. Stages in the development of symbiosis are: preinfection (chemotaxis of bacteria, the exchange of signals, the adsorption of microsymbionts on the root surface), infection of the roots and development of nodules (penetration, formation of infection threads, the transformation of rhizobia into bacteroids), as well as the functioning of the nodules of nitrogen fixation [9, 18, 19].

Plants throughout their lifespan release various matter into the environment. The presence of lectins has been found in the soil [20], which are secreted during the germination of seeds together with other biologically active substances [21]. It is believed that in the first stages of interaction between rhizobia and legumes an important role is played by the chemotactic response to bacteria. Nodule bacteria and other soil microorganisms, in response to plant exudates, stimulate the reproduction and active movement of bacteria towards the roots, colonizing the rhizosphere and rhizoplane. A positive chemotaxis of rhizobia on the root exudates and germinating seeds can be either nonspecific (due to the excretion of organic acids, carbohydrates, vitamins), or specific (induced by flavonoids and lectins) [9, 18, 22].

The initial stage of many plant-microbe interactions is recognition of partners, which is largely due to the exchange of molecular signals [23, 24]. This process begins with the exudation by the host plant of chemicals such as flavonoids and betaines that induce the gene expression of nodulation in rhizobia [25]. Flavonoids induce the transition of bacteria from free-living to those in a symbiotic state [26]. The first chemically-characterized inducer of the expression of nod-gene was luteolin, drawn from the extract of lucern seeds [27]. For each type of rhizobia, plants produce individual flavonoid signals that can stimulate or inhibit the nod-genes of rhizobia [28]. The identification of factors by microorganisms which are released by plants initiates physiological processes required for the infection of the host plant. In turn, the microbial signals induce the plant to express the genes required for the formation of responses [5, 18, 25, 29]. Amongst the signals produced by the bacteria, the most studied are the lipochitooligosaccharide Nod-factors [14, 24]. During the formation of symbiotic relationships lectins on the plant recognize the nod-factor signal released by the bacteria. It is believed that lectins binding to the bacteria whith the surface of root hairs are able to identify the rhizobial signaling molecules, through surface polysaccharides on the rhizobia [14, 30]. It has been shown that the lectin of dolichos roots, which is localized on the surface of root hairs, is able to connect with some of the Nod-factors [14]. Furthermore, from the soybean root tissue there has been found and characterized a chitin-binding protein of the plasma membranes of cells. The presence of this protein and the specificity of the induction of the biological response to the binding of ligand *Bradyrhizobium japonicum* indicates its importance in the initiation of response to the binding of chitin in soybean [31].

Therefore Nod-factors play the role of a trigger mechanism for initiating bacterial invasion and nodulation of the plant [24]. Lipopolysaccharides inform the plant about the transition to a symbiotic interaction and the formation of the so called functional areas [32]. Legume plants are unique in their ability to form a response to Nod-factors, which consist of a preparation for the penetration of bacteria into the roots. Unfortunately, we do not know how the representatives of this family developed the ability to recognize such signals and how they improved upon these mechanisms [26].

An important stage of preinfection is the adsorption of bacterial cells on the surface of root hairs [33, 34]. The process of attachment of the bacteria from the *Rhizobiaceae* family consists of several stages with the participation of bacterial surface proteins in the first stage (rickadhesin, porin) and polysaccharide fibrils in the second stage. The first stage, involving proteins, is more decisive for the success of infection than the second stage, mediated by specific cellulose fibrils. The latter promote the retention of bacterial cells on the surface of the plant, but are not necessary for the infection [35].

The adsorption of rhizobia has also been associated with the ability of legumes to synthesize specific glycoproteins called lectins, which bind to polysaccharides on the surface of rhizobia cells [36, 37]. It is believed [33, 34] that the ability of rhizobia to be adsorbed on the surface of the roots of the host plant (like the chemotaxis) may be either specific or nonspecific. The initial binding of rhizobial cells to the surface of root hairs occurs by the means of exopolysaccharides (EPS). The degree of affinity between EPS microsymbionts and host plant lectin determines the degree of homology between the symbiotic partners and provides an advantage over other homologous strains in the process of plant infection.

Dazo and his co-writer [38] proposed a model of the attachment of rhizobia to the surface of the roots of dicotyledonous plants. The first phase is non-specific attachment which is characterized by the fact that the polyvalent host plant lectin binds carbohydrate receptors on the cell surface of nodule bacteria. This results in an intercellular "bridge" between lectin and polysaccharide. The second phase is the "anchoring" of bacteria on the surface of plant cells. This is the phase of specific attachment. The bacterial cells attach to the plant cell, which is the signal for further stages of infection. At this point microfibrils are formed between the contacting surfaces which consist of cellulose [35].

The "Lectin" hypothesis which explains the specificity for formation of nodules as resulting from the complementary interaction of the surface structures of bacterial and plant cells. This forms an "antigen-antibody" complex, which was first recognized in the early 1980's. Although initially a controversial subject, Diaz and co-writer [39] provided strong genetic evidence that the root lectins of legumes determine the specifics of symbiosis development. They injected special strains of *Agrobacterium* into the roots of clover through transfer of the psl gene (codes for synthesis of pea lectin). This resulted in clover acquiring the ability to form nodules with pea rhizobia. A similar study ten years later had the same result [40]. Transgenic lucerne plants carrying genes encoding soybean or pea lectin generated a structure similar to that of the root nodule in response to incoulation with pea and soybean nodule bacteria producing the specific Nod-factor. However not all of the structures created were able to form complete infection threads. The results confirm the importance of the presence of lectin in establishing contact between the symbionts and demonstrates the unique role of exopolysaccharides (EPS) in the formation of nodules. *Galega officinalis* and *orientalis* are other examples demonstrating the importance of lectins in legume-rhizobial symbiosis. The validity of this result is confirmed by sequencing the amplifiers of DNA extracted from the seeds of the mentioned crops from different geographical locations [41].

Therefore numerous experimental data confirms the important role of lectins in the early stages of the symbiosis and suggests the involvement of these proteins in a variety of physiological processes of plants.

Once they have prepared each other with molecular signals, the partners begin to form the structural basis of symbiosis - a nodule. This leads to a morphophysiological differentiation of bacterial and plant cells [18, 25]. Nod-factors control the phenotypic changes that occur in the roots of the host plant (the initial stage of symbiosis.) In addition there is also the deformation and proliferation of root hairs [25], the expression of early nodule genes [5], the induction of mitotic divisions in the cortex, followed by the onset of the first stages of histogenesis of nodules [42, 43].

Based on the assumption regarding the participation of lectins in various physiological processes in the formation of symbiosis, much attention was paid by researchers to the study of the direct effects of plant lectins on the manifestation of the symbiotic properties of nodule bacteria. In particular, it has been shown that the incubation of nodule bacteria *Bradyrhizobium japonicum* whith soybean root exudates and lectin from its seeds increased the activity of nodule formation in the mutated soybean HS 111, characterized by delayed nodulation, and in doubling the number of nodules formed by wild strain USDA 110 [44]. Later it was shown that treatment of rhizobia by specific plant lectin increases their virulence and competitiveness [34], and also increases the quantity of infectious threads in the roots of leguminous plants [45]. The particular modulating effect of the plant lectin on the formation and functioning of symbiosis has been established, which is manifested in the stimulating effect on partners, homologous data of the proteins, and the neutral or suppressive reaction of lectin which is not in accordance with the symbionts [46]. Moreover it is shown that the incubation of nodule bacteria with homologous lectin has a positive effect on the symbiotic properties of the active strain *Bradyrhizobium japonicum* 634b, and the intensity of the basic physiological processes in plants - nitrogen fixation and photosynthesis. The introduction of the same protein into an inactive strain of rhizobia suspension *Bradyrhizobium japonicum* 604k has a suppressive effect on the symbiotic system which it took part in forming as well as the stated processes [47]. It has also been found that the nature of the influence of homologous lectin on the growth of nodule bacteria in a axenic culture and the biosynthesis of extracellular carbohydrates [48], the ability of rhizobia to form nodules and their nitrogenase activity, as well as its impact on the productivity of host plants depends on the concentration of this protein in the bacterial suspension [49].

Co-incubation of nodule bacteria with lectin enhances soybean growth processes in the early stages of ontogeny as well as in the functioning of the symbiotic system, increases the nitrogen-fixing activity and, consequently, the productivity of plants [47, 49, 50]. This is why in recent years work was done on the selection of the optimal concentration of the homologous lectin, and the length of time of its co-incubation with the culture of *Bradyrhizobium japonicum* in the production of bacterial fertilizers both in liquid and solid form. Lectin is used as a growth stimulating biologically active substance, which is introduced into the culture of bacteria under certain conditions [51].

In the bacterial model *Bacillus subtilis* where antibiotics were used as metabolic inhibitors, which block the processes of replication, transcription or transmission, there was established an ability of carbohydrate-binding proteins, lectins of plant origin, to have various effects on intracellular processes. Among the presumed processes affected by lectins were reparative functions [52]. It was also found that lectins from the seeds of leguminous plants that have a high molecular weight (> 100 000 Da) can stimulate the respiration of some *Rhizobium*. According to the authors [53], this effect is induced by the lectins, and significantly increases the interaction of lectin-*Rhizobium* due to the physiological properties of the bacteria.

The presented data gives cause to consider the homologous lectin, not only as a receptor or a signaling molecule in the initial stages of symbiosis, but also as a molecular signal that changes the metabolism of rhizobia, which significantly affects their symbiotic properties and the physiological status of the host plant.

In the process of development of nodules in legume plants, lectins are localized at least in three different places. Their possible functions are related to areas of infection on the surface of the roots. In the nodule primordia lectins can stimulate mitotic activity reducing the threshold of sensitivity to the rhizobial Nod-factors. At the same time, in the central part of the mature nodule, lectins can function as spare nitrogen compounds [54].

It is thought that legume lectins are involved in the formation of nodules. The activating effect of lectin on the synthesis of extracellular and capsular polysaccharides of rhizobia is shown, which in turn induces the formation of infection threads [55]. Data for the study of lectin gene expression during the development and functioning of root nodules of lucerne provided evidence supporting the involvement of these proteins in the early stages of the ontogenesis of nodules [56]. A study has been made of a number of genes and plant proteins encoded by them, which play a role in the formation and functioning of root nodules. Among them is a lectin related to the pea *ps*1 gene with the presumed function as a mitotic stimulator and soybean lectin *le*1, which takes part in the attachment of cells [19].

It is established that the proteins extracted from lupine and soybean nodules are capable of hemagglutinating activity. A comparative analysis of the nitrogen-fixation of soybean nodules and the lectin activity of the protein extracts from these nodules revealed a link between these rates in ontogenesis [57]. This data suggests that the proteins which are capable of lectin activity may be directly involved in the processes of nodule functioning.

A hypothesis was made by Antoniuk and his co-writer that the lectin of wheat germs acts as a signal for *Azospirillum brasilense,* which changes the metabolism of the bacteria in a direction favorable to the growth and development of the host plant [58, 59]. According to the authors, the level of lectin in the plant depends on a number of conditions and is one of the factors responsible for the variability in the results of wheat inoculation.

Nodule bacteria penetrate the plant cell (as opposed to *Azospirillum*) and transform into bacteroids, which do not divide, but only increase in size. Nonetheless a number of analogies can be made between the influence of the wheat germ lectin on *Azospirillum* and the influence of a specific legume lectin on *Rhizobium*. In the first as well as in the second case, the homologous lectin has a positive effect on the symbiotic properties of bacteria, which promotes a more efficient interaction between the partners in the different symbiotic systems that they create. There remains the little-studied question regarding the participation of lectins in the functioning of the symbiotic apparatus of legumes, but it is possible that the presence of proteins with lectin activity in the nodules is associated with the biosynthesis of proteins (including nitrogenase) in bacteroids and in ensuring conditions for the effective functioning of their symbiotic system.

Legume lectins may influence the receptor and signaling molecules during the stages of the symbiosis. Hemagglutinating activity of the proteins contained in the nodules, varies depending on the efficiency of their functioning. Specific lectins are able to modify the symbiotic properties of the nodule bacteria which positively affect the physiological status of the host plant and, ultimately, the effectiveness of the symbiotic system. Uncovering the role that lectins have in the functioning of the symbiotic apparatus of legumes requires further research. At the same time these proteins can be regarded as one of the factors for the effective symbiosis, which must be considered when developing and implementing new approaches to the management of production process in legume plants.

3. Biotechnology of Diazotrophs

Microorganisms which are the basis of biological preparations must comply with certain requirements, as well as have certain properties including virulence, activity and effectiveness, specificity, competitiveness, and technological properties (the ability to accumulate titer in the standard and production environments). Among the general requirements for the development of new biological preparations are the following: a high titer of active bacteria cells, the desired length of shelf life, transportability, technological properties (solubility, the ability to stay on the seeds, etc.), as well as the economic efficiency of their production.

The efficiency of symbiotic systems, "plant - microorganism" is defined by the virulence and level of activirty of microsymbionts. Virulence of nodule bacteria for example is the ability to penetrate into the soybean root through root hairs and lead to the formation of nodules by complex morphophysiological changes. The first stage of virulence is tumorgenic activity which is the ability to form tumors on the roots. Truly virulent strains are characterized by nodulating activity (nodulation), the ability to form complete nodules. A nodule is a complex organ of the plant, the main structure of which is made up of: tissue infected with bacteria where molecular nitrogen is fixated; conductive tissue through which photosynthates are delivered and products of nitrogen fixation are taken out; and the meristem, due to which the growth of nodules occurs [1, 5, 18].

At certain stages of the formation of associative rhizocenosis and legume-rhizobial symbiosis, other important properties of bacteria develop such as: nitrogen-fixing activity - the recovery rate of which is N_2 in NH_3, and symbiotic effectiveness - the ability of plants to develop intensively, using symbiotrophic feeding on nitrogen. Symbiotic efficiency is to a great extent determined by the nitrogen-fixing activity of nodules, especially in conditions where there is a shortage of fixed nitrogen, but the efficiency of the symbiosis also depends on a number of factors not directly linked to nitrogen fixation. An important role in determining the productivity of plant-bacterial interactions can be played by the compatibility of the metabolic systems of the partners (for example, ways of transporting nitrogen and carbon), and the absence of active defense reactions in plants in response to the presence or the penetration of microorganisms. Bacteria found in the rhizosphere or nodules can synthesize substances that stimulate (phytohormones, vitamins) or inhibit (rhizobiotoxins) the development of the host plant. It has been found that the effective and ineffective strains of nitrogen-fixing bacteria differ on a number of biochemical parameters. Effective strains most probably have a far richer metabolic fund, and their acidifying- restoring processes occur more actively [1, 3, 6, 18].

Specificity is the ability of bacteria to selectively interact with a particular species or group of plants. Specificity is one of the important systematic characteristics of nodule bacteria and is closely linked to their activity. *Rhizobium* for example is divided into: active, low-active, and non-active. It should be noted that the virulence of the bacteria and the activity may depend on the certain strain, species, varietal specificity of the plant, soil, climatic conditions and other factors [5, 6, 18].

Different races of nitrogen-fixing bacteria compete with each other. The more virulent strains will more actively colonize or inoculate relative to other strains, the root system of plants specific to them. Researchers have different interpretations of the concept of bacteria competitiveness. Some consider it to be the ability to compete with the strains which spontaneously inoculate the plants; while others consider it to be resistance to the local saprophyte microflora and to displace the local strains [60, 61].

There are several methods for application of microorganisms to plants: a liquid culture, preparations on gel substrates (bacterial exopolysaccharides, silica gel, highly dispersed materials) and applications on solid carriers (vermiculite, lignin, perlite, peat) [6, 62]. Obtaining the production-strains of nitrogen-fixing bacteria and the creation of biological preparations based on them - is a long process of research and production (Fig. 1), in which we can distinguish the following stages:

1. Research:

- reisolation of the nitrogen-fixing microorganism from its natural habitat;

- an introduction into the culture and analytical selection of perspective strains;

- generation of new highly effective strains using genetic engineering methods;

- the study of physiological and biochemical characteristics, symbiotic properties, competitiveness, efficiency and technological properties of nitrogen-fixing bacteria;

- conducting research trials, registration, depositing and patenting of the strains;

- cultivating of axenic cultures of microorganisms in museum conditions.

2. The preparatory stage of production:

- preparation of nutritional medium, and nutrient supplements;

- revival of the physiological activity of nitrogen-fixing bacteria after storage (reseeding of culture, reactivation on a shaker at constant temperature and aeration);

- establishing a system of air purification, and other activities (if necessary) predetermined by the specifics of the production.

3. Production process:

- the cultivation of bacteria on industrial shakers, in flasks or in fermenters (bioreactors in the form of specially constructed chambers, in which occurs the process of growing microorganisms and fermentation);

- selection of the final product (obtaining a liquid culture of the microorganism);

- preparation of the carrier (packing and sterilization of the substrate, the introduction nutrient supplements), or containers for the liquid substance (sterilization of containers);

- inoculation of the used carrier.

4. Storage or incubation of the biological preparations under certain temperature conditions.

5. Quality control (titer of bacterial cells / 1g of the preparative form, the presence of foreign microflora).

6. Treatment of wastewater and gas emissions, recycling waste.

The general scheme of production (see figure 1) includes the above mentioned stages; however each case has its peculiarities. This is due to the degree of complexity of each separately organized biotechnological process (laboratory conditions, the use of shakers, industrial rockers, different fermenters), and the technological requirements of cultured organisms and choice of the preparative form of the final product (liquid culture, different carriers). For example, a kit for the fermentation of microorganisms in liquid media may consist of inoculation and production fermentors, air purification systems for fermentation, a set of connecting tubes, as well as the compressor and the gas meter.

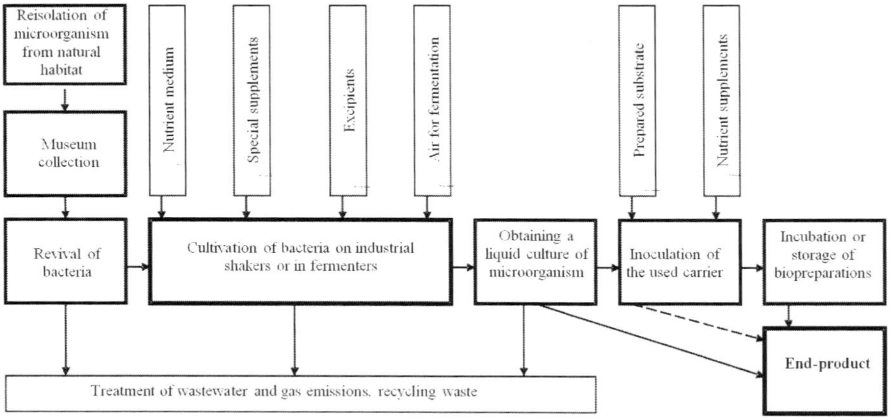

Figure 1. The scheme of creation and production biopreparations, based on nitrogen-fixing bacteria (Sytnikov, 2012).

After establishing a positive impact of the soil microflora on plant productivity, the question arose about the practical application of micro-organisms, in particular those fixing atmospheric nitrogen. The creation of biological products on the basis of nitrogen-fixing bacteria was dictated by the need to preserve their life and functional activity in a specific preparative form (nutrient medium or substrate) aiming to widen its practical application.

The first biological preparation based on the nitrogen-fixing nodule bacteria - Nitragin, was produced in Germany in 1896 [63]. In the Soviet Union the preparation Rizotorfin was created and widely used. It is a peat substrate with nutrient supplements containing highly active and competitive strains of rhizobia for a specific type of legume plants. Bacterial fertilizers for legume plants based on symbiotrophic nitrogen fixers are the most commonly used biological preparation of diazotrophs. For example, in the United States Nitragin and Double-noktin were produced and used on hundreds of thousands of hectares of crops. In other countries the following compounds were used: Nitrosoil and Nitrum in Argentina and Uruguay, Rhizoctonia in New Zealand, Tropical-inoculum, Nodulite and Nitrogerm in Australia, Ariss Agro in India, and Okadin in Egypt [64]. In Ukraine, up to 20% of crop cultures (predominantly soybean) are inoculated by biological preparations [3].

To inoculate the seeds the most commonly used preparations are based on nodule bacteria from the families of *Rhizobium, Bradyrhizobium, Sinorhizobium* for the legume plants on the basis of vermiculite or perlite under the general name of Rizobofit. Symbiotrophic nodule bacteria provide nitrogen fixation of up to 350 kg / ha in lucerne during the second year of vegetation [65], as well as up to 280 kg / ha for soybean [66] and 70 kg / ha in pea [9]. As a result of many years of research it has been found that the inoculation of plants with high efficiency strains of rhizobia increases the productivity of legumes by an average of 10-25% [67]. On the basis of associative nitrogen-fixing bacteria, there has been developed a technology of production of a whole line of biological preparations. The list of biotechnological products and microbial preparations for crop production, has increased considerably over the past decade and includes products that are based on free-living, associative, symbiotrophic nitrogen-fixing bacteria, and phosphate-mobilizing bacteria, as well as products of binary action obtained from a combination of various microorganisms [6, 62, 64].

4. Ways to Improve the Productivity of the Soybean-Rhizobial Symbiosis

One of the objects with prospects in biotechnology are cyanobacteria because of their ability in photosynthesis, nitrogen fixation, the synthesis of biologically active substances and growth activating substances. These substances positively influence on soil fertility and the activity of soil biota. Cyanobacteria are closely linked to the bacteria such as *Rhizobium, Agrobacterium, Pseudomonas,* and are capable of forming a new stable association that opens the perspective for the creation of efficient microbial consortia and preparations based on them. Positive results were obtained when using in agricultural biotechnology artificial algae-rhizobial associations to inoculate the seeds of lotus, peas and clover. To date, it has been proved that under the influence of artificial consortia on the basis of *Nostoc* and various *Rhizobium* the effect of nitrogenation is enhanced on leguminous plants [6, 68-70].

The co-treatment of lucerne seeds by nitrogen-fixing bacteria and cyanobacteria *Nostoc punctiforme,* as well as by their binary compositions stimulates plant growth and development. It was found that the most effective method was co-inoculation by cyanobacteria with certain Tn5-mutants of *Rhizobium* when compared to inoculation by monocultures [71]. We have also studied the reaction of soybean *Glycine max* (L.) Merr. to inoculation by algal-rhizobial compositions based on the nodule bacteria *Bradyrhizobium japonicum,* and cyanobacteria *Nostoc punctiforme* [72]. It is shown that the inoculation of soybean seeds by algal-rhizobial compositions enhances germination and positively influences seedling formation. The inclusion of cyanobacteria into the inoculation suspension of rhizobia and Tn5-mutants in certain combinations can stimulate the growth and development of soybean, the accumulation of photosynthetic pigments in the leaves and the protein content in seeds. However, it does not have a significant effect on the activity of nitrogen-fixing nodules, and plant productivity. This data indicates the need to find effective algal-rhizobial complex compositions to inoculate soybean plants by the optimal selection of the strains of bacteria and the ratio of inoculating agents. In our opinion further study is also required of the algal-rhizobial compositions created on the basis of microorganisms with genetically modified properties.

Creation and selection of compatible algal-rhizobial associations, including the axenic cultures of cyanobacteria and nodule bacteria as well as their Tn5-mutants may be one of the methods of biologically stimulating legume-rhizobial symbiosis, which enhances the significance of the interaction of rhizobia with plants and the efficiency of bacterial preparations based on them.

In the formation of legume-rhizobial symbiosis the essential components of the interaction of the symbiotic partners are polysaccharides which are synthesized by nitrogen-fixing bacteria. Perhaps it was under the influence of these substances, acting as elicitors in the early stages of the morphogenesis of nodules, that gene activation occurrs for a number of plant genes that "sit in silence" in the uninoculated plant roots.

It has been suggested that the polysaccharides of non-rhizobial origin, as well as glycopolymers of rhizobia, are able to mimic the action of phytohormones and stimulate the processes of nodulation and morphogenesis in the legume-rhizobial symbiosis [73]. It is shown that the cells of nodule bacteria during the action of exogenous polysaccharides (Baktozol) increased in growth, produced more biomass and changed the activity of some enzymes of the nitrogen exchange [74, 75]. The stimulating effect of the synthetic polysaccharide (PS MOD-19) on the growth of rhizobia, biomass accumulation and changes in their metabolism during the growth of bacteria on solid and liquid media in the presence of the biopolymer was later revealed. During the processing of the seeds of peas (*Pisum sativum* L.) prior to sowing whith PS MOD-19 there was found an increased rhizogenesis in plants, an increased peroxidase activity in plant cells, as well as the increase of the effectiveness of symbiosis as a whole due to secondary formation of nodules on side roots and the prolonging of the period of their active nitrogen fixation [76]. In this regard, synthetic polysaccharides may be of interest as biologically active compounds for practical application, in particular, for the expansion of the range of substances that can stimulate the growth activity of rhizobia and, to a greater extent, enhance and prolong the activity of nitrogen-fixing nodules formed on roots of leguminous plants. The latter circumstance is of particular importance for legumes with a short vegetation period, of which the most vivid representative is the pea [73]. In our opinion, further study of the stimulating effect of polysaccharides of different origin is a promising direction of improving the efficiency of legume-rhizobial symbiosis.

Our studies also show the effectiveness of biological products of nodule bacteria modified by homologous lectin, and the economic feasibility of their use [47, 49-51, 77-79]. In accordance with the existing concepts about the mechanisms of interaction of plants with rhizobia, polysaccharides of the latter are a factor which provides a "recognition" by the bacteria of the corresponding host plant through complementary binding to the plant lectin. Lectins are proteins that have the ability to reversibly and selectively bind to carbohydrates and carbohydrate parts of biopolymers without changing the covalent structure of the latter [77, 79]. Along with other biologically active substances legume lectins during the germination of seeds secrete themselves into the environment [21]. These proteins stimulate the proliferation and active movement of soil microorganisms to the roots, and influence the growth of microsymbionts and the synthesis by them of exoglycans [48]. Plant lectins are regarded as one of the factors of effective symbiosis, which is also proposed for consideration when developing and implementing new approaches to the management of production processes in legume plants [77]. It is known that treatment of rhizobia by plant lectin specific to them has a positive effect on their virulence and competitiveness [34], and also increases the nitrogen-fixing activity of root nodules. Lectin acts on the biosynthesis of nitrogenase [58, 59] in the bacterial cell. As a consequence, pre-incubation of rhizobia with homologous lectin enhances the growth processes of plants and increases the productivity of the symbiosis [47].

The results of our studies indicate to the prospects of using bacterial agents modified by homologous lectin, both liquid and manufactured on a solid carrier (Table 1).

The use of this protein makes it possible to improve the efficiency of the symbiotic system of soybean *Glycine max* (L.) Merr. and increase its productivity. The tests showed that the concentration of homologous lectin of 100 mcg / ml of bacterial suspension *Bradyrhizobium japonicum* is the optimal dose in both physiological and economical aspects in the manufacture of rhizobia preparations using perlite as a solid carrier (Table 2). In addition, the effectiveness has been established of using bacterial preparations modified by homologous lectin, on the basis of active production-strains of rhizobia and some Tn5-mutants (T66 and T3-11) [78].

Carrier	Lectin concentration, μg/ml	Seed yield, cwt/ha	Yield increment relative to control	
			cwt/ha	%
Liquid	0	29,3 ± 2,7	–	–
	100	36,6 ± 1,5	7,3	24,9
	300	32,5 ± 1,8	3,2	10,9
Perlite	0	34,6 ± 1,2	–	–
	100	43,1 ± 1,1	8,5	24,5
	300	40,3 ± 1,7	5,6	16,4
$LSD_{0,05}$		4,6		

Table 1. Soybean seeds harvest with the application of various preparations forms of nodule bacteria with different homologous lectin concentrations (average of the 2 experiments) [80].

Lectin concentration, µg/ml	Seed yield, cwt/ha				Yield increment relative to control	
	I	II	III	Average of the 3 experiments	cwt/ha	%
0	36,2 ± 0,7	33,1 ± 1,8	22,5 ± 0,7	30,6 ± 1,1	–	–
100	45,1 ± 1,4	41,2 ± 0,8	25,8 ± 1,1	37,4 ± 1,1	6,8	22,2
300	41,9 ± 1,5	38,8 ± 2,0	27,3 ± 1,0	36,0 ± 1,5	5,4	17,6
$LSD_{0,05}$	4,2	5,1	2,5			

Table 2. Productivity of soybean inoculated with biological rhizobial preparation on the solid carrier (perlit) and modified with homologous lectin [80].

Note. I, II - 2005., III - 2006.

The economic efficiency of using bacterial preparations to inoculate the seeds of legume plants before sowing depends on the increase of yield, its cost and additional expenses. The calculation of economic efficiency of soybean production via a typical technological map was shown on a farm in the Kiev region (Table 3). By using bacterial preparations of soybean nodule bacteria (application of inoculation) the yield of the crop increased throughout Ukraine by an average of at least 12%, which makes it possible to increase the profitability of production to 27.8% (see Table 3). The use of bacterial preparations modified by homologous lectin in the conditions of our field trials led to an increase in soybean yield of no less than 9.8% in comparison with conventional inoculation (see Table. 1 and 2). A comparative analysis of various indicators of economic efficiency of soybean cultivation points to the feasibility of using bacterial preparations to inoculate seeds (see Table 3). The profitability of production at the same time increased by 11.7%. The maximum profitability (39.2%) was observed when using bacterial preparations modified by homologous lectin, indicating the benefits of their use. Table 4 also shows that the use of bacterial preparations reduced production costs and increased net income. Thus, the cultivation of soybeans in the conditions and fields of Ukraine can produce about 0.4 hryvnia net profit to every 1 hryvnia spent. Therefore different preparatory forms of nitrogen-fixing microorganisms effectively increase the productivity of plants and can be recommended for agricultural production. The use of bacterial agents leads to a slight increase in the cost of production while the economic effect of using nitrogen-fixing bacteria is achieved due to the additional crop yield, the savings made on fertilizers and the reduction of other production energy costs.

Parameter	Unit of measure	Control	Application of inoculation	Biopreparation with lectin
Seed yield	cwt/ha	20,0	22,4	24,6
Supplementary yield	cwt/ha	–	2,4	4,6
Cost of yield	UAH/ha	2400,0	2688,0	2952,0
Costs of biopreparation	UAH/ha	–	36,0	53,0
Net cost	UAH/cwt	103,34	93,88	86,17
Net profit	UAH/ha	333,13	585,13	832,13
Profitability	%	16,1	27,8	39,2

Table 3. Economic effect of bacterial preparations application while producing soybean seeds (the condition of 2006), the preparations being modified with homologous lectin [80].

Note. UAH - Ukrainian hryvna.

5. Conclusion

Given the uniqueness of the biochemical processes of nitrogen fixation by microorganisms and global importance of biological nitrogen fixation in the nitrogen cycle in the biosphere, this phenomenon requires further study and attention. Scientists still face the problem of finding new nitrogen-fixing communities, the creation of genetically engineered micro-and macro- symbionts, as well as studying specific mechanisms of the relationship between plants and microorganisms.

A comprehensive study of plant-bacterial symbioses and associations made possible the use of biological nitrogen fixation in practice and opened up prospects for the management of this process. Work is presently being done on the generation, testing and introduction of new strains of *Rhizobium* and associative nitrogen fixers for the cultivation of plants in our latitudes. Among the promising works in the creation of biological preparations based on symbiotic nitrogen fixers are those regarding the creation of agricultural consortiums with cyanobacteria, the use of bacterial and synthetic polysaccharides, and plant lectins. The biological preparations created based on beneficial microorganisms are an excellent alternative to mineral fertilizers. However, they have not yet been widely applied, which of course does not facilitate the resolution of a number of existing environmental and economic problems.

Author details

Denis M. Sytnikov[1,2*]

Address all correspondence to: sytnikov@list.ru

1 Department of Phytohormonology of N.G. Kholodnyi Institute of Botany, National Academy of Science of Ukraine, Ukraine

2 Department of Microbiology and Virology, I.I. Mechnikov Odessa National University, Ukraine

References

[1] Kretovich, V. L. (1994). Biokhimia Usvoeniya Azota Vozdukha Rasteniyami (Biochemistry of Plants Nitrogen-Fixation). Moscow: Nauka.

[2] Ignatov, V. V. (1998). Biological Nitrogen-Fixation and Nitrogen Fixers. *Sorosovskii Obrazovatelnii Zhurnal (Soros Education Journal)*, 9, 28-33.

[3] Kots, S., Ya, , Morgun, V. V., Patyka, V. F., et al. (2011). Biologicheskaya Fiksatsiya Azota: Bobovo-Rizobialnyi Simbioz (Biological Nitrogen Fixation: Legume-Rhizobial Symbiosis). Kiev: Logos.

[4] Novikova, N. I. (1996). Modern Views on the Phylogeny and Taxonomy of Nodule Bacteria. *Mikrobiologia*, 65(4), 437-450.

[5] Spaynk, G., Kondoroshi, A., Hukas, P., editors. Tikhonovich, I.A., Provorov, N.A., Engl. translation editors (2002). Rhizobiaceae. Molecular Biology of Bacteria Interacting with Plants. St.-Petersburg.

[6] Patyka, V. P., Kots, S.Ya., Volkogon, V. V., et al. (2003). Biologichnyi Azot (Biological Nitrogen). Kiev: Svit.

[7] Kukk, E.G. (1977). Division of Cyanobacteria. In: Fedorov A.A. (ed.), *Zhyzn Rastenii (The Life of Plants)*, Moscow: Prosveschenie, 78-92.

[8] Patyka, V. P., Krutylo, D. V., & Kovalevska, T. M. (2004). Influence of Aboriginal Populations of Soybean Nodule Bacteria on Symbiotic Activity Introductional Strain Bradyrhizobium japonicum 634b. *Microbiologichnyi Zhurnal*, 66(3), 14-21.

[9] Kots, S.Ya., Malichenko, S. M., Krugova, E. D., et al. (2001). Fiziologo-Biokhimichni Osoblyvosti Zhyvlennya Roslyn Biologichnym Azotom (Physiological and Biochemical Features of Plant Nutrition of Biological Nitrogen). Kiev: Logos.

[10] Antonyuk, V. O. (2005). Lektiny ta ikh Syrovynni Dzherela (Lectins and their Sources of Raw Materials). Lvov: PP «Kvart».

[11] Yamaleyeva, A. A. (2001). Lektiny Rastenii i ikh Biologicheskaya Rol (Plant lectins and their biological role). Ufa: Publisher Bashkir. University.

[12] Aleksidze, G., Ya, , Vyskrebentseva, E. I., & Korolyev, N. P. (1984). Purification and Some Properties of Lectin from the Fraction of the Cell Walls of Sugar Beet. *Fiziologiya Rastenii*, 31(6), 1021-1027.

[13] Vyskrebentseva, E. I., & Borisova, N. N. (1996). Distribution of Lectin Activity in the Mitochondria of Sugar Beet: Lectin Activity of Mitochondrial Membranes and Matrix of Sugar Beet. *Fiziologiya Rastenii*, 43(4), 527-532.

[14] Etzler, M.E. (1998). From Structure to Activity: New Insights into the Functions of Legume Lectins. *Trends in Glycosci. and Glycotechnol*, 10(53), 247-255.

[15] Loris, R., Hamelryck, T., Bouckaert, J., & Wyns, L. (1998). Legume Lectin Structure. *BBA- Protein Str. and Mol. Enzymol.*, 1383(1), 9-36.

[16] Lutsyk, M. D., Panasyuk, E. N., & Lutsyk, A. D. (1981). Lektiny (Lectins). Lvov: Vyscha shkola, Pablisher Lvov University.

[17] Sharon, N., & Lis, H. (1990). Legume Lectins - a Large Family of Homologous Proteins. *FASEB J.*, 4(14), 3198-3208.

[18] Tikhonovich, I. A., Provorov, N. A., editors. (1998). Genetika Simbioticheskoi Azotfiksatsii s Osnovami Selektsii (Genetics of Symbiotic Nitrogen-Fixation and Basis of Breeding). St.-Petersburg: Nauka.

[19] Stougaard, J. (2000). Regulators and Regulation of Legume Root Nodule Development. *Plant Physiol.*, 124(2), 531-540.

[20] Pospelov, S. V., Samorodov, V. M., & Pospelova, G. D. (2001). The Main Directions and Results of Research in Lectin Biology in Poltava State Agricultural Institute. *Bulletin of the Poltava State Agricultural Institute*, 4, 42-47.

[21] Fountain, D. W., Foard, D. E., Replogle, W. D., & Yang, W. K. (1977). Lectin Release by Soybean Seeds. *Science*, 197(4309), 1185-1187.

[22] Chuiko, N. V., Antonyuk, T. S., & Kurdish, I. K. (2002). Chemotaxis of Bradyrhizobium Japonicum to Various Organic Compounds. *Mikrobiologiya*, 71(4), 460-466.

[23] Ovtsyna, A. O., & Tikhonovich, I. A. (2004). Structure, Function, and the Practical Application of Signaling Molecules that Initiate the Development of the Legume-Rhizobial Symbiosis. *Ekologicheskaya Genetika*, 2(3), 14-24.

[24] Geurts, R., & Bisseling, T. (2002). Rhizobium Nod Factor Perception and Signalling. *Plant Cell.*, 14, S239-S249.

[25] Martihus, M. F., Regina, G. S., & Franco, A. A. (2002). Importância Compositos Fenólicos nas Interações Entre Espécies Leguminosas e Rizóbio. *Rev. Univ. rural. Sér. Ciê. vida / Univ. fed. rur. Rio de Janeiro*, 22(1), 65-81.

[26] Hirsh, A. M., Lum, M. R., & Downie, J. A. (2001). What Makes the Rhizobia-Legume Symbiosis so Special? *Plant Physiol.*, 127, 1484-1492.

[27] Peters, N. K., Frost, J. W., & Long, S. R. (1986). A Plant Flavone, Luteolin, Induces Expression of Rhizobium Meliloti Nodulation Genes. *Science*, 233, 977-980.

[28] Parret, X., Staehelin, C., & Broughton, W. J. (2000). Molecular Basis of Symbiotic Promiscuity. *Microbiol. Molecular. Biol. Rev.*, 64(1), 180-201.

[29] Kalsi, G., & Etzler, M. E. (2000). Localization of a Nod Factor-Binding Protein in Legume Roots and Factors Influencing its Distribution and Expression. *Plant Physiol.*, 124, 1039-1048.

[30] Kijne, J. W., Bauchrowitz, M. A., & Diaz, C. L. (1997). Root Lectins and Rhizobia. *Plant Physiol.*, 115(3), 869-873.

[31] Dey, R. B., Okada, M., Ito, Y., et al. (2001). Binding Site for Chitin Oligosaccharides in the Soybean Plasma Membrane. *Plant Physiol.*, 126, 1162-1173.

[32] Mathis, R., Van Gijsegem, F., De Rycke, R., et al. (2005). Lipopolysaccharides as a Communication Signal for Progression of Legume Endosymbiosis. *PNAS*, 102(7), 2655-2660.

[33] Malichenko, S. M., Datsenko, V. K., & Mamenko, P. M. (2001). Adsorption of Nodule Bacteria of Specific and Nonspecific Legume Roots. *In: Plant Ontogenesis, Biological Fixation of Molecular Nitrogen and Nitrogen Metabolism: proceedings of the International Conference*, 1-4 October 2001, Ternopol, Ukraine.

[34] Lodeiro, A. R., Lopez-Garsia, S. L., Vazquez, T. E. E., & Favelukes, G. (2000). Stimulation of Adhesiveness, Infectivity, and Competitiveness for Nodulation of Bradyrhizobium Japonicum by its Pretreatment with Soybean Seed Lectin. *FEMS Microbiol. Lett.*, 188(2), 177-184.

[35] Chumakov, M. I. (1996). Involvement of Surface Polysaccharides and Proteins of Bacteria of the Family Rhizobiaceae in the Adsorption and Attachment to Plant Surfaces. *Mikrobiologiya*, 65(6), 725-739.

[36] Hirsh, A.M. (1999). Role of Lectins (and Rhizobial Exopolysaccharides) in Legume Nodulation. *Current Opinion in Plant Biol.*, 2, 320-326.

[37] Wisniewski, J.-P., Delmotte, F.M. (1996). Modulation of Carbohydrate-Binding Capacities and Attachment Ability of Bradyrhizobium Sp. (Lupinus) to White Lupin Roots. *Can. J. Microbiol.*, 42, 234-242.

[38] Dazzo, F. B., Truchet, G. L., Sherwood, J. E., et al. (1984). Specific Phases of Root Hair Attachment in the Rhizobium Trifolii-Clover Symbiosis. *Appl. and Envir. Microbiol.*, 48(6), 1140-1150.

[39] Diaz, C. L., Hosselet, M., & Logman, G. J. J. (1990). Distribution of Glucose/Mannose-Specific Isolectins in Pea (Pisum Sativum L.) Seedlings. *Planta*, 181, 451-461.

[40] Van Rhijn, P., Fujishige, N. A., Lim, P. O., & Hirsh, A. M. (2001). Sugar-Binding of Pea Lectin Enhances Heterologous Infection of Transgenic Alfalfa Plants by Rhizobium Leguminosarum Biovar Viciae. *Plant Physiol.*, 126, 133-144.

[41] Baimiyev, Al., Ch, , & Gubaidullin, I. I. (2001). Differences in the Carbohydrate-Binding Structure of the Lectin Peptides from Galega Orientalis and Officinalis Determine the Difference in their Choice of a Partner Legume-Rhizobial Symbiosis. *In: Young Scientists of the Volga-Ural region at the turn of the century: proceedings of the Jubilee Scientific Conference of Young Scientists*, 24-26 October, Ufa.

[42] Chimote, Vivek., Kashyap, L.R. (2001). Lipochitooligosaccharides and Legume-Rhizobium Symbiosis. A new concept. *Indian J. Exp. Biol.*, 39(5), 401-409.

[43] D'Haeze, W., & Holsters, M. (2002). Nod-Factor Structures, Responses, and Perception During Initiation of Nodule Development. *Glycobiology*, 12(6), 79R-105R.

[44] Halverson, L. J., & Stacey, G. (1986). Effects of Lectin on Nodulation by Wild-Type Bradyrhizobium Japonicum and a Nodulation-Defective Mutant. *Appl. Environ. Microbiol.*, 51(4), 753-760.

[45] Brelles-Marino, G., Costa, G. A., & Boiardi, J. L. (1996). Enhancement of Infection Thread Formation by Rhizobium Etli Incubated with Bean Seed Lectin. *Microbiol. Res.*, 151, 243-246.

[46] Malichenko, S. M., Datsenko, V. K., Mamenko, P. M., & Kots, S.Ya. (2002). Participation Specific and Nonspecific Lectins to Leguminous Plants Nodule Bacteria in the Formation and Functioning of Nitrogen-Fixing Symbiosis. *Sci. Notes of Ternopol Pedagogical Univ.*, 3(18), 49-57.

[47] Sytnikov, D. M., Kots, S.Ya., & Malichenko, S. M. (2005). Efficiency of Symbiotic System Soybean- Bradyrhizobium Japonicum under the Influence of Homolohous Lectin and Different Mineral Nitrogen Supply. *Fiziol. Biokhim. Kul't. Rast.*, 37(5), 394-401.

[48] Kosenko, L. V., & Mandrovskaya, N. M. (1998). Effect of Pea Lectin on the Growth of Pea Microsymbionts and the Biosynthesis of Exoglycans. *Mikrobiologiya*, 67(5), 626-630.

[49] Sytnikov, D. M., Malichenko, S. M., Yakimchuk, R. A., & Mosendz, G. N. (2006). Incubation Duration Influence of Inoculation Suspension with Lectin on the Growing, Nitrogenase Activity and Productivity of Soybean. *Fiziol. Biokhim. Kul't. Rast.*, 38(4), 310-316.

[50] Sytnikov, D. M., Kots, S.Ya., Malichenko, S. M., & Kiriziy, D. A. (2006). Photosynthetic Rate and Lectin Activity of Soybean Leaves after Inoculation with Rhizobia together with Homologous Lectin. *Russian Journal of Plant Physiology*, 53(2), 169-175.

[51] Sytnikov, D. M., Kots, S.Ya., & Datsenko, V. K. (2007). Efficacy of Biological Preparations of Soybean Root Nodule Bacteria Modified with a Homologous Lectin. *Applied Biochemistry and Microbiology*, 43(3), 274-279.

[52] Karpova, I. S., Koretska, N. V., & Kochubey, T. O. (2005). The Ability of Lectins Modulate the Action of Antibiotics on the Growth of Mutants Bacillus Subtilis. *Ukr. Biochem. J.*, 77(3), 136-141.

[53] Martínez, C. R., Albertini, A. V. P., Figueiredo, M. V. B., et al. (2004). Respiratory Stimulus in Rhizobium Sp. by Legume Lectins. *World J. Microbiol. and Biotechnol.*, 20(1), 77-83.

[54] Brewin, N. J., & Kardailsky, I. V. (1997). Legume Lectins and Nodulation by Rhizobium. *Trend in Plant Sci.*, 2(3), 92-98.

[55] Mody, B., & Mody, V. (1987). Peanut Agglutinin Induced Alterations in Capsular and Extracellular Polysaccharide Synthesis and Ex-Planta Nitrogenase Activity of Cowpea Rhizobia. *J. Biol. Sci.*, 12(3), 289-296.

[56] Bauchrovitz, M. A., Barker, D. G., & Truchet, G. (1996). Lectin Genes are Expressed throughout Root Nodule Development and During Nitrogen-Fixation in the Rhizobium-Medicago Symbiosis. *Plant J.*, 9(1), 31-43.

[57] Sytnikov, D. M., Kots, S.Ya., & Malichenko, S. M. (2006). Lectin Activity of Different Soybean Organs under Conditions of Efficient and Inefficient Symbiosis. *Fiziol. Biokhim. Kul't. Rast.*, 38(1), 53-60.

[58] Antonyuk, L. P., & Ignatov, V. V. (2001). The Role of Wheat Germ Agglutinin in Plant-Bacterial Interactions: a Hypothesis and Experimental Data to Support it. *Fiziologiya Rastenii*, 48(3), 427-433.

[59] Antonyuk, L. P., Fomina, O. R., & Ignatov, V. V. (1997). Effect of the Wheat Lectin on the Metabolism of Azospirillum Brasilense: Induction of Protein Biosynthesis. *Mikrobiologiya*, 66(2), 172-178.

[60] Sharypova, L. A., & Simarov, B. V. (1985). Method of Comparing the cCompetitiveness of Effective Strains Rhizobium Meliloti. *Trudy VNII S.-Ch. Mikrobiologii*, 15, 85-91.

[61] Butovina, O., Yu, , Tolkachev, N. Z., & Knyazev, A. V. (1997). Highly Competitive Strains of Nodule Bacteria- the Basis of the Effectiveness of Biological Preparations. *Mikrobiologichnyi Zhurnal*, 59(4), 123-131.

[62] Sherstoboyeva, E. V., Dudinova, I. A., Kramarenko, S. N., & Sherstoboyev, N. K. (1997). Biological Preparation of Nitrogen-Fixing Bacteria: Problems and Prospects of their Application. *Mikrobiologichnyi Zhurnal*, 59(4), 109-117.

[63] Dorosinskii, L.M. (1970). Klubenkovye Bakterii i Nitragin (Nodule Bacteria and Nitragin). Leningrad: Kolos.

[64] Sherstoboyeva, O.V. (2004). The Role of Microbiological Preparation in Enhancing the Productivity of Plants Environmentally Safe Methods. *Fiziol. Biokhim. Kul't. Rast.*, 36(3), 229-238.

[65] Kots, S., Ya, , & Mykhalkiv, L. M. (2005). Fiziologiya Symbiozu ta Azotne Zhyvlennya Lutserny (Physiology of Symbiosis and Nitrogen Nutrition of Alfalfa). Kiev: Logos.

[66] Tolkachev, N.Z. (1997). The Potential for Symbiotic Nitrogen Fixation by Soybean in the Southern Ukraine. *Mikrobiologichnyi Zhurnal*, 59(4), 34-41.

[67] Kozhemyakov, A.P. (1997). Productivity of Nitrogen Fixation in Agrocenozes. *Mikrobiologichnyi Zhurnal*, 59(4), 22-28.

[68] Gromov, B. V. (1996). Cyanobacteria in Biosphere. *Sorosovskii Obrazovatelnii Zhurnal (Soros Education Journal)*, 9, 33-39.

[69] Parshykova, T. V., Patsko, E. V., & Sirenko, L. A. (2007). The Use of Microscopic Algae of Eutrophic Water to Maintain Soil Fertility and Increase Yields of Agricultural Crops. *Algologiya*, 17(2), 262-272.

[70] Pankratova, E. M., Trefilova, L. V., Zyablykh, R.Yu., & Ustyuzhanin, I. A. (2008). Cyanobacteria Nostoc Paludosum Kütz as the Basis for Agriculturally Useful Microbial Associations on the Example of Bacteria of the Genus Rhizobium. *Mikrobiologiya*, 77(2), 266-272.

[71] Sytnikov, D. M., Vorobey, N. A., & Kots, S.Ya. (2009). Physiological Reaction of Legume Plants to Inoculation with Algal-Rhizobial Association. *Acta Agronomica Hungarica*, 57(2), 239-244.

[72] Sytnikov, D. M., Vorobey, N. A., & Patsko, E. V. (2010). Reaction of Soybean on Inoculation whith Algal-Rhizobial Compositions. *Biotechnology* (Ukraine), 3(6), 42-48.

[73] Krugova, E. D., Kots, S.Ya., & Mandrovskaya, N. M. (2009). Effects the Synthetic Polysaccharide MOD-19 on the Formation and Functioning of the Symbiotic System Pisum Sativum L.- Rhizobium Leguminosarum Bv. Viciae. *Prikladnaya Biokhimiya i Mikrobiologiya*, 45(3), 324-330.

[74] Krugova, E. D., Mandrovskaya, N. M., & Okhrimenko, S. M. (2002). Effect of Bacterial Exopolysaccharides on the Efficiency of Symbiotic Nitrogen Fixation of Pea and Soybean Plants. *Fiziol. Biokhim. Kul't. Rast.*, 34(3), 239-244.

[75] Kosenko, L.V., Mandrovskaya, N.M., Krugova, E.D., Varbanets, L.D. (2003). Effects of Plant Growth Stimulator Bactozol on Rhizobium Leguminosarum Bv. Viciae 250A and Nitrogen Resistant Mutant M-71 under Conditions of Different Nitrogen Supply. *Mikrobiologiya*, 72(1), 40-47.

[76] Mandrovskaya, N. M., Krugova, E. D., Kots, S.Ya., & Dulnyev, P. G. (2005). Effects of Synthetic Polysaccharide on the Growth of Nodule Bacteria and Rhizogenesis of Roots. *Agroekologichnyi Zhurnal*, 4, 47-51.

[77] Kots, S.Ya., & Sytnikov, D. M. (2007). Legume Lectins as a Factor of Efficient Symbiosis. *Fiziol. Biokhim. Kul't. Rast.*, 39(6), 463-475.

[78] Sytnikov, D. M., Vorobey, N. A., & Beregovenko, S. K. (2008). Efficacy of Bacterial Preparations Based on Tn5-mutants with Homologous Lectin. *Bull. Charkov National Agrarian Univ. Series Biology*, 3(15), 46-52.

[79] Sytnikov, D. M., & Kots, S.Ya. (2009). Participation of Lectins in Plant Physiological Processes. *Fiziol. Biokhim. Kul't. Rast.*, 41(4), 279-296.

[80] Sytnikov, D.M. (2012). Economic Effect and Application of Bacterial Preparations Modified with Homologous Lectin. *Microbiologiya i biotekhnologiya*, 1(17), 75-84.

Chapter 5

Inoculation Methods of *Bradyrhizobium japonicum* on Soybean in South-West Area of Japan

Takeo Yamakawa and Yuichi Saeki

Additional information is available at the end of the chapter

http://dx.doi.org/10.5772/52183

1. Introduction

Inoculation with efficient rhizobia at the ordinary dose does not increase appreciably the seed yield of soybean because the occupation ratio of the inoculated rhizobial strains in the nodules is very low due to competition with less efficient indigenous rhizobia [1, 2]. In order to increase the seed yield by rhizobial inoculation, the occupation ratio of the inoculated strains must be increased. The increase of the occupation ratio has been examined from various viewpoints such as improvement of inoculation method [3]. For the screening of efficient and competitive strains, a large number of useful strains had been isolated from mutagenized and recombinant rhizobia [4, 5, 6].

Furthermore, Williams and Lynch [7] who identified a non-nodulating line of soybean among the progenies from the cross between cvs. Lincoln and Richard, showed that the abnormal nodulation reaction was controlled by a single recessive gene of the host plant, rj_1. Thereafter, cv. Hardee was found to nodulate ineffectively with *Bradyrhizobium japonicum* belonging to the strains 3-24-44 and 122 serogroups [8, 9]. It was demonstrated that the ineffective nodulation was controlled by a host dominant gene, Rj_2. Furthermore, this cultivar was found to nodulate ineffectively with the *B. elkanii* strain 33 due to the presence of another *Rj*-gene, Rj_3 [10]. Cvs. Hill and Amsoy 71 harbor a gene (Rj_4) that was responsible for the ineffective nodulation. These Rj_4-cultivars were nodulated ineffectively with *B. elkanii* strain 61 [11]. Soybean plants harboring these *Rj*-genes (*Rj*-cultivars) were considered to restrict effective nodulation with appropriate serogroups of strains and to prefer certain types of rhizobia for nodulation. If this assumption holds true, planting of *Rj*-cultivars could increase the populations of rhizobial strains highly compatible with those cultivars in soils. Therefore, the relationship between the *Rj*-genotypes of soybean and the preference of the *Rj*-cultivars for various types of *Bradyrhizobium* strain was examined [12, 13]. These *Bradyrhizobium* strains were classified into three nodulation types, type A, B, and C, based on the compatibility with *Rj*-cultivars. Nodulation type A strains nodulated with almost all the cultivars except for the rj_1-ones (non-nodulating lines) and were preferred by non-*Rj*-ones. Type B or type C strains nodulated soybean cultivars other than the Rj_2Rj_3-ones or Rj_4-ones, respectively except for rj_1-ones and were preferred by Rj_4-ones or Rj_2Rj_3-ones, respectively.

This chapter deals with the developmental process and experimental trial of inoculation methods using effective *Bradyrhizobium* strains and various *Rj*-genotypes to increase the yield of soybean.

2. Isolation of $Rj_2Rj_3Rj_4$ – Genotypes from progenies of a cross between soybean cvs. IAC-2 (Rj_2Rj_3) and hill (Rj_4)

In order to analyze in more detail the relationship between the *Rj*-genotype of soybean and the preferences of *Rj*-cultivars for various types of rhizobia, the preferences of the soybean cultivars harboring both Rj_2- and Rj_4-genes for nodulation with *Bradyrhizobium* strains should be studied. For this purpose, the isolation of Rj_2Rj_4-lines from the cross between the Rj_2-cultivar IAC-2 and the Rj_4-cultivar Hill was firstly conducted. Secondly, the isolation of $Rj_2Rj_3Rj_4$-lines from progenies of Rj_2Rj_4-lines was conducted [14, 15]. This section deals with the processes of isolation of Rj_2Rj_4-lines and $Rj_2Rj_3Rj_4$-lines. Furthermore, the preference of various *Rj*-genotypes for rhizobial strains was analyzed in greater detail.

2.1. Materials and methods

Detection of Rj_2- and Rj_4-genes in progenies from the cross of cvs. IAC-2 and Hill: Soybean (*Glycine max* L. Merr.) cvs. IAC-2 (Rj_2Rj_3) and Hill (Rj_4) were crossed. Twenty-three seeds were obtained and used to isolate the progenies homozygous for the Rj_2- and Rj_4-genes. *Bradyrhizobium japonieum* strains Is-80 (nodulation type A), Is-1 (type B), and Is-34 (type C), which were isolated and characterized in a previous study, were used to detect the Rj_2- and Rj_4-genes (Ishizuka et al. 1991a). Soybean cv. CNS (Rj_2Rj_3) was used instead of cv. IAC-2 (Rj_2Rj_3) in inoculation tests. Seeds of progenies from the cross were sterilized with sodium hypochlorite solution (30 g L^{-1}), sown in sterilized vermiculite wetted with the nitrogen-free culture solution used in the previous study (Ishizuka et al. 1991a), inoculated with the combined inoculum of broth cultures of *B. japonicum* strains Is-1 and Is-34, and grown in a phytotron at 25°C under natural light conditions. Four weeks after sowing, the plants were harvested and examined for the nodulation reaction. Non-nodulated and ineffectively nodulated plants which formed small nodule-like structures with a white center, were selected as a genotype harboring both Rj_2- and Rj_4-genes at least heterozygously. Although the parent cultivar IAC-2 harbors the Rj_3-gene in addition to Rj_2-gene, it was not used because the isolation of progenies homozygous for the three genes was very complicated.

Detection of Rj_3-gene in Rj_2Rj_4-lines: To detect the Rj_3-gene, Rj_2Rj_4-lines characterized, A250, B340, B345, B346, B349, B410, C242, C244, C247, and C249, and cvs. Bragg (non-*Rj*),

D-51 (Rj_3), IAC-2, CNS, and Hardee (Rj_2Rj_3), and Akisengoku and Hill (Rj_4) were used. The *Rj*-genotypes are indicated in parentheses [12, 14]. To examine the preference of *Rj*-cultivars for *B. japonicum* for nodulation, Bragg and Akisirome (non-*Rj*), IAC-2 (Rj_2Rj_3), and Hill (Rj_4), A250 and C242 (Rj_2Rj_4) were grown in pots filled with gray lowland alluvial soils. Soybean seeds used were harvested at Kyushu University Farm in 1992. *B. japonicum* USDA33 and USDA110 obtained from the USDA Beltsville Rhizobium Culture Collection, USA, and *B. japonicum* Is-1 and Is-34 were used. These strains were stored at 4°C on yeast extract mannitol agar (YMA) plate, and were cultured in yeast extract mannitol broth [16] on a rotary shaker (100 rpm) at 30°C for 7 days. These cultures were diluted to 1/200 (ca. 10^7 cells mL^{-1}) with sterile saline water (9 g NaCl L^{-1}) immediately before inoculation. Is-1, USDA33, or Is-34 cannot form effective nodules on soybean harboring the Rj_2Rj_3-, Rj_3- or Rj_4-gene, respectively [10, 14]. Seeds of the Rj_2Rj_4-lines and various *Rj*-cultivars were sterilized by immersion in a sodium hypochlorite solution (50 g L^{-1} as active chlorine) for 5 min, washed with ethanol, rinsed five times in sterile water. In the Erlenmeyer flask method, two seeds were sown in a sterilized Erlenmeyer flask containing 110 mL of vermiculite and 60 mL of nutrient solution consisting of 1.6 mM K_2HP0_4, 2.0 mM $CaCl_2$, 2.5 mM $MgSO_4$, and 0.5 mM NH_4NO_3. In the porcelaneous pot method, 14 seeds were sown in a porcelaneous pot containing 2.8 L of vermiculite and 1.4 L of nutrient solution. The flasks or pots were autoclaved (121°C, 20 min). Five milliliters of USDA33 cell suspension per seed was inoculated. Plants were grown in a Phytotron at 25°C and 70% relative humidity under natural light. Three or four weeks after planting, the plants were harvested and examined for the number, size, and shape of the nodules to estimate the effectiveness of the nodules. A globular nodule larger than 1 mm in size was referred to as an effective nodule, and a nodule appearing as a protuberance smaller than 1 mm in size was referred to as an ineffective one.

Isolation of *B. japonicum* from nodules: Six weeks after planting in a/5,000 porcelain Wagner pots, plants were harvested and examined for the number of nodules. Approximately 12 to 14 nodules for each independent pot after 4 weeks of plant growth were collected from soybeans grown in soil. The nodules were immersed in 90% ethanol for 1 min. Then the nodules were transferred to a solution of 5% hydrogen peroxide, soaked for 5 min and washed with sterile saline water three times. The nodules were crushed in a sterile test tube with a sterile glass rod and diluted with sterile saline water. The bacteroid suspension was streaked with a sterile bamboo stick on YMA plates containing Congo red (25 mg L^{-1}). The YMA plates were incubated for one week at 30°C. The colonies grown on the plates and the residual bacteroid suspension were subjected to the Indole acetic acid (IAA)-producing ability assay and serological test, respectively.

Serological identification of baeteroids in nodules: The antisera developed against the somatic antigens of the *B. japonicum* strains Is-1, Is-34, USDA110 and *B. elkanii* strain USDA33 used in this study were prepared as in the previous study [13]. The antisera were diluted to 1/200 and 1/400 with saline water immediately before use. For the serological test, the nodules were harvested, washed and put into a test tube containing saline water. The test tube was heated in boiling water for 30 min. After cooling, the nodules were transferred to small test tubes, one nodule in each tube, and crushed with a round-ended glass rod. Then the bacteroids were dispersed using a vibrator with a small amount of saline water containing sodium azide (0.5 g L^{-1}). After precipitation of the nodule debris, each drop of the bacteroid suspension and the diluted antiserum were put into a well of a micro-test assay plate (Becton Dickinson and Co., USA) using fresh Pasteur pipettes, covered with a thin polyethylene film and shaken gently. The plate was incubated at 37°C for 2 h and then stored in a cold room (4°C) overnight. The agglutination reaction was checked on the next day by comparison with the bacteroid-saline control.

IAA-producing ability: The IAA-producing ability was estimated basically by the methods of Minamisawa et al. [17]. Single colony of isolates on the YMA plates was suspended into 5 mL of sterile saline water and a drop of the suspension was poured into 2.5 mL of the Tris-YMRT (pH 6.8) broth medium [18] containing 0.3 mM L-tryptophan. The cultures were incubated for one week at 30°C on a rotary shaker at a rate of 100 rpm, and the IAA concentration was determined colorimetrically by the addition of 5 mL of Salkowski's reagent [19].

2.2. Results and discussion

Isolation of Rj_2Rj_4-lines: For the verification of the above assumption, the nodulation reactions of the F4 progenies from selected F3 plants with the combined inoculum were examined (Table 1). All the F4 progenies from C3-1, -3, and -5 plants nodulated effectively and produced more than five nodules per plant. Within the limits of the authors' preliminary experiments, soybean plants inoculated with incompatible *B. japonicum* strains had never produced more than two nodules under natural light conditions in winter, so that these plants were assumed to harbor either rj_2 or rj_4 homozygously. Since the ratios of effectively nodulated progenies from the F3 plants, lines A2-4 and C3-4 were approximately 0.44, the expected ratio according to Mendel's laws, these F4 plants were assumed to belong to the genotype $Rj_2rj_2Rj_4rj_4$ [14]. Thus, these five lines (F4 generation) were omitted for further selection. In order to predict the genotypes of the remaining lines, A2-5, B3-4, C2-4, and C2-5, the bacteroids present in the nodules were identified serologically. The genotype of A2-5 was considered to be $Rj_2Rj_2Rj_4rj_4$ because the ratio of effectively nodulated plants and the nodule number of the nodulated plants were 0.25 and more than five, respectively, and almost all the nodules were occupied by Is-34 [14]. On the other hand, the ratios of the nodulated progenies of the B3-4 and C2-4 plants were nearly 0.25 also, but the nodule numbers were only four per four and three plants, respectively, and half or three fourths of these were produced by other serotypic rhizobia than the inoculum strains.

Thirty-one plants of the F4 generation were harvested, and the seeds (F6 generation) of 22 plants were examined for their compatibility with *B. japonicum* Is-1 and 1s-34 by the method described above. The plants were classified based on the nodule number (Table 2), and it was eventually concluded that the progenies from line A2-5 segregated into the genotypes $Rj_2Rj_2Rj_4Rj_4$ and $Rj_2Rj_2Rj_4rj_4$. The lines A2-50 and A2-53 belonged to the former and latter, respectively. Although line A2-52 may also belonged to the former genotype, the genotype of line A2-51 was not clearly revealed because five plants out of 20 plants formed more than four nodules and the soybean cvs. CNS (Rj_2Rj_3) and Hill (Rj_4) formed six nodules per plant

http://dx.doi.org/10.5772/52183

in this assay. Therefore, the lines A2-51, A2-52, and A2-53 were omitted. The progenies from line B3-4 could be considered to belong to the genotype $Rj_2Rj_2Rj_4Rj_4$ except for the lines B3-44 and B3-47. Though the lines B3-44 and B3-47 were also expected to belong to the same genotype as the other progenies of line B3-4, these were omitted for further examination to avoid the risk. All the tested progenies from line C2-4 were found to belong to the genotype $Rj_2Rj_2Rj_4Rj_4$ unlike those of line C2-5, based on the results shown in Table 2. The facts described above indicate that 12 soybean lines homozygous for the Rj_2- and Rj_4-genes were isolated from the cross cvs. IAC-2 (Rj_2Rj_3) × Hill (Rj_4). These lines showed some differences in physiological characteristics, for instance, hypocotyl color, stem length and maturity.

F3 plant	No. of plants tested	Nodulation reaction of F. progeny			Ratio of effectively nodulated plants
		E	I	N	
A2-4	16	9	7	0	0.56
A2-5	16	4	9	3	0.25
B3-4	16	4	12	0	0.25
C2-4	14	3	5	6	0.21
C2-5	15	1	2	12	0.07
C3-1	12	12	0	0	1.00
C3-3	13	13	0	0	1.00
C3-4	15	8	7	0	0.53
C3-5	15	15	0	0	1.00

E, I, and N denote effective, ineffective, and non-nodulated, respectively

Table 1. Nodulation reactions of F4 progenies from the cross IAC-2 × Hill with the combined inoculum of *B. japonicum* strains Is-I and -34.

Isolation of $Rj_2Rj_3Rj_4$-lines: Soybean cultivar D-51 harboring the Rj_3-gene frequently formed a small number of effective nodules and a large number of ineffective ones upon the inoculation of *B. elkanii* USDA33. Therefore, the criteria for harboring the Rj_3-gene were determined as follows. When the ratio of ineffective to the total nodule number (I/T) exceeded 0.5, it was assumed that the test plant harbored the Rj_3-gene. Table 3 shows the nodulation of test plants inoculated with *B. elkanii* USDA33. Cvs. Bragg (non-*Rj*), Hill and Akisengoku (Rj_4) formed only effective nodules and not ineffective nodules. However, cvs. IAC-2, CNS, and Hardee (Rj_2Rj_3) did not produce any effective nodules and formed 0.8 to 17.5 ineffective nodules per plant at 3-weeks after the inoculation of USDA33 using an Erlenmeyer flask. In cv. D-51 (Rj_3), 3.7 effective and 30.0 ineffective nodules per plant were counted at 4-weeks after the inoculation using a porcelain pot. I/T ratio of all the cultivars harboring the Rj_3-gene exceeded 0.5 which corresponded to the criterion for the detection of the Rj_3-gene. Also, the I/T ratio of all the Rj_2Rj_4-lines tested here ranged from 0.6 to 0.9 for the plants grown for 3 weeks and a value of 1.0 was recorded at 4 weeks after inoculation. Based on these results, it was concluded that all the Rj_2Rj_4-lines tested here harbored the Rj_3-gene and displayed the $Rj_2Rj_3Rj_4$-genotype. Also it may be necessary for the plant to grow for 4 weeks for the identification of the Rj_3-gene, because the growth of ineffective nodules was delayed compared to that of the effective nodules and counting and discrimination were difficult on plants grown for only 3 weeks (Table 3).

Line	Inoculum	Plant number[a]						Nodule number	
		0	1	2	3	4<	Total	E	I
A2-50	C	18	2	0	0	0	20	0.1	0
A2-51	C	8	5	1	1	5	20	1.7	0.2
A2-52	C	11	5	3	0	1	20	0.8	0
A2-53	C	0	0	1	0	13	14	8.1	0
B3-40	C	10	8	2	0	0	20	0.6	0
B3-44	C	10	6	2	2	0	20	0.8	0
B3-45	C	13	4	3	0	0	20	0.5	0
B3-46	C	8	11	1	0	0	20	0.7	0.1
B3-47	C	7	8	3	1	1	20	1.1	0
B3-49	C	9	8	3	0	0	20	0.2	0
B3-410	C	17	3	0	0	0	20	0.7	1.8
C2-40	C	18	2	0	0	0	20	0.1	2.9
C2-42	C	17	3	0	0	0	20	0.2	1.7
C2-44	C	15	5	0	0	0	20	0.3	1.3
C2-47	C	14	6	0	0	0	20	0.3	0.9
C2-48	C	20	0	0	0	0	20	0	0
C2-49	C	13	6	1	0	0	20	0.4	0
C2-50	C	8	2	1	1	8	20	2.6	0.8
C2-51	C	0	0	2	0	17	19	6.7	0
C2-52	C	1	0	0	2	16	19	5.5	0
C2-53	C	9	2	2	2	5	20	2	0.6
C2-55	C	1	0	1	1	17	20	6.2	0
CNS	C	0	0	0	0	6	6	6.2	0
Hill	C	0	1	0	0	5	6	5.8	0
CNS	Is-1	4	0	0	0	0	3	0	0
Hill	Is-1	0	0	0	0	4	4	5.8	0
CNS	Is-34	0	0	0	0	3	3	9.3	0
Hill	Is-34	4	0	0	0	0	4	0	0

C: combined inoculum of Is-1 and Is-34. Underlined part: line with genotype $Rj_2Rj_2Rj_4Rj_4$. E and I: number of effective and ineffective nodule per plant, respectively. [a] Number of plants which produced nodules with the number shown below.

Table 2. Nodulation of F6 progenies from the cross between soybean cvs. IAC-2 and Hill inoculated with *B. japonieum* Is-l and -34.

Preference of $Rj_2Rj_3Rj_4$-lines for indigenous rhizobia: Table 4 shows the nodule numbers of soybean plants grown in pots containing Chikugo soil. The nodules formed on the roots of all the soybean plants grown in this soil were all effective. Non-*Rj*-cultivars, Akisirome and Bragg formed a relatively large number of nodules on the tap roots and the number was not significantly different from that on the lateral roots. In the case of the *Rj*-cultivars, however, the nodule number of the tap roots was markedly lower than that of the lateral roots, that is, all the *Rj*-cultivars formed 3.0 to 4.3 nodules on the tap roots and 16.3 to 24.3 nodules on the lateral roots per plant. These data clearly indicate that nodule formation in the non-*Rj* cultivars occurs earlier than in the *Rj*-cultivars, because nodule formation occurred first on the tap roots and then gradually developed on the lateral roots with the progression of growth.

http://dx.doi.org/10.5772/52183

The difference in the onset of nodule formation between non-*Rj*- and *Rj*-cultivars may be due to the difference in populations of compatible rhizobial strains with both genotypes. Also, although the $Rj_2Rj_3Rj_4$-genotype was derived from the cross between Rj_2Rj_3- and Rj_4-genotype in order to restrict the nodulation with undesirable strains, the difference in the nodulation ability was not appreciable among the three *Rj*-genotypes. Therefore, it is suggested that the $Rj_2Rj_3Rj_4$-genotype cannot decrease the ability of nodule formation but may increase the preference of certain types of rhizobia for nodulation.

Cultivar or line	Nodule number					I/T	Rj_3-gene
	Effective			Ineffective	Total		
	TR	LR	WR	(I)	(T)		
Rj-cultivar							
Bragg*	1.2	1.5	2.7	0.0	2.7	0.0	−
IAC-2	0.0	0.0	0.0	17.5	17.5	1.0	+
CNS	0.0	0.0	0.0	5.6	5.6	1.0	+
Hardee	0.0	0.0	0.0	0.8	0.8	1.0	+
D-51*	0.1	3.6	3.7	30.0	33.7	0.9	+
Hill	0.5	4.7	5.2	0.0	5.2	0.0	−
Akisengoku	2.2	3.8	6.0	0.0	6.0	0.0	−
$Ri_2 Rj_4$ -line							
A250	0.0	0.6	0.6	3.9	4.5	0.9	+
A2S0*	0.1	0.5	0.6	28.1	28.7	1.0	+
B340*	0.0	0.0	0.0	32.8	32.8	1.0	+
B345*	0.0	0.0	0.0	16.8	16.8	1.0	+
B346*	0.0	0.0	0.0	21.4	21.4	1.0	+
B347*	0.0	0.0	0.0	17.6	17.6	1.0	+
B349*	0.0	0.0	0.0	37.6	37.6	1.0	+
B410*	0.0	0.5	0.5	38.3	38.8	1.0	+
C242	0.1	0.9	1.0	1.2	2.2	0.5	+
C242*	0.0	0.2	0.2	24.9	25.1	1.0	+
C244	0.2	2.2	2.4	3.9	6.3	0.6	+
C244*	0.0	0.1	0.1	28.4	28.5	1.0	+
C247	0.2	0.6	0.8	5.5	6.3	0.9	+
C249	0.2	1.2	1.4	8.6	10.0	0.9	+
C249*	0.1	0.2	0.3	42.1	42.4	1.0	+

TR, LR, and WR: tap root, lateral root, and whole root, respectively. Cultivar* and line* indicate that for the detection of the Rj_3-gene a porcelain pot was used, and the nodule number corresponds to a mean of 14 plants. The other cultivars and lines show a mean of 5-6 plants when Erlenmeyer flasks were used. The * marked plants were assayed 4-weeks after planting and the other after 3 weeks. + or -: presence or absence of Rj_3-gene, respectively.

Table 3. Identification of Rj_3-gene in Rj_2Rj_4-lines and *Rj*-cultivars with inoculum of *B. elkanii* USDA33.

Agglutination reaction with the antiserum against USDA110 was assayed with a bacteroid suspension of nodules from soybean plants grown in soil. Isolates separated from the residual bacteroid suspension were used for the assay of IAA-producing ability. Table 5 shows that the nodule occupancy rate of serotype 110, the bacteroid suspension of which reacted positively with the antiserum against USDA110, was significantly higher in the $Rj_2Rj_3Rj_4$-genotype than in the other genotypes, i.e. non-*Rj*-, Rj_2Rj_3-, and Rj_4-cultivars. Furthermore, C242 in the $Rj_2Rj_3Rj_4$-genotype showed significantly lower ratios of IAA forming isolates than in the other genotypes but A250 in this genotype showed the same ratios as those in other genotypes. Based on the above results, the $Rj_2Rj_3Rj_4$-genotype of soybean was considered to nodulate more actively serogroup USDA 110 of rhizobia which contains Hup$^+$ strains at high rates for nodulation compared with the other genotypes.

Further studies should be carried out to determine whether the $Rj_2Rj_3Rj_4$-gene is able to repress infection with *B. elkanii* which does not contain Hup$^+$ strains but contains rhizobitoxine-producing strains [20]. Nevertheless, the $Rj_2Rj_3Rj_4$-genotype of soybean is superior to non-*Rj*-, Rj_2Rj_3- and Rj_4-genotypes in that it prefers more active rhizobial strains forming efficient nodules for nitrogen fixation. However, it remains to be determined whether the $Rj_2Rj_3Rj_4$-genotype is superior to other *Rj*-genotypes in terms of nodule occupancy of inoculum strain in field experiments using *B. japonicum* USDA 110 as effective inoculum strain.

Cultivar or line	*Rj*-genotype	Tap root	Lateral root	Whole root
Akisirome	non-*Rj*	18.8	26.0	44.8
Bragg	non-*Rj*	8.8	9.8	18.6
IAC-2	Rj_2Rj_3	3.0	16.3	19.3
Hill	Rj_4	4.3	24.3	28.6
A250	$Rj_2Rj_3Rj_4$	3.5	17.5	21.0
C242	$Rj_2Rj_3Rj_4$	4.0	21.0	25.0

Data show a mean of 4-plants

Table 4. Number of nodules in soybean plants grown in pots containing gray soils of Kyushu Agricultural Experiment Station.

Cultivar or line	Serotype 110		IAA formation	
	Nodule occupancy (%)	Statistical analysis (p=0.05)	Nodule occupancy (%)	Statistical analysis (p=0.05)
Akisirome	36	a	24	a
Bragg	56	b	17	abc
IAC-2	39	a	15	abcd
Hill	58	b	19	ab
A250	73	c	11	bcde
C242	81	c	4	e

Each figure in the column of nodule occupancy is significantly different at the 5% level by chi-square test when different letters are indicated in the column of statistical analysis. Twelve to 14 nodules or isolates from the tap roots and lateral roots of two plants in each independent pot were assayed. Data were calculated for the whole root and correspond to the mean of two replicate pots.

Table 5. Positive agglutination reaction with antiserum against USDA110 (serotype 110) of bacteroids in nodules of soybean plants grown in pots containing the soil of Kyushu Agricultural Experiment Station and IAA formation of isolates from the bacteroid suspension.

3. Occupation of serogroup USDA110 in nodules of soybean plants harboring various *Rj*-genes grown in a field

In order to analyze in more detail the relationship between the *Rj*-genotype of soybean and the preferences of *Rj*-cultivars for various types of rhizobia, the preferences of the soybean cultivars harboring both Rj_2-, Rj_3- and Rj_4-genes for nodulation with *Bradyrhizobium* strains should be studied. For this purpose, the isolation of Rj_2Rj_4-lines from the cross between the Rj_2-cultivar IAC-2 and the Rj_4-cultivar Hill was conducted. Furthermore, to analyze in greater detail the preference of various *Rj*-genotypes for rhizobial strains, the isolation of $Rj_2Rj_3Rj_4$-lines from the cross between the Rj_2Rj_3-cultivar IAC-2 and the Rj_4-cultivar Hill was conducted [14, 15]. The $Rj_2Rj_3Rj_4$-lines were predicted to be nodulated only by type A strains, such as *B. japonicum* USDA110. Thus, the competition with less efficient indigenous rhizobia in soils might be reduced by the use of the reciprocal relation between *Rj*-cultivars and rhizobia. Therefore, the relationship between the *Rj*-genotypes of soybean and their preference for various types of rhizobia for nodulation was investigated.

3.1. Materials and methods

Plants: Soybean (*Glycine max* L. Merr.) cultivars Bragg and Akisirome (*non-Rj*), IAC-2, CNS, and Hardee (Rj_2Rj_3), Hill, Fukuyutaka and Akisengoku (Rj_4), and B340, B349 and C242 ($Rj_2Rj_3Rj_4$) were used. The *Rj*-genotypes are indicated in parentheses [14, 15].

Rhizobium:*B. japonicum* USDA110 was stored at 4°C on YMA plate, and was cultured in YMB [16] on a rotary shaker (100 rpm) at 30°C for 7 days. This culture was diluted to 1/200 (ca. 10^7 cells mL^{-1}) with saline water immediately before inoculation.

Plant cultivation: Field experiment was conducted on gray lowland soils of Kyushu University Farm, located in the northern flooded plain of Fukuoka Prefecture. The soil pH (soil: water ratio was 1:2.5) ranged from 6.9 to 7.3. The field was 16 m long and 18 m wide. The row width and intra-row spacing were 70 and 20 cm, respectively. Fused phosphate, calcium superphosphate, and potassium sulfate were applied at the time of plowing over the experimental field at the rates of 450 kg N, 150 kg P_2O_5 and 200 kg K_2O per ha, respectively. The seeds were sown at the rate of two seeds per hill on June 9, 1992. For each cultivar, the entire row (16 m long) was divided into four parts. The individual hills with alternative two parts were inoculated with 50 mL of 1/200 diluted USDA110 culture (1 week old) at the rate of 5×10^8 cells per hill immediately after sowing and the individual hills with the other parts received 50 mL of saline water as control treatment. The experiment was conducted with two replications.

Sampling: In order to evaluate nodulation, the relative efficiency (RE) of nitrogen fixation and occupation ratio of the inoculum strain, two hills per soybean genotype in the inoculated or non-inoculated plots were harvested. The plant roots were washed with tap water. The growth stage corresponded to the vegetative stage at the time of the first sampling conducted during the period from July 28 to August 5, 1992. The second sampling was carried out in order to evaluate the occupation ratio of the inoculum strains on September 2, from the flowering to the pod elongation stages depending on the plant genotypes.

H_2 evolution: Plant roots with nodules were severed at the cotyledonary node and individually placed in 200-mL conical flasks. The flasks were sealed with a serum stopper. One milliliter of air was drawn out with a syringe, and was analyzed for the amount of hydrogen evolved from the nodules at 5 and 30 min after sealing by using a thermal conductivity gas chromatograph (GC-8A, Shimadzu, Kyoto, Japan) equipped with a stainless steel column (3 mm in diameter, 2 m length). The column was filled with molecular sieve 5A, 60-80 mesh (Nacalai Tesque, Inc., Kyoto, Japan). Column and injector temperatures were 50 and 70°C, respectively. Carrier gas was N_2 (flow rate: 36 mL min^{-1}).

Acetylene reduction activity (ARA): After the determination of H_2 evolution, the flask containing roots with nodules was used for the ARA assay. ARA was assayed according to the methods of Haider et al. [21]. After the assay, the nodules were separated from the roots, weighed, and counted.

Relative efficiency (RE): RE of nitrogen fixation was calculated from the amount of H_2 evolved and ARA as follows:

$$RE = 1 - \frac{H_2\,\text{evolution}\ (\mu\text{molg FW}^{-1}\text{h}^{-1})}{\text{ARA}\ (\mu\text{molg FW}^{-1}\text{h}^{-1})}$$

Serological test with antiserum USDA110 and occupation ratio: The nodules on the tap root and lateral roots of two hills of each genotype of soybean plants were separately collected into a test tube, with the addition of saline water (8.5 g L^{-1}) containing 0.5 g L^{-1} of sodium azide and allowed to stand for 30 min in boiling water. The nodules were kept at 4°C until the serological test was performed. One hundred nodules per cultivar were used for the serological test, that is, a maximum number of 30 nodules from the tap root and the other nodules from the lateral roots per cultivar. The serological test was conducted with the antiserum against *B. japonicum* USDA110, according to the method described in a previous paper [15].

Occupation ratios of the tap, lateral and whole roots were expressed in percentages of positive nodules against antiserum USDA110 to all the nodules tested, respectively.

3.2. Results and discussion

The seeds were inoculated or irrigated immediately after sowing with 50 mL of diluted inoculum or saline water, as described above in the inoculation method of rhizobium. As a result, it was considered that the inoculum strain USDA110 was distributed to the cylindrical soil block in which the tap root of soybean seedlings could extend. Therefore, initial nodulation by USDA110 as inoculum was assumed to be limited to the upper part of the tap root. The effects of the inoculation of *B. japonicum* USDA110 on the nodulation of almost all the tested genotypes of soybean were not significant in terms of nodule number and fresh weight (FW) per plant, and nodule size (mg FW nodule^{-1}), except for B349 among the soybeans with $Rj_2Rj_3Rj_4$-genotypes [22]. These results suggest that the population of indigenous rhizobia in this field was not the limiting factor for nodulation.

Consequently, the information acquired from this experiment can be more useful for the cultivation of soybean than that derived from experimental results under controlled conditions and fields, in which the population of indigenous rhizobia is low. ARA did not show any significant difference between the inoculated and non-inoculated plants (Table 6). RE of the inoculated soybean plants tended to be higher than that of the non-inoculated ones in all the cultivars/lines. However, there was no significant difference in the effects among the *Rj*-genotypes of the host plants. Especially, RE of IAC-2, CNS (Rj_2Rj_3) and B349 ($Rj_2Rj_3Rj_4$) increased significantly by USDA110 inoculation, reflecting the reduction in H_2 evolution by Hup^+ of USDA110. RE generally increased by inoculation of USDA110 (Hup^+ strain) and biological nitrogen fixation was considered to be enhanced. In four experiments conducted by Evans et al. [23], it was observed that the content of crude seed protein of soybean plants inoculated with the Hup^+ strain increased by 8.9% compared with the plants inoculated with Hup^- strains and that the REs (calculated from Evans's data) of the plants inoculated with Hup^+ and Hup^- strains were 0.94-1.00 and 0.56-0.79, respectively. As higher RE values were observed in the inoculated plants than in the non-inoculated plants, in spite of the lack of significant difference (Table 6), the occupation ratios of the serogroup USDA110 of rhizobia in the nodules of lateral roots were determined at the time of the first sampling (Table 7).

Rj-genotype	Cultivar and Line	Inoculation	H_2 evol. (μmol gFW^{-1} h^{-1})	ARA (μmol gFW^{-1} h^{-1})	RE
Non-*Rj*	Akisirome	+	3.1 ab	7.1 ab	0.51 bcde
		-	2.3 abcd	3.8 ab	0.4 def
	Bragg	+	2.5 abc	7.4 a	0.66 abc
		-	3.4 a	6.9 ab	0.5 bcde
Rj_2Rj_3	IAC-2	+	1.4 cde	3.4 bc	0.58 bcd
		-	2.6 abc	4 abc	0.35 ef
Rj_2Rj_3	CNS	+	0.6 e	2.2 c	0.67 ab
		-	1.4 cde	2.5 c	0.46 cde
Rj_4	Hill	+	1.3 cde	3.9 abc	0.66 abc
		-	1.8 bcde	4.7 abc	0.63 abc
Rj_4	Fukuyutaka	+	0.6 e	2.4 c	0.78 a
		-	1.2 cde	3 bc	0.62 abc
$Rj_2Rj_3Rj_4$	B340	+	0.9 de	2.6 c	0.67 ab
		-	1.0 de	2.3 c	0.55 bcd
$Rj_2Rj_3Rj_4$	B349	+	0.7 e	2 c	0.63 abc
		-	1.7 bcde	2.2 c	0.24 f
$Rj_2Rj_3Rj_4$	C242	+	0.6 e	1.4 c	0.55 bcd
		-	0.6 e	1.3 c	0.53 bcde

ARA and RE refer to the acetylene reduction activity and relative efficiency, respectively. Planting and inoculation were performed on June 9 in 1992, and sampling of these plants was performed on July 28. Values are means of 4 plants (two hills). Means in the same column followed by the same letter are not significantly different at the 5% level by LSD.

Table 6. Nodule activity of field-grown soybeans inoculated with *B. japonicum* USDA110.

The occupation ratios of serogroup USDA110 in the nodules of the lateral roots were 53-67, 40-58, 63-83, and 62-77% in the inoculated plants with the non-*Rj*-, Rj_2-, Rj_4- and $Rj_2Rj_3Rj_4$-genotypes, respectively. Therefore, these results indicated that infection by USDA110 occurred rapidly after the inoculation. $Rj_2Rj_3Rj_4$- genotypes suppress nodulation by the nodulation types B and C which are specific serogroups of the strains and they can select strains of nodulation type A to form nodules. As a result, they form nodules with only strains of the nodulation type A [14, 15]. However, the $Rj_2Rj_3Rj_4$-genotypes were not always nodulated by only USDA110 belonging to the nodulation type A. Soybean plants may possess a factor concerned with the preference for *B. japonicum* USDA110, except for these *Rj*-genes. The mechanism of preference has not been elucidated, but it is assumed to relate to the increase in the population of the serogroup USDA110 in soils by the root activity. Also,

Cregan and Keyser [24] screened out PI genotypes which excluded the *B. japonicum* strain of serogroup 123 in favor of the inoculated strain, and they suggested that the trait identified in the PI genotypes could exert a beneficial effect on soybean productivity by excluding all or a part of the indigenous serogroup 123 population in favor of more effective strains of *B. japonicum*. Their concept involving the preference of soybean genotypes for rhizobial strains may indicate that the planting of soybean genotypes compatible with efficient rhizobial strains could increase nitrogen fixation and dry matter production, as well as soybean yield. The second sampling was performed to investigate the changes in the occupation ratio of serogroup USDA110 with the progression of growth. The values revealed that the occupation ratios of serogroup USDA110 decreased for all the genotypes compared with those at the first sampling (Table 7). The reduction of the occupation ratio of serogroup USDA110 from the 1st until the 2nd sampling was lowest (0.13-0.16) in the $Rj_2Rj_3Rj_4$-genotypes, excluding B349, followed by the non-*Rj*- and Rj_2Rj_3-genotypes and highest (0.52-0.69) in the Rj_4-genotypes, excluding Hill. Therefore, it was considered that the population of compatible rhizobia with host soybean plants increased in the rhizosphere with the progression of development and growth, based on the following experimental basis.

Rj-genotype	Cultivar and Line	Occupation ratio (%) 1st sampling	2nd sampling	Reduction
Non-*Rj*	Akisirome	67 bcd	35 hi	0.48
Non-*Rj*	Bragg	53 fg	32 hi	0.40
Rj_2Rj_3	IAC-2	58 c	33 hi	0.43
Rj_2Rj_3	Hardee	40 ghi	32 hi	0.20
Rj_4	Hill	83 a	56 c	0.33
Rj_4	Fukuyutaka	83 a	26 i	0.69
Rj_4	Akisengoku	63 cde	30 hi	0.52
$Rj_2Rj_3Rj_4$	B340	74 abc	41 fgh	0.45
$Rj_2Rj_3Rj_4$	B349	62 de	54 ef	0.13
$Rj_2Rj_3Rj_4$	C242	77 ab	65 bcde	0.16

First sampling and second sampling conducted on July 28 and September 2, respectively. The same small letters next to each symbol are not significantly different at the 5 % level by X^2 test. Reduction = 1 - [Occupation ratio of the nodules on the lateral root at the vegetative stage] / [Occupation ratio of the nodules on the lateral root at the reproductive stage].

Table 7. Changes in the occupation ratio of *B. japonicum* belonging to serogroup USDA110 in the nodules on the lateral roots of field-grown soybeans with the progression of growth.

As a result, it was considered that in the rhizosphere of the $Rj_2Rj_3Rj_4$-genotypes, the growth of the type A rhizobia was enhanced, while that of the types B and C rhizobia was repressed. Therefore, with the expansion of the root area of the host plant, the occupation ratio of the type A rhizobia including the serogroup USDA110 was high. Consequently, the $Rj_2Rj_3Rj_4$-genotypes were considered to be superior to other *Rj*-genotypes in terms of the inoculation effects of the nodulation type A of *B. japonicum* USDA110. However, it was estimated that the $Rj_2Rj_3Rj_4$-genotypes were not always nodulated by only USDA110 belonging to the nodulation type A. The mechanism of preference of the $Rj_2Rj_3Rj_4$-genotypes for the USDA110 serogroup has not been elucidated.

The findings in this experiment on nodulation and RE, and the results of the serological tests, indicate that the $Rj_2Rj_3Rj_4$-genotypes of soybean are superior to the non-*Rj*-, Rj_2Rj_3- and Rj_4-genotypes because they prefer rhizobial strains of serogroup USDA110 which are more active and form more efficient nodules in nitrogen fixation, compared with the other genotypes and because they are able to restrict nodulation with certain indigenous rhizobia. However, the preference of the $Rj_2Rj_3Rj_4$-genotypes for serogroup USDA110 is not sufficient to rule out the competition with other serogroups. Therefore, the study should be focused on the isolation of efficient (Hup^+) and highly compatible rhizobial strains with $Rj_2Rj_3Rj_4$-genotypes. Also, the breeding of new $Rj_2Rj_3Rj_4$-genotypes of soybean might be considered using both CNS (Rj_2Rj_3) and Hill (Rj_4) cultivars in which the occupation ratio was found to be high by the inoculation of *B. japonicum* USDA110 (Table 7).

4. Effect of *Rj* genotype and cultivation temperature on the community structure of soybean-nodulating bradyrhizobia

Saeki et al. [25] investigated the genetic diversity and geographical distribution of indigenous bradyrhizobia isolated from five sites in Japan (Hokkaido, Fukushima, Kyoto, Miyazaki, Okinawa) by PCR restriction fragment length polymorphism (PCR-RFLP) analysis of the 16S-23S rRNA gene internal transcribed spacer (ITS) region and revealed that geographical distribution of indigenous bradyrhizobia varied from the northern to southern regions in Japan. As a result, the representative clusters of isolated indigenous bradyrhizobia were in the order of *Bradyrhizobium japonicum* USDA123, 110, and 6 and *Bradyrhizobium elkanii* USDA 76^T clusters from northern to southern regions in Japan. It has been suggested that an environmental factor such as temperature will influence the localization of Japanese indigenous bradyrhizobia. Saeki et al. [26] investigated the occupancy of three *Bradyrhizobium japonicum* strains and one *Bradyrhizobium elkanii* strain under different temperature conditions in soil and liquid media and suggested that temperature is one of the environmental factors that affect the occupancy of indigenous bradyrhizobia in soil.

Minami et al. [27] isolated 260 indigenous bradyrhizobia from 13 soybean cultivars of five *Rj*-genotypes (non-*Rj*, Rj_2Rj_3, Rj_3, Rj_4, and $Rj_2Rj_3Rj_4$) from an Andosol and estimated the nodulation tendency among *Rj*-genotype soybeans. The results showed that indigenous bradyrhizobial communities among the same *Rj*-genotype soybean cultivars were similar to each other, whereas indigenous bradyrhizobial communities between the Rj_2Rj_3-genotype and non-*Rj*-, Rj_3-, or Rj_4-genotype soybean cultivars were significantly different. However, They couldn't investigate the nodulation tendency by indigenous bradyrhizobia under different temperature conditions.

In the present section, to examine the influence of combinations of several cultivation temperatures and *Rj*-genotype soybean cultivars on the nodulation tendency and community structure of indigenous bradyrhizobia, we isolated indigenous bradyrhizobia from an Andosol using soybean cultivars of different *Rj*-genotypes and several cultivation temperatures. The isolates were analyzed by PCR-RFLP of the 16S-23S rRNA gene ITS region, and a dendrogram was constructed to classify the isolates into clusters. The effects of cultivation temperature and *Rj*-genotype on soybean-nodulating bradyrhizobial communities were also estimated.

4.1. Materials and methods

Soybean cultivars and soil samples: We used 13 soybean cultivars of four *Rj*-genotypes to investigate the effect of several cultivation temperatures and *Rj*-genotypes of host soybean cultivars. The soybean cultivars were Akishirome, Bragg and Orhihime as non-*Rj*-genotypes, Bonminori, CNS, Hardee and IAC-2 as Rj_2Rj_3-genotypes, Akisengoku, Fukuyutaka and Hill as Rj_4-genotypes, and A-250-3, B349 and C242 as $Rj_2Rj_3Rj_4$-genotypes [14, 15]. As the soil sample, an Andosol (pH [H_2O] 5.04, electrical conductivity [EC] = 0.05 dS m^{-1}; The National Agricultural Research Center for the Tohoku Region, Arai, Fukushima, Japan [25, 27]) was used for soybean cultivation because a high diversity of indigenous bradyrhizobia has been found in this soil in previous studies [25, 28].

Soybean cultivation: To isolate indigenous bradyrhizobia, we grew soybean cultivars in 1-liter culture pots for 4 weeks. The culture pots were filled with vermiculite with N-free nutrient solution [29] at 40% (vol/vol) water content and then autoclaved at 121°C for 20 min. Soybean seeds were sterilized by soaking them in 70% ethanol for 30 s and in a dilute sodium hypochlorite solution (0.25% available chlorine) for 3 min and then washing them with sterile distilled water. A soil sample (2 to 3 g) was placed in the vermiculite at a depth of 2 to 3 cm, and the soybean seeds were sown on the soil. The plants were grown for 4 weeks in a growth chamber (low: day, 23°C for 16 h, and night, 18°C for 8 h; middle: day, 28°C for 16 h, and night, 23°C for 8 h; and high: day, 33°C for 16 h, and night, 28°C for 8 h) with a weekly supply of sterile distilled water. After cultivating, 20 nodules were randomly collected from among all of the nodules harvested from soybean roots and sterilized by soaking them in 70% ethanol for 3 min and in a diluted sodium hypochlorite solution (0.25% available chlorine) for 30 min and then washing them with sterile distilled water.

DNA samples of indigenous bradyrhizobia: Total DNA for the PCR template was extracted from a nodule directly as described by Hiraishi et al. [30] with slight modifications [29]. Each nodule was homogenized in 50 μL of BL buffer (40 mM Tris-HCl, 1% Tween 20, 0.5% Nonidet P-40 and 1 mmol of EDTA [pH 8.0]), 40 μl of sterile distilled water, and 10 μL of proteinase K (1 mg mL^{-1}) and then incubated at 60°C for 20 min and 95°C for 5 min. After centrifugation, the supernatant was collected and used as the PCR template. A total of 780 DNA samples were obtained for further analysis.

PCR-RFLP analysis of the 16S-23S rRNA gene ITS region: As reference strains, total DNA for PCR template of strains *B. japonicum* USDA4, 6^T, 38, 110, 115, 123, 124, and 135 and *B. elkanii* USDA46, 76^T and 94 [19] was prepared as described previously [31]. PCR was carried out using *Ex Taq* DNA polymerase (TaKaRa Bio, Otsu, Japan). For ITS amplification, we used the ITS primer set BraITS-F (5'-GACTGGGGTGAAGTCGTAAC-3') and BraITS-R (5'-ACGTCCTT CATCGCCTC-3') [25]. The PCR cycle consisted of a pre-run at 94°C for 5 min, denaturation at 94°C for 30 s, annealing at 55°C for 30 s, and extension at 72°C for 1 min. This temperature control sequence was repeated for a total of 30 cycles and was followed by a final post-run at 72°C for 10 min. The RFLP analysis of the 16S-23S rRNA gene ITS region was investigated using restriction enzymes *Hae*III, *Hha*I, *Msp*I, and *Xsp*I (Ta- KaRa Bio) [31]. A 5 μL aliquot of the PCR product was digested with restriction enzyme at 37°C for 16 h in a 20 μL reaction mixture. The restricted fragments were separated by agarose gel electrophoresis and visualized with ethidium bromide.

Cluster analysis: For the cluster analysis, we calculated the genetic distance between pairs of isolates (D). D was calculated from N_{AB} (the number of RFLP bands shared by the two strains) and N_A and N_B (the numbers of RFLP bands in strains A and B, respectively) [32, 33]. The cluster analysis was carried out using the unweighted pair group method with arithmetic average (UPGMA) method. The dendrograms were constructed using the PHYLIP software program v3.69 (J. Felsenstein, University of Washington, Seattle, WA).

Diversity analysis of bradyrhizobial communities: To estimate the diversity of the bradyrhizobial communities isolated from the host soybean cultivars, we used the Shannon-Wiener diversity index [28, 34, 35]. The formula for the diversity index was

$$H' = -\Sigma Pi \ln Pi$$

where Pi is the dominance of the isolates expressed by ni/N, N is the total number of tested isolates (n = 20), and ni is the total number of tested isolates belonging to a particular dendrogram cluster. The indexes of alpha diversity ($H'\alpha$), beta diversity ($H'\beta$), and gamma diversity ($H'\gamma$) were calculated [36, 37]. These diversity indices were used to estimate the differences in the bradyrhizobial communities between cultivation temperature pairs. The $H'\alpha$ index represents a weighted average of the diversity indices of each of the two bradyrhizobial communities, the $H'\beta$ index represents the differences between the two bradyrhizobial communities from the two host soybean cultivars and the $H'\gamma$ index represents the diversity of the total isolate communities from the two host soybean cultivars (n = 40). The relationship among these indices is

$$H'\beta = H'\gamma - H'\alpha$$

We also estimated the differences among the compositions of the bradyrhizobial communities by comparing the ratio of the beta to the gamma index ($H'\beta/H'\gamma$), taking into consideration the difference in gamma diversity in each pairwise comparison of bradyrhizobial communities.

4.2. Results and discussion

The PCR products of amplified 16S-23S rRNA gene ITS region were digested by four restriction enzymes, and the restriction fragments were separated by electrophoresis. The fragment sizes were estimated using a 50-bp ladder marker. A total of 36 operational taxonomic units (OTUs) containing 11 reference strains were detected [38]. The dendrogram was generated using the differences in fragment size and pattern. The maximum similarity among OTUs of the reference strains was 86% and occurred between *B. japonicum* USDA 38 and 115. These results were then applied as the criterion for distinguishing clusters in the dendrogram, which produced 11 clusters, each of which contained 11 reference strains. The indigenous bradyrhizobia isolates in the middle and high cultivation temperatures were classified into seven clusters, Bj6, Bj38, Bj110, Bj115, Bj123, Be76, and Be94, while the indigenous bradyrhizobia isolates in the low cultivation temperature were classified into five clusters, Bj6, Bj38, Bj110, Bj115, and Bj123 [38]. For the low and middle cultivation temperatures, most of the indigenous bradyrhizobia were classified into four major clusters, Bj6, Bj38, Bj110, and Bj123, while most of the indigenous bradyrhizobia in the high cultivation temperature were classified into five major clusters, Bj6, Bj38, Bj110, Be76, and Be94. The indigenous bradyrhizobia belonging to the Bj123 cluster was not a major cluster at the high cultivation temperature.

Cluster analysis provided us with information about the cluster occupancy of each *Rj*-genotype and cultivation temperature. The occupancy rate of the Bj6, Bj38, Bj110, Bj115, Bj123, Be76, and Be94 clusters on the non-*Rj*-, Rj_2Rj_3-, Rj_4-, and $Rj_2Rj_3Rj_4$-genotype soybean cultivars is shown in Table 8. Interestingly, the occupancy rate of Bj123 cluster was significantly decreased with increasing cultivation temperature. On the other hand, the occupancy rate of Bj110 cluster tended to increase with increasing cultivation temperatures. The Be76 and Be94 clusters had the same tendency as Bj110 cluster, but their occupancy rates were lower than that of Bj110 cluster (Table 8).

Diversity analysis of the bradyrhizobial communities at various cultivation temperatures: The differences in bradyrhizobial communities among the cultivation temperatures of each *Rj*-genotype were also estimated by the $H'\beta/H'\gamma$ ratios. There was no significant difference (Tukey-Kramer test) based on the cultivation temperature in each *Rj*-genotype due to the variation of bradyrhizobial communities among each *Rj*-genotype soybean cultivars was large, but the values of $H'\beta/H'\gamma$ between low and high cultivation temperature pairs tended to be higher than those of other cultivation temperature pairs (Fig. 1). In addition, the values of $H'\beta/H'\gamma$ of Rj_2Rj_3- and $Rj_2Rj_3Rj_4$-genotype soybean cultivars between low and high cultivation temperature pairs tended to be lower than the values of non-*Rj*- and Rj_4-genotype soybean cultivars (Fig. 1). The values of $H'\beta/H'\gamma$ of $Rj_2Rj_3Rj_4$-genotype soybean cultivars, tended to be comparatively lower than those of non-*Rj*- and Rj_4-genotype soybean cultivars and were similar to the values of $H'\beta/H'\gamma$ of Rj_2Rj_3-genotype soybean cultivars.

With increasing cultivation temperature from low to high, the occupancy rate of the Bj123 cluster decreased and the occupancy rate of the Bj110 cluster increased. Furthermore, the soybean cultivars with Rj_2Rj_3-genotypes and $Rj_2Rj_3Rj_4$-genotypes showed a higher occupancy rate of the Bj110 cluster (50–73.8%) than other *Rj*-genotype soybean cultivars (Table 8), suggesting that the host soybean *Rj*-genotype affected the infection of some specific bradyrhizobia. Yamakawa et al. [23] reported that $Rj_2Rj_3Rj_4$- genotype cultivars were superior to other *Rj*-genotypes for inoculation with *B. japonicum* USDA110. In the present study, we did not demonstrate the inoculum efficiency of *B. japonicum* USDA110; however, this previous result suggested that Rj_2Rj_3- and $Rj_2Rj_3Rj_4$- genotype soybean cultivars may enhance the occupancy rate of the inoculum of *B. japonicum* USDA110. In addition, the occupancy rate of the Bj110 cluster in $Rj_2Rj_3Rj_4$-genotypes did not show significant differences among the three cultivation temperature conditions tested (Table 8), suggesting that $Rj_2Rj_3Rj_4$-genotype soybean cultivars were unaffected by cultivation temperature changes and may enhance the inoculum efficiency such as that of *B. japonicum* USDA110.

http://dx.doi.org/10.5772/52183

Cluster	Mean nodule occupancy (%) ± SE of clusters on each *Rj* genotype											
	Non-*Rj*			Rj_2Rj_3			Rj_2Rj_3			Rj_2Rj_3		
	Low	Middle	High	Low	Middle	High	Low	Middle	High	Low	Middle	High
Bj6	3.3 ± 1.7 ^{B}	20.0 ± 5.0 ^{A}	10.0 ± 2.9 ^{AB}	12.5 ± 1.4	6.3 ± 2.4	7.5 ± 2.5	8.3 ± 4.4	21.7 ± 4.4 ^{A}	10.0 ± 5.0	10.0 ± 2.9	6.7 ± 4.4	10.0 ± 5.0
BJ38	8.3 ± 1.7	10.0 ± 5.8	28.3 ± 6.0	1.3 ± 1.3	3.8 ± 3.8	–	11.7 ± 3.3	8.3 ± 1.7	3.3 ± 3.3	3.3 ± 3.3	3.3 ± 1.7	3.3 ± 3.3
Bj110	10.0 ± 2.9	38.3 ± 9.3	35.0 ± 14.4	50.0 ± 4.1 ^{B}	63.8 ± 6.3 ^{AB}	73.8 ± 4.3 ^{A}	13.5 ± 8.8	26.7 ± 19.2	68.3 ± 9.3	60.0 ± 2.9	71.7 ± 1.7	68.3 ± 9.3
Bj115	3.3 ± 1.7	3.3 ± 1.7	1.7 ± 1.7	1.3 ± 1.3	1.3 ± 1.3	–	3.3 ± 1.7	1.7 ± 17.0	–	–	–	±
Bj123	75.0 ± 2.9 ^{A}	26.7 ± 9.3 ^{B}	13.3 ± 4.4 ^{C}	35.0 ± 5.4 ^{A}	17.5 ± 5.2 ^{AB}	5.0 ± 3.5 ^{B}	63.3 ± 10.1 ^{A}	41.7 ± 16.4 ^{AB}	5.0 ± 0.0 ^{B}	26.7 ± 3.3 ^{A}	13.3 ± 3.3 ^{B}	5.0 ± 0.0 ^{B}
Be76	–	–	3.3 ± 1.7	–	5 ± 2	7.5 ± 1.4	–	–	5.0 ± 0.0	–	1.7 ± 1.7	6.7 ± 6.7
Be94	–	1.7 ± 1.7	8.3 ± 6.0	–	2.5 ± 1.4	6.3 ± 2.4	–	–	5.0 ± 2.9	–	3.3 ± 1.7	6.7 ± 4.4

[a] The nodule occupancy rate was expressed as the mean (n = 3 or 4) of DNA samples belonging to certain clusters within identical *Rj*-genotypes. Significantly different test results were found for each *Rj*-genotype. Different superscript capital letters indicate significant differences (Tukey-Kramer test) at $P < 0.05$ for different cultivation temperatures in the same *Rj*-genotype. –, not detected.

Table 8. Nodule occupancy rate of soybean-nodulating bradyrhizobia for cluster analysis[a].

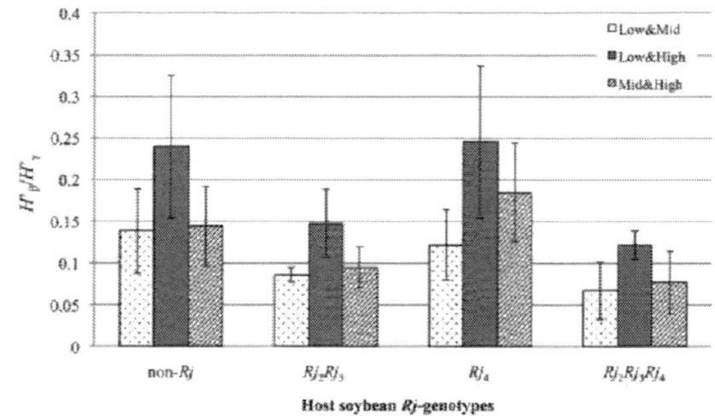

Figure 1. Difference in beta diversity compared to gamma diversity ($H'\beta/H'\gamma$) among pairs of cultivation temperatures. Each value is expressed as the mean ± the standard error (n = 3 or 4].

We also investigated the differences in bradyrhizobial communities for the pairs of cultivation temperature. The nodulation tendencies of soybean cultivars were similar for each cultivation temperature, and differences in the community structures between low and high cultivation temperatures were relatively larger than the other comparisons, although the statistical significant difference was not detected. This possible reason is that responses of soybean cultivars for cultivation temperatures on soybean nodulating bradyrhizobial communities are different among each soybean cultivar even in same *Rj*-genotypes. Therefore, analyses of soybean-nodulating rhizobial communities on not only *Rj*-genotypes but also every soybean cultivars must be conducted for environmental factors affecting soybean-nodulating rhizobial community structures such as cultivation temperature in further studies. The responses of host soybean and soybean-nodulating bradyrhizobia under cultivation conditions such as a suboptimal root zone temperature were reported previously. The lowering of temperature delayed bradyrhizobia infection of soybean roots and lowered the genistein secretion from soybean roots [39, 40]. It also appeared to inhibit the expression of the nodulation (*nod*) gene of soybean-nodulating bradyrhizobia [41]. However, Pan and Smith [42] reported that the concentration of daidzein secreted from soybean roots increased with decreasing root zone temperature. The physiological factors of bradyrhizobia involved in the nodulation are the expression of the nod gene and growth capability in soil and rhizosphere. Yokoyama [43] demonstrated that the expression level of the nod gene of three *Bradyrhizobium* strains, *B. japonicum* USDA110 strain, *B. elkanii* USDA76 strain, and *Bradyrhizobium* sp. TARC 64 strain (isolates from Thailand soil [44]), which were mutants of *nodY-lacZ* fusion, were different depending on the incubation temperature (20, 23, 26, 30, 33, 35, 37, and 40°C) and suggested that the transcriptional responses of the *nod* gene of USDA110 strain and USDA76 strain were distinctly different at 23 to 35°C. Saeki et al. [28] demonstrated that the population occupancy of four *Bradyrhizobium* USDA strains, *B. japonicum* USDA6^T, 38, and 123 and *B. elkanii* USDA76^T, in soil microcosms changed with different

http://dx.doi.org/10.5772/52183

temperature conditions and indicated that USDA76^T was dominant over a wide range of temperature conditions, especially at higher temperature, whereas USDA123 was dominant at low soil temperatures. These results suggested that temperature is one of the environmental factors affecting the infection of soybean and the bradyrhizobia occupancy in soils. Furthermore, Duzan et al. [45] reported that the deformations of soybean root hair decreased with decreasing root zone temperature. The infection and nodulation of soybean by bradyrhizobia under different temperature conditions may be affected by other, as-yet-unknown factors as well.

5. *Rj*-gene specific nodulation genes (*Rj-gsn* genes) of *Bradyrhizobium* strains

Genotype-specific nodulation (*gsn*) genes are reported in *B. japonicum*. The *nolA* gene allowed serogroup 123 strains to nodulate soybean plants having USDA123 restricting Plant Introduction (PI) genotypes [46]. The *noeD* gene restricted nodulation of soybean genotype PI 417566 with USDA110 [47]. There have been, however, no reports of the identification of the *Rj-genotype* specific nodulation (*Rj-gsn*) gene that is related to the incompatible combinations with *Rj* soybeans. This study reports the isolation and characterization of Tn*5* mutants of *B. japonicum* strain Is-1 with the ability to nodulate cv. CNS (Rj_2Rj_3) and shows the candidates of the *Rj-gsn* genes with the identifications of Tn5-flanking sequences in these mutants. As this is the first report to overcome the nodulation restriction conditioned by Rj_2-cultivars, these mutants will be helpful in identifying the *Rj-gsn* gene in further studies.

5.1. Materials and methods

Plants: Soybean (*Glycine max* L. Merr.) cvs. CNS (Rj_2Rj_3), D- 51 (Rj_3), Hardee (Rj_2Rj_3), Hill (Rj_4) and IAC-2 (Rj_2Rj_3) were used. No soybean cultivar harboring only the Rj_2-gene has been reported to date. The *Rj*-genotypes are indicated in parentheses [11, 14].

Bacterial strains and plasmids: The bacterial strains and plasmids used in this study are described in Table 9. The *B. japonicum* strain Is-1 [13] and Tn5 mutants were grown on HM salt medium [48] supplemented with 0.1% arabinose at 30°C. *Escherichia coli* strains were grown on Luria–Bertani medium [49] at 37°C. Antibiotics were added to the media at the following final concentrations: kanamycin at 50 μg mL^{-1} for Tn5 mutant and ampicillin at 100 μg mL^{-1} and kanamycin at 50 μg mL^{-1} for *E. coli* S17-1 harboring pUTKm (Table 9).

Tn5 mutagenesis: Transposon mini-Tn*5* was introduced into *B. japonicum* strain Is-1 by mating with *E. coli* strain S17-1 [50] carrying the suicide vector pUTKm [51]. The *E. coli* donor strain S17-1 (pUTKm) was grown with kanamycin and ampicillin for 12 h at 100 rpm in a rotatory shaker, and then 1 mL of culture was spun down for 2 min at 4,000 × g, and then rinsed three times with HM salts medium [52] to remove the antibiotics, and then suspended into 50 μL of HM salts medium. The *B. japonicum* strain Is-1 was grown for 6 days at 100 rpm in a rotatory shaker, and then 2 mL of the culture was spun down for 3 min at 9,000 × g and then suspended into 50 μL of HM salt medium. Equal volumes (50 μL) of *E. coli* donor and *B. japonicum* recipient cells prepared as described above were mixed in a 1.5-mL eppendorf tube. The mating mixture was spread onto a cellulose acetate membrane filter (Advantec, Tokyo, Japan; pore size 0.45 μm). The mating was carried out on HM plates at 30°C. The incubation time of the mating was 2 days. After mating, cells were suspended in 1 mL of HM salts medium and 100 μL aliquots of this suspension were spread onto 10 HM plates. Kanamycin was added to select for kanamycine-resistant (Km^r) transconjugants.

Strain and plasmid	Relevant chracteristics	Reference or source
Bradyrhizobium japonicum		
Is-1	Incompatible with *Rj*2*Rj*3 soybean and compatible with *Rj*4	Ishizuka et al. (1991a)
1C1	Is-1::Tn*5*, Km^r	This study
1C2	Is-1::Tn*5*, Km^r	This study
5C1	Is-1::Tn*5*, Km^r	This study
6C1	Is-1::Tn*5*, Km^r	This study
7C1	Is-1::Tn*5*, Km^r	This study
7C2	Is-1::Tn*5*, Km^r	This study
10C1	Tn*5*-induced spontaneous mutant, Km^r	This study
10C2	Is-1::Tn*5*, Km^r	This study
Escherichia coli		
S17-1	*pro thi recA hsdR*; chromosomal RP4-2(Tn*1*::ISRI *tet*::Mu Km::Tn7)	Simon et al. (1983)
JM109	*rec*A1; cloning strain	Takara Bio, Shiga, Japan
Plasmid		
pUTKm	Tn5-based delivery plasmid with Km^r, Ap^r	Herrero et al. 1990

Km, kanamycin; Amp, ampicillin.

Table 9. Bacterial strains and plasmids used in this study.

Isolation of Tn5 mutants: All Km^r transconjugant colonies in each HM plate were suspended in 20 mL of HM medium and cultivated for 7 days at 30°C. These 10 cultures containing Km^r transconjugants were separately spun down for 10 min at 6,000 × g and then rinsed twice with half-strength modified Hoagland nutrient (MHN) solution [53] to remove the antibiotic. Each pellet of Km^r transconjugants was suspended in 60 mL of MHN solution. These 10 suspensions were separately inoculated onto CNS (Rj_2Rj_3) at a rate of 5 mL per seed in each pot. The pots and seeds were prepared as follows. Both vermiculite (2.8 L) and MHN solution (1.4 L) were added to a 3.0-L porcelain pot and autoclaved at 121°C for 20 min. The seeds were sterilized with 5% sodium hypochlorite for 5 min and then rinsed with ethanol five times, and then rinsed with MHN solution five times. The surface-sterilized seeds of CNS were planted in a vermiculite medium at a rate of 12 seeds per pot. The inoculated plants were grown in a phytotron controlled at 25 ± 1°C under natural light conditions and sterilized water was supplied to maintain the moisture content. After 4 weeks of cultivation, Tn5 mutants were isolated from nodules produced on the root of CNS. The nodules were washed to remove vermiculite and immersed in 95% ethanol for 1 min, and then in 5% hydrogen peroxide solution for 5 min. The sterilized nodule was crushed and suspended in 5 mL of autoclaved 0.9% saline water. A drop of the turbid suspension was transferred to a yeast extract mannitol agar (YMA) plate [16] containing Congo Red (25 μg mL^{-1}) and streaked. These plates were incubated for 7 days at 30°C.

Amplification of the Tn5-flanking sequence: Kwon and Ricke [54] developed an efficient method to specifically amplify the transposon-flanking sequences with unique Y-shape linkers. The amplification of Tn5- flanking sequence was carried out according to their method. Because pUTKm was used to carry out Tn5 mutagenesis in both our study and the study by Kwon and Ricke [54], the same oligonucleotides are used in these two experiments: Linker 1, 5′- TTTCTGCTCGAATTCAAGCTTCTAACGATGTACGGGGACACATG-3′; Linker 2, 5′-TGTCCCCGTACATCGTTAGAACTACTCGTACCATCCACAT-3′; Y linker primer, 5′-CTGCTCGAATTCAAGCTTCT-3′; Tn5 primer, 5′- GGCCAGATCTGATCAAGAGA-3′. The genomic DNA was prepared using ISOPLANT (Nippon gene, Toyama, Japan). The Y linker was designed to have a 3′ overhang complementary to the sticky end generated by the *Nla* III or *Sph* I. In the present study, *Sph* I (Toyobo, Osaka, Japan) was used for digestion of the genomic DNA. Polymerase chain reactions (PCR) were carried out using the Program Temp Control System (ASTEC, Fukuoka, Japan) and TaKaRa Ex Taq Hot Start Version (Takara Bio, Shiga, Japan). Thermal cycling conditions were as follows: initial denaturing was carried out for 2 min at 95°C followed by 30 cycles of denaturing (30 s at 95°C), annealing (30 s at 60°C) and extension (90 s at 70°C). After the final cycle an extension step of 5 min at 70°C was added. The PCR products were analyzed on a 3% agarose gel (Nacalai Tesque, Kyoto, Japan) and stained with ethidium bromide (Nacalai Tesque, Kyoto, Japan).

Cloning, DNA sequencing and sequence analysis: The PCR products were cloned with pGEM-T Easy Vector System (Promega, Madison, WI, USA) (Table 9). Cloned DNA fragments were sequenced commercially by Macrogen (Seoul, Korea). Homology searches were carried out using the BLASTN program in the RhizoBase, which is a database for the genome of *B. japonicum* USDA110 (http://www.kazusa.or.jp/rhizobase/Bradyrhizobium/).

5.2. Results and discussion

Eight nodules and two popcorn-like nodules were produced on cv. CNS (*Rj*$_2$*Rj*$_3$) inoculated with the Km^r transconjugants. Tn*5* mutants (1C1, 1C2, 5C1, 6C1, 7C1, 7C2, 10C1 and 10C2) were isolated from the eight nodules and no Tn*5* mutants were isolated from the popcorn-like nodules. The nodulation type [13] of Tn*5* mutants was determined with CNS (*Rj*$_2$*Rj*$_3$) and Hill (*Rj*$_4$). All of them nodulated CNS effectively and retained the ability to nodulate Hill effectively [55]. Nodule effectiveness was judged by whether or not the nodule section indicated a section of red color. All Tn5 mutants were classified into nodulation type A. Although all nodules formed by Tn*5* mutants indicated a red color section, the nitrogen fixation activities were different for each one. In particular, the nitrogen fixation activities of 6C1 and 7C1 were drastically low. The nodulation profile test was carried out with D-51 (*Rj*$_3$), Hardee (*Rj*$_2$*Rj*$_3$) and IAC-2 (*Rj*$_2$*Rj*$_3$). All Tn*5* mutants nodulated both the *Rj*$_2$*Rj*$_3$- and the *Rj*$_3$-cultivars effectively.

The results of the electrophoresis of the PCR products showed that all mutants except for 10C1 contained a single copy of Tn*5* and that the fragment size of each PCR product was different. The PCR product amplified from 10C1 was not detected [55]. These results are equal to those obtained by genomic southern blot analysis (data not shown).

Tn5 mutant	Length	Identity	Location[†]	Gene[‡]	Deduced gene product
1C1	773	769/773(99%)	5009532–5010304 b	*bll4521*	Probable rare lipoprotein A
1C2[§]	1484	1478/1484(99%)	8203627–8205110 b	*bsr7468*	Cold shock protein
5C1	756	752/756(99%)	1540572–1541327 b	*blr1414*	Transcriptional regulatory protein ArsR family
6C1	227	227/227(100%)	1308155–1307929 b	*bll1193*	Integral inner membrane metabolite transport protein
7C1	1896	1892/1896(99%)	6149473–6151368 b	*bll5593*	Unknown protein
7C2	2092	2070/2092(98%)	6353378–6355469 b	*blr5786*	Putative integral membrane protein
10C2	2092	2070/2092(98%)	6353378–6355469 b	*blr5786*	Putative integral membrane protein

[†]The location in the genome of USDA110 homologous to the Tn*5* flanking sequence is shown. [‡]The gene of USDA110 that contains the homologous sequence to the Tn*5*-flanking sequence inside or upstream of itself is shown. [§]The location in the genome of USDA110 homologous to the Tn5 flanking sequence of 1C2 is upstream of *bsr7468*.

Table 10. Results of the blast searches of Tn5-flanking sequences based on RhizoBase.

The results of blast searches of cloned PCR products against the RhizoBase revealed that all homologous sequences to Tn*5*-flanking sequences were found in the complete genome sequence of *B. japonicum* USDA110 [56] (Table 10). All of them were located in different regions in the genome of *B. japonicum* USDA110 and all of them except for one in mutant 1C2 are within the open reading frames predicted by the USDA110 genome sequence project (Table 10). Using the "plasmid rescue" method [57], we have already determined Tn*5*-flanking sequences longer than the ones found in this report [58]. These sequences have been deposited into the DNA Data Bank of Japan (DDBJ) database (sequential accession numbers AB243409 through to AB243415].

B. japonicum strain USDA110 has an ability to form root nodules on soybeans (*Glycine max* L. Merr.) and is superior at symbiotic nitrogen fixation with soybeans compared with other strains [2]. The complete genome sequence of this strain has been determined [56]. This sequence information is a useful tool to determine the mechanism of nodulation and nitrogen fixation in *B. japonicum*. The effect of rhizobial inoculation depends largely on the nitrogen fixation ability of the strain that forms most of the nodules. However, in soils containing indigenous bradyrhizobia, the inoculation of highly effective rhizobia does not always result in the formation of effective nodules and in high nitrogen fixation [59]. To improve the effectiveness of the inoculation, the molecular genetic information of various strains, as well as USDA110, is believed to be required. The *B. japonicum* strain Is-1 nodulated Hill (Rj_4) but did not nodulate CNS (Rj_2Rj_3) and was classified into the nodulation type B. Although the *gsn* genes have been reported [46, 60], there have been no reports of the identification of an *Rj-gsn* gene. In this study, Tn*5* mutants were isolated from nodules produced on CNS inoculated with the Km^r transconjugants. It is confirmed that these mutants are derived from Is-1 using amplified fragment length polymorphism (AFLP) analysis [61].

Tn*5*-flanking sequences were specifically amplified from Tn*5* mutants except for 10C1 [58]. These results show that all mutants except for 10C1 contained a single copy of Tn*5* and that each Tn*5* insertion site was different and that 10C1 is probably a spontaneous kanamycin-resistant mutant. Wei and Bauer [62] also reported that Tn*5*-induced mutants with different phenotypes from the wild strain did not contain Tn*5* insertion. The analyses of these sequences showed that all Tn*5* mutants contained a single copy of Tn*5* and that each Tn*5* insertion site was different. There was no clear relationship among the Tn*5*-inserted gene products. However, most of the Tn*5*-inserted gene products related to the cell membrane structure

http://dx.doi.org/10.5772/52183

(e.g. the probable rare lipoprotein A, cold shock protein, the transcriptional regulatory protein ArsR family, integral inner-membrane metabolite transport protein and putative integral membrane protein) (Table 10). The probable rare lipoprotein A, integral inner-membrane metabolite transport protein and putative integral membrane protein are a part of the membrane structure. Meanwhile, it is thought that cold shock protein and the transcriptional regulatory protein ArsR family may be indirectly related to the membrane structure. The membrane-bound pump is regulated by the ArsR family [63] and the relationship between cold shock protein and membrane composition was reviewed by Ulusu et al. [64]. Therefore, changes or damage to the cell membrane structure in mutants may overcome the nodulation restriction conditioned by Rj_2-soybean. Judd et al. [65], however, reported that the transposon was not responsible for host range extension in the Tn*5* mutant of USDA 438. In the future, we must confirm that the Tn*5* insertion is responsible for the acquisition of the ability to nodulate Rj_2-soybean.

6. Inoculation techniques using compatibility between *Rj*-genes and *Bradyrhizobium* strains

In our laboratory, an experiment for testing effects of inoculation into a cylindrical soil block from just above the seed were conducted using bacterial suspension of *Bradyrhizobium japonicum* USDA110 as useful rhizobia in the past. The results showed that the occupation ratio of USDA110 in the tap roots was as high as 70-100% and that in the lateral root was 44-77% lower than that of tap root and decreased in progression of the growth and developing of the host plant. In particular, the extreme decrease of USDA110 occupation was observed in soybean variety Fukuyutakaa, which was cultivated as a recommended variety in the southwest warm region of Kyushu in Japan. It is likely that the cause of reduced occupancy in lateral roots occurred because rhizobia were inoculated in the narrow range around the tap root, while lateral roots continued to grow outside of the inoculation column. If the range of inoculation were enlarged, the occupancy ratio of lateral roots probably would have increased.

In 2005, a field experiment was conducted to clarify the effect of the difference on soybean production between the inoculation of rhizobia on the seed surface compared with the plow layer [66]. Non-inoculated plot (NI), seed coat inoculation (SI) plot, and plow layer rhizobial solution inoculation (RI) plots were tested in three replications. Rhizobial concentration for inoculation was 10^7 cells/seed in both plots. As the result, the number of nodulation and the occupancy ratio of serotype USDA110 were highest in the SI plot, however yield (kg/10a) was higher in the RI plot versus other plots. This occurred is because the inoculum density in the SI plot was high, resulting in many nodules formed. This increased competition of photosynthetic products between the growth of soybean and nodulation, and consequently the initial growth was suppressed. For the SI plot, the effect of inoculation was expected to decline with time. In addition, since the occupancy ratio of inoculum was less than 50%, it is considered likely that the yield would increase if the seed surface inoculation were high.

Based on these results, it is important to examine the relationship between nodulation method and the effective inoculum concentration on the seed.

Therefore, the purpose of this study was to clarify the effect of the difference in inoculation method and inoculum density of *B. japonicum* USDA110 on soybean production.

6.1. Materials and methods

Soybean (*Glycine max* L. Merr.) variety Fukuyutak and *B. japonicum* USDA110 having uptake hydrogenase (Hup^+) were used in this experiment. *B. japonicum* USDA110 was cultured in HM medium for eight days with shaking at 30°C used for inoculation. Fukuyutaka was cultivated in the farm of Kyushu University Faculty of Agriculture, where barley was grown as the previous crop and soybean had not been grown for the past five years. Experimental plots had three replications for the treatments: non-inoculation (NI) plot, seed coat inoculation (SI) plots; 10^5 cells/seed (SI5) plot and 10^7 cells/seed (SI7) plot, and plow layer rhizobial peat-moss inoculation (PI) plots; 10^7 cells/seed (PI7) plot and 10^9 cells/seed (PI9) plot. Peat-moss inoculation was conducted using mixture of BM2 (raw materials: peat-moss, Group Berger Peat Moss Ltd., Canada) USDA110 culture inoculated into the plow layer before seed sowing to be 10^7 or 10^9 cells/seed.

Inoculum for SI plot per 100 seeds was made from 1.5 mL deionized water and 10 ml of 12% aqueous solution of gum arabic, 10 g of BM2 and 0.015 mL (SI5) or 1.5 mL (SI7) of USDA110 culture solution (1×10^9 cells/mL). Inoculum for PI plot per m^2 was mixed well BM2 of 200 g spread on the plastic sheet, tap water of 40% of the maximum water holding capacity of BM2 and USDA110 culture solution of 0.25 mL (10^7 cells/seed) or 25.0 mL (10^9 cells/seed). After spraying the mixture into the row of the test plot, the rows were plowed (approximately 15 cm depth) by a tractor. The amount of USDA110 in the plow layer after inoculation was 10^7 or 10^9 cells per 1,800 cm^3 ($60 \times 20 \times 15$ cm^3) of soil occupying the root zone of one hill (three seeds sown). As a result, the inoculum density of USDA110 in the plow layer of each plot, PI9 and PI7 was estimated to be 1.7×10^3 and 1.7×10^5 cells/g DW soil. The amount of lime carbonate for adjusting to 6.5 soil pH was estimated by the buffer curve method using soil samples collected from 6 points (15 cm soil depth) in the experimental field. After applying the lime carbonate, on July 11, Mame kasei (3.0% ammonia nitrogen, 10.0% acid soluble phosphorus, 10.0% water-soluble potash) was applied at a rate of 80 kg per 10 a to all layers. Inoculation of the PI plot was done on July 18. On July 28, seeds of Fukuyutaka of all plots were sown into hills of either 20 or 60 cm apart, at a depth of approximately 2-3 cm. The seedlings were thinned to two seedlings at the trifoliate stage (V2). At V5 stage, the inter row area was cultivated and ridged. Pesticides were sprayed according to the occurrence of insect pests. Plants of one hill per each plot were sampled for roots at V6.4 and R5.7 stage, and 10 hills were harvested by cutting the stem at R8 stage (November 2 or 8).

To estimate the density of indigenous rhizobia, soils were collected from two locations in the experimental field before fertilization and the rhizobia density was measured by the most probable number (MPN) method using soybean cultivars, Orihime (non-*Rj*-genotype) and Fukuyutaka (Rj_4-genotype). Plants sampled for roots at V6.4 and R5.7 stage [67] were divided into shoots and roots. Roots were used for measurement of acetylene reduction activity (ARA) and counting of nodules. Counting of nodules was done after partitioning the plant into five parts the upper 3 cm of tap root, the lower part of the tap root, the lateral roots generated from the upper tap root, the lateral roots generated from the lower tap root and the superficial roots. Nodules from these samples were freeze-dried. Plant parts were air-dried at 80°C for 24-48 hrs. Also, in order to determine the effect of inoculation, the freeze-dried nodules were evaluated with serological tests performed using USDA110 antiserum produced previously in our laboratory. USDA110 occupancy was analyzed by the χ^2 test.

After examining individual harvest yield components (effective number of pods/m^2, seeds number/pods, complete seed percentage, one hundred seed weight (g), yield (g/m^2) in each plot), plant dry weight of plant part was weighed. For each growth stage, significant differences were determined according to LSD test (Fisher) after ANOVA analysis.

6.2. Results and desucassion

Occupation of useful rhizobium inoculated on the seed coat was known to be low because of their low competition against indigenous rhizobia. This object was to clarify the effect of inoculation method and inoculum density of *B. japonicum* USDA110 on production of soybean. Five experimental plots were designed:, no inoculation (NI), seed coating inoculation with a density of 10^5 cells/seed (SI5) and 10^7 cells/seed (SI7), previous inoculation into the root zone with a density of 1.7×10^3 cells/g soil (PI7) and 1.7×10^5 cells/g soil (PI9). The PI plots were plowed after the BM2 mixture was soaked with rhizobium culture.

Growth Stage	Treatment	Superficical root	Tap root		Lateral roots		Total
			Upper	Lower	Upper	Lower	
V6.4	NI	13.7	32.7	24.3	39.7 b	13.7	124.0
	SI5	18.3	37.7	33.0	38.3 b	16.3	143.7
	SI7	17.3	33.0	18.7	42.7 b	27.7	139.3
	PI7	8.3	33.7	34.7	70.7ab	25.3	160.7
	PI9	16.0	21.0	24.0	90.7 a	18.3	170.0
R5.7	NI	47.7	46.7	28.0 a	214.3	35.3	372.0
	SI5	55.0	30.3	24.7ab	167.0	27.7	304.7
	SI7	61.7	37.3	13.3 b	211.3	34.0	357.7
	PI7	31.3	32.0	15.7ab	237.7	25.3	342.0
	PI9	39.0	29.3	14.0ab	183.3	27.0	292.7

Mean followed by same and without letters within a column of each stage are not significantly different using LSD (P<0.10).

Table 11. Transition of nodule number in each growth stage effected by inoculation methods.

Table 11 shows the effect of inoculation methods on nodulation efficiency. The PI treatment had high efficiency for the lateral roots at the V6.4 stage, but at R5.7, the efficiency was low. For all other root positions and inoculation methods, the efficiency was low. Occupation ratio of serotype USDA110 on the total root was significantly highest for SI7 and PI9 and lowest in PI7 at the V6.4 stage. At the R5.7 stage, the occupation ratio in the superficial and upper lateral roots increased especially by inoculation with the *B. japonicum* USDA110, except the superficial roots of PI7. Also, for whole roots, the occupation ratio was significantly increased by inoculation of *B. japonicum* USDA110, except for PI7 (Table 12). However, the ARA showed a tendency to decrease with high density inoculum (Table 13). These results were complemented by seed yields of Fukuyutaka being highest in SI5 and SI7 (Table 14).

Growth Stage	Treatment	Superficial root	Tap root		Lateral roots		Total
			Upper	Lower	Upper	Lower	
V6.4	NI	12.5	15.0bc	25.0ab	20.5bc	5.6 b	17.3 b
	SI5	16.7	23.1bc	29.7 b	27.5 b	12.5 b	24.0 b
	SI7	0.0	70.3 a	42.9 a	34.8ab	16.7 b	32.7 a
	PI7	0.0	6.3 c	9.1 b	11.8 c	5.0 b	8.7 c
	PI9	6.7	27.8 b	31.8ab	46.3 a	53.3 a	38.7 a
R5.7	NI	10.0 b	31.6	50.0	22.4 b	35.7	26.0 b
	SI5	44.8 a	38.5	23.1	45.7 a	50.0	43.3 a
	SI7	47.8 a	58.8	28.6	53.9 a	28.6	50.0 a
	PI7	27.3ab	25.0	40.0	25.0 a	25.0	24.7 b
	PI9	50 a	53.0	33.3	55.2 a	54.0	53.3 a

Mean followed by same and without letters within a column of each stage are not significantly different Chi squaretest (P<0.10).

Table 12. Occupation ratio of serotype USDA110 in nodule of each growth stage.

Treatment	Growth stage	
	V6.4	R5.7
NI	48.0 a	9.2 c
SI5	50.0 a	16.3 b
SI7	39.1ab	7.9 c
PI7	48.0 a	28.4 a
PI9	33.3 b	10.1 c

Mean followed by same and without letters within a column of each stage are not significantly different using LSD (P<0.10).

Table 13. Acetylene reduction activity (ARA: μmole g^{-1} DW) at each growth stage.

Tratment	Pod number m^{-2}	Seed number pod^{-1}	Full seed ratio	100 seed g DW	Yield g DW m^{-2}	Yield index
NI	477.5	1.81	0.93	21.2	165.5 b	0.57
SI5	545.5	1.85	0.94	21.5	210.2 a	0.58
SI7	527.5	1.80	0.92	22.7	195.9 a	0.55
PI7	486.7	1.84	0.94	20.6	172.7 b	0.57
PI9	577.8	1.79	0.93	21.8	208.8 a	0.56

Mean followed by same and without letters within a column of each stage are not significantly different using LSD (P<0.10).

Table 14. Seed yield and yield components of each treatment.

Consequently, in SI5, SI7, and PI9 plots, occupation of serotype USDA110 was significantly high vs. other treatments. Since greater fixed nitrogen was distributed for these treatments to pods and seeds, yield (g/m^2) significantly increased. It was thought that the density of 10^5 cells/seed was more effective at seed inoculation because there were no effect by increasing the inoculum above this density. And from the results of the PI9 plot, it was thought previ-

http://dx.doi.org/10.5772/52183

ous inoculation using BM2 (1.7×10^5 cells/seed) was effective in competition with the indigenous rhizobia for nodulation.

Previous work showed that to successfully compete with indigenous rhizobium, introduced inoculum must be 1000 times greater [1]. However, in Canada, standard inoculum for soybean, kidney bean, and pea were 10^5 cells/seed [60]. Others have reported unsuccessful nodulation at high concentrations of *B. japonicum* USDA110 [68]. In this study, we demonstrated a significant increase in yield in both SI5 and SI7 treatments with inoculation concentrations of 10^5 cell/seed and 1.7×10^5 cell/seed, respectively. Increased inoculum density above these levels did not increase seed yield. We concluded that it is possible to increase yield through the introduction of rhizobium species having greater nitrogen fixation rates.

Author details

Takeo Yamakawa[1*] and Yuichi Saeki[2]

*Address all correspondence to: yamakawa@agr.kyushu-u.ac.jp

1 Department of Biosciences & Biotechnology, Faculty of Agriculture, Kyushu University, 6-10-1 Hakozaki, Higashi-ku, Fukuoka, Japan

2 Department of Biotechnology and Biosciences, Faculty of Agriculture, University of Miyazaki, Miyazaki, Japan

References

[1] Weaver RW., Frederick LR. Effect of Inoculum Rate on Competitive Nodulation of *Glycine max* L. Merrill. II. Field Studies. Agronomy Journal 1974: 66(2) 233-236.

[2] Kvien CS., Ham GE., Lambert JW. Recovery of Introduced *Rhizobium japonicum* Strains by Soybean Genotypes. Agronomy Journal 1981: 73(5) 900-905.

[3] Takahashi T., Ito A., Suzuki H. Interaction between Effective Bacteria and Host Plant. In: Ishizuka J. (ed.) Studies on Nodule Formation and Nitrogen Fixation in Legume Crops. Tokyo: Ministry of Agriculture, Forestry and Fisheries; 1996. p92-107.

[4] Maier RJ., Brill WJ. Mutant Slrams of *Rhizohium japonicum* with Increased Ability to Fix Nitrogen for Soybean. Science 1978: 201(4354) 448-450.

[5] Williams LE., Phillips DA. Increased Soybean Productivity with a *Rhizobium japonicum* Mutant. Crop Science 1983: 23(2) 246-250.

[6] Maier J., Graham L. Mutant Strains of *Bradyrhizobium japonicum* with Increased Symbiotic N_2 Fixation Rates and Altered Mo Metabolism Properties. Applied Environmental Microbiology 1990: 56(8) 2341-2346.

[7] Williams LF., Lynch LD. Inheritance of a Non-nodulating Character in the Soybean. Agronomy Journal 1954: 46(1) 28-29.

[8] Caldwell BE. Inheritance of a Strain-Specific Ineffective Nodulation in Soybeans. Crop Science 1966: 6(5) 427-428.

[9] Caldwell BE., Hinson K., Johnson HW. A Strain-Specific Ineffective Nodulation Reaction in the Soybean *Glycine max* L. Merrill. Crop Science 1966: 6(5) 495-496.

[10] Vest G. Rj3 A Gene Conditioning Ineffective Nodulation. Crop Science 1970: 10(1): 34-35.

[11] Vest G., Caldwell EB. Rj4 - A Gene Conditioning Ineffective Nodulation in Soybean. Crop Science 1972: 12(5) 692-693.

[12] Ishizuka J., Suemasu Y., Mizogami K. Prefarence of *Rj*-Soybean Cultivars for *Bradyrhizobium japonicum* for Nodulation. Soil Science and Plant Nutrition 1991: 37(1) 15-21.

[13] Ishizuka J., Yokoyama A., Suemasu Y. Relationship between Serotypes of *Bradyrhizobium japonicum* and Their Compatibility with *Rj*-Cultivars for Nodulation. Soil Science and Plant Nutrition 1991: 37(1) 23-30.

[14] Ishizuka J., Kim SD., Hussain AKMA., Yamakawa T. Soybean Preference for *Bradyrhizobium japonicum* for Nodulation. Isolation of Rj_2Rj_4-Lines from the Cross of Soybean Cvs. IAC-2 (Rj_2) and Hill (Rj_4). Soil Science and Plant Nutrition 1993: 39(1) 79-86.

[15] Yamakawa T., Eriguchi M., Hussain AKMA., Ishizuka J. Soybean Preference for *Bradyrhizobium japonicum* for Nodulation. Nodulation by $Rj_2Rj_3Rj_4$-Genotypes Isolated from the Progenies of Cross of Soybean Cvs. IAC-2 (Rj_2Rj_3) and Hill (Rj_4). Soil Science and Plant Nutrition 1999: 45(2) 461-469.

[16] Vincent J.M. A Manual for the Practical Study of the Root-Nodule Bacteria. IBP Handbook, No.15. Oxford; Blackwell Scientific Publications; 1970.

[17] Minamisawa K., Iino H., Saito N., Onodera S., Nakatuka Y., Nakazima S., Kubota M., Asami T. Determination of Genotype Group of *Bradyrhizobium japonicum* by Indole-3-Acetic Production. Japanese Journal of Soil Science and Plant Nutrition 1992: 63(4) 459-462. (in Japanese)

[18] Owens LD., Wright DA. Production of the Soybean-Chlorosis Toxin by *Rhizobium japonicum* in Pure Culture. Plant Physiology 1965: 40(5) 931-933.

[19] Gordon SA., Weber RP. The Colorimetric Estimation of IAA. Plant Physiology 1951: 26(1) 192-195.

[20] Minamisawa K., Fukai K. Production of Indole-3-Acetic Acid by *Bradyrhizobium japonicum*: A Correlation with Genotype Grouping and Rhizobitoxine Production. Plant Cell Physiology 1991: 32(1) 1-9.

[21] Haider J., Hussam AKMA., Ikeda M., Yamakawa T., Ishizuka J. Effects of Nitrate Application on Growth, Nodulation and Nitrogen Fixation of Nitrate-Tolerant Mutant of Soybean. Soil Science and Plant Nutrition 1991: 37(3) 521-529.

[22] Yamakawa T, Hussain AMKA, Ishizuka J. Soybean Preference for *Bradyrhizobium japonicum* for Nodulation. Occupation of Serogroup USDA110 in Nodules of Soybean Plants Harboring Various *Rj*-Genes Grown in a Field. Soil Science and Plant Nutrition 2003: 49(6) 835-841.

[23] Evans HI., Harker AR., Papen H., Russell SA., Hauns FI., Zuber M. Physiology, Biochemistry and Genetics of the Uptake Hydrogenase in Rhizobia. Annual Review of Microbiology 1987: 41 335-361.

[24] Cregan PB., Keyser HH. Host Restriction of Nodulation by *Bradyrhizobium japonicum* Strain USDA 123 in Soybean. Crop Science 1968: 26(5) 911-916.

[25] Saeki Y., Aimi N., Tsukamoto S., Yamakawa T., Nagatomo Y., Akao S. Diversity and Geographical Distribution of Indigenous Soybean-Nodulating Bradyrhizobia in Japan. Soil Science and Plant Nutrition 2006: 52(4) 418-426.

[26] Saeki Y., Ozumi S., Yamamoto A., Umehara Y., Hayashi M., Sigua GC. Change in Population Occupancy of Bradyrhizobia under Different Temperature Regimes. Microbes Environments 2010: 25(4) 309-312.

[27] Minami M., Yamakawa T., Yamamoto A., Akao S., Saeki Y. Estimation of Nodulation Tendency among *Rj*-Genotype Soybeans using the Bradyrhizobial Community Isolated from an Andosol. Soil Science and Plant Nutrition 2009: 55(1) 65-72.

[28] Saeki Y., Minami M., Yamamoto A., Akao S. Estimation of the Bacterial Community Diversity of Soybean-Nodulating Rhizobia Isolated from *Rj*-Genotype Soybean. Soil Science and Plant Nutrition 2008: 54(5) 718-724.

[29] Saeki Y., Akagi I., Takaki H., Nagatomo Y. Diversity of Indigenous *Bradyrhizobium* Strains Isolated from Three Different *Rj*-Soybean Cultivars in Terms of Randomly Amplified Polymorphic DNA and Intrinsic Antibiotic Resistance. Soil Science and Plant Nutrition 2000: 46(4) 917–926.

[30] Hiraishi A., Kamagata Y., Nakamura K. Polymerase Chain Reaction Amplification and Restriction Fragment Length Polymorphism Analysis of 16S rRNA Genes from Methanogens. Journal of Fermentation and Bioengineering 1995: 79(6) 523–529.

[31] Saeki Y., Aimi N., Hashimoto M., Tsukamoto S., Kaneko A., Yoshida N., Nagatomo Y., Akao S. Grouping of *Bradyrhizobium* USDA Strains by Sequence Analysis of 16S rDNA and 16S-23S rDNA Internal Transcribed Spacer Region. Soil Science and Plant Nutrition 2004: 50(4) 517–525.

[32] Nei M., Li HW. Mathematical Model for Studying Genetic Variation in Terms of Restriction Endonucleases. Proceeding of the National Academy of Sciences of the United States of America 1979: 76(10) 5269–5273.

[33] Sakai M., Futamata H., Kim SJ., Matuguchi T. Effect of Soil Salinity on Population Structure of Fluorescent Pseudomonads in Spinach Rhizosphere. Soil Science and Plant Nutrition 1998: 44(4) 701–705.

[34] MacArthur RH. Patterns of Species Diversity. Biological Reviews 1965: 40(4) 510 – 533.

[35] Pielou EC. Ecological Diversity and Its Measurement. In: An Introduction to Mathematical Ecology. New York: John Wiley & Sons Inc; 1969. p221–235.

[36] Kobayashi S. Multivariate Analysis of Biological Communities. Tokyo; Soju Shobo; 1995. (In Japanese.)

[37] Whittaker RH. Evolution and Measurement of Species Diversity. Taxon 1972: 21(2/3) 213–251.

[38] Shiro S., Yamamoto A., Umehara Y., Hayashi M., Yoshida N., Nishiwaki A., Yamakawa T., Saeki Y. Effect of *Rj* Genotype and Cultivation Temperature on the Community Structure of Soybean-Nodulating Bradyrhizobia. Applied Environmental Microbiology 2011: 78(4) 1243-1250.

[39] Zhang F., Smith LD. Effect of Low Root Temperature on the Early Stages of Symbiosis Establishment between Soybean (*Glycine max* (L.) Merr.) and *Bradyrhizobium japonicum*. Journal of Experimental Botany 1994: 45(10) 1467–1473.

[40] Zhang F., Smith LD. Genistein Accumulation in Soybean (*Glycine max* [L.] Merr.) Root Systems under Suboptimal Root Zone Temperatures. Journal of Experimental Botany 1996: 47(6) 785-792.

[41] Zhang F., Charles CT., Pan B., Smith LD. Inhibition of the Expression of *Bradyrhizobium japonicum nod* Genes at Low Temperatures. Soil Biology and Biochemistry 1996: 28(12) 1579-1583.

[42] Pan B., Smith LD. Genistein and Daidzein Concentrations and Contents in Seedling Roots of Three Soybean Cultivars Grown under Three Root Zone Temperatures. Journal of Agronomy and Crop Scence 1998: 180(2) 77-82.

[43] Yokoyama T. Effects of Temperature on Competition for Nodulation in Phylogenetically Different *Bradyrhizobium* Strains. Japanese Journal of Soil Science and Plant Nutrition 2005: 76(5) 599–607. (In Japanese with English summary.)

[44] Yokoyama T., Ando S., Murakami T., Imai H. Genetic Variability of the Common *nod* Gene in Soybean Bradyrhizobia Isolated in Thailand and Japan. Canadian Journal of Microbiology 1996: 42(12) 1209-1218.

[45] Duzan HM., Zhou X., Souleimanov A., Smith LD. Perception of *Bradyrhizobium japonicum* Nod Factor by Soybean [*Glycine max* (L.) Merr.] Root Hairs under Abiotic Stress Conditions. Journal of Experimental Botany 2004: 55(408) 2641-2646.

[46] Sadowsky MJ., Cregan PB., Guottfert M., Sharma A., Gerhold D., Rodriguez-Quinones F., Keyzer HH., Henneke H., Stacy G. The *Bradyrhizobium japonicum nolA* Gene

and Its Involvement in the Genotype-Specific Nodulation of Soybeans. Proceeding of the National Academy of Sciences of the United States of America 1991: 88(2) 637-641.

[47] Lohrke SM., Day B., Kolli VS., Hancock R., Yuen JP., de Souza ML., Stacey G., Carlson R., Tong Z., Hur HG., Orf JH., Sadowsky MJ. The *Bradyrhizobium japonicum noe*D Gene: A Negatively Acting, Genotype Specific Nodulation Gene for Soybean. Molecular Plant-Microbe Interctions 1998: 11(6) 476– 488.

[48] Cole MA., Elkan GH. Transmissible Resistance to Penicillin G, Neomycin, and Chloramphenicol in *Rhizobium japonicum*. Antimicrobal Agents and Chemotherapy 1973: 4(3) 248–253.

[49] Sambrook J., Fritsch EF., Maniatis T. Molecular Cloning: A Laboratory Manual, 2nd edn. Cold Spring Harbor; 1989.

[50] Simon R., Prieffer U., Pühler A. A Broad Host Range Mobilization System for *In Vivo* Genetic Engineering: Transposon Mutagenesis in Gram-Negative Bacteria. Nature Biotechnology 1(9), 784–791, 1983.

[51] Herrero M., de Lorenzo V., Timmis KN. Transposon Vectors Containing Non-Antibiotic Resistance Selection Markers for Cloning and Stable Chromosomal Insertion of Foreign Genes in Gram-Negative Bacteria. Journal of Bacteriology 1990: 172(11) 6557– 6567.

[52] Kuykendall LD. Isolation and Identification of Genetically Marked Strains of Nitrogen-Fixing Microsymbionts of Soybean. In: Elkan GH (ed.) Symbiotic Nitrogen Fixation Technology. New York and Basel: Marcel Dekker, Inc; 1987. p205-220.

[53] Nakano Y., Yamakawa T., Ikeda M., Ishizuka J. Nodulation of *Rj*-Soybean Varieties with *Rhizobium fredii* USDA193 under Limited Supply of Nutrients. Soil Science and Plant Nutrition 1997: 43(4) 929–932.

[54] Kwon YM., Ricke SC. Efficient Amplification of Multiple Transposon-Flanking Sequences. Journal of Microbiological Methods 2000: 41(3) 195–199.

[55] Tsurumaru H., Yamakawa T., Tanaka M., Sakai M. Tn*5* Mutants of *Bradyrhizobium japonicum* Is-1 with Altered Compatibility with Rj_2-Soybean Cultivars. Soil Science and Plant Nutrition 2008: 54(2) 197-203, 2008.

[56] Kaneko T., Nakamura Y., Sato S., Minamisawa K., Uchiumi T., Sasamoto S., Watanabe A., Idesawa K., Iriguchi M., Kawashima K., Kohara M., Matsumoto M., Shimpo S., Tsuruoka H., Wada T., Yamada M., Tabata S. Complete Genomic Sequence of Nitrogen-fixing Symbiotic Bacterium *Bradyrhizobium japonicum* USDA110. DNA Reserch 2002: 9(6) 189-197.

[57] Kiessling U., Platzer M., Strauss M. Plasmid Rescue – A Tool for Reproducible Recovery of Genes from Transfected Mammalian Cells? Molecular and General Genetics 1984: 193(3) 513–519.

[58] Tsurumaru H., Yamakawa T., Tanaka M., Sakai M. The Efficient Strategy of Plasmid Rescue from Tn*5* Mutants Derived from *Bradyrhizobium japonicum* Is-1, Based on Whole Genome Sequence Information of Strain USDA110. Journal of the Faculty of Agriculture, Kyushu University 2008: 53(1) 27-31.

[59] Hardarson G., Golbs M., Danso SKA. Nitrogen Fixation in Soybeans (*Glycine max* L. Merrill) as Affected by Nodulation Patterns. Soil Biology and Biochemistry 1989: 21(6) 783-787.

[60] Lohrke SM., Madrzak CJ., Hur H., Judd AK., Orf JH., Sadowsky MJ. Inoculum Density-Dependent Restriction of Nodulation in the Soybean-*Bradyrhizobium japonicum* Symbiosis. Symbiosis 2000: 29(1) 59-70.

[61] Yamakawa T., Tanaka M., Sakai M., Sarr PS. Genetic Grouping of *Bradyrhizobium* Strains Compatible with Soybean (*Glycine max* L. Merr.) Harboring *Rj*-gene by AFLP-fingerprinting Analysis. Journal of the Faculty of Agriculture, Kyushu University 2011: 56(2) 223-229.

[62] Wei X., Bauer WD. Tn5-Induced and Spontaneous Switching of *Sinorhizobium meliloti* to Faster-Swarming Behavior. Applied and Environmental Microbiology 1999: 65(3) 1228–1235.

[63] Rosen BP., Bhattacharjee H., Shi W. Mechanisms of Metalloregulation of an Anion-Translocating ATPase. Journal of Bioenergetics and Biomembranes 1995: 27(1) 85–91.

[64] Ulusu NN., Tezcan EF. Cold Shock Proteins. Turkish Journal of Medical Sciences 2001: 31(4), 283–290.

[65] Judd AK., Sadowsky MJ., Bhagwat AA., Cregan PB., Liu RL. Isolation of a *Bradyrhizobium japonicum* Serogroup 123 Mutant Which Has an Extended Host Range for Nodulation-Restricting Soybean Genotypes. FEMS Microbiology Letters 1993: 106(2) 205–210.

[66] Fukushima Y., Yamakawa, T. Effect of Difference of Inoculation Method and Inoculum Density of *Bradyrhizobium japonicum* USDA110 on Production of Soybean (*Glycine max* L. Merr.), Science bulletin of the Faculty of Agriculture, Kyushu University 2006: 61(2) 171-176. (in Japanese with English summary)

[67] Fehr WR., Cavines, CE., Burmood DT., Pennington JS. Stage of Development Descriptions for Soybeans, *Glycine max* (L) Merrill. Crop Science 1971: 11(6) 929-931.

[68] Smith RS. Legume Inoculant Formulation and Application. Canadian Journal of Microbiology 1992: 38(6) 485-492.

Chapter 6

Soybean Seed Production and Nitrogen Nutrition

Takuji Ohyama, Ritsuko Minagawa, Shinji Ishikawa,
Misaki Yamamoto, Nguyen Van Phi Hung,
Norikuni Ohtake, Kuni Sueyoshi, Takashi Sato,
Yoshifumi Nagumo and Yoshihiko Takahashi

Additional information is available at the end of the chapter

http://dx.doi.org/10.5772/45867

1. Introduction

The world population is consistently increasing, and it is over 7 billion in 2012, while the land area for agricultural use is limited. Therefore, the increase in crop production per area is very important. Soybean (*Glycine max* (L.) Merr.) originates from East Asia, and soybean seed is one of the most important protein sources for human and livestock all over the world. Annual production of soybean (262 M (million) t in 2010) is the fourth of the major grain crops, after maize (844 M t), paddy rice (672 M t) and wheat (650 M t) [1]. In the whole world, over 85% of soybean is used for oil and the residue is used for animal feed. Annual soybean seed production has been steadily increasing for recent decades (91 M t in 1980, 109 M t in 1990, 161 M t in 2000, 262 M t in 2010) [1]. The cultivation area of soybean is 102 M ha in 2010. Major soybean production countries (annual production in 2010) are USA (90.6 M t), Brazil (68.5 M t), Argentina (52.7 M t), China (15.1 M t), and India (9.8 M t) in this sequence. Soybean production in Japan in 2010 was only 223,000t and it accounted for 5% of the total consumption in Japan. The world average seed yield is 2.56 t ha^{-1} in 2010, and is higher in the USA (2.92 t ha^{-1}), Brazil (2.94 t ha^{-1}), and Argentina (2.90 t ha^{-1}) compared with China (1.77 t ha^{-1}), Japan (1.62 t ha^{-1}) and India (1.07 t ha^{-1}) and other countries.

The nutrient composition of soybean seeds (per 100 g) produced in Japan is as follows [2]; energy 417 kcal (1,745 kJ), water 12.5g, protein 35.3g, lipids 19.0g, carbohydrate 28.2g, minerals 5g. The composition is quite different from the other grain crop seeds, such as "Paddy rice"; energy 350 kcal (1,464 kJ), water 15.5g, protein 6.8g, lipids 2.7g, carbohydrate 73.8g, minerals 1.2g, "Wheat"; energy 337 kcal (1,410 kJ), water 12.5g, protein 10.6g, lipids 3.1g, carbohydrate 72.2g, minerals 1.6g, and "Corn"; energy 350 kcal (1,464 kJ), water 14.5g, protein 8.6g, lipids 5.0g, carbohydrate 70.6g, minerals 1.3g. The protein concentration in soybean seeds is very high about 4-5 times higher than that of rice, wheat and corn, but carbohydrate concentration is lower.

The storage protein of soybean seeds mainly consists of glycinin and β-conglycinin. The β-conglycinin is comprised of three subunits, designated as α', α, and β-subunits. The β-subunit of β-conglycinin is especially low in sulfur amino acids, containing only one cysteine and no methionine residue in its mature form. Soybean seeds contain a large amount of lipids (20%), and about 90% is unsaturated fatty acid (linoleic acid 51-57%, oleic acid 32-36% and linolenic acid 2-10%) and 10% is saturated fatty acid (palmitic acid 4-7%, stearic acid 4-7%) [3]. Linoleic acid and linolenic acid are essential lipids, which cannot be synthesized by ourselves. Although soybean seeds contain about 28% carbohydrates, most of them are structural carbon like cell walls and oligosaccharides (sucrose 5%, stachyose 4%, raffinose 1%). Starch is tentatively accumulated in young immature soybean seed, however, it decreases and converts to lipid and protein at maturity. Soybean seeds contain relatively a high amount of minerals (5%) compared with cereal seeds (about 1%). Soybean seeds contain abundant potassium (1,900 mg), calcium (240 mg), magnesium (220 mg), phosphorous (220 mg), iron (9.4 mg), zinc (3.2 mg) per 100g seeds. Soybean seeds contain vitamins, both lipid soluble vitamins (Vitamin E (1.8 mg)) and water soluble vitamins (V B_1 (0.83 mg) and V B_2 (0.30 mg)). Soybean seeds contain isoflavonoids, daidzein and genistein. These isoflavonoids are expected to play a role as a female hormone or to decrease fat in blood.

2. Soybean seed yield

2.1. Potential of soybean yield

The world average soybean yield has been increasing by about 60% for 30 years from 1980 (1.6 t ha^{-1}) to 2010 (2.6 t ha^{-1}). The highest yield of soybean in Japan was recorded at 7.8 t ha^{-1}, and soybean seed yield can reach 4-6 t ha^{-1} with well-managed fields under good climatic and soil conditions [4]. Recently, an amazing high soybean seed yield over 10 t ha^{-1} was recorded in 2008 and 2010 by a farmer, Mr. Kip Cullers in Missouri, USA [5]. Therefore, the potential productivity of soybean should be much higher than we have thought.

Figure 1 shows the yield components of soybean. Soybean "seed yield" is calculated by multiplying the "seed number" per area and one average "seed weight". Seed number per area is calculated by the "pod number" and the average "seed number per pod". The pod number is decided by "flower number" and the "pod formation rate". The flower number depends on the "node number". The node number per area is decided by "stem number per plant" and "planting density".

Farmers can control the planting density. Planting density is an important factor for soybean growth and seed yield, although the planting density is not directly proportional to the dry matter production and seed yield. When planting density is high, the branching of each plant is depressed and the number of the lateral stems decrease. In addition, under excess planting density the competition for photosynthesis and nutrient absorption among plants become severe and the stems grow tall and thin and plants are proneto lodge.

http://dx.doi.org/10.5772/45867

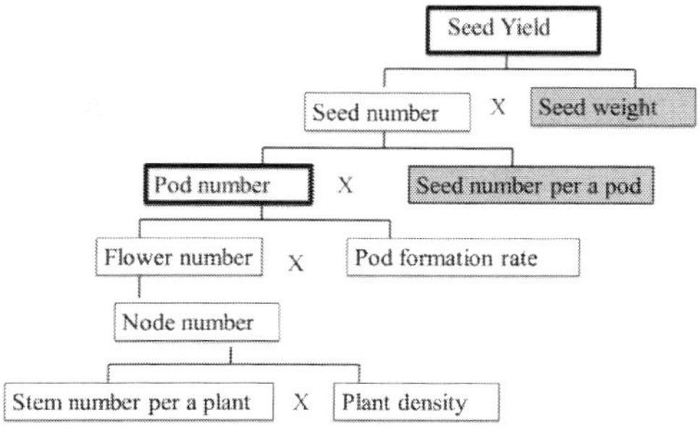

Figure 1. Yield components in soybean cultivation.

Generally, soybean seed yield depends mostly on pod number per area. Generally over 50% of soybean flowers and young pods abort and they don't make mature pods and seeds. Under bad weather and growing conditions, the percentage of flower and pod abortion increases. The average number of seeds in a pod is relatively constant, although the seed number per pod differs from 1 to 4 in soybean. Average seed weight is affected by growing conditions in late growth stages.

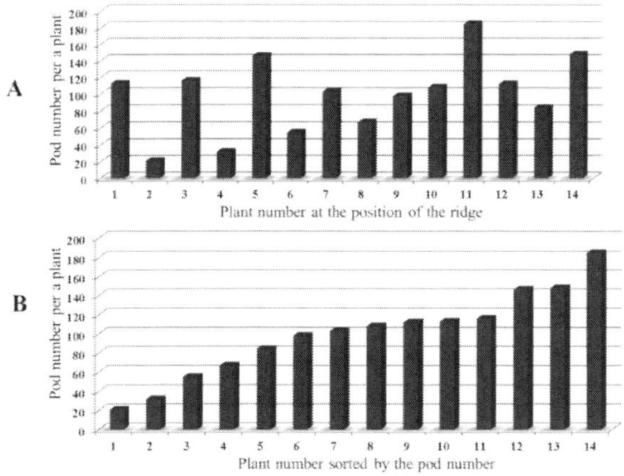

Figure 2. Pod number per a plant grown in the same row in Nagakura field in 2011. A: The number shows the original position in the row. B: The number of the plant is sorted from low to high pod number.

The low average yield compared with potential productivity (10 t ha^{-1}) may be due to several factors that interfere with maximum growth. First, soybean plants are very susceptible to physical, chemical and biological conditions of the soil as well as climatic conditions. Figure 2 shows the pod number per plant of soybean plants grown in a row. Figure 2A shows the data of the plants at the original position of the row, and Figure 2B shows data are sorted from low to high pod number. In this row, the highest pod number was about 180 and the lowest was only 18, and the average pod number was about 100 pods per a plant. As shown in Figure 2A, a plant with many pods tended to neighbor to a plant with low pod number. This may be mainly due to competition for solar radiation, and the plant growth is easily depressed by shading of neighbor bigger plant.

Another example of adaptation of soybean plants to environmental conditions is shown in Figure 3 and 4. When soybean plants are planted in a small pot, the growth was inferior (Figure 3) and the "dwarf" soybean formed only 3-5 pods with normal seeds. Figure 4 shows an example of "giant" soybean cultivated with a low planting density at 2 plants m^{-2} [6]. This plant had a very thick basal stem (25 mm diameter), and the dry weight per a plant was 572g (leaves 100g, stems 204g, pods 220g, roots 41g, and nodules 7g). It had 1,874 nodules on the roots. The plant had 17 lateral branches, 178 stem nodes, 600 pods and 1,687 seeds as shown in Figure 5.

Figure 3. Soybean plants grown in a small flask.

Figure 4. Soybean plants (cv. Williams) cultivated at the density of 2 plants m^{-2} in Ikarashi sandy dune field in 2000. [6]

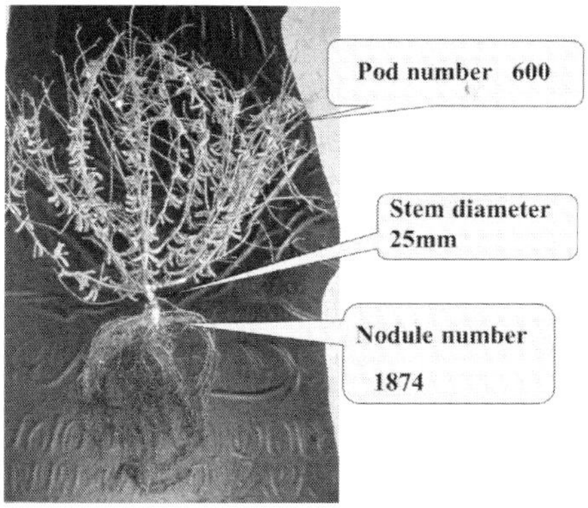

Figure 5. Characteristics of soybean plant (cv. Williams) cultivated at the density of 2 plants m^{-2} in Ikarashi sandy dune field in 2000. [6]

2.2. Characteristics of soybean growth

Optimal planting date for a cultivar in the growing area is very important to get good growth and seed yield. When planting is delayed only a few weeks, the stem length and plant dry matter accumulation at R1 stage may decrease by half. Optimum planting density is also important. In Niigata, the planting density for cultivar "Enrei" is 89,000 plants ha^{-1} by a single seed planting, which means one seed is planted in each seeding spot, with 75 cm row spacing and 15 cm planting distance in a row. When the germination rate of the seeds is not good, the planting density will decrease. Therefore, multiple seeding is sometimes carried out to avoid subnormal plant population. However, the total number of nodes, pods and seeds per area are almost the same between single-seeded and multi-seeded planting.

Figure 6 shows the stage description for the vegetative and reproductive growth of soybean proposed by Fehr and Caviness [8]. Soybean seeds germinate in a few days and emergence of seedling occurs about 7-10 days after sowing (PE stage), if soil moisture and temperature are optimum. After cotyledon leaves appear, a pair of unifoliolate leaves unroll (VC stage). During this stage, the storage compounds in the cotyledons support the nutrition for root and shoot growth. Then trifoliolate leaves appear one by one at V1 (the first trifoliolate leaf), V2 (the second trifoliolate leaf) and so on (Vn stage), and the shoot and roots grow during the vegetative stage.

The reproductive growth starts from beginning bloom (R1 stage). Soybean is short-day plant, and they bloom when the day length become shorter than 14 hrs, although it depends on the varieties and planting date. Bloom period lasts for 15-50 days from the beginning to the end of bloom. Full bloom is described as R2 stage. Pod initiation starts (R3 stage) about one month after the R1 stage. Then about one month after R3 stage, the seed begins to enlarge (R5 stage) to full seed (R6 stage). By one month after R5, the soybean plant starts to become mature (R7 stage). At 2-3 weeks after the R7 stage, the plant dries down to harvest maturity (R8 stage). Vegetative growth continues after R1 in both determinate and indeterminate soybeans. Vegetative growth of stems and leaves does not stop until about the R5 stage.

Vegetative (V) soybean growth stages

PE: Plant emergence (depends on temperature and moisture).
VC: Unifoliolate leaves unrolled in addition to cotyledons. One node.
V1: One unrolled trifoliolate leaf. Two nodes.
V2: Two unrolled trifoliolate leaves. Three nodes.
Vn: (n) number of trifoliolate leaves unrolled; (n) + 1 number of nodes

Reproductive (R) soybean growth stages

R1: Beginning bloom. At least one flower is present on the main stem.
R2: Full bloom. Flowers are found on any of the top two nodes.
R3: Beginning pod. Pods are 3/16 inch (4.8 mm) long on one of the top four nodes.
R4: Full pod. Pods are 3/4 inch (19 mm) long on one of the top four nodes.
R5: Beginning seed. Seeds are 1/8 inch inch (3.2 mm) long on one of the top four nodes.
R6: Full seed. Pods are completely filled by seeds on one of the top four nodes.
R7: Beginning maturity. One mature pod found on the plant.
R8: Full maturity. 95% pods have reached mature pod color.

Figure 6. Stage description for the vegetative and reproductive growth of soybean. (from Clemson Cooperative Extension, Home Page)

Root growth starts with a seminal root, which becomes a primary root. The secondary roots are formed from the primary root. The first root nodules are formed on the basal part of the primary roots, and they become visible at about 10 days after planting. They start to fix nitrogen (N_2) at about 15-20 days after planting when the diameter reaches about 2 mm [9]. In the later stage, the nodules formed at the basal part of primary roots degrade, and a large number of new nodules form on the lateral roots near the soil surface, and they play an important role for supplying N during the pod filling stage.

2.3. Factors affecting soybean yield

Soybean plants are very susceptible to environmental conditions, such as climatic conditions (solar radiation, day length, temperature, rain fall etc), soil conditions (drought, excess water, pH, soil fertility, mineral nutrition, etc). Secondly, soybean seed yield often severely declines with pests, such as insects, weeds, diseases, and nematodes. Third, nitrogen fixation by the root nodules (Figure 7) with the soil microorganism bradyrhizobia is very important for soybean production [10,11], however, it is difficult to obtain the optimum condition of nodulation and nitrogen fixation. The nodule formation and nitrogen fixation is sensitive to the external factors such as climate, soil properties, pests etc, and internal factors such as competition among plants or competition among organs, pods, leaves, roots and nodules. Therefore, many stress conditions, such as drought stress, decrease in oxygen supply, a high or low pH, nutrient imbalance etc., may depress nodule formation and nitrogen fixation activity. In addition, low population of compatible bradyrhizobia or the dominance of inefficient strains of indigenous bradyrhizobia in the field may decrease nitrogen fixation activity. The inoculation of efficient strains of bradyrhizobia may promote soybean growth and seed yield.

Figure 7. A photograph of the nodulated roots of soybean plant (cv. Williams) inoculated with *Bradyrhizobium japonicum* USDA110 and cultivated in a glass bottle with culture solution.

In Japan, over 80% of soybean cultivation is carried out in rotated paddy fields by block rotation with rice, because rice has been over produced in Japan. For example, one block of rice field is drained and soybean cultivation continued there for 3 years. Then this field is returned to rice cultivation for the next 3 years. When drainage of water is good and the ground-water level is maintained at lower than 30cm, soybean cultivation will be successful. However, bad drainage of water like in heavy clay soil near Niigata will depress root and nodule development. Therefore, soybean growth and seed yield is very poor. Hosokawa developed a new method of a raised planting bed cultivation by changing the blades of a reverse rotary tiller. Soybean plants grow very well in a raised bed due to efficient drainage of water especially after heavy rain fall. Nagumo et al. reported that a higher seed yield has been obtained by raised bed tillage with sigmoidal releasing-type coated urea fertilizer in the rotated paddy field under poor drainage conditions. Respiration of nodules and roots is severely depressed by excess water in soil due to oxygen deficiency. Soybean plant can be cultivated in water culture (Figure 7) when aeration is good. However, under water logging conditions, oxygen deficiency in soil occurs, because diffusion of oxygen through water is very slow compared with gas diffusion through soil, and soil microorganisms respire O_2 actively under high temperature conditions. It is known that soybean nodule respiration is about 4 times higher than that in roots in order to support nitrogen fixation activity and nitrogen assimilation to ureides. Therefore, low oxygen supply is fatal for nodules.

2.4. Nitrogen assimilation and seed yield

Soybeanplants assimilate a large amount of nitrogen during both vegetative and reproductive stages, and the total amount of N assimilated in a plant is highly correlated with the soybean seed yield. One t of soybean seed requires about 70-90 kg N, which is about four times more than in the case of rice [13]. Soybean plants assimilate the N from three sources, 1) N derived from symbiotic N_2 fixation by root nodules (Ndfa), 2) absorbed N from soil mineralized N (Ndfs), and 3) N derived from fertilizer when applied (Ndff) (Figure 8). For the maximum seed yield of soybean, it is necessary to use both N_2 fixation and absorbed N from roots [14-15]. When only N_2 fixation is available to the plant vigorous vegetative growth does not occur, which results in reduced seed yield. On the other hand, a heavy supply of N often depresses nodule development and N_2 fixation activity and induces nodule senescence, which also results in reduced seed yield. In addition, a heavy supply of N from fertilizer or from the soil causes luxuriant shoot growth, which result in lodging and poor pod formation. Therefore, for soybean cultivation no nitrogen fertilizer is applied or only a small amount of N fertilizer is applied as a "starter N" to promote the initial growth.

Figure 8. Three sources of nitrogen for soybean plants, nitrogen fixed in root nodules, nitrogen absorbed by roots from fertilizer and from soil nitrogen.

In Niigata fields, about 60-75% of N assimilation in soybean was estimated to derive from N_2 fixation [16,17]. Figure 9 shows the growth of cultivar "Enrei" and the non-nodulated isogenic line "En1282" planted in the same Nagakura field. It is obvious that non-nodulated soybean grew very poor with pale leaf color due to N deficiency by the lack of nitrogen fixation. It is said that the legume nitrogen fixation is variable, but it is a valuable process in world agriculture [18-20].

3. Characteristics of nitrogen nutrition of soybean

3.1. Characteristics of nitrogen nutrition related to seed yield

Soybean seed contains an extraordinary high concentration of protein about 35-40% based on the seed weight. Many field researches showed the soybean seed yield is proportional to the total assimilated N in plants. Figure 10 shows the relationship between total amounts of N in soybean shoot at the R7 stage and seed yield in rotated paddy field in Nagakura from 1989-1991 [21]. The seed yield exhibited a linear correlation (r=0.855) with the amount of nitrogen accumulation.

Figure 9. Comparison of the growth of nodulating soybean cv. "Enrei" (left) and the non-nodulated mutant "En1282" (right) cultivated in Nagakura field.

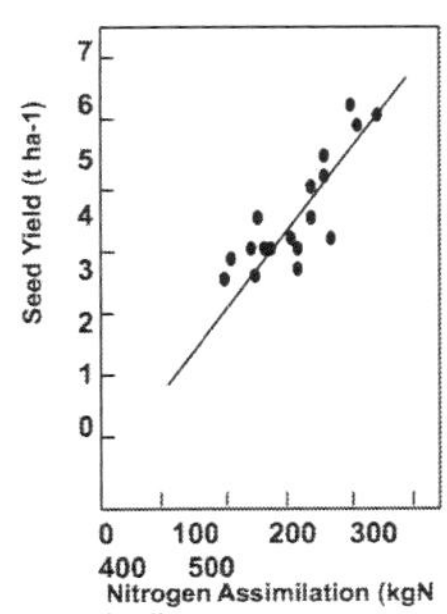

Figure 10. Relationship between amount of nitrogen accumulated in soybean shoot at R7 stage and seed yield of soybean cv. "Enrei". In Nagakura field with various fertilizer treatments.[21]

The protein concentration in soybean seeds is about 4 times higher than cereal seeds such as rice grain (Figure 11) [12]. Due to a high concentration of seed protein, 1t of soybean seed production requires about 70-90 kg of N, while 1t of rice grain requires only 20 kg of N. Soybean plants assimilate about 20 % of total N until initial flowering stage (R1 stage), and 80% of N during the reproductive stage. On the other hand, rice assimilates about 80% of N until flowering. Therefore, the continuous assimilation of nitrogen after initial flowering stage is essential for good growth and high seed yield in soybean cultivation.

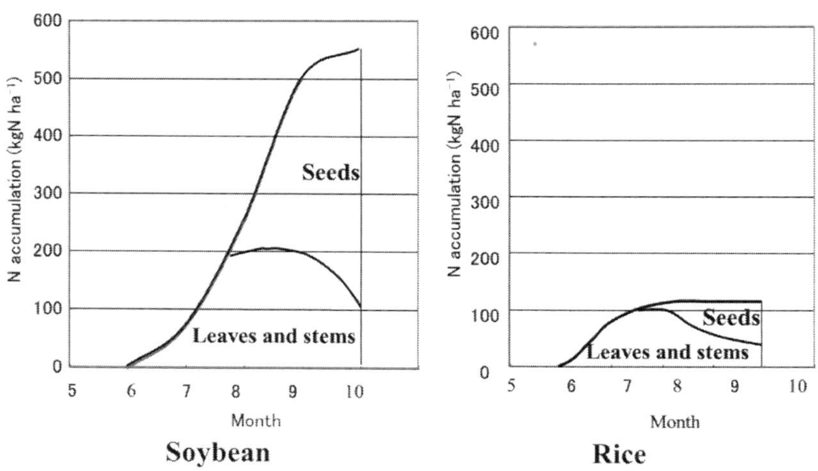

Figure 11. Comparison of the nitrogen assimilation and distribution pattern of soybean and rice.

To obtain high seed yield of soybean, good nodulation and high and long lasting nitrogen fixation activity are very important (Figure 12). Nodule formation and nodule growth are influenced by various soil conditions (water content, pH, nutrition) and climatic conditions (solar radiation, temperature, rain fall etc). Soybean can fix atmospheric N_2 by their root nodules associated with soil bacteria, bradyrhizobia. In addition, soybean can absorb inorganic nitrogen, such as nitrate and ammonia from soil or fertilizer. Usually a high yield of soybean was obtained in a field with high soil fertility. By supplying a constant but low concentration of nitrogen either from soil or organic manure, soybean growth will occur without depressing nodulation and nitrogen fixation activity. However, it is well known that a high concentration of mineral N depresses nodule formation and nitrogen fixation activity. Especially, nitrate, the most abundant inorganic nitrogen in upland fields, severely inhibits nodulation and nitrogen fixation of soybean, when nodulated roots are in direct contact with the soil solution containing nitrate [22-24].

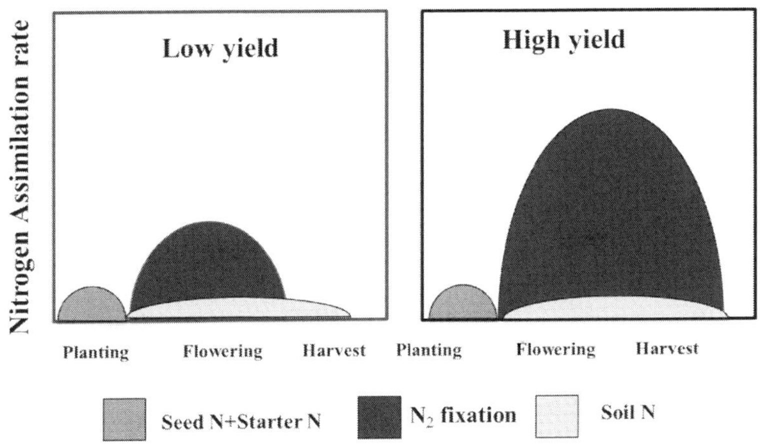

Figure 12. Comparison of the time course of nitrogen assimilation in soybean plants with a low yield (left) and a high yield (right).

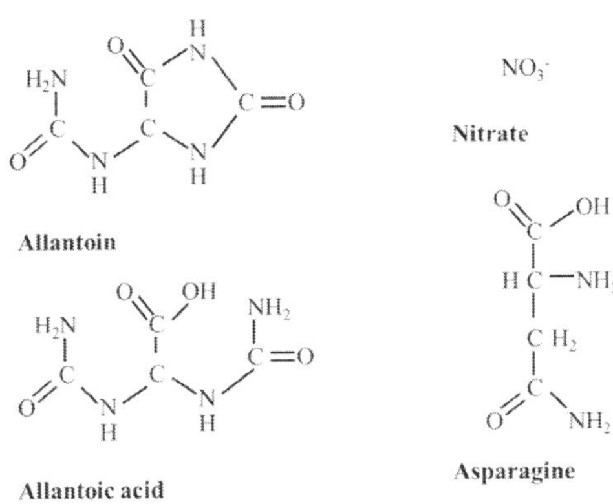

Figure 13. Chemical formula of ureides (allantoin and allantoic acid), nitrate and asparagine.

3.2. Nitrogen assimilation in nodules

Ammonia is known to be the initial product of biological nitrogen fixation by the enzyme nitrogenase [25]. After discovering an enzyme glutamate synthase (GOGAT) in *Aerobacter aerogenes* [26], it was confirmed that ammonia can be assimilated via glutamine synthetase (GS) and glutamate synthase (GOGAT) pathway in soybean nodules by ^{15}N tracer experiments [27,28]. The ^{15}N assimilation was investigated in the cytosol (plant cytoplasm) and bacteroid fractions of soybean nodules [29]. The result suggested that most of the fixed N is immediately exported from the bacteroid to the plant cytosol and assimilated via GS/GOGAT pathway into various amino acids via transamination from glutamate. Ureides, allantoin and allantoic acid are synthesized from amino acids and amides in the cytosol (Figure 13,14). Kushizaki et al. discovered that nodulated soybean plants contain a large amount of ureides in their stems, while non-nodulating isolines contain much less [30]. Reviews on ureide biosynthesis in legume nodules were published [31,32].

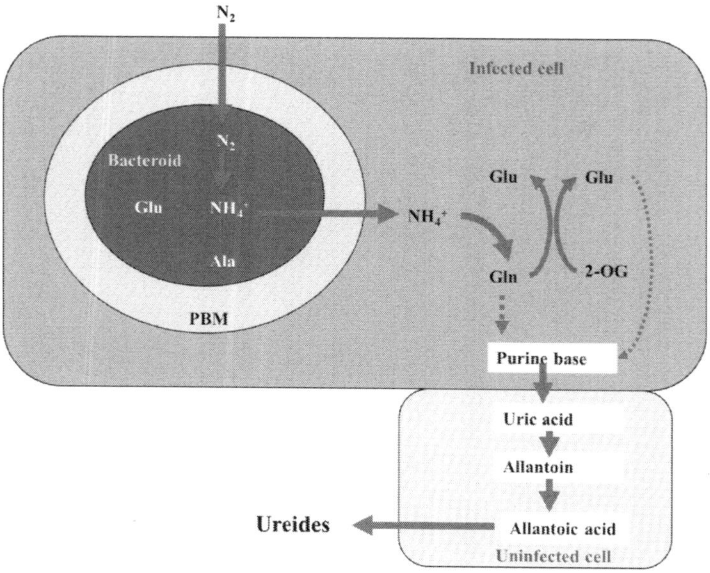

Figure 14. A model of the flow of fixed N_2 in infected cell and uninfected cell of soybean nodule.

3.3. Nitrogen absorption and assimilation in soybean roots

The outlines of absorption and metabolism of ammonium and nitrate in plant cells are shown in Figure 15. Ammonium (NH_4^+) and nitrate (NO_3^-) are major sources of inorganic nitrogen in soil. Ammonium is the most reduced form of nitrogen and nitrate is the most oxidized form. The NH_4^+ ion is absorbed through the membrane bound protein, ammonium transporter. The NO_3^- ion is absorbed through the nitrate transporter with $2H^+$ co-transport. There are two types of nitrate transporter, a high affinity nitrate transporter system (HATS) and a low affinity nitrate transporter system (LATS) [33]. The kinetics of the absorption rate versus nitrate concentration indicated the presence of only one HATS, having a Km value of 19 μ mole in soybean roots.

http://dx.doi.org/10.5772/45867

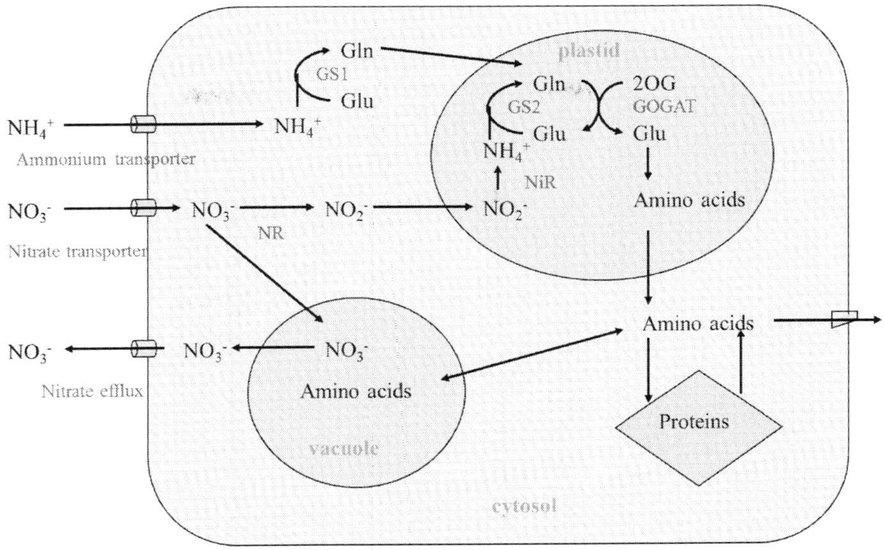

Figure 15. A model of absorption and metabolism of ammonium and nitrate in plant cell.

The diurnal rhythm in NO_3^- absorption by intact soybean plants was investigated by sampling the culture solution every 15 min [34]. The NO_3^- absorption rate was different between day (1.10 mgN L^{-1} h^{-1}) and night period (0.77 mgN L^{-1} h^{-1}), and the nitrate absorption rate at night was about 60-75% of that in the daytime. The temporary interruption of NO_3^- absorption was observed twice a day at dawn and dusk. The changes in NO_3^- absorption rate were not controlled by the shoots, because the rhythm continued under the extended dark period or by cutting the shoots [34]. When the roots were put in the water bath at a constant temperature of 30°C, the rhythm of NO_3^- absorption disappeared. This suggests that the nitrate absorption rate of soybean roots may be controlled by monitoring temperature changes by the root.

Some parts of the NO_3^- absorbed in the root cell is reduced to nitrite (NO_2^-) by nitrate reductase (NR) in the cytosol, and NO_2^- is reduced to ammonia by nitrite reductase (NiR) in the plastids. Then amino acids are formed followed by the assimilation via GS/GOGAT pathway. When a high concentration of NO_3^- is supplied, a part of NO_3^- is temporarily stored in the vacuoles of cortex cells in roots. Some parts of NO_3^- are transported cell to cell via the symplast pathway and effluxed in the stele and transported via the xylem with the transpiration stream in the form of NO_3^-. Plant NR requires NADH or NAD(P)H which means both NADH and NADPH act as electron donors. In soybean, there are two types of NAD(P)H-NR and one type of NADH-NR [15].

After adding $^{15}NO_3^-$ in the solution, the ^{15}N concentration of asparagine increased markedly, indicating that asparagine is a major assimilatory compound of NO_3^- in soybean roots. Nitrogen assimilation and transport of plants supplied with $^{15}NO_3^-$ was investigated by analyzing xylem sap collected from decapitated soybean plants [35]. Some part of the NO_3^- absorbed in the roots was immediately exported to the shoots, whereas another parts of the NO_3^- was temporarily stored in the vacuoles of root cells and then gradually released to the xylem. On the other hand, some other parts of the NO_3^- was reduced and assimilated in the roots and synthesized into asparagine.

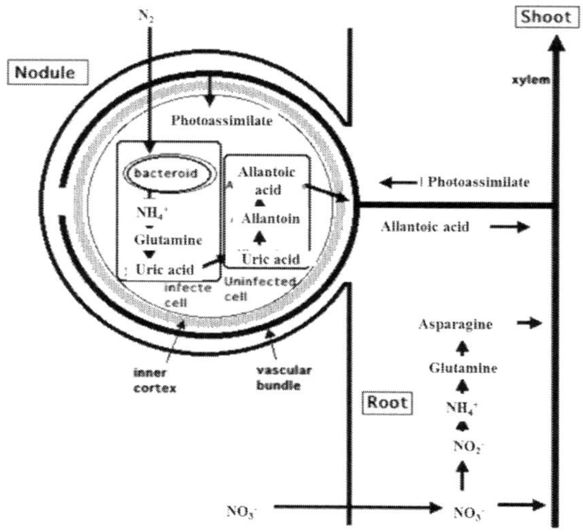

Figure 16. Comparison of the fate of fixed N_2 in nodules and absorbed nitrate in roots.

3.4. Comparison of the fate of fixed N_2 and absorbed NO_3^- in soybean plant

The labeling patterns of ureides and amino acids were compared from the data of $^{15}N_2$ and $^{15}NO_3^-$ feedings [36]. The result proved that the ureides in stems are mainly derived from fixed N_2, and only a small amount of ureides is synthesized in the root. Figure 16 shows the metabolic pathways and transport of N derived from N_2 fixation and NO_3^- absorption in soybean plants.

3.5. Effect of combined nitrogen on nodule growth and nitrogen fixation activity

It is well known that a heavy supply of nitrogen fertilizer causes the inhibition of nodulation and nitrogen fixation. The inhibitory effect of combined nitrogen depends on the forms, concentration, application site, and soybean growth stage. The inhibitory effect of nitrate is stronger than urea or ammonium. Direct and indirect effects of nitrate have been known. Direct or local effect of nitrate is the effect of nitrate in direct contact with the nodulated part of roots. When nitrate was supplied to the hydroponically grown soybean roots, the nitrate inhibition on nodule growth and nitrogen fixation was shown to be rapid and reversible [22-24]. On the other hand, indirect or systemic effect of nitrate, which means the effect of nitrate absorbed from distant part of the roots, depended on the site, duration and concentrations of nitrate supply [37,38]. The indirect effect of nitrate was investigated by a two-layered pot system separating the upper roots and lower roots. A continuous high concentration of nitrate (5mM) supply in the lower roots depressed the nodulation and nitrogen fixation of the upper roots. However, a continuous supply of a low concentration of nitrate (1 mM) resulted in the promotion of nodulation and nitrogen fixation activity in the upper roots [38].

3.6. Nitrogen metabolism in soybean leaves

Plant leaves are the important organ for nitrogen metabolism, in addition to photosynthetic activity. The N absorbed from roots or fixed in root nodules is transported via the xylem in stems and petioles, and the N in leaves is translocated to the sink organs such as pods and seeds via the phloem. The flow of N in leaves was investigated by petiole girdling and $^{15}N_2$ or $^{15}NO_3^-$ treatment [39]. By petiole girdling treatment, the accumulation of amino acids (x 2.5) especially asparagine (x 8.8) in leaf blades was observed, indicating that these compounds are the transport forms from leaves to sink organ via the phloem. However, nitrate and ureides are not accumulated in girdled leaves compared with intact leaves, suggesting that nitrate and ureides are not transported from leaves to sinks via the phloem (Figure 17). There are two different ureide degradation pathways in soybean leaves, either by allantoate amidinohyfrolase or by allantoate amidohydrolase [40].

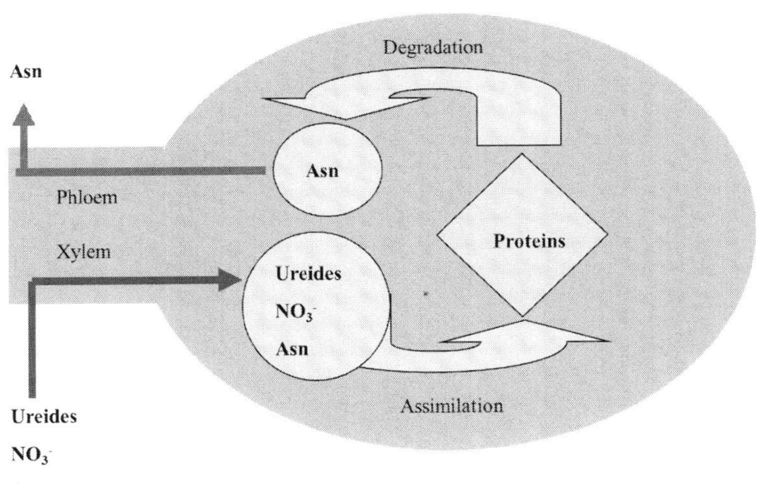

Figure 17. A model of N flow in a soybean leaf.

A comparative study on the nitrogen metabolism and transport was done at the pod filling stage by $^{15}N_2$ or $^{15}NO_3^-$ treatment [41]. Based on the results obtained, we proposed a model of N flow derived from N_2 and NO_3^- in soybean plants as shown in Figure 18 [42]. The N derived from N_2 fixed by the root nodules is rapidly assimilated into ureides (allantoin and allantoate), and some ureides are directly transported to pods and used for seed development. Ureides are also used for leaf protein synthesis, but the contribution is relatively lower than N derived from NO_3^- absorbed from the roots. On the other hand, some part of NO_3^- absorbed from the roots is immediately reduced in the roots, and transported in the form of amino acids, especially asparagine. Another part of NO_3^- is transported to the leaf blades via transpiration, and assimilated into leaf protein. The remobilization of storage protein in leaves and roots may be a major source for seed N source in the case of NO_3^- nutrition.

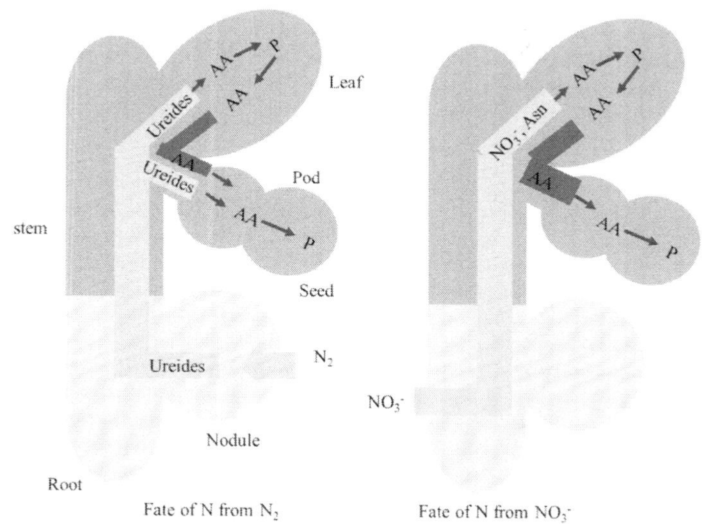

Figure 18. A model of N flow in soybean plant originated from N_2 (left) and NO_3^- (Right). AA: amino acids, P: protein

3.6. Assimilation of nitrogen in pods and seeds

Figure 19 shows the outline of the N flow in soybean pod and seed (cotyledon). Ureides are transported from the root nodules via the xylem and are accumulated in the pod. Allantoin and allantoic acids are metabolized into amino acids in the pod or seed coat and excreted to the inside of seed coat. Asparagine from roots via the xylem or from leaves via phloem is also metabolized to amino acids and then transported into the apoplast space between the seed coat and cotyledon. The cotyledon cells absorb amino acids from the apoplast and they synthesize storage proteins and accumulate them into protein bodies.

http://dx.doi.org/10.5772/45867

Rainbird et al. reported that glutamine is the principal N supply to the cotyledon, contributing 55% of the embryo nitrogen requirement, and 20% comes from asparagine, with negligible amounts from ureides, allantoin and allantoic acid. Haga and Sodek also reported that glutamine was the most efficient source in terms of protein accumulation in the cultured soybean cotyledons, while asparagine was less efficient and allantoin was a poor source of nitrogen. Ohtake et al. reported that the rapid N transport to pods and seeds in N-deficient soybean plants were faster compared with N-sufficient plants.

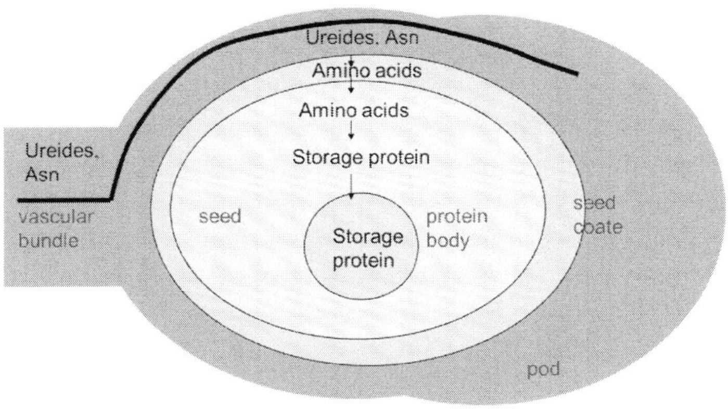

Figure 19. The model of nitrogen assimilation in soybean pod and seed.

The storage protein of soybean seeds mainly consists of glycinin and β-conglycinin. β-conglycinin is comprised of three subunits, designated as α', α, and β–subunits [46]. We happened to discover the lack of β-subunit of β-conglycinin in several non-nodulated soybean lines, although an electrophoretic protein band due to this protein was clearly detected in the corresponding nodulated isolines [47]. The suppression of the β–subunit in the non-nodulating isoline T201 is regulated at the level of mRNA accumulation. The α'–and α–subunits mRNAs were actively expressed in both line. Nitrogen regulation for storage protein in soybean seeds was evaluated using T201 and T202 with solution culture in the greenhouse [48]. The results indicated that non-nodulated T201 has a normal, non-defective, β–subunit genes, and that limited N availability decreases the accumulation of β–conglycinin.

4. Inoculation of efficient strains of bradyrhizobia to improve soybean seed yield

4.1. Inoculation of bradyrhizobia and indigenous strains in fields

Soybean plants form root nodules by symbiosis with soil bacteria *Bradyrhizobium* (*B. japonicum, B. elkani and B. lianingense*), *Sinorhizobium* (*S. fredii, S. xinjiangense*) and Mesorhizobium (*M. tianshanense*) [49]. Only *B. japonicum, B.* elkani, and *S. fredii* are used as commercial inoculants for soybean with *B. japonicum* being the most widely employed [49]. Genus *Bradyrhizobium* belongs to α-Proteobacteria, family Bradyrhizobiaceae. *Bradyrhizobium* species are rod shaped Gram negative bacteria with a single subpolar or polar flagellum. They can be isolated from nodules and grown on a Yeast Extract Mannitol agar plate (Figure 20) [50]. Rhizobia and bradyrhizobia can exist either as a free living organism in soil or a symbiotic state in the infected cells of root nodules. When bradyrhizobia inhabit a soil, they do not fix N_2 and they depend on organic nitrogen. After they infect soybean roots and form root nodules, they become "bacteroids" in a symbiotic state and fix N_2.

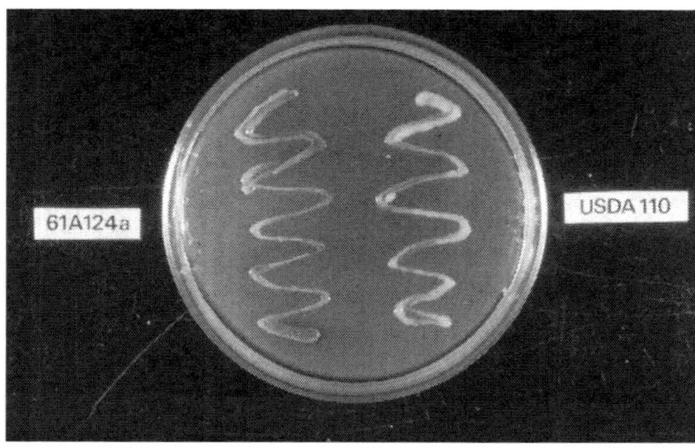

Figure 20. A culture of *Bradyrhizobium japonicum* USDA110 (right) and the *gus* mutant line 61A124a (left) on a Yeast Extract Mannitol agar plate which contain GUS substrate X-Gluc.

When soybean plants are cultivated in a new field where soybeans are cultivated for the first time, the inoculation of compatible strains of bradyrhizobia may significantly promote plant growth and seed yield. However, after cultivation of soybean, rhizobia will predominate throughout the soil. Therefore, soybean plants are usually nodulated with indigenous rhizobia in most fields of Japan without inoculation. Tewari et al. reported that the inoculation of the efficient strain of bradyrhizobia USDA110 in a paper pot filled with vermiculite was very effective in the first year of a soybean crop after 30 cm layer of mountain soil was dressed, where population of bradyrhizobia was very low [51]. Figure 21 shows the plant with uninoculated paper pot (left) and that with inoculated paper pot (right) cultivated in the mountain soil. The average seed weight per plant was 8.7 g in uninoculated plants and 22.3 g in inoculated plants, respectively. In a rotated paddy field in Nagakura where indigenous rhizobia had been established, Tewari et al. inoculated USDA110 in a paper pot, and the plant growth and seed yield was about 10-20% higher than the uninoculated paper pot (Figure 22). Good nodulation occurred in the roots of the plant in the uninoculated paper pot, because indigenous bradyrhizobia already inhabited the field.

Figure 21. The effect of inoculation of *Bradyrhizobium japonicum* USDA110 on soybean growth cultivated in montain soil where the population of indigenous bradyrhizobia was very low. Left plant was cultivated with a paper pot without inoculation. Right plant was cultivated with a paper pot with inoculation of bradyrhizobia [51].

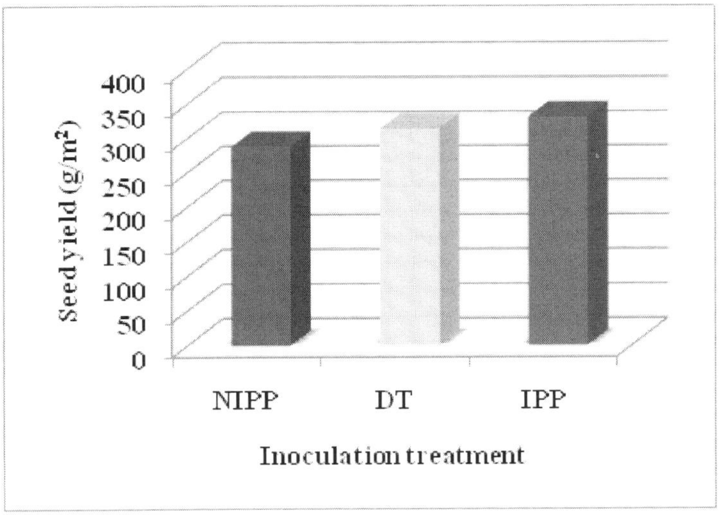

Figure 22. The effect of inoculation method for seed yield of soybean plants in a rotated paddy field in Nagakura. NIPP; Non-inocultated paper pot.DT; Direct inoculation of bradyrhizobia to soybean seed. IPP; Inoculated paper pot [52].

Because the soybean plant can form nodules in most of the fields including a rotated paddy rice field in Japan, farmers generally do not inoculate bradyrhizobia, except in the Tokachi area in Hokkaido. The Tokachi Federation of Agricultural Cooperatives in Hokkaido provides biofertilizer inoculants for soybean seeds [50]. Although many types of bradyrhizobia exist in a field [53,54], not all are efficient strains. Minamisawa et al. reported that 44 isolates of *Bradyrhizobium japonicum* from Nakazawa field in Niigata were divided into 33 genetically different groups by genetic analysis using a repeated sequence specific hybridization method. The similar diversity of indigenous bradyrhizobia was also shown in 6 sites in Japan, including the Nakazawa and Nagakura fields in Niigata [54]. The Nakazawa and Nagakura fields are very near to one another, but bradyrhizobia types differ between the fields. In Nagakura, *B. japonicum* hup^+ (uptake hydrogenase positive) and hup^- (uptake hydrogenase negative) groups made up about 80% and 20% of bradyrhizobium, respectively. On the other hand, in the Nakazawa field, *B. japonicum* hup^+, hup^- and *B. elkanii* made up about 50%, 20% and 30%, respectively, of the local bradyrhizobium. Significant diversities and site-dependent variations were observed, and the fingerprints at Ishigaki island with no history of soybean cultivation were less diverse than the other sites. They suggested that soybean bradyrhizobia might be diversified in individual fields by association with host plants and local soil conditions [54].

4.2. Use of marker strain as an inoculant

The ecological study of inoculated strains in soil is important to establish the efficient way of inoculation. Minagawa et al. used *gus* (β-glucuronidase gene)-marked *Bradyrhizobium* strain to estimate the number of inoculated strains in the soil. The *gus* gene from *Eschrichia coli* was introduced into *Bradyrhizobium japonicum*. This strain absorbs and hydrolyzes GUS(β-glucuronidase)-substrate (X-Gluc; 5-bromo-4-chloro-3-indolyl-β-glucuronide) and precipitates an indigo pigment in the cell (Figure 20). The accumulated indigo blue metabolite in *gus*-marked strain can be determined for the population of liquid cultured rhizobia using optical density at 645 nm. However, when the *gus*-marked strain was inoculated in soil, the stained bradyrhizobia were difficult to separate from the soil. Therefore, we extracted the blue pigment by phenol-water after they are incubated with X-Gluc for 4 days. The absorbance of the extracted blue pigment in the phenol-layer was measured optically at 645 nm. The initial number of bradyrhizobia in culture media or soil and the absorbance of the phenol extract of GUS metabolite was positively correlated. In addition, the occupancy by *gus*-strain in each nodule can be determined by staining the nodule slice with substrate X-Gluc (Figure 23). The nodule occupancy by *gus*-marked strain was 50% when the same population of *gus*-marked strain and USDA110 were inoculated at the same time [55].

The *gus*-marked strains were inoculated in five different types of soil (Nagakura and Sonoki; two types of Alluvial soil from rotated paddy fields), volcanic ash soil of upland field (Nakazawa), sandy dune soil (Ikarashi), and calcinic vermiculite. Soybean cultivation was frequent in Nagakura and Nakazawa fields, but soybean plants have not been grown for a long time in Sonoki and Ikarashi fields. The number of indigenous bradyrhizobia was estimated by MPN method, and the populations were as follows; Nagakura (6×10^5 cells g^{-1} soil), Nakazawa (3×10^5), Sonoki (2×10^4), Ikarashi (8), and Vermiculite (0), respectively. The population of *gus*-marked strains increased over 10 times in all types of soils for 1 week after

http://dx.doi.org/10.5772/45867

inoculation of 9 x10^6 cells g^{-1} soil (Figure 24). The population was higher in Nakazawa and Nagakura than that in Sonoki, Ikarashi, and vermiculite at 1 week after inoculation. From this result it was indicated that the population of indigenous bradyrhizobia may not restrict the growth of inoculated strains.

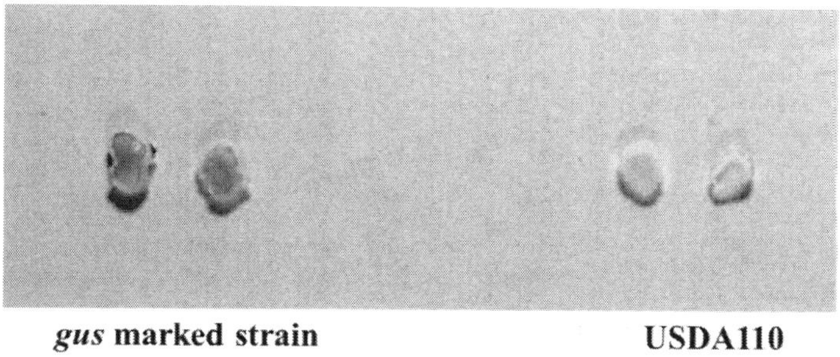

Figure 23. Detection of *gus*-marked strain in the nodule by X-Gluc treatment.

The population of inoculated strains was determined in rhizosphere soil, non-rhizosphere soil and roots (Figure 25) [51]. The population increased about 10 times in a week both in rhizosphere and non-rhizosphere soils at one week after inoculation.

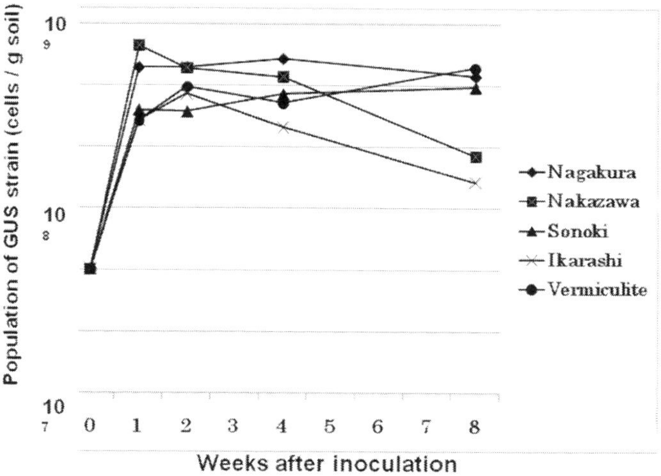

Figure 24. Changes in population of inoculated *gus*-marked strain in various types of soils in Niigata.

The proliferation and mobility of inoculated *gus*-marked strains were examined in a rhizobox containing various soil types (Figure 26), where a soybean plant was cultivated in the center of the box. Inoculation and watering were supplied from one side of the box. The proliferation rates of the *gus*-marked strain in a whole box at 25 days after planting were different in various types of soils; Nagakura (x 1,218), Nakazawa (x 538), Sonoki (x 513), Ikarashi, (x173) and vermiculite (x 98). In all the soil types, bradyrhizobia were distributed in many compartments of the rhizobox (Figure 27). It was observed that most of bradyrhizobia attached to the soil particle, however, some of them moved by water flow through soil apertures or along with the root elongation.

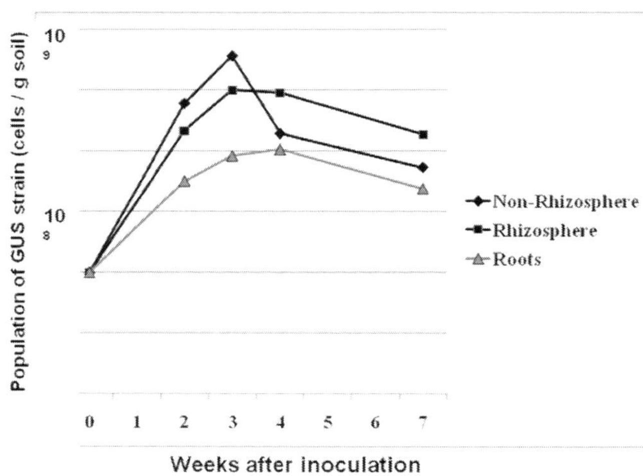

Figure 25. Changes in population of inoculated *gus*-marked strains in non-rhizosphere soil, rhizosphere soil and roots.

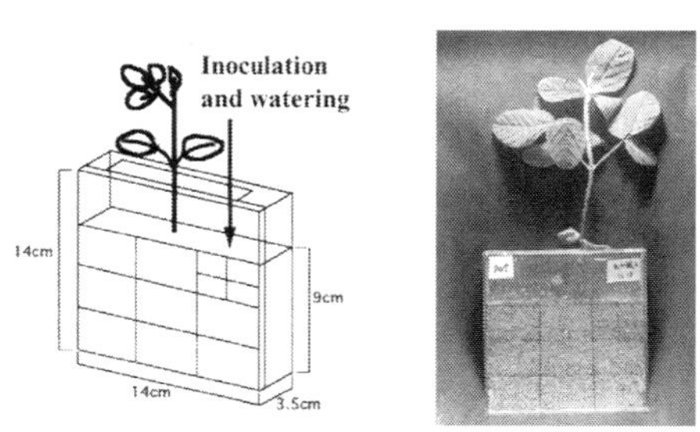

Figure 26. Rhizobox experiment system.

A soybean plant was grown in the center of the box, and inoculation of the *gus*-marked strain and the watering was carried out from one side of the box (arrow).

For soils lacking indigenous bradyrhizobia (vermiculite) or having a very low density of indigenous *Bradyrhizobia* (Sonoki and Ikarashi), the nodules were formed almost exclusively by the inoculated *gus*-marked strain Sonoki (100%; occupancy rate by *gus*-strain), Ikarashi (98%) and vermiculite (100%). However, the major nodules were formed by indigenous strains in the soil types, which contain a high population of indigenous *Bradhrhizobia,* in Nagakura (25% occupancy rate by *gus*-marked strain), Nakazawa (35%*gus*-marked strain), although the inoculated strain proliferated very well. The results suggested that nodule occupancy may be simply related to the population rate of inoculated strains vs. indigenous strains rather than the competition between inoculated and indigenous strains.

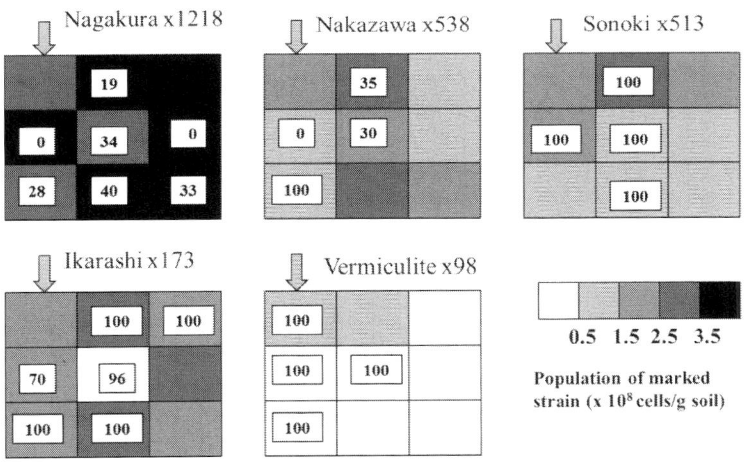

Figure 27. Population of *gus*-marked strains in 9 compartments of each rhizobox (color of each compartment), and the percentage of nodule occupancy by *gus*-marked strains (number in the center of each compartment) [55]. Arrow indicates the site of inoculation and watering.

4.3. Survival of inoculated *gus*-marked strain

The survival of the inoculated strain is very important to improve indigenous strains in a field. We investigated the survival of *gus*-strains in the first soybean cultivation just after inoculation of the *gus*-strain, and in the second year in the pot filled with five soil types. The nodule occupancy rate (%) by the *gus*-marked strain in the first and second year are shown in Figure 28. In the Ikarashi and vermiculite soils with low densities of indigenous bradyrhizobia, the nodule occupancy rates by the *gus*-marked strains were high (about 80-100%) both in the first and second year. However, in Nagakura, Nakazawa and Sonoki soils, the nodule occupancy rates were lower in the second year than the first year. This result suggests that the survival ability of the inoculated single strain may be inferior to the indigenous strains. The genetic diversity of the indigenous strain may be related to the competition between inoculated and indigenous strains.

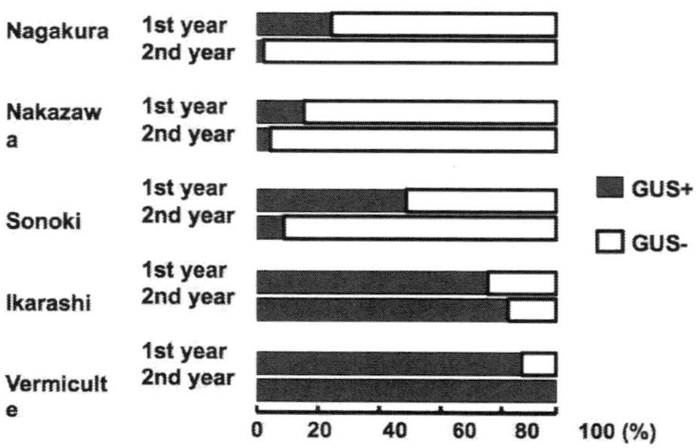

Figure 28. Percentage of nodules occupied by *gus*-marked strain (GUS+) and indigenous strain (GUS-) in various soils in Niigata in the first and second cropping years after inoculation.

5. Field estimation of nitrogen fixation activity and nitrogen absorption rate of soybean by relative ureide method

5.1. Concept of relative ureide method

Many tropical grain legumes, such as soybean, common bean, cowpea, pigeon pea, and mung bean that have the spherical determinate type of nodule, transport the bulk of fixed N as ureides (allantoin and allantoic acid). On the other hand, nitrate and amino acids (especially amide, asparagine) are the major transport forms of N derived from the soil and fertilizer N absorbed by soybean roots [36, 56]. Herridge et al. [57,58] developed the "Relative ureide method" for evaluation of % Ndfa by analyzing the concentration of nitrogen compounds in xylem sap obtained from the bleeding sap from a cut stump, or vacuum collection from stems. The concentrations of ureide-N, nitrate-N and α-amino-N can be determined by colorimetry.

Figure 29 shows a concept of relative ureide mathod for estimation of nitrogen fixation activity and Figure 30 shows the equation of %Ndfa by this method. The xylem sap obtained from soybean plants, which depend solely on nitrogen fixation contains about 80-90% of N in the form ureides (allanotate and allantoin), in addition to some amino acids such as asparagine. On the other hand, in the xylem sap from soybean plants depending only on nitrate, nitrate and amino acids (asparagine) are the major N compounds accounting for 80-90%, and they contain about 10-20% as ureides. When the xylem sap in N_2 grown soybean contains about 10-20% of N as amino acids, and nitrate grown soybean contains the same percentage of N as ureides, the relative dependence on N_2 fixation is estimated by the relative ureide content as shown in Figure 29.

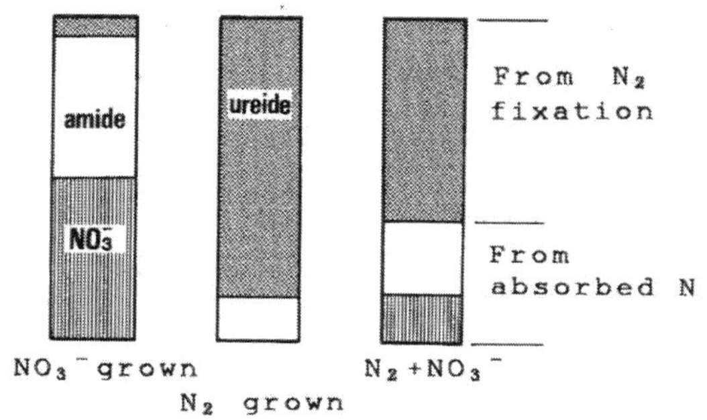

Figure 29. Concept of the relative ureide method using the concentration of xylem solute ureide-N, nitrate-N and amino-N in soybean. NO_3^- grown; Plants depend on sole NO_3^- absorption. N_2 grown; Soybean plants depend on sole N_2 fixation, $N_2 + NO_3^-$; Soybean plants depend on both N_2 fixation and NO_3^- absorption. [16,50]

This method is reliable in soybean field experiments, without any requirement of reference plants. It is the easiest way to measure the percentage of fixed N in any fields, because no preparation is necessary before sampling. This method is also applicable for experiments with variable N fertilizer application. In field conditions, the simple equation can be adapted for the estimation of % Ndfa (nitrogen derived from atmospheric dinitrogen) by the equation in Figure 30. The original equation proposed by Herridge used "α-Amino-N" instead of "2 X α-Amino-N" in our equation. Based on the analysis of the animo acid compositions in root bleeding xylem sap, asparagine (2N amide) was the major amino acid throughout the stages and the average N number in amino acids was 1.7, so we use "2 X α-amino-N" for this equation.

$$\%\text{Ndfa} = \frac{\text{Ureide-N}}{\text{Ureide-N} + \text{Nitrate-N} + 2 \times \alpha\text{-amino-N}}$$

Figure 30. Equation for calculating percentage of nitrogen derived from atmospheric N_2 (%Ndfa).

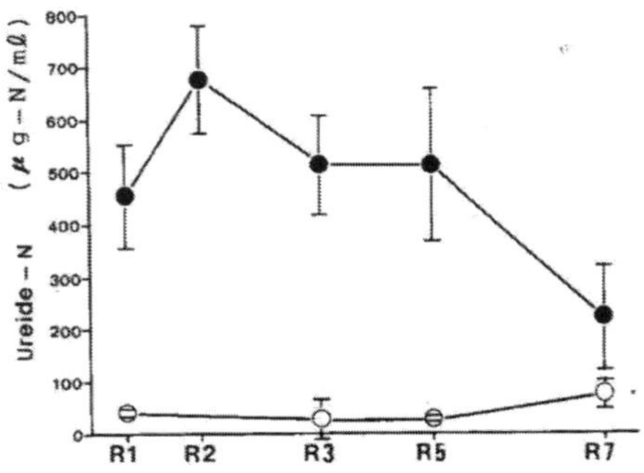

Figure 31. Changes in the concenration of ureide-N in nodulated T202 (closed circle) and non-nodulated T201 (open circle) [16].

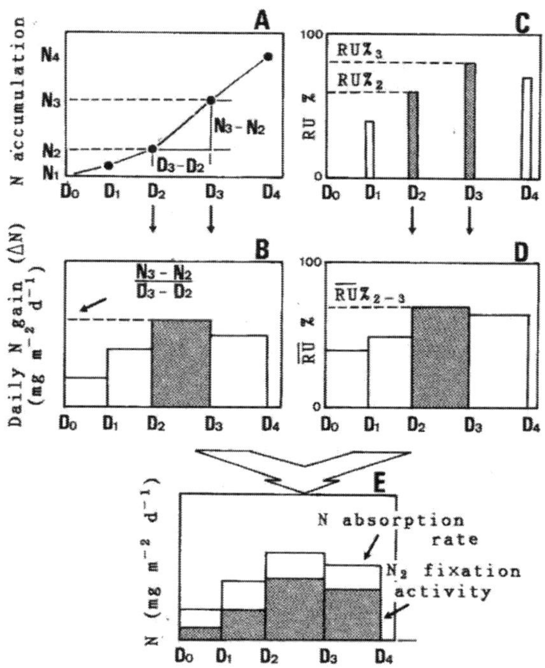

Figure 32. Outline of estimation of daily N_2 fixation activity and N absorption rate based on the data obtained by relative ureide-N and total N analysis of the shoot [16,50].

http://dx.doi.org/10.5772/45867

Figure 31 shows the changes in the concentration of ureide-N in xylem sap of nodulated (T202) and non-nodulated (T201) soybean plants [16]. The concentration of ureide-N in T202 was significantly higher than that in T201. In the field estimation, the relative ureide N percent is used as an indicator of relative dependence for nitrogen fixation at the sampling time.

5.2. Quantitative estimation of daily N_2 fixation and N absorption rate

By periodic sampling of soybean shoots and xylem sap, a quantitative estimation of the seasonal changes in N_2 fixation activity and N absorption rate is possible as shown in Figure 32 [16]. We usually sample soybean plants four or three times at R1, R3, R5 and R7 stages or three times at R1, R5, and R7 stages for xylem sap and plant N analyses. The examples of the evaluation of Ndfa by relative ureide method are shown in Figure 33 [4].

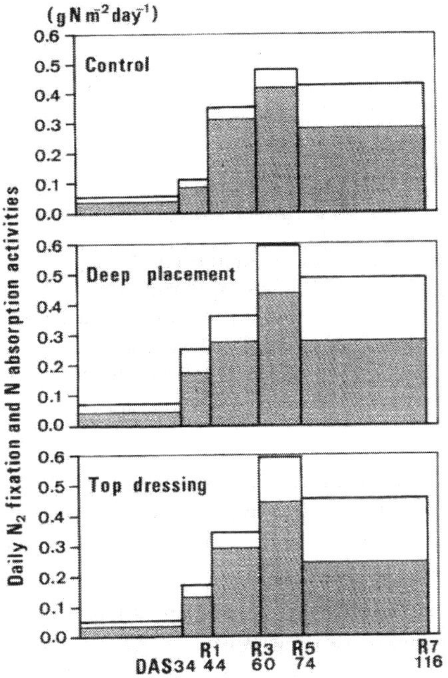

Figure 33. Example of the estimation of daily N_2 fixation activity and N absorption rate based on the data obtained by relative ureide-N and total N analysis of the shoot. Three fertilizer treatments were used: control; without additional fertilizer, Deep placement; deep placement of 100 day type coated urea, Top dressing; top dressing of 70 day type coated urea [4].

D_1, D_2, D_3, D_4 indicate the sampling date (Days after planting) of xylem sap and shoots. We usually sample at R1, R3, R5 and R7 stage for D_1, D_2, D_3, D_4. D_0 means the planting date. RU $\%_n$ indicates the relative ureide N percent in xylem sap at sampling time at D_n. $RU\%_{n-n+1}$ means average of RU% at D_n and D_{n+1}. We use $RU\%_{1-0}$ as $RU\%_1$, because we cannot measure the RU% at planting.

6. Possiblity of the agronomical use of hypernodulation mutant lines of soybean for promoting yield

Soybean plants can control the nodule number by auto-regulation of nodulation. Supernodulation or hypernodulation mutant lines of soybean have been isolated from wild type soybean, and it was expected that these mutants would produce higher seed yield. However, most of these mutants showed reduced vegetative growth and seed yield, possibly due to a heavy burden of excess nodules. Hypernodulation or supernodulation mutant lines of soybeans have been selected from several different cultivars [59-61]. A genetic defect in the autoregulatory control of nodulation causes more profuse nodulation than the wild type (Figure 34). The nodulation trait depends on the shoot genotype and not on the root genotype. In the wild type parents, some shoot-derived signal (autoregulation signal) arrests nodule primordia and suppresses nodule development in response to some signals (infection signal) derived from nodulated roots after infection. Sato et al. reported that a rooted single leaf of soybean retains the autoregulation trait.

Recently, the genes were identified for the hypernodulation lines of lotus (*HAR1*) and soybean (*GmNARK*), which play important roles in the autoregulation of nodulation, and they were shown to encode a receptor-like kinase protein that contains a leucine-rich repeat [64-66]. These legume genes are homologous to *Arabidopsis CLAVATA1* (*CLV1*), which is involved in the control of cell proliferation in the shoot apical meristem [67]. The results by Ito et al. suggest that the protein coded by GmNARK may play some roles on leaf growth as well.

In spite of profuse nodulation in the hyprnodulation mutant lines, the root and shoot growth is inferior in most of these lines compared with the wild type with or without nitrate supply. All the hypernodulation mutant lines are partially tolerant to NO_3^-. The supernodulation line first reported was nominated "nts", which means "nitrate-tolerant symbiosis" mutant [60]. The labeling experiments using $^{14}CO_2$ or $^{13}CO_2$ indicated that the hypernodulating mutant NOD1-3 supplied a larger amount of photoassimilate to the nodules than to the roots under nitrogen free conditions, and that the photoassimilate transport to the nodules was less sensitive to nitrate than that of the parent line [69].

Assimilation of $^{15}N_2$ and $^{15}NO_3^-$ was compared among hypernodulation mutant lines, NOD1-3, NOD2-4, and NOD3-7 and the parent Williams [70]. The 5 mM NO_3^- treatment resulted in a 95 to 97% decrease in nodule mass and $^{15}N_2$ fixation by Williams, while the three mutant lines retained 30 to 40% of the nodule mass and 17 to 19% of the $^{15}N_2$ fixation of control Williams. The hypernodulation mutant lines, which had restricted root growth, absorbed less $^{15}NO_3^-$ than Williams. These results confirmed that nodule formation and development are less sensitive to external NO_3^- in mutant lines than in the Williams parent. The partial tolerance of nodulation for nitrate in mutant lines may be partly due to less NO_3^- absorption activity and smaller roots.

Figure 34. Nodulated roots of Williams (left) and the hypernodulation mutant NOD1-3 (right).

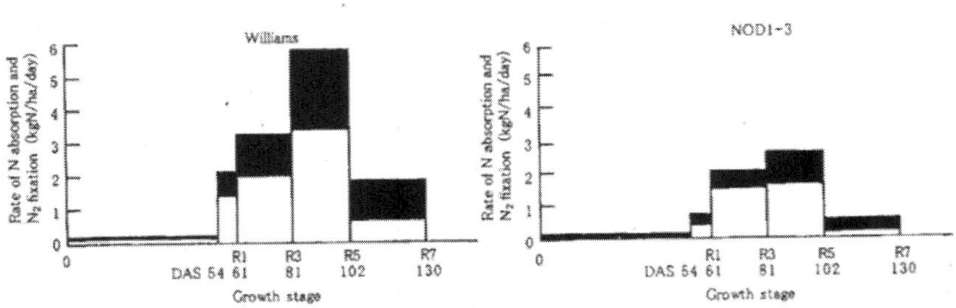

Figure 35. Changes in daily N_2 fixation activity and N absorption rate by Williams and the hypernodulation line NOD1-3 cultivated in a sandy dune field in Ikarashi [6].

The hypernodulation mutant lines of soybean may have some advantages for promoting seed yield, due to higher N_2 fixation activity or the nitrate tolerant trait to nodulation. Wu and Harper evaluated the N_2 fixation potential and yield of hypernodulating soybean NOD1-3, NOD2-4 and NOD3-7 compared with the parent Williams. In the absence of N fertilizer, all hypernodulation mutants had greater N_2 fixation potential than did Williams in early growth stages. However, the seed yields from the hypernodulation mutants were 10 to 30% less than that for Williams. Suganuma et al. also compared the growth and N_2 fixation activity of NOD1-3 and Williams in a sandy dune field. Figure 35 shows the daily rate of N_2 fixation and N absorption by Williams and the hypernodulation mutant NOD1-3. The % Ndfa was higher in NOD1-3 (65 %) than Williams (58 %), however, the rates of N_2 fixation and N absorption were lower in NOD1-3 than in Williams. The hypernodulation mutant lines have not been used for cultivar improvement, but recently "Sakukei no. 4" bred from En6500 (hypernodulating line from "Enrei") and "Tamahomare" in Japan may be useful in agricultural production by increasing the planting density.

7. A new fertilization technique to promote nitrogen fixation and seed yield

7.1. Nitrogen fertilization in soybean cultivation

Nitrogen fertilization usually depresses nodulation and nitrogen fixation activity. It often results in the same or less yield than the control treatment without fertilizer. Top dressing of N fertilizer sometimes gives positive effects on seed yield, but results are not consistent. Takahashi et al. reported the effect of basal side-dressings of various types of controlled release nitrogen fertilizer (coated urea) on shoot growth and seed yield of soybean [73]. The yield was significantly higher in all the coated urea basal side-dressing treatments compared with control, particularly in CUS120 which releases urea from 60 to 120 days after planting with a sigmoid pattern of N release.

We have developed a new fertilization technique for soybean cultivation to supplement N during seed filling stage, without the concomitant depression of N_2 fixation, by deep placement (20 cm depth from soil surface) of slow release N fertilizer coated urea and lime nitrogen. We analyzed the beneficial effects from both plant nutrition and soil analysis aspects.

7.2. Deep placement of controlled release nitrogen fertilizer coated urea

Takahashi et al. [4,16, 74-77] developed a new fertilization technique for soybean to supplement nitrogen during the seed filling stage without concomitant depression of symbiotic N_2 fixation by deep placement of coated urea, a slow release N fertilizer. They applied 100 kg N ha^{-1} coated urea by deep placement (20 cm depth from soil surface) using a fertilizer injector devised by Shioya [78]. They used CU-100, a 100-day type coated urea, the commercial name "LP-100" produced by Chisso-Asahi Fertilizer Co. Ltd, Tokyo (JCAM AGRI Co. Ltd., Tokyo at present). CU-100 linearly releases urea and 80 % of which is released in 100 days in water at 25°C.

A polymer coated controlled release nitrogen fertilizer (commercial name LP in Japan or MEISTER outside Japan) has been invented by Fujita and coworkers [79]. Linear types of coated urea were first marketed in 1982. This type of fertilizer has a spherical shape of about 3mm diameter with 50-60 μm coat thickness which consists of polyolefin (polyethylene), ethylene vinyl acetate and talc mineral. Different from chemically synthesized slow release N fertilizers such as IBDU (Isobutylidene diurea) and CDU (Crotonylidene diurea), the N release rate from the coated urea is temperature dependent and not affected by other chemical, physical and biological conditions. Therefore, the release pattern of urea can be predicted as a function of temperature and time period after application. Since the release of N from the fertilizer meets the plant N demand, and the fertilizer efficiency (recovery rate of N in plants from fertilizer N) is high, the use of coated urea can reduce environmental problems by decreasing nitrate accumulation and leaching in the soil. Also the use of coated urea saves the labor of farmers by eliminating top dressing or split dressing of fertilizer to supply N during late growth stages.

Fertilizer experiments were carried out from 1989 to 1991 in the fields, which had been converted from a paddy rice field the previous year in Nagakura. [21]. As shown in Figure 36 the seed yield was significantly higher in the plants with the deep placement of CU-100 than the control in each year. The seed yield was from 10 (1991) to 23 % (1990) higher in deep placement than the control treatments. The promotion of leaf growth and retardation of leaf senescence were observed at the maturing stage by deep placement of CU. In 1990, the seed yield was very high, about 6 t ha^{-1} in deep placement, due to favorable climatic conditions. The absorption efficiency of fertilizer N determined by ^{15}N labeled fertilizers was calculated from recovery of ^{15}N in the shoots at the R7 stage. In 1990, the absorption efficiency at R7 from the deep placement of CU-100 was 62 %, which was much higher than the top dressing of CU-70 (33 %) and basal application of ammonium sulfate (9 %). It was observed that the CU-100 deep placement increased root growth and water and nutrient absorption activity revealed by the uptake of rubidium tracer in the field [75]. Owing to the promotion of subsoil root growth and N absorption activity with supplementing N fertilizer without depression of N_2 fixation, plant growth was promoted from the early vegetative stage until late maturing. Leaf area index (LAI) and chlorophyll content were always higher in CU-100 deep placement compared with the control, and the leaf senescence was retarded at R7 stage [21].

Takahashi et al. analyzed the concentration of urea, ammonium and nitrate in the upper 0-10cm and lower 15-25cm layers of the soil in control and deep placement of CU-100 treatments. In the upper layer, the concentration of urea and nitrate was very low (less than 10 mg N kg^{-1} soil) both in control and deep placement of CU-100 treatments. However, the accumulation of ammonium (up to 150 mg N kg^{-1}) and nitrate (up to 50 mg N kg^{-1}) was observed in the lower layer of deep placement of CU100 in August. Although the urea released from the coated urea was rapidly hydrolyzed to ammonia, NH_4^+-N could not be easily nitrified in the deep soil layers of the converted rice field owing to the low activity of nitrification and restricted O_2 supply. As a result, the nodulation and N_2 fixation near the surface layer were not depressed, and instead were promoted through the improvement of plant growth and photosynthetic activity. The mechanism of promotion of deep placement of coated urea for soybean growth and seed yield is summarized in Figure 37.

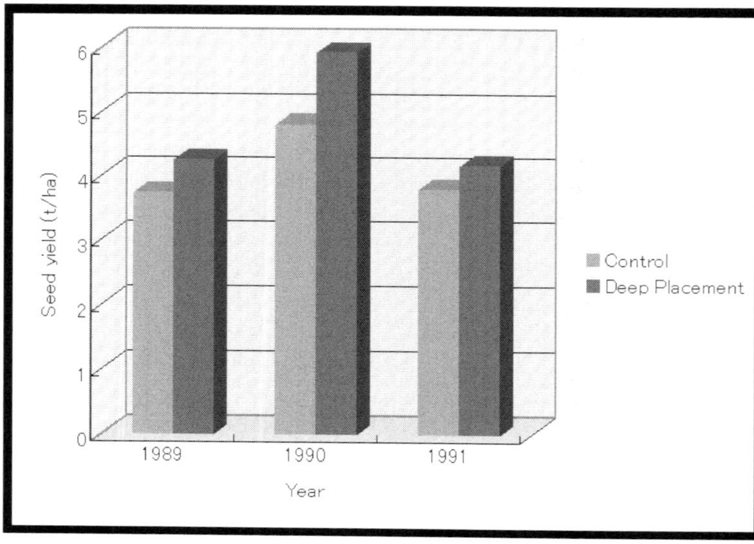

Figure 36. Comparison of the seed yield of control and deep placement of coated urea in rotated paddy fields in Nagakura in 1989, 1990 and 1991. [4]

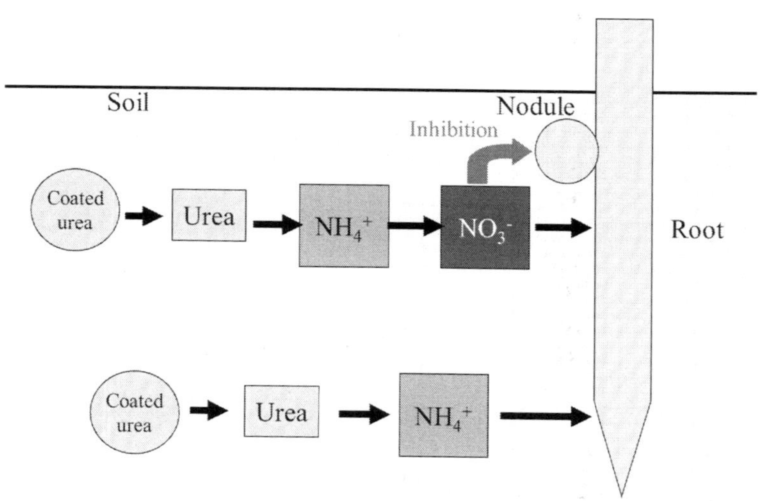

Figure 37. A model of the promotive effect of deep placement of coated urea on nitrogen fixation and seed yield of soybean.

a. Deep placement of coated urea slowly releases urea inside. The urea is rapidly hydrolyzed to ammonium and the ammonium does not leach out from the fertilization sites at 20cm depth.

b. The abundant supply of N in the deep layer promotes the root growth, and water and nutrients absorption activity and fertilizer N is efficiently absorbed from the lower roots.

c. The abundant supply of N from the lower part of roots promotes leaf growth and extends the photosynthetic activity until maturity. The leaf area and chlorophyll content were higher in the leaves of deep placement than those in control ones.

d. An abundant supply of photoassimilate to nodules supports the nodule growth and N_2 fixation activity for an extended period during the seed filling stage.

e. The continuous supply of N from nodules and roots with increased photo assimilate supply promotes seed yield with good quality.

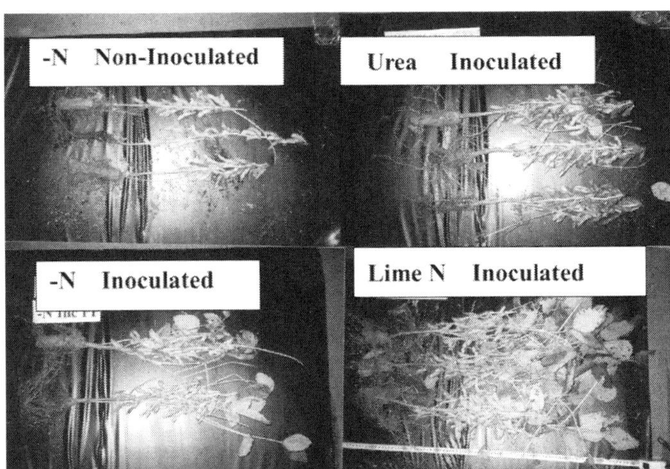

Figure 38. Soybean plants at R7 cultivated in a newly reclaimed field with a 30 cm depth of mountain soil. Plants were planted with inoculated or non-inoculated paper pot. –N: without N deep placement, Urea: deep placement of urea at 20cm depth. Lime N: deep placement of lime nitrogen. [51]

7.3. Deep placement of lime nitrogen

Recently, Tewari et al. [51,52,80,81] investigated the effects of deep placement of lime nitrogen (calcium cyanamide, $CaCN_2$) in comparison with coated urea. The fertilizer experiments were combined with a new inoculation method of bradyrhizobia using a paper pot transplantation. All the experiments were carried out in 2001 at three different sites in Niigata

Prefecture. The tests were initiated in a field reclaimed with the application of mountain soil about 30 cm depth without indigenous bradyrhizobia (Figure 38) [51], a rotated paddy field in Nagakura in Niigata Agricultural Research Institute (Figure 39) [52], and a sandy dune field of Niigata University in Ikarashi [80].

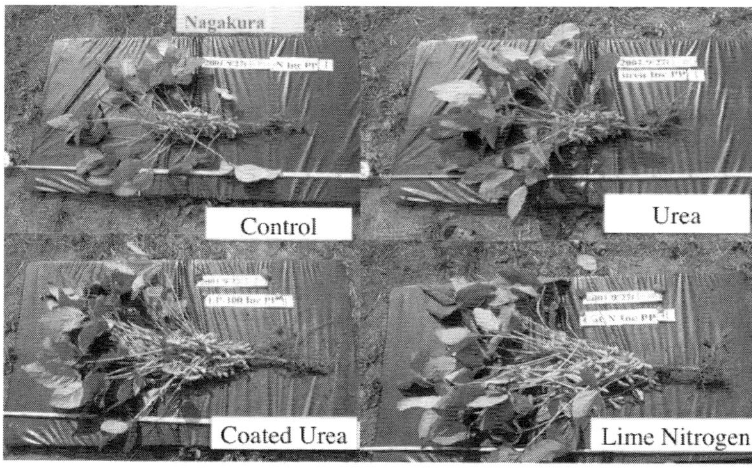

Figure 39. Soybean plants cultivated in a rotated paddy field in Nagakura. Plants were planted with inoculated paper pots. Control: without N deep placement, Urea: deep placement of urea at 20cm. Coated Urea: deep placement of coated urea. Lime Nitrogen: deep placement of lime nitrogen. [52]

Lime nitrogen is composed of about 60% calcium cyanamide ($CaCN_2$) with calcium oxide and carbon, and the N content is about 20-23%. After application to the soil, the calcium cyanamide is converted to urea, which is again degraded into NH_3 and CO_2. Dicyandiamide contained in lime nitrogen or formed during the degradation of calcium cyanamide is a potent nitrification inhibitor, which retards the oxidation of NH_4^+ to NO_3^-. Therefore the ammonium produced by $CaCN_2$ decomposition persists for a longer period of time and the nitrate concentration remains low in the soil. It is expected that the inhibition of nodulation and of the N_2 fixation activity may be alleviated by low a level of nitrate accumulation. Also this fertilizer exerts some hormonal effects on plants and is used for controlling soil diseases caused by bacteria and fungi.

In each of these fertilizer treatments, IPP (Inoculated paper pot), DT (Direct transplantation of inoculated seedlings without paper pot) and NIPP (Non-inoculated paper pot) seedlings were transplanted in separate plots. Paper pots (height 13.5 cm, diameter 3 cm) were made of a biodegradable paper designed to break down in the field. The pots were open at the bottom to allow root expansion below the pot. A paper pot was filled with vermiculite and a seed was planted in each pot, and followed by inoculation of one ml suspension of *Bradyrhizobium japonicum* USDA110 of about 10^8 cells ml^{-1}. Since the bradyrhizobium population in-

http://dx.doi.org/10.5772/45867

creases about 100 times in vermiculite for a few weeks [55], efficient infection of inoculated bradyrhizobia can also be expected by paper pot inoculation with vermiculite.

In regard to the inoculation method, IPP (inoculated paper pot) tended to show the highest seed yield than DT and NIPP treatments. Especially in the newly reclaimed field having mountain soil with no indigenous rhizobia at the 30 cm depth, the inoculation by IPP or DT promoted seed yield to a level more than twice as much as the control treatment. Among the inoculation methods, the IPP and DT seedlings showed a higher seed yield than the NIPP seedlings.

Significantly higher seed yields in the rotated paddy field were obtained with the deep placement of $CaCN_2$ IPP (6.12 t ha^{-2}) and CU-100 IPP (6.04 t ha^{-2}), compared with the Urea IPP (4.67 t ha^{-2}) and Control IPP (3.31 t ha^{-2}) treatments (Figure 39) [51]. A similar effect was observed in a reclaimed field and sandy dune field, where deep placement of lime nitrogen gave the same or better seed yields compared with coated urea. Recently, Sakashita et al. reported the promotive effect of deep placement of lime nitrogen in 8 sites of farmers' field in 2008, 2009 and 2010. Seed yields increased about 30% on average by deep placement of lime nitrogen.

The mechanism of yield promotion by deep placement of lime nitrogen for soybean growth and seed yield is summarized in Figure 40.

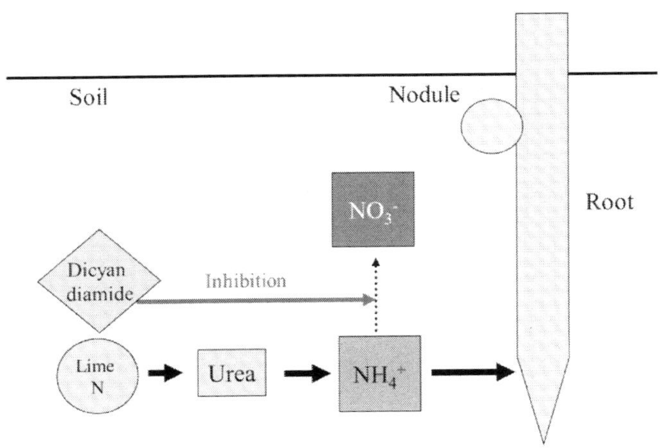

Figure 40. A model for the promotive yield effect of deep placement of lime nitrogen on nitrogen fixation and seed yield of soybean.

a. Deep placement of lime nitrogen is hydrolyzed to urea, then to ammonium and carbon dioxide. The ammonium does not easily leach out from the fertilization sites at 20 cm depth. Dicyandiamide contained in lime nitrogen or formed in the soil from cyanamide depresses nitrification to prevent ammonium oxidation to nitrate. As a result, nitrate leaching and denitrification is reduced and the ammonium can be sustained in the subsoil for a long time.

b. The abundant supply of N in the lower layer promotes the lower root growth, and water and nutrient absorption activity and fertilized N is efficiently absorbed from the lower roots.

c. The abundant supply of N from lower parts of roots promotes leaf growth and extends the photosynthetic activity until maturity.

d. An abundant supply of photoassimilate to nodules supports nodule growth and N_2 fixation activity for an extended period until maturity is reached.

e. The continuous supply of N from nodules and roots with increased photoassimilate supply promotes seed yield without decreasing the quality.

8. Conclusion

Potential soybean seed yield may over 10 ton ha^{-1} based on recent world record soybean production in the USA. However, the average soybean seed yield is only 2.56t. Soybean seed yield is proportional to the total assimilation of nitrogen. Soybean seeds contain an extraordinary high concentration of protein (about 35%), therefore, one t of seeds requires about 70-90 kg of N. To get a high seed yield, soybean plants need to assimilate a sufficient amount of nitrogen not only during vegetative stages but also during reproductive stages. At first, a high and long lasting activity of nodules during the pod filling stage is the most important factor to get high seed yield. To obtain optimum nodulation and nitrogen fixation activity, physical, chemical and biological conditions in the soil are very important in addition to appropriate cultivation method (fertilization, planting date, planting density, weed and pest management etc.). Inoculation of effective bradyrhizobium strains may improve soybean growth and seed yield. Although it is well known that a heavy supply of nitrogen fertilizer severely depresses nodulation and nitrogen fixation of soybean, a continuous supply of a low level of combined nitrogen from the subsoil (eg. soil organic matter) may support the plant vigor and photosynthetic activity and promote nitrogen fixation. The use of coated urea and lime nitrogen for deep placement has been successful in various types of soil in Japan.

Author details

Takuji Ohyama[1], Ritsuko Minagawa[1], Shinji Ishikawa[1], Misaki Yamamoto[1],
Nguyen Van Phi Hung[1], Norikuni Ohtake[1], Kuni Sueyoshi[1], Takashi Sato[2],
Yoshifumi Nagumo[3] and Yoshihiko Takahashi[1]

1 Faculty of Agriculture and Graduate School of Science and Technology, Niigata University, Niigata, Japan

2 Faculty of Bioresource Sciences, Akita Prefectural University, Japan

3 Niigata Agricultural Research Institute, Niigata, Japan

References

[1] FAOSTAT crop production, http://faostat.fao.org/site/567/default.aspx#ancor

[2] All Guide, Standard Table of Food Composition in Japan, Jikyo Syupan Ltd. Tokyo: 2009

[3] Yamauchi F., Ohkubo K. (Ed.) Science of soybean: Series Science of Food. Tokyo: Asakura Syupan Ltd. 1992.

[4] Takahashi Y., Chinushi T., Nakano T., Ohyama T. Evaluation of N2 fixation and N absorption activity by relative ureide method in field grown soybean plants with deep placement of coated urea. Soil Sci. Plant Nutr., 1992:38; 699-708.

[5] Schafer S., Kip Cullers Sets New Soybean World Record. AgWeb: http://www.agweb.com/article/kip_cullers_sets_new_soybean_world_record/ (accessed 18, October 2010)

[6] Suganuma T., Fujikake H., Ohtake N., Sueyoshi K., Ohyama T. Comparison of the growth and nitrogen fixation activity of the hypernodulatin soybean mutant NOD1-3 and its parent cv. Williams in field cultivation. Bull. Facul. Agric. Niigata Univ., 2001:53; 123-132.

[7] Wiatrak, P. Soybean vegetative and generative growth stages. Clemson Cooperative Extension Home Page http://www.clemson.edu/extension/rowcrops/soybeans/guide/growth_stages.html

[8] Fehr WR., Caviness CE., Burmood DT., Pennington JS. Stage development description for soybean Glycine max (L.) Merill. Crop Sci., 1971:11; 929-931.

[9] Sato T., Onoma N., Fujikake H., Ohtake N., Sueyoshi K., Ohyama T. Changes in four leghemoglobin components in nodules of hypernodulating soybean (*Glycine max* [L] Merr.) mutant and its parent in the early nodule developmental stage. Plant and Soil, 2001: 237; 129-135

[10] Bohlool BB., Ladha JK., Garrity DP., George T. Biological nitrogen fixation for sustainable agricultre: A perspective. Plant and Soil, 1992;141:1-11.

[11] Keyser HH., Li F. Potential for increasing biological nitrogen fixation in soybean. Plant and Soil, 1992: 141;119-135

[12] Nagumo Y., Sato T., Hattori M., Tsuchida T., Hosokawa H., Takahashi Y., Ohyama T. Effect of sigmoidal releasing-type coated urea fertilizer and ridge tillage on nitrogen accumulation and rate of side-wrinkled seeds in soybean cultivated in rotated paddy fields under poor drainage conditions. Jpn. J. Soil Sci. Plant Nutr., 2010: 81; 360-366

[13] Hoshi S. Nitrogen fixation, growth and yield of soybean. In: Nitrogen fixation in Root, (Ed.) Japanese Society of Soil Science and Plant Nutrition, Hakuyusha Publishers, Ltd. Tokyo, 1982. p 5-33.

[14] Harper JE. Soil and symbiotic nitrogen requirements for optimum soybean production. Crop Sci., 1974: 14; 255-260.

[15] Harper JE. Nitrogen metabolism. (In) Soybeans: Improvement, production and Uses. 2nd ed. Agronomy Monograph no.16. ASA-CSSA-SSSA, 1987: 497-533.

[16] Takahashi Y., Chinushi T., Ohyama T. Quantitative estimation of N2 fixation and absorption rate in field grown soybean plants by relative ureide method. Bull. Fac. Agric. Niigata Univ., 1993:45; 91-105.

[17] Ohyama T., Takahashi Y., Chinushi T., Nakano T. Evaluation of N2 fixation activity and nitrogen absorption rate in field grown soybean plants by simple relative ureide method. Agriculture and Horticulture, 1992: 67; 1157-1164.

[18] Harderson G. Method for enhancing symbiotic nitrogen fixation. Plant and Soil, 1993: 152; 1-17.

[19] Herridge DF., Danso SKA. Enhancing crop legume N2 fixation through selection and breeding. Plant and Soil, 1995:174;51-82

[20] Herridge D., Rose I. Breeding for enhanced nitrogen fixaion in crop legumes. Field Crop Research, 2000:65; 229-248.

[21] Takahashi Y., Chinushi T., Nakano T., Ohyama T. Yield components of soybean plants with deep placement of N fertilizer, related to high productivity. J. Niigata Agric. Exp. Stn., 1994:40; 7-15.

[22] Fujikake H., Yashima H., Sato T., Ohtake N., Sueyoshi K., Ohyama T. Rapid and reversible nitrate inhibition of nodule growth and N2 fixation activity in soybean (*Glycine max* (L.) Merr.). Soil Sci. Plant Nutr. 2002: 48; 211-217.

[23] Fujikake H., Yamazaki A., Ohtake N., Sueyoshi K., Matsuhashi S., Ito T., Mizuniwa C., Kume T., Hashimoto S., Ishioka N-S., Watanabe S., Osa A., Sekine T., Uchida H., Tsuji A. Ohyama T. Quick and reversible inhibition of soybean root nodule growth by nitrate involves a decrease in sucrose supply to nodules. J. Exp. Bot., 2003:54; 1379-1388.

[24] Ohyama T., Fujikake H., Yashima H., Tanabata S., Ishikawa, S., Sato, T., Nishiwaki, T., Ohtake, N., Sueyoshi, K., Ishii, S., Fujimaki, S., Effect of nitrate on nodulation and nitrogen fixation of soybean, In: El-Shemy HA. (ed.) Soybean, Physiology and Biochemistry, Rijeka: InTech; 2011. p333-364

[25] Bergersen FJ. Ammonia-an early stable product of nitrogen fixation by soybean nodules. Aust. J. Biol. Sci., 1965; 18: 1-9.

[26] Tempest DW., Meers, JL., Brown CM. Synthesis of glutamate in *Aerobacter aerogenes* by a hitherto unknown route. Biochem. J., 1970: 117; 405-407

[27] Ohyama T., Kumazawa K. Incorporation of 15N into various nitrogenous compounds in intact soybean nodules after exposure to 15N2 gas. Soil Sci.Plant Nutr., 1978:24; 525-533.

[28] Ohyama T., Kumazawa K. Nitrogen assimilation in soybean nodules I. The role of GS/GOGAT system in the assimilation of ammonia produced by N2 fixation. Soil Sci.Plant Nutr., 1980: 26; 109-115.

[29] Ohyama T., Kumazawa K. Nitrogen assimilation in soybean nodules II. 15N2 assimilation in bacteroid and cytosol fractions of soybean nodules. Soil Sci.Plant Nutr., 1980: 26; 205-213.

[30] Kushizaki M, Ishizuka J, Akamatsu F. Physiological studies on the nutrition of soybean plants. 2. Effect of nodule formation on nitrogenous constituents of soybeans. J. Sci. Soil Manure, Japan, 1964:35;323-327.

[31] Schubert KR. Products of biological nitrogen fixation in higher plants; symbiosis, transport and metabolism. Annu. Rev. Plant Physiol., 1986:37;539-574.

[32] Tajima S., Nomura M., Kouchi H. Ureide biosynthesis in legume nodules,. Frontiers in Bioscience, 2004:9; 1374-1381.

[33] Crawford NM, Glass ADM. Molecular and physiological aspects of nitrate uptake in plants. Trends in Plant Sci., 1998: 3; 389-395.

[34] Ohyama T., Saito K., Kato N. Diurnal rhythm in nitrate absorption by roots of soybeans (*Glycine max*). Soil Sci.Plant Nutr., 1989:35;33-42.

[35] Ohyama T., Kato N., Saito K. Nitrogen transport in xylem of soybean plant supplied with 15NO3-. Soil Sci.Plant Nutr., 1989:35; 131-137.

[36] Ohyama T., Kumazawa K. Assimilation and transport of nitrogenous compounds originated from 15N2 fixation and 15NO3- absorption. Soil Sci.Plant Nutr., 1979: 25; 9-19.

[37] Yashima H., Fujikake H., Sato T., Tewari K., Ohtake N., Sueyoshi K., Ohyama T. Systemic and local effects of long term application of nitrate on nodule growth and N2 fixation in soybean (*Glycine max* (L.) Merr.). Soil Sci. Plant Nutr., 2003:21; 981-990

[38] Yashima H, Fujikake H, Yamazaki A, Ito S, Sato T., Tewari K, Ohtake N, Sueyoshi K, Takahashi Y, Ohyama, T. Long-term effect of nitrate application from lower part of roots on nodulation and N2 fixation in upper part of roots of soybean (*Glycine max* (L.) Merr.) in two-layered pot experiment. Soil Sci. Plant Nutr., 2005:21; 981-990

[39] Ohyama T., Kawai, S. Nitrogen assimilation and transport in soybean leaves: Investigation by petiole girdling treatment. Soil Sci. Plant Nutr., 1983:29; 227-231.

[40] Vadez V., Sinclair TR. Ureide degradation pathways in intact soybean leaves. J. Exp. Bot., 2000:51;1459-1465.

[41] Ohyama T. Comparative studies on the distribution of nitrogen in soybean plants supplied with N2 and NO3- at the pod filling stage. Soil Sci. Plant Nutr., 1983:29; 133-145.

[42] Ohyama T. Comparative studies on the distribution of nitrogen in soybean plants supplied with N2 and NO3- at the pod filling stage. II. Assimilation and transport of nitrogenous constituents. Soil Sci.Plant Nutr., 1984:30; 219-229.

[43] Rainbird RM., Thorne JH., Hardy RWF. Role of amides, amino acids and ureides in the nutrition of developing soybean seeds. Plant Physiol., 1984:74; 329-334.

[44] Haga KI., Sodek L. Utilization of nitrogen sources by immature soybean cotyledons in culture. Ann. Bot., 1987: 59; 597-601.

[45] Ohtake N., Sato A., Fujikake H., Sueyoshi K., Ohyama T., Ishioka NS., Watanabe S., Osa A., Sekine T., Matsuhashi S., Ito T., Mizuniwa C., Kume T., Hashimoto S., Uchida H., Tsuji A., Rapid N transport to pods and seeds in N-deficient soybean plants. J. Exp. Bot., 2001:52; 277-283.

[46] Shuttuck-Eidens DM., Beachy RN. Degradation of b-conglycinin in early stage of soybean embryogenesis. Plant Physiol., 1985:76; 301-306.

[47] Ohtake N, Nishiwaki T., Mizukoshi K, Chinushi T, Takahashi Y, Ohyama T, Lack of β-subunit of β–conglycinin in non-nodulating isolines of soybean. Soil Sci. Plant Nutr., 1994:40;345-349.

[48] Ohtake N., Yamada S., Suzuki M., Takahashi N., Takahashi Y., Chinushi T. and Ohyama T. Regulation of accumulation of β-subunit of β–conglycinin in soybean seeds by nitrogen. Soil Sci. Plant Nutr., 1997:43;247-253.

[49] Perez-Gimenez J., Quelas JI., Lodeiro AR. Competition for nodulation. In: El-Shemy HA. (ed.) Soyean, Physiology and Biochemistry, Rijeka: InTech; 2011. 139-166

[50] Biofertilizer Manual, FNCA (Forum for Nuclear Cooperation in Asia): 2006; http://www.fnca.mext.go.jp/english/bf/bfm/e_bfm.html

[51] Tewari K., Minagawa R., Suganuma T., Fujikake H., Ohtake N., Sueyoshi K., Takahashi Y., Tsuchida T. Ohyama T. Effect of deep placement of slow release nitrogen fertilizers and inoculation of Bradyrhizobia on the first cropping of soybean in the field dressed with mountain soil. Jpn. J. Soil Sci. Plant Nutr., 2003:74;183-189.

[52] Tewari K., Suganuma T., Fujikake H., Ohtak N., Sueyoshi K., Takahashi Y., Ohyama T. Effect of deep placement of calcium cyanamide, coated urea, and urea on soybean (*Glycine max* (L.) Merr.) seed yield in relation to different inoculation methods. Soil Sci. Plant Nutr., 2002:48;855-863.

[53] Minamisawa K., Seki T., Onodera S., Kubota M., Asami T. Genetic relatedness of *Bradyrhizobium japonicum* field isolates as revealed by repeated sequences and various other characteristics. Appl. Env. Microbiol., 1992:58(9);2832-2839.

[54] Minamisawa K., Nakatsuka Y., Isawa T. Diversity and field site variation of indigenous populations of soybean bradyrhizobia in Japan by fingerprints with repeated sequences RSa and RSb. FEMS Microbiol. Ecol., 1999:29; 171-178.

[55] Minagawa R., Isoda K., Ohtake N., Yamada S., Ikarashi T., Minamisawa K., Ohyama T. Estimation of gus marked bradyrhizobial number by phenol extraction of GUS metabolite. Jpn. J. Soil Sci. Plant Nutr., 1997:68;148-155.

[56] Ohtake N., Nishiwaki T., Mizukoshi K., Minagawa R., Takahashi Y., Chinushi T., Ohyama T. Amino acid composition in xylem sap of soybean related to the evaluation of N2 fixation by the relative ureide method. Soil Sci. Plant Nutr., 1995:41;95-102.

[57] Herridge DF., People MB. Ureide assay for measuring nitrogen fixation by nodulated soybean calibrated by 15N methods. Plant Physiol., 1990: 93; 495-503.

[58] Herridge DF., Bergersen FJ., People MB. Measurement of nitrogen fixation in by soybean in the field using the ureide and natural 15N abundant methods. Plant Physiol., 1990: 93; 708-716.

[59] Akao S., Kouchi H. A supernodulating mutant isolated from soybean cultivar Enrei. Soil Sci. Plant Nutr., 1992; 38: 183-187.

[60] Carroll BJ., McNeil DL., Gresshoff PM. A supernodulation and nitrate-tolerant symbiotic (*nts*) soybean mutant. Plant Physiol., 1985: 78; 34-40.

[61] Gremaud MF., Harper JE. Selection and initial characterization of partially nitrate tolerant nodulation mutants of soybean. Plant Physiol., 1989: 89;169-173.

[62] Gerahty N., Caetano-Anolles G., Joshi PA., Gresshoff PM. Anatomical analysis of nodule development in soybean reveals an additional autoregulatory control point. Plant Sci., 1992:85;1-7.

[63] Sato T., Yashima H., Harper JE., Akao S., Ohyama T. Nodule formation and N2 fixation traits of the rooted-single leaf isolated from hypernodulating mutants and the wild type. Jpn J. Soil Sci. Plant Nutr., 1997: 68; 444-447.

[64] Krusell L., Madsen LH., Sato S., Aubert G., Genua A., Scyglowski K, Duc G, Kaneko T, Tabata S, de Bruijn F, Pajuelo E, Sandal N Stougaard J. Shoot control of root development and nodulation is mediated by a receptor-like kinase. Nature, 2002:420; 422-426.

[65] Nishimura R., Hayashi M., Wu GJ., Kouchi H., Imaizumi-Anraku H., Murakani Y., Kawasaki S., Akao S., Ohmori M., Nagasawa M., Harada K., Kawaguchi M. HAR1 mediates systemic regulation of symbiotic organ development. Nature, 2002:420; 426-429.

[66] Searle IR., Men AE., Laniya TS., Buzas DM., Iturbe-Ormaetxe I., Carroll BJ., Gresshoff PM. Long-distance signaling in nodulation directed by a CLAVATA1-like receptor kinase. Science, 2003:299; 109-112.

[67] Clark SE., Williams RW., Meyerowits EM. The *CLAVATA1* gene encodes a putative receptor kinase that controls shoot and floral meristem size in *Arabidopsis*. Cell, 1996:89; 575-585.

[68] Ito S., Ohtake N., Sueyoshi K. Ohyama T. The autoregulation of nodulation mechanism is related to leaf development. Plant Cell Phisiol., 2008:49; 121-125.

[69] Fujikake H, Tamura Y, Ohtake N, Sueyoshi K, Ohyama T, Photoassimilate partitioning in hypernodulation mutant of soybean (*Glycine max* (L.) Merr.) NOD1-3 and its parent Williams in relation to nitrate inhibition of nodule growth. Soil Sci. Plant Nutr., 2003:49;583-590

[70] Ohyama T., Nicholas JC., Harper JE. Assimilation of 15N2 and 15NO3- by partially nitrate-tolerant nodulation mutants of soybean. J. Exp. Bot., 1993:44; 1739-1747.

[71] Wu S., Harper JE. Dinitrogen fixation potential and yield of hypernodulating soybean mutants: a field evaluation. Crop Sci., 1991:31; 1233-1240.

[72] Nakamura T., Nakayama N., Yamamoto R., Shimamura S., Kim Y., Hiraga S., Ohyama T., Komatsu S., Shimada S. Nitrogen utilization in the supernodulating soybean variety "Sakukei 4" and its parental varieties, "Enrei" and "Tamahomare". Plant Prod. Sci., 2010:13(2);123-131.

[73] Takahashi Y, Ohtake N, Hottori M, Nagumo Y, Ohyama T. Effect of basal side-dressing of various types of coated urea fertilizer on shoot growth, yield components and seed composition of soybean (*Glycine max* (L.) Merr.), Soil Sci. Plant Nutr., 2006:52; 264-273.

[74] Takahashi Y., Chinushi T., Nagumo Y., Nakano T., Ohyama T. Effect of deep placement of controlled release nitrogen fertilizer (coated urea) on growth, yield and nitrogen fixation of soybean plants. Soil Sci. Plant Nutr., 1991:37; 223-231.

[75] Takahashi Y., Chinushi T., Nakano T., Hagino K., Ohyama T. Effect of placement of coated urea on root growth and rubidium uptake activity in soybean plants. Soil Sci. Plant Nutr., 1991:37; 735-739.

[76] Takahashi Y., Chinushi T., Nakano T., Ohyama T. Behavior of fertilized N from top-dressed or deep placed coated urea in soil of soybean field. Jpn J. Soil Sci. Plant Nutr., 1993:64; 338-340.

[77] Takahashi Y., Ohyama, T. Technique for deep placement of coated urea fertilizer in soybean cultivation. JARQ, 1999:33; 235-242.

[78] Shioya T. Abstract, 1985 Meeting, Farm Work Research Congress, Japan, 1985: p47

[79] Fujita T., Shoji S. Meister: Controlled relrease fertilizer. (ed.) Shoji S, Sendai: Konno Printing Company Ltd.;1999.

[80] Tewari K., Suganuma T., Fujikake H., Ohtake N., Sueyoshi K., Takahashi Y., Ohyama T. Effect of deep placement of N fertilizers and different inoculation methods of bradyrhizobia on growth, N2 fixation activity and N absorption rate of field grown soybean plants. J. Agronomy and Crop Science, 2004:190; 46-58.

[81] Tewari K., Nagumo Y., Takahashi Y., Sueyoshi K., Ohtake N., Ohyama, T. A New Tecnology of Deep Placement of Slow Release Nitrogen Fertilizers for Promotion of Soybean Growth and Seed Yield. New York: NOVA Science Publishers. 2011.

[82] Sakashita H., Nosaka Y., Hosokawa H. Effect of deep placement of lime nitrogen in high ridge tillage cultivation. Agriculture and Horticulture, 2011: 86: 981-986

Section 2

Soybean Agricultural Economics

Chapter 7

The Comparative Advantage of Soybean Production in Vietnam: A Policy Analysis Matrix Approach

Huynh Viet Khai and Mitsuyasu Yabe

Additional information is available at the end of the chapter

http://dx.doi.org/10.5772/51000

1. Introduction

Vietnam is a country with a long agricultural tradition in which highly-skilled farmers make up 68% of the population. Annually, Vietnam has been exporting a lot of agricultural products, which contributes a large amount of foreign currency to the governmental budget. Recently, Vietnam has become the second largest exporter of rice after Thailand, of cashew after Indonesia and of coffee after Brazil. Moreover, Vietnam is one of the top-leading exporters for rubber, and black pepper.

Although the contribution of farmers is very important for agriculture as well as for the economy of Vietnam, farmers' earnings are much lower than the rest of the population. This fact has created an income gap between rural and urban people. To solve this problem, the government has applied a variety of policies to increase the farmers' income, for instance, policies related to import, export, agricultural outputs and inputs etc. However, studies on the effectiveness and impacts of such policies are still limited. How have the policies undertaken by the government had an impact on farmers? Are farmers receiving any benefits from policies provided by the government? These are the questions that the study partly seeks to answer. In this study, the method of policy analysis matrix (PAM) was used to analyze the effect of policies on soybean production in Vietnam. Moreover, the analysis of PAM also answers whether the development of soybean production has potential in Vietnam, in other words, whether soybean cultivation has a comparative advantage in Vietnam compared to the other soybean producers in the world. The results of the study are useful to policy-makers, and also help us better understand the effects of the current policies undertaken by the government.

The first section of the study discusses data collection, the theoretical framework and some useful policy parameters of PAM. The second section reviews the current situation of soybean production in the study area by doing the cost - benefit analysis (CBA) of soybean production. The third section presents the establishment of, results of, some sensitivity analyses of PAM, and policy simulations to help determine likely changes on PAM's parameters. This will be followed with some conclusions.

2. Data

The primary data of this study was collected in a field survey in two agro-ecological areas of the Mekong Delta of Vietnam in 2004. In the Mekong Delta, rice is the main crop. Farmers often apply mixed farming systems such as one-rice and one-fish crop, or two-rice and one-vegetable crop to improve income and soil conditions. Consequently, farmers grow soybean once a year. The soybean crop is usually cultivated in January and February after the Winter-Spring rice crop and harvested in March and April. In this study, farmers who grow two-rice and one-soybean crop were selected for interviews.

The interviews were conducted in Can Tho Province, representing the lower reaches of the Mekong Delta, and An Giang Province, representing the upper one. The total sample of 113 farmers, of which 58 farmers were in Can Tho and 55 farmers in An Giang, was interviewed following a stratified random sampling procedure. The survey was conducted by the staffs of Can Tho University, Vietnam.

3. Methodology

The approach of Policy Analysis Matrix (PAM) is applied to estimate the comparative advantage of soybean production. The PAM is a tool to create policy indicators for which values can be estimated, notable among which are the nominal protection coefficient, effective protection coefficient, private cost ratio and domestic resource cost ratio.

Regarding practical issues addressed by the PAM, Monke and Pearson (1989, p.17) proposed that the application of PAM approach is suitable for three areas of economic analysis:

- The impact of policies on the competitiveness of commodity systems;

- The impact of investment policy on economic efficiency and comparative advantage;

- The effects of agricultural research policy on steering the processes of technological change in desirable directions.

The PAM is built through double entry book-keeping, with the purpose of ensuring complete and consistent coverage of all policy influences on the returns to, and costs of, agricultural production or marketing. Indicators of the economic consequences of policies could be derived from the parameters in the matrix. The main empirical task is to construct accounting matrices of revenues, costs and profits. A PAM is constructed for each commodity system to be analyzed. Thus, the impact of commodity and macroeconomic policies is measured by comparing results in the presence and the absence of policy.

The PAM is comprised of revenues, costs and profits, at private and social (often called 'shadow') prices (Table 1). The top row of the matrix is a budget showing costs of production and marketing at market prices, the only unusual aspect being the division of cost elements into two categories: tradable and non-tradable inputs (usually defined as domestic resources – the immovable domestic factors of production).

The second row in the matrix shows the same cost elements expressed at social prices, i.e. social opportunity cost. For tradable products, adjusted world prices are normally taken as social prices, applying import or export parity measures as appropriate. The social price of domestic resources is taken as their opportunity cost, in other words the return at the margin in the best available alternative.

	Revenues	Tradable input costs	Domestic factor costs	Profits
Private accounts	A	B	C	D
Social accounts	E	F	G	H
Divergences	I	J	K	L

Note: Private Profits: D = A – B – C; Social Profits: H = E – F – G; Output Transfers: I = A – E; Input Transfers: J = B – F; Factor Transfers: K = C – G; Net Transfers: L = D – H *Source: Monke and Pearson 1989.*

Table 1. The Policy Analysis Matrix (PAM)

An important general point about the PAM is that the opportunity costs of domestic resources will be a function of current policy. Thus, strictly, these opportunity costs are only relevant under a particular set of policy constraints, i.e. they are constrained second best equilibrium values. If policy was to change, so would opportunity costs. For this reason, the PAM is not completely satisfactory in terms of economic theory, being based on a partial equilibrium rather than general equilibrium approach. It is a practical, indicative approach to policy, which recognizes that practitioners of policy analysis will only rarely have the data or the time to construct a fully specified general equilibrium model capable of generating useful estimates of opportunity costs under different policy scenarios.

The third row of the PAM is simply the first row minus the second. It shows the net impact of: market failure; distorting policies; and efficient policies (those which correct market failure). The signs of the revue and cost terms in the third row indicate whether the net effects of policy and market imperfections for these categories amount to an implicit subsidy or tax. If for example, I letter were positive, the net effect of policy or market failure is that the market price paid to the system is in excess of the social opportunity cost, i.e. output prices are subsidized. The right-hand entry in the third row, L, summarizes the net effect of polices or market failures on the profitability of the system, known as 'net transfers'. If D > H, then the net effect of policy is to subsidize the system. In this case, policy reforms to bring about greater economic efficiency will reduce the gap between D and H, and this will induce adjustments in the commodity system in question, which may involve changes in the proportions in which resources are used and, at least in the short term, some contraction in the scale of operation.

Regarding indicators in the PAM, the basis PAM permits twelve indicators of economic efficiency, six of which are non-ratio indicators and six are ratio-indicators. Ratio measures are more useful for comparison of commodity systems which are dissimilar in the relative proportions in which they use inputs.

The primary objective of constructing a PAM is to derive a few important policy parameters for policy analysis. Seven of the most commonly used parameters are private cost ratio (PCR), domestic resource cost (DRC), nominal protection coefficient output (NPCO), nominal protection coefficient input (NPCI), effective protection coefficient (EPC),Profitability Coefficient (PC) and Subsidy Ratio to Producers (SRP) explained as followings:

- PCR is the ratio of factors costs (C) to value added in private prices (A-B). In Table 1, PCR = C/(A-B). This ratio measures the competitiveness of a commodity system. This system is competitive if PCR is less than 1.

- DRC is the ratio of domestic factor cost valued at social prices to the value-added created by the same resources at social prices. In Table 1, DRC = G/(E-F). It is, in fact, a social cost-benefit ratio which helps determine the desirability of certain domestic production systems relative to the international market in terms of economic efficiency. The social cost is the opportunity cost of domestic resources involved in the production process. The social benefit is the value-added generated by the resources measured at social prices. If the cost is greater than the benefit (DRC>1), the production of the product is not desirable from the social point of view. On the other hand, if the cost is less than the benefit (DRC<1), the production of that product is socially desirable. If the cost is equal to the benefit (DRC=1), it is just worthwhile to produce the commodity. It also implies that in regard to the commodity in question, the allocation of productive resources is such that domestic resources are being used in a way that reduces the country's welfare.

- NPCO is the ratio of domestic market price of a product to its parity price at the farm-gate. In Table 1, NCPO=A/E. If NPCO>1, it indicates that the private price of output is greater than its parity price and hence producers are positively protected for the product. If NPCO<1, it indicates that producers are implicitly taxed on the product. If NCPO=1, it indicates a neutral situation.

- NPCI is the ratio of the private to the social values of all the tradable inputs (or input components). In Table 1, NPCI=B/F. If NPCI>1, it indicates that producers are taxed when they buy tradable inputs. If NPCI<1, it indicates that they are subsidized. NPCI=1 represents a neutral situation.

- NPCO and NPCI consider the distortion of government policy in the product and tradable input markets respectively in isolation. EPC measures the total effects of intervention in both markets. It is defined as the ratio of value-added measured at private prices to that at social prices, or EPC=(A-B)/(E-F). If EPC>1, it implies that the overall impact of the existing policy results in a net positive incentive to produce the commodity. EPC<1 represents a net disincentive. EPC=1 implies either no intervention or impact of various distortions in both the input and product markets results in a neutral effect on value-added.

- PC measures the impact of all transfers on the private profits. It equals the ratio of private profits to social profits or PC = D/H.

- SRP is a single measure of all transfer effects. In Table 1, SRP = L/E = (D-H)/E. It indicates the extent to which the system revenues are increased or decreased because of transfers. If the market failures are insignificant, the SRP shows the net impact of distorting policies on the system revenues.

5. Results and discussion

5.1. Cost and benefit analysis of soybean production

The income is estimated by multiplying yield with price and the profit is calculated by using income minus the total cost of soybean production presented in Table 2.

	Can Tho	An Giang	Overall
Yield (kg/10a)	277.01	259.04	267.79
Price (VND/kg)	4,974.55	5,333.62	5,158.85
Income (VND/10a)	1,378,006.98	1,381,598.04	1,381,462.72
Total costs (VND/10a)	936,978.29	800,466.64	860,258.04
Profit (VND/10a)	441,028.69	581,131.40	521,204.68
Family labor (days/10a)	4.87	4.89	4.88
Profit-income ratio	0.3	0.4	0.4
Profit-cost ratio	0.5	0.7	0.6
Income-cost ratio	1.5	1.7	1.6
Profit-family labor ratio	90,544.25	118,730.95	106,738.45

Note: 10a (are) = 1,000m2. *Source: computed from the survey data*

Table 2. Yield, income and profit of soybean production

Table 2 shows that harvesting the average yield of soybean production of 268kg per 10a, soybean farmers sell their product with the price of 5,200VND per kg. However, there is a big variation in soybean price among farmers in the sample. The minimum price is only 1,000VND per kg, while the maximum is 7,000VND per kg. The reasons for this may come from the differences in output quality, the time of selling and unstable markets. Normally, the soybean price is fixed by purchasers while it is sometimes formed by the negotiation between buyers and sellers. At harvesting time, the buyers come to the farm to buy soybean at the farm-gate. Very few farmers could sell their product at the market price because of the lack of market information. They even sell their soybean at a low price compared to the market price due to poor storage facilities and the lack of savings for daily living.

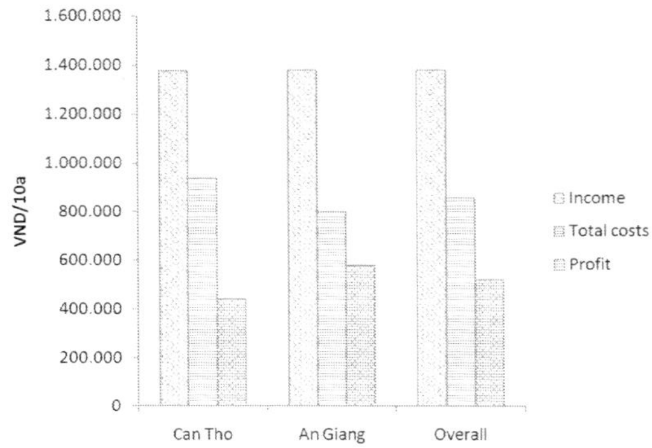

Figure 1. Yield, income and profit of soybean production

With an average soybean income of around 1,400,000VND per 10a, farmers could receive a profit of 521,000VND per 10a from their soybean cultivation. There is a big difference in profit among the surveyed sample. Some farmers could obtain the maximum profit of around 1,767,000VND per 10a while 20 percent of farmers had negative profits from soybean cultivation.

In rural areas in Vietnam, family labor is considered as the source of income. The ratio of profit to family labor means the earning of farmers received in their own land. This ratio of 106,000VND per day is three times higher than the hired labor price of 35,000VND per day, indicating that it is more profitable to do their own farming compared to working for other farmers as hired laborers.

The income-cost ratio of 1.6 indicates that farmers spent 1VND on soybean production costs and received an income of 1.6VND resulting in a profit of 0.6VND for soybean production performed by the profit to cost ratio. Although these ratios are rather high, the absolute amount of cash profit received by farmers is relatively low because of their small-scale production of soybean (the average soybean area of 0.68ha). In addition, the poor farming techniques and lack of market information result in unstable yields and prices that may not ensure sustainable income for soybean farmers.

5.2. Establishing the Policy Analysis Matrix

The basic information needed for compiling a PAM are yields, input requirements, the actual market prices and social prices of inputs and outputs. The major sources of data used for the private account in the PAM are from the data of the soybean farm household survey.

The first step is to establish a table of physical input-output relationships for soybean production. This numerical description of the soybean production function summarizes the technology used in this system. In this illustrative system, the inputs used by the representative farmers are about 18kg Urea, 18kg NPK and 3.3 labor days for crop care per 10 a. The average yield of soybean is around 268kg per 10a. These input-output coefficients are drawn from the synthetic budgets and the farm interviews presented in Table 3.

I-O	Quantities	An Giang	Can Tho	Overall
Tradable	Fertilizer			
	Urea (kg/10a)	18.5	17.0	17.8
	NPK (kg/10a)	16.4	19.4	17.9
	DAP (kg/10a)	12.8	13.3	13.1
	Other fertilizer (VND)	13,461.8	4,205.1	8,710.6
	Pesticide (VND/10a)			
	Herbicide	45,664.8	22,230.9	33,636.8
	Fungicide	17,754.4	6,887.3	12,176.6
	Insecticide	151,777.7	122,718.5	136,862.4
	Other pesticide	17,425.5	14,860.3	16,108.8
	Seed (kg/10a)	13.9	9.6	11.7
	Fuel of Irrigation (liters/10a)	4.7	2.7	3.6
Factors	Labor (days/10a)			
	Hired labor	4.8	1.9	3.3
	Machinery (days/10a)			
	Land preparation	1.1	3.0	2.1
	Irrigation	-	2.4	1.2
	Harvest	4.6	4.2	4.4
	Other machinery (VND/10a)	363.6	8,298.2	4,401.8
Output	(kg/10a)	277.0	259.0	267.8

Source: Own estimates, 2006; data appendix available from authors

Table 3. Physical Input-Output of soybean

The second step is to compile a table of private (actual market) prices for each of the inputs used and output produced in the system. These prices should be representative of the base year of the study. The private prices for the soybean system are presented in Table 4.

P-Price	Quantities	An Giang	Can Tho	Overall
Tradable	Fertilizer			
	Urea (VND/kg)	3,302	3,401	3,352
	NPK (VND/kg)	3,573	3,601	3,588
	DAP (VND/kg)	3,990	4,101	4,049
	Other fertilizer (VND)	-	-	-
	Pesticide (VND)			
	Herbicide	-	-	-
	Fungicide	-	-	-
	Insecticide	-	-	-
	Other pesticide	-	-	-
	Seed (VND/kg)	6,463	6,562	6,514
	Fuel of Irrigation (VND/liters)	5,190	5,624	5,373
Factors	Labor (VND/day)			
	Hired labor	37,368	29,628	34,159
	Machinery (VND/day)			
	Land preparation	24,000	27,179	25,272
	Irrigation	-	26,277	26,277
	Harvest	43,784	38,783	41,012
	Other machinery (VND)	-	-	-
Output	(VND/kg)	4,975	5,334	5,159

Source: Own estimates, 2006; data appendix available from authors

Table 4. Private Prices of soybean

The most difficult task for constructing a PAM is the estimation of social prices and the separation of inputs into their tradable and non-tradable components. We use the world price as a reference price. In the study, the social price of soybean is the respective import parity price of soybean equivalents at the farm gate as an imported commodity. The CIF import price of soybean is listed on the website of the Ministry of Agriculture and Rural Development. The CIF price is adjusted at the farm gate by adding to it the transportation cost from the port of Ho Chi Minh City, one of the biggest ports in Vietnam. The value of transportation is assumed to be equal to 1 percent of the CIF price of soybean. The social price of soybean at the farm gate is then calculated by subtracting the distribution costs to farm.

As for tradable inputs, Vietnam has been importing chemical fertilizer, pesticides, fuel and other major farm inputs from international markets. Thus, the tradable inputs for fertilizer, fuel, etc. are the respective import parity prices at the farm gate and the social price of soybean seed is assumed to be equal to 20 percent of the social soybean price. However, because of complications for pesticide price, the social price of pesticides was not calculated in the study. Regarding domestic factors, since these factors are not tradable internationally and thus do not have world price, their social opportunity costs are estimated through observations of rural factor markets. The domestic inputs in the study are hired labor, machinery etc. They are assumed to be equal to the maximum prices in the sample.

When the parity prices of soybean and inputs are estimated, the shadow exchange rate instead of the official exchange rate is used to convert the international prices in US$ into VND. In the study, shadow exchange is assumed to be 16,000VND/US$ for calculating the social prices of soybean and its inputs. The social prices of soybean is calculated by adjusting form the international price to farm-gate level presented in Table 5. Similarly, the social price of tradable inputs such as fertilizers and diesel are also estimated in Table 6 and Table 7.

Import parity prices	**Soybean**
CIF Vietnam (US$/ton) *	366.3
Exchange rate (VND/US$)	16,000.0
Exchange rate premium (%)	0.10%
Equilibrium exchange rate (VND/US$)	16,016.0
CIF Vietnam in domestic currency (VND/ton)	5,866,700.0
Weight conversion factor (kg/ton)	1,000.0
CIF Vietnam in domestic currency (VND/kg)	5,866.7
Transportation and handing costs (VND/kg)	58.7
Value before processing (VND/kg)	5,925.4
Processing conversion factor (%)	1
Import parity value (VND/kg)	5,925.4
Distribution costs to farm (VND/kg)	200.0
Import parity value at farm gate (VND/kg)	5,725.4

* The source from the website of Ministry of Agriculture and Rural Development, 6/4/2006 *Source: Own estimates, 2006; data appendix available from authors*

Table 5. Adjustment of International Price of soybean to Farm-gate Level

Import Parity Prices	Urea	NPK	DAP
CIF HCM port (US$/ton) *	147.19	161.77	199.33
Exchange rate (VND/US$)	16,000	16,000	16,000
Exchange rate premium (%)	0.10%	0.10%	0.10%
Equilibrium exchange rate (VND/US$)	16,016	16,016	16,016
CIF in domestic currency (VND/ton)	2,357,352.0	2,590,913.3	3,192,442.8
Weight conversion factor (kg/ton)	1,000	1,000	1,000
CIF in dom.currency and weigh units (VND/kg)	2,357.35	2,590.91	3,192.44
VAT	5%	5%	5%
CIF and VAT in domestic currency (VND/kg)	2,475.22	2,720.46	3,352.06
Transportation and handing costs (VND/kg)	230.0	230.0	230.0
Value before processing (VND/kg)	2,705.2	2,950.5	3,582.1
Processing conversion factor	1	1	1
Import parity value at wholesale (VND/kg)	2,705.2	2,950.5	3,582.1
Distribution costs to farm (VND/kg)	200.0	200.0	200.0
Import parity value at farm gate (VND/kg)	2,505.2	2,750.5	3,382.1

* The source from the website of Ministry of Agriculture and Rural Development, 6/4/2006
Source: Own estimates, 2006; data appendix available from authors

Table 6. Adjustment of International Prices of fertilizers to Farm-gate Level

Import Parity Prices	Diesel
CIF HCM port (US$/ton) *	359
Exchange rate (VND/USD)	16,000
Exchange rate premium (%)	0.10%
Equilibrium exchange rate (VND/US$)	16,016
CIF in domestic currency (VND/ton)	5,749,744.00
Weight conversion factor (kg/ton)	1000
CIF in domestic currency and weigh units (VND/kg)	5,749.74
Luxury tax and transportation	10%
CIF including luxury tax	6,324.72
VAT	10%
CIF and VAT in domestic currency (VND/kg)	6,957.19
Transportation and handing costs (VND/kg) **	500

http://dx.doi.org/10.5772/51000

Import Parity Prices	Diesel
Value before processing (VND/kg)	7,457.19
Processing conversion factor	1
Import parity value at wholesale (VND/kg)	7,457.19
Distribution costs to farm (VND/kg)	200
Import parity value at farm gate (VND/kg)	7,257.19

* The source from the website of Labor newspaper, 19/10/2006**The source from the website of Viet Nam trade promotion Agency, 21/4/2006 *Source: Own estimates, 2006; data appendix available from authors*

Table 7. Adjustment of International Prices of Diesel to Farm-gate Level

S-Price	Quantities	An Giang	Can Tho	Overall
Tradable	Fertilizer			
	Urea (VND/kg)	2,505.22	2,505.22	2,505.22
	NPK (VND/kg)	2,750.46	2,750.46	2,750.46
	DAP (VND/kg)	3,382.06	3,382.06	3,382.06
	Other fertilizer (VND)	-	-	-
	Pesticide (VND)			
	Herbicide	-	-	-
	Fungicide	-	-	-
	Insecticide	-	-	-
	Other pesticide	-	-	-
	Seed (VND/kg)	6,870.44	6,870.44	6,870.44
	Fuel of Irrigation (VND/liters)	7,257.19	7,257.19	7,257.19
Factors	Labor (VND/day)			
	Hired labor	73,000.00	73,000.00	73,000.00
	Machinery (VND/day)			
	Land preparation	35,000.00	35,000.00	35,000.00
	Irrigation	-	83,000.00	83,000.00
	Harvest	80,000.00	80,000.00	80,000.00
	Other machinery (VND)	-	-	-
Output	(VND/kg)	5,725.37	5,725.37	5,725.37

Source: Own estimates, 2006; data appendix available from authors

Table 8. Social prices of soybean

Unit: VND/10a

Quantities	Private values			Social values		
	An Giang	Can Tho	Overall	An Giang	Can Tho	Overall
Tradable factors						
Fertilizer						
Urea	61,188	57,957	59,543	46,425	42,690	44,508
NPK	58,730	69,740	64,373	45,213	53,264	49,346
DAP	51,210	54,472	52,898	43,410	44,923	44,187
Other fertilizer	13,462	4,205	8,711	13,462	4,205	8,711
Pesticide						
Herbicide	45,665	22,231	33,637	45,665	22,231	33,637
Fungicide	17,754	6,887	12,177	17,754	6,887	12,177
Insecticide	151,778	122,718	136,862	151,778	122,718	136,862
Other pesticide	17,426	14,860	16,109	17,426	14,860	16,109
Seed	89,970	62,821	76,018	95,643	65,773	80,175
Fuel of Irrigation	24,327	14,944	19,589	34,018	19,285	26,456
Domestic factors						
Labor						
Hired labor	179,692	56,251	113,236	351,033	138,596	241,994
Machinery						
Land preparation	26,142	80,851	51,996	38,124	104,116	72,012
Irrigation	-	62,748	32,207	-	198,198	101,730
Harvest	199,272	161,483	178,500	364,102	333,103	348,191
Other machinery	364	8,298	4,402	364	8,298	4,402
Output						
Total Revenue	1,378,007	1,381,598	1,381,463	1,585,993	1,483,074	1,533,168
Total costs (excluding land)	936,978	800,467	860,258	1,264,416	1,179,150	1,220,495
Profit (excluding land)	441,029	581,131	521,205	321,578	303,925	312,672

Source: Own estimates; data appendix available from authors.

Table 9. Private and social revenues, costs and profits of soybean.

http://dx.doi.org/10.5772/51000

Getting the results from these above tables (Table 5, 6, 7), the full set of social prices for the illustrative soybean system is presented in the following Table 8.

After the calculation of private and social prices for tradable, non-tradable inputs and soybean, Table 9 shows a farm budget. The table is divided into two blocks. The first block records private prices, calculates the costs of inputs, and separates these costs into their tradable and non-tradable components. The second block is similar to the first block but all the values are calculated in social prices.

Unit: VND/10a

	Revenues	Costs		Profits
		Tradable	Factors	
An Giang				
Private	1,378,006.98	531,508.89	405,469.40	441,028.69
Social	1,585,993.28	510,793.46	753,622.20	321,577.62
Divergences	-207,986.30	20,715.43	-348,152.80	119,451.06
	DRC = 0.70; PCR = 0.48			
Can Tho				
Private	1,381,598.04	430,835.33	369,631.31	581,131.40
Social	1,483,074.31	396,837.71	782,311.92	303,924.67
Divergences	-101,476.27	33,997.61	-412,680.61	277,206.73
	DRC = 0.72; PCR = 0.39			
Overall				
Private	1,381,462.72	479,916.49	380,341.55	521,204.68
Social	1,533,167.61	452,166.02	768,329.20	312,672.40
Divergences	-151,704.89	27,750.48	-387,987.65	208,532.28
	DRC = 0.71; PCR = 0.42			

Source: Own estimates; data appendix available from authors.

Table 10. Results of the PAM analyses of soybean

The summary information from Table 9 is extracted to form a PAM of soybean production as shown in Table 10.

The principal determinant of transfers to farm production activity is the difference between world and domestic prices. The study shows that farmers received the private soybean price of 5,159VND per kilogram. The CIF price of soybean is 5,725VND per kilogram equivalent to a farm-gate social price (after converting to social costs and subtracting the social value of transport costs). For 10a of soybean, the private profit is 521,205VND, while the social profit is only 312,672VND.

5.3. Comparative and competitive advantage of the soybean farming system

The ability of an agricultural system to compete without distorting government policies can be strengthened or eroded by changes in economic conditions. Dynamic comparative advantage refers to shifts in competitiveness that occur over time because of changes in three categories of economic parameters – long-run world prices of tradable outputs and inputs, social opportunity costs of domestics factors of production (labor, capital and land), and production technologies used in farming or marketing. Collectively, these three parameters determine comparative advantage.

	Total Revenue	NPCO
An Giang		
Private	1,378,006.98	0.87
Social	1,585,993.28	
Divergences	-207,986.30	
Can Tho		
Private	1,381,598.04	0.93
Social	1,483,074.31	
Divergences	-101,476.27	
Overall		
Private	1,381,462.72	0.90
Social	1,533,167.61	
Divergences	-151,704.89	

Table 11. Output transfer of soybean farming system

Comparative advantage of an agricultural system, in the PAM table, is indicated by the value of the Domestic Resources Cost Ratio (DRC). The DRC serves as a proxy measure for social profits. Minimizing the DRC is equivalent to maximizing social profits. Comparative advantage is an indicator of potential advantage and will be fully received if there is no policy distortion in the system. If a commodity has comparative advantage, its production is economically efficient.

Based on information provided in Table 10, the DRC of soybean-farming system is 0.71. This result indicates that the soybean system has a comparative advantage. Growing Soybean in Can Tho is as efficient as that in An Giang because DRC of the two provinces are nearly the same. In other words, soybean production in An Giang has the same comparative advantage to that in Can Tho.

The determination of profit actually received by farmers is a straightforward and important initial result of the PAM approach. The results indicate which farmers are currently competitive. In the PAM table, the competitiveness of a system is measured by the private profitability (D) or Private Cost Ratio (PCR). Based on information given in Table 10, the PCR of soybean production is 0.42. This result indicates that soybean cultivation is profitable and thus competitive.

5.4. Transfers and impacts of government policies

In the Policy Analysis Matrix (PAM), impacts of government policies can be identified by the divergences identity in the third row of the PAM table. Divergences cause private prices to differ from their social counterparts. A divergence arises either because a distorting policy intervenes to cause a private market price to diverge from an efficient price or because underlying market forces have failed to provide an efficient price. Divergences in PAM can also be indicated by the ratio between the values in the first row (private prices) and the values in the second row (social prices). The ratio's indicators are more frequently used because of their ability to compare different systems producing unlike outputs.

	Tradable input costs							
	Urea	**NPK**	**DAP**	**Other fer.**	**Pes.**	**Seed**	**Fuel**	**Total**
An Giang								
Private	61,188	58,730	51,210	13,462	232,622	89,970	24,327	531,509
Social	46,425	45,213	43,410	13,462	232,622	95,626	34,018	510,777
Divergences	14,763	13,516	7,800	0	0	-5,656	-9,691	20,732
	NPCI = 1.04							
Can Tho								
Private	57,957	69,740	54,472	4,205	166,697	62,821	14,944	430,835
Social	42,690	53,264	44,923	4,205	166,697	65,762	19,285	396,826
Divergences	15,267	16,476	9,549	0	0	-2,941	-4,341	34,009
	NPCI = 1.09							
Overall								
Private	59,543	64,373	52,898	8,711	198,785	76,018	19,589	479,916
Social	44,508	49,346	44,187	8,711	198,785	80,161	26,456	452,152
Divergences	15,035	15,028	8,712	0	0	-4,142	-6,868	27,764
	NPCI = 1.06							

Table 12. Tradable input transfers of soybean farming system

Table 11 shows output transfers of soybean production. The ratio formed to measure output transfers is called the Nominal Protection Coefficient on Output (NPCO). The NPCOs of soybean in Can Tho and An Giang are slightly different. The NPCOs of Can Tho and An Giang are 0.93 and 0.87, respectively. Both values of NPCO are less than 1. This result indicates that soybean farmers received slightly lower prices than they would have received facing world prices or that systems are receiving very slight protection. The positive output transfers are caused mainly by indirect quantitative restriction (quotas) on soybean imports.

Moreover, the value of NPCI is 1.06. This result indicates that soybean farmers are taxed when they buy tradable inputs. Details of tradable input transfers of soybean production are presented in Table 12.

	Revenues	Tradable input costs	EPC
An Giang			
Private	1,378,006.98	531,508.89	0.79
Social	1,585,993.28	510,793.46	
Divergences	-207,986.30	20,715.43	
Can Tho			
Private	1,381,598.04	430,835.33	0.88
Social	1,483,074.31	396,837.71	
Divergences	-101,476.27	33,997.61	
Overall			
Private	1,381,462.72	479,916.49	0.83
Social	1,533,167.61	452,166.02	
Divergences	-151,704.89	27,750.48	

Source: Own estimates; data appendix available from authors.

Table 13. Effective Protection Coefficients for soybean farming system

The EPCs of the soybean farming systems are depicted in Table 13. Regarding the total effects of government intervention in the output of soybean and tradable input markets, the study estimates the value of EPC=0.83. It indicates that there is no subsidy of soybean production in the soybean output and tradable input markets from government policies. The costs or profits of soybean producers are 17 percent less than they would have been in the absence of policy on output and tradable inputs.

5.5. Sensitivity analysis of soybean production

The aim of sensitivity analysis in this section is to examine whether soybean production will have comparative advantage or not when the key factors vary and change in the future. The expected results will help answer the question "How sensitively are ratios of a PAM subject to the changes of the key factors?" In the study, based on the unstable market of inputs, and the changes of government policy, three possible scenarios are assumed as: the reduction of soybean tariff from 15 percent to 5 percent, an increase of 10 percent in fertilizer prices and in exchange rate of VND/US$ of 10 percent. The results of sensitivity analysis of a PAM are performed in Table 14.

- *The decrease of the soybean tariff:* Vietnam officially jointed the WTO in November 2006. In compliance with WTO rules, the import tariff for soybean has to decrease from 15 percent to 5 percent. Table 14 shows that when the soybean tariff is reduced from 15 percent to 5 percent, soybean production in Vietnam still has a comparative advantage since the DRC becomes 0.83 less than 1. Moreover, the NPCO = 1.01, indicating a neutral situation, meaning that there is almost no intervention of the Government in the soybean market.

	DRC	NPCI	NPCO	EPC
Basic scenario	0.71	1.06	0.9	0.83
A decrease from 15% to 5% in the soybean tariff	0.83	1.08	1.01	0.97
An increase of 10% in the fertilizer prices	0.72	1.07	0.9	0.83
An increase of 10% in the exchange rate	0.72	1.02	0.9	0.85

Source: Own estimates; data appendix available from authors.

Table 14. Sensitivity analysis of a PAM

- *The increase of fertilizer prices:* According to the annual statistics, the prices of fertilizer have been increasing steadily and it is likely that the upward trend will continue in the future. Thus, we assume that the price of fertilizers will go up about 10 percent in the future and then we investigate how the change of the PAM ratios is. The result shows that soybean farmers in Vietnam still have comparative ability when the prices of fertilizer increase 10 percent.

- *An increase in the exchange rate:* Recently, since the Vietnamese currency gradually has been losing value compared with the US$ currency, the study assumes the US$ currency increases about 10 percent compared with the currency of VND and considers what happens to the PAM ratios. The analysis indicates that there are no big changes for Vietnamese soybean in terms of competitiveness and soybean cultivation and that Vietnam would still have a comparative advantage when the exchange rate increases 10 percent.

6. Conclusion

Calculating the costs, income and profit of soybean production, we described and estimated briefly the current situation of soybean production in Vietnam. It revealed that with the average soybean yield of 268kg per 10a, farmers obtained income of 1.38 million VND. After excluding input costs such as fertilizers, pesticides, hired labors, machinery service etc., the soybean farm could receive a profit of 521,000VND. The financial ratios of soybean production showed if the farmer invested 1VND into soybean production, they could earn 1.6VND for income, thus receive a profit of 0.6VND. In addition, the ratio of profit to family labor indicated it was more profitable for Vietnamese soybean farmers to do their own farming than to work as hired laborers for other farmers. In other words, they could obtain the much higher amount of opportunity cost of around 107,000VND per day from soybean cultivation in comparison with the average hired labor of 35,000VND per day in the Vietnamese rural areas.

By applying the approach of policy analysis matrix (PAM) to analyze the competitiveness of soybean production in Vietnam, the study showed that soybean production had a comparative advantage since DRC was less than 1. Moreover, we determined that government policy had almost no positive impacts on the soybean farmers. It even reduced the competitiveness of Vietnamese soybean. Because PAM analysis could not capture the potential changes in prices and productivity, the results of scenarios in Table 14 are subject to changes in market conditions. Some sensitivity analyses were estimated to catch some potential changes. The simulation results showed that if any of the following forecasts became true, soybean production still had a comparative advantage: the decrease of soybean tariff from 15 percent to 5 percent, the increase of fertilizer prices of 10 percent, an increase of 10 percent in the exchange rate.

Author details

Huynh Viet Khai[1] and Mitsuyasu Yabe[2]

1 Can Tho University, Vietnam

2 Kyushu University, Japan

References

[1] Adesina, A. A., & Coulibaly, O. N. (1998). Policy and Competitiveness of Agroforestry-Based Technologies for Maize Production in Cameroon: An Application of Policy Analysis Matrix, *Agricultural Economics*, 19(1-2).

[2] Fang, C., & John, C. B. (2000). Food Seft-Suffciency, Comparative Advantage, and Agricultural Trade: A Policy Analysis Matrix for Chinese Agriculture Center for Agricultural and Rural Development Publications 99wp223, *Center for Agricultural and Rural Development (CARD) at Iowa State University.*

[3] Gotsch, C. G., Pearson, S. R., & Bahri, S. (2003). Computer Tutorial for Policy Analysis Matrix in Indonesia, *Agriculture Stanford University.*

[4] Harrigan, K., Loader, R., & Thirtle, C. (1992). Agricultural Price Policy: Government and the Market, *FAO, Rome.*

[5] Khai, H. V., Yabe, M., Yokogawa, H., & Sato, G. (2008). The Situation and Comparative Advantage of Soybean Production in the Mekong River Delta of Viet Nam, *Journal of the Faculty of Agriculture, Kyushu University*, 53(2).

[6] Kydd, J., Pearce, R., & Stockbridge, M. (1996). The Economic Analysis of Commodity Systems: Extending the Policy Analysis Matrix to Account for Environmental Effects and Transactions Costs, *Agricultural Systems*, 55(2).

[7] Mohanty, S., Fang, C., & Chaudhary, J. (2003). Assessing the Competitiveness of Indian Cotton Production: A Policy Analysis Matrix Approach, *Journal of Cotton Science*, 7(3).

[8] Monke, E. A., & Pearson, S. R. (1989). *The Policy Analysis Matrix for Agricultural Development Stanford University.*

[9] Nelson, G. C., & Panggabean, M. (1991). The Costs of Indonesian Sugar Policy: A Policy Analysis Matrix Approach, *American Journal of Agricultural Economics*, 73(3).

[10] Pearson, S. R., Gotsch, C. G., & Bahri, S. (2003). Applications of the Policy Analysis Matrix in Indonesian, *Agriculture Stanford Unversity.*

[11] Scarborough, V., & Kydd, J. G. (1992). Economic Analysis of Agricultural Markets: a Manual, *NRI Marketing Series, 5, Chatham.*

[12] Staal, S. J., Shapiro, B. I. (., & D., A. The Effects of Recent Price Liberalization on Kenyan Peri-Urban Dairy, *Food Policy*, 19(6).

Section 3

Soybean Agronomy and Physiology

Chapter 8

Molecular Design of Soybean Lipoxygenase Inhibitors Based on Natural Products

Isao Kubo, Tae Joung Ha and Kuniyoshi Shimizu

Additional information is available at the end of the chapter

http://dx.doi.org/10.5772/52703

1. Introduction

Lipoxygenases (EC 1.13.11.12) are non-heme iron-containing enzymes that catalyze the site-specific oxygenation of polyunsaturated fatty acids to produce hydroperoxides. Lipoxygenases are suggested to be involved in the early event of atherosclerosis by inducing plasma low-density lipoprotein (LDL) oxidation [1, 2]. Lipoxygenase inhibitors have also been suggested to be potential cancer chemopreventives [3,4]. On the other hand, lipid peroxidation is well known as one of the major factors in deterioration during the storage and processing of foods, because it can lead to the development of unpleasant rancid or off flavors as well as potentially toxic end products [5]. Hence, lipoxygenase inhibitors should have broad applications [6].

Lipoxygenase inhibitors act in different ways; by chelating the iron of the active site of the enzyme [7, 8] and/or by reducing the ferric form of the enzyme to an inactive ferrous form [9, 10] (Fig. 1). Resorcinol exhibits no effect on this enzyme, indicating that the pentadeca(en)yl tail portion seemed an essential element in eliciting the inhibitory activity. The available information now demonstrates that lipoxygenase inhibitors can be designed when appropriate head portions are selected. 5-Pentadeca(en)yl salicylic acids, commonly known as anacardic acids, were previously reported to inhibit the linoleic acid **(30)** peroxidation catalyzed by soybean lipoxygenase-1 [11, 12]. This inhibition comes from chelating the iron of the active site of the enzyme but not by reducing the ferric form of the enzyme to an inactive ferrous form. In our structure and activity relationships (SAR) studies with lipoxygenase inhibitors, lipoxygenase inhibitors can be designed mainly by selecting appropriate head portions (See chemical structures, Fig. 2). Based on this concept, the experiment has been extended to search for lipioxygenase inhibitors acting both by chelating the iron of the active site of the enzyme and by reducing the ferric form of the enzyme to an inactive ferrous form.

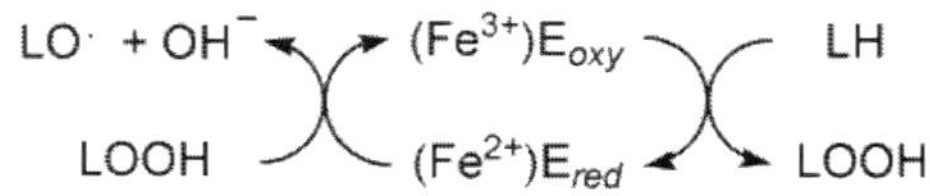

Figure 1. Commercial lipoxygenase contains a non-hemeferrous ion (E_{red}) that must be oxidized to yield the catalytically active ferric enzyme (E_{oxy}) and therefore a catalytic amount of LOOH (13-HPOD, **27**) is usually added as a cofactor to LH (linoleic acid, a substrate, **30**).

2. Results

To begin with, ethylenediaminetetraacetic acid (EDTA, **28**), a well known chelate agent was tested butit did not show any lipoxygenase inhibitory activity, indicating that the chelation ability alone is not enough in eliciting the inhibitory activity. For example, it appears that hydrophilic ligands with a longer alkyl chain length tend to be more potent inhibitors. The molecular lipophilicity has a critical impact on the ability of this class of chelator to inhibit lipoxygenase. Nordihydroguaiuretic acid (NDGA, **26**) is known to inhibit lipoxygenase by reducing the ferric form of the enzyme to an inactive ferrous form but not by chelating the iron of the active site of the enzyme. Hence, various phenolic acids were first tested for their radical scavenging activity using 1,1-diphenyl-2-picrylhydrazyl (DPPH, **29**). This activity can be measured as decolorizing activity following the trapping of the unpaired electron of DPPH. In general, DPPH radical scavenging activity does not take into account the reaction time, and hence, the activity was first observed to be 20 min and then extended to 24 h for comparison. In preliminary experiments, their hexyl, nonyl and dodecyl (lauryl) esters were tested to see if the alkyl group affects scavenging activity. All the esters tested regardless of their alkyl chain length, scavenged six molecules of DPPH. In addition, one molecule of alkyl protocatechuates, regardless of their alkyl chain length, scavenges six molecules of DPPH radical and the alkyl chain length is not associated with the scavenging activity. Thus, alkyl protocatechuatesare oxidized three times by DPPH within 20 min. The scavenging activity does not correlate with the hydrophobic alkyl chain length. This is consistent with the previous report with alkyl gallates [13, 14]. Hence, all the alkyl protocatechuates can be used as scavenging antioxidants and selection of the tail portion is flexible. Alkyl protocatechuates can be expected to inhibit lipoxygenase by chelating the iron of the active site of theenzyme and by reducing the ferric form of the enzyme to an inactive ferrous form. Both 2,3-dihydroxybenzoic acid (**24**) and 2,5-dihydroxybenzoic acid (**25**) are known to scavenge DPPH [15]. In addition to their hexyl, nonyl and dodecyl esters, the same esters of 3,4-dihydroxyphenyl alkanoates and 2,3-dihydroxyphenyl alkanoates were synthesized and tested for their effect on DPPH. All the esters tested, regardless of their alkyl chain length, were found to scavenge DPPH radicals. In a preliminary test, a series of alkyl 2,3-dihydroxybenzoates and 3,4-dihydroxybenzoates also scavenge DPPH.In addition, 2,3- and 3,4-dihydroxybenzoic acids(**24** and **23**) were previously reported to interact with Fe(II) and inhibit iron-induced oxidative DNA damage [16].

http://dx.doi.org/10.5772/52703

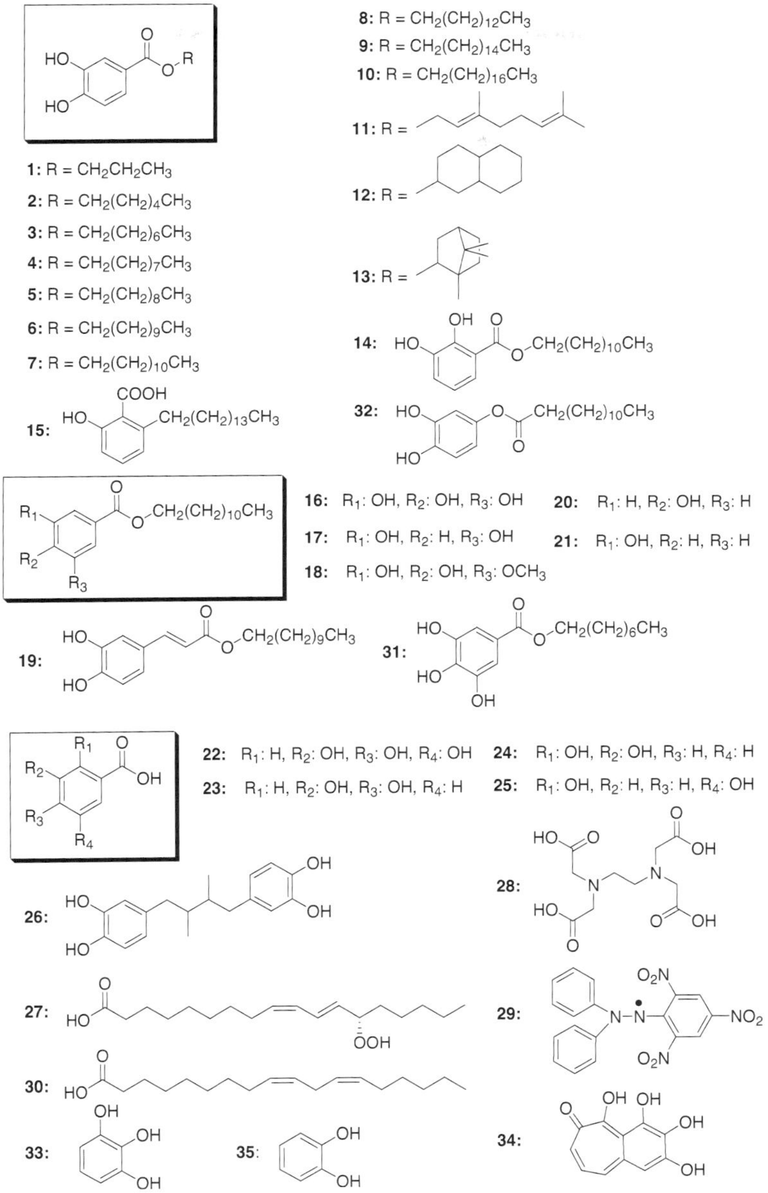

Figure 2. Chemical structures.

A homologous series of alkyl (C_1-C_{20}) protocatechuates (3,4-dihydroxybenzoates) and their related derivatives were previously synthesized by one-step esterification utilizing DCC (*N,N'*-dicyclohexylcarbodiimide) as an activating agent [17]. The head and tail structures are synthetically easily accessible as esters and therefore the construction of a wide range of structurally diverse multifunctional mimics are available for evaluation. Thus, there are unlimited choices from common natural products for both head and tail portions. In contrast to their DPPH scavenging activity, the lipoxygenase inhibitory activity of alkyl protocatechuates was found to correlate with their alkyl chain lengths and require a certain head portion. This indicates that lipoxygenase inhibitors can be designed when appropriate head portions are selected.

The synthesized alkyl protocatechuates (Fig. 2)were tested for their effects on the soybean lipoxygenase-1 catalyzed oxidation of linoleic acid (**30**). Results are listed in Table 1. It appears that the inhibitory activity of alkyl protocatechuates was a parabolic function of their lipophilicity and maximized with alkyl chain lengths between C_{12} and C_{14}. Among the compounds tested, tetradecanyl (C_{14}) protocatechuate (**8**) showed the most potent inhibition with the concentration leading to 50% activity loss (IC_{50})of 0.05μM, followed by dodecyl (C_{12}) protocatechuate (**7**) with an IC_{50} of 0.06μM. On the other hand, their parent compounds, protocatechuic acid did not show any activity up to 200 μM, indicating that the alkyl chain was an essential element in eliciting the inhibitory activity. It thus appears that hydrophilic ligands with a longer alkyl chain length up to C_{16} tend to be the more potent inhibitors. Since their head portions are the same, the data are interpreted to mean that changes in the hydrophobic tail portions correlated to the activity. Apparently a small change in chemical structure affects activity to a large extent.

Compounds	IC_{50} (μM)	Inhibition Type	K_i (μM)
1 (C_3)	39.4	Competitive	19.7
2 (C_6)	2.3	Competitive	0.98
3 (C_8)	0.55	Competitive	0.23
4 (C_9)	0.25	-	-
5 (C_{10})	0.15	-	-
6 (C_{11})	0.09	-	-
7 (C_{12})	0.06	Slow-binding	-
8 (C_{14})	0.05	-	-
9 (C_{16})	0.31	-	-
10 (C_{18})	0.42	-	-
11 (Geranyl)	0.32	-	-
12 (Decahydro-2-naphthyl)	0.1	Slow-binding	-
13 (Bornyl)	0.09	-	-
14	>100	-	-

- Not tested.

Table 1. IC_{50}andK_i values and inhibition type of alkyl protocatechuates against soybean lipoxygenase-1

The inhibitory activity of soybean lipoxygenase-1 was measured by two methods for comparison, because this enzyme seems to be sensitive to assay conditions. In the current study, linoleic acid (**30**) was used as a substrate. Soybean lipoxygenase-1 is known to catalyze the dioxygenation of the (1*Z*, 4*Z*)-pentadiene moiety of linoleic acid (**30**). In plants, the primary dioxygenation product is 13(*S*)-hydroperoxy-9*Z*,11*E*-octadienoic acid (13-HPOD, **27**) [18]. Hence, the enzyme assay was usually performed using a UV spectrophotometer to detect the increase at 234 nm associated with the (2*Z*, 4*E*)-conjugated double bonds newly formed in the product but not in the substrate. In previous reports, the data were obtained at pH 9 since soybean lipoxygenase-1 has its optimum activity at pH 9.0 [19], but the absorption at 234 nm suffered from unstable baseline activity of unknown origin attributable to the presence of the linoleic acid (**30**) substrate. This pseudoactivity of the blank control had to be substracted from activity of the enzyme assay, making precise measurements difficult. Moreover, this basic pH value may not be practical to use for food protection. Since the stable data were obtained at pH 8.0, the evaluation was performed at pH 8.0 [20]. This was also compared with that of NADG (**26**) used as a reference compound. Its IC_{50} was obtained as 82 μM when the experiment was performed at pH 9.0 but 0.2 μM at pH 8.0. As a result, octyl protocatechuate(3,4-dihydroxybenzoate) (**3**) showed a dose-dependent inhibitory effect on this oxidation as shown in Figure 3. As the concentration of octyl protocatechuate (**3**) increased, the enzyme activity was rapidly decreased with eventual complete suppression. The inhibitory concentration leading to 50% activity loss(IC_{50}) was estimated to be 0.55 μM. As the need arises, the assay was also monitored by using polarography (oxygen consumption) for comparison. The IC_{50} of octyl protocatechuate (**3**) obtained was 1.53 μM, which is almost comparable with that measured by the spectrophotometric method. The difference of IC_{50}values of these methods is due to a K_m for linoleic acid (**30**) almost 3-fold larger than using the oxygen monitor. This is in good agreement with previously reported observations [20, 21].

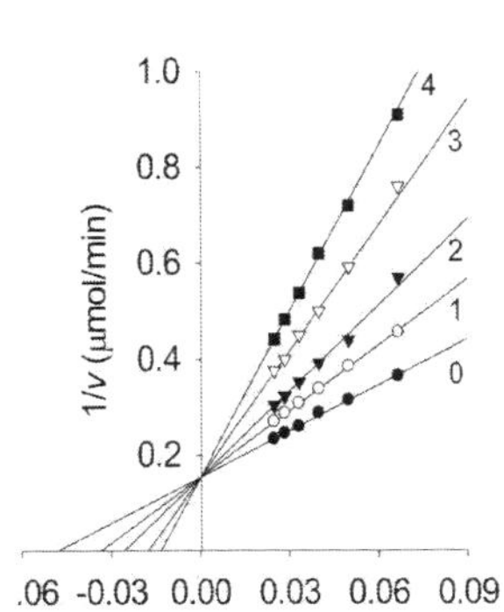

Figure 3. Lineweaver-Burk plots of 13-HPOD (**27**) generation by soybean lipoxygenase-1 in the presence of octyl protocathechuate (**3**) at 25 °C, pH 8.0. Concentrations of octyl protocathechuate (**3**) for curves 0-4 were 0, 0.1, 0.2, 0.4, and 0.6 μM, respectively.

Subsequently, the inhibition kinetics of soybean lipoxygenase-1 by octyl protocatechuate (**3**) was investigated as the representative molecule. The kinetic behavior of soybean lipoxygenase-1 during the oxidation of linoleic acid (**30**) was studied first. Under the conditions employed in the present investigation, the oxidation of linoleic acid (**30**) catalyzed by soybean lipoxygenase-1 follows Michaelis–Menten kinetics. The kinetic parameters for this oxidase obtained from a Lineweaver–Burk plot show that K_m is equal to 20.9μM and V_{max} is equal to 6.5μmol/min. The estimated value of K_m obtained with a spectrophotometric method is in good agreement with the previously reported value (20, 21). As illustrated in Figure 3, the inhibition kinetics analyzed by Lineweaver–Burk plots show that octyl protocatechuate (**3**) is a competitive inhibitor because increasing octyl protocatechuate (**3**) resulted in a family of lines with a common intercept on the $1/v$ axis but with different slopes. This may suggest that octyl protocatechuate (**3**) displaces linoleic acid (**30**) from the enzymatic site of oxidation. The equilibrium constant for inhibitor binding, K_I, was obtained from a plot of the apparent Michaelis–Menten constant versus the concentration of octyl protocatechuate (**3**), which is linear.

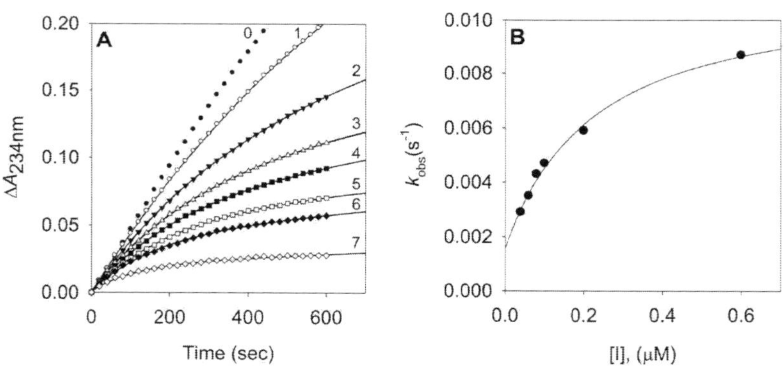

Figure 4. Time-dependent inhibition of soybean lipoxygenase-1 in the presence of dodecyl protocathechuate (**7**). (A) Conditions were as follows: 40 μM linoleic acid (**30**), 1.3 nM lipoxygenase-1, and concentrations of dodecyl protocathechuate (**7**) for curves 0-7 were 0, 0.02, 0.04, 0.06, 0.08, 0.1, 0.2, and 0.6 μM, respectively. The k_{obs} values at each inhibitor concentration were determined. (B) Dependence of the values for k_{obs} on the concentration of dodecyl protocathechuate (**7**). The k_{obs} values, determined on panel A

To investigate the inhibitory effect of dodecyl protocatechuate (**7**) on dioxygenase enzyme, we assayed soybean lipoxygenase-1 activity with the inhibitor. Soybean lipoxygenase-1 showed time-dependent inhibition in the presence of dodecyl protocatechuate (**7**) (Figure 4, panel A). Increasing dodecyl protocatechuate (**7**) concentrations led to the decrease in both the initial velocity (v_i) and the steady-state rate (v_s) in the progress curve. The progress curves obtained using various concentrations of the inhibitors were fitted to eq 1 to determine v_i, v_s, and k_{obs}.

$$A = v_s t + (v_i - v_s)[1 - \exp(-k_{obs}t)] / k_{obs} + A_0 \tag{1}$$

The plot for k_{obs} versus [I] are shown in panel B in Figure 4. That plot showed a hyperbolic dependence on the concentration of the dodecyl protocatechuate (**7**), so the inhibition of lipoxygenase-1 by dodecyl protocatechuate (**7**) followed mechanism A (Figure 5). The kinetic parameters, k_5, k_6, and K_i^{app} were derived from the plots by fitting the results to following eq 2.

$$k_{obs} = k_6 + [(k_5 \times [I]) / (K_i^{app} + [I]) \quad (2)$$

Thus, analysis of data yielded the following values: k_5 = 9.8x10^{-3} s^{-1}, k_6 = 1.5x10^{-3} s^{-1}, K_i^{app} = 0.23μM.The kinetic model can be written as:

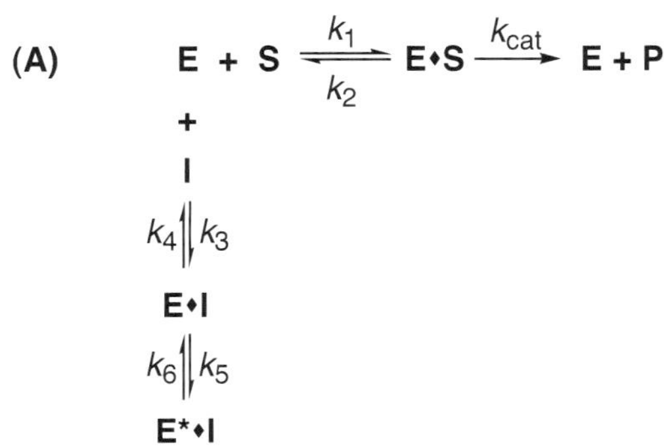

Figure 5. The kinetic model of mechanism A

where E, S, I, and P denote enzyme, substrate, inhibitor (dodecyl protocatechuate) (**7**) and product (13-HPOD, **27**), respectively. ES and EI are respective complexes. Because k_5 is greater than k_6 the enzyme first quickly and reversibly binds with dodecyl protocatechuate (**7**) and then undergoes a slow interaction of dodecyl group with the hydrophobic portion near the active site. In the case of octyl protocatechuate (**3**) [14], conversely k_6 is greater than k_5, and hence, it was observed as a competitive inhibitor. The effect of dodecyl protocatechuate (**7**) on the soybean lipoxygenase-1 catalyzed oxidation of linoleic acid (**30**) is similar to octyl protocatechuate (**3**) in many aspects but different to some extents.

Subsequently, the progress curves of 13-HPOD (**27**) generation show thatdecahydro-2-naphthylprotocatechuate (**12**) also inhibited soybean lipoxygenase-1 by slow-binding inhibition mechanism A (Figure 5). The k_{obs} values for the decahydro-2-napthyl protocatechuate (**12**) inhibition of lipoxygenase-1 at different concentrations of decahydro-2-napthyl protocatechuate (**12**) were determined by fitting data to the slow-binding equation (eq1). The k_{obs} values were plotted as a function ofdecahydro-2-napthyl protocatechuate (**12**) concentration. The results indicated that decahydro-2-napthyl protocatechuate (**12**) inhibits soybean lipoxygenase-1 by slow enzyme isomerization. This was evidenced by the observation that the k_{obs} values exhibited a hyperbolic dependence on the inhibitor concentration as shown in Figure

6B. Thus, analysis of data according to eq2 yielded the following values: k_5 = 8.8x10^{-3} s^{-1}, k_6 = 1.7x10^{-3} s^{-1}, K_i^{app} = 0.12μM.

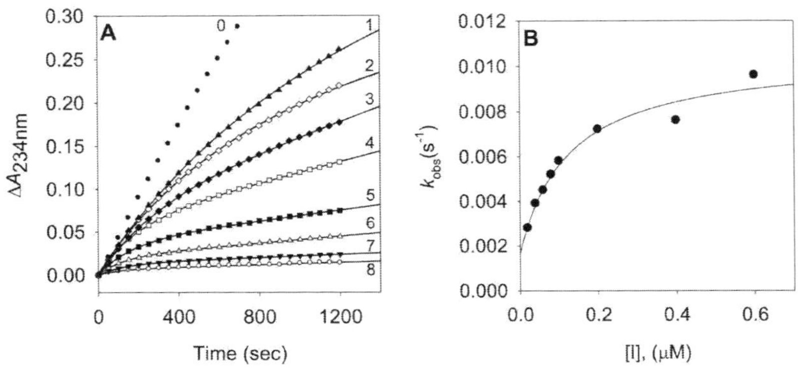

Figure 6. Time-dependent inhibition of soybean lipoxygenase-1 in the presence of decahydro-2-napthylprotocathechuate (**12**). Conditions were as follows: 40 μM linoleic acid, 1.3 nM lipoxygenase-1, and concentrations of decahydro-2-naphtyl protocathechuate for curves 0-8 were 0, 0.02, 0.04, 0.06, 0.08, 0.1, 0.2, 0.4, and 0.6 μM, respectively. The k_{obs} values at each inhibitor concentration were determined by fitting the data to eq1. (B) Dependence of the values for k_{obs} on the concentration of decahydro-2-naphtyl protocathechuate (**12**).

Soybean lipoxygenase-1 was used if dodecyl protocatechuate (**7**) inhibited the linoleic acid (**30**) peroxidation by this enzyme. The result is shown in Figure 4. Dodecyl protocatechuate (**7**) inhibited the linoleic acid (**30**) peroxidation 80% at 30 μg/mL (88.8 μM) and IC_{50} was established as 3.5 μg/mL (10.3 μM). Soybean lipoxygenase-1 inhibition mechanism of dodecyl protocatechuate (**7**) first competes with the active site, and then the hydrophobic dodecyl side chain interacts with the hydrophobic domain close to the active site in the enzyme. The inhibition of the enzyme by dodecyl protocatechuate (**7**) is a reversible reaction without residual activity. The hydrophobic interaction with the enzyme increases with increasing the alkyl chain length. The results imply that the inhibitory activity of soybean lipoxygenase-1 increases with increasing the alkyl chain length.

Some lipoxygenases reduce the ferric form of the enzyme to an inactive ferrous form [9,10]. This enzyme reduction mechanism is directly measurable by electron spin resonance(ESR)and fluorescence and also is indirectly observable when the inhibitor-dependent decrease in 13-HPOD (**27**) concentration, catalyzed by lipoxygenase-1, using the variation in A_{234} nm. Nordihydroguaiuretic acid (NDGA, **26**) is a well known soybean lipoxygenase-1 inhibitor. It acts by reducing the catalytically active ferric enzyme to the catalytically inactive ferrous enzyme [9]. As a result, Figure 7 demonstrates the fast decrease in $A_{234,}$ observed with addition of NDGA (**26**) in the pseudoperoxidase activity assay system. However, dodecyl protocatechuate (**7**) did not have a significant difference between controls. It should be noted that the inhibitor concentration used in this assay (10 μM) was 150-fold higher than the IC_{50} value for the inhibition of the lipoxygenase reaction. These results indicate that do-

http://dx.doi.org/10.5772/52703

decyl protocatechuate (**7**) did not convert the catalytically active ferric form to the inactive ferrous form, despite the fact that dodecyl protocatechuate (**7**) inhibition potencies are within the range of those of reducing inhibitors.

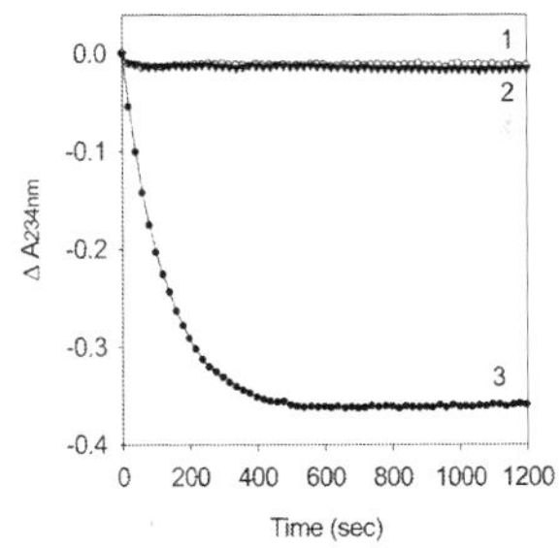

Figure 7. Effects of NDGA (**26**) and dodecyl protpcatechuate (**7**) on soybean lipoxygenase-1 catalyzed consumption of 13-HPOD (**27**). The conditions of incubation were 0 (1), 10 μM dodecyl protocatechuate (2), and 10 μM NDGA (3) with 13-HPOD (15 μM, **27**) at 25°C in Tris-HCl buffer, pH 8.0. Dodecyl protocatechuate (**7**) did not reduce the catalytically active ferric form to the inactive ferrous form.

Since the head portion of alkyl protocatechuates is the same, the data are interpreted to mean that changes in the hydrophobic tail portions correlate and are responsible for this specific activity. As far as protocatechuic acid (**23**) is concerned, it remains the head portion. Their lipoxygenase inhibitory activity can be designed by selecting a side chain length to give the appropriate partition coefficient (log *P*) as a standard and the maximum activity was observed to be around 6. However, it should be noted that other factors are also possibly associated with the activity. For example, the introduction of branching or unsaturation into the hydrophobic moiety is known to increase the solubility of the molecules in water and consequently may be associated with the activity [23]. Since geranyl protocatechuate (**11**), decahydro-2-naphthyl protocatechuate (**12**) and bornylprotocatechuate (**13**) showed potent inhibitory activity, the hydrophobic portion can be flexible. In brief, the hydrophilic head moiety requires a certain specific structural feature but the hydrophobic tail portion is flexible. As mentioned earlier, the head and tail structures are synthetically easily accessible as esters and therefore the design of a wide range of structurally diverse lipoxygenase inhibitors can be available for evaluation. Thus, there are unlimited choices from common natural products for both head and tail portions. The target molecules can be designed using their log *P* values as a standard. Based on this concept, bornylprotocatechuate (**13**) was synthesized by connecting two common natural products, protocatechuic acid (**23**) and borneol, as an example. Thus, bornylprotocatechuate (**13**) inhibited the lipoxygenase-catalyzed peroxidation of linolenic acid (**30**) with an IC_{50} of 0.09 μM.

Based on the head and tail concept, octylgallate(3,4,5-trihydroxybenzoate) (**31**) [13] and dodecyl gallate (**16**) [14] were found to be potent lipoxygenase inhibitors. Since alkyl gallates are known to have iron binding properties, their galloyl moiety initially and rapidly binds with the active site as a chelator and then the alkyl group undergoes a slow interaction with the hydrophobic environment. The same head and tail concept has been extended to evaluate octylprotocatechuate (**3**) and dodecyl protocatechuate (**7**), and both were found to show even slightly more potent inhibitory activity than that of the corresponding alkyl gallates. On the other hand, dodecyl 3,5-dihydroxybenzoate (**17**) which does not have an ability to chelate ironwas alsoevaluated for comparison. This resorcinol derivative was also found to inhibitthe enzymatic linoleic acid (**30**) peroxidation, but to a much lesser extent compared to dodecyl gallate (**16**) or dodecyl protocatechuate (**7**). It appears that the activity of dodecyl gallate (**16**), which has the ability to chelate iron, is conspicuously potent. Compared to an acardic acid ($C_{15:0}$) (**15**), dodecylgallate (**16**) is nearly 200-fold more potent (Table 2). In the case of alky gallates, the hydrophilic ligands with a longer alkyl side chain of up to C_{14} in length tend to be the more potent inhibitors [14]. Since their hydrophilic head portions are the same, the data are consequently interpreted to mean that changes in the hydrophobic tail portions correlated to the activity.On the one hand, dodecyl 3,4-dihydroxy-5-methoxybenzoate (**18**) still exhibited this inhibitory activity but to a much lesser extent compared to that of dodecyl gallate (**16**). This noticeable inferior activity can be explained by a steric hindrance to its approach to the active site in the enzyme, but dodecyl gallate (**16**) still reduces the ferric form of the enzyme to an inactive ferrous form. It would thus seem that a relatively small change in chemical structure of these molecules affects their biological activity to a large extent. The activity of alkyl 3,4-dihydroxybenzoates was compared with those of alkyl 2,3-dihydroxybenzoates, and thus the differences should be due to imply the hydrophilic head portions. Noticeably, dodecyl protocatechuate (**7**) showed potent lipoxygenase inhibitory activity with an IC_{50} value of 0.06 μM, whereas 3,4-dihydroxyphenyltridecanoate (**32**) also exhibited the inhibitory activity but to a much lesser extent. In these catechol (**35**) structures, the head and tail portions are connected as esters, either -COOR or -OCOR. Thus, the ester groups, either an electron withdrawing (-COOR) or an electron donating (-OCOR), were connected at the C-5 position to the catechol (**35**) moiety. The results obtained indicate that –COO Resters exhibited more potent inhibitory activity compared to the corresponding -OCOR esters, although the reason for this still remains obscure.

Compounds Tested	**IC_{50} (μM)**
Anacardic acid ($C_{15:0}$) **(15)**	14.3
Dodecyl 3,4,5-trihydroxybenzoate **(16)**	0.07
Dodecyl 3,5-dihydroxybenzoate **(17)**	60.3
Dodecyl 3,4-dihydroxy-5-methoxybenzoate **(18)**	10.5
Dodecyl 3,4-dihydroxycinnamate **(19)**	3.3
Dodecyl 4-hydroxybenzoate **(20)**	200
Dodecyl 3-hydroxybenzoate **(21)**	200

Table 2. Lipoxygenase-1 inhibitory related activity of selected compounds

The data obtained so far indicates that alkyl 3,4-dihydroxybenzoates can be expected as superior as lipoxygenase inhibitors similar to the alkyl 3,4,5-trihydroxybenzoates. It should be noted, however, that alkyl 3,4,5-trihydroxybenzoates were oxidized in part in basic condition. The lipoxygenase assay was performed using a UV spectrophotometer to detect the increase at 234 nm associated with the (2*Z*, 4*E*)-conjugated double bonds newly formed in the product but not the substrate. In previous reports, the data were obtained at pH 9 since soybean lipoxygenase-1 had its optimum pH at 9.0 [19]. Alkyl 3,4,5-trihydroxybenzoates were oxidized in part at pH 9 and therefore their data, obtained with the lipoxygenase inhibition,may not be their effects. Pyrogallol (1,2,3-benzentriol, **33**) was previously described to react with dioxygen in weakly alkaline solution to form purpurogallin (2,3,4,6-tetrahydroxy-5*H*-benzocycloheptene-5-one, **34**) [24]. Notably, alkyl gallates were oxidized in part at pH 9 but alkyl protocatechuates were not.

The data obtained so far indicates that alkyl 3,4-dihydroxybenzoates can be expected as superior as lipoxygenase inhibitors similar to the alkyl 3,4,5-trihydroxybenzoates. It should be noted, however, that alkyl 3,4,5-trihydroxybenzoates were oxidized in part in basic condition. The lipoxygenase assay was performed using a UV spectrophotometer to detect the increase at 234 nm associated with the (2*Z*, 4*E*)-conjugated double bonds newly formed in the product but not the substrate. In previous reports, the data were obtained at pH 9 since soybean lipoxygenase-1 had its optimum pH at 9.0 [19]. Alkyl 3,4,5-trihydroxybenzoates were oxidized in part at pH 9 and therefore their data obtained with the lipoxygenase inhibition may not be their effects. The data obtained so far indicates that alkyl 3,4-dihydroxybenzoates can be expected as superiorlipoxygenase inhibitors. Hence, a series of alkyl 3,4-dihydroxybenzoates was assayed in detail for their effects on the linoleic acid (**30**) peroxidation catalysed by the soybean lipoxygenase-1.

In contrast to their DPPH scavenging activity, the lipoxygenase inhibitory activity of alkyl protocatechuates was found to correlate with their alkyl chain lengths and require a certain head portion. This indicates that lipoxygenase inhibitors can be designed when appropriate head portions are selected. For example, dodecyl 3,5-dihydroxybenzoate (**17**) exhibited any lipoxygenase inhibitory activity up to 200 μg/mL.

3. Discussion

Alkyl protocatechuate antioxidants may act in a variety of ways including direct quenching of reactive oxygen species, inhibition of enzymes involved in the production of the reactive oxygen species, and chelation of low valent metal ions such as Fe^{2+}, Fe^{3+} or Cu^{2+}. In connection with this, both pyrogallol (**33**) and catechol (**35**) groups have iron binding properties *in vitro*. Alkyl gallates and protocatechuates may suppress the superoxide- driven Fenton reaction, which is currently believed to be the most important route to active oxygen species [25]. More specifically, alkyl gallates and protocatechuates may prevent cell damage induced by H_2O_2 since this can be converted to the more reactive oxygen species, hydroxy radicals, in the presence of these metal ions. Thus, metal chelation may play a large role in determining the antioxidant activity [26]. Interestingly, the inhibition of iron absorption *in vivo* has been positively correlated with the presence of galloyl groups but not catechol (**35**) groups [27].

Alkyl protocatechuates act as both lipoxygenase inhibitors and scavengers. Safety is a primary consideration for antioxidants in food products. After ingestion, the alkyl protocatechuates are likely hydrolyzed, at least in part, to protocatechuic acid and the corresponding alcohols that are common in edible plants. For example, pecan nuts contain protocatechuic acid as a one of predominant phenolic acids [28]. If alkyl protocatechuates can reach the sites where antioxidants are needed, a more lipophilic alkyl side chain may partition into lipophilic membranes of cells and organelles, where it presumably exerts its antioxidant activity, similar to the phytyl chain in tocopherols and tocotrienols [29]. The site of antioxidant location is known to be important, however, it is not clear if alkyl protocatechuates can reach, without being metabolized, the sites where antioxidants are needed for protection from oxidative damages. It should be noted, however, that the role of alkyl protocatechuates in the human body is unknown when orally ingested. It is not clear if alkyl protocatechuates are absorbed into the system through the intestinal tract and delivered to the places where lipoxygenase inhibitors are needed without being metabolized.

4. Materials and methods

Chemicals. A series of alkyl protocatechuates was available from our previous study [17].

Enzyme Assay. The assay was performed as previously reported [11]. The soybean lipoxygenase 1 (EC 1.13.11.12, Type I) used for the bioassay was purchased from Sigma Chemical Co. Linoleic acid (**30**) was used as a substrate. The reaction mixture consisted of 1.2 mM linoleic acid (**30**), 60 mM phosphate buffer (pH 7.0) and different concentrations of sample. At zero time, the enzyme solution (100 units) was added to the reaction mixture. The lipoxygenase activity was measured polarographically with an oxygen electrode(YSI 53, Yellow Springs Instrument Co., Yellow Springs, OH) at 25 °C. Calibration of a Clark-type oxygen electrode was performed by using 4-tert-butylcatechol.The native enzyme is quite capable of producing 13-HPOD (**27**) directly from linoleic acid (**30**) without any prior activation.

Acknowledgements

The work was presented in part at the Symposium of Diet and the Prevention of Gender Related Cancers in Division of Agricultural and Food Chemistry for the 222nd ACS National Meeting in Chicago, Il.

Abbreviations

LDL, low-density lipoprotein; SAR, structure and activity relationships; NDGA, nordihydroguaiuretic acid; DPPH, 1,1-diphenyl-2-picrylhydrazyl; EDTA, ethylenediaminetetraacetic acid; IC_{50}, the half maximal inhibitory concentration; DCC, *N,N'*-dicyclohexylcarbodiimide; gallic acid, 3,4,5-trihydroxybenzoic acid; gallate, 3,4,5-trihydroxybenzoate; protocatechuic acid, 3,4-dihydroxybenzoic acid; protocatechuates, 3,4-dihydroxybenzoate; 13-HPOD, 13(*S*)-hydroperoxy-9*Z*,11*E*-octadienoic acid; ESR, electron spin resonance; K_i, inhibition constant; K_m, Michaelis constant; V_{max}, maximum velocity; log *P*, partition coefficient

Author details

Isao Kubo, Tae Joung Ha and Kuniyoshi Shimizu

Department of Environmental Science, Policy and Management, University of California, Berkeley, California, USA

References

[1] Cornicelli JA, Trivedi BK. 15-Lipoxygenase and its Inhibition: a Novel Therapeutic Target for Vascular Disease. Current Pharmaceutical Design 1999;5(1) 11-20.

[2] Kris-Etherton PM, Keen CL. Evidence that the Antioxidant Flavonoids in Tea and Cocoa are Beneficial for Cardiovascular Health. Current Opinion in Lipidology. 2002;13(1):41-9.

[3] Steele VE, Holmes CA, Hawk ET, Kopelovich L, Lubet RA, Crowell JA, Sigman CC, Kelloff GJ. Lipoxygenase Inhibitors as Potential Cancer Chemopreventives. Cancer Epidemiology, Biomarkers & Prevention. 1999;8(5):467-83.

[4] Fischer SM & Klein RD. Lipoxygenases as Targets for Cancer Prevention. Totowa: Humana Press Inc; 2004

[5] Shibata D, Axelrod B. Plant Lipoxygenases. Journal of Lipid Mediators and Cell Signalling 1995; 12 (2-3): 213-228.

[6] Richard-Forget F, Gauillard F, Hugues M, Jean-Marc T, Boivin P, Nicolas J. Inhibition of Horse Bean and Germinated Barley Lipoxygenases by Some Phenolic Compounds. Journal of Food Science 1995; 60, 1325-1329.

[7] Clapp CH, Banerjee A, Rotenberg SA. Inhibition of Soybean Lipoxygenase 1 by n-Alkylhydroxylamines. Biochemistry 1985; 24, 1826-1830.

[8] Nelson MJ. Catecholate Complex of Ferric Soybean Lipoxygenase 1. Biochemistry 1988; 27, 4273-4278.

[9] Kemal C, Louis-Flamberg P, Krupinski-Olsen R, Shorter A. Reductive Inactivation of Soybean Lipoxygenase-1 by Catechols: a Possible Mechanism for Regulation of Lipoxygenase Activity. Biochemistry 1987; 26 (22) 7064–7072

[10] Mansuy D, Cucurou C, Biatry B, Battioni JP. Soybean Lipoxygenase-catalyzed Oxidations by Linoleic Acid Hydroperoxide: Different Reducing Substrates and Dehydrogenation of Phenidone and BW 755C. Biochemical and Biophysical Research Communications 1988; 151, 339-346.

[11] Ha TJ, Kubo I. Lipoxygenase Inhibitory Activity of Anacardic Acids. Journal of Agricultural and Food Chemistry 2005; 53, 4350-4354.

[12] Kubo I, Ha TJ, Shimizu K, Lipoxygenase Inhibitory Activity of 6-Pentadecanylsalicylic Acid without Prooxidant Effect. Natural Product Communications 2010; 5 (1) 85-90

[13] Ha TJ, Nihei K, Kubo I, Lipoxygenase Inhibitory Activity of Octyl Gallate. Journal of Agricultural and Food Chemistry 2004; 52, 3177-3181.

[14] Ha TJ, Kubo I. Slow-binding Inhibition of Soybean Lipoxygenase-1 by Dodecyl Gallate. Journal of Agricultural and Food Chemistry 2007; 55, 446-451.

[15] Sroka Z, Cisowski W. Hydrogen Peroxide Scavenging, Antioxidant and Anti-radical Activity of Some Phenolic Acids. Food and Chemical Toxicology 2003; 42, 753-758.

[16] Lodovici M, Guglielmi F, Meoni M, Dolara P. Effect of Natural Phenolic Acids on DNA Oxidation In Vitro. Food and Chemical Toxicology 2001; 39(12) 1205-1210

[17] Nihei K, Nihei A, Kubo I. Rational Design of Antimicrobial Agents: Antifungal Activity of Alk(en)yl Dihydroxybenzoates and Dihydroxyphenyl Alkanoates. Bioorganic & Medicinal Chemistry Letters 2003; 13, 3993-3996.

[18] Grechkin A. Recent Developments in Biochemistry of the Plant Lipoxygenase Pathway. Progress in Lipid Research 1998; 37(5) 317-52.

[19] Axelrod B, Cheesbrough TM, Laakso S. Lipoxygenase from Soybeans. Methods in Enzymology 1981; 71, 441-457.

[20] Berry H, Debat H, Larreta-Garde V. Excess Substrate Inhibition of Soybean Lipoxygenase-1 Is Mainly Oxygen-Dependent. FEBS Letters 1997;408(3) 324-6.

[21] Glickman MH, Klinman JP. Lipoxygenase Reaction Mechanism: Demonstration That Hydrogen Abstraction from Substrate Precedes Dioxygen Binding during Catalytic Turnover. Biochemistry, 1996, 35 (39) 12882–12892.

[22] Schilstra MJ, Veldink GA, Johannes JV, Vliegenthart FG. Effect of Lipid Hydroperoxide on Lipoxygenase Kinetics, Biochemistry 1992, 31 (33), 7692–7699

[23] Rosen MJ. Surfactants and Interfacial Phenomena, 2nd ed., New York: Wiley Interscience; pp. 1-32; 1989.

[24] Abrash HI, Elias DSW, Malekmehr F. A Kinetic Study of the Air Oxidation of Pyrogallol and Purpurogallin. International Journal of Chemical Kinetics 1989; 21(6) 465–476

[25] Afanas'ev IB, Dcrozhko AI, Brodskii AV, Kostyuk VA, Potapovitch AI. Chelating and Free Radical Scavenging Mechanisms of Inhibitory Action of Rutin and Quercetin in Lipid Peroxidation. Biochemical Pharmacology 1989; 38(11)1763-1769

[26] Arora A, Nair MG, Strasburg GM. Structure-activity Relationships for Antioxidant Activities of a Series of Flavonoids an a Liposomal System. Free Radical Biology and Medicine 1998; 24, 1355-1363.

[27] Brune M, Rossander L, Hallberg L. Iron Absorption and Phenolic Compound: Importance of Different Phenolic Structures. European Journal of Clinical Nutrition 1989; 43, 547-558.

[28] Senter SD, Horvat RJ, Forbus WRJ. Relation between Phenolic Acid Content and Stability of Pecans in Accelerated Storage. Journal of Food Science 1980; 45(5) 1380-1382.

[29] Burton GW, Joyce A, Ingold KU. Is Vitamin E the Only Lipid-soluble, Chain-breaking Antioxidant in Human Blood Plasma and Erythrocyte Membranes? Archives of Biochemistry and Biophysics 1983; 221(1) 281-90

Chapter 9

Challenges to Increased Soybean Production in Brazil

Hilton S. Pinto, Ana Maria H. de Avila and
Andrea O. Cardoso

Additional information is available at the end of the chapter

http://dx.doi.org/10.5772/52647

1. Introduction

Soybean (*Glycine max* (L.) Merril) is one the of the main sources of protein and vegetable oil of the world being grown commercially and used in animal and human feed for hundreds of years. Originally from Asia where the soybean was considered as sacred grain, next to rice, wheat, barley, millet to be essential to the stability of Chinese civilization.

According to EMBRAPA SOJA (2008) soybean is the main oleaginous grain cultivated in the world. In 2006/07 it contributed about 60% of the total 385 million tons of grain produced globally by major oil seeds (soybean, sunflower, canola, peanut, cotton and castor). According to Dall'agnoll and Hirakuri (2008) the high rates of increase in soybean world wide production (more than five million tons/year in the period 1970 to 2007) should continue due to the growth of the global population (70 million/year) and to the change in food consumption habits of the population.

On the other hand, the intense search for alternative sources of energy also stimulates consumption of soybean oil as a fuel. Thus, in the future soybean may also become a major energy crop. Although use rates are not increasing as much as for oil, demand for soybean as a livestock meal is also increasing. Since the early 1980sthe total consumption of soybean mealon a worldwide basishas grown about three times from the original 127.2 million tons. The expansion of income, especially in Asia, with the effect on the consumption of meats, especially poultryands wine, was the driver for the robust growth rate of soybean for the animal feed industry (Pinazza, 2007). Also according to Pinazza (2007), the long-term projections for soybeansin a global context shows that the total demand for soybeansis expected to grow around 3% in the next ten years. The crushing world will grow from the current level of 174.8 million tons to253.9 million tons.

Projections to 2015 indicate that from a world production of 71.5 million tons of soybeans, 66.4 million tons comes from the exporting countries Brazil, Argentina and the USA with 4.9 million from other countries. According to the Baseline Projections of the USDA, world GDP will continue growing at an annual average rate of 3.2%. At the same time the world population is expected to grow in the same period at an annual rate of 1.1%, to 6.87 billion people in 2015.

Soybean has enormous economical significance due to its high food value and its role in the vegetable oil industry as well in bio-diesel production. Leguminous plants are sources of various industrial enzymes, protein, fiber and unsaturated fat, and rich in vitamins and minerals, a nutritious meal with high food value. It has been explored worldwide for its inclusion in a variety of foods and is also an important constituent for animal feed. It is the source of various enzymes, particularly important in seed imbibitions and germination. Many of these enzymes have industrial significance and due to their plant origin are more acceptable, particularly by the food industry (Dwevedi and Kayastha, 2011).

According to FAPRI (2009), estimates for soybean grain indicate a Brazilian production of 86.5 million tons in 2020/2021. Soybean is expected to expand through a combination of initiating production in new areas where land is available and the substitution of soybean for crop land that is already in pasture land or other crops. The annual growth rate for soybean production land is expected to be 2.3% from now until 2020/2021 (Figure 1). This rate is close to the global rate for the next ten years, estimated by FAPRI (2009) at 2.30% per year. Domestic consumption of soybeans is expected to reach 45.6 million tons by 2020/2021, which represent 52.7% of Brazil's production. Consumption is projected to grow at an annual rate of 1.9%.

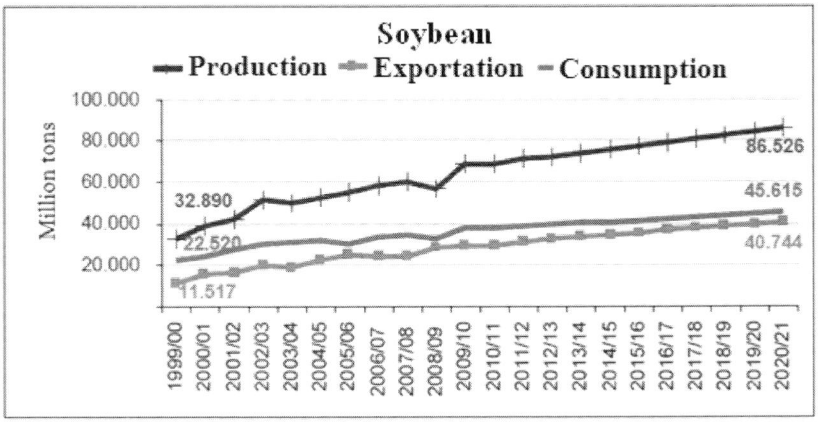

Figure 1. Production, exportation and consumption of soybean from Brazil (Source: Brasil Projeções do Agronegócio 2010/2021 – Ministério da Agricultura Pecuária e Abastecimento).

Considering plant breeding technology the productive potential of a cultivar is obtained only if the culture is under optimal environmental conditions. However, in natural conditions for farming, a number of biotic and non biotic factors acting individually or interactively may prevent optimum conditions from occurring. Therefore, soybean can be exposed to many different stresses. To avoid reductions in crop yields it is essential to use management practices ranging from the choice of the variety best suited to your region, the proper management of soil and crops, rationalizing the use of lime, fertilizer and pesticides, to knowledge and use of market information that will allow you to maximize profits in the marketing of the crop.

Brazilian agriculture has shown a very high performance from 1995 to 2011 and is considered the largest worldwide exporter of soybeans. It accounts for 39% of the total grain sold in the world. Brazil has potential to increase soybean production by conversion of up to 50 million hectares of degraded pasture land. Thus, biofuel production can be expanded without affecting either food production or the rainforest areas. Soybean is an ideal crop to restore fertility to degraded pastures in an integrated agriculture and livestock system due to the capacity to fix nitrogen from the air using bacteria in the root nodules.

Brazilian territory is located between the latitudes of 5º N and 33º S and consequently 93% of the country is located in the southern hemisphere. This region has a large variety of climates favorable to agricultural development that can be affected by global warming. A study done by Pinto et all, 2005 and 2008, based on the IPCC (2001, 2007) estimated that climate change would reduce the Brazilian soybean production area by 23% until 2020 unless measures were taken to develop adaptable varieties that are tolerant to drought and heat stress. The objective of this study is to predict the effects of climate change on the soybean industry of Brazil. The study follows the methodology used by the Ministries of Agriculture and Agrarian Development to outline production risks in different climatic zones of Brazil. Climatic risk zoning in the country is a public policy for farmers and establishes climatic risk levels for the different regions indicating optimal crops and planting dates for given locations.

Climatic risk zoning was determined for the period from 2030 through 2040. Based on climatic scenarios for given regions it was determined that in order to reduce crop risk, two main research areas needed to be addressed: adaptation of varieties through plant breeding research and improvement of the management system through new technologies in the field. The strategy is to develop plants adapted to high temperatures and water deficit. Cost estimates for the research project were based on an average period of 8-12 years for breeding and release of improved cultivars. The average total investment for the development of each cultivar was designed around US$ 5 million or an average cost of 1 million Reais a year (1 US$ = 2 Reais)

Assad et al. (2007a) stated that for soybean production at higher elevations, the size of low climatic risk areas will decrease as the mean temperatures and rainfall increase for the main planting period. In Brazil, the total area with soybean cultivations in 2010/2011 was 24,2 million hectares with a total potential for future production of 86 million tons in 2020, without considering any climate change (CONAB, 2011). The biggest soybean production area in Brazil is the municipality of Sorriso, in Mato Grasso State, located in the Midwest region. A description of current areas suitable and unsuitable for soybean production is given in Figure 2.

In Brazil, soybean is the crop that will probably suffer most from global warming. According to studies by Pinto et al. (2008), this will create a different geographic distribution for the crop (Fig. 3), if current varieties and cultural practices are not changed. Financial losses from a reduction in production area could reach R$7 billion/year by 2070.

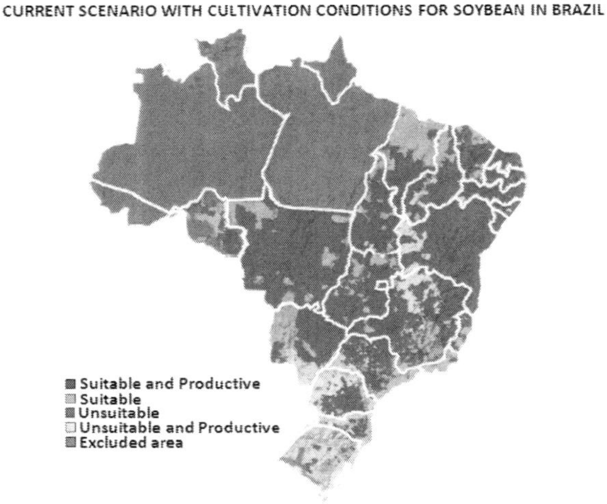

Figure 2. Geographical Distribution of soybean in Brazil (Source Pinto et al., 2008) showing the actual areas where the grain is already cultivated with low risk due to soil and climate conditions (blue), suitable but with no crop cultivation (green), unsuitable due to drought (red), not indicated by the risk zoning, but with soybean cultivation with low productivity (yellow) and not allowed due to permanent protection regions (brown).

By 2070 the low risk area for cultivation in Brazil is set to be 60% lower than that of today due to the decreased rainfall and the likelihood of more intense short summer droughts. The South of Brazil and the North-eastern Cerrado region will be affected more severely. The crop that currently has the highest production value in Brazilian agriculture – US$17 billion (according to 2010 data) and is also the country's main agricultural export product, may suffer a R$ 3.9 billion to R$ 4.3 billion annual loss by 2020 (in scenarios IPCC A2 and B2, respectively) caused by reduction in the low risk area of 21.62% and 23.59% for A2 and B2 scenarios, respectively. In 2050 the loss may increase to somewhere between R$ 5.47 billion (B2) and R$ 6.3 billion (A2), reflecting a reduction in low risk area of 29.6% to 34.1% compared with the 2010 production area. For 2070 in the more optimistic outlook the loss will be R$ 6.4 billion (- 34.86% of suitable land) and in the worst case scenario it will stand at R$ 7.6 billion (-41.39%).

http://dx.doi.org/10.5772/52647

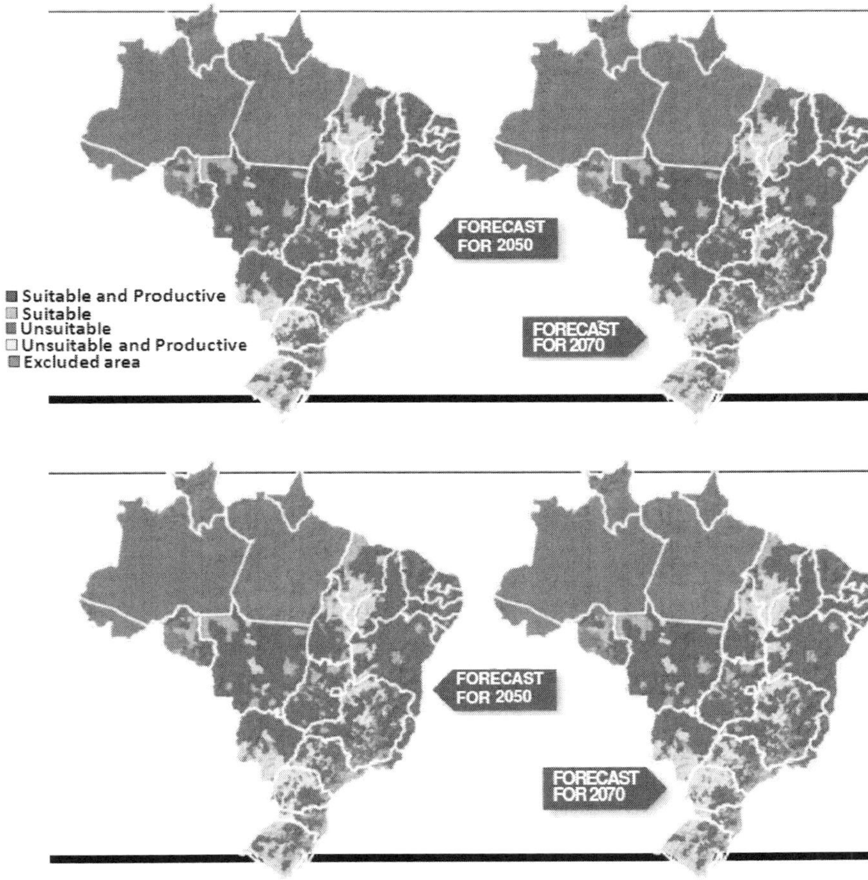

Figure 3. The geographical distribution of soybean in Brazil in 2050 and 2070 for the optimistic and pessimistic scenarios - B2 and A2 (Adapted from Pinto et al., 2008).

Despite the high vulnerability to climatic change detected in this study, agriculture has a great capacity of adaptation to new climatic conditions, depending on the technology available to deal with the problems. Assad et al. (2007b) highlighted the factors required for the adaptation of the main Brazilian crops: heat tolerance for the whole country, drought tolerance for the Southern and Northern regions and soil management to increase water conservation capacity. The biodiversity of the Cerrados and Amazon regions may contain genes that facilitate the adaptation of present crops to environmental stress and tolerate drought and heat. The substitution of crops or the use of more resistant varieties or species may be another method of adapting agriculture to climatic changes. Papers issued by Pinto et al.

(2005, 2007, 2008), Zullo et al. (2006) and Assad et al. (2004) indicate that the higher elevation regions with colder and milder climate may become low climate risk areas for most part of crop production as temperatures increase with global warming.

According to Pinto and Assad (2008) the worst prediction for soybean in 2020, considering the climate scenario A2 made by the IPCC (2001) showed a 24% reduction in the low risk area for soybean cultivation in Brazil. A more recent study (Pinto and Assad, 2012) using a combination of 23 Global Climate Models (GCMs) indicated by AR4 WGI Report IPCC (2007) and three Regional Climate Models (RCMs) - PRECIS (Jones et al. 2004), ETA (Pisnichenko and Tarasova 2009) and BRAMS (Freitas et al. 2009) showed a decrease in the low risk area of the country also of 24% in the pessimistic scenario and of 13% in the optimistic one. The Brazilian government estimates a production of 86 million tons of soybean in 2020 without taking into account the reduction of low risk areas due to climate change. In this case, close to 20 million tons will be lost if the temperature continues to increase, or, in other words, at actual prices of US $500/ton, Brazil will lose close to US$ 10 billion/year.

However, at the same time, scientists are developing new soybean genotypes and cultivars which are more tolerant to high temperatures and droughts. The studies are at an advanced stage, but it should be underlined that, even if they result in more resistant plants, there is a limit to how far genetic improvement can go. The alterations are capable of handling the problem with up to a 2°C temperature rise. Above that mark, the plants begin to have difficulties in photosynthesizing, demanding the adoption of other measures. One alternative to this limitation could be the so-called second generation crops of Genetically Modified Organism (GMO).Instead of just being herbicide-tolerant or pest-resistant, these would be more suited to severe environmental conditions.

This proposal aims to find out plants that are naturally more tolerant for high temperatures and water shortages and use their genes to produce more resistant farm crops. Embrapa Cerrados is analyzing typical species of the biome which are more adapted to the characteristic variations in temperature and rainfall in the region.

The researchers have already identified five plants (broadleaf pauterra, small-leaf pau-terra, pacari, faveiro and sucupirapreta) which occur in over 80% of the biome, suggesting a high adaptive capacity. The next step is to isolate the genes which give them these characteristics. Similar studies are also being planned for the Caatinga region. The value of the biodiversity of the two biomes is yet another argument for the prevention of its deforestation.

Plant breeding research developed by Embrapa Soja at Londrina, Parana has already showed a prospective tolerant cultivar (Figure 4). Using the same water content in the soil (2.5%) the modified cultivar shows a normal development when compared to a normal commercial soybean.

The general purpose of this study is to show the current potential of national agriculture to mitigate some problems related to global warming and to adapt the soybean research program accordingly. The purpose is to aid researchers in avoiding the potential problems that could occur as global warming continues to increase at its current pace.

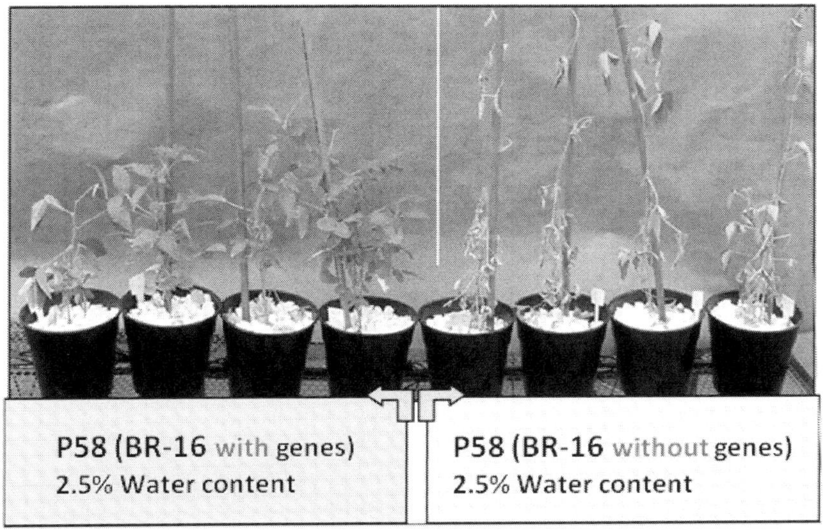

Figure 4. Gene expression drought-tolerance in soybeans (Source: Nepomuceno, A. L.-Embrapa Soja).

List of acronyms

BRAMS: Brazilian Developments on the Regional Atmospheric Modelling System)

CONAB: National Supply Company

EMBRAPA: Brazilian Agricultural Research Corporation

ETA: Greek Letter - Weather Forecast Model

FAPRI: Food and Agricultural Policy Research Institute

GCMs: Global Climate Models

GDP: Gross Domestic Product

GMO: Genetically Modified Organisms

IPCC: Intergovernmental Panel on Climate Change

PRECIS: Providing Regional Climates for Impacts Studies

RDMs: Regional Climate Models

WGI: Working Group I

Author details

Hilton S. Pinto[1], Ana Maria H. de Avila[1] and Andrea O. Cardoso[2]

1 CEPAGRI, UNICAMP, Campinas – SP, Brazil

2 UFABC, Universidade Federal do ABC, Santo Andre, SP, Brazil

References

[1] Assad, E. D., Pinto, H. S., Zullo, J., Jr , , Ávila, e., & de , A. M. H. (2004). Impacto das MudançasClimáticas no ZoneamentoAgroclimático do Café no Brasil. Pesq.Agrop.Bras.PAB.39(11)., 1057-1064.

[2] Assad, E. D., Pinto, H. S., Zullo, J., & Jr 2007a, . (2007a). Impacts of Global Warming in the Brazilian Agroclimatic Risk Zoning. In: Silva Dias, P. L. S.; Ribeiro, W. C.; e Nunes, L. H. A Contribution to Understanding the Regional Impacts of Global Change in South America.IEA/USP. São Paulo, SP. http://www.iea.usp.br/iea/artigos/globalchangeinsouthamerica.pd, 175-182.

[3] Assad, E. D., Pinto, H. S., Zullo, J. R. e., & Marin, F. R. (2007). b MudançasClimáticas e Agricultura: umaAbordagemAgroclimatológica. Ciência e Ambiente 34. , 169-182.

[4] CONAB,(2011). Sojaemnúmeros (safra 2010/2011).http://www.cnpso.embrapa.br/index.php?cod_pai=2&op_page=294 Accessed in 13 Jul 2012.

[5] Dall'agnol, A., & Hirakuri, M. H. (2009). Realidade e Perspectivas do BrasilnaProdução de Alimentos e Agroenergia, com ÊnfasenaSoja.Acessed in 12 march. 2012: http://www.redeagroenergia.cnpm.embrapa.br.

[6] Alka, Dwevedi., Arvind, M., & Kayastha, (2011. (2011). Soybean: a Multifaceted Legume with Enormous Economic Capabilities, Soybean- Biochemistry, Chemistry and Physiology, Prof. Tzi-Bun Ng (Ed.), 978-9-53307-219-7InTech, Available from: http://www.intechopen.com/books/soybean-biochemistry-chemistry-and-physiology/soybean-a-multifaceted-legume-with-enormous-economic-capabilities

[7] EMBRAPA SOJA.(2008). EmpresaBrasileira de PesquisaAgropecuária.Disponívelem: http://www.cnpso.embrapa.br.Accessed in: 16 march. 2012.

[8] FAPRI.(2009). U.S and world Agricultural Outlook. FAPRI Staff Report FSR 1 1534-4533Food and Agricultural Policy Research Institute- Iowa State University University of Missouri-Columbia Ames, Iowa. U.S.A., 09.

[9] Freitas, S. R., Longo, K. M., Silva, Dias. M. A. F., Chatfield, R., Silva, Dias. P., Artaxo, P., Andrea, M. O., Grell, G., Rodrigues, L. F., Fazenda, A., & Panetta, J. (2009). The Coupled Aerosol and Tracer Transport Model to the Brazilian Development on the

Regional Atmospheric Modeling System (CATT-BRAMS). Part 1: Model Description and Evaluation. Atmospheric Chemistry and Physics 9., 2843-2861.

[10] Hirakuri, M. H. Estimativa de custo de produção e lucratividade da soja, safra (2008/09). para o Paraná e Santa Catarina.Londrina: EmbrapaSoja, 2008. 15 p. (EmbrapaSoja.Circular Técnica, 65).

[11] IPCC.(2001). Third Assessment Report. Climate Change 2001.The Physical Science Basis.Summary for Policymakers.

[12] IPCC.(2007). Fourth Assessment Report. Climate Change.The Physical Science Basis.Summary for Policymakers.

[13] Jones, R. G., Noguer, M., Hassell, D. C., Hudson, D., Wilson, S. S., Jenkins, G. J., & Mitchell, J. F. B. (2004). Generating Hight Resolution Climate Change Scenarios Using PRECIS.Met Office Hadley Centre. Exeter, UK, 40 pp.

[14] MAPA- Ministério da Agricultura, Pecuária e Abastecimento.(2007). Secretaria de Política Agrícola. Instituto Interamericano de Cooperação para a Agricultura- IICA. Série Agronegócios. Cadeia Produtiva da Soja. , 2

[15] MAPA- Ministério da Agricultura, Pecuária e Abastecimento- Assessoria de GestãoEstratégica.(2011). Brasilprojeções do Agronegócio 2010/2011 a 2020/2021.

[16] Pinazza, L. A. ., & Coord, . Cadeia produtiva da soja. Brasília, DF: Ministério da Agricultura, Pecuária e Abastecimento: IICA, (2007). p (MAPA.Agronegócios, , 1

[17] Pinto, H. S., Assad, E. D., & 201, . (2012). Impacts of Climate Change on Brazilian Agriculture. In: Brazil: Assessment of the Vulnerability and Impacts of Climate Change on Brazilian Agriculture. Development report for World Bank Project Coord. Erick C.M. Fernandes. May, 25- 2012., 118037.

[18] Pinto, H. S. e., & Assad, E. D. (2008). Global Warming and The New Geography of Agricultural Production in Brazil. The British Embassy. Brasília, DF. 42 pp. Disponívelem: http://www.cpa.unicamp.br/aquecimento_agricola_en.html.

[19] Pinto, H. S., Zullo, J. R. J., Assad, E. D. e., Avila, A. M. H., & de , . (2005). th International Congress of Biometeorology. Annalen der Meteorologie, , 1, 223-226.

[20] Pinto, H. S., Zullo, J. J. R., Assad, E. D. e., & Evangelista, B. A. (2007). O Aquecimento Global e a CafeiculturaBrasileira. Boletim da SociedadeBrasileira de Meteorologia., 65-72.

[21] Pisnichenko, I. A., Tarasova, T. A., & 200, . (2009). Climate Version of the ETA Regional Forecast Model.TheorApplClimatol. Spring-Verlag. 18 pp. Published online 05 May, 2009.

[22] Zullo, J. R. J., Pinto, H. S. e., & Assad, E. D. (2006). Impact Assessment Study of Climate Change on Agricultural Zoning.MeteorolAppl 13(S1)., 69-80.

Chapter 10

Drought Stress and Tolerance in Soybean

Yee-Shan Ku, Wan-Kin Au-Yeung, Yuk-Lin Yung,
Man-Wah Li, Chao-Qing Wen, Xueyi Liu and
Hon-Ming Lam

Additional information is available at the end of the chapter

http://dx.doi.org/10.5772/52945

1. Introduction

Agriculture requires ~70% of the total fresh water resources [1]. Agricultural drought refers to the shortage of precipitation that causes deficit in soil water and reduction of ground water or reservoir levels, which will hamper farming and crop production [2]. To assess the severity of agricultural drought, different indices based on various parameters have been adopted. An expert meeting of the World Meteorological Organization was held in 2010 in Geneva to discuss and categorize the recently in-use indices into 7 types (Table. 1) [3]. With the establishment of networks under the cooperation of authorities at different administrative levels, one purpose is to develop monitoring tools and early warning systems for droughts. Nevertheless, a universal agricultural index satisfying all common interests has not yet arrived.

The effects of drought on soybean have been extensively reported, including morphological changes of the vegetative plant and the reduction in seed quantity and quality. Methods for assessing both quantitative and qualitative morphological parameters have been reported.

Tremendous efforts have been placed on the enhancement of drought tolerance of soybean, with a primary goal of enhancing yield under drought. Traditional breeding is a widely accepted strategy which will combine desirable agronomic traits from soybean germplasms, via repeated crossing and selection processes. The recent advances in genomics, genetics, and molecular biology facilitate the identification of molecular markers and functional genes that are related to drought tolerance in soybean. Therefore, the ideas of enhancing drought tolerance by marker-assisted breeding and genetic modification have gained growing attention.

Types of indices	Examples	Advantages
Precipitation-based	Standardized Precipitation Index; Days without Rainfall	Simple; required data can be easily accessed
Temperature-based	Cold Spell Duration Index; Warm Spell Duration Index	Straightforward monitoring of trends in the frequency or intensity of events
Precipitation- and temperature-based	Palmer Drought Severity Index; Standardized Precipitation Evapotranspiration Index	Simple to apply; more comprehensive
Precipitation, temperature, and soil moisture/ soil characteristics based	Relative Soil Moisture; Accumulated Drought Index	Better understanding about the interaction between crops and the environment during drought; help determine the influence of drought on the crop growth and yield
Temperature, relative humidity, solar radiation, wind speed, and soil moisture/ soil characteristics based	Aridity Anomaly Index; Soil Moisture Anomaly and Relative Soil Moisture Index	Fully consider both crop and soil water supply and demand
Remote sensing-based	Normalized Difference Vegetation Index; Temperature Condition Index	Required spectral data from instruments are easily available for most parts of the world; can monitor vegetation conditions
Composite	Utilize and incorporate a consolidation of indices into a comprehensive one	Comprehensive and flexible for monitoring different types of drought in different places

Table 1. Types of commonly used drought indices.

In addition to the genetic improvement programs for soybean, agronomic practices aimed at minimizing water input, reducing water loss, and increasing plant water usage efficiency have also been developed to cope with the problem of water scarcity. Some of these can be applied for soybean cultivation.

In this chapter, we will summarize the understandings of drought stress and drought tolerance in soybean from available literatures. We have integrated information from traditional breeding and agronomic measures to molecular aspects of this subject, and highlighted unsolved problems and possible strategies to tackle them.

2. Effects of drought on soybean production

Soybean is among the top 10 of the most widely grown crops, with a total production of over 260 million tonnes in 2010 (FAO data). The cultivated area of soybean occupies more than 100 million hectares worldwide, with about half in the U.S.A. and Brazil (FAO data). Greenhouse and field studies showed that drought stress led to significant reduction in seed yield (24~50%) from distinct locations and time [4, 5].

Numerous efforts have been put to examine the effects of drought and irrigation at various vegetative stages on soybean production. A 2-year field experiment by Brown et al (1985) on 4 determinate cultivars Davis, Lee 74, Sohoma and Centennial demonstrated that mositure stress initiated at R2 or R4 reduced yield significantly [6].

An in-depth analysis of the effects of drought at various growth stages on seed yield of soybean cultivar Douglas was reported by Eck et al (1987) [7]. In their study, yield loss was the most severe when drought stress was applied throughout the seed development period (R5-R7), resulting in a reduction of 45% and 88% respectively in two consecutive years [7]. Besides, Desclaux et al (2000) conducted a comprehensive analysis of yield components when drought stress was applied to soybean cultivar Weber at different developmental stages [8]. In this experiment, the stress condition was attained by temporally withholding irrigation for 4 to 5 days until the plant available water reduced to 50% or 30% of the normal conditions. The major findings showing various adverse effects of drought were summarized in Table 2. The most severe effect of this treatment was observed during the seed filling period [8].

On the other hand, Korte et al (1983) conducted a 3-year study on 8 soybean cultivars to assess yield enhancement by irrigation, using non-irrigated soybean plants as the control group [9]. The experimental groups were irrigated at different developmental stages (one stage or different stages in combination), including the flowering stage (R1-R2), the pod elongation stage (R3-R5), and the seed enlargement stage (R5-R6) [9]. Results of factorial analysis indicated that the yield was sensitive to the enhancement by irrigation, at pod elongation stage (R3-R4) and the seed enlargement stage (R5-R6) [9]. For 5 cultivars, the enhancement effect by irrigation followed the order: seed enlargement stage (R5-R6) > pod elongation stage (R3-R4) > flowering (R1-R2) [9]. A separate experiment by Kadhem et al (1985) supported the sensitivity toward irrigation at the pod elongation stage in determinate cultivars (R3.7 and R4.7) [10].

Traits			Growth stage with drought stress application			
Vegetative	During Flowering	Pod Lengthening	Seed Filling			
Dry matter value	Main-stem height	Internode length	*	*		
		Number of nodes	*			
	Stem diameter		*			
	Leaf surface area		*			
Yield components	Pods per dry matter				*	
	Seeds per pod					*
	Individual seed weight				*	*

* indicates significant effect of drought on the trait. Growth stages were characterized by the Fehr and Caviness scale [11]. Experiments were carried on the indeterminate cultivar, Weber.

Table 2. Effects of drought at different developmental stages on different agronomic traits (Modified from [8])

It has been clearly demonstrated that water availability will affect seed yield, though the growth stages that are most sensitive to drought stress vary among reports on different cultivars. In contrast, there are controversial reports on the effects of drought on soybean seed quality. Germination rate is a crucial criterion for assessing seed quality. A 2-year field study conducted on 3 soybean cultivars of Maturity Group (MG) IV, V and VI respectively in the southern U.S.A. reported a reduction of seed germination to less than 80% of the control, when drought stress was applied at any of the tested reproductive stages [12]. This observation is supported by a greenhouse experiment reporting that the germination rate was reduced in medium seeds from plants subjected to drought during seed filling period [13].

On the contrary, in a greenhouse experiment using the cultivar Gnome [14], drought stress led to a reduction of seed yield mainly due to the reduction of seed number. Nevertheless, there were only slight reductions in standard germination percentage and seedling axis dry weight of the harvested seeds. The authors suggested that drought stress affects the seed yield to a larger extend than seed quality. This result is supported by a separate experiment using other determinate and indeterminate cultivars (Essex, Union, Harper and McCall), in which drought did not result in production of seeds with reduced germination rate or vigor, except for those shriveled, flat, and underdeveloped seeds [15].

The study by Dornbos and Mullen (1991) further showed that the effect of drought on the germination rate of seeds from stressed plants was more significant when the air temperature reached 35°C. The authors also reported an increase in the percentage of hard seeds with increased duration of drought stress, and a negative relationship between seed weight and the percentage of hard seeds [16]. Hard seeds possess impermeable seed coats that will impede germination. In conclusion, drought clearly affects seed quality on some soybean cultivars. However, the discrepancies among different reports suggest that such effects are not universal to all cultivars under different stress conditions.

The contents of seed protein and oil are major parameters determining the nutritional value of soybean. Soybean seed protein content in general is negatively correlated with the amount of seed oil [17]. A differential irrigation experiment performed on soybean cultivars Gnome and Hodgson 78 in a greenhouse setting reported a 4.4% increase in protein content and 2.6% decrease in oil content under severe drought [18]. Furthermore, a 6-year field experiment was conducted using 60 soybean cultivars and breeding lines (Figure 1). The results confirmed both the negative correlation between seed protein and seed oil contents as well as the effect of drought on seed protein and seed oil contents [19]. The variations in contents of seed protein and oil were attributed largely to the differential rainfall during the seed filling stage [19].

Soybean seeds are also rich in isoflavones, a group of secondary metabolites exhibiting estrogenic, antifungal, and antibacterial activities [20]. The level of isoflavones is affected by drought during seed development [21]. While drought stress reduced the total content of isoflavones in soybean seeds under 28°C and 700 ppm CO_2, an increase was observed when the drought stress was applied at 23°C and 700 ppm CO_2 [21]. The results implied that the isoflavone content in soybean seeds is responsive to drought but also to other environmental factors including temperature and CO_2 level.

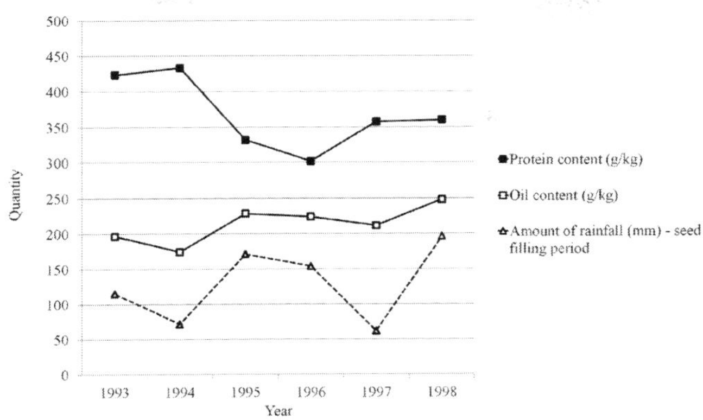

Figure 1. Variations of the average seed protein and oil contents of 60 soybean germplasms and breeding lines, and the amount of rainfall at the experimental field during the reporting year (based on data from [19]).

3. Parameters for measuring the degree of drought stress in soybean

3.1. Parameters related to seed

Seed weight can be evaluated using 100-seed weight or seed weight distribution. To eliminate the effects of the large measurement errors on the weight of a single seed, the weights of batches of 100 seeds are measured instead. Despite the general decreasing trend of seed weight under drought, the seed weight may not reduce uniformly as a function of drought intensity [16]. Therefore, seed weight distribution has become another parameter employed to evaluate the effect of drought on seed weight, through the assessment of weight of seeds of different sizes. Dornbos and Mullen (1991) reported that under severe drought, the proportion of seeds of diameter larger than 4.8mm was reduced by 30%-40% while the proportion of seeds of diameter smaller than 3.2mm was increased by 3%-15%. Under drought, soybean plants continued to produce heavy seeds. However, a greater portion of seeds were of low weight [16].

3.2. Parameters related to vegetative tissues

Drought stress reduced the number of nodes which is a result due to the reduction of main stem height and the decreased node emergence rate [8]. Length of internode is also a parameter for evaluating drought stress. However, the change in internode length is dependent on the timing of drought. In the experiment reported by Desclaux et al (2000), only the internodes which initiated during drought stress showed reduction in length [8].

Reduction in leaf area is a convenient morphological parameter for measuring drought stress experienced by the plant. Commercially available leaf area meters provide a non-destructive means to measure leaf area in the field. Alternately, the area of detached leaves can be measured simply by creating a digital image of the leaf using desk-top scanners followed by image analysis by computer software [22].

Drought stress also leads to a reduction in leaf relative growth rate [23], which can be calculated using the following formula:

$$RGR = \ln(FDW) - \ln(IDW) / (t_2 - t_1)$$

where, FDW refers to the final dry weight; IDW refers to the initial dry weight; t_2 refers to the time in days at the end of the experiment; t_1 refers to the time in days at the beginning of the experiment.

The degree of chlorophyll reduction in soybean leaves was correlated with the strength of drought treatments [24]. Chlorophyll can be simply extracted by immersing the plant tissue in N,N-dimethylformamide (DMF) [25]. After incubation and mixing, the DMF is subjected to OD determination. The total chlorophyll content is calculated as $C_t = 8.24\ A_{664} + 23.97\ A_{647} - 16.64\ A_{603}$, where C_t is the total chlorophyll content in μg/ml of the DMF subjected to measurement [25].

4. Accessing drought tolerance of soybean

To facilitate cultivation of soybean in semi-arid and arid regions, it is important to rank soybean cultivars according to their drought tolerance. Various parameters have been adopted to assess drought tolerance (Table 3). Due to different environmental and temporal factors, the results of assessment may be varied. Therefore, experiments have to be conducted consecutively for a few years in the same regions with large sampling size using various assessment parameters in order to achieve a more reliable classification.

Index	Description	Refs
A. Direct yield scoring methods		
1. Average yield	Average yield of certain germplasm in arid region or in the same region for several years.	[26]
2. Water usage efficiency	Unit yield on the expense of a unit of water.	[26]
3. Environmental index	Yield at targeted region relative to the average yield of all regions tested.	[26]
B. Drought tolerance coefficient-based methods		
1. Drought tolerance coefficient (I)	Yield at water deficit year/region relative to water sufficient year/region.	[26]

Index	Description	Refs
2. Drought tolerance index (DI)	$DI = (I \times YD)/AVER_{yield}$ where I is the drought tolerance coefficient, YD is the yield of targeted germplasm in arid region, $AVER_{yield}$ is the average yield of all germplasms tested in the same field. This method accounts for the fact that a germplasm that has a higher drought tolerance coefficient can actually have a lower yield.	[26]
C. Others		
1. Germination stress index (GSI)	Soybean seeds are germinated in water or hypertonic solution. $GSI = PI_{treat}/PIcontrol$ where PI is the summation of the germination rate at day 2, 4, 6, 8 in a ratio of 1, 0.75, 0.5 and 0.25 respectively.	[27]
2. Survival rate	Water is withdrawn during the 3rd – 5th trifoliate stage. Time needed for 50% of plant to die is determined.	[26]
3. Repeated drought during seedling stage	Water is withdrawn from seedlings at 1st trifoliate stage. The seedlings are re-watered when half of the seedlings are suffering from permanent wilting. This process is repeated 2-3 times to determine the amount of seedlings that can recover from the treatment. Switching between hypertonic solution and water can also be used as treatment.	[26]
4. Membership function method	To compare performance of different germplasms under the same standard, each agronomic trait of each germplasm was converted to membership function using the principle of fuzzy logic. Germplasms can be ranked using the membership function into 5 levels at which 1 is the most tolerant and 5 is the most sensitive.	[28]
5. Canopy wilting index	Degree of wilting of the canopy is determined during the mid-day of fine weather. Degree of wilting can be ranked into 6 levels from 0 to 5, where 0 and 5 represent no wilting and completely wilted, respectively.	[26]
D. Methods used officially in China		
1. Tolerance during germination	Seeds are air-dried at 20℃ before experiment. 100 seeds are germinated either in water-soaked (control) or 40% PEG-soaked (treatment) cotton at 25℃. Experiment is conducted with triplication. Germination rates are determined 7 days later. Relative germination rate (RGI) is calculated by dividing the treatment germination rate with control germination rate. Level 1: RGI >95% Level 2: RGI between 80% and 95% Level 3. RGI between 65% and 80% Level 4: RGI between 35% and 65% Level 5: RGI < 35%	[29]

Index	Description	Refs
2. Whole-growth-stage tolerance	Test should be carried out in field with precipitation less than 50mm. Seeds of each germplasm in each treatment are sown in single row of 1.5 m. The control field is irrigated (7 times) to maintain the field soil moisture. In the treatment field, irrigation is only applied before sowing to ensure the germination of seeds. Plant height, number of branching, number of pods per plant and yield per plant of 10 plants are determined upon harvest. Drought tolerance coefficients of each trait are calculated. Average of coefficients (RI) of all traits will be used to rank the germplasm into 5 levels. Level 1: RI > 0.6500 Level 2: RI between 0.5000-0.6500 Level 3: RI between 0.3500-0.5000 Level 4: RI <0.3500 Level 5: Plant died or cannot reproduce	[29]

Table 3. Common parameters for assessing drought tolerance of soybean cultivars.

5. Morphological and physiological adjustments of soybean under drought stress

For ease of discussion, we define the term drought tolerance loosely to include all mechanisms that allow soybean to survive better under drought. Soybean cultivars of different drought tolerance exhibit a spectrum of differential morphological and physiological changes under drought stress, presumably due to the differences in their genotypes.

5.1. Morphological and growth adjustments

Morphological adjustments are sometimes effective means to avoid drought stress. A number of root-related traits have been proposed as indicators of drought tolerance in soybean [30-34]. Root distribution, which is measured in terms of horizontal and vertical root length density or dry matter in soil of different depth [34, 35], will change in drought tolerant soybean cultivars under drought stress [36]. It was reported that under seasonal drought, there is a low root density in the dry surface soil but a high root density in the deeper region of the soil where the water content is higher [34]. Moreover, using data from drought tolerant soybean cultivars, it was found that there is a positive correlation between drought tolerance and dry root weight/ plant weight; total root length/ plant weight, and root volume/ plant weight [30].

Root to shoot ratio increases under water deficit conditions [37]. It has been proposed that the cessation of shoot but not root growth can be explained by the higher sensitivity to water deficit of shoot than root [37]. The differential growth is closely related to the differential change in cell wall composition, which involves the thickening of shoot cell wall and relaxing of the expansion of root cell wall by certain catalytic enzymes and stiffening agents [37]. There are only limited reports on related studies in soybean. The study on GmRD22 from soybean suggested a relationship between osmotic stress and cell wall metabolism. GmRD22 is a BURP-domain containing protein localized in the apoplast, which may play a role in stress tolerance by regulating lignin content of cell wall under stress, presumably through interacting with cell wall peroxidases [38].

The adjustments of leaf morphology may play a role in drought tolerance. Some cultivars take advantage from the maintenance of leaf area which provides a possible benefit for the growth of soybean plant after the stress is relieved [39]. Under stress, drought tolerant soybean cultivars exhibited a larger leaf area when compared with less tolerant cultivars [23, 35]. This phenomenon was associated with the larger extent of reduction in stomatal conductance and yet a smaller extent of reduction in photosynthetic rate in the tolerant cultivar [23]. In this case, the drought tolerant cultivar may benefit from the reduction of water loss while minimizing the cost of reduction of photosynthesis.

5.2. Physiological and biochemical adjustments

To survive over an extended drought period, it is important for the soybean leaves to adjust its stomatal conductance to prevent excessive water loss. For example, after 30 days of water stress, the drought tolerant soybean variety MG/BR46 exhibited a higher degree of reduction in stomatal conductance when compared to the drought sensitive cultivar BR16 (65% versus 50% reduction) [23]. After 45 days of stress, the reduction in stomatal conductance was no longer detectable in the sensitive cultivar while it had reached 79% in the tolerant cultivar [23].

Another important adjustment under drought stress is to maintain cell turgidity. In a field test conducted using the drought tolerance soybean cultivar PI 416937 and the sensitive cultivar Forrest, it was found that PI 416937 maintained a lower solute potential yet a higher water potential and water use efficiency. As a result, PI 416937 gave a higher seed weight and yield than Forrest under drought. This report provided evidence on the positive correlation between turgor maintenance of leaves and drought tolerance [40].

To maintain cell turgidity under stress, osmotic adjustment is a common mechanism which involves active accumulation of solutes in cells [39]. In soybean, drought stress up-regulates the expression of the soybean *P5CS* gene which encodes the enzymeΔ1-pyrroline-5-carboxylate synthase, a key enzyme in proline biosynthesis [41]. When the expression of the soybean *P5CS* gene was knocked-down, survival under drought stress was hampered [42]. However, a recent study comparing a drought tolerant and a drought sensitive soybean did not reveal an increase in proline level under stress, although the proline level of the tolerant cultivar was higher than that in the sensitive cultivar [43]. The involvement of proline accumulation in drought stress adjustment in soybean awaits further confirmation.

The cellular biochemical adjustment under drought stress involves the scavenging of reactive oxygen species (ROS). Under normal situation, ROS including singlet oxygen, superoxide radical, hydrogen peroxide, and hydroxyl radical are continuously synthesized and eliminated in plant cells as "by-products" of photosynthesis, photorespiration, and respiration in chloroplast and mitochondria [44]. Under drought stress, ROS accumulates when the production outweighs the removal [45]. The over-produced ROS will attack cellular components including nucleic acids, protein, and lipid and eventually leads to cell death [46].

ROS scavenging enzymatic activities of superoxide dismutase, catalase, and glutathione peroxidase increased in 5 soybean germplasms under drought stress [24]. The tested germplasms displayed different basal and treatment-induced level of ROS scavenging enzymatic activities, which were correlated positively to the final seed yield [24]. The study on GmPAP3 from soybean provides another example for the correlation between enhanced ROS scavenging activity and the adaptation to osmotic stress. GmPAP3 is a mitochondria localized purple acid phosphatase [47]. Ectopic expression of the *GmPAP3* gene significantly reduces ROS accumulation and thereby alleviates osmotic stress [48].

Adverse environmental conditions can bring forth the misfolding of proteins that will accumulate in endoplasmic reticulum (ER) [49]. The resulting ER stress will activate unfolded protein response [49]. By global expression-profiling analyses on soybean leaves exposed to ER stress inducers and polyethylene glycol, a number of genes were identified as candidate regulatory components integrating ER stress signaling and osmotic stress responses [50]. Moreover, overexpression of soybean BiP (binding protein), an ER-resident molecular chaperone, can enhance drought tolerance in soybean [51]. This evidence tightens the link between ER stress and drought response through the activity of chaperones.

6. Molecular mechanisms of drought tolerance in soybean

In higher plants, the drought stimuli are presumably perceived by osmosensors (that are yet to be identified) and then transduced down the signaling pathways, which activate downstream drought responsive genes to display tolerance effects [52]. The tolerance involves not only the activities of protein receptors, kinases, transcription factors, and effectors but also the production of metabolites as messengers for transducing the signals. Drought tolerance is of multigenic nature, involving complex molecular mechanisms and genetic networks.

The signaling pathway of drought stress is largely overlapping with the signaling pathway of osmotic stresses which has been extensively reviewed [52]. Here, we provide a summary directly related to drought tolerance and include updated information when appropriate.

6.1. Searching for osmosenors

The perception of drought stimulus is presumably via unknown osmosensors. It is speculated that these sensors are associated with alterations in membrane porosity, integrity [53], and turgor pressure [54]. From the spatial perspective, membrane proteins, cell wall receptors, and cytosolic enzymes are all potential sensors for osmotic stress [55, 56]. For example, the families of THESEUS 1 and FERONIA receptor-like kinases (RLKs) in *A. thaliana* are putative stress sensors in cell wall to perceive changes in cell wall integrity and turgor pressure

http://dx.doi.org/10.5772/52945

[57-59]. On the other hand, from the functional point of view, calcium ion (Ca^{2+}) channels, Ca^{2+} binding proteins, two-component histidine kinases, receptor-like protein kinases, G-protein coupled receptors are also potential candidates of osmosensors [60-63]. For instance, AHK1 has been postulated as a cell surface sensor that activates the high-osmolarity glycerol response 1 (HOG1) mitogen-activated kinase (MAPK) cascade in transgenic yeast. [64]

In soybean, two-component histidine kinases (GmHK07, GmHK08, GmHK09, GmHK14, GmHK15, GmHK16 and GmHK17) and receptor-like protein kinases (GmCLV1A, GmCLV1B, GmRLK1, GmRLK2, GmRLK3 and GmRLK4) have been identified as candidates of osmosensors [65-67]. However, direct evidence for their functions to perceive stress signals in soybean is still missing.

6.2. Signal transduction under drought stress

Abscisic acid (ABA) regulates the physiology (e.g. closure of stomata) and metabolism of plants (e.g. expression of enzymes) to rapidly cope with environmental challenges [68]. Biosynthesis, accumulation, and catabolism of ABA are all crucial for the transduction of ABA-mediated signals. The accumulation of ABA in response to drought is associated with the changes in the levels of Ca^{2+} and ROS [60, 69]. *In planta*, ABA is synthesized in various cell types including root cells, parenchyma cells, and mesophyll cells. Under drought stress, ABA is transported to guard cells to control stomatal aperture [70]. ABA reaching the target tissues and cells will be recognized and the signals will be transduced down the ABA signalosome [71], including ABA receptors (PYR/PYL/RCAR), negative regulators (e.g. group A protein phosphatases 2C), and positive regulators (e.g. SnRK-type kinases).

Components of this system have been discovered in soybean. For example, GsAPK is a SnRK-type kinase from wild soybean that is up-regulated by drought stress in both leaves and roots, but down-regulated by ABA treatment in roots [72]. *In vivo* assay revealed that the phosphorylation activities of GsAPK is activated by ABA in a Ca^{2+}-independent manner, suggesting that GsAPK may play a role in the ABA-mediated signal transduction [72]. Activated SnRK-type kinases in rice and *A. thaliana* will phosphorylate target proteins including bZIP transcription factors and membrane ion channels [72].

Perceived stress signals may trigger transient changes in the cytosolic Ca^{2+} level which acts as a second messenger [73]. Ca^{2+} sensors in turn transmit and activate the signaling pathways for downstream stress responses [60]. Ca^{2+} sensors include various types of Ca^{2+}-binding proteins: CaMs (calmodulins), CMLs (CaM-like proteins), CDPKs (Ca^{2+}-dependent protein kinases), and CBLs (calcineurin B-like proteins) [74]. Among these Ca^{2+} sensors, all are plant and protist-specific with the exception of CaM.

Expression of the soybean CaM (GmCaM4) in transgenic *A. thaliana* activated a R2R3 type MYB transcription factor which in turn up-regulated several drought-responsive genes, including *P5CS* (encoding a proline anabolic enzyme) [75]. While the application of Ca^{2+} affects the nodulation of soybean [76], the gene encoding a soybean CaM binding protein was found to be differentially expressed in soybean nodules under drought stress [77].

The drought tolerance related CDPK family is well-studied in rice and *A. thaliana* [78, 79]. In isolated soybean symbiosome membrane, a CDPK was demonstrated to phosphorylate an aquaporin called nodulin 26 and hence enhance the water permeability of the membrane. I was hypothesized that this is an integral part of the drought tolerance mechanism [80, 81].

Besides Ca^{2+}, phosphatidic acid (PA) and the intermediates of inositol metabolism are also second messengers for signal transduction [82-84]. However, there are only very limited evidence supporting the involvement of phospholipid signaling in drought stress response of soybean. The soybean nodulin gene *G93* encoding a ZR1 homologue was down-regulated under drought stress [85]. Plant ZR1 homologue such as RARF-1 in *A. thaliana* may involve in lipid signaling via interaction with phosphatidylinositol 3-phosphate [86].

When plants are subjected to drought stress, accumulation of cellular ROS will trigger the generation of hydrogen peroxide, a signaling molecule that will activate ROS scavenging mechanisms [87]. In soybean, exogenous application of hydrogen sulphide alleviates symptoms of drought stress, probably via triggering an antioxidant signaling mechanism [88].

Many studies support the roles of protein kinases in stress signaling [89, 90]. In plants, the drought responsive signal transduction of the MAPK family (MAPK, MAPKK/MEKK, MAPKKK/MKK) as well as the MAPK phosphatases (MKP) family have been relatively well-studied in *A. thaliana* and rice [89], but remained under-explored in soybean, although a PA-responsive MAPK has been identified in soybean [91].

On the other hand, some non-MAPK type protein kinases found in soybean may be related to drought responses. The soybean gene encoding a serine/threonine ABA-activated protein kinase was found to be up-regulated by ABA, Ca^{2+}, and polyethylene glycol treatments [92]. The With No Lysine protein kinase 1 of soybean is another serine/threonine protein kinase that is a putative osmoregulator [93].

The ubiquitin-mediated protein degradation pathway is also an integral part of the signal transduction network [94]. This pathway directs the degradation of target proteins by the 26S proteasome and is responsive to drought stress. Two ubiquitin genes and one gene encoding ubiquitin conjugating enzyme were identified as differentially expressed genes in nodulated soybean under drought stress [77]. Overexpression of the ubiquitin ligase gene *GmUBC2* enhances drought tolerance in *A. thaliana*, via up-regulating the expression of genes encoding ion transporters (AtNHX1 and AtCLCa), a proline biosynthetic enzyme (AtP5CS), and a copper chaperone (AtCCS) [94].

6.3. Drought-responsive transcription factors

Transcription regulation plays an important role in drought stress response. For instance, using oligo microarray analysis, transcriptions of 4,433 and 5,098 soybean genes were found to be significantly up-regulated and down-regulated respectively when subjected to a no-irrigation period for 4 days [95]. The signal transduction pathways can ultimately regulate the expression of drought-responsive genes through diverse transcription factors. Transcription factors often target the corresponding *cis*-acting promoter elements, such as the drought stress related elements DRE, ABRE, Gbox, and T/Gbox [95, 96].

In the soybean genome, ~500 transcription factors were *in silico* annotated [96]. Increasing efforts have been placed to characterize their importance and functions in relation to drought [97-101]. Soybean transcription factors that confer drought tolerance are summarized in Table 4.

Transcription factor	Expression under drought [a]	Plant system used	Refs
	AP2/ERF		
GmDREB	+	Wheat	[102]
GmDREB2	+	*A. thaliana*	[103]
GmDREB3	nc	*A. thaliana*	[104]
GmERF3 [b]	+	Tobacco	[105]
GmERF4	+	Tobacco	[106]
GmERF089	+	Tobacco	[107]
	bZIP		
GmbZIP1	+	*A. thaliana*, wheat	[108]
	GT		
GmGT-2A	+	*A. thaliana*	[109]
GmGT-2B [b]	+	*A. thaliana*	[109]
	Zinc finger		
GsZFP1	+	*A. thaliana*	[110]
	Zinc finger—WRKY		
GmWRKY54	+	*A. thaliana*	[111]
GmWRKY57B	nt	Tobacco	[112]

[a]+: up-regulated; -: down-regulated; nc: no change; nt: not tested

[b] The transcription activity was verified by transactivation tests in yeast.

Table 4. Soybean transcription factors that exhibit protective function against drought in transgenic plant systems.

7. Strategies for breeding drought tolerant soybean cultivars

To combat water deficit, one of the most effective ways is to breed for new cultivars that exhibit durable drought tolerance. A combination of conventional breeding, marker-assisted breeding, and transgenic approaches will shed light on the crop improvement program of drought tolerance in soybean.

7.1. Conventional breeding

The high biodiversity nature of soybean allows the stacking of desirable traits through breeding. Since the genetic background of soybean germplasms varies due to spatial adaptations to diverse habitats, breeding with soybean germplasms from different origins can effectively accelerate crop improvement. A recent study suggested that wild soybean exhibited higher allelic diversity compared to cultivated soybean [113]. Since they are sexually compatible, wild soybean can potentially serve as a good genetic source in the breeding programs.

Conventional breeding could be a long and tedious process. For example, the breeding of the drought tolerant soybean cultivar Jindou 21 started by breeding Lín Xiàn White Soybean (an old cultivar of higher drought tolerance but lower yield) against Jindou 2 (drought tolerant, high yield, and early maturation). After the selective breeding for six generations, the resulting drought tolerance line was used as a parent of the next selection breeding and crossed with Jindou 14. The final selection breeding of Jindou 21 was carried out in the arid region of western Shanxi for seven years (1987 – 1993). Comparing to its parent Jindou 14, Jindou 21 exhibited increased yield and enhanced drought tolerant. Since then, Jindu 21 has become one of the most popularized drought tolerant soybean cultivars grown in semi-arid regions of Gansu, Ningxia, and Shanxi Provinces of China, particularly in regions where irrigated agriculture is not practical. The total cultivation area is over 3.75 million hectare [114].

7.2. Marker-assisted breeding

Drought tolerance in crops may involve different mechanisms depending on the nature of drought, making it difficult for phenotypic selection and screening through conventional breeding. A recent genomic study showed that soybean is a species of exceptionally high linkage disequilibrium (low recombination frequency) and hence marker-assisted breeding is a promising approach. The same study also identified more than 200,000 tagged SNPs for this purpose [113].

Marker-assisted breeding makes use of DNA markers that are closely linked to the target QTLs, to expedite the selection of progeny lines by replacing some time consuming phenotypic characterizations [115]. For example, delayed wilting response of canopy is associated with drought tolerance [116, 117]. Four QTLs that are associated with this trait were mapped [118], which are significantly associated with 16 SSR markers. One of the identified QTLs was identified in all tested environments which is therefore a promising candidate for marker-assisted breeding for delayed canopy wilting trait in different environments, including those with the soil type and moisture level inadequately characterized [119, 120].

7.3. Genetic engineering

With the advancement of biotechnology and availability of genomic sequence information, germplasm resources, and increasing genomic tools available for soybean research, transgenic approach has become an attractive alternative strategy in breeding. One critical hurdle of this approach is to identify ideal candidate genes that can improve drought tolerance but do not have a yield penalty when introduced into the soybean genome.

Rapid gain-of-function experiments using heterologous model plant systems (tobacco, *A. thaliana* and rice) have been employed to screen for potential candidates. Although of lower efficiency, there are established systems of soybean transformation [121-124], allowing direct assessments of the protective functions of both native and heterologous genes in soybean.

Some promising results using this approach have been obtained, although they are all at the experimental stage. For example, AtMYB44 is a R2R3-type MYB transcription factor from *A. thaliana* that participates in the ABA-mediated abiotic stress signaling [125]. Ectopic expression of AtMYB44 in soybean led to improved drought tolerance and yet suffered from reduced growth phenotype under normal conditions [126]. Transgenic soybean expressing the *AtP5CR* gene (encoding L-Δ1-Pyrroline-5-carboxylate reductase) resulted in enhanced tolerance toward drought stress with significantly higher relative water content [127]. Introducing the *NTR1* gene from *Brassica campestris* (encoding a jasmonic acid carboxyl methyltransferase) into soybean led to increased accumulation of methyl jasmonate and enhanced tolerance toward dehydration during seed germination [128]. Overexpression of the soybean gene *GmDREB3* (encoding a dehydration-responsive element-binding transcription factor) also enhances drought tolerance, in parallel to the accumulation of proline [129].

8. Agronomic practices to alleviate the impacts of drought

While breeding programs often take a long time to complete, agronomic practices aiming at efficient use of limited water resources will give immediate results. These measures include: (1) minimizing water input; (2) reducing water loss from irrigation system and the field; and (3) increasing crop water use efficiency (WUE). In agriculture, WUE is defined as the yield of irrigated plant per total water in actual evapotranspiration (ET). A higher WUE value usually suggests a better use of water though not necessarily a higher yield [26].

Traditional irrigation systems involve open and unlined ditches channeling water from uncovered sources like wells and rivers to the fields. Besides, irrigation by flooding furrows or the whole field is common in many regions simply due to low cost [130]. A loss of more than 50% of irrigated water happens in these irrigation systems through evaporation, leakage, seepage, and percolation especially when the water source is far away from the field [131]. A well-managed pipe system can achieve 90-100% conveyance efficiency [131]. Pressurized water application methods such as advanced sprinkler and dripping systems at the terminal of the closed irrigation channels help further reduce water input [130, 132]. Sprinklers can evenly spray desirable amounts of water onto the field such that water loss through seepage and percolation can be reduced. Dripping can deliver water precisely to the root zone of the plant. This can reduce the loss of water in barren areas or consumption by weeds.

The plant at different growth stages requires different amount of water to grow and survive. ET accounts for both the evaporation and transpiration and is a measure of the amount of water used by the crop. ET by the soybean plant roughly appears as a bell shaped curve during its life cycle. It gradually increases from the germination stage through the vegetative stage to a maximum at the early reproductive stage (R1-R2); then reduces continuously until the maturation stage [133]. The yield of soybean grown in arid regions without irrigation exhibits significant yield reduction, compared to those grown on fully irrigated land [134]. Nevertheless, delayed irrigation at flowering and early podding stages can effectively regain most of the yield as the fully irrigated plant [134-136]. Limiting irrigation to growth stages critical to the final yield can be an effective mean to reduce the input of water while water resources are scarce [137, 138].

The drought stress response of the plant involving ABA can also be used in formulating effective water saving agricultural strategies. ABA reduces stomatal aperture and hence reduces water loss through transpiration [139-142]. On the other hand, soil water deficit and water replenishment induce root growth in crops such as maize, corn [143, 144], and some soybean varieties [35]. The outgrowth of roots benefits both water and nutrient absorption upon water replenishment [145]. Based on these researches, regulated deficit irrigation (RDI) has been developed to save agricultural water by improving WUE as a result of supplying water less than the full ET of the plant. A recent study in soybean showed that compared with the fully irrigated control, irrigating with 75% water of the fully irrigated treatment could maintain over 90% of the yield while increasing the water productivity from 0.44 to 0.56 kg/m^3 [146].

Controlled alternate partial root-zone irrigation (CAPRI) or so-called partial root-zone drying (PRD) is a derivative of RDI. Instead of just reducing the amount of irrigation, the strategy of CAPRI is to supply water only to spatially separated parts of the root system while keeping the unirrigated parts dry [147]. The drought stress signal will be generated in the dry parts of the root system to induce growth of the whole root system and reduce stomatal aperture. On the other hand, the irrigated half of the root system will continue to absorb water to support the growth of the whole plant [145]. To prevent undesirable anatomical changes and severe damages to the root, different parts of the root system will be irrigated in turn [145]. Dripping irrigation has played an important role in CAPRI as it can precisely irrigate the desired part of the root system. Application of alternate partial root-zone drip irrigation (APRDI) has achieved promising water saving effects on different crops like cotton, grapes, and potato [148-151]. Similar strategies can be applied in soybean cultivation.

Traditional mulching involves covering of the field with straw or other harvest left-overs. The mulch can trap moisture and hence retain soil water. The degrading organic mulch also adds humus to the soil and improves the water holding capacity of the soil. In China, plastic mulch has been widely used on soybean interplanted with maize, potato or cucumber. For example, a study conducted in Shouyang County of the Shanxi Province, China suggested that mulching cultivation with hole-sowing or row-sowing techniques can increase soybean yield up to 23.4% and 50.6%, respectively [152]. Ridge-furrow mulching and whole year mulching cultivation could increase WUE by 37.3% - 58.0% and yield by 40.8% - 41.9%, respectively, in the Loess Plateau of China, compared to traditional open field cultivation [153, 154].

9. Conclusion

Soybean is nutritionally and economically important. Due to the adverse effects of agricultural drought on soybean production, drought stress in soybean has become a hot research topic. From measurement of the effect of drought on soybean to the studies of drought responsive mechanisms at morphological, physiological, and molecular level, the knowledge on drought stress and tolerance in soybean has been accumulated rapidly. With the advancement of breeding programs and agronomic practices, the production of soybean under drought can be improved by integrating all technologies and knowledge involved.

Acknowledgement

This work is supported by the Hong Kong RGC Collaborative Research Fund CUHK3/CRF/11G.

Author details

Yee-Shan Ku[1], Wan-Kin Au-Yeung[1], Yuk-Lin Yung[1], Man-Wah Li[1], Chao-Qing Wen[1], Xueyi Liu[2] and Hon-Ming Lam[1,2*]

*Address all correspondence to: honming@cuhk.edu.hk

1 State Key Laboratory of Agrobiotechnology and School of Life Sciences, The Chinese University of Hong Kong, Shatin, Hong Kong

2 Institute of Economic Crops, Shanxi Academy of Agricultural Sciences, Fengyang, Shanxi, People's Republic of China

References

[1] Cosgrove W. Water in a changing world, in The United Nations World Water Development Report, Programme W.W.A., Editor 2009.

[2] Motha R. Monitoring, assessment and combat of drought and desertification, in Commission for Agricultural Meteorology Reports 1992, World Meteorological Organization: Geneva.

[3] Sivakumar M.V.K., Motha R.P., Wilhite D.A., Wood D.A. Agricultural Drought Indices. in WMO/UNISDR Expert Group Meeting on Agricultural Drought Indices. 2010. Murcia, Spain: Geneva, Switzerland: World Meteorological Organization.

[4] Frederick J.R., Camp C.R., Bauer P.J. Drought-stress effects on branch and mainstem seed yield and yield components of determinate soybean. Crop Science 2001;41(3) 759-763.

[5] Sadeghipour O., Abbasi S. Soybean response to drought and seed inoculation. World Applied Sciences Journal 2012;17(1) 55-60.

[6] Brown E., Brown D., Caviness C. Response of selected soybean cultivars to soil moisture deficit. Agronomy Journal 1985;77(2) 274-278.

[7] Eck H.V., Mathers A.C., Musick J.T. Plant water stress at various growth stages and growth and yield of soybeans. Field Crops Research 1987;17(1) 1-16.

[8] Desclaux D., Huynh T.T., Roumet P. Identification of soybean plant characteristics that indicate the timing of drought stress. Crop Science 2000;40(3) 716-722.

[9] Korte L.L., Williams J.H., Specht J.E., Sorensen R.C. Irrigation of soybean genotypes during reproductive ontogeny. I. Agronomic responses. Crop Science 1983;23(3) 521-527.

[10] Kadhem F.A., Specht J.E., Williams J.H. Soybean irrigation serially timed during stages Rl to R6. I. Agronomic responses. Agronomy Journal 1985;77(2) 291-298.

[11] Fehr W.R., Caviness C.E. Stages of Soybean Development: Agriculture and Home Economics Experiment Station, Iowa State University of Science and Technology; 1977.

[12] Heatherly L.G. Drought stress and irrigation effects on germination of harvested soybean seed. Crop Science 1993;33(4) 777-781.

[13] Samarah N.H., Mullen R.E., Anderson I. Soluble sugar contents, germination, and vigor of soybean seeds in response to drought stress. Journal of New Seeds 2009;10(2) 63-73.

[14] Dornbos D.L., Mullen R.E., Shibles R.E. Drought stress effects during seed fill on soybean seed germination and vigor. Crop Science 1989;29(2) 476-480.

[15] Vieira R.D., TeKrony D.M., Egli D.B. Effect of drought and defoliation stress in the field on soybean seed germination and vigor. Crop Science 1992;32(2) 471-475.

[16] Dornbos D.L., Mullen R.E. Influence of stress during soybean seed fill on seed weight, germination, and seedling growth rate. Canadian Journal of Plant Science 1991;71(2) 373-383.

[17] Chung J., Babka H.L., Graef G.L., Staswick P.E., Lee D.J., Cregan P.B., Shoemaker R.C., Specht J.E. The seed protein, oil, and yield QTL on soybean linkage group I. Crop Science 2003;43(3) 1053-1067.

[18] Dornbos D.L., Mullen R.E. Soybean seed protein and oil contents and fatty acid composition adjustments by drought and temperature. Journal of the American Oil Chemists' Society 1992;69(3) 228-231.

[19] Vollmann J., Fritz C.N., Wagentristl H., Ruckenbauer P. Environmental and genetic variation of soybean seed protein content under Central European growing conditions. Journal of the Science of Food and Agriculture 2000;80(9) 1300-1306.

[20] Eldridge A.C., Kwolek W.F. Soybean isoflavones: effect of environment and variety on composition. Journal of Agricultural and Food Chemistry 1983;31(2) 394-396.

[21] Caldwell C.R., Britz S.J., Mirecki R.M. Effect of temperature, elevated carbon dioxide, and drought during seed development on the isoflavone content of dwarf soybean [Glycine max (L.) Merrill] grown in controlled environments. Journal of Agricultural and Food Chemistry 2005;53(4) 1125-1129.

[22] O'Neal M.E., Landis D.A., Isaacs R. An inexpensive, accurate method for measuring leaf area and defoliation through digital image analysis. Journal of Economic Entomology 2002;95(6) 1190-1194.

[23] Stolf-Moreira R., Medri M.E., Neumaier N., Lemos N.G., Pimenta J.A., Tobita S., Brogin R.L., Marcelino-Guimarães F.C., Oliveira M.C.N., Farias J.R., Abdelnoor R.V., Nepomuceno A.L. Soybean physiology and gene expression during drought. Genetics and Molecular Research 2010;9 1946-1956.

[24] Masoumi H., Masoumi M., Darvish F., Daneshian J., Nourmohammadi G., Habibi D. Change in several antioxidant enzymes activity and seed yield by water deficit stress in soybean (Glycine max L.) cultivars. Notulae Botanicae Horti Agrobotanici Cluj-Napoca 2010;38(3) 86-94.

[25] Moran R. Formulae for determination of chlorophyllous pigments extracted with N, N-dimethylformamide. Plant Physiology 1982;69(6) 1376.

[26] Liu X. Drought. In: Lam H.M., Chang R., Shao G.,Liu Z., Editors. (ed.) Research on tolerance to stresses in chinese soybean. China agricultural press: Beijing. 2009.

[27] Bouslama M., Schapaugh W.T. Stress tolerance in soybeans. I. Evaluation of three screening techniques for heat and drought tolerance. Crop Science 1984;24(5) 933-937.

[28] Liu X. Discusing assessment methods of drought resistance of osybean. Chinese Journal of Oil Crop Sciences 1986;4(2) 23-26.

[29] Qiu L., Chang R. Descriptors and data standard for soybean (Glycine spp.): China agricultural press; 2006.

[30] Liu Y., Gai J.Y., Lu H.N., Wang Y.J., Chen S.Y. Identification of drought tolerant germplasm and inheritance and QTL mapping of related root traits in soybean (Glycine max (L.) Merr.). Acta Genetica Sinica 2005;32(8) 855-863.

[31] Wang M., Cheung C.Y., Ma T.F., Yao W.C. Studies on the drought resistance of seedling in soybean. Chinese Journal of Oil Crop Science 2004;26(3) 29-32.

[32] Jia Q.S., Wei L., Yang H.F. Primary report on the relationship between the root system and drought resistance in soybean seedlings. Shaanxi Journal of Agricultural Sciences 2006;(2) 12-13.

[33] Lu G.H. Studies on root properties and drought-resistance for different types of drought. Journal of Shanxi Agricultural Sciences 2000;28(2) 37-40.

[34] Garay A.F., Wilhelm W. Root system characteristics of two soybean isolines undergoing water stress condition. Agronomy Journal 1983;75 973-977.

[35] Benjamin J.G., Nielsen D.C. Water deficit effects on root distribution of soybean, field pea and chickpea. Field Crops Research 2006;97(2-3) 248-253.

[36] Tzenova V., Kirkova Y., Stoimenov G. Methods for plant water stress evaluation of soybean canopy, in Balwois 2008 - Water Observation and Information System for Decision Support 2008: Ohrid, Republic of Macedonia.

[37] Wu Y., Cosgrove D.J. Adaptation of roots to low water potentials by changes in cell wall extensibility and cell wall proteins. Journal of Experimental Botany 2000;51(350) 1543-1553.

[38] Wang H., Zhou L., Fu Y., Cheung M.Y., Wong F.L., Phang T.H., Sun Z., Lam H.M. Expression of an apoplast-localized BURP-domain protein from soybean (GmRD22) enhances tolerance towards abiotic stress. Plant, Cell & Environment 2012.

[39] Manavalan L.P., Guttikonda S.K., Tran L.S.P., Nguyen H.T. Physiological and molecular approaches to improve drought resistance in soybean. Plant and Cell Physiology 2009;50(7) 1260-1276.

[40] Sloane R.J., Patterson R.P., Carter Jr T.E. Field drought tolerance of a soybean plant introduction. Crop Science 1990;30(1) 118-123.

[41] Porcel R., Azcón R., Ruiz-Lozano J.M. Evaluation of the role of genes encoding for Δ-pyrroline-5-carboxylate synthetase (P5CS) during drought stress in arbuscular mycorrhizal and plants. Physiological and Molecular Plant Pathology 2004;65(4) 211-221.

[42] de Ronde J.A., Spreeth M.H., Cress W.A. Effect of antisense L-Δ1-pyrroline-5-carboxylate reductase transgenic soybean plants subjected to osmotic and drought stress. Plant Growth Regulation 2000;32(1) 13-26.

[43] Silvente S., Sobolev A.P., Lara M. Metabolite adjustments in drought tolerant and sensitive soybean genotypes in response to water stress. PLoS ONE 2012;7(6) e38554.

[44] Foyer C.H., Noctor G. Redox sensing and signalling associated with reactive oxygen in chloroplasts, peroxisomes and mitochondria. Physiologia Plantarum 2003;119(3) 355-364.

[45] Agarwal S., Sairam R., Srivastava G., Meena R. Changes in antioxidant enzymes activity and oxidative stress by abscisic acid and salicylic acid in wheat genotypes. Biologia Plantarum 2005;49(4) 541-550.

[46] Mittler R. Oxidative stress, antioxidants and stress tolerance. Trends in plant science 2002;7(9) 405-410.

[47] Liao H., Wong F.L., Phang T.H., Cheung M.Y., Li W.Y.F., Shao G., Yan X., Lam H.M. GmPAP3, a novel purple acid phosphatase-like gene in soybean induced by NaCl stress but not phosphorus deficiency. Gene 2003;318 103-111.

[48] Li W.Y.F., Shao G., Lam H.M. Ectopic expression of GmPAP3 alleviates oxidative damage cuased by salinity and osmotic stresses. New Phytologist 2008;178(1) 80-91.

[49] Liu J.X., Howell S.H. bZIP28 and NF-Y transcription factors are activated by ER stress and assemble into a transcriptional complex to regulate stress response genes in Arabidopsis. The Plant Cell 2010;22(3) 782-796.

[50] Irsigler A., Costa M., Zhang P., Reis P., Dewey R., Boston R., Fontes E. Expression profiling on soybean leaves reveals integration of ER-and osmotic-stress pathways. BMC Genomics 2007;8(1) 431.

[51] Valente M.A.S., Faria J.A.Q.A., Soares-Ramos J.R.L., Reis P.A.B., Pinheiro G.L., Piovesan N.D., Morais A.T., Menezes C.C., Cano M.A.O., Fietto L.G. The ER luminal binding protein (BiP) mediates an increase in drought tolerance in soybean and delays drought-induced leaf senescence in soybean and tobacco. Journal of experimental botany 2009;60(2) 533-546.

[52] Ahuja I., de Vos R.C.H., Bones A.M., Hall R.D. Plant molecular stress responses face climate change. Trends in Plant Science 2010;15(12) 664-674.

[53] Mahajan S., Tuteja N. Cold, salinity and drought stresses: an overview. Archives of Biochemistry and Biophysics 2005;444(2) 139-158.

[54] Reiser V., Raitt D.C., Saito H. Yeast osmosensor Sln1 and plant cytokinin receptor Cre1 respond to changes in turgor pressure. Journal of Cell Biology 2003;161(6) 1035-1040.

[55] Kader M.A., Lindberg S. Cytosolic calcium and pH signaling in plants under salinity stress. Plant Signaling & Behavior 2010;5(3) 233.

[56] Humphrey T.V., Bonetta D.T., Goring D.R. Sentinels at the wall: cell wall receptors and sensors. New Phytologist 2007;176(1) 7-21.

[57] Hématy K., Höfte H. Novel receptor kinases involved in growth regulation. Current Opinion in Plant Biology 2008;11(3) 321-328.

[58] Hématy K., Sado P.E., Van Tuinen A., Rochange S., Desnos T., Balzergue S., Pelletier S., Renou J.P., Höfte H. A receptor-like kinase mediates the response of Arabidopsis cells to the inhibition of cellulose synthesis. Current Biology 2007;17(11) 922-931.

[59] Cheung A.Y., Wu H.M. THESEUS 1, FERONIA and relatives: a family of cell wall-sensing receptor kinases? Current Opinion in Plant Biology 2011;14(6) 632-641.

[60] Xiong L., Zhu J.K. Molecular and genetic aspects of plant responses to osmotic stress. Plant, Cell & Environment 2002;25(2) 131-139.

[61] Kacperska A. Sensor types in signal transduction pathways in plant cells responding to abiotic stressors: do they depend on stress intensity? Physiologia Plantarum 2004;122(2) 159-168.

[62] Grene R., Vasquez-Robinet C., Bohnert H.J. Molecular biology and physiological genomics of dehydration stress. Plant Desiccation Tolerance 2011; 255-287.

[63] Huang G.T., Ma S.L., Bai L.P., Zhang L., Ma H., Jia P., Liu J., Zhong M., Guo Z.F. Signal transduction during cold, salt, and drought stresses in plants. Molecular Biology Reports 2011; 1-19.

[64] Urao T., Yakubov B., Satoh R., Yamaguchi-Shinozaki K., Seki M., Hirayama T., Shinozaki K. A transmembrane hybrid-type histidine kinase in Arabidopsis functions as an osmosensor. The Plant Cell 1999;11(9) 1743-1754.

[65] Yamamoto E., Karakaya H.C., Knap H.T. Molecular characterization of two soybean homologs of Arabidopsis thaliana CLAVATA1 from the wild type and fasciation mutant. Biochimica et Biophysica Acta (BBA) 2000;1491(1) 333-340.

[66] Yamamoto E., Knap H.T. Soybean receptor-like protein kinase genes: paralogous divergence of a gene family. Molecular Biology and Evolution 2001;18(8) 1522-1531.

[67] Le D.T., Nishiyama R., Watanabe Y., Mochida K., Yamaguchi-Shinozaki K., Shinozaki K., Tran L.S.P. Genome-wide expression profiling of soybean two-component system genes in soybean root and shoot tissues under dehydration stress. DNA Research 2011;18(1) 17-29.

[68] Zhang J., Jia W., Yang J., Ismail A.M. Role of ABA in integrating plant responses to drought and salt stresses. Field Crops Research 2006;97(1) 111-119.

[69] Zhao Z., Chen G., Zhang C. Interaction between reactive oxygen species and nitric oxide in drought-induced abscisic acid synthesis in root tips of wheat seedlings. Functional Plant Biology 2001;28(10) 1055-1061.

[70] Wilkinson S., Davies W.J. Drought, ozone, ABA and ethylene: new insights from cell to plant to community. Plant, Cell & Environment 2010;33(4) 510-525.

[71] Umezawa T., Nakashima K., Miyakawa T., Kuromori T., Tanokura M., Shinozaki K., Yamaguchi-Shinozaki K. Molecular basis of the core regulatory network in ABA responses: sensing, signaling and transport. Plant and Cell Physiology 2010;51(11) 1821-1839.

[72] Yang L., Ji W., Gao P., Li Y., Cai H., Bai X., Chen Q., Zhu Y. GsAPK, an ABA-activated and Calcium-Independent SnRK2-Type kinase from G. soja, mediates the regulation of plant tolerance to salinity and ABA stress. PLoS ONE 2012;7(3) e33838.

[73] McAinsh M.R., Pittman J.K. Shaping the calcium signature. New Phytologist 2009;181(2) 275-294.

[74] DeFalco T., Bender K., Snedden W. Breaking the code: Ca2+ sensors in plant signalling. Biochemical Journal 2010;425 27-40.

[75] Yoo J.H., Park C.Y., Kim J.C., Do Heo W., Cheong M.S., Park H.C., Kim M.C., Moon B.C., Choi M.S., Kang Y.H. Direct interaction of a divergent CaM isoform and the transcription factor, MYB2, enhances salt tolerance in Arabidopsis. Journal of Biological Chemistry 2005;280(5) 3697-3706.

[76] Bell R.W., Edwards D.G., Asher C.J. External calcium requirements for growth and nodulation of six tropical food legumes grown in flowing solution culture [peanut; pigeon pea; guar; soybean; cowpea cv Vita 4 and CPI 28215]. Australian Journal of Agricultural Research 1989;40.

[77] Clement M., Lambert A., Herouart D., Boncompagni E. Identification of new up-regulated genes under drought stress in soybean nodules. Gene 2008;426(1-2) 15-22.

[78] Zhu S.Y., Yu X.C., Wang X.J., Zhao R., Li Y., Fan R.C., Shang Y., Du S.Y., Wang X.F., Wu F.Q. Two calcium-dependent protein kinases, CPK4 and CPK11, regulate abscisic acid signal transduction in Arabidopsis. The Plant Cell 2007;19(10) 3019-3036.

[79] Saijo Y., Hata S., Kyozuka J., Shimamoto K., Izui K. Over-expression of a single Ca2+-dependent protein kinase confers both cold and salt/drought tolerance on rice plants. The Plant Journal 2000;23(3) 319-327.

[80] Guenther J.F., Chanmanivone N., Galetovic M.P., Wallace I.S., Cobb J.A., Roberts D.M. Phosphorylation of soybean nodulin 26 on serine 262 enhances water permeability and is regulated developmentally and by osmotic signals. The Plant Cell 2003;15(4) 981-991.

[81] Rivers R.L., Dean R.M., Chandy G., Hall J.E., Roberts D.M., Zeidel M.L. Functional analysis of nodulin 26, an aquaporin in soybean root nodule symbiosomes. Journal of Biological Chemistry 1997;272(26) 16256-16261.

[82] Munnik T. Phosphatidic acid: an emerging plant lipid second messenger. Trends in Plant Science 2001;6(5) 227-233.

[83] Testerink C., Munnik T. Phosphatidic acid: a multifunctional stress signaling lipid in plants. Trends in Plant Science 2005;10(8) 368-375.

[84] Xue H.W., Chen X., Mei Y. Function and regulation of phospholipid signalling in plants. Biochemical Journal 2009;421(Pt 2) 145.

[85] Clement M., Boncompagni E., de Almeida-Engler J., Herouart D. Isolation of a novel nodulin: a molecular marker of osmotic stress in Glycine max/Bradyrhizobium japonicum nodule. Plant, Cell & Environment 2006;29(9) 1841-1852.

[86] Drøbak B.K. PARF-1: an Arabidopsis thaliana FYVE-domain protein displaying a novel eukaryotic domain structure and phosphoinositide affinity. Journal of Experimental Botany 2002;53(368) 565-567.

[87] Cruz C.M.H. Drought stress and reactive oxygen species: Production, scavenging and signaling. Plant Signaling & Behavior 2008;3(3) 156.

[88] Zhang H., Jiao H., Jiang C.X., Wang S.H., Wei Z.J., Luo J.P., Jones R.L. Hydrogen sulfide protects soybean seedlings against drought-induced oxidative stress. Acta Physiologiae Plantarum 2010;32(5) 849-857.

[89] Boudsocq M., Laurière C. Osmotic signaling in plants. Multiple pathways mediated by emerging kinase families. Plant Physiology 2005;138(3) 1185-1194.

[90] Bartels S., Besteiro M.A.G., Lang D., Ulm R. Emerging functions for plant MAP kinase phosphatases. Trends in Plant Science 2010;15(6) 322-329.

[91] Lee S., Hirt H., Lee Y. Phosphatidic acid activates a wound-activated MAPK in Glycine max. The Plant Journal 2001;26(5) 479-486.

[92] Luo G.Z., Wang Y.J., Xie Z.M., Gai J.Y., Zhang J.S., Chen S.Y. The putative Ser/Thr protein kinase gene GmAAPK from soybean is regulated by abiotic stress. Journal of Integrative Plant Biology 2006;48(3) 327-333.

[93] Wang Y., Suo H., Zheng Y., Liu K., Zhuang C., Kahle K.T., Ma H., Yan X. The soybean root-specific protein kinase GmWNK1 regulates stress-responsive ABA signaling on the root system architecture. The Plant Journal 2010;64(2) 230-242.

[94] Zhou G.A., Chang R.Z., Qiu L.J. Overexpression of soybean ubiquitin-conjugating enzyme gene GmUBC2 confers enhanced drought and salt tolerance through modulating abiotic stress-responsive gene expression in Arabidopsis. Plant Molecular Biology 2010;72(4-5) 357-367.

[95] Maruyama K., Todaka D., Mizoi J., Yoshida T., Kidokoro S., Matsukura S., Takasaki H., Sakurai T., Yamamoto Y.Y., Yoshiwara K. Identification of cis-acting promoter elements in cold-and dehydration-induced transcriptional pathways in Arabidopsis, rice, and soybean. DNA Research 2012;19(1) 37-49.

[96] Mochida K., Yoshida T., Sakurai T., Yamaguchi-Shinozaki K., Shinozaki K., Tran L.S.P. In silico analysis of transcription factor repertoire and prediction of stress responsive transcription factors in soybean. DNA Research 2009;16(6) 353-369.

[97] Le D.T., Nishiyama R., Watanabe Y., Mochida K., Yamaguchi-Shinozaki K., Shinozaki K., Tran L.S.P. Genome-wide survey and expression analysis of the plant-specific NAC transcription factor family in soybean during development and dehydration stress. DNA Research 2011;18(4) 263-276.

[98] Mizoi J., Shinozaki K., Yamaguchi-Shinozaki K. AP2/ERF family transcription factors in plant abiotic stress responses. Biochimica et Biophysica Acta (BBA)-Gene Regulatory Mechanisms 2011.

[99] Nakashima K., Takasaki H., Mizoi J., Shinozaki K., Yamaguchi-Shinozaki K. NAC transcription factors in plant abiotic stress responses. Biochimica et Biophysica Acta (BBA)-Gene Regulatory Mechanisms 2011.

[100] Pereira S.S., Guimarães F.C.M., Carvalho J.F.C., Stolf-Moreira R., Oliveira M.C.N., Rolla A.A.P., Farias J.R.B., Neumaier N., Nepomuceno A.L. Transcription factors expressed in soybean roots under drought stress. Genetics and Molecular Research 2011;10(4) 3689-3701.

[101] Phang T.H., Li M.W., Cheng C.C., Wong F.L., Chan C., Lam H.M. Molecular responses to osmotic stresses in soybean. In: Sudaric A., Editor (ed.) Soybean - Molecular Aspects of Breeding. InTech: Rijeka, Croatia. 2011. p 215-240.

[102] Gao F., Xiong A., Peng R., Jin X., Xu J., Zhu B., Chen J., Yao Q. OsNAC52, a rice NAC transcription factor, potentially responds to ABA and confers drought tolerance in transgenic plants. Plant Cell, Tissue and Organ Culture 2010;100(3) 255-262.

[103] Chen M., Wang Q.Y., Cheng X.G., Xu Z.S., Li L.C., Ye X.G., Xia L.Q., Ma Y.Z. GmDREB2, a soybean DRE-binding transcription factor, conferred drought and high-salt tolerance in transgenic plants. Biochemical and Biophysical Research Communications 2007;353(2) 299-305.

[104] Chen M., Xu Z.S., Xia L.Q., Li L.C., Cheng X.G., Dong J.H., Wang Q.Y., Ma Y.Z. Cold-induced modulation and functional analyses of the DRE-binding transcription factor gene, GmDREB3, in soybean (Glycine max L.). Journal of Experimental Botany 2009;60(1) 121-135.

[105] Zhang G., Chen M., Li L., Xu Z., Chen X., Guo J., Ma Y. Overexpression of the soybean GmERF3 gene, an AP2/ERF type transcription factor for increased tolerances to salt, drought, and diseases in transgenic tobacco. Journal of Experimental Botany 2009;60(13) 3781-3796.

[106] Zhang G., Chen M., Chen X., Xu Z., Li L., Guo J., Ma Y. Isolation and characterization of a novel EAR-motif-containing gene GmERF4 from soybean (Glycine max L.). Molecular Biology Reports 2010;37(2) 809-818.

[107] Liao Y., Zhang J.S., Chen S.Y., Zhang W.K. Role of soybean GmbZIP132 under abscisic acid and salt stresses. Journal of Integrative Plant Biology 2008;50(2) 221-230.

[108] Gao S.Q., Chen M., Xu Z.S., Zhao C.P., Li L., Xu H., Tang Y., Zhao X., Ma Y.Z. The soybean GmbZIP1 transcription factor enhances multiple abiotic stress tolerances in transgenic plants. Plant Molecular Biology 2011;75(6) 537-553.

[109] Xie Z.M., Zou H.F., Lei G., Wei W., Zhou Q.Y., Niu C.F., Liao Y., Tian A.G., Ma B., Zhang W.K. Soybean trihelix transcription factors GmGT-2A and GmGT-2B improve plant tolerance to abiotic stresses in transgenic Arabidopsis. PLoS ONE 2009;4(9) e6898.

[110] Luo X., Bai X., Zhu D., Li Y., Ji W., Cai H., Wu J., Liu B., Zhu Y. GsZFP1, a new Cys2/His2-type zinc-finger protein, is a positive regulator of plant tolerance to cold and drought stress. Planta 2011; 1-15.

[111] Zhou Q.Y., Tian A.G., Zou H.F., Xie Z.M., Lei G., Huang J., Wang C.M., Wang H.W., Zhang J.S., Chen S.Y. Soybean WRKY-type transcription factor genes, GmWRKY13,

GmWRKY21, and GmWRKY54, confer differential tolerance to abiotic stresses in transgenic Arabidopsis plants. Plant Biotechnology Journal 2008;6(5) 486-503.

[112] Zhang L., Wang X.P., Bi Y.D., Zhang C.Y., Fan Y.L., Wang L. Isolation and functional analysis of transcription factor GmWRKY57B from soybean. Chinese Science Bulletin 2008;53(22) 3538-3545.

[113] Lam H.M., Xu X., Liu X., Chen W., Yang G., Wong F.L., Li M.W., He W., Qin N., Wang B. Resequencing of 31 wild and cultivated soybean genomes identifies patterns of genetic diversity and selection. Nature Genetics 2010;42(12) 1053-1059.

[114] Chang R., Qiu L. Evaluation and utilization of soybean germplasm in China. In: Lam H.M., Chang R., Shao G.,Liu Z., Editors. (ed.) Research on tolerance to stresses in chinese soybean. China agricultural press: Beijing. 2009.

[115] Kumar P., Gupta V.K., Misra A.K., Modi D.R., Pandey B.K. Potential of molecular markers in plant biotechnology. Plant Omics 2009;2(4) 141-162.

[116] Carter T.E., DeSouza P.I., Purcell L.C. Recent advances in breeding for drought and aluminum resistance in soybean: Superior Printing; 1999.

[117] Carter Jr T.E., Orf J., Purcell L., Specht J., Chen P., Sinclair T., Rufty T. Tough times, tough plants - new soybean genes defend against drought and other stresses. in American Seed Trade Association Conference Proceedings. 2006. Alexandria, VA.

[118] Bhatnagar S., King C.A., Purcell L., Ray J.D. Identification and mapping of quantitative trait loci associated with crop response to water-deficit stress in soybean [Glycine Max (L.) Merr.]. 2005.

[119] Du W., Wang M., Fu S., Yu D. Mapping QTLs for seed yield and drought susceptibility index in soybean (Glycine max L.) across different environments. Journal of Genetics and Genomics 2009;36(12) 721-731.

[120] Mian M.A.R., Carter T.E., Parrott W.A., Wells R., Bailey M.A., Ashley D.A., Boerma H.R. Molecular markers associated with water use efficiency and leaf ash in soybean. Crop Science 1996;36(5) 1252-1257.

[121] Xue R.G., Xie H.F., Zhang B. A multi-needle-assisted transformation of soybean cotyledonary node cells. Biotechnology Letters 2006;28(19) 1551-1557.

[122] Dang W., Wei Z. An optimized Agrobacterium-mediated transformation for soybean for expression of binary insect resistance genes. Plant Science 2007;173(4) 381-389.

[123] Rech E.L., Vianna G.R., Aragao F.J.L. High-efficiency transformation by biolistics of soybean, common bean and cotton transgenic plants. Nature Protocols 2008;3(3) 410-418.

[124] Liu M., Yang J., Cheng Y., An L. Optimization of soybean (Glycine max (L.) Merrill) in planta ovary transformation using a linear minimal gus gene cassette. Journal of Zhejiang University-Science B (Biomedicine & Biotechnology 2009;10(12) 870-876.

[125] Jung C., Seo J.S., Han S.W., Koo Y.J., Kim C.H., Song S.I., Nahm B.H., Do Choi Y., Cheong J.J. Overexpression of AtMYB44 enhances stomatal closure to confer abiotic stress tolerance in transgenic Arabidopsis. Plant physiology 2008;146(2) 623-635.

[126] Seo J.S., Sohn H.B., Noh K., Jung C., An J.H., Donovan C.M., Somers D.A., Kim D.I., Jeong S.C., Kim C.G. Expression of the Arabidopsis AtMYB44 gene confers drought/ salt-stress tolerance in transgenic soybean. Molecular Breeding 2012; 1-8.

[127] de Ronde J.A., Laurie R.N., Caetano T., Greyling M.M., Kerepesi I. Comparative study between transgenic and non-transgenic soybean lines proved transgenic lines to be more drought tolerant. Euphytica 2004;138(2) 123-132.

[128] Xue R.G., Zhang B., Xie H.F. Overexpression of a NTR1 in transgenic soybean confers tolerance to water stress. Plant Cell, Tissue and Organ Culture 2007;89(2) 177-183.

[129] Qi Q. Effect of transgenic DREB3 drought resistant soybean on soil enzyme activity and soil functional microorganism, in Agricultural Science 2012, Northeast Agricultural University. p. 183.

[130] Cooley H., Christian-Smith J., Gleick P.H. More with less: Agricultural water conservation and efficiency in California, Ross N., Editor 2008, Pacific Institute: Oakland, CA.

[131] Barta R., Broner I., Schneekloth J., Waskom R. Colorado high plains irrigation practices guide - water saving options for irrigators in Eastern Colorado, 2004, Colorado Water Resources Research Institute.

[132] Peterson J.M., Ding Y. Economic adjustments to groundwater depletion in the high plains: Do water-saving irrigation systems save water? American Journal of Agricultural Economics 2005;87(1) 147-159.

[133] Specht J.E., Elmore R.W., Eisenhauer D.E., Klocke N.W. Growth stage scheduling criteria for sprinkler-irrigated soybeans. Irrigation Science 1989;10(2) 99-111.

[134] Klocke N.L., Eisenhauer D.E., Specht J.E., Elmore R.W., Hergert G.W. Irrigation soybeans by growth stages in Nebraska. Applied Engineering in Agriculture 1989;5(3) 361-366.

[135] Sweeney D.W., Granade G.V. Effect of a single irrigation at different reproductive growth stages on soybean planted in early and late June. Irrigation Science 2002;21(2) 69-73.

[136] Sweeney D.W., Long J.H., Kirkham M.B. A single irrigation to improve early maturing soybean yield and quality. Soil Science Society of America Journal 2003;67(1) 235-240.

[137] FAO. Crop water information: soybean. 2012 [cited 2012 May]; Available from: http://www.fao.org/nr/water/cropinfo_soybean.html.

[138] Rogers D.H. Soybean Production Handbook: Kansas State University; 1997.

[139] Zhang J., Davies W.J. Changes in the concentration of ABA in xylem sap as a function of changing soil-water status can account for changes in leaf conductance and growth. Plant Cell and Environment 1990;13(3) 277-285.

[140] Zhang J.H., Davies W.J. Does ABA in the xylem control the rate of leaf growth in soil-dried maize and sunflower plants? Journal of Experimental Botany 1990;41(230) 1125-1132.

[141] Zhang J.H., Davies W.J. Sequential response of whole plant water relations to prolonged soil drying and the involvement of xylem sap ABA in the regulation of stomatal behavior of sunflower plants. New Phytologist 1989;113(2) 167-174.

[142] Zhang J.H., Schurr U., Davies W.J. Control of stomatal behavior by abscisic-acid which apparently originates in the roots. Journal of Experimental Botany 1987;38(192) 1174-1181.

[143] Liang J., Zhang J., Wong M.H. Effects of air-filled soil porosity and aeration on the initiation and growth of secondary roots of maize (Zea mays). Plant and Soil 1996;186(2) 245-254.

[144] Skinner R.H., Hanson J.D., Benjamin J.G. Root distribution following spatial separation of water and nitrogen supply in furrow irrigated corn. Plant and Soil 1998;199(2) 187-194.

[145] Kang S.Z., Zhang J.H. Controlled alternate partial root-zone irrigation: its physiological consequences and impact on water use efficiency. Journal of Experimental Botany 2004;55(407) 2437-2446.

[146] Tabrizi M.S., Parsinejad M., Babazadeh H. Efficacy of partial root drying technique for optimizing soybean crop production in semi-arid regions. Irrigation and Drainage 2012;61(1) 80-88.

[147] Sepaskhah A.R., Ahmadi S.H. A review on partial root-zone drying irrigation. International Journal of Plant Production 2010;4(4) 241-258.

[148] Du T., Kang S., Zhang J., Li F., Yan B. Water use efficiency and fruit quality of table grape under alternate partial root-zone drip irrigation. Agricultural Water Management 2008;95(6) 659-668.

[149] Du T.S., Kang S.Z., Zhang J.H., Li F.S. Water use and yield responses of cotton to alternate partial root-zone drip irrigation in the arid area of north-west China. Irrigation Science 2008;26(2) 147-159.

[150] Huang Z.D., Qi X.B., Fan X.Y., Hu C., Zhu D.H., Li P., Qiao D.M. Effects of alternate partial root-zone subsurface drip irrigation on potato yield and water use efficiency. Ying Yong Sheng Tai Xue Bao 2010;21(1) 79-83.

[151] Li F.S., Wei C.H., Zhang F.C., Zhang J.H., Nong M.L., Kang S.Z. Water-use efficiency and physiological responses of maize under partial root-zone irrigation. Agricultural Water Management 2010;97(8) 1156-1164.

[152] Guo Z.-L., Sun C.-Q., Liang N. Impacts of plastic mulching on water saving and yield increasing of dry land spring soybean and its density effect. Chinese Journal of Eco-Agriculture (in Chinese) 2007;15(1) 205-206.

[153] Wang L., Chen G., Zhang G., Li X., Ni S., Yang R. Water use efficiency of soybean under different mulching in dryland. Soybean Science 2010;29(5) 767-771.

[154] Wang L., Chen G., Zhang G., Li X., Ni S., Yang R. Effects of mulching and water conservation technology for soybean in the rainfed highland of the Loess Plateau. Crops 2011;6 95-98.

Chapter 11

Biologically Active Constituents of Soybean

Tzi Bun Ng, Randy Chi Fai Cheung and Jack Ho Wong

Additional information is available at the end of the chapter

http://dx.doi.org/10.5772/52526

1. Introduction

A diversity of bioactive biomolecules present in soybean is covered in this chapter such as e-Phe-Leu and Trp-Leu, and Val-Leu-Ile-Val-Pro derived from glycinin; Bowman-Birk inhibitor, Kunitz trypsin inhibitor, hemagglutinin, isoflavone-deprived soy peptide, neutral PR-5 protein, SbPRP protein, ferritin, peroxisomal proteins, defense proteins such as calmodulin, disease resistance protein, beta-glucan-binding protein, l unasin, and glysojanin; enzymes such as phenylalanine ammonia-lyase and tyrosine ammonia-lyase, defense-related enzymes, cysteine proteinase, isoflavone synthase, isocitrate lyase, vestitone reductase and chalcone reductase, UDP-glucose: flavonoid 3-O-glucosyltransferase, beta-glucosidase, isoflavone conjugate-hydrolyzing beta-glucosidase, genes related to 2-oxo-glutarate-dependent dioxygenases, 5'-adenylylsulfate reductase, and ATP sulfurylase, anticarcinogenic daidzein-rich fraction, polysaccharides, glyceollins, and isoflavones. The aforementioned soybean constituents manifest many uses and consumption of soybean promotes health.

Soybean represents a rich and yet relatively inexpensive source of proteins. Hence it is a common dietary component in many countries. It is well known that some of its constituents like isoflavones are beneficial to health, and regular consumption of soybean has been associated with a reduced incidence of diseases such as osteoporosis, cancer and cardiovascular disease. A wealth of information pertaining to the various bioactive constituents of soybean has been accumulated. The aim of this article is to review the information that is available.

2. Storage proteins

A combined proteomic approach was employed for the separation, identification, and comparison of two major storage proteins, beta-conglycinin and glycinin, from cultivated (*Glycine max*) and wild (*Glycine soja*) soybean seeds. Two-dimensional polyacrylamide gel electrophoresis with three different immobilized pH gradient strips effectively resolved many storage proteins. The pH range 3.0 - 10.0 was good for separating most of the beta-conglycinin subunits while pH ranges 4.0 - 7.0 and 6.0 - 11.0 were satisfactory for separating acidic and basic glycinin polypeptides, respectively. Although the distribution pattern of the protein spots was in general alike in both genotypes by employing pH 3.0 - 10.0, variations in number and intensity of protein spots were better resolved when a combination of pH 4.0 - 7.0 and pH 6.0 - 11.0 was utilized. The total number of storage protein spots detected in wild and cultivated genotypes was approximately 44 and 34, respectively [1].

Krishnan et al have developed a fast and simple fractionation technique using 10 mM Ca^{2+} to precipitate soybean seed storage globulins, glycinin and beta-conglycinin. This method eliminates over 80% of the highly abundant seed proteins from the extract, facilitating detection of previously inconspicuous proteins in soybean seed [2].

3. β-conglycinin

Ferreira et al conducted a study to investigate the effect of daily oral administration of soybean β-conglycinin (7S protein) (300 mg/kg/day) alone and in conjunction with the hypolipidemic drugs fenofibrate (30 mg/kg/day) and rosuvastatin (10 mg/kg/day) in hypercholesterolemic rats for 28 days. The data reveal that conglycinin exhibits effects similar to those of fenofibrate and rosuvastatin in the control of plasma cholesterol, HDL-C and triglycerides, and suggests that the association of soybean β-conglycinin with rosuvastatin changes the action of the drug in cholesterol homeostasis [3].

4. Glycinin

Glycinin was isolated and characterized from soybeans stored under different ambient conditions (i) in 84% relative humidity at 30 °C (adverse conditions) for 9 months and (ii) in 57% relative humidity at 20 °C, (iii) in the cold (4 °C), and (iv) in an uncontrolled ambient garage for 18 months.. Glycinin purified from soybean exposed to the adverse conditions possessed a significant amount of sugar and demonstrated a decrease in hydrophobic interactions after 3 months; the total free sulfhydryl content in glycinin dropped, but the intramolecular disulfide bonds increased; the alpha-helix content of secondary structure underwent a slight rise, but the beta-sheet content fell. The structure of glycinin purified from and subjected to the other three aforementioned conditions remained unaltered after 18 months of storage when compared to the control. The molecular mass of glycinin remained in the range of 313-340 kDa throughout the storage period for all four conditions [4].

Addition of 1% cholesterol and 0.5% cholic acid induced hypercholesterolemia in Male Wistar rats without altering weight gain. Daily administration (300 mg/kg/day) of glycinin (11S globulin) for 28 days led to a 2.8% increase of dietary protein intake and resulted in elevated HDL-C, reduced hepatic triglycerides and improved atherogenic index in hypercholesterolemic animals [5].

Sitohy et al [6] isolated glycinin, basic subunit and β-conglycinin from soybean protein isolate and tested them for antimicrobial activity against pathogenic and spoilage bacteria as compared to penicillin. All three of them demonstrated antibacterial activity identical to or higher than penicillin in the order; basic subunit > glycinin > β-conglycinin with minimal inhibitory concentration (MIC) of 50, 100 and 1000 μg/ml, respectively. The half maximal inhibitory concentration (IC_{50}) values of the basic subunit, glycinin and β-conglycinin were 15, 16 and 695 μg/ml against *Listeria monocytogenes*; 17, 20, and 612 μg/ml against *Bacillus subtilis*; and 18, 21 and 526 μg/ml against *S. enteritidis*, respectively. Transmission electron microscopy images of *L. monocytogenes* and *S. enteritidis* revealed larger sizes and separation of cell wall from cell membrane following treatment with glycinin or basic subunit. Scanning electron microscopy of *B. subtilis* disclosed an irregular wrinkled external surface, fragmentation, adhesion and aggregation of damaged cells or cellular debris following treatment with glycinin or the basic subunits but not with penicillin. All tested substances, especially the basic subunit, manifested enhanced concentration-dependent cell permeation as indicated by crystal violet uptake. Glycinin and basic subunit demonstrated a more rapid antimicrobial action compared with penicillin. The cell killing efficiency was in the following descending order; basic subunit > glycinin > penicillin > β-conglycinin and the susceptibility of the bacteria to the antimicrobial agents was in the order: *L. monocytogenes* > *B. subtilis* > *S. enteritidis*. Addition of glycinin and the basic subunit to pasteurized milk inoculated with the three bacteria; *L. monocytogenes, B. subtilis* and *S. enteritidis* (approximately 5 log CFU/ml) impeded their proliferation after storage at 4 °C for 16-20 days [6].

5. Soybean 7S globulin

In rats fed a hypercholesterolemic diet soybean 7S globulin reduced plasma cholesterol and triglycerides and upregulated liver beta- very-low-density lipoprotein (VLDL) receptors [7].

6. Basic 7S globulin (Bg7S)

Basic 7S globulin (Bg7S) is expressed by soybeans in response to biotic or abiotic stress. It is capable of binding to a 4-kDa protein involved in cell proliferation. Bg7S is detected in legumes, and other plants. However, its function has not been elucidated. Bg7S has been successfully crystallized. Orthorhombic and monoclinic crystals were procured under various conditions and belonged to space groups P2(1)2(1)2, with unit-cell parameters a = 111.9, b = 130.1, c = 287.8 Å, and P2(1), with unit-cell parameters a = 85.3, b = 137.6, c = 162.1 Å, β = 91.2°, respectively [8].

7. Soybean food allergens

As part of the safety assessment of genetically modified (GM) soybean, 2-dimensional gel electrophoresis analyses were performed with the isoxaflutole and glyphosate-tolerant soybean FG72, its non-GM near-isogenic counterpart (Jack) and three commercial non-GM soybean lines. Rouquié et al [9] compared the known endogenous human food allergens in the seeds of the five different soybean lines to assess any potential unanticipated consequences of the genetic modification. Thirty-seven protein spots representing five well known soybean food allergen groups were quantified in each genotype. Among the various protein spots, the levels of accumulation of three allergens were slightly attenuated in GM soybean than in non-GM counterparts although all allergenic proteins were detected in the different genetic backgrounds. There was no significant rise in the level of allergens in FG72 soybean seeds which can thus be considered as safe as its non-GM counterpart [9].

8. Immunodominant soybeanallergen P34

Protein P34 (Gly m Bd 30K) is the immunodominant allergen in soybean. Among commercial soy ingredients, soy flour displayed the highest P34 antigenicity (32 mg/g extracted protein) followed by soy protein isolate (29 mg/g extracted protein) and soy protein concentrate (24 mg/g extracted protein). Among soy consumer products, soymilk exhibited the highest P34 antigenicity (from 7 to 23 mg/g extracted protein), followed by tempeh (8 mg/g extracted protein), soy infant formula (3.4 mg/g extracted protein), soy powder (2 mg/g extracted protein), and soy cheese products (0.50 mg/g extracted protein). Korean miso, soy sauce, soy chili mix, soy nuts, soy cream cheese, soy meat patty, texturized soy protein, and soy cereal exhibited undetectable P34 antigenicity (detection limit = 0.45 ng). Selecting soybean varieties low in this allergen, or via processing, potentially can render soybean products less antigenic [10].

The sera from 3 of 9 (33%) patients with outgrown soybean allergy and 6 of 9 (66%) patients with soybean allergy were classified as positive. Sodium dodecyl sulfate (SDS)-treated recombinant P34 retained its structure and biological activity. Recombinant P34 is a useful tool for the analysis of antigen-specific response in soybean allergy. A modified form of recombinant P34 for the diagnosis or treatment of soybean allergy using specific immunotherapy techniques may be available in the future [11].

9. GmPep914

The peptide DHPRGGNY from soybean leaves, at a concentration of 0.25 nM, alkalinizes the media of soybean suspension cells, a response associated with defense peptides. Only 5 to 10 min are needed to achieve a maximal pH change. The peptide is present at the carboxy-terminal end of a 52-amino acid precursor protein (Glyma12g00990) deduced from the soybean genome project. A search of the soybean databank discloses a homolog (Glyma09g36370) that contains a similar peptide, DLPRGGNY. The synthetic peptide exhibits the same activity. The peptides, designated GmPep914 (DHPRGGNY) and GmPep890 (DLPRGGNY), can induce the expression of both Glyma12g00990 (GmPROPEP914) and Glyma09g36370 (GmPROPEP890) in cultured soybean suspension cells within 1 h. Both peptides induced the expression of defense genes such as CYP93A1(a cytochrome P450 gene implicated in phytoalexin synthesis), chitinase b1-1(a chitinase implicated in pathogen defense), and soybean chalcone synthase1 (Gmachs1), and chalcone synthase, implicated in phytoalexin production. Both GmPROPEP914 and GmPROPEP890 were highly expressed in the roots, relative to the aerial portions of the plant. However, treatment of the aerial portion of soybean plants with hormones involved in elicitation of defense responses revealed augmented expression levels of GmPROPEP914 and GmPROPEP890 [12].

10. Cyclo(his-pro)

Cyclo(His-Pro),(CHP),a naturally occurring, cyclic dipeptide structurally related to thyrotropin-releasing hormone, was obtained from soybean meal by hydrolysis with flavourzyme and alcalase. When rat insulinoma cells (RINm5F) were exposed to 2 mM streptozotocin (STZ), insulin secretion declined to approximately half. CHP treatment restored the insulin-secreting activity of RINm5F cells to approximately 71% of that of untreated control cells. CHP significantly protected the cells from STZ -mediated cytotoxicity via reduction of STZ-induced nitric oxide production (2.3-fold) and lipid peroxidation (1.9-fold). STZ-induced apoptotic events, including activation of caspase-3, poly(ADP-ribose) polymerase (PARP) cleavage, and DNA fragmentation in RINm5F cells were attenuated by CHP. These results suggest that CHP could be employed as a protective and therapeutic agent against STZ-mediated cytotoxicity and apoptosis [13].

11. Immunostimulating glutamine-rich peptide

An immunostimulating glutamine-rich peptide, isolated from the soybean protein fraction digested with *Rhizopus oryzae* Peptidase R, was located at or in the vicinity of the glutamine-rich region 202 to 222 of the glycinin G4 subunit. The peptide significantly upregulated the number of CD8(+), CD11b(+), and CD49b(+) cells in C3H/HeN mouse spleen cell cultures. Two chemically synthesized glutamine-rich peptides (QQQQQQKSHGGR and

KQGQHQQEEEEE), corresponding to residues 202 to 213 and residues 214 to 225 of the glycinin G4 subunit, elevated the number of interleukin (IL)-12(+)CD11b(+) cells. Peptide 202-213 increased the number of CD49b(+), IL-2(+)CD4(+), and interferon-gamma(+)CD4(+) cells, and stimulated the cytotoxic activity of splenocytes toward K562 human erythroleukemia cells. The data suggest that the glutamine-rich region of the soybean glycinin G4 subunit enhances the cellular immune system in murine splenocyte cultures [14].

12. Soy protein

Soy protein mitigates lipotoxicity in the liver and pancreas in various rodent models of obesity. Dietary soy protein can suppress cardiac cholesterol and triglyceride concentrations and reduce cardiacceramide concentration by downregulating the expression of serine palmitoyl transferase, a key enzyme in sphingolipid biosynthesis in the heart of obese rodents, and by reducing lipid accumulation. Thus, soy protein consumption may be a useful dietary therapeutic strategy for l prevention of lipotoxic cardiomyopathy [15].

13. Soy protein-derived angiotensin I-converting enzyme (ACE)inhibitory hydrolysate

Cha and Park [16] produced a soy protein-derived ACE inhibitory hydrolysate and characterized its activity, physicochemical, and biochemical properties. The competitive ACE inhibitory activity remained stable after incubation with gastric enzymes for 2 hours

14. Angiotensin I-converting enzyme inhibitory peptides

Two non-competitive ACE inhibitors, Ile-Phe-Leu (IC_{50}, 44.8 M) and Trp-Leu (IC_{50}, 29.9 M), were isolated from the tofuyo (fermented soybean food) extract. The first sequence is contained in the alpha- and beta-subunits of beta-conglycinin, while the second sequence exists in the B-, B1A- and BX-subunits of soybean glycinin. The inhibitory activity of Trp-Leu was unaffected by the gastrointestinal proteases pepsin, chymotrypsin and trypsin. About 30% of the activity remained after successive digestion by these proteases [17].

15. Angiotensin I-converting enzyme inhibitory peptide derived from glycinin

The pentapeptide Val-Leu-Ile-Val-Pro derived from glycinin has an IC_{50} of 1.69 +/- 0.17 M and a K_i of 4.5 +/- 0.25 x 10^{-6} M for ACE inhibition. This peptide was resistant to digestion by gastrointestinal proteases [18].

16. Bowman-Birk Inhibitor (BBI)

The low cancer incidence associated with soybean consumption has been related to the presence of Bowman-Birk inhibitor (BBI) and lectin [19]. Two different strategies were used for the extraction of BBI and lectin. The first method entailed extraction of soybean proteins using Tris-HCl buffer followed by isolation of BBI and lectin by isoelectric precipitation of other soybean proteins. The second method involved direct extraction of BBI and lectin using an acetate buffer. The effect of previous delipidation on the extraction of BBI and lectin was examined. The possibility of using a high-intensity focalized ultrasonic probe for expediting the extraction was investigated and the extraction time and ultrasound amplitude were optimized. After evaluation of its analytical characteristics (linearity, precision, and recovery), the method was used for quantitating BBI and lectin in different varieties of soybean [19].

17. Bowman-Birk (BBI) isoinhibitors

Two major soybean isoinhibitors, IBB1 and IBBD2, were found in soybean. IBB1 inhibited both trypsin and chymotrypsin while IBBD2 was trypsin-specific. The median inhibitory concentration values of IBB1 and IBBD2 on HT29 colon cancer cell growth were not significantly different (39.9+/-2.3 and 48.3+/-3.5 M, respectively). The distribution pattern of HT29 colon cancer cells was affected by BBI: cells remained in the G0-G1 phase of the cell cycle [20].

18. Kunitz trypsininhibitor

Duranti et al described a method for the extraction and purification of Kunitz trypsin inhibitor from soybean seeds. The first step involved heat treatment of whole soybean seeds in water at 60 °C for 90 min. It was found that 8.4% of total trypsin inhibitory activity of the seeds was secreted during heat treatment. The aqueous medium was applied on an affinity chromatography column with immobilized trypsin. The method is readily scalable to pilot plant or industrial preparations [21].

19. Hemagglutinin

Soybean agglutinin (SBA) is a specific N-acetylgalactosamine-binding lectin that can agglutinate a number of cell types. SBA has potential for utilization as an affinity tag for high-quality purification of tagged proteins, isolation of fetal cells from maternal blood for genetic screening, screening and treatment of breast cancer, and use as a carrier system for oral drug delivery. Tremblay et al employed transient and stable expression systems in potato and *Nicotiana benthamiana*, respectively, for the production of recombinant SBA, with the transgenic protein accumulated to 0.3% and 4% of total soluble protein in potato tubers and *Nicotiana benthamiana* leaves, respectively. Both rSBAs were able to resist degradation in simulated gastric and intestinal fluids. High yield and purity of rSBA can be obtained by affinity column chromatography using N-acetylgalactosamine as a specific ligand [22].

20. Glyceollins

Glyceollins produced from daidzein in soybeans cultured with fungi exert antifungal and cancer preventive actions [23]. Isoflavones and metabolites serve as a mixed agonist/antagonist for estrogen. Glyceollins suppress some cancer cells via anti-estrogenic activity. Estrogenic effects of glyceollins have been demonstrated using E-screen assay and pS2 expression. Glyceollins exhibited higher affinity for estrogen receptor beta than estrogen receptor alpha. After *Aspergillus sojae* infection glyceollins were more efficiently produced de novo in minced soybean than in half-sliced soybean. Hence glyceollins may be useful for preventing or alleviating postmenopausal complications due to potent estrogenic activity, and their production could be varied depending upon processing prior to fungal inoculation [23].

21. EAR-motif-containing gene GmERF4

The ERF-associated amphiphilic repression (EAR) motif is vital for transcriptional repression. The soybean protein GmERF4 contains one AP2/ERF domain, two putative nuclear localization signal regions and one EAR motif. It is preferentially localized in the nuclei of onion epidermis cells and bound specifically to the GCC box and DRE/CRT element *in vitro*. Its expression was inducible by soybean mosaic virus, salt, cold, drought, ethylene, jasmonic acid, and salicylic acid, and inhibited by abscisic acid. Constitutive expression of GmERF4 in transgenic tobacco plants enhanced tolerance to salt and drought stresses compared with wild-type plants, but did not elicit detectable resistance against bacterial infection [24].

22. Isoflavones

Contradictory reports are available on the influence of source of dietary protein and soy isoflavones on blood pressure. Therefore, Gilani et al. conducted a study to examine whether the source of dietary protein (casein vs. soybean protein, washed by alcohol to eliminate most isoflavones), dietary extracted soybean isoflavones and anthocyanins modulate the lifespan of Stroke-prone Spontaneously Hypertensive rats. The survival times and survival rates of animals fed casein and soybean protein diets were not statistically different. However, there was a significant effect of isoflavone or anthocyanins supplementation on survival times and survival rates. Death occurred significantly earlier in the isoflavones- or anthocyanins-supplemented groups [25].

23. Soybean resistant protein

The effect of soybean resistant protein on cholesterol levels in serum and the liver and fecal steroid excretion was examined. The protein downregulated hepatic cholesterol, probably because of elevated excretion of cholesterol and bile acids. The serum cholesterol level was not different as between resistant protein and other soy-derived proteins [26].

24. Isoflavone-deprived soy peptide

NF-gammaB mediates processes of carcinogenesis related to deregulation of the normal control of proliferation, angiogenesis, and metastasis. NF-gammaB suppression is associated with cancer chemoprevention. Heat shock protein 90 is important to NF-gammaB activation and stabilizes key proteins implicated in cell cycle control and apoptosis signaling. Administration of isoflavone-deprived soy peptide (3.3 g/rat/day) in the diet reduced the incidence of 7,12-dimethylbenz[alpha]anthracene (DMBA)-induced rat mammary ductal carcinomas, the number of tumors, and prolonged the latency period of tumor development compared to control diet animals. Isoflavone-deprived soy peptide downregulates the expression of heat shock protein 90, hence thwarting the signaling pathway that brings about NF-gammaB activation.It upregulates the expression of p21, p53, and caspase-3 proteins; and downregulates the expression of VEGF. In keeping with in vivo, treatment with the soy peptide inhibited growth and induced apoptosis in human breast MCF-7 tumor cells. [27].

25. Neutral PR-5protein

A gene designated as *Glycine max* osmotin-like protein, b isoform (GmOLPb, accession no. AB370233) encodes a putative protein with striking resemblance (89% identity) to chickpea PR-5b. In contrast to the two acidic isoforms of soybean PR-5 protein (GmOLPa and P21), GmOLPb possesses a C-terminal extension for possible vacuolar targeting and after maturation displays a calculated molecular mass of 21.9kDa and a pI of 6.0. Under conditions of high-salt stress, GmOLPb was highly induced in the leaves 18-72h after onset of the stress. Gene expression of P21e (a homolog of P21) was transiently induced by high-salt stress, but this occurred prior to the gene expression of GmOLPa and GmOLPb. Differential expression also was noted in studies employing methyl jasmonate and salicylic acid. These results suggested that each soybean PR-5 might have a unique function in the defensive system to protect the soybean plant from stress imposed by high salt, especially in the leaves [28].

26. New defense gene

The SbPRP gene putatively encodes a bimodular protein with 126 amino acids similar to developmentally regulated proteins in other plants. It has a distinct proline-rich domain (17 amino acids) and a long hydrophobically cysteine-rich domain (84 amino acids). Its steady-state mRNA level accumulated differentially in response to salicylic acid, and to the inoculation of soybean mosaic virus Sa strain. It responded to drought treatment and salt stress. It appears that SbPRP plays a defense role in soybean [29].

27. SBTX, a new toxic protein distinct from soyatoxin

SBTX is a 44-kDa basic glycoprotein composed of a 27-kDa polypeptide chain and a 17-kDa polypeptide chain linked by a disulfide bond. The N-terminal sequences of the 44 and 27 kDa chains were identical. The secondary structure content was 35 % alpha-helix, 13% beta-strand and beta-sheet, 27 % beta-turn, 25 % unordered, and 1 % aromatic residues. SBTX lacked protease-inhibitory and hemagglutinating activities, but was immunologically related to other toxic proteins, such as soyatoxin and canatoxin, and cross-reacted weekly with soybean trypsin inhibitor and agglutinin. It inhibited the growth of *Cercospora sojina,* a fungus causing frogeye leaf spot in soybeans at a concentration much lower than the lethal dose to mice. Thus it may be useful for the development of transgenic plants with augmented resistance to pathogens [30].

28. Ferritin

Soybean seed ferritin contains a great deal of bioavailable iron and is important for human iron supplementation and prevention of anemia caused by iron deficiency. Dong et al [31] employed a rapid and simple *Escherichia coli* expression system for producing the soybean seed ferritin complex. The two subunits, H-2 and H-1, were encoded in a single plasmid, and optimal expression was accomplished by coexpressing in addition a team of molecular chaperones, trigger factor and GroEL-GroES. The His-tagged ferritin complex was purified by Ni^{2+} affinity chromatography. An intact ferritin complex was obtained following His-tagged enterokinase digestion [31].

29. Peroxisomal proteins

Peroxisomal proteins comprise enzymes for metabolite transport, stress response, fatty acid beta-oxidation, the glyoxylate cycle, and photorespiratory glycolate metabolism, an enoyl-CoA hydratase/isomerase family protein, a short-chain dehydrogenase/reductase family protein, 3-hydroxyacyl-CoA dehydrogenase-like protein, and a voltage-dependent anion-selective channel protein [32].

30. Calmodulin

Calmodulin (CaM) plays a role in defense responses in plants. In soybean, induction of transcription of calmodulin isoform 4 (GmCaM4) occurred within half an hour following pathogen stimulation. . Park et al utilized the yeast one-hybrid system to isolate two cDNA clones encoding proteins (GmZF-HD1 and GmZF-HD2) which bind to a 30-nt A/T-rich sequence in the GmCaM4 promoter, a region with two repeats of a conserved homeodomain binding site, ATTA. The two proteins are members of the zinc finger homeodomain (ZF-HD) transcription factor family. A homeodomain motif, but not the two zinc finger domains, is capable of binding to the 30-nt GmCaM4 promoter sequence. The interaction between GmZF-HD1 and two homeodomain binding site repeats is subject to regulation by pathogen stimulation. GmZF-HD1 can activate the expression of GmCaM4 through interaction with the two repeats. Thus GmZF-HD1 and GmZF-HD-2 proteins are ZF-HD transcription factors that activate GmCaM4 gene expression upon encounter with a pathogen [33].

31. Soybean disease resistance protein RHG1-LRR domain

RHG1 encoded by rhg1 was a soybean transmembrane receptor-like kinase (EC 2.7.11.1) with an extracellular leucine-rich repeat (LRR) domain. The LRR of RHG1 was believed to be involved in elicitor recognition and interaction with other plant proteins. Afzal and Lightfoot expressed the LRR domain in Escherichia coli (RHG1-LRR) and produce refolded protein [34].

32. ABC-transporter

A new ATP-binding cassette (ABC) transporter was identified as a salicylic acid-induced gene from soybean in a subtractive suppression hybridization approach. The encoded 1447-amino acid protein has two similar repeat units typical of full-size ABC transporters and a close relationship to plant pleiotropic drug resistance (PDR)-type transporters. GmPDR12 responds quickly to salicylic acid, functional analogues of salicylic acid and methyl jasmonate. Salicylic acid is required for the hypersensitive reaction in soybean cell suspension cultures inoculated with *Pseudomonas syringae* pv. glycinea [35].

33. Beta-glucan-binding protein

Soybean beta-glucan-binding protein (GBP) binds a microbial cell wall elicitor, triggering the activation of defense responses. It hydrolyzes beta-1,3-glucans present in the cell walls of potential pathogens [36].

34. Lunasin

Lunasin, a chemopreventive soybean peptide, suppresses transformation of mammalian cells induced by chemical carcinogens and inhibits skin carcinogenesis in mice. All lunasin fractions partially purified by anion exchange and immunoaffinity column chromatography inhibit colony formation induced by the ras-oncogene and inhibit core H3-histone acetylation. The peptide is found in different soybean varieties and commercially available soy proteins by Western blot analysis [37].

35. Pti1 homologue

A full-length Pti1-like gene GmPti1, encoding a protein of 366 amino acids with a Ser/Thr/Tyr kinase domain, was identified from soybean in an EST sequencing project by its homology to tomato Pti1. GmPti1 expression was inducible by wounding and salicylic acid. GmPti1 protein was expressed in *E. coli* as an MBP fusion, purified by amylase resin and examined for its autophosphorylation ability. GmPti1 exhibited kinase activity in the presence of Mn^{2+} ions [38].

36. Glysojanin

A monomeric 25-kDa antifungal protein designated as glysojanin, with an N-terminal sequence resembling a segment of chitin synthase, was isolated from the seeds of the black soybean Glycine soja. The protein demonstrated potent antifungal activity against the fungi *Fusarium oxysporum* and *Mycosphaerella arachidicola*. It inhibited [methyl-^{3}H]thymidine incorporation by mouse spleen cells with an IC_{50} of 175 M, translation in the rabbit reticulocyte lysate with an IC_{50} of 20 M, and HIV-1 reverse transcriptase with an IC_{50} of 47 M [39].

37. Syringolide-induced/ HCD associated proteins

Syringolide elicitors produced by bacteria expressing *Pseudomonas syringae* pv. glycinea avirulence gene D (avrD) induce hypersensitive cell death (HCD) only in soybean (*Glycine max* [L.] Merr.) plants carrying the Rpg4 disease resistance gene. Employing a differential display method, we isolated 13 gene fragments induced in cultured cells of a soybean cultivar Harosoy (Rpg4) treated with syringolides. Several genes for isolated fragments were induced by syringolides in an rpg4 cultivar Acme as well as in Harosoy; however, the genes for seven fragments designated as SIH (for syringolide-induced/HCD associated) were induced exclusively or strongly in Harosoy. cDNA clones for SIH genes were obtained from a cDNA library of Harosoy treated with syringolide. Several sequences are homologous to proteins associated with plant defense responses. The SIH genes did not respond to a non-specific beta-glucan elicitor, which induces phytoalexin accumulation but not HCD, suggesting that the induction of the SIH genes is specific for the syringolide-Harosoy interaction. HCD and the induction of SIH genes by syringolides were independent of H_2O_2. On the other hand, Ca^{2+} was required for HCD and the induction of some SIH genes. These results suggest that the induction of SIH genes by syringolides could be activated through the syringolide-specific signaling pathway and the SIH gene products may play an important role(s) in the processes of HCD induced by syringolides [40].

38. Anticarcinogenicdaidzein-rich fraction

Partial purification of the methanol extract of soybean powder, by employing solvent fractionation and silica gel chromatography, yielded an active fraction rich in daidzein with an active component(s) that enhanced HL-60 cell differentiation. The fraction appeared to exert cytotoxic activity via an apoptotic pathway as evidenced by DNA fragmentation and caspase-3 induction. It also upregulated TGF-beta2 expression, but did not affect the expression of other members of the TGF-beta family of cytokines and their receptors, or on the expression of the vascular endothelial growth factor gene [41].

39. Phenylalanine ammonia-lyase and tyrosineammonia-lyase

Key enzymes of the phenylpropanoid pathway including phenylalanine ammonia-lyase and tyrosine ammonia-lyase are induced in response to biotic (such as chitin from fungal cell walls) and abiotic cues. The elevation of enzyme activity enzymes following application of chitin and chitosan to soybean leaf tissues was dependent on the chain length of the oligomers and time post-treatment. Treatment with chitin hexamer and chitosan pentamer produced the maximal activities at 36 h. Total phenolic content of soybean leaves rose, indicating a positive correlation between enzyme activity and total phenolic content [42].

40. Defense-related enzymes

A new class of bacteriocin (class IId) stimulates plant growth in a way analogous to Nod factors which have been shown to provoke aspects of plant disease resistance. Cut stems of soybean seedlings at the first trifoliate stage were exposed to the bacteriocin solutions Phenylalanine ammonia lyase (PAL) activity, superoxide dismutase (SOD) activity, and one APX isozyme (28kDa isoform) level in bacteriocins thuricin 17 (T17)-treated leaves peaked at 72h after treatment. At 72h post-treatment, compared with the control, guaiacol peroxidase (POD) activity in leaves exposed to T17 and bacteriocin bacthuricin F4 (BF4) rose by 72.7 and 91.3%, respectively; ascorbate peroxidase (APX) activity increased by 52.3 and 49.6%; and SOD activity was elevated by 70.5 and 60.2%, respectively. The 33-kDa polyphenol oxidase isozyme was induced strongly by both bacteriocins. Thus class IId bacteriocins can act as an inducer of plant disease defense-related enzymes and may employ mechanisms similar to Nod factors [43].

41. Cysteine proteinase

Two cDNAs, were isolated from the cotyledons of growing soybean seedlings by cDNA representational difference analysis (cDNA RDA) and rapid amplification of cDNA ends (RACE). Both CysP1 and CysP2 encode a cysteine proteinase (CPR) with a C-terminal KDEL motif. CysP1 and CysP2 were expressed from 6 days to 13 or 14 days after germination in the cotyledons of growing seedlings and in the root, flower and pod of soybean plants [44].

Two types of cysteine proteases, low-specificity enzymes from the papain family and Asn-specific enzymes from the legumin family, are endopeptidases that play a role in the degradation of seed storage proteins during early growth of seedlings. The action of the enzymes (CPPh1 and LLP, respectively) from the common bean *(Phaseolus vulgaris L.)* on the common bean storage protein phaseolin, and on the homologous soybean storage protein beta-conglycinin, has been examined. The two most active proteinases detected in common bean seedlings individually cannot bring about extensive degradation of phaseolin. However, the successive action of LLP and CPPh1 leads to extensive hydrolysis of phaseolin. CPPh1 accomplished extensive hydrolysis of beta-conglycinin [45].

42. Isoflavone synthase

Isoflavonoids are secondary metabolites important in nodulation and defense responses. They have a common occurrence in leguminous plants. Isoflavone synthase (IFS) catalyzes the key entry point step of isoflavone biosynthesis from the general phenylpropanoid pathway. IFS transcripts were detected in seeds and roots. The transcript levels of the isoforms IFS1 and IFS2 are altered in response to *Bradyrhizobium japonicum* or salicylic acid. The expression pattern of IFS1 in soybean is indicative of the roles of isoflavonoids in defense against pathogens and signal molecules to symbiotic bacteria [46].

43. UDP-glucose:Flavonoid 3-O-glucosyltransferase

The seed coats of black soybean accumulate red (cyanidin-based), orange (pelargonidin-based) blue (delphinidin-based), purple (petunidin-based), and anthocyanins almost exclusively as 3-O-glucosidesThe full-length cDNA encoding UDP-glucose:flavonoid 3-O-glucosyltransferase (UGT78K1), the enzyme which catalyzes the last step in anthocyanin biosynthesis, was isolated from black soybean seed coat using rapid amplification of cDNA ends (RACE). The recombinant enzyme glucosylated only anthocyanidins and flavonols with 3-OH regiospecificity. Galactose could also be transferred, albeit with relatively low activity, to the 3-position of cyanidin or delphinidin *in vitro*. The results are consistent with the presence of mainly 3-O-glucosylated and minor amounts of 3-O-galactosylated anthocyanins in the black soybean seed coat. The recombinant enzyme exhibited pronounced substrate inhibition by cyanidin at 100 M acceptor concentration. The accumulation of anthocyanins and flavonols was reinstated by transfer of UGT78K1 into the *Arabidopsis* T-DNA mutant (ugt78d2) lacking in anthocyanidin and flavonol 3-O-glucosyltransferase activity, signifying the role of the enzyme as a flavonoid 3-O-glucosyltransferase. Genomic and phylogenetic analyses suggest the presence of three additional soybean sequences with remarkable homology to UGT78K1. RT-PCR confirmed the co-expression of one of these genes (Glyma08g07130) with UGT78K1 in the black soybean seed coat, suggesting possible functional redundancies in anthocyanin biosynthesis in this tissue [47].

44. Beta-glucosidase

A novel 75-kDa beta-glucosidase with strict specificity toward glucosyl isoflavones but not malonylglucosidic conjugates of soybean isoflavones, was isolated from soybean. Its N-terminal amino acid sequence EYLKYKDPKA closely resembled those of maize and wheat glycosidases. Its optimal temperature and pH were 45 degrees C and 4.5, respectively. It was completely inhibited by 1 mM Hg^{2+} or 10 mM Al^{3+} ion, and glucose and mannose also affected the activity [48].

45. Isoflavone conjugate-hydrolyzing beta-glucosidase

Soybeans excrete from the roots isoflavones, which are involved in plant-microbe interactions such as symbiosis and as a defense against infections. The release of free isoflavones from their conjugates, the latent forms, is catalyzed by an isoflavone conjugate-hydrolyzing beta-glucosidase. The purified enzyme, isoflavone conjugate-hydrolyzing beta-glucosidase from (GmICHG), is a homodimeric 116-Da glycoprotein capable of directly hydrolyzing genistein 7-O-(6 ''-O-malonyl-beta-d-glucoside) to yield free genistein. GmICHG cDNA is highly expressed in the seedling roots but only very weakly in cotyledons and hypocotyls. The enzyme is exclusively localized in the cell wall and intercellular space of seedling roots, especially in the cell wall of root hairs. GmICHG is a member of glycoside hydrolase family 1 [49].

46. Isocitratelyase

A four-stage purification procedure comprising ammonium sulfate precipitation and ion exchange chromatography on DEAE cellulose has been elaborated for isolation of isocitrate lyase (EC 4.1.3.1) isoforms from soybean cotyledons, the isoform that migrated quickly in PAAG had a much lower affinity to isocitrate (K_M - 50 M) than the slowly migrating form (K_M - 16 M). The conservation of activity of the isoforms obtained depends on the presence of divalent cations (Mn^{2+} and Mg^{2+}) in the medium. It is suggested to use the isoforms of isocitrate lyase isolated from soybeans for the development of biosensors for biochemical and kinetic assays [50].

47. Vestitonereductase and chalconereductase

The complete mRNA sequences of two soybean genes-vestitone reductase and chalcone reductase, were amplified using the rapid amplification of cDNA ends methods. Soybean vestitone reductase gene encodes a 327 amino-acid protein with pronounced resemblance to *Medicago sativa* vestitone reductase. The soybean chalcone reductase gene encodes a 314 amino-acid protein highly homologous to the kudzu vine and medicago sativa chalcone reductases. The two genes were differentially expressed in soybean embryos and endosperm, flowers, leaves, stems, and roots [51].

48. Genes related to 2-oxoglutarate-dependentdioxygenases

A screening for genes involved in root nodule senescence has resulted in the isolation of the senescence-associated nodulin 1 (SAN1) multigene family from soybean. The SAN1-encoded proteins display sequence resemblance and highly conserved motifs with plant 2-oxoglutarate-dependent dioxygenases (2-ODDs). SAN1A is downregulated whereas SAN1B is upregulated during senescence induced by treatment with fixed nitrogen or darkness. The expression of the SAN1genes is not found only in nodules, signifying a more general role in plant metabolism [52].

49. 5'-adenylylsulfate reductase

Soybean seeds contain only low levels of sulfur-containing amino acids. Although expressed in various tissues throughout the plant, there is an abundant expression of the APS reductase gene and activity of the encoded protein in the early developmental stages of soybean seed, which decreases as the seed matures. Sulfur and phosphorus deprivation and cold-treatment elevate the expression level, while nitrogen starvation reduces APS reductase mRNA transcript and protein levels. This study allows an insight into the sulfur assimilation pathway [53].

50. Soybean ATP sulfurylase

Analysis of the ATP sulfurylase ATPS clone isolated from a soybean seedling cDNA library revealed an open-reading frame, encoding a 52-kDa polypeptide with an N-terminal chloroplast/plastid transit peptide. ATP sulfurylase mRNA was present with the highest abundance in root tissue. mRNA accumulated and the specific activity of ATP sulfurylase in root tissue was enhanced by cold treatment. The transcript levels and specific activity of the enzyme dropped during the later stages of seed development. Augmenting the expression levels of this key enzyme during soybean seed de-

http://dx.doi.org/10.5772/52526

velopment could increase the availability of sulfur amino acids and elevate the nutritional value of the crop [54].

51. Polysaccharides

Soluble soybean polysaccharide (SSPS) was fractionated into two sub-fractions, a high-molecular-weight fraction (HMF) and a low-molecular-weight fraction (LMF) by the ethanol-extraction method. The major constituent of HMF was a large polysaccharide with covalently-attached peptides, probably corresponding to the intact SSPS molecule. LMF was composed of free peptides and saccharides of small size, which might have occurred as by-products during the process of SSPS production. HMF exhibited high ability to emulsify oil droplets and stabilize alpha-casein dispersions in an acidic pH region. LMF showed lower activity in this regard although but it had higher potency in preventing the oxidation of emulsified lipids. The data indicate that HMF and LMF possess dissimilar structural and functional characteristics, and that the combination of the two sub-fractions displays the multiple functions of commercial SSPS [55].

52. Conclusion

The foregoing account has shown a constellation of bioactive constituents in soybean. They include enzymes that catalyze crucial steps in metabolic pathways, and also nonenzymatic proteins with important activities such as defense. Storage proteins in soybean like glycinin and conglycinin exhibit health promoting activities. Other proteins and peptides have blood pressure lowering and anticarcinogenic activities. It is anticipated that more components with potentially exploitable activities will be discovered and more applications of these soybean components to the benefit of mankind will be found.

Author details

Tzi Bun Ng*, Randy Chi Fai Cheung and Jack Ho Wong

*Address all correspondence to: b021770@mailserv.cuhk.edu.hk, jack1993@yahoo.com

School of Biomedical Sciences, Faculty of Medicine, The Chinese University of Hong Kong, Hong Kong, China

References

[1] Natarajan SS, Xu C, Bae H, Caperna TJ, Garrett WM. Characterization of storage proteins in wild (Glycine soja) and cultivated(Glycine max) soybean seeds using proteomic analysis. J Agric Food Chem. 2006;54(8)3114-20.

[2] Krishnan HB, Oehrle NW, Natarajan SS.A rapid and simple procedure for the depletion of abundant storage proteins from legume seeds to advance proteome analysis: a case study using Glycine max. Proteomics 2009;9(11)3174-88.

[3] Ferreira ES, Silva MA, Demonte A, Neves VA. β-conglycinin combined with fenofibrate or rosuvastatin have exerted distinct hypocholesterolemic effects in rats. Lipids Health Dis. 2012;1111.

[4] Hou DH, Chang SK. Structural characteristics of purified glycinin from soybeans stored undervarious conditions. J Agric Food Chem. 2004;52(12)3792-800.

[5] Fassini PG, Noda RW, Ferreira ES, Silva MA, Neves VA, Demonte A.Soybean glycinin improves HDL-C and suppresses the effects of rosuvastatin onhypercholesterolemic rats. Lipids Health Dis. 2011;10:165.

[6] Sitohy MZ, Mahgoub SA, Osman AO. In vitro and in situ antimicrobial action and mechanism of glycinin and its basicsubunit. Int J Food Microbiol. 2012;15419-29.

[7] Duranti M, Lovati MR, Dani V, Barbiroli A, Scarafoni A, Castiglioni S, Ponzone C,Morazzoni P.The alpha' subunit from soybean 7S globulin lowers plasma lipids and upregulates liver beta-VLDL receptors in rats fed a hypercholesterolemic diet. J Nutr. 2004;134(6)1334-9.

[8] Yoshizawa T, Hashimoto H, Shimizu T, Yamabe M, Shichijo N, Hanada K, Hirano H,Sato M.Purification, crystallization and X-ray diffraction study of basic 7S globulinfrom soybean. ActaCrystallogr Sect F StructBiolCrystCommun. 2011;6787-9.

[9] Rouquié D, Capt A, Eby WH, Sekar V, Hérouet-Guicheney C. Investigation of endogenous soybean food allergens by using a 2-dimensional gelelectrophoresis approach. RegulToxicolPharmacol. 2010;3SupplS47-53.

[10] Wilson S, Martinez-Villaluenga C, De Mejia EG. Purification, thermal stability, and antigenicity of the immunodominant soybeanallergen P34 in soy cultivars, ingredients, and products J Food Sci. 2008;73(6)T106-14.

[11] Morita H, Kaneko H, Ohnishi H, Kato Z, Kubota K, Yamamoto T, Matsui E, TeramotoT, Fukao T, Kasahara K, Kondo N. Structural property of soybean protein P34 and specific IgE response torecombinant P34 in patients with soybean allergy. Int J Mol Med. 2012;29153-8.

[12] Yamaguchi Y, Barona G, Ryan CA, Pearce G.GmPep914, an eight-amino acid peptide isolated from soybean leaves, activatesdefense-related genes. Plant Physiol. 2011;156(2)32-42.

[13] Koo KB, Suh HJ, Ra KS, Choi JW. Protective effect of cyclo(his-pro) on streptozotocin-induced cytotoxicity andapoptosis in vitro. J MicrobiolBiotechnol. 2011;21(2)218-27.

[14] Egusa S, Otani H. Characterization of a cellular immunostimulating peptide from a soybean proteinfraction digested with peptidase R. J NutrSciVitaminol (Tokyo). 2009;55(5)428-33.

[15] Torre-Villalvazo I, Gonzalez F, Aguilar-Salinas CA, Tovar AR, Torres N. Dietary soy protein reduces cardiac lipid accumulation and the ceramideconcentration in high-fat diet-fed rats and ob/ob mice. J Nutr. 2009;139(12)2237-43.

[16] Cha M, Park JR. Production and characterization of a soy protein-derived angiotensin I-convertingenzyme inhibitory hydrolysate. J Med Food. 2005;8(3)305-10.

[17] Kuba M, Tanaka K, Tawata S, Takeda Y, Yasuda M.Angiotensin I-converting enzyme inhibitory peptides isolated from tofuyofermented soybean food. BiosciBiotechnol-Biochem. 2003;67(6)1278-83.

[18] Mallikarjun Gouda KG, Gowda LR, Rao AG, Prakash V. Angiotensin I-converting enzyme inhibitory peptide derived from glycinin, the 11Sglobulin of soybean (Glycine max). J Agric Food Chem. 2006;54(13)4568-73.

[19] Anta L, Luisa Marina M, García MC.Simultaneous and rapid determination of the anticarcinogenic proteins Bowman-Birkinhibitor and lectin in soybean crops by perfusion RP-HPLC. J Chromatogr A. 2010;1217(45)7138-43.

[20] Clemente A, Moreno FJ, Marín-ManzanoMdel C, Jiménez E, Domoney C. The cytotoxic effect of Bowman-Birkisoinhibitors, IBB1 and IBBD2, from soybean(Glycine max) on HT29 human colorectal cancer cells is related to their intrinsicability to inhibit serine proteases.Mol NutrFood Res. 2010;54(3):396-405.

[21] Duranti M, Barbiroli A, Scarafoni A, Tedeschi G, Morazzoni P. One-step purification of Kunitz soybean trypsin inhibitor. Protein ExprPurif. 2003;30(2)167-70.

[22] Tremblay R, Feng M, Menassa R, Huner NP, Jevnikar AM, Ma S. High-yield expression of recombinant soybean agglutinin in plants using transientand stable systems. Transgenic Res. 2011;20(2)345-56.

[23] Kim HJ, Suh HJ, Kim JH, Kang SC, Park S, Lee CH, Kim JS. Estrogenic activity of glyceollins isolated from soybean elicited withAspergillussojae. J Med Food. 2010;13(2)382-90.

[24] Zhang G, Chen M, Chen X, Xu Z, Li L, Guo J, Ma Y. Isolation and characterization of a novel EAR-motif-containing gene GmERF4 fromsoybean (Glycine max L.). MolBiol Rep. 2010;37(2)809-18.

[25] Gilani GS, NimalRatnayke WM, Mueller R, Mazza G. Effects of source of protein and supplementary extracted isoflavones andanthocyanins on longevity of Stroke-prone Spontaneously Hypertensive (SHRSP)rats. J Toxicol Sci. 2009;34(3)335-41.

[26] Yamatani K, Saeki T, Iwami K, Suzuka T, Kanamoto R.Soybean resistant protein elevates fecal excretion of cholesterol and bile acids and decreases hepatic cholesterol content in comparison with soybean proteinisolate. BiosciBiotechnolBiochem. 2009;73(4)921-2.

[27] Park K, Choi K, Kim H, Kim K, Lee MH, Lee JH, Kim Rim JC.Isoflavone-deprived soy peptide suppresses mammary tumorigenesis by inducingapoptosis. ExpMol Med. 2009;41(6)371-81.

[28] Tachi H, Fukuda-Yamada K, Kojima T, Shiraiwa M, Takahara H. Molecular characterization of a novel soybean gene encoding a neutral PR-5protein induced by high-salt stress. Plant PhysiolBiochem. 2009;47(1)73-9.

[29] He C, Wu X, Dongfang Y, Du B, Zhang J, Chen S. Isolation and characterization of a new defense gene from soybean. Sci China C Life Sci. 2001;44(4)409-20.

[30] Vasconcelos IM, Morais JK, Siebra EA, Carlini CR, Sousa DO, Beltramini LM, MeloVM, Oliveira JT. SBTX, a new toxic protein distinct from soyatoxin and other toxic soybean[Glycine max] proteins, and its inhibitory effect on Cercosporasojina growth. Toxicon. 2008;51(6)952-63.

[31] Dong X, Tang B, Li J, Xu Q, Fang S, Hua Z. Expression and purification of intact and functional soybean (Glycine max) seedferritin complex in Escherichia coli. J MicrobiolBiotechnol. 2008;18(2)299-307.

[32] Arai Y, Hayashi M, Nishimura M. Proteomic analysis of highly purified peroxisomes from etiolated soybeancotyledons. Plant Cell Physiol. 2008;49(4)526-39.

[33] Park HC, Kim ML, Lee SM, Bahk JD, Yun DJ, Lim CO, Hong JC, Lee SY, Cho MJ, Chung WS.Pathogen-induced binding of the soybean zinc finger homeodomain proteins GmZF-HD1and GmZF-HD2 to two repeats of ATTA homeodomain binding site in the calmodulinisoform 4 (GmCaM4) promoter. Nucleic Acids Res. 2007;35(11)3612-23.

[34] Afzal AJ, Lightfoot DA.Soybean disease resistance protein RHG1-LRR domain expressed, purified andrefolded from Escherichia coli inclusion bodies: preparation for a functionalanalysis. Protein ExprPurif. 2007;53(2)346-55.

[35] Eichhorn H, Klinghammer M, Becht P, Tenhaken R. Isolation of a novel ABC-transporter gene from soybean induced by salicylic acid. J Exp Bot. 2006;57(10)2193-201.

[36] Fliegmann J, Montel E, Djulić A, Cottaz S, Driguez H, Ebel J. Catalytic properties of the bifunctional soybean beta-glucan-binding protein, amember of family 81 glycoside hydrolases. FEBS Lett. 2005;579(29)6647-52.

[37] Jeong HJ, Park JH, Lam Y, de Lumen BO. Characterization of lunasin isolated from soybean. J Agric Food Chem. 2003;51(27)7901-6.

[38] Tian AG, Luo GZ, Wang YJ, Zhang JS, Gai JY, Chen SY. Isolation and characterization of a Pti1 homologue from soybean. J Exp Bot. 2004;55(396)535-7.

[39] Ngai PH, Ng TB.Purification of glysojanin, an antifungal protein, from the black soybean Glycinesoja. Biochem Cell Biol. 2003;81(6)387-94.

[40] Hagihara T, Hashi M, Takeuchi Y, Yamaoka N. Cloning of soybean genes induced during hypersensitive cell death caused bysyringolide elicitor. Planta. 2004;218(4)606-14.

[41] Lee JS, Son KH, Sung MK, Kim YK, Yu R, Kim JS. Anticarcinogenic properties of a daidzein-rich fraction isolated from soybean. J Med Food. 2003;6(3)175-81.

[42] Khan W, Prithiviraj B, Smith DL.Chitosan and chitin oligomers increase phenylalanine ammonia-lyase and tyrosineammonia-lyase activities in soybean leaves. J Plant Physiol. 2003;160(8)859-63.

[43] Jung WJ, Mabood F, Souleimanov A, Smith DL.Induction of defense-related enzymes in soybean leaves by class IIdbacteriocins (thuricin 17 and bacthuricin F4) purified from Bacillus strains.Microbiol Res. 2011;167(1)14-9.

[44] Ling JQ, Kojima T, Shiraiwa M, Takahara H. Cloning of two cysteine proteinase genes, CysP1 and CysP2, from soybeancotyledons by cDNA representational difference analysis. BiochimBiophysActa. 2003;1627(2-3)129-39.

[45] Zakharov A, Carchilan M, Stepurina T, Rotari V, Wilson K, Vaintraub I. A comparative study of the role of the major proteinases of germinated commonbean (Phaseolus vulgaris L.) and soybean (Glycine max (L.) Merrill) seeds in the degradation of their storage proteins. J Exp Bot. 2004;55(406)2241-9.

[46] Subramanian S, Hu X, Lu G, Odelland JT, Yu O. The promoters of two isoflavone synthase genes respond differentially tonodulation and defense signals in transgenic soybean roots. Plant Mol Biol. 2004;54(5)623-39.

[47] Kovinich N, Saleem A, Arnason JT, Miki B. Functional characterization of a UDP-glucose:flavonoid 3-O-glucosyltransferasefrom the seed coat of black soybean (Glycine max (L.) Merr.). Phytochemistry. 2010;71(11-12)1253-63.

[48] Chiou TY, Lin YH, Su NW, Lee MH. Beta-glucosidase isolated from soybean okara shows specificity toward glucosylisoflavones. J Agric Food Chem. 2010;58(15)8872-8.

[49] Suzuki H, Takahashi S, Watanabe R, Fukushima Y, Fujita N, Noguchi A, Yokoyama R, Nishitani K, Nishino T, Nakayama T.An isoflavone conjugate-hydrolyzing beta-glucosidase from the roots of soybean(Glycine max) seedlings: purification, gene cloning, phylogenetics, and cellular localization. J Biol Chem. 2006;281(40)30251-9.

[50] Eprintsev AT, Diachenko EV, Lykova TV, Kuen CT, Popov VN. Preparation and properties of isocitratelyase isoforms from the cotyledons ofGlycine max L. PriklBiokhimMikrobiol. 2010;46(1)103-8.

[51] Liu GY. Isolation, sequence identification and tissue expression profile of two novelsoybean (glycine max) genes-vestitonereductase and chalconereductase. MolBiol Rep. 2009;36(7)1991-4.

[52] Webb CJ, Chan-Weiher C, Johnson DA.Isolation of a novel family of genes related to 2-oxoglutarate-dependentdioxygenases from soybean and analysis of their expression during root nodulesenescence. J Plant Physiol. 2008;165(16)1736-44.

[53] Phartiyal P, Kim WS, Cahoon RE, Jez JM, Krishnan HB. The role of 5'-adenylylsulfate reductase in the sulfur assimilation pathway ofsoybean: molecular cloning, kinetic characterization, and gene expression. Phytochemistry 2008;69(2)356-64.

[54] Phartiyal P, Kim WS, Cahoon RE, Jez JM, Krishnan HB.Soybean ATP sulfurylase, a homodimeric enzyme involved in sulfur assimilation, isabundantly expressed in roots and induced by cold treatment. Arch BiochemBiophys. 2006;450(1)20-9.

[55] Li J, Matsumoto S, Nakamura A, Maeda H, Matsumura Y. Characterization and functional properties of sub-fractions of soluble soybean.BiosciBiotechnolBiochem. 2009;73(12)2568-75.

Chapter 12

Cell Death Signaling From the Endoplasmic Reticulum in Soybean

Pedro A.B. Reis and Elizabeth P. B. Fontes

Additional information is available at the end of the chapter

http://dx.doi.org/10.5772/52711

1. Introduction

Plants are constantly subjected to adverse environmental conditions, such as extremes of temperature, cold, salinity and drought. As a consequence, plant cells have developed coordinated and integrated mechanisms that respond to these injuries and are immediately activated upon stresses. To cope with the stress, cell signaling pathways are activated and promote up or down regulation of specific genes, which minimize the deleterious effect of stresses within the cell. The endoplasmic reticulum (ER) is a key signaling organelle involved in the activation of cellular stress responses in eukaryotic cells. One such well-characterized signaling event is the unfolded protein response (UPR), which is activated to cope with the disruption of ER homeostasis that results in the accumulation of unfolded or misfolded proteins in the lumen of the organelle. Upon disruption of ER homeostasis, plant cells activate at least two branches of the unfolded protein response (UPR) through IRE1-like (Inositol-Requiring Enzyme 1) and ATF6-like (Activating Transcription factor 6) transducers, resulting in the up-regulation of ER-resident molecular chaperones and the activation of the ER-associated degradation protein system. However, if ER stress is sustained, an apoptotic pathway is activated. Persistent ER stress has been shown to trigger both ER-stress specific apoptotic pathways and shared PCD (programmed cell death) signaling pathways elicited by other death stimuli. One plant-specific, ER stress-shared response is the ER and osmotic stress-integrated signaling, which converges on N-rich proteins (NRPs) to transduce a cell death signal. NRP-mediated cell death signaling is a distinct, plant-specific branch of the ER stress pathway that has been shown to integrate the ER and osmotic stress signals into a full response. This ER- and osmotic-stress induced cell death signaling pathway has been uncovered in soybean and constitutes the major focus of this chapter. A second cell death pathway induced by ER stress has been shown to be mediated by the G protein in Arabidopsis, but it remains to be determined whether it operates in soybean as well.

2. ER stress response: The cytoprotective unfolded protein response

The ER is a highly dynamic organelle, which mediates several cellular functions, such as the folding and post-translational modification of secretory proteins and protein quality control in addition to maintaining Ca^{2+} homeostasis (Schröder 2008). The loading of unfolded protein in the lumen of ER for maturation is tightly controlled and dependent on the cellular requirements. Under stress conditions, the folding capacity of the ER can be overloaded causing the accumulation of unfolded proteins and disruption of cellular homeostasis (Xu, Bailly-Maitre & Reed 2005). To cope with this stress condition, eukaryotic cells evolved a sophisticated signaling mechanism referred to as unfolded protein response (UPR; Malhotra and Kaufman., 2007). In mammalian cells, the UPR is transduced through three distinct ER-transmembrane sensors: PERK (protein kinase RNA-like ER kinase), Ire1 (inositol-requiring enzyme-1) and the basic leucine zipper transcription factor ATF6 (activating transcription factor-6; Ron and Walter, 2007; Malhotra and Kaufman, 2007, Kapoor and Sanyal, 2009). The activation of the UPR allows the ER processing and folding capacities to be balanced with protein loading into the lumen of the organelle under conditions of ER stress (Malhotra and Kaufman, 2007). This balance is achieved by (i) shutting down protein synthesis via PERK activation, (ii) up-regulating the expression of ER-resident processing proteins, such as molecular chaperones and foldases, via activation of Ire1 and ATF6, and (iii) inducing the ER-associated protein degradation (ERAD) machinery, through activation of Ire1, which mediates the targeting and subsequent degradation of unfolded proteins by the proteosome. However, if the ER stress is sustained, multiple apoptotic pathways can be activated in mammalian cells.

In plants, the UPR seems to operate as a bipartite module, as the ER stress signal is transduced through homologs of the Ire1 and ATF6 transducers, but a PERK-mediated branch of the UPR has not been shown (Urade 2009; Chen and Brandizzi., 2012). Two components of the Ire1-mediaded branch of the UPR is known. The first one is the Ire1 ortholog that is represented by two copies in the Arabidopsis genome, Ire1a and Ire1b, and one copy, OsIre1, in the rice genome. Like the mammalian counterpart, plant Ire1 is associated with the ER membrane and exhibits ribonuclease activity and autophosphorylation activities, as shown for Ire1a, Ire1b and OsIre1(Koizumi et al., 2001; Okushima et al., 2002). The second component is ER membrane-associated transcription factor bZIP60. Upon ER stress, bZIP60 mRNA is spliced in an IRE1-mediated process to generate an alternatively spliced transcript that lacks the transmembrane domain-encoding sequences (Liu et al., 2007 e 2008 ; Deng et al., 2011; Nagashima et al., 2011). This splicing leads to the synthesis of a soluble and functional bZIP60 transfactor that can be translocated to the nucleus, where it activates ER stress inducible promoters, such as the BiP3 promoter. Likewise, OsbZIP74 or OsbZIP50 from rice, an ortholog of Arabidopsis AtbZIP60, is regulated through the IRE1-mediated splicing of its

http://dx.doi.org/10.5772/52711

RNA to render the activation of ER stress-inducible promoters (Hayashi et al., 2011; Lu et al., 2011).

The second branch of UPR in plants mechanistically resembles the ATF6-mediated transduction of the ER stress signal. Upon ER stress, the membrane-associated Arabidopsis ATF6 homologs bZIP17 and bZIP28 are relocated to the Golgi, where their transcriptional domains are proteolytically released from the membrane by SP2 (Tajima et al., 2008: Che et al., 2010). The released bZIP domain of these transfactors is then translocated to the nucleus, where it acts in concert with the heterotrimeric NF-Y complex to activate UPR genes (Liu e Howel., 2010). The NF-Y complex is composed the transcriptional factors NF-YA4, NF-YB3 and NF-YC2.

Comprehensive genome-wide evaluations of ER stress-induced changes in gene expression have provided evidence that the UPR operates in a similar fashion in both soybean and Arabidopsis (Irsigler et al., 2007). Inducers of ER stress, such as tunicamycin and AZC, promote the up-regulation of a class of genes that functions in protein folding and ERAD. In the protein folding category, the up-regulated genes include ER-resident molecular chaperones such as BiP, calreticulin, calnexin, and the folding catalyst protein disulfide isomerase (PDI). ERAD-associated genes that are up-regulated by ER stress in soybean include those encoding polyubiquitin, ubiquitin conjugating enzyme, the alpha subunit of the proteasome, CDC48 and Derlin. These genomic analyses suggested that soybean, like Arabidopsis, have evolved at least two different mechanisms that mediate UPR: (i) transcriptional induction of genes encoding chaperones and vesicle trafficking proteins and (ii) upregulation of the ER-associated protein degradation (ERAD) system for rapid disposal of unfolded proteins in the ER as a protective measure.

In addition to the cytoprotective bipartite response to ER stress in plants, two apparently distinct branches of the ER stress-induced pathways have been shown to transduce a cell death signal: (i) the ER membrane associated Gβ-Gγ heterodimer-mediated signaling events that trigger UPR-associated cell death in *Arabidopsis* (Wang et al., 2007) and (ii) the ER stress-induced NRP-mediated cell death response that has been uncovered in soybean (Reis and Fontes, 2012).

3. The ER-stress-induced NRP-mediated cell death response

NRP-mediated cell death signaling is a distinct, plant-specific branch of the ER stress pathway that has been uncovered in soybean and has been shown to integrate the ER and osmotic stress signals into a full response. This integrative pathway was first identified through genome-wide approaches and expression profiling, which revealed the existence of a modest overlap of the ER and osmotic stress-induced transcriptomes in soybean seedlings treated with PEG (an inducer of osmotic stress) or tunicamycin and AZC (potent inducers of ER stress; Irsigler et al., 2007). The co-regulated genes were first considered to be downstream targets of the integrated pathway based on similar induction kinetics and a synergistic response to the combination of osmotic and ER stress-inducing treatments. Based on these criteria, the selected downstream components of this ER and osmotic stress response-integrating pathway encode proteins with diverse roles, such as plant-specific development and cell death (DCD) domain-containing proteins (NRP-A and NRP-B), an ubiquitin-associated (UBA) protein homolog and NAC domain-containing proteins (GmNAC6). Among them, NRP-A and NRP-B were the first ones to be characterized and to show to induce a cell death response when ectopically expressed in tobacco leaves or soybean protoplasts (Costa et al., 2008). As a consequence, the ER and osmotic stress response-integrating pathway has been designated as the NRP-mediated cell death response.

An upstream component of the NRP-mediated cell death response, GmERD15 (*Glycine max* Early Responsive to Dehydration 15), has been recently identified using one-hybrid screening that targeted the NRP-B promoter in yeast (Alves et al., 2011). GmERD15 is induced by ER and osmotic stress to activate the expression of *NRP* genes (NRP-A and NRP-B). Up-regulation of NRP-B leads to the induction of an NAC domain-containing protein, GmNAC6, which is a critical mediator of stress-induced cell death in plants (Faria et al., 2011). These components of the ER stress-induced NRP-mediated cell death signaling pathway, GmERD15, NRPs and GmNAC6, have been further characterized.

3.1. GmERD15 is a ssDNA binding transcriptional activator

The Early Responsive Dehydration (ERD) genes are rapidly induced in response to water deficit and form a family comprised by ERD1 to ERD16 representatives. The ERD encoded proteins exhibit diverse and heterogeneous biochemical functions and fall into different classes of proteins, such as chloroplast ATP-dependent protease (ERD1), cytosolic HSP70 (ERD2), glutationa-S-transferases (ERD9,ERD11, ERD13) among others (Soitano et al., 2008; Kiyosue et al., 1994; Kiyosue et al., 1993). ERD15 was first identified in Arabidopsis as a hydrophilic protein that possesses a PAM2 domain that interacts with polyA-binding proteins (PABP11; Kiyosue et al., 1994; Kariola et al., 2006)., ERD15 has been shown to function as a negative regulator of the abcisic acid (ABA)-mediated response (Kariola et al., 2006). Overexpression of ERD15 reduces the ABA sensitivity of Arabidopsis, whereas silencing of ERD15 by RNAi promotes hypersensitivity to the hormone. The negative effect of ERD15 on ABA signaling enhances salicylic acid-dependent defense because overexpression of ERD15 was associated with increased resistance to the bacterial necrotroph Erwiniacarotovora, and the enhanced induction of marker genes for systemic acquired resistance. These results are consistent with the observed antagonistic effect of ABA on salicylic acid-mediated defense and may implicate ERD15 as a shared component of these responses.

The soybean GmERD15 homolog has been described as a new ER stress- and osmotic stress-induced transcription factor that binds to the promoter and induces the expression of the NRP-B gene. In fact, GmERD15 was isolated by its capacity to associate stably with the promoter of NRP-B in yeast cells using the one-hybrid system (Alves et al., 2011). The GmERD15 binding site in the NRP-B promoter was mapped to a 12-bp palindromic sequence (511 AGCAnnnnTGCT -500) that resembles binding sites for ssDNA binding proteins, such as NF1C and PBF2 that recognize the sequences -TTGGCnnnnnGCCAA-3' and 5- TGACAnnnnTGTCA-3′, respectively (Wang and Kiledjian., 2000). Furthermore, GmERD15 is located in the nucleus, and chromatin immunoprecipitation (ChIP) assays revealed that it binds to the NRP-B promoter in vivo (Alves et al., 2011). The ectopic expression of GmERD15 in soybean cells activates the NRP-B promoter and induces NRP-B expression. Collectively, these results indicate that GmERD15 functions as an upstream component of the NRP-mediated cell death signaling pathway that is induced by ER stress and osmotic stress

3.2. NRPs: Molecular and functional characterization

The N Rich Protein (NRP) gene was first identified by its rapid induction in response to pathogen incompatible interactions in soybean (Ludwig and Tenhaken, 2001). The NRP designation was derived from its high content of asparagine residue, about 25 %. NRP is represented in the soybean genome by a small family of three genes: NRP-A, NRP-B and NRP-C. The encoded proteins share a highly conserved development and cell death (DCD) domain at the C-terminal portion in addition to a high content of asparagine residues at their more divergent N termini. The asparagine rich domain is not well characterized but harbors putative glycosylation and myristoylation sites that may be relevant for function. The DCD domain is found exclusively in plant proteins and it is composed of about 130 amino acid residues, organized into several conserved motifs: FGLP and LFL in the N-treminal region of the domain, PAQV and PLxE at its C-terminus (Tenhaken et al; 2005). DCD domain-containing proteins may be subdivided into four groups, according to the localization of the DCD domain in the primary structure. NRPs belong to the the subgroup I of DCD domain-containing family of proteins, as their domains are located at the C-terminal portion of the protein (Tenhaken et al; 2005).

NRPs are critical mediators of ER and osmotic stress-induced cell death in soybeans (Costa et al., 2008). The cell death response mediated by NRPs resembles a programmed cell death event. The overexpression of NRPs in soybean protoplasts induces caspase-3-like activity and promotes extensive DNA fragmentation. Furthermore, the transient expression of NRPs in plants causes leaf yellowing, chlorophyll loss, malondialdehyde production, ethylene evolution and the induction of senescence marker genes, which are hallmarks of leaf senescence.

NRPs are up-regulated by ER or osmotic stress but need both stress signals for full induction (Isrigler et al., 2007). This synergistic interaction of both signals upon NRP induction indicates that the ER stress and osmotic stress responses converge at the level of gene expression to potentiate a NRP-mediated cell death response (Costa et al., 2008). NRPs are also up-regulated by other abiotic and biotic signals, such as salt stress, oxidative stress and pathogens. Because the NRP-mediated cell death signaling pathway represents a shared response to multiple stress signals in plants, it might permit coordinate adaptive cellular responses under a large array of stress conditions

3.3. GmNAC6 as a downstream component of the NRP-mediated cell death response

NAC domain-containing proteins are plant-specific transcriptional factors that are expressed in several tissues and developmental stages. The NAC transfactors are organized into a general structure that consists of a highly conserved N-terminal domain involved in

DNA binding (called NAC domain) and a C-terminal region highly divergent in sequence and length that functions as the activation domain. The NAC domain was derived from comparison of consensus sequences among NAM from Petunia, ATAF1/2 and CUC2 from Arabidopsis (Souer et al., 1996.). It comprises nearly 160 amino acid residues, divided into five subdomains (A–E) exhibiting a negative net charge and a nuclear localization signal (Xie et al., 1999; Seoet al., 2008). The subdomains A, C and D are conserved among plant species whereas B and E subdomains are variable (Ooka et al., 2003). The C- terminus harbors a protein-protein interaction domain in some NAC-containing proteins while a transmembrane domain is present in other transcriptional factors (Seo et al., 2008). Therefore, the NAC family is comprised by both soluble, nuclear transactivators and membrane proteins.

The members of the NAC gene family are involved in a variety of developmental events and defense responses, such as shoot apical meristem formation and maintenance (SAM; Aida et al., 1997; Souer et al.,1996; Weir et al., 2004), hormone signaling (Fujita et al., 2004; Xie et al., 2000), response to pathogen infection (Ren et al., 2000; Selth et al., 2005; Xie et al., 1999), leaf senescence (John et al., 1997) and response to different abiotic stresses (Hegedus et al., 2003; Tran et al.,2004).

The soybean NAC family is comprised by 180 putative sequences of NAC domain-containing proteins, which display different expression profiles in response to distinct environmental stress conditions and developmental signals (Mochida et al.,2009; Mochida et al., 2010; Wang et al., 2010). Frequently, the stress-induced expression profile of the soybean NAC genes reflects the functional profile of the encoded protein (Pinheiro et al., 2009). GmNAC6 was identified by its synergistic induction in response to a combined treatment of inducers of osmotic stress (polyethylene glycol) and ER stress (tunicamycin) and was functionally linked to the NRP-mediated cell death response (Faria et al., 2011). Transient expression of GmNAC6 promotes cell death and hypersensitive-like responses *in planta*. GmNAC6 and NRPs also share overlapping responses to biotic signals, but the induction of NRPs peaks before the increased accumulation of GmNAC6 transcripts. Consistent with the delayed kinetics of GmNAC6 induction, increased levels of NRP-A and NRP-B transcripts induce promoter activation and the expression of the GmNAC6 gene. Therefore, GmNAC6 is biochemical and functionally linked to the ER stress- and osmotic stress-integrating cell death response, in which it acts downstream of the NRPs.

GmNAC6 encodes a 33kDa protein that belongs to the TERN (Tobacco elicitor-responsive gene encoding NAC domain protein) group of the NAC family, which is induced by elicitors of the pathogen response (Ooka et al.,2003). Likewise, GmNAC6 is induced by the pathogenic bacteria *Pseudomonas syringaepatovar tomato* which elicits an incompatible interaction in soybean and by cell wall-degrading enzymes, which mimic bacterial pathogen attack. Like GmNAC6, the other components of the ER- and osmotic-stress induced cell death signaling pathway, GmERD15 and NRPs, are also induced by other biotic and abiotic signals, such drought and pathogen incompatible interactions. Therefore, the activation of the NRP-mediated senescence-like response is not specific to ER stress or osmotic stress but is, rather, a shared branch of general environmental adaptive pathways.

Acknowledgement

This research was supported by the Brazilian Government Agencies CNPq grants 559602/2009-0, 573600/2008-2 and 470878/2006-1 (to E.P.B.F.), the FAPEMIG grant CBB-APQ-00070-09, and the FINEP grant 01.09.0625.00 (to E.P.B.F.). P.A.B.R. is supported by CNPq graduate fellowships.

Author details

Pedro A.B. Reis and Elizabeth P. B. Fontes

Departamento de Bioquímica e Biologia Molecular/ Bioagro, Universidade Federal de Viçosa, Viçosa, MG, Brazil

References

[1] Aida, M., Ishida, T., Fukaki, H., Fujisawa, H., & Tasaka, M. (1997). Genes involved in organ separation in Arabidopsis, an analysis of the cup-shaped cotyledon mutant. The Plant Cell, 0153-2298X, 9, 841-857.

[2] Alves, M. S., Reis, P. A. B., Dadalto, S. P., Faria, J. A. Q. A., Fontes, E. P. B., & Fietto, L. G. (2010). A novel transcription factor, early responsive to dehydration 15, connects ER stress with an osmotic stress-induced cell death signal. Journal of Biological Chem. istry, 0108-3351X, 286, 20020-20030.

[3] Che, P.; Bussell, J.D.; Zhou, W.; Estavillo, G.M.; Pogson, B.J.; Smith, S.M. (2010) Signaling from the Endoplasmic Reticulum Activates Brassinosteroid Signaling and Promotes Acclimation to Stress in Arabidopsis. Science Signaling, 3: ra69, ISSN 1945-0877

[4] Chen, Y., & Brandizzi, F. (2012). AtIRE1A/AtIRE1B and AGB1 independently control two essential unfolded protein response pathways in Arabidopsis. The Plant Journal0136-5313X, 69, 266-277.

[5] Costa, M. D. L., Reis, P. A. B., Valente, M. A. S., Irsigler, A. S. T., Carvalho, C. M., Loureiro, M. E., Aragaão, F. J. L., Boston, R. S., Fietto, L. G. ., & Fontes, E. P. B. (2008). A new branch of endoplasmic reticulum stress signaling and the osmotic signal converge on plant-specific asparagine-rich proteins to promote cell death. Journal of Biological Chemistry0108-3351X, 20209-20219.

[6] Deng, Y., Humbert, S., Liu, J. X., Srivastava, R. J., Rothstein, J. S., & Howell, S. H. (2011). Heat induces the splicing by IRE1 of a mRNA encoding a transcription factor involved in the unfolded protein response in Arabidopsis. Proceedings of the National Academy of Sciences, 1091-6490, 108, 7247-7252.

[7] Faria, J. A. Q. A., Reis, P. A. B., Reis, M. T. B., Rosado, G. L., Pinheiro, G. L., Mendes, G. C., & Fontes, E. P. B. (2011). The NAC domain-containing protein, GmNAC6, is a downstream component of the ER stress- and osmotic stress-induced NRP-mediated cell-death signaling pathway. BMC Plant Biology, 11: 129, 1471-2229, 1471-2229.

[8] Fujita, M., Fujita, Y., Maruyama, K., Seki, Motoaki., Hiratsu, K., Ohme-Takagi, M., Tran-S, L., , P., Yamaguchi-Shinozaki, K., & Shinozaki, K. (2004). A dehydration-induced NAC protein, RD26, is involved in a novel ABA-dependent stress-signaling pathway. The Plant Journal0136-5313X, 39, 863-876.

[9] Hayashy, S., Wakasa, Y., Takahashi, H., Kawakatsu, T., & Takaiwa, F. (2012). Signal transduction by IRE1-mediated splicing of bZIP50 and other stress sensors in the endoplasmic reticulum stress response of rice. The Plant Journal, 0136-5313X, 69, 946-956.

[10] Hegedus, D.., Yu, M., Baldwin, D., Gruber, M., Sharpe, A., Parkin, I., Whitwill, S., & Lydiate, D. . (2003). Molecular characterization of Brassica napus NAC domain transcriptional activators induced in response to biotic and abiotic stress. *Plant Molecular Biology*, 383-397, 1572-9818.

[11] Irsigler, A. S. T., Costa, M. D. L., Zhang, P., Reis, P. A. B., Dewey, R. E., Boston, R. S. ., & Fontes, E. P. B. (2007). Expression profiling on soybean leaves reveals integration of ER- and osmotic-stress pathways. *BMC Genomics*, 1471-2164, 1471-2164.

[12] John, I., Hackett, R., Cooper, W., Drake, R., Farrell, A., & Grierson, D. (1997). Cloning and characterization of tomato leaf senescence-related cDNAs. *Plant Molecular Biology*, 33, 641-651, 1572-9818.

[13] Kapoor, A., & Sanyal, A. J. (2009). Endoplasmic Reticulum Stress and the Unfolded Protein Response. *Clinics in Liver Disease*, 581-590, 1089-3261.

[14] Kariola, T., Brader, G., Helenius, E., Li, J., Heino, P., & Palva, E. T. (2006). EARLY RESPONSIVE to DEHYDRATION 15, a negative regulator of abscisic acid responses in Arabidopsis. Plant Physiology, 1532-2548, 142, 1559-73.

[15] Kiyosue, T., Yamaguchi-Shinozaki, K., & Shinozaki, K. (1994). Cloning of cDNAs for genes that are early-responsive to dehydration stress (ERDs) in Arabidopsis thaliana L.: identification of three ERDs as HSP cognate genes. *Plant Molecular Biology*, 25, 791-8, 1572-9818.

[16] Kiyosue, T., Yamaguchi-Shinozaki, K., & Shinozaki, K. (1993). Characterization of two cDNAs (ERD11 and ERD13) for dehydrationinducible genes that encode putative glutathione S-transferases in Arabidopsis thaliana L. FEBS Letters, 0014-5793, 335, 189-92.

[17] Koizumi, N., Martinez, I. M., Kimata, Y., Kohno, K., Sano, H., & Chrispeels, M. J. (2001). Molecular characterization of two Arabidopsis Ire1 homologs, endoplasmic reticulum-located transmembrane protein kinases. Plant Physiology, 1532-2548, 127, 949-962.

[18] Liu, J. X., Srivastava, R. ., , P. ., & Howell, S. H. (2007). Salt stress responses in Arabidopsis utilize a signal transduction pathway related to endoplasmic reticulum stress signaling. The Plant Journal, 0136-5313X, 51, 897-909.

[19] Liu, J. X., Srivastava, R., & Howell, S. H. (2008). Stress-induced expression of an activated form of AtbZIP17 provides protection from salt stress in Arabidopsis. Plant-Cell & Environment. 1365-3040, 31, 1735-1743.

[20] Liu, J. X., & Howell, S. H. (2010). Endoplasmic Reticulum Protein Quality Control and Its Relationship to Environmental Stress Responses in Plants. The Plant Cell, 0153-2298X, 22, 1-13.

[21] Lu-J, S., Yang-T, Z., Sun, L., Sun, L., Song-T, Z., & Liu-X, J. (2011). Conservation of IRE1 -Regulated bZIP74 mRNA Unconventional Splicing in Rice (Oryza sativa L.) Involved in ER Stress Responses. *Molecular Plant*, 1752-9867.

[22] Ludwig, A. A., Tenhaken, R., & (2001, . (2001). A new cell wall located N-rich protein is strongly induced during the hypersensitive response in Glycine max L. *European Journal of Plant Pathology*, 107, 323-336, 1573-8469.

[23] Malhotra, J. D. ., & Kaufman, R. J. (2007). The Endoplasmic Reticulum and the Unfolded Protein Response. Semininars in Cell and Developmental Biology, 1084-9521, 18, 716-73.

[24] Mochida, K., Yoshida, T., Sakurai, T., Yamaguchi-Shinozaki, K., Shinozaki, K., & Tran, L. S. (2009). In silico analysis of transcription factor repertoire and prediction of stress responsive transcription factors in soybean. DNA Research., 1756-1663, 16, 353-69.

[25] Mochida, K., Yoshida, T., Sakurai, T., Yamaguchi-Shinozaki, K., Shinozaki, K., & Tran, L. S. (2010a). LegumeTFDB: an integrative database of Glycine max, Lotus japonicus and Medicagotruncatula transcription factors. *Bioinformatics*, 26, 290-1, 1460-2059.

[26] Nagashima, Y., Mishiba, K., Suzuki, E., Shimada, Y., Iwata, Y., & Koizumi, N. (2011). Arabidopsis IRE1 catalyses unconventional splicing ofbZIP60 mRNA to produce the active transcription factor. Scientific Reports, 1:29, 2045-2322, 2045-2322.

[27] Okushima, Y., Koizumi, N., Yamaguchi, Y., Kimata, ., Kohno, K., & Sano, H. (2002). Isolation and characterization of a putative transducer of endoplasmic reticulum stress in Oryza sativa. Plant Cell Physiology, 1471-9053, 43, 532-539.

[28] Ooka, H., Satoh, K., Doi, K., et al., et al. (2003). Comprehensive analysis of NAC family genes in Oryza sativa and Arabidopsis thaliana. DNA Research. 10; 1756-1663, 239-247.

[29] Pinheiro, G. L., Marques, C. S., Costa, M. D. B. L., Reis, P. A. B., Alves, M. S., Carvalho, C. M., Fietto, L. G. ., & Fontes, E. P. B. (2009). Complete inventory of soybean NAC transcription factors: Sequence conservation and expression analysis uncover their distinct roles in stress response. *Gene*, 444, 10-23, 0378-1119.

[30] Reis, P.A.B and Fontes, E.P.B.(2012). N-rich protein (NRP)-mediated cell death signaling: a new branch of the ER stress response with implications for plant biotechnology. Plant Signaling & Behavior. 1559-2316, 7

[31] Ren, T., Qu, F., & Morris, T. J. (2000). HRT gene function requires interaction between a NAC protein and viral capsid protein to confer resistance to turnip crinkle virus. The Plant Cell 0153-2298X, 12, 1917-1926.

[32] Ron, D., & Walter, P. (2007). Signal integration in the endoplasmic reticulum unfolded protein response. Nature ReviewsMolecular Cell Biology 1471-0072, 8, 519-529.

[33] Schröder, M. (2008). Endoplasmic reticulum stress responses. Cellular and Molecular Life Sciences0142-0682X, 65, 862-894.

[34] Selth, L. A., Dogra, S. C., Rasheed, M. S., Healy, H., Randles, J. W., & Rezaian, M. A. (2005). A NAC domain protein interacts with tomato leaf curl virus replication accessory protein and enhances viral replication. The Plant Cell 0153-2298X, 17, 311-325.

[35] Seo, P. J., Kim, S. G., & Park, C. M. (2008). Membrane bound transcription factors in plants. Trends in Plant Science1360-1385, 13, 550-6.

[36] Souer, E., van Houwelingen, A., Kloos, D., Mol, J., & Koes, R. (1996). The no apical meristem gene of Petunia is required for pattern formation in embryos and flowers and is expressed at meristem and primordial boundaries. Cell, 0092-8674, 85, 159-70.

[37] Tajima, H., Iwata, Y., Iwano, M., Takayama, S., & Koizumi, N. (2008). Identification ofan Arabidopsis transmembranebZIP transcription factor involved in the endoplasmicreticulum stress response. Biochemical and Biophysical Research Communications, 0000-6291X, 374, 242-247.

[38] Tenhaken, R., Doerks, T., & Bork, P. (2005). DCD- a novel plant specific domain in proteins involved in development and programmed cell death. BMC Bioinformatics 6:169, 1471-2105, 1471-2105.

[39] Tran, L. S., Nakashima, K., Sakuma, Y., et al. (2004). Isolation and functional analysis of Arabidopsis stress-inducible NAC transcription factors that bind to a drought-responsive cis-element in the early responsive to dehydration stress 1 promoter. The Plant Cell0153-2298X, 16, 2481-2498.

[40] Udvardi, M. K., Kakar, K., Wandrey, M., et al. (2007). Legume transcription factors: global regulators of plant development and response to the environment. *Plant Physiology*, 144, 538-49, 1532-2548.

[41] Wang, S., Narendra, S., & Fedoroff, N. (2007). Heterotrimeric G protein signaling in the Arabidopsis unfolded protein response. Proceedings of the National Academy of Sciences, ISSN 1091-6490, 104, 3817-3822.

[42] Wang, Z., & Kiledjian, M. (2000). The Poly(A)-Binding Protein and an mRNA Stability Protein

[43] Jointly Regulate an Endoribonuclease Activity.Molecular and Cellular Biology, 1098-5549, 20, 6334-6341.

[44] Wang, Z., Libault, M., Joshi, T., et al. (2010). SoyDB: a knowledge database of soybean transcription factors. BMC Plant Biology, 10: 14, 1471-2229, 1471-2229.

[45] Weir, I., Lu, J., Cook, H., Causier, B., Schwarz-Sommer, Z., & Davies, B. (2004). CUPULIFORMIS establishes lateral organ boundaries in Antirrhinum. *Development*, 131, 915-922, 1011-6370.

[46] Xie, Q., Sanz-Burgos, A. P., Guo, H., Garcia, J. A., & Gutierrez, C. (1999). GRAB proteins, novelmembers of the NAC domain family, isolated by their interaction with a geminivirus protein. Plant Molecular Biology. 1572-9818, 39, 647-656.

[47] Xu, C., Bailly-Maitre, B., & Reed, J. C. (2005). Endoplasmic reticulum stress: cell life and death decisions. Journal of Clinical Investigation. 0021-9738, 115, 2656-2664.

Chapter 13

Soybean Under Water Deficit: Physiological and Yield Responses

Gustavo M. Souza, Tiago A. Catuchi,
Suzana C. Bertolli and Rogerio P. Soratto

Additional information is available at the end of the chapter

http://dx.doi.org/10.5772/54269

1. Introduction

The worldwide importance of soybean and the main limitations to crop yields

Because of its potential for large-scale production, soybean (*Glycine max* (L.) Merrill) has excelled in the world agricultural economy as a major oilseed crop. At present, soybeans are grown primarily for oil extraction and for use as a high protein meal for animal feed (Singh & Shivakumar, 2010). According to Li-Juan & Ru-Zhen (2010), soybean has a protein content of approximately 40% and an oil content of approximately 20%. In 2010, the area planted with soybeans worldwide was 102.4 million hectares, with total production of 261.6 million tons in the same year (Faoestat, 2012). This crop is currently being produced around the world, including in much of North America, South America and Asia. The U.S. and Brazil are the world's largest producers and exporters of soybean (Kumudini, 2010).

According to Mutei (2011), with the increasing world population, which is expected to reach nine billion people by 2050, the worldwide production of food should increase by 70% to ensure food security, which should be achieved through growth in productivity, rather than simply through the expansion of cultivation areas. Câmara & Heiffing (2000) indicate that to obtain increases in soybean yields, it is necessary to understand the interaction between cultivars and the production environment. Based on these factors, crop management can be adjusted to achieve proper development of plants in each production environment. Soybean is very responsive to environmental conditions, and the main climatic factors affecting its crop yields include the photoperiod, which influences the availability of full light, temperature and water availability (Mundstock & Thomas, 2005).

Although the effects of various environmental factors interfere with the performance of crops, water restriction is the main limiting environmental factor that contributes to the failure to obtain maximum soybean yields (Casagrande et al, 2001), influencing the use of other environmental resources. According to Confalone & Navarro Dujmovich (1999), the efficiency of the use of solar radiation by soybean remains relatively constant at different development stages. When there is moderate water stress, soybean tends to maximize the efficiency of radiation utilization and reduce the efficiency of the interception of photosynthetically active radiation, while under severe water deficits, there is a reduction of the efficiency of radiation utilization.

Lisar et al. (2012) report that the impacts of water stress in crop plants can reduce productivity by 50% in various parts of the world. Under stress conditions, the plants present a series of changes in their morphology, physiology and biochemistry, negatively affecting their growth and productivity. According to Gerten & Rost (2010), two-thirds of world food production through cultivation occurs under water stress. In this context and because of the prospect of global climate change, most crops will be exposed to negative impacts caused by drought.

2. Effects of water deficits on soybean: Crop yields and general responses

Water participates in nearly all physiological and biochemical processes in plants, comprising approximately 90% of their mass (Farias et al, 2007). It is responsible for the thermal regulation of the plant, acting both to maintain the cooling and heat distribution and to promote mechanical support of the plant (Taiz & Zeiger, 2009). It also functions as a solvent, through which gases, minerals and other solutes enter cells and move within plants (Nepomuceno et al, 1994).

The need for water in soybean increases with plant development, peaking during the flowering-grain filling stages (7-8 mm day^{-1}) and decreasing thereafter. The total water requirement for maximum productivity varies between 450 and 800 mm, depending on weather conditions, crop management practices and cycle timing (Embrapa, 2011, Farias et al, 2007). The loss of productivity under water deficit conditions depends on the soybean phenological stage, duration and intensity of water shortages (Doss & Thurlow, 1974). Kron et al. (2008) evaluated the responses of soybean to water stress induced in different phases in the plants and concluded that plants subjected to water stress during the V4 stage showed an increased tolerance to water shortages in later stages. This stage was considered to represent a "developmental window" in soybean, characterized as a specific period during plant development when environmental disturbances can be embodied, thereby improving subsequent plant resistance to environmental changes (Kron et al., 2008).

Desclaux et al. (2000) evaluated the effects of water stress at various stages of development in soybean plants and found the average length of the internodes to be the most sensitive feature to drought imposed during the vegetative stages (V4) and flowering (R1-R3), and a reduction in plant height was associated with water stress induced in the V4 stage. The number of pods per unit of shoot dry matter was significantly affected by water deficits in the reproductive stages (R3-R5). When stress occurred during grain filling (R5), the characteristics of the plant that were most affected were the number of grains per pod and the grain weight. Rosolem (2005) notes that the water demand of soybean is highest at the initiation of flowering, but a water deficit from pod initiation (R3) until 50% yellow leaves (R7) is the most critical stage for productivity. In a study performed by the same author correlating rainfall with grain yields, it was found that when water restriction occurred between flowering and the emergence of pods, the grain yield of soybean was 1,275 kg ha^{-1}, but under no water limitation at this stage, there may be an increase in productivity of 3.8 kg ha^{-1} for each mm of rain. When water restriction occurred during grain filling, the yield was 878 kg ha^{-1}, and for each mm of rain, there was an increase in productivity of 13 kg ha^{-1}, indicating the greater susceptibility of the soybean to water stress during grain filling, although the highest water demand in the crop occurred at the beginning of flowering (Desclaux et al., 2000). These results are in agreement with those of Nogueira& Nagai (1988). However, other studies observed that seed filling is not the most drought prone period of soybean development. When the water deficit starts during R1 (early flowering) and R4 growth stage, the seed yield may be significantly reduced (Eck et al., 1987, Brown et al., 1985, Hoogenboom et al., 1987) compared to R6-R7 growth stage.

According to Santos & Carlesso (1998), the most prominent responses of plants to water deficits in terms of morphological processes are decreases in leaf area and acceleration of the senescence and abscission of leaves. Catuchi et al. (2011) studied the conventional cultivar CD 220 and the transgenic cultivar CD 226RR under water deficits induced at the V4 stage, and they observed a reduction of approximately 40% of leaf area per plant compared to control plants and decreasing shoot dry mass of approximately 50% for both cultivars. Akinci & Losel (2012) reported that water stress drastically decreased root elongation and the expansion of leaf area in soybean, though these two processes were not equally affected because leaf expansion is usually reduced by a greater proportion than root growth, and carbon partitioning shifts to increase the root/shoot ratio. Catuchi et al. (2012) studied the responses of biomass and leaf area in plants of two soybean cultivars, BR 48 and EMBRAPA 16, which are considered tolerant and susceptible to water restriction, respectively, grown under water deficits. The authors observed a reduction of all traits in both cultivars due to a water deficit imposed in the reproductive stage of culture, with the exception of root dry mass in Embrapa 48, which remained unchanged, even under conditions of water restriction (Table 1). This phenomenon may occur because drought can promote the expansion of the root system to reach additional deeper moisture zones in the soil profile, a process that begins gradually after drying of the soil surface (Santos & Carlesso, 1998). The reduction of the other biomass parameters under conditions of water deficits is related to decreased photosynthetic rates, and biomass accumulation and translocation to grain are consequently impaired (Neumaier et al., 2000).

One of the most important processes of nitrogen nutrition of soybean, which results in improvements of productivity and profitability of the crop, is the symbiotic nitrogen fixation. Nevertheless, this process is negatively influenced by low moisture (Purcell et al, 2004; Purcell & Specht, 2004). Decreased nitrogen fixation starts when water potential of root nodules starts falling below -0.2 to -0.4 MPa (Pankhurst & Sprent, 1975). According to the Purcell et al. (2000), the water deficit promotes the accumulation of products of N_2 fixation (ureides) in the shoot of soybean plant, causing a feedback reduction in fixation of N_2. Thus, the authors report that proper nutrition with manganese (Mn^{+2}) promotes the breaking of ureides and extends N_2 fixation in plants under water deficit. Furthermore, soybean plants that produce larger nodules are less susceptible to reduction of nitrogen fixation in water deficit conditions (King & Purcell, 2001).

Attributes	BR-16		Embrapa 48	
	Control	Water deficits	Control	Water deficits
SM (g)	26.3$^{A(1)}$	13.9^{B}	23.9^{A}	11.4^{B}
SD_m (g)	57.1^{A}	29.1^{B}	51.8^{A}	25.2^{B}
RD_m (g)	8.3^{A}	5.1^{B}	5.6^{A}	4.6^{A}
OD_m (g)	65.4^{A}	34.3^{B}	57.4^{A}	29.8^{B}
A_l(cm^2)	1637.7^{A}	756.7^{B}	1356.1^{A}	724.3^{B}

$^{(1)}$Means followed by the same letters between the levels of water reposition in each cultivar not differ by Tukey test (P = 5%).

Table 1. Seed mass per plant (SM), shoot dry mass (SD_m), root dry mass (RD_m), overall dry mass (OD_m) and overall area leaf (A_l)of BR-16 and Embrapa 48 soybean cultivars with 100% and 40% water reposition (adapted from Catuchi et al., 2012)

Water restriction may be caused by several factors in plants, with the principal cause being an absence or an irregular distribution of rainfall during the crop cycle (Gopefert et al., 1993). In recent years, due to global climate change, climate stability, which allows the cultivation of crops to be planned, has been more limited. For plants to withstand periods of water restriction, they should be able to maintain their water status at normal turgor pressure during the hottest hours of the day, when the water vapor atmospheric demand is greater. This requires that the plant have a well-developed root system allowing it to reach water in deeper layers in the soil profile (Farias et al, 2007). In some cases, the limited extent of the root system reduces the water supply to plants. These responses are typical of soils containing toxic aluminum (Al^{3+}) combined with low rainfall during the crop cycle. The presence of Al^{3+} could limit the development of the root system due to inhibition of DNA synthesis and cell division, limiting the elongation of the roots and thus, the absorption of water from the deeper layers of soil. It also causes changes in nutrient uptake and in the overall nutritional balance of plants (Machado, 1997). Mascarenhas et al. (1984) observed a reduction of primary roots of plants of two cultivars of soybeans due to increased levels of Al^{3+} in a nutrient solution. Nolla et al. (2007) assessed the root development of soybean seedlings grown in solution with various concentrations of Al^{3+} (0.0, 0.30, 0.60, and 1.20 mmol L^{-1} Al) and observed a significant reduction of the root dry mass due to the increased concentration of Al^{3+} at pH = 4 (Figure 1).

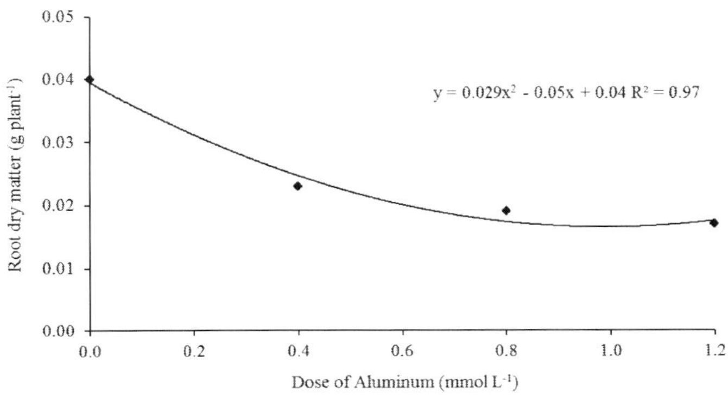

Figure 1. Root dry mass of soybean seedlings grown under different concentration of aluminum. (adapted from Nolla et al, 2007).

Moreover, soil compaction caused by the pressure of agricultural implements on farms is another limiting factor for deep root development. In the work carried out by Cardoso et al. (2006), there was a significant correlation found between the root volume of two soybean cultivars and the resistance to penetration (RP), with the root volume decreasing linearly in both cultivars associated with increases in RP. According to Beutler & Centurion (2003), the growth of the soybean root system is limited when the RP is greater than 2 MPa. Beutler et al. (2006) note that soybean yield decreases in RPvalues from 2.24 to 2.97 MPa.

In this sense, the use of implements that do not cause soil compression and cultivation techniques such as "no-tillage systems" (NTS) that result in better soil physics, promoting better root development at depth, are extremely important to avoid loss of productivity due to water restriction. Furthermore,some agricultural practices, such as lime and gypsum application may promote the correction of the soil profile (Santos et al, 2010).

According to Franchini et al. (2009), under NTS, the maintenance of the soil covering reduces evaporative water loss due to the formation of a physical barrier and reduces the temperature of the soil and runoff because of the increased capacity of water infiltration associated with protection of the surface of the soil against the impact of raindrops, thus preventing crusting. Similarly, increasing the percentage of soil organic matter (SOM), which is associated with a reduced intensity of soil cultivation, substantially improves the soil structure, which favors the development of soybean roots and thus increases the size of the water reservoir available. In addition, improvements in the soil structure provided by NTS increase infiltration and the water retention of the soil, thus favoring the upward flow of water from deeper layers to upper layers, where the majority of the soybean root system develops. The effects of SOM are connected with the hygroscopicity and high specific surface area of this type of material, which promotes increased water retention (Braida et al, 2011). In a long-term study, Franchini et al. (2009) reported that during the first four seasons after the adoption of aNTS, the soybean yield was similar or slightly lower than that obtained under conventional tillage (CT). However, from the fifth year onward, when the system had matured and consolidated, the soybean yield was higher under the NTS than the CT (Figure 2).

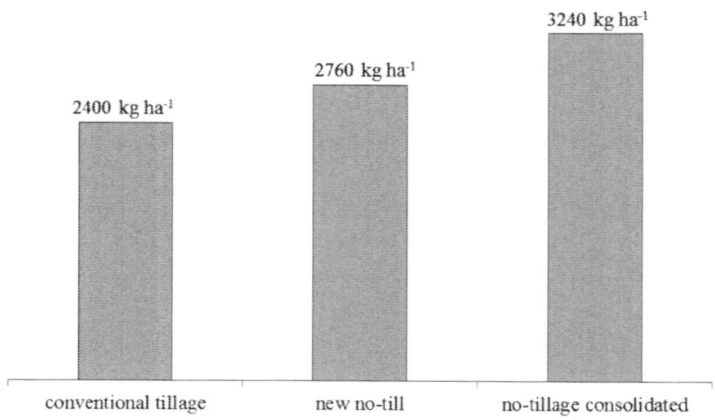

Figure 2. Soybean yield in different management systems: conventional tillage (13 years); new no-tillage (3 years) and no-tillage consolidated (13 years), (adapted from Franchini et al., 2009).

Sowing according to agroclimatic zoning for each agricultural environment is another strategy to avoid productivity losses due to water restriction. The cultivar must be adapted to the region considering the climate and soil type that determine water retention. In regions with higher occurrences of drought, it is essential to cultivate material that is more tolerant to water restriction. Thus, when the chemical and physical conditions of soil are suitable, allowing good root development at depth, and the cultivar and sowing time are selected to minimize the effects of water restriction, it is possible to obtain a high productivity soybean grain yield.

Overall, to achieve productivity under any conditions, it is essential that the process of photosynthesis, which is responsible for all carbon assimilated for the production of biomass, has a minimum efficiency. Under water deficit conditions during the soybean cycle, photosynthesis is one of the main physiological processes affected.

3. Influence of water deficiency on plant physiology: An overview

Drought stress is a factor that occurs when little water is available in an appropriate thermodynamic state. The term "dry" indicates a period without appreciable precipitation during which the water content in the soil is reduced (Larcher, 2000). This situation can be exacerbated when atmospheric conditions cause continuous water loss via transpiration or evaporation (Jaleel et al., 2009).

Depending on environmental conditions, different levels of water loss can occur. While water deficiency is a state caused by a moderate loss of water, which causes stomatal closure and reduction of gas exchange (Jaleel et al., 2009), desiccation is caused by much more extensive losses of water, which can potentially lead to major disruption of cellular structure and metabolism and eventually to an interruption of enzyme-catalyzed reactions (Smirnoff, 1993, Jaleel et al. 2009).According to Jaleel et al. (2009), water stress is characterized by reductions in water content and leaf water potential and loss of cellular turgor, which decreases stomatal conductance and cellular expansion, consequently constraining plant growth. Severe water stress can result in impairment of photosynthesis, metabolic disorders and ultimately death of the plant. In addition, other factors, such as high temperatures, soil salinity, pathogen attack and mechanical damage, may also contribute to differences in the amount of water in plants by interfering with their development and reducing their productivity (Larcher, 2000).

Once successful stand establishment is achieved, one of the most sensitive physiological processes to drought stress is reduced cell expansion resulting from decreased turgor pressure (Raper & Kramer, 1987). As leaf water potential falls, cell and leaf expansion are affected before photosynthesis. Bunce (1977) reported a linear relationship between soybean leaf elongation rate and turgor pressure. Decreasing leaf water potential to -0.80 MPa reduced leaf elongation rate by 40% relative to greater values of water potential. Consequently, leaf area and plant dry matter were reduced 60% and 65%, respectively. These results were subsequently confirmed in field experiments (Muchow et al., 1986). Thus, occurrence of water deficit during vegetative growth (emergence to R5) can reduce the leaf area indices (LAI) and the interception of photosynthetically active radiation by the total leaf area to levels insufficient for optimal crop growth rate (CGR) and yield. The effects of water stress on photosynthetic rates of soybean leaves are readily detectable at leaf water potentials about -1.0 to -1.2 MPa (Raper& Kramer, 1987). The rate starts declining more rapidly as water potential falls below -1.8MPa. Plants suffering this level of drought would have greater reductions of CGR and yield because not only would LAI be reduced, but the net assimilation rate (photosynthetic rate per unit LAI) would also be reduced. Drought stress effects on photosynthesis become irreversible once water potential falls below -1.6 MPa.

The stress induced by water deficits in plants depends on the conditions provided by the environment, varying according to the intensity and duration of water deficits, the rate of water deficit induction and the stage of plant development when drought occurs (Pinheiro & Chaves, 2011, Bertolli et al., 2012). Moreover, these factors can strongly influence the process of mitigation associated with acclimation to conditions after water shortages (Chaves et al., 2009). Acclimation responses in plants under water stress generally include responses related to growth inhibition or leaf shedding, reducing the water consumption by these tissues and contributing to the maintenance of water balance and the assimilation of carbon (Chaves et al., 2009, Pinheiro & Chaves, 2011). Osmoregulatory elements produced in response to slow dehydration also have the function of maintaining metabolic activity, improving cell osmotic adjustment (Pang et al., 2011). The increase in sugar levels during leaf dehydration appears to be the result of coordinated regulation of the synthesis and translocation of sucrose, contributing to osmotic adjustment and enabling the maintenance of turgor in meristematic regions. These responses may eventually lead to restoration of cellular homeostasis, thereby increasing plant survival under stress (Chaves, 1991, Chaves et al., 2009).

A central question in plant biology is related to the problem of the optimization of CO_2 fixation in environments with limited water. In environments where there is water restriction caused by a lack of water from the soil or by a high atmospheric water demand plants tend to close their stomata to conserve water via reducing transpiration losses, which may limit the CO_2 intake into the leaf for photosynthesis (Kaizer, 1987, Flexas et al., 2009, Pinheiro & Chaves, 2011). Control of the entry of CO_2 and water loss is performed by continuously setting the opening of the stomata distributed across the leaves. This dynamic of opening and closing of stomata is heterogeneous and can be particularly enhanced under conditions of water stress, causing the patchy stomatal conductance phenomenon to occur (Mott & Buckley, 1998, Flexas et al., 2009).

The response of photosynthesis under water stress has been the focus of study and debate for decades, particularly with respect to what the most limiting factors for photosynthesis are (Lawlor & Cornic, 2002, Lawlor & Tezara, 2009, Flexas et al., 2009). However, there is still some controversy regarding the importance of the main physiological parameters and the time period over which they limit photosynthesis (Pinheiro & Chaves, 2011).

The decrease in CO_2 diffusion from the atmosphere to the carboxylation site of Rubisco (ribulosebiphosphate carboxylase/oxygenase) is generally considered to be the main cause of reduced photosynthesis under conditions of mild and moderate water deficits (Grassi & Magnani, 2005, Chaves et al., 2003, 2009, Pinheiro & Chaves, 2011). Greater resistance to the diffusion of CO_2 may be caused by both stomatal limitation and by lower conductance in the mesophyll (Flexas et al., 2009). The stomata act in regulating the entry of CO_2 (an essential substrate for photosynthesis) into the leaves and the loss of water vapor to the atmosphere via transpiration. The control of gas exchange between the leaf and the atmosphere through the stomata is essential for the maintenance of photosynthetic activity and tissue hydration (Reynolds-Henne, et al. 2010, Aasamaa & Sõber, 2011). When the decrease in stomatal conductance is combined with high light levels, the leaves are subjected to an excessive amount of incident energy in relation to the amount of intercellular CO_2 available for photosynthesis, and the rate of reductant energy production can therefore overlap with the rate of its consumption by the Calvin cycle. Under these circumstances, down-regulation of photosynthesis or even photoinhibition can become a powerful defense mechanism for plants (Pinheiro & Chaves, 2011, Sanda et al. 2011). This protection can be achieved via thermal dissipation occurring in the light harvesting complex of the photochemical apparatus, involving the xanthophyll cycle (Demmig-Adams et al. 2006, Sarlikioti et al. 2010), and through alternative sinks for the excess energy, such as photorespiration (Lawlor & Cornic, 2002, Miyake, 2010) or the Mehler reaction, now referred to as the water-water cycle

(Miyake, 2010). Because there is dependence on the fixation of CO_2 and photochemical efficiency (ATP generation and NADPH) and vice-versa (Miyake et al., 2009), any stress or situation that reduces the consumption of photochemical products through decreased carboxylation efficiency can generate excessive excitation energy (Baker et al., 2004, Sarlikioti et al., 2010). When the non radioactive de-excitation energy (NPQ) is insufficient to address this excess energy, alternative electron sinks, such as the water-water cycle and photorespiration, are activated, which is crucial for maintaining photosynthetic activity (Miyake et al., 2009, 2010, Lawlor & Tezara, 2009).

Part of the reduction of photosynthesis can be attributed to non-stomatal effects at the chloroplast level, and electron transport and photophosphorylation are the main targets of inhibition (Sanda et al., 2011). The reduction in ATP synthesis that is an initial response to water deficits can lead to reduction of the capacity for Ribulose-1,5-*bis*phosphate (RuBP) regeneration, reducing potential photosynthesis (Lawlor & Tezara, 2009, Pinheiro & Chaves, 2011). Furthermore, Rubisco activity may be impaired by Rubisco activase activity and the reduction of ATP. Lawlor & Tezara (2009) claim that inhibitors such as RuBP analogues bind to the active site of Rubisco, decreasing its activity, especially when the concentration of RuBP is under saturated due to water deficiency. The regulation and restoration of Rubisco are mediated by Rubisco activase and require a high rate of ATP/ADP conversion. Thus, because phosphorylation is reduced under water stress, the activity of Rubisco is depressed (Parry et al., 2002).

In addition to solar energy, water availability and air temperature are elements that show seasonal variation (Rossato et al. 2009) and can affect photochemical activity. High temperatures may also affect photochemical activity directly through the inactivation of photosystem II and structural disorganization of the thylakoids or indirectly by the reduction of the activity of the Calvin cycle (Zhang & Sharkey, 2009). Plants subjected to water deficits usually exhibit reduced transpiration and therefore show a low capacity for dissipation of latent heat, increasing the temperature of the leaves. This condition can decrease the fixation of CO_2 because under increasing temperatures, Rubisco activity tends to be reduced (Kumar et al., 2009).

3.1. Effects of water stress on photosynthesis in soybean plants

Studies on the soybean crop have been focused on describing the impacts caused by the imposition of water stress on physiological parameters, particularly on photosynthesis in different genotypes (Desclaux & Roumet, 1996, Kron et al., 2008, Firmano et al., 2009, Catuchi et al., 2011, Bertolli et al., 2012, Catuchi et al., 2012). In the cultivars 'New Tanbakuro' and 'Tamanisiki' (Ohashi et al. 2009), 'CD 202' and 'CD 226RR' (Catuchi et al., 2011), 'BR 16' and 'Embrapa 48' (Catuchi et al., 2012) and 'MG/BR-46' (Conquista) (Stolf-Moreira et al., 2011) and wildtype plants (Kao & Tsai, 1998), it has been observed that the photosynthetic rate, stomatal conductance and transpiration were reduced by water deficits. Moreover, there was a reduction in the intercellular CO_2 concentration (Ci) observed in the cultivars 'BR 16' (Catuchi et al., 2012) and 'Embrapa 48' and wildtype species. Additionally, in the cultivars 'New Tanbakuro' and 'Tamanisiki', Ci was altered slightly under early water stress treatments, while for the cultivars 'MG/BR-46' (Conquista) and 'BR 16' (Stolf-Moreira et al., 2011), Ci increased as the water deficit progressed, indicating different physiological responses for different soybean cultivars. Furthermore, an increase in the intrinsic efficiency of water use was observed when the cultivars 'CD 202', 'CD 226RR', BR 16 and EMBRAPA 48 were subject to episodes of water restriction (Catuchi et al., 2011 and 2012). According to Manavalan et al. (2009), this increase may indicate better control of water loss via transpiration, contributing to the productivity of soybean. However, it is important to take into account that the studies discussed herein used different methods of water stress induction, which could interfere with making more reliable and suitable comparisons among cultivars (Bertolli et al., 2012).

Reduction of net photosynthesis in soybean plants can be induced by both stomatal and non-stomatal factors (of both biochemical and photochemical origin). When a crop is subjected to a water deficit, the plants can reduce their stomatal conductance (gs), limiting the entry of CO_2 into the substomatal chambers and thus reducing the diffusion of carbon to the site of carboxylation, resulting in significant decreases in carbon assimilation (Yu et al., 2004). Furthermore, Flexas et al. (2006a) report that the effects of water stress on the initial activity of Rubisco may be reproduced by induction of stomatal closure, independent of the reduction in the relative water content in the leaves of soybean plants. Thus, we can expect a lower regulation of photochemical and biochemical processes when the availability of CO_2 is the most limiting component for photosynthesis in plants under severe water stress (Flexas et al. 2006b).

Although Pankovic et al. (1999) report that the content of Rubisco in soybean leaves increases as plants acclimate to water stress, other authors have observed that under the imposition of water deficits in pine, tobacco and soybean, there is reduced transcription of the subunits of this enzyme (Pelloux et al., 2001, Kawaguchi, et al., 2003, Majumdar et al., 1991, respectively). Reduction of the Rubisco activity in soybean plants under drought stress can be induced by reducing the content of the enzyme itself and possibly through increases in strongly binding inhibitors, as reported by Flexas et al. (2006a). When stomatal closure occurs for a period of several days, another mechanism involving gene expression can operate, resulting in a decrease in the total amount of Rubisco and/or an increase in the content of inhibitors that bind strongly to this enzyme (Flexas et al., 2006a).

Moreover, when the availability of CO_2 and biochemical activity are reduced due to water deficits, the excess reductants in the photochemical apparatus must be dissipated as heat or drained through alternative electron sinks (Miyake et al., 2009, 2010) to reduce photoinhibition and the production of reactive oxygen species (ROS). Studies have revealed that PSII of soybean plants is resistant to moderate water stress (Kirova et al., 2008), and the potential quantum efficiency of PSII (F_v / F_m) and electron transport rate (ETR) are not altered by the imposition of water stress (Ohashi et al., 2006). Bertolli et al. (2012) reported that the decrease in the ETR was more sensitive than the decrease of F_v / F_m when the relative water content declined in soybean plants (cv. CD202), supporting the idea that the potential photochemical efficiency would not be readily affected by water deficiency. However, the reduction of the ETR could be due to a lower energy (ATP /

http://dx.doi.org/10.5772/54269

NADPH) demand being required for carbon fixation metabolism under water deficit. Because the stomatal conductance was decreased as water deficiency progressed, the reduction of CO_2 diffusion from the substomatic chamber to the active site of Rubisco may have influenced the efficiency of the Calvin cycle, which, in turn, may have reduced the consumption of ATP / NADPH from the photochemical apparatus, reducing the efficiency of the electron transport chain (Bertolli et al., 2012).

Depending on the intensity and duration of drought stress, metabolic limitations are frequently observed to be correlated with decreases in ATP, which reduces the capacity for RuBP turnover (Parry et al., 2002). This reduction in ATP synthesis is due to the decrease in electron transport and photophosphorylation caused by the reduction of the reactions associated with the chloroplast membranes (Catuchi et al., 2011). These membranes may suffer structural damage caused by decreasing the content of free water and ROS overproduction (Lawlor & Cornic, 2002), reducing the efficiency of the photochemical apparatus. Kao & Tsai (1998) also reported that the amount of quinone (QA) in the reduced state in a wild type soybean plant under water stress is greater than in plants irrigated under high light. The increase in the reduced state of QA is associated with the amount of inactivated and damaged D1 protein in the PSII reaction centers, indicating susceptibility of the photochemical apparatus to water deficits.

4. Aspects of mineral nutrition in the relationship between water deficits and plant physiology

Under water stress, plants develop various physiological and molecular mechanisms to maintain productivity. Among these mechanisms, Nepomuceno et al. (2001) highlight the activation of genes induced by drought to promote cell tolerance to dehydration and osmotic adjustment to maintain the water potential and turgor close to optimum levels. Additionally, to minimize the oxidative damage to cells generated by reactive oxygen species (ROS), plants develop antioxidant systems (Apel & Hirt, 2004).

In addition to the internal mechanisms in plants, the negative effects of water stress can be minimized through a balanced supply of nutrients (Waraich et al., 2011). Among the nutrients classified as essential (Dechen & Nachtigall, 2006), potassium (K), phosphorus (P) and calcium (Ca) are the most studied in relation to their roles in reducing the effects of water stress on the physiology of soybean (Waraich et al, 2011).

4.1. Effects of potassium

Potassium is considered to be the second most extracted element in soybean crops (Malavolta, 2006). From 1,000 kg of seeds produced by a soybean crop, 20 kg of K_2O can be extracted (Mascarenhas, 2004). In this context, in some agricultural production systems, particularly tropical systems, K application is required to ensure soil productivity. More than 60 enzymes involved in cell metabolism are K dependent for normal activity because this nutrient is an important enzymatic activator (Prado, 2008). Moreover, K plays an important role in cell expansion, which involves the formation of a large central vacuole occupying 80% to 90% of cell volume.

Under water deficit conditions, stomatal conductance is reduced (Oliveira et al, 2005) consequently decreasing the intercellular CO_2 concentration (Kaiser, 1987, Lawlor & Tezara, 2009). Thus, the light energy used for the fixation of CO_2 is diverted to O_2, generating high accumulation of ROS in the chloroplast (Pitzschke et al, 2006). According to Cakmak (2005), when plants are grown under low K availability, the production of free radicals may be increased because the lack of this nutrient disturbs the opening and closing mechanism of stomata, causing a reduction of photosynthesis, and consequently, the excess electrons are diverted to the production of ROS. Therefore, under conditions of water stress, the plant exhibits an increased demand for K to maintain photosynthesis and protect the chloroplasts from oxidative damage. This author also stresses the importance of K in the translocation of assimilates. Under K deprivation, there is reduced exportation of the products of photosynthesis to the drain region of the plant. Thus, the accumulation of photoassimilates in the chloroplast can decrease the fixation of CO_2 through down-regulation, thus increasing the generation of ROS.

According to Prado (2008), K promotes maintenance of the turgor of guard cells, allowing better opening and closing dynamics of the stomatal pores. Sangakkara et al. (2000) evaluated the effect of moisture and K fertilization on the physiology of two common bean cultivars and observed that the addition of K to the system via a nutrient solution promoted an increased photosynthetic rate under conditions of water stress in both cultivars. Catuchi et al. (2012) evaluated the net CO_2 assimilation rate (A) in two soybean cultivars under water deficit conditions as well as 12 hours after rehydration and following supplementation with 0, 90 and 180 mg dm^{-3} K. The authors concluded that in general, the A values (Figure 3) in both cultivars decreased by 50% under water deficit, irrespective of the K level. In contrast, after rehydration, the cultivar BR 16 showed A values that were 27% higher in plants without the addition of K and 42% higher in plants supplied with 90 mg dm^{-3} K compared to the values in plants under drought. However, the higher dose of K did not allow the recovery of A after rehydration. Moreover, Embrapa 48 responded positively to supplementation with two doses of K in terms of the recovery of A. While in plants without added K, there was no recovery of A observed. Plants that received doses of 90 and 180 mg dm^{-3} K showed A values that were 57 and 38% higher, respectively, than those in plants under water stress. These responses indicate that K may promote greater recovery of photosynthesis in soybean after a period of water restriction. According to Flexas et al. (2004), the intensity and duration of water restriction are key factors that define the speed and rate of recovery of plants after rehydration. In general, plants subjected to severe drought stress exhibit recovery of only 40-60% of the maximum photosynthetic rate on the next day. In a study performed by Catuchi et al. (2012), these values were only achieved in plants that were supplemented with K. The response of the recovery of plants supplied with K via fertilization may be related to the influence of this nutrient on the repair of oxidative damage to cells under conditions of water stress (Soleimanzadeh et al., 2010). The higher photosynthetic rates of plants supplied with K after recovery could provide greater restoration of plant growth, minimizing productivity losses.

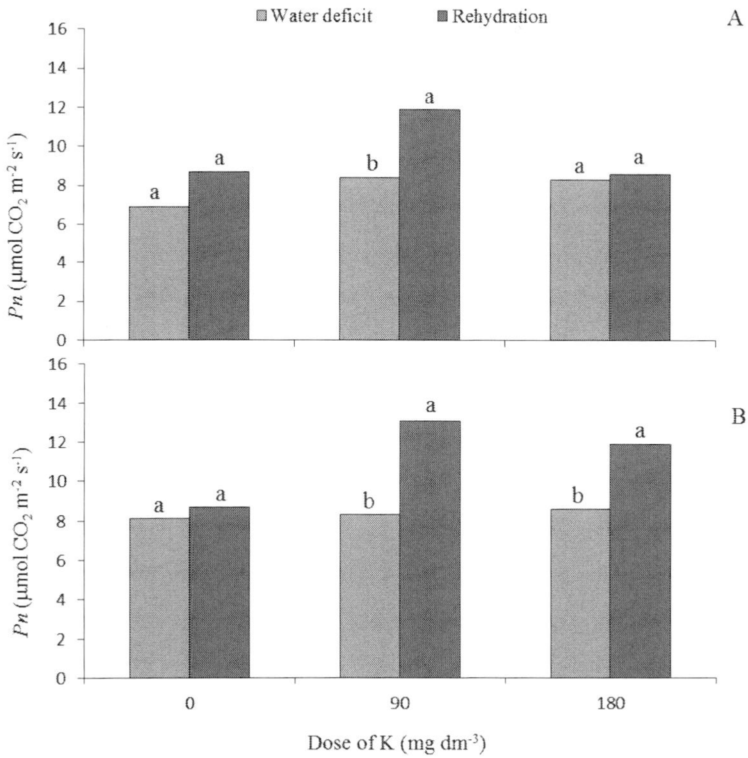

Figure 3. Average of net photosynthesis (Pn) of cultivars BR-16 (A) and Embrapa 48 (B) grown under water stress (40%) and 12 hours after rehydration. The letters above the bars indicate the statistical difference ($p < 0,005$) between the water levels in each dose of potassium (adapted from Catuchi et al, 2012).

4.2. Effects of phosphorus

Because of its role in the formation of adenosine triphosphate (ATP), phosphorus (P) plays key roles in the production of energy necessary for photosynthesis, the translocation of assimilates and many other metabolic processes. In its inorganic form, P is the substrate or end product in many enzymatic reactions, including photosynthesis and carbohydrate metabolism, and it is essential for the regulation of metabolic pathways in the cytoplasm and chloroplast, sucrose and starch synthesis, triose phosphate transport, translocation of sucrose and hexose synthesis (Araújo & Machado, 2006).

According to Lantmann & Castro (2004), for each ton of soybeans produced, the plant consumes 15 kg of P_2O_5 on average. Under conditions of low soil water availability, there is a marked reduction in P uptake by plants (Santos et al., 2006). When there is a lack of inorganic P (Pi) in the chloroplast, decreases in the production of ATP and NADPH may occur, resulting in a decrease in the regeneration of ribulose-1,5-biphosphate, which is crucial in the photosynthetic assimilation of CO_2 (Lawlor & Cornic, 2002).

The decrease in ATP synthesis in the chloroplast may be caused by low availability of free cytoplasmic Pi, which is exchanged for triose phosphate in the chloroplast by phosphate transporters that use Pi as a substrate (Flügge et al., 2003). The carbon partitioning between starch and sucrose is dependent on the concentration of cytoplasmic Pi, which regulates the export of triose-P from the chloroplast to the cytosol, and a decrease in the recycling of P between the cytoplasm and chloroplasts can generate inhibition of photosynthesis via carbohydrate accumulation (Foyer, 1988). During drought periods lasting approximately ten days, the diffusive flux of P from the soil to plants stops almost completely, causing a significant loss of productivity (Novais & Smyth, 1999). Thus, there is a need for a stock of P under optimal conditions of water availability to reduce the effects of the lack of P during water stress (Prado, 2008). Furthermore, after rehydration, the absorption and uptake of P should begin rapidly to restore the diffusion flow.

The direct role of P in the maintenance of plant productivity under low water availability is also related to the maintenance of stomatal conductance (Waraich et al., 2011). This function of P is associated with the osmotic regulation of stomatal guard cells because P supplementation can be related to the accumulation of proline, which is an important regulator of cell osmolarity (Al-Karaki et al., 1996). Firmano et al. (2009) evaluated the effects of P on photosynthesis in soybean plants grown under a water deficit and observed (Figure 4) that fertilization with 200 kg ha^{-1} P maintained net photosynthesis under water stress better in comparison to what was observed in plants that were not supplemented. According to these authors, these results were due to increased stomatal conductance promoted by P under conditions of water restriction.

Santos et al. (2006) evaluated the effect of foliar supplementation with inorganic phosphate (Pi) in two common bean genotypes, A320 and Ouro Negro, under water deficit conditions for 7 days. They observed that the rates of net photosynthesis and stomatal conductance were not affected by supplementation of Pi during dehydration in either genotype. However, after rehydration, stomatal conductance and photosynthesis were increased associated with foliar Pi being supplied in relation to the plants without Pi supplementation. Likewise, as noted by Firmano et al. (2009) in soybean, the role of Pi in the regulation of photosynthesis recovery after a water deficit appears to be important in reducing the deleterious effects of a temporary lack of water.

http://dx.doi.org/10.5772/54269

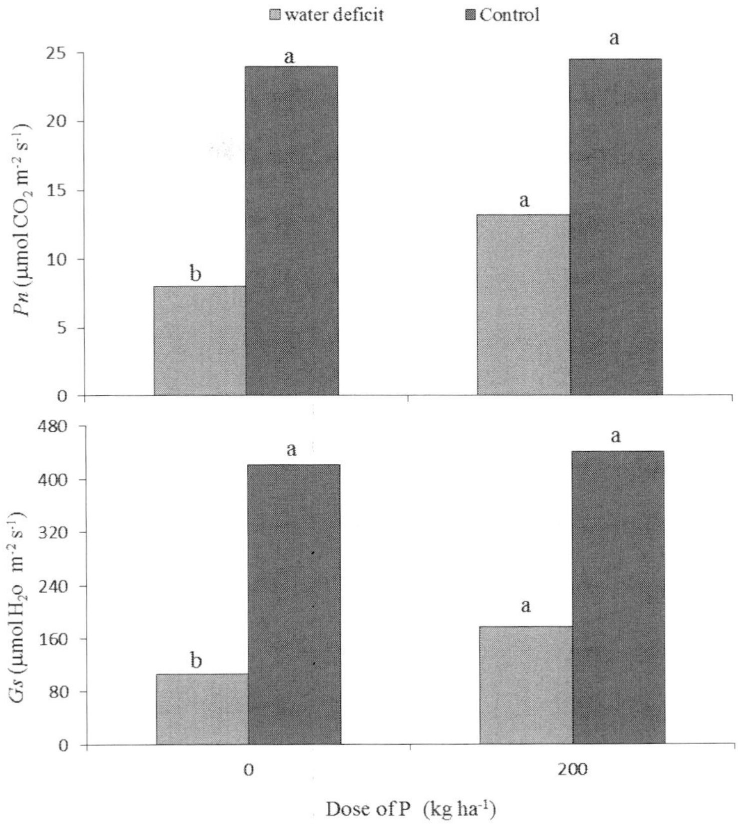

Figure 4. Average of net photosynthesis (Pn) and stomatal conductance (Gs) of soybean plants grown with and without water restriction. The lowercase letters above the bars indicates the statistical difference ($p < 0,05$) between the P doses in each water level (adapted from Firmano et al, 2009).

4.3. Effects of calcium

The main functions of Ca in plants are acting as a component of the cell wall and as a second messenger in signaling associated with different processes in the cell. This nutrient plays an important role in ion uptake, root development and the germination of pollen grains (Vitti et al., 2006). During stress, Ca plays an important role in the regulation of plant metabolism, along with the calmodulin protein, which can promote the maintenance of cellular metabolism under water deficit conditions (Waraich et al, 2011). Ca assists the plant in its recovery after a water shortage because this nutrient functions in the activation of the ATPase enzyme in the cell membrane, promoting the pumping back to the cell of electrolytes that were lost because of membrane damage caused by water deficit (Palta, 1990, Waraich et al., 2011).

In addition to the direct effects of K, P, and Ca on the maintenance of plant metabolism under water deficit conditions, balanced nutrition regarding all essential elements (both macro- and micronutrients) can support plant development under limiting conditions by improving the initial steps of vegetative growth, such as leaf area expansion. This improved growth will allow the achievement of high photosynthetic rates and, hence, good root development, thereby improving the absorption of water into deeper layers and allowing the plants to survive water deficit periods.

4.4. Future directions

Breeders and geneticists involved in soybean breeding are interested in consolidating the current knowledge about physiology and functional genomics to improve crop breeding programs (Manavalan et al., 2009), especially based on studies aimed at providing the information needed to improve the resistance / tolerance of cultivars to a multitude of stress factors (Kulcheski et al., 2011, Makbul et al., 2011). Through proteomic analysis, 145 genes that are differentially expressed according to the imposition of water stress were identified in two soybean cultivars, MG/BR46 [Conquista] and BR 16, that are considered tolerant and sensitive to water deficits, respectively (Stolf-Moreira et al., 2011). These genes were classified into nine functional categories: energy, transcription factors, metabolism, stress responses, protein synthesis, cell communication, the cell cycle, cellular transport, and other unknown functions. Additionally, 11 micro-RNAs that show different expression patterns during the imposition of biotic and abiotic stress were identified in the cultivars 'Embrapa 48' (tolerant to drought stress) and 'BR 16' (sensitive to water stress) (Kulcheski et al., 2011), and the transcription of several other proteins related to oxidative damage, isoflavonoids and lignin synthesis was detected in soybean under water stress (Yamaguchi et al., 2010). Furthermore, Alam et al. (2010) reported that there are two enzymes involved in carbohydrate metabolism (UDP-glucose pyrophosphorylase and 2.3-biphosphoglycerate independent phosphoglyceratemutase) that are suppressed after exposure to a water deficit. The levels of these enzymes tended to revert to the basal level after rehydration of the plants, suggesting that the change in the allocation of carbon in soybean plants under drought may indicate an adaptive response. According to the authors of this report, the metabolism of carbohydrates is one of the processes that are most susceptible to water stress, after photosynthesis. Other studies have identified several soybean wildtypes that can be specifically adapted to adverse conditions, such as wind, water logging, salinity and water deficits, and may be useful for identifying genes related to tolerance / resistance to a variety of biotic and abiotic stresses (Lee et al., 2010). Such studies are required because genetic diversity has been lost in the process of domestication of *G. max* (Hyten et al., 2006), and wildtype soybean have been useful for contributing new and unique genes to increase yields under different worldwide crop conditions (Wang et al., 2004).

Moreover, as discussed in previous sections of this chapter, to improve soybean growth under water deficit conditions, the application of additional strategies is necessary, such as constant development of new soil management techniques, allowing the development of root systems to increase the water intake capacity as well as provide balanced nutrition to the crop, supporting adequate development of the plants throughout their life cycle.

Author details

Gustavo M. Souza[1], Tiago A. Catuchi[2], Suzana C. Bertolli[1] and Rogerio P. Soratto[2]

1 Laboratório de Ecofisiologia Vegetal, Universidade do Oeste Paulista, Presidente Prudente, SP, Brazil

2 Departamento de Produção Vegetal, Faculdade de Ciências Agronômicas, Universidade Estadual Paulista, Botucatu, SP, Brazil

References

[1] Aasamaa, K., Sõber, A. (2011). Responses of stomatal conductance to simultaneous changes in two environmental factors. *Tree Physiology*. v. 31, p. 855-864.

[2] Alam, I., Sharmin, S. A., Kim, K-H., Yang, J. K., Choi, M. S., Lee, B-H. (2010). Proteome analysis of soybean roots subjected to short-term drought stress. *Plant Soil*. v. 333, p. 491-505.

[3] Araújo, A. P. and Machado, C. T. T. (2006). Fósforo. In: Fernandes, M. S. *Nutrição mineral de plantas*. SBCS, p. 253-280.

[4] Al-Karaki, G. N., Clark, R. B. and Sullivan, C. Y. (1996). Phosphorus nutrition and water effects on proline accumulation in sorghum and bean. *Journal of Plant Physiology*, Leipzig, Germany, v. 148, p. 745-751.

[5] Apel, K, Hirt H. (2004). Reactive oxygen species: Metabolism,oxidative stress, and signal transduction. *Ann Rev Plant Bio*, v. 55, p. 373-399.

[6] Akýnci, S. & Lösel, D. M. (2012). Plant Water-Stress Response Mechanisms. In:Rahman, I. M. M. and Hasegawa, H. *Water Stress*, InTech, Croatia, p. 15-42.

[7] Baker, N. R., Rosenqvist, E. (2004). Applications of chlorophyll fluorescence can improve crop production strategies: an examination of future possibilities. *Journal of Experimental Botany*. v. 55, p. 1607–1621.

[8] Beutler, A. N. & Centurion, J. F. (2003). Efeito do conteúdo de água e da compactação do solo na produção de soja. *Pesq. agropec. bras.*, v. 38, n. 7, p. 849-856.

[9] Beutler, N. A., Centurion, J. F. Centurion, M. A. P. C. & da Silva, A. P. (2006). Efeito da compactação na produtividade de cultivares de soja em latossolo vermelho. *R. Bras. Ci. Solo*, v. 30, p. 787-794.

[10] Bertolli, S. C., Rapchan, G. L., Souza, G. M. (2012). Photosynthetic limitations caused by different rates of water-deficit induction in Glycine max and Vignaunguiculata. *Photosynthetica.* DOI: 10. 1007/s11099-012-0036-4.

[11] Braida, J. A., Bayer, C., Albuquerque, J. A. & Reichert, J. M. (2011). Matéria Orgânica e seu Efeito na Física do Solo. In:Klauberg Filho, O., Mafra, A. L. andGatiboni, L. C. *Tópicos em Ciência do Solo.* SBCS, v. 7, p. 221-278.

[12] Brown, E. A., Caviness, C. E., Brown, D. A. (1985). Response of selected soybean cultivars to soil moisture deficit. Agron. J., v. 77, pp. 274-278.

[13] Bunce, J. A. (1977). Leaf elongation in relation to leaf water potential in *soybean*. Exp. Bot., v. 28, pp. 158-63.

[14] Cakmak, I. (2005). The role of potassium in alleviating detrimental effects of abiotic stresses in plants. *J. Plant Nutr. Soil Sci.* Turquia, v. 168, p. 521-530.

[15] Câmara, G. M. S. & Heiffig, L. S. (2000). Fisiologia, ambiente e rendimento da cultura da soja. In: Câmara, G. M. S. *Soja: tecnologia da produção.* Piracicaba: ESALQ/LPV, p. 81 – 120.

[16] Casagrande, E. C., Farias, J. R. B., Neumaier, N., Oya, T., Pedroso, J., Martins, P. K., Breton, M. C. & Nepomuceno, A. L. (2001). Expressão gênica diferencial durante déficit hídrico em soja. *R. Bras. Fisiol. Veg.* v. 13, n. 2, p. 168-184.

[17] Catuchi, T. A., Guidorizzi, F. V. C., Guidorizi, K. A., Barbosa, A. M. & Souza, G. M. (2012). Respostas fisiológicas de cultivares de soja à adubação potássica sob diferentes regimes hídricos. *Pesq. agropec. bras.,* v. 47, n. 4, p. 519-527.

[18] Catuchi, T. A., Vítolo, H. F., Bertolli, S. S. & Souza, G. M. (2011). Tolerance to water deficiency between two soybean cultivars: transgenic *versus* conventional. *Ciência Rural,* Santa Maria, v. 31, n. 3, p. 373-378.

[19] Cardoso, E. C., Zotarelli, L., Piccinin, J. L., Torres, E., Odilon Ferreira Saraiva, O. F. & Guimarães, M. F. (2006). Sistema radicular da soja em função da compactação do solo no sistema de plantio direto. *Pesq. agropec. bras.,* v. 41, n. 3, p. 493-501.

[20] Carmo-Silva, A. E., Keys, A. J., Andralojc, P. J., Powers, S. J., Arrabaça, M. C., Parry, M. A. J. (2010). Rubisco activities, properties, and regulation in three different C4 grasses under drought. *Journal of Experiemental Botany.* v. 61, p. 2355-2366.

[21] Chaves, M. M., Flexas, J., Pinheiro, C. (2009). Photosynthesis under drought and salt stress: regulation mechanisms from whole plant to cell. *Annals of Botany.* v. 103, p. 551–560.

[22] Confalone, A. & Navarro Dujmovich, M. (1999). Influência do "déficit" hídrico sobre a eficiência da radiação solar em soja. *Rev. Bras. de Agrociência,* v. 5, n. 3, p. 195-198.

[23] Dechen, A. R. & Nachtigall, G. R. (2006). Elementos essências e benéficos às plantas superiores. In:Fernandes, M. S. *Nutrição mineral de plantas.* SBCS, p. 01-06.

[24] Demmig-Adams, B., Adams, Iii W. W., Mattoo A. (2006). Photoprotection, photoinhibition, gene regulation and environment. In: *Advances in photosynthesis and respiration.* v. 21. Dordrecht: Springer. eds.

[25] Desclaux, D., Huynh, T. & Roumet, P. (2000). Identification of soybean plant characteristics that indicate the timing of drought stress. *Crop Science.* v. 40, p. 716-722.

[26] Desclaux, D., Roumet, P. (1996). Impact of drought stress on the phenology of two soybean (Glycine max L. Merr) cultivars. *Field Crops Research.* v. 46, p. 61-70.

[27] Doss, B. D. & Thulow, D. L. (1974). Irrigation, row width and plant population in relation to growth characteristics of two soybean varieties. *Agronomy Journal,* Madison, v. 65. p. 620-623.

[28] Eck, H. V., Mathers, A. C. & Musick, J. T. (1987). Plant water stress at various growth stages and growth and yield of soybeans. *Field Crops Res.,* v. 17, pp. 1-16.

[29] Embrapa. (2011). Sistema de Produção 15. Exigências Climáticas. In: *Tecnologia de produção de soja - Região central do Brasil – 2012 e 2013.* Londrina: Embrapa Soja, p. 11-12. <http://www. cnpso. embrapa. br/download/SP15-VE. pdf>. [Accessed on 09/06/2012].

[30] FAOSTAT – Food And Agriculture Organization Of The United Nations (2012). *Production - crops.* Roma, Food And Agriculture Organization Of The United Nations. Disponível em: < http://faostat. fao. org/site/567/default. aspx#ancor >. [Accessed on 05/06/2012].

[31] Farias, J. R. B, Nepomuceno, A. L. & Neumaier, N. (2007). *Ecofisiologia da soja.* Londrina: EMBRAPA-CNPSo, 9 p. (EMBRAPA-CNPSo. Circular técnica, 48).

[32] Firmano, R. S Kuwahara, F. A. & Souza, G. M. (2009). Relação entre adubação fosfatada e deficiência hídrica em soja. *Ciência Rural,*v. 39, n. 7, p. 1967-1973.

[33] Flexas, J., Barón, M., Bota, J., Ducruet, J., Gallé, A., Galmés, J., Jiménez, M., Pou, A., Ribas-Carbó, M., Sajnani, C., Tomás, M., Medrano, H. (2009). Photosynthesis limitations during water stress acclimation and recovery in the drought-adapted Vitis hybrid Richter-110 (V. berlandieri x V. rupestris). *Journal of Experimental Botany.* v. 60, p. 2362-2377.

[34] Flexas, J., Ribas-Carbó, M. Bota, J., Galmés, J., Henkle, M., Martínez-Cañellas, S., Medrano, H. (2006a). Decrease Rubisco activity during water stress is not induced by decreased relative water content but related to conditions of low stomatal conductance and chloroplast CO2 concentration. *New Phytologist.* v. 172, p. 73-82.

[35] Flexas, J., Bota, J., Galmés, J., Medrano, H., Ribas-Carbó. (2006b). Keeping a positive carbon balance under adverse conditions: responses of photosynthesis and respiration to water stress. *PhysiologiaPlantarum.* v. 127, p. 343-352.

[36] Flexas, J., Bota, J., Loreto, F., Cornic, G. & Sharkey, T. D. (2004). Diffusive and metabolic limitations to photosynthesis under drought and salinity in C3 plants. *Plant Biology,* v. 6, p. 269-279.

[37] Flügge, U. I., Häusler, R. E., Ludewig, F. & Fischer, K. (2003). Functional genomics of phosphate antiport systems of plastids. *PhysiologiaPlantarum,* Hoboken, v. 118. p. 475-482.

[38] Foyer, C. H. (1988). Feedback inhibition of photosynthesis through source-sink regulation in leaves. *Plant Physiology and Biochemistry*, Paris, v. 26, p. 483-492.

[39] Franchini, J. C., Debiasi, H., Sacoman, A. Nepomuceno, A. L. & Farias, J. R. B. (2009). Manejo do solo para redução das perdas de produtividade pela seca. In: Embrapa-Cnpso, 39p. (Embrapa-Cnpso. Circular técnica, 314).

[40] Grassi, G., Magnani, F. (2005). Stomatal, mesophyll conductance and biochemical limitations to photosynthesis as affected by drought and leaf ontogeny in ash and oak trees. *Plant, Cell and Environment.* v. 28, p. 834–849.

[41] Gerten, D. & Rost, S. (2010). *Development and Climate Change: Climate change impacts on agricultural water stress and impact mitigation potential.* Potsdam Institute for Climate Impact Research (PIK), Germany, 8p.

[42] Hyten, D., Song, Q., Zhu, Y., Choi, I. Y., Nelson, R. L., Costa, J. M., Specht, J. E., Shoemaker, R. C., Cregan, P. B. (2006). Impacts of genetic bottlenecks on soybean genome diversity. *Proceedings of the National Academy of Sciences.* v. 103, p. 16666-16671.

[43] Gopefert, H., Rossetti, L. A., Souza, J. (1993). *Eventos generalizados e seguridade agrícola.* Brasília: IPEA, Ministério do Planejamento, 78p.

[44] Hoogenboom, G, Peterson, C. M., Huck, M. G. (1987). Shoot growth rate of soybean as affected by drought stress. *Agron. J.*, v. 79, pp. 598-607.

[45] Jaleel, C. A., Manivannan, P., Wahid, A., Farooq, M., Al-Juburi, H. J., Somasundaram, R., Panneerselvam, R. (2009). Drought stress in plants: a review on morphological characteristics and pigments composition. *International journal of agriculture & biology.* v. 11, p. 100-105.

[46] Kaiser, W. M. (1987). Effects of water deficit on photosynthetic capacity. *Physiologia-Plantarum.* Copenhagen, v. 71, p. 142-149.

[47] Kao, W. -Y, Tsai, T. -T. (1998). Tropic leaf movements, phosynthetic gas exchange, leaf $\delta^{13}C$ and chlorophyll a fluorescence of three soybean species in response to water availability. *Plant, Cell and Environment.* v. 21, p. 1055-1062.

[48] Kawaguchi, R., Williams A. J., Bray, E. A., Bailey-Serres, J. (2003). Water-deficit-induced translational control in Nicotianatabacum. *Plant, Cell and Environment.* v. 26, p. 221-229.

[49] King, C. A. & Purcell, L. C. (2001). Soybean Nodule Size and Relationship to Nitrogen Fixation Response to Water Deficit. *Crop Science,* v. 41, p. 1099 - 1107.

[50] Kirova, E., Tzvetkova, N., Vaseva, I., Ignatov, G. (2008). Photosynthetic Responses of Nitrate-Fed and Nitrogen-Fixing soybeans to progressive water stress. Journal of Plant Nutrition. v. 31, p. 445-458.

[51] Kumar, A., Li, C., Portis A. R. Jr. (2009). Arabidopsis thaliana expressing a thermostable chimeric Rubisco activase exhibits enhanced growth and higher rates of photosynthesis at moderately high temperatures. *Photosynth Res.* v. 100, p. 143-153.

[52] Kulcheski, F. R., Oliveira, L. F. V., Molina, L., Almerão, M. P., *et al.* (2011). Identification of novel soybean microRNAs invoved in abiotic and biotic stresses. *BMC Genomics.* v. 12, p. 307.

[53] Kron, A. P., Souza, G. M. & Ribeiro, R. V. (2008). Water deficiency at different developmental stages of *Glycine max* can improve drought tolerance. *Bragantia,* v. 67, n. 1, p. 43-49.

[54] Kumudini, S. (2010). Soybean Growth & Development. In: B Singh, (Ed.). *The Soybean: Botany, Production and Uses.* CAB International, Oxfordshire, UK, p. 48–73.

[55] Lantmann, A. F. & Castro, C. (2004). Resposta da soja à adubação fosfatada. In: Yamada, T. & Abdalla, S. R. S. *Fósforo na agricultura brasileira,* p. 223-239.

[56] Larcher, W. (2000). *Ecofisiologia Vegetal.* Traduzido por PRADO, C. H. B. A. RiMa Artes e textos, São Carlos. p. 531.

[57] Lawlor, D. W. & Cornic, G. (2002). Photosynthetic carbon assimilation and associated metabolism in relation to water deficits in higher plants. *Plant, Cell and Environment,* v. 25. p. 275-294.

[58] Lawlor, D. W., Tezara, W. (2009). Causes of decreased photosynthetic rate and metabolic capacity in water-deficient leaf cells: a critical evaluation of mechanisms and integration of processes. *Annals of Botany.* v. 103, p. 561–579.

[59] Lee, J. -D., Shannon J. G., Vuong, T. D., Moon, H., Nguyen, H. T., Tsukamoto, C., Chung, G. (2010). Genetic diversity in wild soybean (Glycine sojaSieb. and Zucc.) accessions from southern island of Korean peninsula. *Plant Breeding.* v. 129, p. 257-263.

[60] Li-Juan, Q. & Ru-Zhen, C. (2010). The Origin and History of Soybean. In: B Singh, (Ed.). *The Soybean: Botany, Production and Uses.* CAB International, Oxfordshire, UK, p. 01–23.

[61] Lisar, S. Y. S., Motafakkerazad, R., Hossain, M. M. & Rahman, I. M. M. (2012). Water Stress in Plants: Causes, Effects and Responses. In:Rahman, I. M. M. & Hasegawa, H. *Water Stress,* InTech, Croatia, p. 01-14.

[62] Machado, P. L. O. A. (1997). Considerações gerais sobre a toxicidade do alumínio nas plantas. EMBRAPA-CNPS, 22p. (EMBRAPA-CNPS. Documentos , 2).

[63] Majumdar, S., Ghosh, S., Glick, B. R., Dumbroff, E. B. (1991). Activities of chlorophyllase, phosphoenolpyruvate carboxylase and ribulose-1,5-bisphosphate carboxylase in

the primary leaves of soybean during senescence and drought. *PhysiologiaPlantarum.* v. 81, p. 473-480.

[64] Makbul, S., SaruhanGüler, N., Durmus, N., Güven, S. (2011). Changes in anatomical and physiological parameters of soybean under drought stress. *Turk J. Bot.* v. 35, p. 369-377.

[65] Malavolta, E. *Função dos Macros e Micronutrientes:* Manual de Nutrição Mineral de Plantas. São Paulo. Editora Agronômica Ceres, 2006. 638p.

[66] Manavalan, L. P., Guttikonda, S. K., Tran, L-S. P., Nguyen, H. (2009). Physiological and molecular approaches to improve drought resistance in soybean. *Plant and Cell Physiology.* v. 50, p. 1260-1276.

[67] Mascarenhas, H. A. A., Camargo, C. E. de O. & Falivene, S. M. P. (1984). Efeito do alumínio sobre o crescimento de raízes, peso seco da parte aérea e raízes de diferentes cultivares de soja. Bragantia, v. 43, n. 1, p. 191-200.

[68] Mascarenhas, H. A. A., Tanaka,R. T., Wutke, E. B., Braga, N. R., Miranda. M. A. C. (2004). Potássio para a soja. *Informações Agronômicas* Nº105, 2p.

[69] Miyake, C. (2010). Alternative electron flow (water-water cycle and cyclic electron flow around PSI) in photosynthesis: molecular mechanisms and physiological function. *Plant and Cell Physiology.* v. 51, p. 1951-1963.

[70] Miyake, C., Amako, K., Shiraishi, N., Sugimoto, T. (2009). Acclimation of tobacco leaves to high light intensity drives the plastoquinone oxidation system – relationship among the fraction of open PSII centers, non-photochemical quenching of chl fluorescence and the maximum quantum yield off PSII in the dark. *Plant and Cell Physiology.* v. 50, p. 730-743.

[71] Mott, K A., Buckley, T. N. (1998). Stomatal heterogeneity. *Journal of Experimental Botany.* v. 49, p. 407-417.

[72] Muchow, R. C., Sinclair, T. R., Bennett, J. M. & Hammond, L. C. (1986). Response of leaf growth, leaf nitrogen, and stomatal conductance to water deficits during vegetative growth of field-grown soybean. *Crop Sci.,* v. 26, pp. 1190–1195.

[73] Mundstock, C. M. & Thomas, A. L. (2005). *Soja: fatores que afetam o crescimento e rendimento de grãos.* Porto Alegre: Universidade Federal do Rio Grande do Sul, 31 p.

[74] Mutei, H. (2011). Papel do Brasil no combate a fome no mundo. In: Siqueira, F., Caju, J. & Moreira, M. *Boletim de pesquisa da soja.* Fundação MT, Mato Grosso, Brazil, n. 15, p. 45-48.

[75] Nepomuceno, A. L., Farias, J. R. B., Neumaier, N. (1994). Efeitos da disponibilidade hídrica no solo sobre a cultura da soja. In: EMBRAPA-CNPSo, Ata - Documentos 72. *Reunião de Pesquisa de Soja da Região Central do Brasil.* Londrina, PR, v. 15, p. 42-43.

[76] Neumaier, N., Nepomuceno, A. L., Farias, J. R. B. (2000). Estresses de ordem ecofisiológica. In: Bonato, E. R. (ed.). *Estresses em soja.* Passo Fundo: EMBRAPA Trigo, 254p.

[77] Nepomuceno, A. L., Neumaier, N., Farias, J. R. B., Oya, T. (2001). Tolerância à seca em plantas. *Biotecnologia Ciência e Desenvolvimento.* v. 23, p. 12-18.

[78] Nogueira, S. S. S. & Nagai, V. (1988). Deficiência hídrica simulada nos diferentes nos diferentes estádios de desenvolvimento de uma cultura precoce de soja. *Bragantia.* v. 47, n. 1, p. 9-14.

[79] Nolla, A., Schlindwein, J. A. & Anghinoni, I. (2007). Crescimento, morfologia radicular e liberação de compostos orgânicos por plântulas de soja em função da atividade de alumínio na solução do solo de campo natural. *Ciência Rural,*v. 37, n. 1, p. 97-101.

[80] Novais, R. F., Smyth, T. T. (1999). *Fósforo em solo e planta em condições tropicais.* Viçosa: Universidade Federal de Viçosa, 399p.

[81] Ohashi, Y., Nakayama, N., Saneoka, H., Mohapatra, P. K., Fujita, K. (2009). Differences in the responses of stem diameter and pod thickness to drought stress during the grain filling stage in soybean plants. *ActaPhysiol Plant.* v. 31, p. 271-277.

[82] Ohashi, Y., Nakayama, N., Saneoka, H., Fujita, K. (2006). Effects of drought stress on photosynthetic gas exchange, chlorophyll fluorescence and stem diameter of soybean plants. *Biologia Plantarum.* v. 50, p. 138-141.

[83] Oliveira, A., Fernandes, E. J., Rodrigues, T. J. D. (2005). Condutância estomática como indicador de estresse hídrico em feijão. *Eng. Agríc., Jaboticabal,* v. 25, n. 1, p. 86-95.

[84] Palta, J. P. (1990). Stress Interactions at the Cellular & Membrane Levels. *Hort. Sci.* v. 25, n. 11, p. 1377-1381.

[85] Pang, J., Yang, J., Ward P., Siddique K. H. M., Lambers, H., Tibbett, M., Ryan, M. (2011). Contrasting responses to drought stress in herbaceous perennial legumes. *Plant Soil.* v. 348, p. 299-314.

[86] Pankhurst, C. E. & Sprent, J. I. (1975). Effects of water stress on respiratory and nitrogen-fixing activity of soybean root nodules. *J. Exper. Bot.,* v. 26, pp. 287-304.

[87] Pankovic, D., Sakac, D., Kevresan, S., Plesnicar, M. (1999). Acclimation to long-term water deficit in the leaves of two sunflower hybrids: photosynthesis, electron transport and carbon metabolism. *Journal of Experimental Botany.* v. 50, p. 127-138.

[88] Parry, M. A. J., Andralojc, P. J., Khan, S., Lea, P., Keys, A. J. (2002). Rubisco activity: effects of drought stress. *Annals of Botany.* v. 89, p. 833–839.

[89] Pelloux, J., Jolivet, Y., Fontaine, V., Benvoy, J., Dizengremel, P. (2001). Changes in Rubisco and Rubisco activase gene expression and polypeptide content in Pinushalepensis M. subjected to ozone and drought. *Plant, Cell and Environment.* v. 24, p. 123-131.

[90] Pinheiro, C., Chaves, M. M. (2011). Photosynthesis and drought: can we make metabolic connections from available data? *J. Experimental Botany.* v. 62, p. 869-882.

[91] Pitzschke, A., Forzani, C. & Hirt, H. (2006). Reactive Oxygen Species Signaling in Plants. *Antioxidants and Redox Signaling,* v. 8, n. 9, p. 1757-1764.

[92] Prado, R. M. (2008). *Nutrição de Plantas.* (Ed.)., São Paulo. Unesp, 2008. p. 161-181.

[93] Purcell, L. C., King, C. A., Ball, R. A. (2000). Soybean Cultivar Differences in Ureides and the Relationship to Drought Tolerant Nitrogen Fixation and Manganese Nutrition. *Crop Science,* v. 40, v. 1062 - 1070.

[94] Purcell, L. C, Serraj, R., Sinclair, T. R., De, A. (2004). Soybean N2 Fixation Estimates, Ureide Concentration, and Yield Responses to Drought. *Crop Science,* v. 44, p. 484-492.

[95] Purcell, L. C. & Specht, J. E. (2004). Physiological traits for ameliorating drought stress. In: *Soybeans: improvement, production, and uses (third ed.),*Boerma, H. R., Specht, J. E. (Eds.), pp. 569-620, American Society of Agronomy, Madison, WI.

[96] Raper, C. D. & Kramer, P. J. (1987). Stress Physiology, In: *Soybeans: Improvement, production, and uses,* Wilcox, J. R. (Ed.)., pp. 589-641. American Society of Agronomy, Madison, WI.

[97] Reynolds-Henne, C. E., Langenegger, A., Mani, J., Schenk, N., Zumsteg, A., Feller, U. (2010). Interactions between temperature, drought and stomatal opening in legumes. *Environmental and Experimental Botany.* v. 68, p. 34-43.

[98] Rosolem, C. A. (2005). Papel do Brasil no combate a fome no mundo. In: Suzuki, S., Yuyama, M. M. & Camacho, S. A. *Boletim de pesquisa da soja.* Fundação MT, Mato Grosso, Brazil, n. 09, p. 95-102.

[99] Rossatto, D. R., Hoffmann, W. A., Franco, A. C. (2009). Differences in growth patterns between co-occurring forest an savanna trees affect the forest-savanna boundary. *Functional Ecology.* v. 23, p. 689-698.

[100] Sanda, S., Yoshida, K., Masayoshi, K., Kawamura, T., Munekage, Y. N., Akashi, K., Yokota, A. (2011). Responses of the photosynthetic electron transport system to excess light energy caused by water deficit in wild watermelon. *PhysiologiaPlantarum.* 142: 247-264.

[101] Sangakkara, U. R., Frehner, M. & Nösberger, J. (2000). Effect of Soil Moisture and Potassium Fertilizer on Shoot Water Potential, Photosynthesis and Partitioning of Carbon in Mungbean and Cowpea. *J. Agronomy e Crop Science.* v. 185, p. 201-207.

[102] Santos M. G., Ribeiro, R. V., Oliveira, R. F., Machado, E. C. & Pimentel, C. (2006). The role of inorganic phosphate on photosynthesis recovery of commom bean after a mild water deficit. *Plant Science,* v. 170, p. 659-664.

[103] Santos, R. F., Carlesso, R. (1988) Déficit hídrico e os processos morfológicos e fisiológicos das plantas. *R. Bras. Eng. Agríc. Ambiental,* v. 2, n. 3, p. 287-294, 1998.

[104] Santos, A. C., Vilela, L. C. Barreto, P. M., Castro, G. J. D. & J. E C. Silva. (2010). Alterações de atributos químicos pela calagem e gessagem superficial com o tempo de incubação. *Revista Caatinga,* v. 23, n. 1, p. 77-83.

[105] Sarlikioti, V., Driever, S. M., Marcelis, L. F. M. (1993). Photochemical reflectance index as a mean of monitoring early water stress. *Annals of Applied Biology.* 157: 81-89. 2010.

[106] Singh, G. & Shivakumar, B. G. (2010) The role of soybean in agriculture. In: B Singh, (Ed.). *The Soybean: Botany, Production and Uses.* CAB International, Oxfordshire, UK, pp. 24–47.

[107] Smirnoff, N. (1993) The role of active oxygen in the response of plants to water deficit and desiccation. *New Phytol.* v. 125, p. 27–58.

[108] Soleimanzadeh, H., Habibi, D., Ardakani, M. R., Paknejad, F. & Rejali, F. (2010). Effect of Potassium Levels on Antioxidant Enzymes and Malondialdehyde Content under Drought Stress in Sunflower (*Helianthus annuus*L.). *American Journal of Agricultural and Biological Sciences,* v. 5, n. 1, p. 56-61.

[109] Sprent, J. I. (1971). The effects of water stress on nitrogen-fixing root nodules. *New Phytologist,* v. 70, pp. 9-17.

[110] Stolf-Moreira, R., Lemos, E. G. M., Carareto-Alves, L, et al. (2011). Transcriptional profiles of roots of different soybean genotypes subjected to drought stress. *Plant Mol. Biol. Rep.* v. 29, p. 19-34.

[111] Taiz, L. & Zeiger, E. (2009). *Fisiologia vegetal.* 4º ed. Artmed Editora, Porto Alegre, 848 p.

[112] Vitti, G. C., Lima, E. & Cicarone, F. (2006). Cálcio, magnésio e nxofre. In: Fernandes, M. S. *Nutrição mineral de plantas.* SBCS, p. 299-326.

[113] Zhang, R., Sharkey, T. D. (2009). Photosynthetic electron transport and photon flux under moderate heat stress. *Photosynth Res.* v. 100, p. 29-43.

[114] Wang, D., Graef, G. L., Procopiuk, A. M., Diers, B. W. (2004). Identification of putative QTL that underline yield in interspecific soybean backcross populations. *Theor. Appl. Genet.* v. 108, p. 458-467.

[115] Waraich, E. A., Ahmad, R., Saifullah., Ashraf, M. Y., & Ehsanullah. (2011). Role of mineral nutrition in alleviation of drought stress in plants. *Australian Journal of Crop Science,* v. 5, n. 6, p. 764-778.

[116] Yamaguchi, M., Valliyodan, B., Zhang, J., Lenoble, M. E., Yu, O., Rogers, E. R., Nguyen, H. T., Sharp, R. E. (2010). Regulation of growth response to water stress in the soybean primary root. I. Proteomic analysis reveals region-specific regulation of phenylpropanoid metabolism and control of free iron in the elongation zone. *Plant, Cell and Environment.* v. 33, p. 223-343.

[117] Yu, G-R., Wang, Q-F., Zhuang, J. (2004). Modeling the water use efficiency of soybean and maize plants under environmental stress: application of a synthetic model of photosynthesis-transpiration based on stomatal behavior. *J. Plant Physiol.* v. 161, p. 303-318.

Chapter 14

Interaction of Photosynthetic Source-Sink Balance and Activities of Membrane H⁺ Pumps in Soybean

Minobu Kasai and Wataru Takahashi

Additional information is available at the end of the chapter

http://dx.doi.org/10.5772/50544

1. Introduction

There is evidence suggesting that in plants photosynthetic matter production is regulated by photosynthetic source-sink balance, i.e., the ratio of photosynthetic source organs (e.g., leaves) to non-photosynthetic sink organs (e.g., roots) and/or the balance of supply and demand of photosynthetic carbohydrate(s) within the plant (Kasai, 2008, 2011). Plant photosynthetic dry matter production is the source of a variety of metabolic and structural compounds. Because of increasing population, shortages of energy and food may become more severe (von Caemmerer & Evans, 2010; Raines, 2011). Plant photosynthetic dry matter production is also essential for maintaing environmental quality. For example, a well-known environmental problem is climatic warming of the earth, which mainly comes from deforestation (Brovvkin et al., 2004). Improvement of plant dry matter productivity may be an effective way for solving the problems of energy, foods and climatic warming. Thus, it is important to elucidate the mechanism(s) of regulation of plant photosynthetic matter production through photosynthetic source-sink balance.

Data from a number of studies including field investigations implicate that in plants, accumulation of photosynthetic carbohydrate(s) in leaves, which occurs when photosynthetic source capacity exceeds sink capacity, can regulate leaf photosynthetic rate (Sawada et al., 1999; Kasai, 2008, 2011; Kasai et al., 2012). In soybean a significant negative correlation exists between leaf photosynthetic carbohydrate (sucrose or starch) content and photosynthetic rate (Sawada et al., 1986, 2001; Kasai, 2008). There have also been findings of photosynthetic carbohydrate-mediated decrease in the activity or the amount of Rubisco, the CO_2-fixing enzyme in leaves (Sage et al., 1989; Xu et al., 1994; Martin et al., 2002; Paul & Pellny, 2003), although the detailed mechanism(s) is still unclear. To date, many studies have focused on photosynthetic carbohydrate-mediated inhibition of leaf photosynthesis to elucidate the mechanism(s) of regulation of photosynthetic matter production through photosynthetic source-sink balance. However, in contrast, there is also evidence suggesting that leaf photosynthetic rate is not necessarily affected by accumulated photosynthetic carbohydrate(s) in leaf (Nebauer et al., 2011). Apart from the regulation of leaf photosynthesis through levels of photosynthetic carbohydrate(s), it is important to examine the mechanism(s) of regulation of photosynthetic dry matter production through photosynthetic source-sink balance by focusing on new enzyme(s) thought to be important.

Data from recent studies implicate that in plants, activity(ies) of membrane H^+ pump(s) such as tonoplast H^+ pump(s) can be important in the regulation of photosynthetic dry matter production through photosynthetic source-sink balance (Kasai &Muto, 1990.; Schumacher et al., 1999; Li et al., 2005; Yang et al., 2007; Wang et al., 2011). However, the effect of photosynthetic source-sink balance on the activity(ies) of membrane H^+ pump(s) has not been investigated. We show here experimental data of our recent study relating to this subject. We investigated in soybean plants how removal of pods, which decreases the ratio of sink to source organs, affects various characteristics related to photosynthetic dry matter production. Factors studied were leaf photosynthetic rate, stomatal conductance, transpiration rate and intercellular CO_2 concentration, initial and total activities of Rubisco, chlorophyll, total protein, inorganic phosphate, photosynthetic carbohydrates (sucrose and starch), and dry weights of source and sink organs. We also investigated the effect of pod removal on activities of the H^+ pumps of the leaf plasma membrane (H^+-ATPase) and tonoplast (H^+-ATPase and H^+-PPase). It is now well known that soybean is one of the most important crops grown in the world (Board & Kahlon, 2011; Ainsworth et al., 2012). On the basis of our experimental data and the other relevant information, we also consider how membrane H^+ pump(s) can be important in the regulation of photosynthetic dry matter production through photosynthetic source-sink balance.

2. Materials and methods

2.1. Plant materials

Soybean (*Glycine max* L. Merr. cv. Tsurunoko) seeds were sown in plastic pots (13.5 cm in height, 8.5 cm in diameter) containing mixed vermiculite and sand (1:1 in volume) and grown in growth chambers (Koitotoron, HNL type; Koito Industries Ltd., Tokyo, Japan) under daily light/dark periods of 10/14 h, day/night temperatures of 25/17°C and relative humidity of 60 %. After 51 days, pods were all removed together with small floral organs from half of the plants, and the depodded plants were grown with the remaining plants (controls) for 3 days under same growth conditions. Nutrients were supplied twice a week with a 1000-fold diluted solution of Hyponex [6-10-5 type (N:P:K = 6:10:5); Hyponex Co., Osaka, Japan], and tap water was supplied in sufficient amounts. Intensity of light, which was supplied with incandescent lamps, was 80 μmol photons m^{-2} s^{-1} (400-700 nm) on pots.

2.2. Leaf photosynthetic rate, transpiration rate, stomatal conductance and intercellular CO_2 concentration

Leaf photosynthetic rate, transpiration rate, stomatal conductance and intercellular CO_2 concentration were determined on day 3 after pod removal in fully expanded middle trifoliate leaves at a light intensity of 800 μmol photons m^{-2} s^{-1}, air flow rate of 200 ml min^{-1}, air temperature of 25 °C, relative humidity of 60 % and CO_2 concentration of 350 ppm using a portable photosynthetic analyzer (Cylus-1; Koito Industries Ltd.). After measurements, leaf disks (1.79 cm^2) were taken from the middle trifoliate leaves for the other analyses (see 2.3), as described previously (Kasai, 2008).

2.3. Rubisco activity, chlorophyll, protein, phosphate, sucrose and starch

Initial and total activities of Rubisco in leaf extract were determined at 25 °C as described previously (Kasai, 2008). Leaf chlorophyll content was determined according to the method of Mackinney (1941). Leaf total protein content was determined by quantifying protein included in the leaf extract that had been prepared for determination of Rubisco activity by the method of Bradford (1976). Leaf inorganic phosphate content was determined according to the method of Saheki et al. (1985). Leaf sucrose and starch contents were determined as described by Sawada et al. (1995).

2.4. Dry weight

For determination of dry weights of source (leaves) and sink organs (stems, floral organs including pods, and roots), organs were separated from plants on day 3 after pod removal and dried at 70°C for a week.

2.5. Plasma membrane, tonoplast and H^+ pump activity

The activities of H^+ pumps of leaf plasma membrane and tonoplast were determined using plasma membrane vesicles and tonoplast vesicles prepared from leaves. Plasma membrane vesicles and tonoplast vesicles were prepared from leaves (25 g in fresh weight) of plants on day 3 after pod removal essentially as described by Nouri & Komatsu (2010) and Maeshima & Yoshida (1989), respectively. For preparation of plasma membrane vesicles, leaf-homogenizing medium consisted of 0.3 M sucrose, 50 mM Tris, 8 mM EDTA (acid form), 2 mM PMSF, 4 mM DTT and 0.2 % (w/v) BSA, and its volume was 200 ml. After homogenization, the medium was filtered through four layers of gauze and the filtrate was centrifuged at 10,000 g for 20 min. After supernatant was centrifuged at 80,000 g for 40 min, the pellets were suspended with a sucrose-containing medium [0.3 M sucrose, 5 mM KH_2PO_4, 5 mM KCl, 0.1 mM EDTA and 0.1 mM DTT (pH 7.8)], of which volume was 10 ml. The plasma membrane vesicles were prepared from the suspension by using aqueous two-phase partitioning method. Dilution of the upper layers that had been obtained was conducted with a sorbitol-containing medium [0.25 M sorbitol, 5 mM HEPES-BTP and 0.1 mM DTT (pH 7.0)]. The final pellets of the plasma membrane vesicles after centrifugation (80,000 g, 40 min) were suspended with another sorbitol-containing medium [1 M sorbitol, 5 mM HEPES-BTP and 0.1 mM DTT (pH 7.0)] and stored at -80 °C until uses. For preparation of tonoplast vesicles, leaf-homogenizing medium consisted of 0.25 M sorbitol, 50 mM HEPES-KOH, 5 mM EGTA, 1 mM PMSF, 2.5 mM $Na_2S_2O_5$ and 1.5% (w/v) PVP (pH 7.6), and its volume was 200 ml. After homogenization, the medium was filtered through four layers of gauze and the filtrate was centrifuged at 4000 g for 10 min. After supernatant was centrifuged at 80,000 g for 60 min, the pellets were suspended with a sucrose-containing medium [0.3 M sucrose, 10 mM KH_2PO_4, 1 mM EGTA and 2 mM DTT (pH 7.8)], of which volume was 5 ml, and 3 ml of a sorbitol-containing medium [0.25 M sorbitol, 50 mM HEPES-KOH, 1 mM EGTA and 2 mM DTT (pH 7.3)] was put on the suspension and the solution was centrifuged (120,000 g, 40 min). The resulting middle layer was diluted with the same sorbitol-containing medium. The final pellets of the tonoplast vesicles after centrifugation (150,000 g, 20 min) were suspended with another sorbitol-containing medium [0.25 M sorbitol, 5 mM HEPES-BTP, 2 mM DTT (pH 7.5)] and stored at -80°C until uses.

The activity of H^+ pump, i.e., H^+-ATPase of leaf plasma membrane was determined at 30°C as vanadate-sensitive ATP-hydrolytic activity (kasai & Sawada, 1994). The activities of H^+ pumps, i.e., H^+-ATPase and H^+-PPase of leaf tonoplast were determined at 30°C as nitrate-sensitive ATP-hydrolytic activity and Na^+-sensitive PPi-hydrolytic activity (Kasai et al., 1993; Kasai & Sawada, 1994), respectively. Reaction medium (500 μl) for the activity of plasma membrane H^+ pump consisted of 50 mM HEPES-BTP (pH 7.0), 3 mM $MgSO_4$, 3 mM ATP, 1 mM EGTA, 50 mM KCl, ± 0.1 mM Na_3VO_4, 0.02% (w/v) Triton X-100 and membrane vesicles (10 μg). Reaction medium for the activity of tonoplast H^+-ATPase consisted of ±50 mM HEPES-BTP (pH 7.5), 3 mM $MgSO_4$, 3 mM ATP, 1 mM EGTA, ±50 mM KCl, ± 50 mM KNO_3, 0.02% (w/v) Triton X-100 and membrane vesicles (10 μg). Reaction medium for the activity of tonoplast H^+-PPase consisted of 50 mM HEPES-BTP (pH 7.5), 5 mM $MgSO_4$, 0.5 mM PPi, 1 mM EGTA, 50 mM KNO_3, ± 50 mM $NaNO_3$, 0.02% (w/v) Triton X-100 and membrane vesicles (10 μg). Phosphate liberated from substrate ATP or PPi was determined according to the method of Saheki et al. (1985).

3. Results

Analyzed leaf photosynthetic rate and transpiration rate were significantly lower in depodded plants than in control plants (Fig. 1). Leaf stomatal conductance was also lower in depodded plants than in control plants, while leaf intercellular CO_2 concentration did not differ significantly between control and depodded plants (Fig. 2). Initial and total activities of Rubisco in leaf extract did not differ significantly between control and depodded plants (Fig. 3). Contents of chlorophyll, total protein and inorganic phosphate in leaves were all significantly higher in depodded plants than in control plants (Fig. 4). Contents of sucrose and starch in leaves did not differ significantly between control and depodded plants (Fig. 5). Activity of H^+ pump (H^+-ATPase) of leaf plasma membrane and activities of H^+ pumps (H^+-ATPase and H^+-PPase) of leaf tonoplast were all significantly lower in depodded plants than in control plants (Fig. 6). Dry weights of leaves, stems and roots did not differ significantly between control and depodded plants (Fig. 7). When the ratio of sink (stems + floral

http://dx.doi.org/10.5772/50544

organs including pods + roots) to source organs (leaves) was calculated, those in control and depodded plants were on the average 1.25 (100%) and 0.70 (56%), respectively.

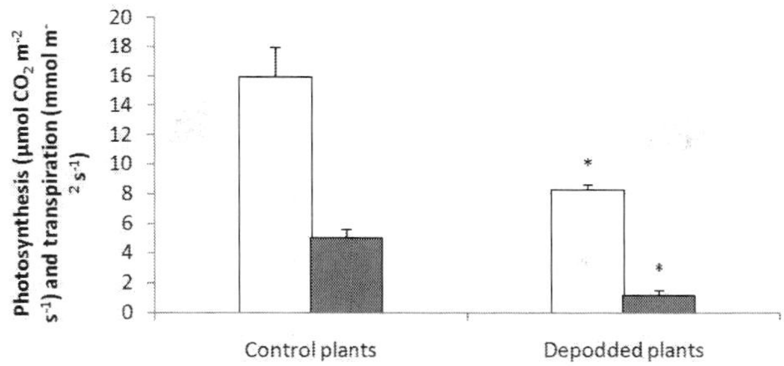

Figure 1. Leaf photosynthetic rate and transpiration rate of soybean plants on day 3 after pod removal. Open bar, leaf photosynthetic rate; gray bar, leaf transpiration rate. Vertical bars indicate S.D. (n=3). **P*<0.01 (*t*-test) when compared with control plants.

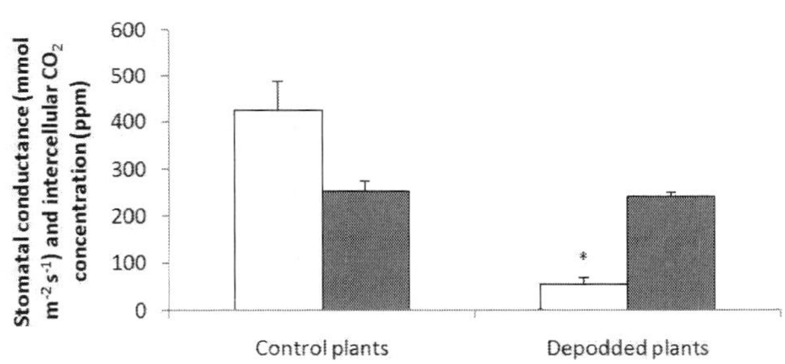

Figure 2. Leaf stomatal conductance and intercellular CO_2 concentration of soybean plants on day 3 after pod removal. Open bar, leaf stomatal conductance; gray bar, leaf intercellular CO_2 concentration. Vertical bars indicate S.D. (n=3). **P*<0.01 when compared with control plants. The leaf intercellular CO_2 concentration did not differ significantly (*P*>0.05) between control and depodded plants.

4. Discussion

As described in the Introduction, it is important to examine the mechanism(s) of regulation of plant photosynthetic dry matter production through photosynthetic source-sink balance by focusing on a new enzyme(s). We focused on H^+ pumping enzymes of leaf plasma membrane (H^+-ATPase) and tonoplast (H^+-ATPase and H^+-PPase), and investigated in soybean plants how removal of pods, which decreases the ratio of sink to source organs, affects various characteristics related to photosynthetic dry matter production and the activities of H^+ pumps. Pod removal was shown to decrease largely leaf photosynthetic rate, transpiration rate and stomatal conductance without affecting significantly leaf intracellular CO_2 concentration (Fig. 1 and 2). These results imply that pod removal decreased equally the rate of CO_2 diffusion via leaf stomata and the rate of CO_2 fixation in leaf photosynthetic cells. In plants, Rubisco is a major protein in leaves (Furbank et al., 1996; von Caemmerer et al., 2005), and there is evidence from studies altering the expressions of Rubisco or its activation enzyme, Rubisco activase, that changes in the activity or the amount of Rubisco in leaves significantly affect leaf photosynthetic rate (Furbank et al., 1996; von Caemmerer et al., 2005). There is also a report demonstrating that a rough and positive correlation exists between leaf chlorophyll content and photosynthetic rate (Arp, 1991). Therefore, it is speculated that the pod removal-induced decrease in leaf photosynthetic rate might have resulted from a decrease in the activity or the amount of Rubisco in the leaf or the content of leaf chlorophyll. However, data of Figure 3 and 4 indicate that pod removal did not significantly affect potential activity of Rubisco in the leaf and could not decrease the content of Rubisco or chlorophyll, suggesting that the pod removal-induced decrease in leaf photosynthetic rate did not result from either of these factors.

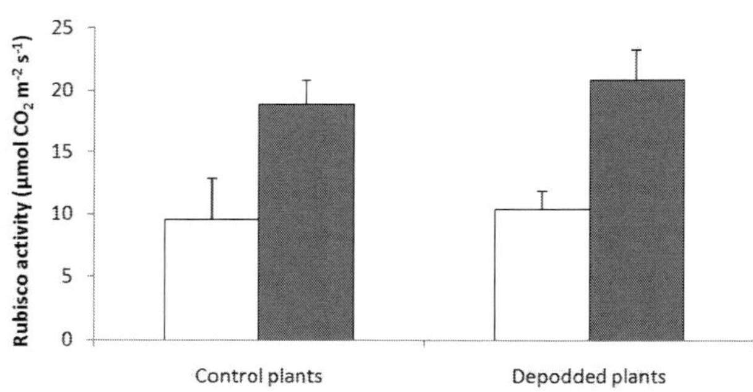

Figure 3. Initial and total activities of Rubisco in leaf extract from soybean plants on day 3 after pod removal. Open bar, initial activity; gray bar, total activity. Vertical bars indicate S.D. (n=3). Both initial and total activities did not differ significantly (*P*>0.05) between control and depodded plants.

Previously, in single-rooted soybean leaves that are the same species as we used in our pod removal study, it was demonstrated that a decrease in leaf inorganic phosphate content can result in a decrease in leaf Rubisco activity in vivo (Sawada et al., 1990, 1992). In vitro, inorganic phosphate has been found to promote the binding of activator CO_2 to uncarbamylated inactive Rubisco (Bhagwat, 1981; McCurry et al., 1981; Anwaruzzaman et al., 1995). Data of Figure 4 indicate that pod removal did not decrease leaf inorganic phosphate content. This result indicates that the pod removal-induced decrease in leaf photosynthetic rate did not result from a decrease in leaf inorganic phosphate content.

http://dx.doi.org/10.5772/50544

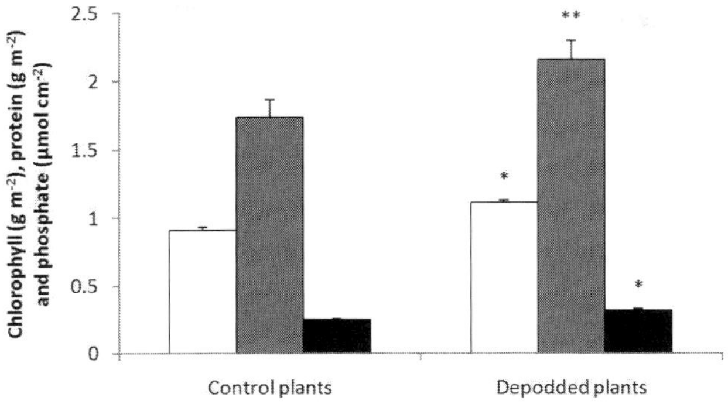

Figure 4. Leaf chlorophyll, total protein and inorganic phosphate contents in soybean plants on day 3 after pod removal. Open bar, chlorophyll; gray bar, total protein; black bar, inorganic phosphate. Vertical bars indicate S.D. (n=3). $^*P<0.01$ ($^{**}P<0.05$) when compared with control plants.

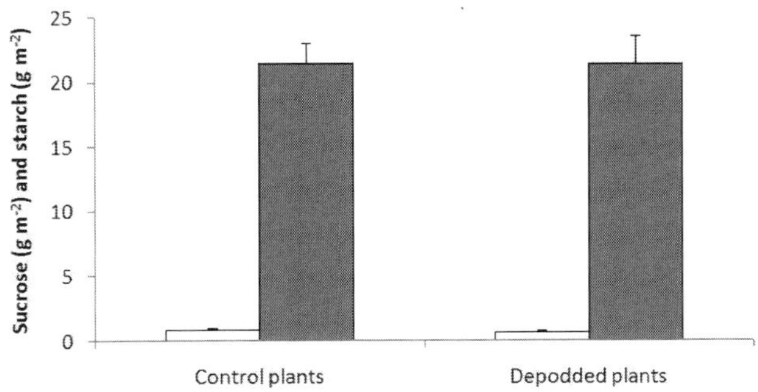

Figure 5. Leaf sucrose and starch contents in soybean plants on day 3 after pod removal. Open bar, sucrose content; gray bar, starch content. Vertical bars indicate S.D. (n=3). Both sucrose and starch contents did not differ significantly ($P>0.05$) between control and depodded plants.

There is a hypothesis of inhibition of photosynthesis through accumulation of sucrose in the leaf, although the detailed mechanism(s) is still unclear (Kasai, 2008). For example, in a study having continuous exposure to light of single-rooted soybean leaves, a significant negative correlation was shown between leaf sucrose content and photosynthetic rate (Sawada et al., 1986). It is thought that in both control and depodded plants, sucrose-induced inhibition of leaf photosynthesis was, if any, very small. Leaf sucrose content of control plants, which was higher on the average than that of depodded plants (Fig. 5), corresponded with a content that led to a very small decrease in leaf photosynthetic rate of single-rooted soybean leaves (Sawada et al., 1986). Leaf sucrose content of control plants did not seem to decrease leaf photosynthetic rate in our previous pod removal study using soybean plants (Kasai et al., 2008). There is also a hypothesis of inhibition of photosynthesis through accumulation of starch in leaves (Kasai, 2008). In the same study, continuous exposure to light of single-rooted soybean leaves resulted in a significant negative correlation between leaf starch content and photosynthetic rate (Sawada et al., 1986). Another study using single-rooted soybean leaves demonstrated that accumulation of starch decreases the rate of CO_2 diffusion (Sawada et al., 2001; Kasai et al., 1996). In our pod removal study, pod removal did not significantly affect leaf starch content (Fig. 5). Therefore, it is suggested that the pod removal-induced decrease in leaf photosynthetic rate did not result from accumulation of sucrose or starch in leaves.

It is believed that in plants, P-type H^+ pump (H^+-ATPase) exists in the plasma membrane, and V-type H^+ pumps (H^+-ATPase and H^+-PPase) exist in the tonoplast (Hall & Williams, 1991; Barkla et al., 2008). In plant leaves, a decrease in H^+ pump activity of the guard cell plasma membrane can induce decreases of stomatal conductance and transpiration rate by inducing a decrease in stomatal pore size (Tominaga et al., 2001). Although we did not analyze the H^+ pump activity of the guard cell plasma membrane, it was shown that pod removal largely decreased the H^+ pump activity of the leaf plasma membrane (Fig. 6). Essentially, almost the same method is used for isolation of the plasma membrane from leaves and guard cells (Becker et al., 1993). Therefore, it is suggested that a large decrease in H^+ pump activity of the guard cell plasma membrane could cause the pod removal-induced decreases of leaf stomatal conductance and transpiration rate. In plant cells, a decrease in H^+ pump activity of the plasma membrane can induce a decrease in the electrochemical potential difference of H^+ across the plasma membrane (Hall & Williams, 1991; Barkla et al., 2008). There is increasing evidence that depolarization-activated Ca^{2+} channel, Ca^{2+}-activated anion (e.g., Cl^-) channel, depolarization-activated anion channel (e.g., HCO_3^-), electrogenic Ca^{2+}/H^+ antiporter (which has a stoichiometry higher than $2H^+/Ca^{2+}$), and CO_2-transportable and Ca^{2+}-inhibitable water channel are present in the plant plasma membrane (Thuleau et al., 1994; Roberts, 2005; Frachisse et al., 1999, Kasai et al., 1990; Song et al., 2011; Chaumont et al., 2005). Therefore, it is speculated that the observed decrease in H^+ pump activity of the leaf plasma membrane could cause the decrease in the rate of CO_2 transport in leaf photosynthetic cells by inducing a depolarization of the plasma membrane, a decrease in the proton motive force across the plasma membrane and a rise of Ca^{2+} concentration inside the plasma membrane in the leaf photosynthetic cells. The suggestion drawn on the basis of data of Figure 1 and 2 that pod removal decreased equally the rate of CO_2 diffusion via leaf stomata and the rate of CO_2 fixation in leaf photosynthetic cells is roughly consistent with the above-mentioned suggestion and speculation proposing the regulation of leaf stomatal conductance, transpiration rate and CO_2 transport in leaf photosynthetic cells by activity of H^+ pump of leaf plasma membrane. As shown in Figure 6, pod removal also greatly decreased activities of the H^+ pumps (H^+-ATPase and H^+-PPase) of the leaf tonoplast. There is increasing evidence that electrogenic Ca^{2+}/H^+ antiporter (which has a stoichiometry higher than $2H^+/Ca^{2+}$) is also present in the plant tonoplast (Blackford et al., 1990; Mei et al., 2007). Therefore, it is speculated that a large decrease in H^+ pump activity of the leaf plasma membrane and a large decrease in activities of H^+ pumps of the leaf tonoplast could cause cooperatively the pod removal-induced decrease in leaf photosynthetic rate by inducing equal

http://dx.doi.org/10.5772/50544

decreases in the rate of CO_2 diffusion via leaf stomata and the rate of CO_2 fixation in leaf photosynthetic cells. To verify our speculations, more evidence is needed. We emphasize, however, that until now, in similar studies other than our study, activities of the H^+ pumps of plasma membrane and tonoplast have not been analyzed.

With respect to the mechanism(s) of why pod removal decreased the H^+ pump activity of the leaf plasma membrane and activities of the H^+ pumps of the leaf tonoplast, it is inferred that plant hormones abscisic acid and cytokinin could be involved in the mechanism(s). In general, cytokinin is known to have positive effect in synthesizing chlorophyll and protein, and its content in plant cells is known to decrease under deficiency of mineral nutrients such as P and N. In contrast, abscisic acid antagonizes the effects of cytokinin, and its content in plant cells increases under conditions of mineral nutrient deficiencies (Pozsar et al., 1967; Kusnetsov et al., 1998; Salama & Wareing, 1979; Mizrahi & Richmond, 1972; Battal et al., 2003). In our pod removal study, it was shown that pod removal, which decreased the ratio of sink to source organs (Fig. 7), increased significantly the contents of chlorophyll, total protein and inorganic phosphate in the leaf (Fig. 4), implicating that pod removal might have increased cytokinin content relative to abscisic acid content in the leaf by influencing the partitioning of mineral nutrients such as P and N within the plant. In barley, it was demonstrated that abscisic acid has stimulatory effects on activities of tonoplast H^+ pumps, whereas cytokinin has opposite effects antagonizing the effects of abscisic acid (Kasai et al., 1993; Fukuda & Tanaka, 2006). In *Phaseolus vulgaris*, excessive levels of cytokinin were demonstrated to decrease leaf stomatal conductance, transpiration rate and photosynthetic rate (Pospisilova, 2003).

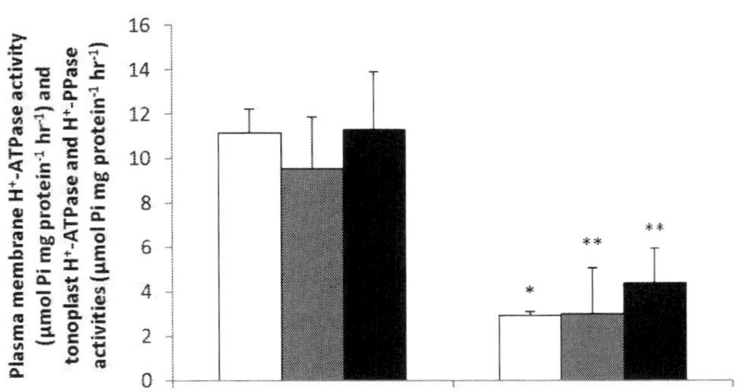

Figure 6. Activities of H^+ pumps of leaf plasma membrane and tonoplast from soybean plants on day 3 after pod removal. Open bar, plasma membrane H^+-ATPase; gray bar, tonoplast H^+-ATPase; black bar, tonoplast H^+-PPase. Vertical bars indicate S.D. (n=3). *$P<0.01$ (**$P<0.05$) when compared with control plants.

Data from recent studies using transgenic plants and those from physiological studies implicate that in plants, activity(ies) of membrane H^+ pump(s) can be important in the regulation of photosynthetic dry matter production through photosynthetic source-sink balance. For example, in a study using *Arabidopsis* plants, overexpression of tonoplast H^+-PPase was shown to result in an increase in whole plant growth, in particular, growth of sink organ roots (Li et al., 2005). In another study using *Arabidopsis* plants, it was shown that overexpression of tonoplast H^+-PPase resulted in a 1.9-fold increase in root dry weight and a 1.5-fold increase in shoot dry weight (Yang et al., 2007). There is also a report that transgenic manipulation increasing tonoplast H^+-ATPase resulted in an increase in root length (Wang et al., 2011). In a study using rye plants, increases of root/shoot ratio and activities of tonoplast H^+ pumps (H^+-PPase and H^+-ATPase) were shown to occur under growth conditions inducing deficiency of mineral nutrients (Kasai et al., 1998). Whereas conditions inducing deficiency of mineral nutrients can increase abscisic acid level in plant cells, it can also decrease cytokinin level (Salama & Wareing, 1979; Mizrahi & Richmond, 1972; Battal et al., 2003). In a recent study using *Arabidopsis* plants, transgenic manipulation decreasing cytokinin level was shown to enhance root/shoot ratio (Werner et al., 2003). In plant cells, levels of abscisic acid and cytokinin can also be affected by levels of photosynthetic carbohydrates, which can interact with the levels of mineral nutrients (Rolland et al., 2002). Interestingly, a more recent study using *Arabidopsis* plants showed that overexpression of tonoplast H^+-PPase resulted in an increase in cell abscisic acid level and a decrease in cell cytokinin (biologically active cytokinin) level (Gonzalez et al., 2010). In the other study using *Arabidopsis* plants, it was shown that a mutant losing about 60% of tonoplast H^+-ATPase activity had a morphology resembling cytokinin-treated plants (Schumacher et al., 1999). As already mentioned, abscisic acid had stimulatory effects on activities of tonoplast H^+ pumps, whereas cytokinin had opposite effects antagonizing the effects of abscisic acid in barley (Kasai et al., 1993; Fukuda & Tanaka, 2006). In our pod removal study using soybean plants, decreasing the ratio of sink to source organs by conducting pod removal was shown to result in a large decrease in activities of H^+ pumps of the leaf plasma membrane and tonoplast. Although characteristics such as leaf photosynthetic rate related to photosynthetic dry matter production were not analyzed in the above studies other than our study, information from these studies and the other information of our data from pod removal study implicate that in plants, changes in activity(ies) of membrane H^+ pump(s) can actually play key roles in the regulation of photosynthetic matter production through photosynthetic source-sink balance. It is suggested from experimental evidence that abscisic acid and cytokinin can be involved at least in the regulation of activities of tonoplast H^+ pumps. Although no direct evidence exists, data from more recent studies let us suppose that now well-known 2C-type protein phosphatase and salt-overly-sensitive 2 protein kinase may be involved (in part) in the regulation of activities of tonoplast H^+ pumps by abscisic acid and cytokinin (Batelli et al., 2007; Huertas et al., 2012). In plants, abscisic acid, cytokinin and membrane H^+ pump(s) are speculated to be symbolic biomolecules that are able to induce or respond to a variety of internal and external environmental changes that cause the regulation of photosynthetic dry matter production through photosynthetic source-sink balance. Many experimental data suggest that interaction actually exists between environmental change and cell abscisic acid and cytokinin levels and environmental change and, for example, activities of tonoplast H^+ pumps and environmental change and photosynthetic matter production and environmental change and photosynthetic source-sink balance (Board & Kahlon, 2011; Peleg & Blumwald,

2011; Kasai, 1999; Rolland et al., 2002). On the basis of a lot of evidence, it seems evident that photosynthetic source-sink balance is an essential factor regulating photosynthetic dry matter production (Kasai, 2008, 2011; Rolland et al., 2002). As described in the Introduction, plant photosynthetic dry matter production is essential for all living organisms and is also essential for creating sound environments. Therefore, further studies are important for elucidation of the detailed mechanism(s) of how membrane H^+ pump(s) are involved in the regulation of photosynthetic dry matter production through photosynthetic source-sink balance. We emphasize that for improvement of plant productivity and for creating sound environments, well-balanced improvement of source and sink would be essential.

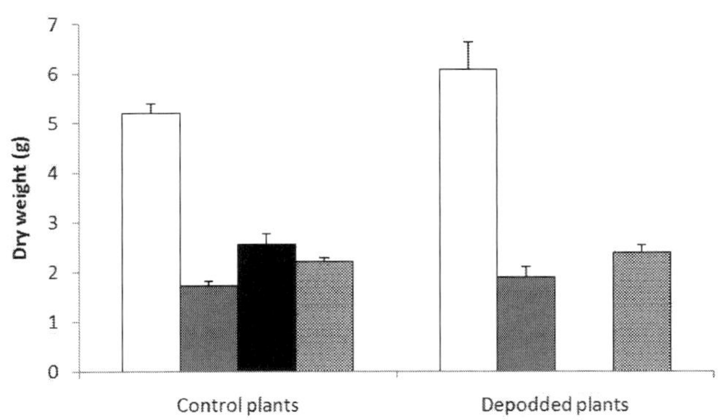

Figure 7. Dry weights of leaves, stems, floral organs including pods, and roots in soybean plants on day 3 after pod removal. Open bar, leaves; gray bar, stems; black bar, floral organs including pods; dotted bar, roots. Vertical bars indicate S.D. (n=3). Dry weights of leaves, stems and roots did not differ significantly (P>0.05) between control and depodded plants.

5. Conclusion

Data from recent studies implicate that in plants, activity(ies) of membrane H^+ pump(s) can be important in regulation of photosynthetic dry matter production through photosynthetic source-sink balance. However, the effect of photosynthetic source-sink balance on the activity(ies) of membrane H^+ pump(s) has not been investigated. In our recent study, we investigated in soybean plants how pod removal, which decreases the ratio of sink to source organs, affects various characteristics related to photosynthetic matter production. We also investigated, for the first time, the effect of pod removal on activities of H^+ pumps of leaf plasma membrane and tonoplast. From the data obtained and the other relevant information, it was concluded that in plants, changes in activity(ies) of membrane H^+ pump(s) can actually play key roles in the regulation of photosynthetic dry matter production through photosynthetic source-sink balance, and that hormones abscisic acid and cytokinin may be involved in regulation of activities of tonoplast H^+ pumps. Plant photosynthetic dry matter production is essential for all living organisms and is also essential for creating sound environments. Therefore, further studies are important to elucidate the detailed mechanism(s) of how membrane H^+ pump(s) are involved in the regulation of photosynthetic dry matter production through photosynthetic source-sink balance.

Author details

Minobu Kasai[1*] and Wataru Takahashi[1]

*Address all correspondence to: minobu@cc.hirosaki-u.ac.jp

1 Department of Biology, Faculty of Agriculture and Life Science, Hirosaki University,, Japan

References

[1] Ainsworth, E. A., Yendrek, C. R., Skoneczka, J. A., & Long, S. P. (2012). Accelerating yield potential in soybean: potential targets for biotechnological improvement. *Plant, Cell & Environment*, 35, 38-52.

[2] Anwaruzzaman, Sawada. S., Usuda, H., & Yokota, A. (1995). Regulation of ribulose-1,5-bisphosphate carboxylase/oxygenase activation by inorganic phosphate through stimulating the binding of the activator CO_2 to the activation sites. *Plant & Cell Physiology*, 36, 425-433.

[3] Arp, W. J. (1991). Effects of source-sink relations on photosynthetic acclimation to elevated CO2. Plant, Cell and Environment , 14, 869-875.

[4] Barkla, B. J., Hirschi, K. D., & Pittman, J. K. (2008). Exchangers man the pumps. Functional interplay between proton pumps and proton-coupled Ca^{2+} exchangers. *Plant Signaling Behavior*, 3, 354-356.

[5] Battal, P., Turker, M., & Tileklioglu, B. (2003). Effects of different mineral nutrients on abscisic acid in maize (Zea mays). *Annals Botanici Fennici*, 40, 301-308.

[6] Batelli, G., Verslues, P. E., Agius, F., Qiu, Q., Fujii, H., Pan, S., Schumaker, K. S., Grillo, S., & Zhu, J. K. (2007). SOS2 promotes salt tolerance in part by interacting with the vacuolar H^+-ATPase and upregulating its transport activity. *Molecular and Cellular Biology*, 27, 7781-7790.

[7] Becker, D., Zeilinger, C., Lohse, G., Depta, H., & Hedrich, R. (1993). Identification and biochemical characterization of the plasma-membrane H^+-ATPase in guard cells of Vicia faba L. *Planta*, 190, 44-50.

[8] Bhagwat, A. S. (1981). Activation of spinach ribulose-1,5-bisphosphate carboxylase by inorganic phosphate. *Plant Science Letters,*, 23, 197-206.

[9] Blackford, S., Rea, P. A., & Sanders, D. (1990). Voltage sensitivity of H^+/Ca^{2+} antiport in higher plant tonoplast suggests a role in vacuolar calcium accumulation. *The Journal of Biological Chemistry*, 265, 9617-9620.

[10] Board, J. E., & Kahlon, C. S. (2011). Soybean yield formation: What controls it and how it can be improved. In Soybean Physiology and Biochemistry, H. A. El-Shemy, Ed., InTech Open Access Publisher, Rijeka, Croatia. , 1-36.

[11] Bradford, M. M. (1976). A rapid and sensitive method for the quantitation of microgram quantities of protein utilizing the principle of protein-dye binding. *Analytical Biochemistry*, 72, 248-254.

[12] Chaumont, F., Moshelion, M., & Daniels, M. J. (2005). Regulation of plant aquaporin activity. Biology of the Cell ., 97, 749-764.

[13] Brovvkin, V., Sitch, S., von, Bloh. W., Claussen, M., Bauer, E., & Cramer, W. (2004). Role of land cover changes for atmospheric CO_2 increase and climate change during the last 150 years. *Global Change Biology*, 10, 1253-1266.

[14] Frachisse, J. M., Thomine, S., Colcombet, J., Guern, J., & Barbier-Brygoo, H. (1999). Sulfate is both a substrate and an activator of the voltage-dependent anion channel of Arabidopsis hypocotyl cells. *Plant Physiology*, 121, 253-261.

[15] Fukuda, A., & Tanaka, Y. (2006). Effects of ABA, auxin, and gibberellin on the expression of genes for vacuolar H^+inorganic pyrophosphatase, H^+ATPase subunit A, and Na^+/H^+ antiporter in barley. *Plant Physiology& Biochemistry*, 44, 351-358.

[16] Furbank, R. T., Chitty, J. A., von, Caemmerer. S., & Jenkins, C. L. D. (1996). Antisense RNA inhibition of RbcS gene expression reduces Rubisco level and photosynthesis in the C_4 plant Flaveria bidentis. *Plant Physiology*, 111, 725-734.

[17] Gonzalez, N., Bodt, S. D., Sulpice, R., Jikumaru, Y., Chae, E., Dhondt, S., Daele, T. V., Milde, L. D., Weigel, D., Kamiya, Y., Stitt, M., Beemster, G. T. S., & Inze, D. (2010). Increased leaf size: different means to an end. *Plant Physiology*, 153, 1261-1279.

[18] Hall, J. L., & Williams, L. E. (1991). Properties and functions of proton pumps in higher plants. *Pest Management Science*, 32, 339-351.

[19] Huertas, R., Olias, R., Eljakaoui, Z., Galvez, F. J., Li, J., Morales, P. A. D., Belver, A., & Rodriguez-Rosales, M. P. (2012). Overexpression of SlSOS2 (SlCIPK24) confers salt tolerance to transgenic tomato. *Plant, Cell & Environment, early view*, 1-16.

[20] Kasai, M. (1999). Environmental stresses and plant vacuolar H^+pumps. In Current Topics in Plant Biology, T. W. Ashenden et al., Eds., Research Trends, Trivandrum, India. , 97-104.

[21] Kasai, M. (2008). Regulation of leaf photosynthetic rate correlating with leaf carbohydrate status and activation state of Rubisco under a variety of photosynthetic source/sink balances. *Physiologia Plantarum*, 134, 216-226.

[22] Kasai, M. (2011). Regulation of leaf photosynthesis through photosynthetic source-sink balance in soybean plants. In Soybean Physiology and Biochemistry H. A. El-Shemy, Ed., InTech Open Access Publisher, Rijeka, Croatia. , 443-460.

[23] Kasai, M., & Muto, S. (1990). Ca^{2+} pump and Ca^{2+}/H^+ antiporter in plasma membrane vesicles isolated by aqueous two-phase partitioning from corn leaves. *The Journal of Membrane Biology,*, 114, 133-142.

[24] Kasai, M., Nakamura, T., Kudo, N., Sato, H., Maeshima, M., & Sawada, S. (1998). The activity of the root vacuolar H^+-pyrophosphatase in rye plants grown under conditions deficient in mineral nutrients. *Plant & Cell Physiology*, 39, 890-894.

[25] Kasai, M., Nakata, H., Seino, H., Kamata, D., & Tsukiyama, T. (2008). Effect of sink-limitation on leaf photosynthetic rate and related characteristics in soybean plants. *Plant Production Science*, 11, 223-227.

[26] Kasai, M., & Sawada, S. (1994). Evidence for decrease in vanadate-sensitive Mg^{2+}-ATPase activity of higher plant membrane preparations in sucrose solution. *Plant & Cell Physiology*, 35, 697-700.

[27] Kasai, M., Yamamoto, Y., Maeshima, M., & Matsumoto, H. (1993). Effects of in vivo treatment with abscisic acid and/or cytokinin on activities of vacuolar H^+ pumps of tonoplast-enriched membrane vesicles prepared from barley roots. *Plant & Cell Physiology*, 34, 1107-1115.

[28] Kasai, M., Koide, K., & Ichikawa, Y. (2012). Effect of pot size on various characteristics related to photosynthetic matter production in soybean plants. *International Journal of Agronomy*, 751731, 7.

[29] Kasai, M., Yamaguchi, A., & Sawada, S. (1996). The effects of CO_2 on the photosynthetic fixation of CO_2 and the activity of ribulose-1,5-bisphosphate carboxylase in single-rooted soybean leaves under sink-limited conditions. *Plant & Cell Physiology*, 37, 1193-1196.

[30] Kusnetsov, V., Herrmann, R. G., Kulaeva, O. N., & Oelmuller, R. (1998). Cytokinin stimulates and abscisic acid inhibits greening of etiolated Lupinus luteus cotyledons by affecting the expression of the light-sensitive protochlorophyllide oxidoreductase. *Molecular & General Genetics*, 259, 21-28.

[31] Li, J., Yang, H., Peer, W. A., Richter, G., Blakeslee, J., Bandyopadhyay, A., Titapiwantakun, B., Undurraga, S., Khodakovskaya, M., Richards, E. L., Krizek, B., Murphy, A. S., Gilroy, S., & Gaxiola, R. (2005). Arabidopsis H^+-PPase AVP1 regulates auxin-mediated organ development. *Science*, 310, 121-125.

[32] Mackinney, G. (1941). Absorption of light by chlorophyll solutions. *The Journal of Biological Chemistry*, 140, 315-322.

[33] Maeshima, M., & Yoshida, S. (1989). Purification and properties of vacuolar membrane proton-translocating inorganic pyrophosphatase from mung bean. *The Journal of Biological Chemistry*, 264, 20068-20073.

[34] Martin, T., Oswald, O., & Graham, I. A. (2002). Arabidopsis seedlings growth, storage mobilization, and photosynthetic gene expression are regulated by carbon: nitrogen availability. *Plant Physiology*, 128, 472-481.

[35] Mc Curry, S. D., Pierce, J., Tolbert, N. E., & Orme-Johnson, W. H. (1981). On the mechanism of effector-mediated activation of ribulose bisphosphate carboxylase/oxygenase. *The Journal of Biological Chemistry*, 256, 6623-6628.

[36] Mei, H., Zhao, J., Pittman, J. K., Lachmansingh, J., Park, S., & Hirschi, K. D. (2007). In planta regulation of the Arabidopsis Ca^{2+}/H^{+} antiporter CAX1. *The Journal of Experimental Botany*, 58, 3419-3427.

[37] Mizrahi, Y., & Richmond, A. E. (1972). Abscisic acid in relation to mineral deprivation. *Plant Physiology*, 50, 667-670.

[38] Nebauer, S. G., Renau-Morata, B., Guardiola, J. L., & Molina, R. V. (2011). Photosynthesis down-regulation precedes carbohydrate accumulation under sink limitation in Citrus. *Tree Physiology*, 31, 169-177.

[39] Nouri, M. Z., & Komatsu, S. (2010). Comparative analysis of soybean plasma membrane proteins under osmotic stress using gel-based and LC MS/MS-based proteomics approaches. *Proteomics*, 10, 1930-1945.

[40] Paul, M. J., & Pellny, T. K. (2003). Carbon metabolite feedback regulation of leaf photosynthesis and development. *The Journal of Experimental Botany*, 54, 539-547.

[41] Peleg, Z., & Blumwald, E. (2011). Hormone balance and abiotic stress tolerance in crop plants. *Current Opinion in Plant Biology*, 14, 290-295.

[42] Pospisilova, J. (2003). Participation of phytohormones in the stomatal regulation of gas exchange during water stress. *Biologia Plantarum*, 46, 491-506.

[43] Pozsar, B. I., El Hammady, M., & Kiraly, Z. (1967). Cytokinin effect of benzyladenine: increase of nucleic acid and protein synthesis in bean leaves. *Nature*, 214, 273-274.

[44] Raines, C. A. (2011). Increasing photosynthetic carbon assimilation in C_3 plants to improve crop yield: Current and future strategies. *Plant Physiology*, 155, 36-42.

[45] Roberts, S. K. (2006). Plasma membrane anion channels in higher plants and their putative functions in roots. *New Phytologist*, 169, 647-666.

[46] Rolland, F., Moore, B., & Sheen, J. (2002). Sugar sensing and signaling in plants. *The Plant Cell*, S185-S205.

[47] Sage, R. F., Sharkey, T. D., & Seemann, J. R. (1989). Acclimation of photosynthesis to elevated CO_2 in five C_3 species. *Plant Physiology*, 89, 590-596.

[48] Saheki, S., Takeda, A., & Shimazu, T. (1985). Assay of inorganic phosphate in mild pH range, suitable for measurement of glycogen phosphorylase activity. *Analytical Biochemistry*, 148, 277-281.

[49] Salama, A. M. S., El -D, , & Wareing, P. F. (1979). Effects of mineral nutrient on endogenous cytokinins in plants of sunflower (Helianthus annuus L.). *The Journal of Experimental Botany*, 30, 971-981.

[50] Sawada, S., Enomoto, S., Tozu, T., & Kasai, M. (1995). Regulation of the activity of ribulose-1,5-bisphosphate carboxylase in response to changes in the photosynthetic source-sink balance in intact soybean plants. *Plant & Cell Physiology*, 36, 551-556.

[51] Sawada, S., Harada, A., Asari, Y., Asano, S., Kuninaka, M., Kawamura, H., & Kasai, M. (1999). Effects of micro-environmental factors on photosynthetic CO_2 uptake and carbon fixation metabolism in a spring ephemeral, Erythronium japonicum, growing in native and open habitats. *Ecological Research*, 14, 119-130.

[52] Sawada, S., Hayakawa, T., Fukushi, K., & Kasai, M. (1986). Influence of carbohydrates on photosynthesis in single, rooted soybean leaves used as a source-sink model. *Plant & Cell Physiology*, 27, 591-600.

[53] Sawada, S., Kuninaka, M., Watanabe, K., Sato, A., Kawamura, H., Komine, K., Sakamoto, T., & Kasai, M. (2001). The mechanism to suppress photosynthesis through end-product inhibition in single-rooted soybean leaves during acclimation to CO_2 enrichment. *Plant & Cell Physiology*, 42, 1093-1102.

[54] Sawada, S., Usuda, H., Hasegawa, Y., & Tsukui, T. (1990). Regulation of ribulose-1,5-bisphosphate carboxylase activity in response to changes in the source/sink balance in single-rooted soybean leaves: the role of inorganic orthophosphate in activation of the enzyme. *Plant & Cell Physiology*, 31, 697-704.

[55] Sawada, S., Usuda, H., & Tsukui, T. (1992). Participation of inorganic orthophosphate in regulation of the ribulose-1,5-bisphosphate carboxylase activity in response to changes in the photosynthetic source-sink balance. *Plant & Cell Physiology*, 33, 943-949.

[56] Schumacher, K., Vafeados, D., Mc Carthy, M., Sze, H., Wilkins, T., & Chory, J. (1999). The Arabidopsis det3 mutant reveals a central role for the vacuolar H^+-ATPase in plant growth and development. *Genes & Development*, 13, 3259-3270.

[57] Song, W. Y., Choi, K. S., Alexis, D. A., Martinoia, E., & Lee, Y. (2011). Brassica juncea plant cadmium resistance 1 protein (BjPCR1) facilitates the radial transport of calcium in the root. *Proceedings of the National Academy of Sciences of the United States of America*, 108, 19808-19813.

[58] Thuleau, P., Ward, J. M., Ranjeva, R., & Schroeder, J. I. (1994). Voltage-dependent calcium-permeable channels in the plasma membrane of a higher plant cell. *The ENBO Journal*, 13, 2970-2975.

[59] Tominaga, M., Kinoshita, T., & Shimazaki, K. (2001). Guard-cell chloroplasts provide ATP required for H^+ pumping in the plasma membrane and stomatal opening. *Plant & Cell Physiology*, 42, 795-802.

[60] von, Caemmerer. S., & Evans, J. R. (2010). Enhancing C_3 photosynthesis. *Plant Physiology*, 154, 589-592.

[61] von, Caemmerer. S., Hendrickson, L., Quinn, V., Vella, N., Millgate, A. G., & Furbank, R. T. (2005). Reductions of Rubisco activase by antisense RNA in the C_4 plant Flaveria bidentis reduces Rubisco carbamylation and leaf photosynthesis. *Plant Physiology*, 137, 747-755.

[62] Wang, L., He, X., Zhao, Y., Shen, Y., & Huang, Z. (2011). Wheat vacuolar H^+-ATPase subunit B cloning and its involvement in salt tolerance. *Planta*, 234, 1-7.

[63] Werner, T., Motyka, V., Laucou, V., Smets, R., Onckelen, H. V., & Schmulling, T. (2003). Cytokinin-deficient transgenic Arabidopsis plants show multiple development alterations indicating opposite functions of cytokinins in the regulation of shoot and root meristem activity. *The Plant Cell*, 15, 2532-2555.

[64] Xu, D. Q., Gifford, R. M., & Chow, W. S. (1994). Photosynthetic acclimation in pea and soybean to high atmospheric CO_2 partial pressure. *Plant Physiology*, 106, 661-671.

[65] Yang, H., Knapp, J., Koirala, P., Rajagopal, D., Peer, W. A., Silbart, L. K., Murphy, A., & Gaxiola, R. A. (2007). Enhanced phosphorus nutrition in monocots and dicots over-expressing a phosphorus-responsive type I H^+-pyrophosphatase. *Plant Biotechnology Journal*, 5, 735-745.

Chapter 15

Soybean Urease: Over a Hundred Years of Knowledge

Rafael Real-Guerra, Fernanda Stanisçuaski and
Célia Regina Carlini

Additional information is available at the end of the chapter

http://dx.doi.org/10.5772/52106

1. Introduction

Ureases are metalloproteins responsible for the one step hydrolysis of urea into ammonia and carbamate [1], the later then rapidly and spontaneously decomposes to form carbon dioxide and a second molecule of ammonia [2]. Plant ureases hold a special place in science history, participating on some important landmarks of biochemistry. For instance, it was by the analysis of *Canavalia ensiformis* urease crystals that the proteinaceous nature of enzymes could be demonstrated [3], rendering a Chemistry Nobel Prize to James Sumner in 1946. Also by the studies carried out with the same protein it was obtained the first proof that Ni^{2+} actually exerts a biological role in living organisms [4]. Ureases usually present two Ni^{2+} in their active site, with a few exceptions reported [5, 6]. However, all this was only possible after Takeuchi's observation that the crude extracts of soybean (*Glycine max*) seeds present high amounts of urease [7]. In those days, urease had been observed only in microorganisms and in algae, being Takeuchi's finding the first evidence of the presence of ureases in higher plants. The importance of this discovery relays on the fact that it made urease largely available for any researcher on the globe, and many more works on ureases followed, utilizing the soybean seed urease with the objective of understanding enzyme functioning. Soybean urease was also one of the key players on the development of enzymology, with studies on this enzyme leading to hypothesis that were essential to confirm the observations of Michaelis and Menten on the rate of reaction of enzymes and substrates [8].

From those first studies more than one century ago until today, soybean ureases continued to be the focus of researchers around the world, in the fields of genetic, biochemistry and physiology. This review will deal with the many faces of these proteins, trying to summarize the great amount of information gathered over time, and to point the many doors that continue to be opened by the studies with this enzyme.

2. Soybean urease – more than one enzyme

During the course of urease history, there were several reports on the presence of nongenetic isoenzymes for *C. ensiformis,* which represented different multimeric interconversible forms of the same enzyme. As many as 12 different forms of the protein could be observed, depending on environmental factors such as temperature, pH, salt concentration, reagents, etc [9-13]. Also, multimeric forms of the soybean urease were observed [14, 15]. By the comparison of the purified urease from soybean seeds and a partially purified urease from a soybean shoot cell culture, it was not possible to differentiate them on basis of electrophoretic species, size exclusion chromatography and immunoinhibition, but they seemed to behave differently on a immunoaffinity chromatography, where 100% of the seed enzyme would be retained in contrast with only 65 % of the culture urease [15]. Although the authors postulated that this behaviour may indicate the existence of two different enzymes, at the time they could not reject the hypothesis of differentially processed forms of the seed enzyme (glycosylated protein, etc).

The next indication of the existence of more than one genetic isoform of soybean urease emerged a few years later, in 1982. After screening a collection of over 6,000 lines of soybean seeds from the United States Department of Agriculture (USDA) Polacco and co-workers were able to identify one urease negative soybean variety (Itachi) [16]. Interestingly, even though no measurable amounts of urease could be detected in the seeds of the Itachi variety, when cells from many parts of the plant were cultured, urease was produced in equivalent amounts to the wild type soybean. Making use of immunoaffinity chromatography with monospecific antibodies that retained 100% of the soybean seed urease, it was observed that none of the cell culture ureases could be retained completely by the antibodies (70 and 45 % for the wild type and Itachi, respectively). Some differences in the inhibition profile by hydroxyurea were also observed, the ureases from both cultures (wild type and Itachi) being less susceptible than the urease purified from the seeds. All these facts pointed to a similar but still different urease synthesized in cell cultures as compared to the one present in the seeds. However, the authors seemed to battle a little with the idea of a second genetic isoform of urease, considering that the observed "heterogeneity" could be due to a glycosylated seed urease.

More definitive proofs of the existence of a second and distinct isoenzyme appeared when Kerr and co-workers [17] carried out biochemical characterizations of the seed and the leaf ureases, showing that the enzymes differed on several characteristics, such as optimum pH, apparent K_m, inhibition profile by hydroxyurea and cross-reactivity against soybean seed urease antibodies. These observations were further extended by Polacco and Winkler [18] who found that the previously observed urease production by the cell cultures of a supposedly urease negative line actually corresponded to the production of a second isoform of urease present in the tested tissues (leaf, callus, seed and cotyledon). The hypothesis of the existence of more than one non-seed isoenzyme, very similar to each other but specific for each tissue, was considered at the time. Since then, this second isoform was denominated

http://dx.doi.org/10.5772/52106

ubiquitous soybean urease (herein referred as ub-SBU) while the seed isoenzyme would later be denominated embryo-specific urease (herein referred as es-SBU) [19].

The data gathered in these studies set ground for investigations in different fields and to the growing understanding of ureases, mainly in genetics, in the search for definitive answers regarding the diversity of this enzyme.

3. The discovery of soybean urease genes

Buttery and Buzzell [14] can be considered the pioneers on genetic studies involving soybean urease. In their work they identified in the soybean seed two variants of urease showing distinct electrophoretical mobility, which they assigned as fast- and slow-moving forms. These forms were later on characterized as the trimeric and hexameric forms of es-SBU, respectively [15, 19]. In their study, Buttery and Buzzell found the slow-moving form to be recessive, while the fast-moving form was considered dominant, and they concluded that a single locus (named *Eu*, after enzyme urease) controlled the expression of these electrophoreticaly variant forms. Many years later, Kloth and Hymowitz [20] advanced the investigation on the two multimeric forms and, by performing a series of crosses between urease null and wild type soybeans, proposed the existence of two codominant alleles, named *Eu1-a*, for the hexamer and *Eu1-b*, for the trimer. The *Eu1* gene, which encodes the es-SBU, was the first urease related gene described for a plant.

Probably the biggest milestone on soybean urease genetics research was the identification of null variants of es-SBU. First obtained by Polacco and co-workers [16], the variety Itachi was found to lack both the es-SBU protein [21] and mRNA [22]. Later on, four more mutants were identified [23], two lacking detectable amounts of es-SBU mRNA, and two producing very low protein levels. One of these later mutants produced a much altered protein. All those mutants were related to the locus *Sun* (for soybean urease null), and the *Sun* allele (normal es-SBU) was dominant over the *sun* allele (null es-SBU activity) [23-25]. The fact that allelic *sun* mutations affected urease transcript as well as es-SBU structure led to the conclusion that *Sun* and *Eu1* describe the same locus, and that the *Eu1-Sun* is the single functional es-SBU structural locus. Thus any mutant defective only on es-SBU is considered a *Eu1* or class I mutant. [25]. A genomic clone for urease was recovered from a soybean library and it was shown that this urease coding sequence was absent from seed urease-null mutants [22]. This would be confirmed as the first partial sequence obtained for the es-SBU years later.

Mutations obtained from ethyl methyl sulfonate treatment of soybean seeds [23] revealed a new class of mutants. Class II mutants produce normal levels of es-SBU and ub-SBU mRNA and protein, however their enzymatic activity is completely absent [23, 26]. Mutants of this class carry damage in one or both loci *Eu2* and *Eu3*, which are unlinked to each other or to *Eu1*. Studies on the urease profile during soybean development have shown that maximal synthesis of es-SBU in the developing cotyledon lags the maximal urease activity, suggesting a very slow maturation process of the enzyme *in vivo*. The facts that es-SBU is a Ni^{2+} containing enzyme [15] and the presence of Ni^{2+} is absolutely required for *in vivo* urease activity [27, 28] but not for urease synthesis [18, 21] are compatible with the assumption that Ni^{2+} emplacement into the active site could be a limiting process for urease activation. Holland and Polacco [29] have shown that class II mutants present normal Ni^{2+} uptake and translocation, thus eliminating the possibility that lack of access to Ni^{2+} was the reason for the absence of activity of both ureases. These observations pointed to *Eu2* and *Eu3* having a direct role on enzyme maturation probably by codifying proteins responsible for the Ni^{2+} emplacement, and that they are common for both es-SBU and ub-SBU [30]. As discussed later, Eu2 and Eu3 were found to encode two different urease accessory proteins, involved in the enzyme maturation process.

After those findings, the remaining question was: is ub-SBU codified by a single gene or are there many tissue specific ureases? This question was clarified when mutants were obtained that presented normal levels of es-SBU activity, but no ub-SBU activity in any tissue. Those mutants were classified as class III mutants. Crosses between Class I and Class III mutants are devoid of ub-SBU even in embryonic tissues. The presence of the protein was detected in all tested tissues of class III mutants despite the lack of urease activity, showing that the lesions affected directly the structural gene of ub-SBU resulting in the production of an inactive protein [26]. The lesions causing this effect were attributed to a new locus named *Eu4* [26, 31-33]. This finding demonstrated that the ub-SBU was actually a single protein synthesized in all tissues, and confirmed the existence of two genetic isoforms of soybean urease. A final blow on this question came when Goldraij and co-workers [34] finally obtained and sequenced the cDNA of both isoforms showing that they are very similar, sharing 87% of identity and 92% of similarity.

The genomic era brought with it the last pieces of the puzzle. The sequencing of the soybean genome [35] confirmed the presence of the two previously described genes of urease, and also revealed a third one (Glyma08g10850), which is believed to be inactive due to a high number of deleterious mutations. Nevertheless, a residual activity was observed in double mutants, lacking both es-SBU and ub-SBU, and accounted for 2 to 10% of activity compared to the wild ub-SBU activity. This activity was designated as "background" and it was attributed to microbial commensals in soybean tissues [26, 31, 36]. Alternatively, complementation between the defective *Eu4* gene and this third urease-like gene may cause the residual activity [37]. The genome sequencing also brought light over the urease accessory proteins. A single gene for UreG, two for UreF and two for UreD were identified in the soybean genome. Table 1 summarizes the data derived from the soybean genome available at the Phytozome databank [38].

Phytozome accession id	Gene (locus)	Gene size (bp)	Protein codified	Protein size (aminoacids)	References
Glyma05g27840	*Eu1*	7736	Embryo-specific urease	839	[23, 33, 34]
Glyma11g37250	*Eu4*	7287	Ubiquitous urease	837	[23, 33, 34]

Phytozome accession id	Gene (locus)	Gene size (bp)	Protein codified	Protein size (aminoacids)	References
Glyma08g10850	Not described	5849	Urease-like protein	713	-
Glyma02g20690	Not described	3473	Accessory protein UreD	308	[39]
Glyma20g17990	Not described	2786	Accessory protein UreD	256	-
Glyma02g44440	*Eu2*	717	Accessory protein UreF	238	[40]
Glyma14g04380	Not described	2032	Accessory protein UreF	238	[40]
Glyma08g08970	*Eu3*	4000	Accessory protein UreG	285	[41]

Table 1. Urease-related genes in soybean

4. The Intricate Process of Urease Activation – Much More Than Structural Genes

The biosynthesis of metalloenzymes usually depends on the participation of several dedicated proteins that are essential for the correct assembly of their active sites, and ureases are no exception. The role of these accessory proteins consist on the stabilization of the apoenzyme in a certain conformation that allows the correct insertion of the metal ion in the active site, dissociating afterwards and releasing the mature enzyme [42]. This process has been fairly studied for bacterial ureases, but the activation of plant ureases still demands more attention. Thus it is described here based mostly on what is known for bacterial ureases.

The activation of ureases require two essential steps: the carbamylation of a lysine residue, that will be responsible for bridging and, consequently, holding the two Ni^{2+} ions into place within the active site; and the actual incorporation of the two Ni^{2+} ions in the active site. The best characterized system so far is that of *Klebsiella aerogenes* urease, for which four accessory proteins, namely UreD, UreE, UreF and UreG, are required for complete activation [2]. Although many efforts have been put on characterizing these proteins, the precise role of each one in the complex process of urease activation is still not clearly understood. These proteins appear to be well conserved among urease producing bacterial organisms and even for higher organisms as plants. In *K. aerogenes,* the accessory proteins bind consecutively to the apourease (apU) forming the following complexes: apU-UreD, apU-UreDF, apU-UreDFG [2, 43]. UreE, a Ni^{2+} binding protein, is the last accessory protein proposed to bind to the complex apU-UreDFG and deliver the metal to the enzyme active site [44]. The process of urease activation requires the hydrolysis of GTP by UreG, after which the accessory proteins dissociate from the active enzyme. A preformed UreDFG complex could be isolated from bacteria lacking UreE [45]. If the accessory proteins act on apU as a preformed complex or in a sequential binding fashion, is a matter that demands further investigation. However, the sequential binding model is to date the most accepted and the one supported by most evidence.

Although the exact role of each accessory protein has not been clearly assigned, some general lines can be traced for their individual actions. The current sequential model assumes that UreD is the first accessory protein to interact with urease. UreD is yet the least characterized protein and it seems to serve as an adapter for the other accessory proteins since neither UreF or UreG are able to bind urease directly in the absence of UreD [2]. UreF interacts directly with UreD and it has been proposed that UreF would be responsible for promoting a conformational change in apU, providing better access to the active site of the protein and allowing the next steps of the process to take place [46]. Recently, a structural model of UreF has indicated that this protein shares structural similarities to some GTP activating proteins (GAP) [47]. UreG is an intrinsically disordered GTPase, as reported for organisms such as *Helicobacter pylori, Bacillus pasteurii, Micobacterium tuberculosis* and also *Glycine max* [48-53]. UreG binds to UreF and, through the cleavage of GTP, provides the necessary energy for the activation process. It is also postulated that GTP cleavage in the presence of CO_2 could form carboxyphosphate, an excellent carbamylation agent [46, 54]. UreG GTPase activity, when detectable, seems to be very low in comparison to other GTPases. It has been proposed [46] that the intrinsically disordered structure of this enzyme serves as a regulatory mechanism of its activity, which would be only maximized when, inserted in the activating complex, it acquires its fully folded state. It is also hypothesized that UreF would act on UreG as a GAP, enhancing its enzymatic activity, although, this hypothesis still waits to be tested.

Available sequences of bacterial urease accessory proteins led to the search of potential orthologs in plants, and the identification of UreG (Eu3) in soybean. This was the first evidence of accessory proteins in plants [41]. Soybean cDNAs for UreD and UreF proteins were also identified later [39], but none of them were assigned as the *Eu2* gene. To date, no UreE equivalent in plants has been identified and sequence analyses supports the hypothesis that its functionality is incorporated into UreG which, in plants but not in bacteria, presents an extended N-terminal rich in aspartic acid and histidine residues. The property of soybean UreG to bind metal ions has been demonstrated [41, 53]. A study in *A. thaliana* [55] has shown that UreD, UreF and UreG are necessary and sufficient to activate urease, since knockout plants for any of the accessory proteins genes lack urease activity. They also showed that simultaneous co-expression of all accessory proteins together with *A. thaliana* urease structural gene was able to generate urease activity on *Escherichia coli* cells. Also the *Oriza sativa* urease could only be activated when all of its three accessory proteins were co-transformed in tobacco [56].

With the exception of dimeric ureases described for some plants, such as canatoxin from *C. ensiformis* [5], and the ureases from *Morus alba* [57] and *Momordica charantia* [58], all reported ureases share a basic trimeric state that may aggregate to form larger oligomers, as hexamers in the case of most plant ureases, in which each monomer carries one active site. For bacteria, each unit of this trimer is usually itself a heterotrimer composed by the subunits UreA, UreB and UreC which co-align with plant ureases with over 50% identity. Thus the single type polypeptide chain of plant ureases corresponds to the collinear fusion of the bacterial subunits (UreA-UreB-UreC). It is curious to notice that despite the high similarity between plant and bacterial ureases, some of the accessory proteins do not share the same degree of sequence identity. Plant and bacterial UreG are very similar, with about 40% of sequence identity, but plant UreD and UreF share only about 20% of sequence identity with their bacterial counterparts [56]. Considering this discrepancy, a recent work using *in silico* structure prediction tools, has shown that despite the sequence disparities, UreF proteins are very conserved at the structural level [56]. Plant UreD, although sharing some structural similarities, possess some marked differences when compared to their bacterial counterparts. This different overall structure of plant and bacterial UreD is hypothesized to reflect distinct requirements for interaction with the bacterial heterotrimeric or the plant single polypeptide urease unit [56].

Despite the fact that the presence of the set of accessory proteins is enough to get ureases activated, there seems to be more to it concerning regulation. In bacteria, UreF and UreD are expressed in very low levels, and it has been shown that over expression of these proteins can hamper urease activation [59, 60]. It has been proposed that differential splicing generating aberrant mRNA could reduce UreD production in plants [55]. Cao and co-workers [56] reported for *ureF* an intron in the 5'leader conserved among 16 plant genomes and they noticed that in almost every case the spliced transcript would be free of AUG codons upstream of the start codon. Special attention was given to the two *ureF* genes from soybean which, although both are spliced at the 5' leader, only the paralog in chromosome 2 (Glyma02g44440) has all the nonstart AUG codons removed. The transcript of the paralog from chromosome 14 (Glyma14g04380) has an out of frame AUG codon upstream of the start codon and therefore was postulated to be ineffectively translated and consequently non functional. In the same work, a low splicing efficiency of *AtureF* was observed, which led to the conclusion that for some plants the 5' leader sequence may have a regulatory role in reducing the amount of *ureF* mRNA either by differential splicing (*Arabidopsis*) or translational inhibition (soybean). Limited expression may be required to ensure that UreF and UreD dissociate from urease after activation to release the active enzyme and that the putative GTPase activating protein UreF does not trigger UreG activity in the absence of urease [56]. Both UreD and UreF have been observed to be very unstable proteins either in plants [56] or bacteria [59], and this intrinsic instability possibly contributes to the regulation of their activities *in vivo* [56].

As mentioned above, soybean genome contains two UreF genes. The one in chromosome 14 UreF (Ch14UreF) has previously been characterized and demonstrated to activate the urease from *S. pombe* [39], but the UreF encoded by chromosome 2 (Ch02UreF) remains to be characterized. Until recently, the product of *Eu2* gene was not identified, and it had been proposed that *Eu2* could represent the functional paralog of UreD or UreF. Analysis of soybean *Eu2* mutants revealed that missense mutations resulted in the expression of an altered form of Ch02UreF protein [40]. Interestingly, these mutants presented no urease activity, even though the Ch14UreF is present and is supposed to activate urease. Another mutant, in which the expression of Ch02UreF was impaired, presented 5-10 % of the wild type urease activity. The authors presented two possible explanations for the results. The first one considered that Ch02UreF could spoil activation by Ch14UreF because of a higher affinity for the activation complex. The second explanation, favoured in their work, proposed that Ch02UreF is more abundant than Ch14UreF, which would be less efficiently translated. Therefore, in *Eu2* mutants, the missense mutants of Ch02UreF block the access of Ch14UreF to the urease activation complex, preventing activation [40].

As pointed out here, although accessory proteins differ widely according to their source, the process of urease activation seems to be very well conserved. Among plants, the urease activation complex seems to be structurally very similar, since accessory proteins from different plants are able to functionally complement each other. Rice urease, for instance, can be activated by *Arabidopsis* bulk of accessory proteins, and UreD and UreG from rice can replace the native accessory protein in mutants of Arabidopsis [56]. The similarities of the activation process seem also to break the kingdom barrier. Soybean UreF has been shown to complement the *Saccharomyces pombe* accessory set [39], and UreG from potato (*Solanum tuberosum*) complements the *K. aerogenes* operon [61].

5. The Physiological Role of Soybean Ureases

After carbon, nitrogen is the main limiting element for plant performance [62], and there is a constant pressure on plants for efficient use of N leading to the development of efficient mechanisms for N uptake and metabolic pathways for N remobilization [37, 56]. Such a pressure even led to a reduction of N content of plant proteins [63]. Urea is an important primary source of N for plants. The action of arginase is the only confirmed pathway for urea generation *in vivo*; urea could also be generated by the degradation of purines and ureides [64], although this later pathway is very controversial. Urea can only be assimilated after its hydrolysis into ammonia and carbon dioxide by urease [65], and that is the main physiological role attributed to ureases in plants [37]. Ammonia will then be re-assimilated by glutamine synthetase using glutamate as substrate [66]. Urease activity is present virtually in all plant species and is ubiquitously distributed in all plant tissues [18, 31, 67], which is indicative of the great importance of its physiological role for the whole plant. For a long time, the relevance of plant urease-mediated metabolism of urea was considered not significant, since it was assumed that urea was hardly taken up by the plants, but instead degraded by microorganisms in the soil and then the ammonia and nitrate were absorbed. Today, it is well known that plants can actively import urea from the soil, through the activity of dedicated urea transporters [68] and can also efficiently process soil-imported urea, even at high concentrations. These findings point at urease as a target for studies on improving plant N metabolism based on urea, the most used N fertilizer in the world [69].

Soybean makes a very interesting model for studies on the physiological role of urease in plants, since it is so far the only genome-sequenced plant that presents more than one isoform of the enzyme. ub-SBU has long been known to be the isoform responsible for recycling all metabolically derived urea [19, 70, 71]. This has been demonstrated since mutants lacking es-SBU do not accumulate urea in any tissue and do not have any impairment on the use of urea as sole nitrogen source, even though ub-SBU is present at levels only 0.1 to 0.3% that of es-SBU [19, 72]. Soybean mutants lacking ub-SBU activity present a characteristic phenotype consisting of necrosis of leaf tips, due to urea "burn", and accumulation of urea in many tissues [32, 36]. Urease is the only Ni^{2+} dependent enzyme yet identified in plants and the same phenotype, early mentioned, is observed for plants grown under Ni^{2+} deprivation [27].

Interestingly, no physiological role, being it assimilatory or of any other nature, could be demonstrated for the very abundant es-SBU. In fact, wild-type cultured cotyledons could not grow in the presence of urea, due to a sudden pH increase resultant of an uncontrolled ammonia release. The same effect was not observed for mutants that have only ub-SBU [36]. It was inferred that es-SBU could be involved in plant defense against predators. A chemical protection was postulated for the case of microbial or insect attack. By this model, wounding or infection of the immature embryo would lead to arginase release from ruptured mitochondria which would generate urea from the large pool of arginine and cytoplasmic urease would rapidly convert urea to ammonia [32]. This hypothesis still waits demonstration, but it has been reported that mutants lacking urease activity were more susceptible to microbial infections [73]. As it will be discussed in the next section, es-SBU can be involved in plant defense not only by conferring chemical protection, but also ureolysis-independent mechanisms, including the generation of toxic peptides. On the other hand, it is tempting to propose that the third urease found in the soybean genome (Glyma08g10850), that apparently has no enzymatic activity, can also be involved in other physiological roles in the plant, such as plant defense, and some indications of that have already been reported (see the next section).

6. New features of an old protein – activities unrelated to the enzymatic one

As stated above, no definite answer to the question of the physiological relevance of es-SBU has been given, since the demonstration that this enzyme plays no role in nitrogen assimilation from urea [32, 65, 73]. During the course of the last decade, a number of biological properties unrelated to the enzymatic activity were described for plant ureases, launching a new look over these proteins and their physiological roles. Table 2 summarizes some of these overlooked biological activities of plant ureases. The main discoveries were made for the jackbean ureases (CNTX and JBU), revealing several interesting properties, such as entomotoxicity [74-77], fungitoxicity [78, 79] and secretory activity [5, 80].

Some of the biological properties described for jackbean's ureases, such as the entomotoxic and fungitoxic activities, are shared by soybean urease. es-SBU displays toxicity toward insects, as demonstrated by [75]. es-SBU is toxic to *Dysdercus peruvianus*, a cotton culture pest, at doses as low as 0.05% (w/w), causing a decrease on insect body weight, delayed development and death. This toxicity was maintained after treating es-SBU with *p*-hydroxymercuribenzoate (an irreversible urease inhibitor), confirming that the entomotoxic property was independent of the enzymatic activity. For the jackbean ureases, it has been demonstrated that this entomotoxic effect is a highly complex event, involving the intact protein as well as the release of a toxic peptide [82, 89, 90]. Jackbean ureases and their derived peptides affect several physiological processes in the insects, including the fluid transport across membranes [83, 91, 92]. So far, no studies regarding es-SBU mechanism of action in insects were performed, but it reasonable to assume that, due to their very similar sequences (92% of similarity), es-SBU and JBU would have the same targets in these organisms. The sequence of the entomotoxic peptide identified in CNTX and JBU, is present in es-SBU and ub-SBU [90]. These findings also support the hypothesis of a similar mode of action in insects.

Activity	CNTX	JBU	es-SBU	GHU
Entomotoxic	*Rhodnius prolixus*, *Callosobruchus maculatus* [74], *Nezara viridula* [81], *Dysdercus peruvianus* [75, 76]	*Dysdercus peruvianus* [75, 76, 82], *Rhodnius prolixus* [83] *Oncopeltus fasciatus* [77]	*Dysdercus peruvianus* [75]	N.D.
Fungitoxic	*Macrophomina phaseolina*, *Colletotrichum gloesporioides*, *Sclerotium rolfsii* [78]	*Fusarium solani*, *Colletotrichum musae*, *Curvularia lunata*, *Penicillium herguei*, *Fusarium oxysporum* [79].	*Colletotrichum musae*, *Penicillium herguei*, *Curvularia lunata*, *Fusarium oxysporum* [79].	*Colletotrichum musae*, *Penicillium herguei*, *Curvularia lunata* [84].
Secretory	Rabbit platelets [85, 86], rat brain synaptosomes [86], rat pancreatic cells [87], rat mast cells [88]	Rabbit platelets [5, 75]	Rabbit platelets [75]	N.D.
Toxicity to mammals	2 mg/kg (LD_{50}) [5]	Not toxic [75]	Not toxic [75]	N.D.

Table 2. Properties of plant ureases unrelated to the enzymatic activity; N.D. = not determined; GHU = *Gossypium hirsutum* urease

Another very interesting property presented by es-SBU is its fungitoxic activity. es-SBU suppressed mycelial growth and/or inhibited spore germination of a series of fungi species and, as demonstrated for the entomotoxic property, this effect also does not require the protein's ureolytic activity [79]. The precise mechanism of action of ureases on fungi has not been elucidated so far, being proposed that ureases may interfere with the cellular osmotic balance. Recently, evidences of the participation of ub-SBU in soybean resistance to fungi were reported [93]. Soybeans mutants, lacking ub-SBU, were more susceptible to necrotrophic fungi, such as *Penicillium herguei*, *Phomopsis* sp and *Rhizoctonia solani*, and to the biotrophic pathogen *Phakopsora pachyrhizi*, responsible for the Asian soybean rust disease. In accordance to a defense role of ub-SBU against fungi infection, a rust-resistant soybean cultivar (PI561356) had a higher level of expression of ub-SBU after infection with *P. pachyrhizi*, in comparison with a susceptible cultivar (Embrapa-48) [93].

These toxic activities, unrelated to the ureolytic one, are interesting findings that point to a possible role of ureases in plant defense against insects and fungi. However, plant ureases also have others bioactivities, that seem not related to plant physiology. JBU and es-SBU were shown to activate exocytosis in blood platelets causing them to aggregate, an effect shown to be independent of their enzymatic activity [75]. This exocytosis-inducing property may be relevant to some urease-producing microorganisms such as *Bacillus pasteurii* and *Helicobacter pylori* [94, 95] but it is unlikely to have a more specific significance for plants. Nevertheless, Carlini and Polacco [73] hypothesized that the secretion-properties of bacterial and plants ureases may play a role in rhizosphere relationships.

7. Soybean as animal feed – what urease has got to do with it?

It is estimated that soybean meal accounts for around 67% of all protein sources used in animal feeds around the world [96], due mainly to its high protein concentration (44 to 48%) [97]. Nevertheless, soybean contains an unusually large number of bioactive compounds with antinutritional and/or toxic properties, which have a negative effect on body metabolism of animals [98]. Urease is one of these factors. Urease content was not evidently different among 11 soybean cultivars tested [99]. In contrast, urease content was found very variable among several other soybean cultivars [100-102], and the levels of urease correlated positively with antinutritional effects in rats [101].

The negative effects of using urease-containing meals as animal feed are reported in the literature. Urea is frequently added to animal feed and, when unprocessed soybeans are mixed with urea, ammonia will be released by the action of urease, which is an undesired effect in a mixed feed [103]. In ruminants, ammonia rapidly enters the blood and can cause adverse affects ranging from depressed feed intake and animal performance, to death from ammonia toxicity [104]. In dairy cows, the liver, responsible for removing potentially toxic ammonia from circulation, was able to remove ammonia added to portal blood until the supply reached 182 mg/min but, at higher infusion rates, peripheral blood ammonia concentrations increased, supporting the assessment that a rapid hydrolysis of dietary urea can exceed the liver's capacity to remove it [105]. In chickens, it was demonstrated that soybean meals from one particular source consistently produced a high incidence of tibial dyschondroplasia (TD) and the most striking difference between the meals was the high antitrypsin and urease values in those that induced the disease [106]. The incidence of TD was demonstrated previously to be increased in broilers when ammonium chloride (1.5 or 30%) was added to the diet [107], but not when calcium chloride was used [108]. These may be indications that the release of ammonia by urease could play a role in the incidence of TD in soybean meal fed chickens.

In order to allow the addition of supplemental nitrogen to the animal feed, while protecting the animals against the production of toxic levels of ammonia, pre-treatment of the soybean meal is necessary. Heat treatment is the main method used to abolish or decrease the effects of the antinutritional and/or toxic factors in soybean, including urease [109, 110], but these treatments should be kept to a minimum, due to the possibility of destroying important seed constituents [98]. To abolish urease activity, several treatments are effective, including steam-heating at 102 °C for 40 min or at 120 °C for 7.5 min [109], boiling at 92 °C for 60 min [101], and dry-heating at 100 °C and 2 kgf/cm^2 for 60 min [111]. All those treatments abolished urease activity, along with a decrease of several antinutritional factors.

The best way to evaluate the adequacy of the processing and final quality of soybean meals is conducting biological tests. However, the cost, time requirement and complexity of these tests impair their use. Since the 1940's, the urease test is used as an indirect way to evaluate the adequacy of heat processing of soybeans due to its rapidity, low skill and minimum amount of laboratory equipment requirements. One research study [109] showed a high correlation among the activities of trypsin inhibitors, urease and lectins, indicating that the adequacy of soybean processing can be estimated to a considerable extent by these analytical criteria. Over the years, many protocols were developed to facilitate the measurement of urease activity. These protocols quantify the released ammonia directly or indirectly. One of the first to be developed, in the Caskey-Knapp method [112] the meal is incubated with urea in a buffered solution and then phenol red is added. After incubation, insufficiently processed meals will cause an increase in the pH of the solution, indicated by a change in color (from red-orange to pink), while adequately processed meals produce little or no color change. One study [113] proposed an alternative method, with the potential to differentiate between meals with low levels of urease activity, based on the incubation of the meal with urea in a buffered solution and the colorimetric determination of the residual urea with *p*-dimethylaminobenzaldehyde. A method for direct titration of ammonia as a measure of urease activity was proposed [114] and adapted [115], in which the incubation of urea and the meal is performed, maintaining the pH of the solution by slowly adding HCl. The system is then titrated with NaOH. The difference in titration between a control (urease inactivated) and the sample is taken as the urease activity of the meal. Two other methods were developed to determine ammonia directly, based on the phenol-hypochlorite reaction [116, 117].

Several modifications and adaptations of these methods were developed during time. But, regardless of the method chosen, urease activity is a very good indicator of under processing of soybean meals. It is worth noting, however, that this activity is not a good indicator of over processing of soybean meals.

8. The biotechnological potential of soybean ureases

The questions of why are ureases so large, and why are they multimeric have been raised, and a possible explanation is that a "primordial" enzyme could have acquired other "traits" under the evolutionary pressure of competition in an increasingly complex biosphere [73].

In the view of these "extra traits" discovered for ureases, some biotechnological applications can now be proposed.

Soybean can be attacked by many different organisms, including fungi, insects, virus and nematodes. These pathogens and pests can cause damage in seeds, roots, leaves, stems and pods, and usually are tissue-specific [118]. And, despite control measures, pests reduce worldwide soybean production by almost 28% [119]. The development of new technologies to control these pests is urgent, and exploring natural plant compounds is a major strategy.

Plant ureases and their derived peptides have a great biotechnological potential. Ureases are abundant in many edible sources, including legumes and potatoes, and even eaten raw in cucumbers, or fruits such as melon and watermelon [73]. Thus, possible biosafety issues could be more manageable. Since JBU, es-SBU and the derived peptide Jaburetox seem non toxic to mammals [75, 90], the entomotoxic and fungitoxic properties of these molecules are relevant when considering biotechnological strategies aiming to protect commercially-relevant crops against natural enemies. The evidences of an *in vivo* effect of soybean urease in protecting the plant from fungi [93] are exciting. The possibility of selecting soybean cultivars with higher urease content or increasing the production of these proteins in the plant through genetic manipulation, in order to increase the resistance against insects and fungi, is very promising. Also, the premise of using plant naturally occurring proteins to improve resistance seems much more appealing to the general public than the alternative of inserting foreign genes (from microorganisms or animals) into crops.

As pointed out a long time ago [15, 120], soybeans with a high content of ureases could also be agronomically valuable, regardless of the defense role, for permitting more efficient assimilation of urea fertilizer by the plant. Also, considering the wide use of soybean meal as animal feed (as discussed above) and the potential of being a protein source for humans, a higher urease content in soybean could be interesting for improving soybean nutritional quality, after the appropriated processing, since urease is richer in methionine than many others soybean seed proteins. Soybean has a limited amount of sulfur aminoacids, almost half of which are considered ideal for animal feed. Although this problem can be overcome by feed supplementation with free methionine, there are problems associated with the supplementation, such as leaching of methionine during processing and bacterial degradation leading to formation of undesirable volatile sulfides [121]. Improving the content of methionine in soybean through the increase of the biosynthesis of endogenous proteins, such as ureases, is a very interesting approach.

9. Conclusion

Soybean ureases were undoubtedly landmarks in science, being the subject of investigations since the beginning of the 1900's. But, despite the more than one century of studies, we still have a long way until fully understanding the complexity of such a striking molecule. The many properties of these proteins revealed that ureases are much more than urea hydrolyzing enzymes, and present a vast array of interesting biotechnological applications. Exploring the toxic properties of plant ureases can be of great interest for the development of alternative strategies to protect agricultural relevant crops against several natural enemies.

Acknowledgements

The authors are thankful for the Brazilian funding agencies Conselho Nacional de Desenvolvimento Científico e Tecnológico (CNPq), Coordenação de Aperfeiçoamento de Pessoal de Nível Superior (CAPES) and Fundação de Amparo à Pesquisa do Estado do Rio Grande do Sul (FAPERGS).

Author details

Rafael Real-Guerra[1*], Fernanda Stanisçuaski[2] and Célia Regina Carlini[1,3]

*Address all correspondence to: rafael.guerra@ufrgs.br

1 Center of Biotechnology, Federal University of Rio Grande do Sul, Porto Alegre, Brazil

2 Department of Molecular Biology and Biotechnology, Biosciences Institute, Federal University of Rio Grande do Sul, Porto Alegre, Brazil

3 Department of Biophysis, Biosciences Institute, Federal University of Rio Grande do Sul, Porto Alegre, Brazil

References

[1] Blakeley, R. L., Hinds, J. A., Kunze, H. E., Webb, E. C., & Zerner, B. (1969). Jack Bean Urease (Ec 3.5.1.5). Demonstration of a Carbamoyl-Transfer Reaction and Inhibition by Hydroxamic Acids. *Biochemistry*, 8(5), 1991-2000.

[2] Carter, E. L., Flugga, N., Boer, J. L., Mulrooney, S. B., & Hausinger, R. P. (2009). Interplay of Metal Ions and Urease. *Metallomics*, 1(3), 207-221.

[3] Sumner, J. B. (1926). The Isolation and Crystallization of the Enzyme Urease. *Journal of Biological Chemistry*, 69-435.

[4] Dixon, N. E., Gazzola, T. C., blakeley, R. L., & Zermer, B. (1975). Jack Bean Urease (Ec 3.5.1.5). A Metalloenzyme. A Simple Biological Role for Nickel? *Journal of the American Chemical Society*, 97(14), 4131-4133.

[5] Follmer, C., Barcellos, G. B., Zingali, R. B., Machado, O. L., Alves, E. W., Barja-Fidalgo, C., Guimaraes, J. A., & Carlini, C. R. (2001). Canatoxin, a Toxic Protein from Jack

Beans (Canavalia Ensiformis), Is a Variant Form of Urease (Ec 3.5.1.5): Biological Effects of Urease Independent of Its Ureolytic Activity. *Biochemical Journal* [Pt 1], 217 .

[6] Carter, E. L., Tronrud, D. E., Taber, S. R., Karplus, P. A., & Hausinger, R. P. (2011). Iron-Containing Urease in a Pathogenic Bacterium. *Proceedings of the National Academy of Sciences of the United States of America*, 108(32), 13095-13099.

[7] Takeuchi, T. (1909). On the Occurence of Urease in Higher Plants. *Journal of the College of Agriculture, Imperial University of Tokyo*, 1.

[8] Van Slyke, D. D., & Cullen, G. E. (1914). The Mode of Action of Urease and Enzymes in General. *Journal of Biological Chemistry*, 19-141.

[9] Fishbein, W. N., Spears, C. L., & Scurzi, W. (1969). Spectrum of Urease Isozymes: Genetic, Polymeric and Conformeric. *Nature*, 223(5202), 191-193.

[10] Fishbein, W. N., & Nagarajan, K. (1971). Urease Catalysis and StructureVii. Factors Involved in Urease Polymerization and Its Kinetic Pattern. *Archives of Biochemistry and Biophysics*, 144(2), 700-714.

[11] Fishbein, W. N., Nagarajan, K., & Scurzi, W. (1976). Urease Catalysis and StructureX. Alternate Bonding-Site Isozymes of Jackbean Urease. *Archives of Biochemistry and Biophysics*, 172(2), 726-733.

[12] Creeth, J. M., & Nichol, L. W. (1960). Evidence for the Chemical Interaction of Urease in Solution. *Biochemical Journal*, 77-230.

[13] Lynn, K. R. (1971). Urease: Two Active Forms. *Enzymologia*, 40(2), 145-150.

[14] Buttery, B. R., & Buzzell, R. I. (1971). Properties and Inheritance of Urease Isozymes in Soybean Seeds. *Canadian Journal Of Botany*, 49, 1101-1105.

[15] Polacco, J. C., & Havir, E. A. (1979). Comparisons of Soybean Urease Isolated from Seed and Tissue Culture. *Journal of Biological Chemistry*, 254(5), 1707-1715.

[16] Polacco, J. C., Thomas, A. L., & Bledsoe, P. J. (1982). A Soybean Seed Urease-Null Produces Urease in Cell Culture. *Plant Physiology*, 69(5), 1233-1240.

[17] Kerr, P. S., Blevins, D. G., Rapp, B. J., & Randall, D. D. (1983). Soybean Leaf Urease: A Comparison with Seed Urease. *Physiologia Plantarum*, 57, 339-345.

[18] Polacco, J. C., & Winkler, R. G. (1984). Soybean Leaf Urease: A Seed Enzyme? *Plant Physiology*, 74(4), 800-803.

[19] Polacco, J. C., Krueger, R. W., & Winkler, R. G. (1985). Structure and Possible Ureide Degrading Function of the Ubiquitous Urease of Soybean. *Plant Physiology*, 79(3), 794-800.

[20] Kloth, R. H., & Hymowitz, T. (1985). Re-Evaluation of the Inheritance of Urease in Soybean Seed. *Crop Science*, 25(2), 352-354.

[21] Winkler, R. G., Polacco, J. C., Eskew, D. L., & Welch, R. M. (1983). Nickel Is Not Required for Apourease Synthesis in Soybean Seeds. *Plant Physiology*, 72(1), 262-263.

[22] Krueger, R. W., Holland, MA, Chisholm, D., & Polacco, J. C. (1987). Recovery of a Soybean Urease Genomic Clone by Sequential Library Screening with Two Synthetic Oligodeoxynucleotides. *Gene*, 54(1), 41-50.

[23] Meyer-Bothling, L. E., Polacco, J. C., & Cianzio, S. R. (1987). Pleiotropic Soybean Mutants Defective in Both Urease Isozymes. *Molecular and general genetics*, 209(3), 432-438.

[24] Kloth, R. H., Polacco, J. C., & Hymowitz, T. (1987). The Inheritance of a Urease-Null Trait in Soybeans. *Theoretical and Applied Genetics*, 73(3), 410-418.

[25] Holland, M. A., Griffin, J. D., Meyerbothling, L. E., & Polacco, J. C. (1987). Developmental Genetics of the Soybean Urease Isozymes. *Developmental Genetics*, 8(5-6), 375-387.

[26] Polacco, J. C., Judd, A. K., Dybing, J. K., & Cianzio, S. R. (1989). A New Mutant Class of Soybean Lacks Urease in Leaves but Not in Leaf-Derived Callus or in Roots. *Molecular and General Genetics*, 217(2-3), 257-262.

[27] Eskew, D. L., Welch, R. M., & Cary, E. E. (1983). Nickel: An Essential Micronutrient for Legumes and Possibly All Higher Plants. *Science*, 222(4624), 621-623.

[28] Eskew, D. L., Welch, R. M., & Norvell, W. A. (1984). Nickel in Higher Plants: Further Evidence for an Essential Role. *Plant Physiology*, 76(3), 691-693.

[29] Holland, M. A., & Polacco, J. C. (1992). Urease-Null and Hydrogenase-Null Phenotypes of a Phylloplane Bacterium Reveal Altered Nickel Metabolism in Two Soybean Mutants. *Plant Physiology*, 98(3), 942-948.

[30] Polacco, J. C., Freyermuth, S. K., Gerendas, J., & Cianzio, S. R. (1999). Soybean Genes Involved in Nickel Insertion into Urease. *Journal of Experimental Botany*, 50(336), 1149-1156.

[31] Torisky, R. S., & Polacco, J. C. (1990). Soybean Roots Retain the Seed Urease Isozyme Synthesized During Embryo Development. *Plant Physiology*, 94(2), 681-689.

[32] Polacco, J. C., & Holland, M. A. (1993). Roles of Urease in Plant-Cells. *International Review of Cytology- a Survey of Cell Biology*, 145, 65-103.

[33] Torisky, R. S., Griffin, J. D., Yenofsky, R. L., & Polacco, J. C. (1994). A Single Gene (Eu4) Encodes the Tissue-Ubiquitous Urease of Soybean. *Molecular & General Genetics*, 242(4), 404-414.

[34] Goldraij, A., Beamer, L. J., & Polacco, J. C. (2003). Interallelic Complementation at the Ubiquitous Urease Coding Locus of Soybean. *Plant Physiology*, 132(4), 1801-1810.

[35] Schmutz, J., Cannon, S. B., Schlueter, J., Mitros, J., Nelson, T., Hyten, W., Song, D. L., Thelen, Q., Cheng, J. J., Xu, J., Hellsten, D., May, U., Yu, G. D., Sakurai, Y., Umezawa,

T., Bhattacharyya, T., Sandhu, M. K., Valliyodan, D., Lindquist, B., Peto, E., Grant, M., Shu, D., Goodstein, S., Barry, D., Futrell-Griggs, K., Abernathy, M., Du, B., Tian, J., Zhu, Z., Gill, L., Joshi, N., Libault, T., Sethuraman, M., Zhang, A., -C, X., Shinozaki, K., Nguyen, H. T., Wing, R. A., Cregan, P., Specht, J., Grimwood, J., Rokhsar, D., Stacey, G., Shoemaker, R. C., & Jackson, S. A. (2010). Genome Sequence of the Palaeopolyploid Soybean. *Nature*, 463(7278), 178-183.

[36] Stebbins, N., Holland, MA, Cianzio, S. R., & Polacco, J. C. (1991). Genetic Tests of the Roles of the Embryonic Ureases of Soybean. *Plant Physiology*, 97(3), 1004-1010.

[37] Witte, C. P. (2011). Urea Metabolism in Plants. *Plant Science*, 180(3), 431-438.

[38] Joint Genome Institute- University of California Regents. (2011). Phytozome. http://www.phytozome.net/.

[39] Bacanamwo, M., Witte, CP, Lubbers, M. W., & Polacco, J. C. (2002). Activation of the Urease of Schizosaccharomyces Pombe by the Uref Accessory Protein from Soybean. *Molecular Genetics and Genomics*, 268(4), 525-534.

[40] Polacco, J. C., Hyten, D. L., Medeiros-Silva, M., Sleper, D. A., & Bilyeu, K. D. (2011). Mutational Analysis of the Major Soybean Uref Paralogue Involved in Urease Activation. *Journal of Experimental Botany*, 62(10), 3599-3608.

[41] Freyermuth, S. K., Bacanamwo, M., & Polacco, J. C. (2000). The Soybean Eu3 Gene Encodes an Ni-Binding Protein Necessary for Urease Activity. *Plant Journal*, 21(1), 53-60.

[42] Kuchar, J., & Hausinger, R. P. (2004). Biosynthesis of Metal Sites. *Chemical Reviews*, 104(2), 509-525.

[43] Park, I. S., & Hausinger, R. P. (1995). Evidence for the Presence of Urease Apoprotein Complexes Containing Ured, Uref, and Ureg in Cells That Are Competent for in Vivo Enzyme Activation. *Journal of Bacteriology*, 177(8), 1947-1951.

[44] Soriano, A., Colpas, G. J., & Hausinger, R. P. (2000). Uree Stimulation of Gtp-Dependent Urease Activation in the Ured-Uref-Ureg-Urease Apoprotein Complex. *Biochemistry*, 39(40), 12435-12440.

[45] Moncrief, M. B., & Hausinger, R. P. (1997). Characterization of Ureg, Identification of a Ured-Uref-Ureg Complex, and Evidence Suggesting That a Nucleotide-Binding Site in Ureg Is Required for in Vivo Metallocenter Assembly of Klebsiella Aerogenes Urease. *Journal of Bacteriology*, 179(13), 4081-4086.

[46] Zambelli, B., Musiani, F., Benini, S., & Ciurli, S. (2011). Chemistry of Ni2+ in Urease: Sensing, Trafficking, and Catalysis. *Accounts of Chemical Research*, 44(7), 520-530.

[47] Salomone-Stagni, M., Zambelli, B., Musiani, F., & Ciurli, S. (2007). A Model-Based Proposal for the Role of Uref as a Gtpase-Activating Protein in the Urease Active Site Biosynthesis. *Proteins*, 68(3), 749-761.

[48] Zambelli, B., Stola, M., Musiani, F., De Vriendt, K., Samyn, B., Devreese, B., Van Beeumen, J., Turano, P., Dikiy, A., Bryant, D. A., & Ciurli, S. (2005). Ureg, a Chaperone in the Urease Assembly Process, Is an Intrinsically Unstructured Gtpase That Specifically Binds Zn2+. *Journal of Biological Chemistry*, 280(6), 4684-4695.

[49] Neyroz, P., Zambelli, B., & Ciurli, S. (2006). Intrinsically Disordered Structure of Bacillus Pasteurii Ureg as Revealed by Steady-State and Time-Resolved Fluorescence Spectroscopy. *Biochemistry*, 45(29), 8918-8930.

[50] Zambelli, B., Musiani, F., Savini, M., Tucker, P., & Ciurli, S. (2007). Biochemical Studies on Mycobacterium Tuberculosis Ureg and Comparative Modeling Reveal Structural and Functional Conservation among the Bacterial Ureg Family. *Biochemistry*, 46(11), 3171-3182.

[51] Zambelli, B., Turano, P., Musiani, F., Neyroz, P., & Ciurli, S. (2009). Zn2+-Linked Dimerization of Ureg from Helicobacter Pylori, a Chaperone Involved in Nickel Trafficking and Urease Activation. *Proteins*, 74(1), 222-239.

[52] Zambelli, B., Cremades, N., Neyroz, P., Turano, P., Uversky, V. N., & Ciurli, S. (2012). Insights in the (Un)Structural Organization of Bacillus Pasteurii Ureg, an Intrinsically Disordered Gtpase Enzyme. *Molecular BioSystems*, 8(1), 220-228.

[53] Real-Guerra, R., Staniscuaski, F., Zambelli, B., Musiani, F., Ciurli, S., & Carlini, C. R. (2012). Biochemical and Structural Studies on Native and Recombinant Glycine Max Ureg: A Detailed Characterization of a Plant Urease Accessory Protein. *Plant Molecular Biology*, 78(4-5), 461-475 .

[54] Soriano, A., & Hausinger, R. P. (1999). Gtp-Dependent Activation of Urease Apoprotein in Complex with the Ured, Uref, and Ureg Accessory Proteins. *Proceedings of the National Academy of Sciences of the United States of America*, 96(20), 11140-11144.

[55] Witte, C. P., Rosso, M. G., & Romeis, T. (2005). Identification of Three Urease Accessory Proteins That Are Required for Urease Activation in Arabidopsis. *Plant Physiology*, 139(3), 1155-1162.

[56] Cao, F. Q., Werner, A. K., Dahncke, K., Romeis, T., Liu, L. H., & Witte, C.P. (2010). Identification and Characterization of Proteins Involved in Rice Urea and Arginine Catabolism. *Plant Physiology*, 154(1), 98-108.

[57] Hirayama, C., Sugimura, M., Saito, H., & Nakamura, M. (2000). Purification and Properties of Urease from the Leaf of Mulberry, Morus Alba. *Phytochemistry*, 53(3), 325-330.

[58] Krishna, B. L., Singh, A. N., Patra, S., & Dubey, V. K. (2011). Purification, Characterization and Immobilization of Urease from Momordica Charantia Seeds. *Process Biochemistry*, 46-1486.

[59] Lee, M. H., Mulrooney, S. B., Renner, MJ, Markowicz, Y., & Hausinger, R. P. (1992). Klebsiella Aerogenes Urease Gene Cluster: Sequence of Ured and Demonstration

That Four Accessory Genes (Ured, Uree, Uref, and Ureg) Are Involved in Nickel Metallocenter Biosynthesis. *Journal of Bacteriology*, 174(13), 4324-4330.

[60] Park, I. S., Carr, M. B., & Hausinger, R. P. (1994). In Vitro Activation of Urease Apoprotein and Role of Ured as a Chaperone Required for Nickel Metallocenter Assembly. *Proceedings of the National Academy of Sciences of the United States of America*, 91(8), 3233-3237.

[61] Witte, C. P., Isidore, E., Tiller, S. A., Davies, H. V., & Taylor, M. A. (2001). Functional Characterisation of Urease Accessory Protein G (Ureg) from Potato. *Plant Molecular Biology*, 45(2), 169-179.

[62] Vitousek, P. M., & Howarth, R. W. (1991). Nitrogen Limitation on Land and in the Sea- How Can It Occur. *Biogeochemistry*, 13(2), 87-115.

[63] Elser, J. J., Fagan, W. F., Subramanian, S., & Kumar, S. (2006). Signatures of Ecological Resource Availability in the Animal and Plant Proteomes. *Molecular Biology and Evolution*, 23(10), 1946-1951.

[64] Polacco, J. C., & Holland, M. A. (1994). Genetic Control of Plant Ureases. *Genetic engineering; principles and methods*, 16-33.

[65] Sirko, A., & Brodzik, R. (2000). Plant Ureases: Roles and Regulation. *Acta Biochimica Polonica*, 47(4), 1189-1195.

[66] Lam, H. M., Coschigano, K. T., Oliveira, I. C., Melo, Oliveira. R., & Coruzzi, G. M. (1996). The Molecular-Genetics of Nitrogen Assimilation into Amino Acids in Higher Plants. *Annual Review of Plant Physiology and Plant Molecular Biology*, 47, 569-593.

[67] Witte, C. P., & Medina-Escobar, N. (2001). In-Gel Detection of Urease with Nitroblue Tetrazolium and Quantification of the Enzyme from Different Crop Plants Using the Indophenol Reaction. *Analytical Biochemistry*, 290(1), 102-107.

[68] Liu, L. H., Ludewig, U., Frommer, W. B., & von, Wiren. N. (2003). Atdur3 Encodes a New Type of High-Affinity Urea/H+ Symporter in Arabidopsis. *The Plant Cell*, 15(3), 790-800.

[69] Food and Agriculture Organization of the United States. (2011). Faostat. http://faostat.fao.org.

[70] Stebbins, N. E., & Polacco, J. C. (1995). Urease Is Not Essential for Ureide Degradation in Soybean. *Plant Physiology*, 109(1), 169-175.

[71] Witte, C. P., Tiller, S. A., Taylor, M. A., & Davies, H. V. (2002). Leaf Urea Metabolism in Potato. Urease Activity Profile and Patterns of Recovery and Distribution of (15)N after Foliar Urea Application in Wild-Type and Urease-Antisense Transgenics. *Plant Physiology*, 128(3), 1129-1136.

[72] Polacco, J. C., & Sparks, R. B. (1982). Patterns of Urease Synthesis in Developing Soybeans. *Plant Physiology*, 70(1), 189-194.

[73] Carlini, C. R., & Polacco, J. C. (2008). Toxic Properties of Urease. *Crop Science*, 48(5), 1665-1672.

[74] Carlini, C. R., Oliveira, A. E., Azambuja, P., Xavier-Filho, J., & Wells, MA. (1997). Biological Effects of Canatoxin in Different Insect Models: Evidence for a Proteolytic Activation of the Toxin by Insect Cathepsinlike Enzymes. *Journal of Economic Entomology*, 90(2), 340-348.

[75] Follmer, C., Real-Guerra, R., Wasserman, G. E., Olivera-Severo, D., & Carlini, C. R. (2004). Jackbean, Soybean and Bacillus Pasteurii Ureases. Biological Effects Unrelated to Ureolytic Activity. European Journal of Biochemistry , 271(7), 1357-1363.

[76] Staniscuaski, F., Ferreira-Dasilva, C. T., Mulinari, F., Pires-Alves, M., & Carlini, C. R. (2005). Insecticidal Effects of Canatoxin on the Cotton Stainer Bug Dysdercus Peruvianus (Hemiptera: Pyrrhocoridae). *Toxicon*, 45(6), 753-760.

[77] Defferrari, M. S., Demartini, D. R., Marcelino, T. B., Pinto, P. M., & Carlini, C. R. (2011). Insecticidal Effect of Canavalia Ensiformis Major Urease on Nymphs of the Milkweed Bug Oncopeltus Fasciatus and Characterization of Digestive Peptidases. *Insect Biochemistry and Molecular Biology*, 41(6), 388-399.

[78] Oliveira, A. E. A., Gomes, V. M., Sales, M. P., Fernandes, K. V. S., Carlini, C. R., & Xavier-Filho, J. (1999). The Toxicity of Jack Bean [Canavalia Ensiformis (L.) Dc.] Canatoxin to Plant Pathogenic Fungi. *Revista Brasileira de Biologia*, 59(1), 59-62.

[79] Becker-Ritt, A. B., Martinelli, A. H., Mitidieri, S., Feder, V., Wassermann, G. E., Santi, L., Vainstein, M. H., Oliveira, J. T., Fiuza, L. M., Pasquali, G., & Carlini, C. R. (2007). Antifungal Activity of Plant and Bacterial Ureases. *Toxicon*, 50(7), 971-983.

[80] Ghazaleh, F. A., Francischetti, I. M. B., Gombarovits, M. E. C., & Carlini, C. R. (1997). Stimulation of Calcium Influx and Platelet Activation by Canatoxin: Methoxyverapamil Inhibition and Downregulation by Cgmp. *Archives of Biochemistry and Biophysics*, 339(2), 362-367.

[81] Carlini, C. R., & Grossi-de-Sa, M. F. (2002). Plant Toxic Proteins with Insecticidal Properties. A Review on Their Potentialities as Bioinsecticides. *Toxicon*, 40(11), 1515-1539.

[82] Piovesan, A. R., Staniscuaski, F., Marco-Salvadori, J., Real-Guerra, R., Defferrari, M. S., & Carlini, C. R. (2008). Stage-Specific Gut Proteinases of the Cotton Stainer Bug Dysdercus Peruvianus: Role in the Release of Entomotoxic Peptides from Canavalia Ensiformis Urease. *Insect Biochemistry and Molecular Biology*, 38(11), 1023-1032.

[83] Staniscuaski, F., Te, Brugge. V., Carlini, C. R., & Orchard, I. (2010). Jack Bean Urease Alters Serotonin-Induced Effects on Rhodnius Prolixus Anterior Midgut. *Journal of Insect Physiology*, 56(9), 1078-1086.

[84] Menegassi, A., Wassermann, G. E., Olivera-Severo, D., Becker-Ritt, A. B., Martinelli, A. H., Feder, V., & Carlini, C. R. (2008). Urease from Cotton (Gossypium Hirsutum)

Seeds: Isolation, Physicochemical Characterization, and Antifungal Properties of the Protein. *Journal of Agriculture and Food Science*, 56(12), 4399-4405.

[85] Carlini, C. R., Guimaraes, J. A., & Ribeiro, J. M. (1985). Platelet Release Reaction and Aggregation Induced by Canatoxin, a Convulsant Protein: Evidence for the Involvement of the Platelet Lipoxygenase Pathway. *British Journal of Pharmacology*, 84(2), 551-560.

[86] Barja-Fidalgo, C., Guimaraes, J. A., & Carlini, C. R. (1991). Lipoxygenase-Mediated Secretory Effect of Canatoxin the Toxic Protein from Canavalia Ensiformis Seeds. *Toxicon*, 29(4-5), 453-459.

[87] Barja-Fidalgo, C., Guimaraes, J. A., & Carlini, C. R. (1991). Canatoxin, a Plant Protein, Induces Insulin Release from Isolated Pancreatic Islets. *Endocrinology*, 128(2), 675-679.

[88] Grassi-Kassisse, D. M., & Ribeiro-Da, Silva. G. (1992). Canatoxin Triggers Histamine Secretion from Rat Peritoneal Mast Cells. *Agents Actions*, 37(3-4), 204-209.

[89] Ferreira-Da, Silva. C. T., Gombarovits, ME, Masuda, H., Oliveira, C. M., & Carlini, C. R. (2000). Proteolytic Activation of Canatoxin, a Plant Toxic Protein, by Insect Cathepsin-Like Enzymes. *Archives of Insect Biochemistry and Physiology*, 44(4), 162-171.

[90] Mulinari, F., Staniscuaski, F., Bertholdo-Vargas, L. R., Postal, M., Oliveira-Neto, O. B., Rigden, D. J., Grossi-de-Sa, M. F., & Carlini, C. R. (2007). Jaburetox-2ec: An Insecticidal Peptide Derived from an Isoform of Urease from the Plant Canavalia Ensiformis. *Peptides*, 28(10), 2042-2050.

[91] Staniscuaski, F., Brugge, V. T., Carlini, C. R., & Orchard, I. (2009). In Vitro Effect of Canavalia Ensiformis Urease and the Derived Peptide Jaburetox-2ec on Rhodnius Prolixus Malpighian Tubules. *Journal of Insect Physiology*, 55(3), 255-263.

[92] Staniscuaski, F., & Carlini, C. R. (2012). Plant Ureases and Related Peptides: Understanding Their Entomotoxic Properties. *Toxins Basel*, 4(2), 55-67.

[93] Wiebke-Strohm, B., Pasquali, G., Margis-Pinheiro, M., Bencke, M., Buecker-Neto, L., Becker-Ritt, A. B., Martinelli, A. H. S., Rechenmacher, C., Polacco, J. C., Stolf, R., Marcelino, F. C., Abdelnoor, R. V., Homrich Del, M.S, Ponte, . E. M., Carlini, C. R., De Carvalho, M. C. C. G., & Bodanese-Zanettini, M. H. (2012). Ubiquitous Urease Affects Soybean Susceptibility to Fungi. *Plant Molecular Biology*, 79(1-2), 75-87.

[94] Olivera-Severo, D., Wassermann, G. E., & Carlini, C. R. (2006). Bacillus Pasteurii Urease Shares with Plant Ureases the Ability to Induce Aggregation of Blood Platelets. *Archives of Biochemistry and Biophysics*, 452(2), 149-155.

[95] Wassermann, G. E., Olivera-Severo, D., Uberti, A. F., & Carlini, C. R. (2010). Helicobacter Pylori Urease Activates Blood Platelets through a Lipoxygenase-Mediated Pathway. *Journal of Cellular and Molecular Medicine*, 14(7), 2025-2034.

[96] American Soybean Association. Soystats. (2011). http://www.soystats.com/.

[97] Cromwell, G. L. (1999). Soybean Meal- the "Gold Standard". *The Farmer's Pride*, 11(20).

[98] Liener, I. E. (1994). Implications of Antinutritional Components in Soybean Foods. *Critical Reviews in Food Science and Nutrition*, 34(1), 31-67.

[99] Gandhi, A. P., Nenwani, M. M., & Ali, N. (1984). Investigations on the Trypsin-Inhibitor, Urease and Cooking Behavior of Soybean Glycine-Max Merr. *Food Chemistry*, 15(3), 215-218.

[100] Vasconcelos, I. M., Siebra, E. A., Maia, A. A. B., Moreira, R. A., Neto, A. F., Campelo, G. J. A., & Oliveira, J. T. A. (1997). Composition, Toxic and Antinutritional Factors of Newly Developed Cultivars of Brazilian Soybean (Glycine Max). *Journal of the Science of Food and Agriculture*, 75(4), 419-426.

[101] Vasconcelos, I. M., Maia, A. A., Siebra, E. A., Oliveira, J. T., Carvalho, A. F., Melo, V. M., Carlini, C. R., & Castelar, L. I. (2001). Nutritional Study of Two Brazilian Soybean (Glycine Max) Cultivars Differing in the Contents of Antinutritional and Toxic Proteins. *The Journal of Nutritional Biochemistry*, 12(1), 55-62.

[102] Becker-Ritt, A. B., Mulinari, F., Vasconcelos, I. M., & Carlini, C. R. (2004). Antinutritional and/or Toxic Factors in Soybean (Glycine Max (L) Merril) Seeds: Comparison of Different Cultivars Adapted to the Southern Region of Brazil. *Journal of the Science of Food and Agriculture*, 84(3), 263-270.

[103] Food and Agriculture Organization of the United States. (2011). Fao. http://www.fao.org/.

[104] Huntington, G. B., Harmon, D. L., Kristensen, N. B., Hanson, K. C., & Spears, J. W. (2006). Effects of a Slow-Release Urea Source on Absorption of Ammonia and Endogenous Production of Urea by Cattle. *Animal Feed Science and Technology*, 130(3-4), 225-241.

[105] Symonds, H. W., Mather, D. L., & Collis, K. A. (1981). The Maximum Capacity of the Liver of the Adult Dairy-Cow to Metabolize Ammonia. *British Journal of Nutrition*, 46(3), 481-486.

[106] Edwards, H. M. Jr. (1985). Effects of Different Soybean Meals on the Incidence of Tibial Dyschondroplasia in the Chicken. *Journal of Nutrition*, 115(8), 1005-1015.

[107] Veltmann, J. R., & Jensen, L. S. (1981). Tibial Dyschondroplasia in Broilers- Comparison of Dietary Additives and Strains. *Poultry Science*, 60(7), 1473-1478.

[108] Edwards, H. M. (1984). Studies on the Etiology of Tibial Dyschondroplasia in Chickens. *Journal of Nutrition*, 114(6), 1001-1013.

[109] Qin, G., Elstter, E. R., Bosch, M. W., & Poel vander, A. F. B. (1996). Thermal Processing of Whole Soya Beans: Studies on the Inactivation of Antinutritional Factors and Effects on Ileal Digestibility in Piglets. *Animal Feed Science and Technology*, 57(4), 313-324.

[110] Quedraogo, C. L., Combe, E., Lalles, J. P., Toullec, R., Treche, S., & Grongnet, J. F. (1999). Nutritional Value of the Proteins of Soybeans Roasted at a Small-Scale Unit Level in Africa as Assessed Using Growing Rats. *Reproduction Nutrition Development*, 39(2), 201-212.

[111] Vasconcelos, I. M., Brasil, I. C., Oliveira, J. T., Campello, C. C., Maia, F. M., Campello, M. V., Farias, D. F., & Carvalho, A. F. (2009). Combination of Chemical Analyses and Animal Feeding Trials as Reliable Procedures to Assess the Safety of Heat Processed Soybean Seeds. *Journal of Agriculture and Food Science*, 57(11), 4668-4673.

[112] Caskey, C. D., & Knapp, F. C. (1944). Method for Detecting Inadequately Heated Soybean Oil Meal. *Industrial and Engineering Chemistry*, 16(10), 640-641.

[113] Schramm, G., & Aines, P. D. (1959). Colorimetric Determination of Urease Activity in Soybean Meals. *Journal of the American Oil Chemists Society*, 36(1), 1-3.

[114] Croston, C. B., Smith, A. K., & Cowan, J. C. (1955). Measurement of Urease Activity in Soybean Oil Meal. *Journal of the American Oil Chemists Society*, 32(5), 279-282.

[115] Smith, A. K., Belter, P. A., & Anderson, R. L. (1956). Urease Activity in Soybean Meal Products. *Journal of the American Oil Chemists Society*, 33(8), 360-363.

[116] Weatherburn, M. W. (1967). Phenol-Hypochlorite Reaction for Determination of Ammonia. *Analytical Chemistry*, 39(8), 971.

[117] Kaplan, A. (1969). The Determination of Urea, Ammonia, and Urease. *Methods of Biochemical Analysis*, 17, 311-324.

[118] Grossi-de-Sa, M. F., Pelegrini, P. B., & Fragoso, R. R. (2011). Genetically Modified Soybean for Insect-Pests and Disease Control. In: Sudaric A (ed.) Soybean- Molecular Aspects of Breeding. *Rijeka: InTech*, 429-452.

[119] Oerke, E. C., & Dehne, H. W. (2004). Safeguarding Production- Losses in Major Crops and the Role of Crop Protection. *Crop Protection*, 23(4), 275-285.

[120] Polacco, J. C. (1976). Nitrogen Metabolism in Soybean Tissue Culture: I. Assimilation of Urea. *Plant Physiology*, 58(3), 350-357.

[121] George, A. A., & Delumen, B. O. (1991). A Novel Methionine-Rich Protein in Soybean Seed- Identification, Amino-Acid-Composition, and N-Terminal Sequence. *Journal of Agriculture and Food Science*, 39(1), 224-227.

Chapter 16

Explanations for the Rise of Soybean in Brazil

Eduardo Antonio Gavioli

Additional information is available at the end of the chapter

http://dx.doi.org/10.5772/51678

1. Introduction

1.1. Historical developments of Soybean Culture

Early reports indicate that the soybean culture came to Brazil around 1882, to the state of Bahia where the Teacher Gustavo Dutra conducted the first studies to evaluate cultivars. Between 1900 and 1901, the Agronomic Institute (IAC), in Campinas, São Paulo state, promoted the first distribution of soybean seeds to producers in the state.

In this same period, soybean culture reached the state of "Rio Grande do Sul", where climatic conditions are similar to the southern U.S. The original introductory varieties came from the southern U.S. Soybean production first occurred in the city of Santa Rosa in the state of "Rio Grande do Sul", in 1914 [1]. Research began during the 1930's with the following breeding objectives: increased productivity; greater plant height and appropriate pod height to facilitate mechanization; development of lodging and pest resistance, and increased seed quality with high oil yield and protein [3].

The first national farm statistics, the Agricultural Yearbook of the "Rio Grande do Sul" state, was published in 1941 and indicated that the production area of only 640 hectares generated 450 tons. By 1949, the production in Brazil had grown to 25,000 tons. This was the first time that Brazilian production figures were recorded in international statistics. By 1970, soybean production had spread throughout the temperate and sub-tropical latitudes (near or above the 30ºS Lat).

During the 1970's, soybean was established as the main crop of the Brazilian agribusiness, rising from 15 million metric tons to over 15million metric tons. There was an increase of the cultivated areafrom 1.3 to 8.8 million hectares, and of the productivity from 1.14 to 1.73 t / ha. Soybean culture was concentrated in the southern region of the country, with more than 80% of the total production [1].

Several factors contributed to the establishment and development of the soybean culture in southern Brazil: similarity with southern U.S, the region from which soybeans were introduced to Brazil; introduction of liming and correction of soil fertility; tax incentives; increased use of vegetable oil vs. animal fats; establishment of a significant industrial soybean processing infrastructure; crop mechanization; emergence of dynamic and efficient cooperatives; establishment of a well-coordinated network of research; and improvements in roads, ports and communications [1].

Another important factor that explains the rise of soybean production in southern Brazil was that prevailing photoperiod and temperature characteristics directly influenced phenological development and yield. Spread of soybeans out of this region to central and northern Brazil and lower latitudes would depend on the development of new phonologically adapted varieties to these areas [2].

Brazil is located in the eastern part of South America between 5°16' N and 33º44' S latitudes (Figure 1). Photoperiod differences between the southern and northern portions of the country have inhibited the expansion of soybean from the original southern base to central and northern Brazil.

Figure 1. Map of Brazil, with its divisions into states, positioned in the South American continent between the parallels of 5°16' N and 33º44' S latitudes.

Research development of varieties adapted to these new areas of production began at the Agronomic Institute (IAC) in Campinas, "São Paulo" State, and the National Center for Soybean Research. In the 1970s, breeding studies were initiated from crosses of North American cultivars, which had the long juvenile trait. Thousands of genotypes are maintained in the Embrapa soybean (CNPSo) germplasm bank in Londrina, "Paraná" State. The Brazilian Agricultural Research Corporation (Embrapa) is under the Ministry of Agriculture. Its mission is to facilitate solutions for research, development and innovation for sustainable agriculture for the benefit of the Brazilian society. The National Center of Soybean Research (CNPSo) is a unit of Embrapa.

The entire germplasm collection is maintained at the Embrapa Genetic Resources in Brasília (Brazil's capital). Most of these accessions are plant introductions from North America which are derived from original plant introductions from China, Japan and other countries with wide genetic diversification [4].

1.2. Expansion of Culture

Because of Brazil's research efforts, developed cultivars were adapted to the short photoperiods of central and northern Brazil. This allowed the soybean expansion to the Cerrado region of Brazil, an area of more than 200 million hectares of undeveloped potential crop land. Brazil is now the second biggest soybean producer in the world with an average yield close to 3,000 kg/ha.

The world and Brazilian production, supply, and trade of soybeans are presented in Tables 1 and 2, respectively, from the 2006/2007 to 2009/2010 growing seasons. Brazil is established as the second largest producer, only behind the United States, and now contributes with about 33% of the world's exported soybean. In the case of end-of-year stocks, the Brazilian product represented approximately 29% of the world supply. During this four-year period, Brazil produced an annual average of 232,550 million metric tons of soybeans, which accounted for 26.50% ofthe world production (61.70 million tons).

Balance	2006/2007	2007/2008	2008/2009	2009/2010
Initial stock	53.206	63.155	52.886	43.973
Production	237.126	221.006	211.964	259.896
Imports	69.066	78.118	77.168	86.661
Total supply	359.398	362.279	342.018	390.530
Exports	71.315	79.589	77.253	89.575
Consumer	224.928	229.804	220.792	237.435
Final stock	63.155	52.886	43.973	63.520

Table 1. World balance of supply and demand for soybeans in thousands of metric tons.Source: USDA. Preparation: [5].

In the 2010/2011 season, Brazilian soybean production rose to 75.0 million metric tons, covering a cultivated area of 24.2 million hectares and an average yield of 3,106 kg/ha. "Mato Grosso" was the state with the greatest production (20.4 million metric tons), cultivated area (6.4 million hectares), and yield (3,190 kg/ha)."Paraná" state was the second largest producer, with production of 15.4 million metric tons, cultivated area of 4.6 million ha and yield of 3,360 kg/ha [6].

Balance	2006/2007	2007/2008	2008/2009	2009/2010
Initial stock	16.641	18.189	18.898	12.037
Production	59.000	61.000	57.800	69.000
Imports	53.000	150.000	44.000	185.000
Total supply	75.694	79.339	76.742	81.222
Exports	23.485	25.364	29.986	28.350
Consumer	34.020	35.077	34.719	36.175
Final stock	18.189	18.898	12.037	16.697

Table 2. Brazilian balance of supply and demand for soybeansin thousands of metric tons.Source: USDA. Preparation: [5].

In the most recent growing season (2011/2012), Brazil produced around 66.5 million tons froman area of 24.7 million hectares, with an average yield of 2,692 kg/ha [6]. The decrease on the production was due to a severe drought that occurred throughout the soybean-growing region. Soybean occupied about 48.9% of Brazil's cropland in that growing season and contributed to about 42% of the country's agricultural output. The central states of"Mato Grosso, Mato Grosso do Sul and Goiás", with more than 10 million hectares, represent approximately half of all Brazil's soybean cultivated area.

Many factors contributed to the establishment of soybean, firstly in southern Brazil (in the 1960's and 1970's) and later in the Cerrado region of central Brazil (in the 1980's and 1990's) [1]. With respect to the central region of Brazil, the wide and rapid soybean acceptance can be attributed to:

Transfer of the national capital from Rio de Janeiro to Brazil's interior in the 1960's, what resulted in a great deal of infrastructure construction including roads, communication, and economic development;

Tax incentives were made available for open new areas of agricultural production, as well as for the acquisition of machinery and construction of silos and warehouses;

Establishment of agro-industries in the region, stimulated by the tax incentives that expanded the agricultural frontier;

Low land values in the central region compared to the southern region during the 1960-1980 period, encouraging the purchase of new farms;

The relatively plain topography of the Cerrado, which was highly favorable to agricultural mechanization;

Good physical soil condition in the region; Improvements in the transport system, such as road, railway, and water transport that aided the marketing of soybean and other crops grown in the region;

New farmers, from southern Brazil, moving into the area, who already had a high technical knowledge of soybean production;

Rainfall in the region, which is highly favorable for summer crops, in contrast to the frequent dry spells occurring in the South, notably in "Rio Grande do Sul" [1].

Currently, soybean (*Glycine max* (L.) Merrill) is cultivated in almost all regions of the country. The biggest innovation that has aided the extension of soybean production across this vast north/south expanse was the development of the long juvenile trait.

The creation of CNPSo - Embrapa in 1975 consolidated the soybean research and greatly enhanced its production and quality. The transfer of this new technology to the farmers aided the expansion of soybean in Brazil. During the 1980's and 1990's soybean expanded again, this time into the tropical region in central Brazil.

From only 2% of the national production in 1970, soybean production in Mato Grosso in central Brazil expanded to 20% of the production in 1980, and then to 40% and 58% in 1990 and 2002, respectively. This transformation has promoted the state of "Mato Grosso" to the national leader on the production and yield of soybean [1].

Data from the Brazilian Institute of Geography and Statistics (IBGE) demonstrates the expansion of cultivation into the interior of the country in the period between 1976 and 2012 (Figure 2).

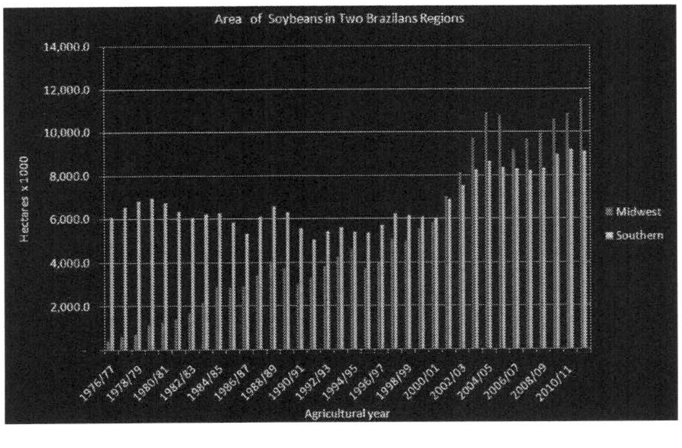

Figure 2. Brazilian soybean area in two regions in the period between 1976 and 2012. Source: [6].

Data for soybean production in two regions are shown in Figure 3 for the period between 1976 and 2012. The production increase in central Brazil made it the largest soybean producer in the country.

The southern region, which consists of "Rio Grande do Sul", "Paraná", and "Santa Catarina", is now the second biggest producer. The stunning increase in soybean production during this period is similar to the rise of sugar cane during the colonial period and the rise of coffee during the Empire era.

The explosive growth of soybean production in Brazil (30 fold increase across a 30-year period) has profoundly changed the Brazilian agriculture. It has boosted farming activities; modernization of the transport system; expansion of the agricultural frontier; professionalization and expansion of the international trade; modification and enrichment of the Brazilian diet; acceleration of the country'surbanization;and population movement from coastal to the interior areas [1].

By the 2010/2011 growing season, soybean production had reached the equatorial region of northern Brazil. Thus, a crop that originally was grown only in southern Brazil became well established in central Brazil and continued to advance into northern Brazil.

Data in Figure 4 shows that 82% of production come from the states of "Mato Grosso", "Paraná", "Rio Grande do Sul' and "Goiás".

However, 13% of soybean production come from the northern state "Tocantins" and the northeast states "Maranhão", "Piauí" and"Bahia".

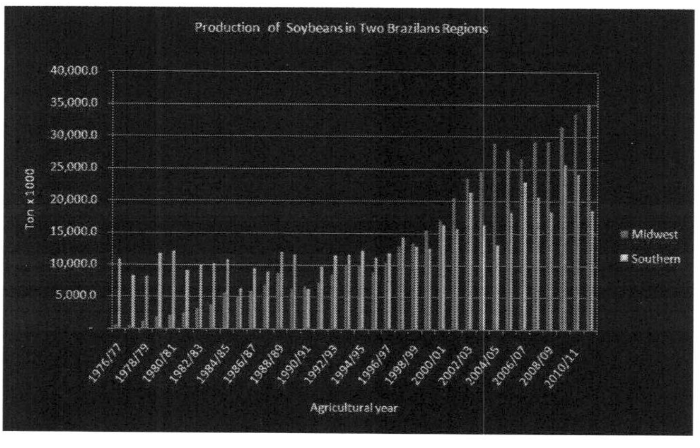

Figure 3. Brazilian soybean production in two regions, in the period between 1976 and 2012. Source: [6].

http://dx.doi.org/10.5772/51678

Figure 4. Brazilian soybean production in 2010/2011 harvest, divided among the states with the highest percentage of harvested grains. Source: [7].

2. Flowering of soybean

2.1. Photoperiod and Photoperiodism

Soybean culture is sensible to photoperiod and temperature and, due to the great diversity among cultivars, problems of adaptation to certain areas may occur [2]. In environments with constant photoperiods, temperature greatly influences flowering time [8]. There is an inverse relationship between temperature and the average number of days to first flower [9]. Days to first flowering are minor when over night temperatures range from 21 to 27°C. As temperatures fall below this range, first flowering is delayed. Above 27°C, flowering is largely inhibited [10].

The length of a day is known as photoperiod and plant developmental responses (i.e. phenology) to photoperiod are called photoperiodism [11]. Photoperiod affects not only days to first flowering, but also lengths of subsequent developmental stages. Variations in the day length are determined by latitude and planting date and both affect photoperiod becauseof tilting of the earth's axis. Plants may respond differently to these changes, as they cause modifications on some processes such as seed germination, inhibition of stem elongation, synthesis of chlorophyll and anthocyanin, leaf expansion, flowering and tuberization. The process by which light regulates plant development is called photomorphogenesis [12].

Soybean is strongly influenced by photoperiod so the culture grows and develops according to which photoperiod is subjected. Soybean is classified as a quantitative short-day plant, which means developmental timing is greatly speeded up and reproductive growth enhanced when day length falls below a critical level. This critical level (called the critical photoperiod) differs with cultivar and maturity group. Soybean responds differently to day length. Differences were observed during the flowering period of soybean when grown on different dates. The discovery of the importance of photoperiod on soybean flowering enabled the soybean classification as short-day plants [8].

Another important concept is the meaning of critical photoperiod, which is related to the quantity of light hours that can cause flowering. The time interval, in number of days between emergence and flowering, is influenced by temperature and photoperiod. There is a limit of the short-day length necessary to induce or to stop flowering. This period is characterized as critical photoperiod [13]. The length of the critical photoperiod also varies among soybean cultivars [14].

The soybean, classified as a short-day plant, only flowers, or flowers more rapidly, when the number of light hours does not exceed the critical period of the considered cultivar for each 24-hour cycle [15].

When undergoing photoperiodic induction, leaf buds are transformed into flower buds. The development of the flower primordial in the Biloxi cultivar started under short days and the flowers opened after three weeks, showing that there is a period between the received induction and anthesis [16]. The authors concluded that initiation of floral induction in soybean occurred with the expansion of the first primary leaves. Further research with Biloxi demonstrated that floral buds were initiated when there was a minimum night length of 10 and a half hours along two or three consecutive photoperiods [17].

When soybean is grown in its adapted area, floral initiation occurs approximately three weeks after germination. Flowering will occur three to five weeks later. Thus, there is a period of approximately three weeks between both developmental stages [18]. Therefore, days to first flower can range from 45 to 50 days depending on the prevailing photoperiod/ temperature and genotype. The minimal period for optimal yield is 45 days from emergence to first flower [19].

Floral induction occurs during the night. It is determined by the duration of darkness and not the number of light hours. This has been demonstrated by studies in which flowering occurred as a result of changes in the night length but not in the day length; and by other studies in which interruption of night time by light breaks altered the flowering response [20]. For example, soybean flowered under either short or long days, as long as nights were short [21].

Several researches in soybean characterized influences of photoperiod in the sub period between emergence and flowering plants [22-25].

Once soybean ends the juvenile phase and is able to perceive the stimulus, and photoperiod conditions are inductive, the plant enters the inductive phase and vegetative meristems change and start producing floral primordials. After the inductive phase is complete and floral organogenesis starts, the plant is in the post-inductive phase until flowering occurs [26].

The sub-period between emergence and when soybean responds to the photoperiod stimulus is termed as the juvenile phase. Recent research showed that soybean has little sensitivity to photoperiod during the juvenile phase [27].

The lengths of such sub-periods are determined by the degree of photoperiod sensitivity of the genotype. Thus, under long days and/or low temperatures, the rate of floral induction and flower development is minor. Developmental rate is important for yield determination, because if the plant develops too rapidly towards first flowering and seed initiation, there will not be enough time to build enough dry matter for optimal yield. Vegetative dry matter accumulation stops at the start of seed filling [26].

2.2. Phytochrome

Promotion or inhibition of the rate of phenological development in soybean is regulated by the phytochrome pigment in the plant. This has been amply demonstrated by night-break studies in which the effect of a long night (or short day) on promotion of flowering is inhibited when a light flash is given early in the night period [28]. Period from emergence to first flower is not only controlled by this mechanism, but also by the rate of phenological development for later reproductive periods [14].

Phytochromeis a blue pigment consisting of an apoprotein, which in turn is connected to a tetrapyrrole fitocromobilina, which serves as achromophore. The chromophore is synthesized in the plastid and is the unused portion of the phytochrome protein responsible for light absorption. The combination of the chromophore with the apoprotein occurs in the cytoplasm.

The phytochrome is found throughout the plant, but the highest concentration is found in the apical meristem of the stem. This is a plant pigment associated with membranes. The phytochrome molecule has two forms, one more stable and inactive and other more unstable and active, working to activate or inactivate reactions, respectively.

Both forms can be transformed into one and another. One form of the phytochrome pigment absorbs far red light (Pfr) at a wavelength of about 730 nm, while the other form of the pigment (Pr) absorbs light in the red range of about 660 nm (Figure 5).

During the day, plants have both forms, with a predominance of Pfr since normal daylight typically has a ratio of red/far red light of about 1.20. During the night, the Pfr form spontaneously converts into Pr. This reversal is essential for the measurement of time by plants as it determines how phenological developmental rate is affected.

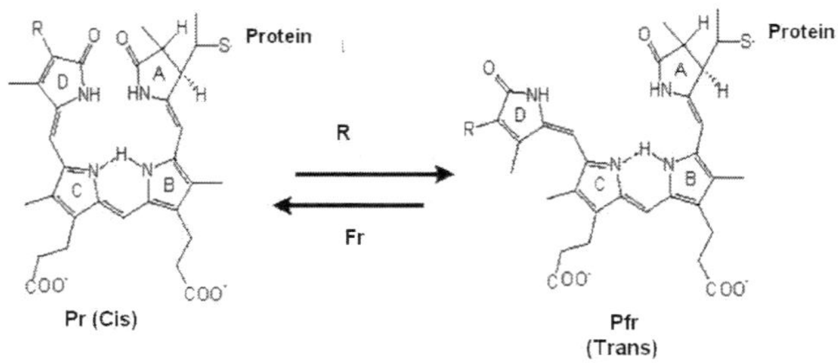

Figure 5. Photoisomerization between C and D rings of the chromophore. The absorption of red for Pr, resulting in the change of the ring D of the cis form (inactive) to the trans form (active) characteristic of Pfr. The protein bound to the chromophore is also changed in its shape.

Temperatures during the night affect the rate of this dark reversion of Pfr to Pr [29]. Application of a red flash of light in the night period inhibits the dark reversion of Pfr to Pr and prevents the effect on the developmental rate induced by normal dark reversion of Pfr to Pr.

Research in the late 1980's identified genes in the *Arabidopsis thaliana* plant that are related to the phytochrome encoding. Five phytochrome genes were isolated from this species: PHYA, PHYB, PHYC, PHYD and PHYE that encode the PHYA, PHYB, PHYC, PHYD and PHYE apoproteins. These proteins constitute the chromophore of the phytochrome [30].

In tomato plants (*Lycopersicumesculentum* Mill.) five genes that encode apoproteins were identified: PHYA, PHYB1, PHYB2, PHYE and PHYF [31]. When a phytochrome has the PHYA apoprotein, it is called type 1 phytochrome. All others are called type 2 phytochromes.

The difference between the two types is that the first one is accumulated mainly in plants grown in the dark and is easily degraded by light. The mechanisms that contribute to the abundance of the type 1 phytochrome in the dark is that the PHYA gene is preferentially transcribed under these conditions and its expression is inhibited by light [32].

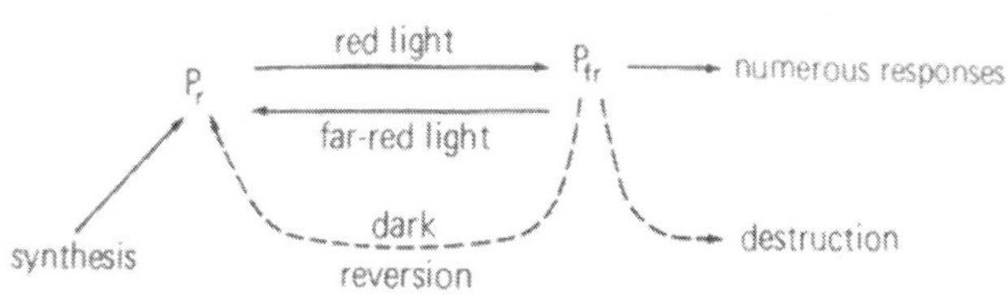

Figure 6. A summary of some transformations of phytochrome. Dashed lines indicating dark reversion and destruction do not seem to occur with type 2 Pfr molecules [32].

The phytochrome forms Pr and Pfr interconvert as shown in Fig. 5. (type 1 phytochrome). The second type of phytochrome may be more stable under conditions of darkness [32].The

http://dx.doi.org/10.5772/51678

mode of action of photoreceptors in the photomorphogenesis process is still unknown [32]. There are two hypothesis:

1. The photoperiodic response is pereceived in the leaf and has a rapid effect on plasma membrane permeability, which sends the flowering response to the apes of the stem.

2. Reduced effect on gene expression.

2.3. Gibberellin

The conversion of Pfrphytochrome to Pr occurs slowly under absence of light. In this condition, the synthesis of the enzyme gibberellin 20 oxidase and 3β-hydroxylase is reduced. They are responsible for turning gibberellin 12 (20 carbons) to gibberellin 1 (19 carbons). Under longer periods of darkness, the following occurs: low concentration of the far red phytochrome; reduced synthesis of gibberellin 20 oxidase and 3β-hydroxylase; higher concentration of gibberellin 12 and a lower concentration of gibberellin 1. This low concentration of gibberellin 1 is responsible for flowering in soybean [33]. The authors described the steps related to Figure 7 as follows:

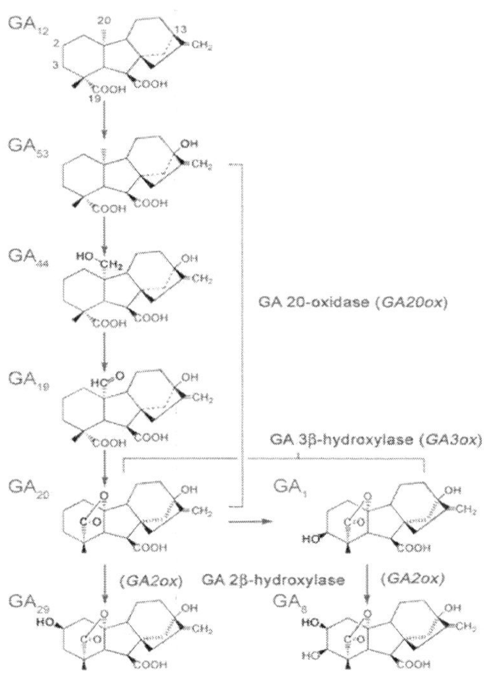

Figure 7. Pathway responsible for the production of the pea plants in GA_1 [33].

GAs in pea pericarps (ovaries) are synthesized mainly via the early 13-hydroxylation pathway. GA_{12} is a 13-hydroxylated to GA_{53}; Carbon 20 (noted as 20 in the figure) is sequentially oxidized by a GA 20-oxidase from GA_{53} to GA_{44}, to GA_{19}, and finally to GA_{20}. GA_{20} is then oxidized by a 3β-hydroxylase to GA_1 (a growth-active GA). Both GA_{20} and GA_1 can be oxidized by a 2β-hydroxylase to GA_{29} and GA_8, respectively. The latter conversion inactivates GA_1. In Figure 8, constructed from previous data [34, 35], an increase occurs in the levels of GA_1 gibberellin in spinach plants submitted to long days.

Morphologically, the end of the juvenile period occurs when soybean becomes responsive to photoperiodically-induced reproductive growth. The internal metabolism which leads to plant blooming seems to be influenced by several factors such as concentration of carbohydrates and gibberellin. It is difficult to exactly determine what regulates this stage of plant development [33]. Studies on maize plants indicated that gibberellin-deficient mutants showed a delayed transition from the juvenile to adult stage. This fact may be associated with a long juvenile period. The application of endogenous gibberellin regulated time in this transition phase [36]. Juvenile plants cannot be induced to flower even under appropriate photoperiod. At this time, the buds of the apical meristem do not respond to the floral stimulus, or the young leaves cannot produce enough stimulus for the induction of floral buds [33]. Research has confirmed the second hypothesis.

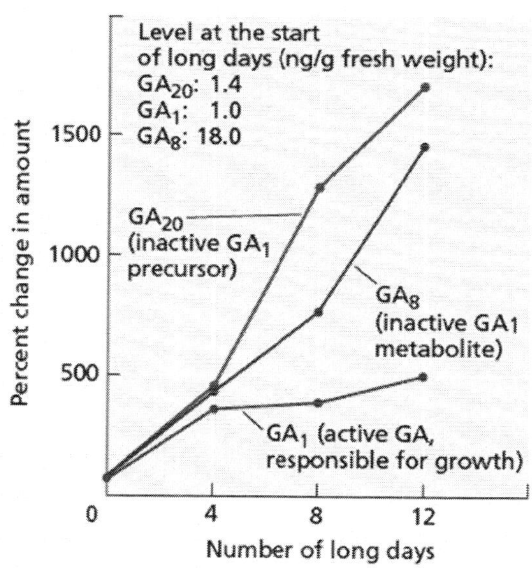

Figure 8. The fivefolfd increase in GA1 is what causes growth in spinach exposed to an increasing number of long days but before stem elongation starts at about 14 days. After [34]; redrawn from data in [35].

http://dx.doi.org/10.5772/51678

Buds of juvenile *Bryophyllum* species were grafted on to adult plants with flowers and they produced flowers. The author concluded that the meristems were competent to flower, but the young leaves produced insufficient amounts of the floral stimulus [37]. In a work with *Perilla*, a short-day plant, it was shown that the second node of the young leaves produce less floral stimulus, and therefore require more inductive photoperiods to induce flowering, as compared to fully mature leaves [38]. The most recent studies on the flowering control of the *Arabidopsis* plant state that the process is regulated by four separate ways:

1. the photoperiodic (long day) pathway, which operates in the leaves;
2. the convergent autonomous (leaf number)/vernalization (low temperature) pathway;
3. the carbohydrate (sucrose) pathway; and
4. the gibberellin pathway.

The latter three pathways all operate in the shoot apical meristem. The four pathways converge on a number of floral pathway integrators that together regulate floral initiation [33].

Recently, the T locus was identified which contains the FT gene related to flowering. It is expressed in leaves, encoding products that fit the description of a universal flowering stimulus. This finding comes against the research carried out for decades which sought that signal [39]. More studies are still seeking to define the paths of integration that exist among the different routes.

Figure 9 provides a complete consideration on this subject. This topic briefly addresses the GA pathway, when it operates, and what is known about its integration with the other pathways.

The photoperiodic pathway is located in the leaves and involves the production of a transmissible floral stimulus, the FTprotein [33]. The gene flowering locus T (FT) is a major output of both the photoperiod and the vernalization pathways controlling the floral transition. FT protein acts at the shoot apex of the plant in concert with a transcription factor, flowering locus D (FD).

In long-day plants (LDPs) such as *Arabidopsis,* the FT protein is produced in the phloem in response to CO (Constans) protein accumulation under long days (LD). It is then translocated via sieve tubes to the apical meristem.In short-day plants (SDPs) such as rice, the transmissible floral stimulus, the Hd3a protein (*Hd3a – heading date gene*), accumulates when the repressor protein, Hd1(*Hd1 – heading date gene*), is not produced under short days (SD), and the Hd3a protein is translocated via the phloem to the apical meristem [33].

A major quantitative trait locus (QTL) controlling response to photoperiod, *Hd1,* was identified by means of a map-based cloning strategy. High-resolution mapping using 1505 segregants enabled us to define a genomic region of ~12 kb as a candidate for *Hd1*. Further analysis revealed that the *Hd1* QTL corresponds to a gene that is a homolog of *Constans* in Arabidopsis.

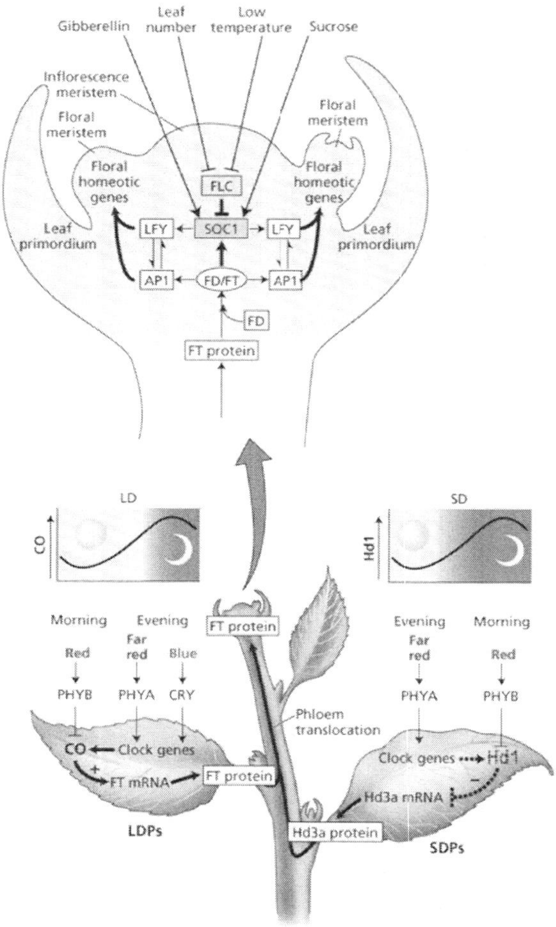

Figure 9. Multiple developmental pathways for floweringin*Arabidopsis:*(a) the photoperiodic (long day) pathway, which operates in the leaves; (b) the convergent autonomous (leaf number)/vernalization (low temperature) pathway; (c) the carbohydrate (sucrose) pathway; and (d) the gibberellin pathway. Fonte: [33].

In *Arabidopsis,* FT binds to FD, and the FT/FD protein complex activates the *AP1* (*Apetala1*) and *SOC1* genes (suppressor of over expression of*CO1*), which trigger the *LFY* (Leafy) gene expression. *LFY* and *AP1* then trigger the expression of the floral homeotic genes. The autonomous (leaf number) and vernalization (low temperature) pathways act in the apical meristem to negatively regulate *FLC*- flowering locus C, a negative regulator of *SOC1*. The sucrose and gibberellin pathways, also located in the meristem, promote *SOC1* expression [33].

2.4. Juvenile Period - Floral Induction in Soybeans

Some plants are indifferent to photoperiod, i.e., flowering and other developmental events are independent of photoperiod. In Brazil, this phenomenon was observed in the "Santa Maria" soybean variety and it was concluded that it was indifferent to day length [40].

Studies with soybean defined four stages of development related to flowering [41]:

Phase I - Juvenile - short days do not induce flowering;

Phase II - Inductive – flowering is induced by a minimum number of short days;

Phase III - Regulation - the number of flowers increases with the continuous conditions of induction and

Phase IV - Post-regulation - there is no effect of day length on flowering.

There is a stage in soybean developmentcalled the juvenile period. Juvenility is the name given to the initial phase of vegetative growth when soybean is not responsive to short-day-induced reproductive development. Until that period is completed, the plant is not cannot start floral initiation, even if it is grown under short days [42]. A determination of the juvenile period of one genotype can be carried out using the technique described in the literature [43]. A plant that flowers later, even under conditions of short days, has a long juvenile period relative to other soybeans. Such genotypes are described as having the long juvenile character [44].

Considering growth and flowering, it can be observed that each cultivar has a typical response in relation to the sowing date [45]. When sowing is early, there is also early flowering and lower plant heights in the most photoperiod sensitive cultivars. Research has shown that when the photoperiod is favorable, there is a combination of two or more endogenous hormones in the plant that produces biochemical changes in the meristematic cells of vegetative nodes. These cells begin to multiply and differentiate into flower buds [46]. After the juvenile period, a sequence of two short days sensitize the soybean leaves through phytochrome [46]. During the day, plants have both forms of phytochrome, with a predominance of Pfr. During the night, the Pfr form converts spontaneously to Pr. This reversal is essential for the measurement of time by plants and for the way they respond to photoperiod.

When cultivars having similar maturity are sown at the same time, they may bloom at different times. Thisis attributed to different juvenile periods [47]. During this period, some metabolic pathways, which are necessary for flowering initiation, are not triggered.The beginning of the studies concentrated on the phenotypic aspects of flowering, relating the effects and not the causes ofthe observed morphological changes. More complex research opened new perspectives to understand such process. There are two hypothesis for the fact that during the juvenile period, plants are not induced to flower even under inductive photoperiod:

1. The buds of the apical meristem are not competent to flowering.
2. The young leaves are still unable to produce enough stimulus for the induction of floral buds [33].

Phenotypic observations reported in many articles have characterizedplant response to photoperiod. Cultivar "Doko" was observed to have a long juvenile period [48], based on a long time to flower under short and long daylengths. Late flowering from several sowings was related to the possible existence of a long juvenile period [49]. The "Doko" cultivar was obtained from a program of selections which were sowedin summer and wintet and has long juvenile period [50].

2.5. Genetic Inheritance of Flowering

Genes affecting flowering response, have been studied in *Pisumsativum* and *Arabidopsis thaliana* [51, 52]. The results show that mutations in these species can change various aspects of the photoperiodic control of flowering. Some of these mutations can eliminate the photoperiodic responses, which are responsible for flowering induction. Others may simply slow down or speed up responses to photoperiod [33]. Much of the regulatory systems of flowering are under either positive or negative control and the presence of mutant plants with changes in photoperiodic responses is very common.

Most mutations result in the loss or alteration of the gene activity. After the mutation, the genes that promoteflowering are changed. Mutations that eliminate plant response to photoperiod can block the production of floral stimulus or may interfere with the ability of the meristem to receive the message [33]. Grafting studies have identified genes in *Pisumsativum* that promote or inhibit flowering and control the sensitivity of the apical meristem signals [52]. Most cultivars of soybean, respond to photoperiod as follows: when the number of daylight hours is below the critical photoperiod, there is flowering induction. Only few cultivars have the long juvenile character in which the effect of genes to promote flowering is reduced.

Early studies showed that a long juvenile character in soybean is genetically controlled and can be transferred in a breeding program [53]. Under short days, the authors identified recessive genes that control the trait. The literature on the subject shows that the long juvenile character is conditioned by recessive genes which can be pleiotropically influenced by other genes in the plant [42, 54-59]. Research conducted under "long-day conditions" indicates that the dominant alleles are responsible for the late cycle: E1/e1 and E2/e2 described by [60], E3/e3 reported by [61], E4/e4 described by [62] and E5/e5 by [63]. In the "short-day conditions", the opposite occurs [53-55]. Under these conditions, the gene J1/j1 was described [64]. Other studies were conducted to determine the type of inheritance. Research was performed under "conditions of short days" with the genotypes "Hill", "Bragg", "UFV-1", "IAC 73-2736" and "PI 159925" [65]. It was observed that the"long juvenile" characteris controlled by one, two or more recessive genes [3]. These and other studies were fundamental to our understanding of theflowering process of soybean plants grown in locations with different latitudes such as occurs in Brazil.

2.6. Long Juvenile Period in Soybean - Practical Application

The possibility of using plants exhibiting the long juvenile character was the solution found by some soybean breeders to delay flowering in short day conditions [42, 54, 58, 66]. Re-

http://dx.doi.org/10.5772/51678

search on the adaptation of soybeans to the tropics began at the Agronomic Institute (IAC) and the National Center for Soybean Research, in the 1970s. There were crosses among American cultivars which had the long juvenile character. Several genotypes with this trait were identified and used in breeding programs: "Santa Maria", "PI 159925" and "PI 240664" [67]. Identification of the character was done through research by EMBRAPA where long juvenile genotypes were planted from September 20 to October 10. By this method, genotypes were identified that had a sufficient delay in days to first flower to optimize dry matter accumulation and yield [66, 68]. The first cultivars developed and recommended for these areas were "Tropical", "Timbira", "BR-10 (Teresina)" and "BR-11 (Carajás)" [68]. Later the following cultivars were released: "BR-27 (Seridó)", "BR-28 (Cariri)", "Embrapa 9 (Bays)", "Embrapa 30 (CVRD)", "Embrapa 31 (Mina)", "Embrapa 32 (Itaqui)", "Embrapa 33 (Cariri RC)", "Embrapa 34 (Teresina RC)", "Embrapa 63 (Mirador)", "MA/BRS-64 (Parnaíba)", "MA/BRS-65 (Sambaíba)", "MA/BRS-163 (Pati)" and "MA/BRS-164 (Seridó RCH)"

The most commonly used cultivars as sources of the long juvenile character are: "Doko", "Doko RC", "Garimpo RCH", "BR/IAC-21", "UFV-16", "UFV-17", "UFV-18", "CAC-1", "CS 301", "MG/BR-46", "MT/BR-45", "BR-9", "FT-Cristalina", "Cristalina FT-RCH", "Tropical", "BR-10", "BR-11" and "Embrapa-33". They allow for a wider sowing time, planting during the offseason, and planting at low latitudes (short-day conditions) [3]. Currently the soybean crop in Brazil has been attacked by the Asian soybean rust, which leads the country to adopt a control measure called fallowing. In such areas, soybean cannot be planted during the off season. The adopted measure is a protection against the Asian soybean rust which has led the country to adopt a control measure called fallowing. In order to stop the rust, production areas are left vacant for part of the year. Asian soybean rust is a disease caused by *Phakopsorapachyrhizi* Sydow which caused a loss of two billion dollars to the Brazilian soybean crop in the 2005/2006 harvest.

3. Conclusions and Prospects of Soybean in Brazil

The achievement of Brazilian research in the development of soybean cultivars adapted to low latitudes has allowed expansion into the central and northern areas of the country. Until 1970, commercial cultivation of soybeans in the world was restricted to regions of temperate and subtropical latitudes which were near or higher than 30º Lat.Brazilian researchers were able to break this barrier by developing cultivars adapted to the short days of the tropics, enabling soybean cultivation anywhere in the country. In the Cerradoregion, more than 200 million hectares were converted into cultivated areas of soybeans and other grains [1].

Currently, states with the highest soybean production are: "Mato Grosso", "Paraná", "Rio Grande doSul" and "Goiás". They produce 82% of Brazil's soybeans. Soybean production is also progressing into new areas in "Maranhão", "Tocantins", "Piauí" and "Bahia", which account for 13.0% of Brazilian production.

The projections of the Ministry of Agriculture, Livestock and Supply (MAPA) show that Brazil is a major supplier of food. The region of Matopiba (Figure 10), an area including the states of "Maranhão", "Tocantins"", Piauí" and "Bahia", has potential for growth of grain production and will stand out in the Brazilian agricultural landscape for years to come. The trend shown in the study (Brazil - Projections of Agribusiness 2010/2011 to 2020/2021), released by the MAPA [69].

Increased soybean production area can occur through a combination of expansion into new areas or replacement of other crops. The production of sugar cane and soybeans are two activities that compete for land in Brazil. The two together will create an increase of 7.4 million hectares, 5.3 million hectares for soybeans and 2.1 hectares for sugarcane. The soybean plants that have the long juvenile characteristic can also be used in crop rotations, particularly in areas of "Sao Paulo" state that have previously been grown to sugar cane. Cultivars have been developed which are adapted to 1.2 to 1.4 million hectares of this area [70].

Most expansion should occur in areas of high yield potential, such as included in the region that is now called "Matopiba". "Mato Grosso" is not expected to have a large increase in arable land, mainly because land prices in the state are more than double that for land in the "Matopiba" region. Since agricultural expansion into these new areas includes large tracts of farm land, land price is a deciding factor.

	Production			Consumption			Exportation		
Year	Projection	*L. Lim.	**U. Lim.	Projection	*L. Lim.	**U. Lim.	Projection	*L. Lim.	**U. Lim.
2012/13	72,764.7	61,176.9	84,352.4	39,349.0	33,725.9	44,972.2	32,985.5	26,829.9	39,141.1
2013/14	74,531.2	61,196.3	87,866.0	40,140.8	33,748.6	46,533.1	33,573.3	25,990.3	41,156.3
2014/15	76,241.0	61,344.5	91,137.4	40,921.0	33,839.2	48,002.7	34,262.6	24,693.8	43,831.4
2015/16	77,958.6	61,650.0	94,267.2	41,703.6	33,994.5	49,412.7	35,705.5	24,497.3	46,913.7
2016/17	79,672.0	62,062.8	97,281.3	42,485.8	34,196.5	50,775.0	36,972.7	24,235.3	49,710.1
2017/18	81,385.8	62,565.6	100,206.0	43,268.0	34,436.6	52,099.4	37,671.2	23,258.6	52,083.8
2018/19	83,099.3	63,141.3	103,057.2	44,050.2	34,708.1	53,392.3	38,595.2	22,438.5	54,751.9
2019/20	84,812.7	63,778.5	105,847.0	44,832.4	35,006.1	54,658.7	39,774.6	22,022.7	57,526.5
2020/21	86,526.2	64,468.1	108,584.2	45,614.6	35,326.9	55,902.3	40,744.4	21,430.0	60,058.8

Table 3. Projections for the production, consumption and trade of soybean (thousand tons).Adapted from [7]. *Lower Limit **Upper Limit

The estimates for soybean indicate a future Brazilian production of 86.5 million metric tons in 2020/2021 (Table 3) . The annual growth rate is expected to be 2.3% from 2010/11 to 2020/2021. This rate is close to the global rate for the next ten years [7]. The domestic consumption of soybeans is expected to reach 45.6 million metric tons at the end of the 2020/2021 season, representing 52.7% of the production. The projection is for an annual rate of increase of 1.9%. As it is known, soybean is an essential component in the manufacture of animal feed and is a gaining importance as human food [7]. The projected expansion for the-area with soybeans in Brazil should exceed 30.0 million hectares in by 2020/2021. This is an increase of more than 5.3 million hectares from the current level. The expansion of soybean production in the country comes from a combination of expanding areas andincreased yield. Production is forecast to increase at a rate of 2.0% per year with area of production expanding at the annual rate of 1.9% [7]

Soybean meal and oil will have a moderate increase in future years. Bran exports shall grow at 1.1% per year and the soybean oil exports at 0.5% per year. Domestic consumption for both is expected to grow at high rates. The consumption of soybean oil is expected to grow at an annual rate of 2.2% between 2010/11 and 2020/2021, while the soybean meal consumption is expected to grow at 2.5% per year.

Figure 10. Matopiba.The new agricultural frontier in Brazil.Fonte: [69].

These data reflect the dynamism of the internal market for these products, given the human and animal consumption.The relationship between consumption and production of soybean oil in future years is around 78%. Most of the oil is for human consumption and the other part has been used for the production of biodiesel. About 22% of the production will be exported. For soybean meal, between 47.0 and 49.0% should be directed to domestic consumption, and about 50% exported. Thisbrief account of the soybean in Brazil demonstrates the importance of this crop to the national economy. The expansion of the area for production of soybean was due largely to basic and applied research involving genetic and physiological mechanisms affectingthe timing of flowering and other developmental events. Thus, much of soybean's expansion has been due to the quality of Brazil's national agricultural research.

Acknowledgements

I especially thank my wife Rita de Cássia and daughter Mariana for the encouragement and support so that I could write this chapter.

I thank my family for their confidence in my work.

I thank the authors cited in the chapter, because without their research, it would be impossible to accomplish this work.

Author details

Eduardo Antonio Gavioli[1*]

Address all correspondence to: eduardogavioli@terra.com.br

1 Taquaritinguense Institute of Higher Education - ITES, Department of Agronomy, Brazil

References

[1] Embrapa- Empresa Brasileira de Pesquisa Agropecuária. (2004). Tecnologias de Produção de Soja Região Central do Brasil. http://www.cnpso.embrapa.br/producaosoja/SojanoBrasil.htm, accessed 12 mar 2012.

[2] Sediyama, T., Santos, O. S., & Escolha de Cultivares. (1998). In: Santo O.S. (coord.). *A Cultura da Soja- 1. Rio de Janeiro: Publicações Globo Rural*, 91-108.

[3] Sediyama, T., et al. (2005). Melhoramento da Soja. In: Borém A. (coord.). Melhoramento de Espécies Cultivadas. *Viçosa: UFV*, 553-603.

[4] Almeida, L. A., et al. (1997). Melhoramento da soja. In: Abreu A.F.B., Gonçalves F.M.A., Marques Jr. O.G., Ribeiro P.H.E. (eds.). Simpósio sobre Atualização em Genética e Melhoramento de Plantas. Lavras, MG: UFLA-GEN , 09-55.

[5] Deser- Departamento de Estudos Sócio-Econômicos Rurais. (2010). Boletim Eletrônico. *Conjuntura Agrícola. Nº 169. m.*, http://www.deser.org.br/documentos/imagem/Soja.pdf, accessed 02 may 2012.

[6] IBGE- Instituto Brasileiro de Geografia e Estatística. (2012). http://www.ibge.gov.br/home/estatistica/indicadores/agropecuaria/lspa/lspa_201203.pdf, accessed 02 may 2012.

[7] Mapa- Ministério da Agricultura Pecuária e Abastecimento- Assessoria de Gestão Estratégica. (2011). Projeções do Agronegócio Brasil 2010/11 a 2020/21.2011. http://www.agricultura.gov.br/arq_editor/file/Ministerio/gestao/projecao/PROJECOES

%20DO%20AGRONEGOCIO%202010-11%20a%202020-21%20-%202_0.pdf, accessed 02 may 2012.

[8] Garner, W. W., & Allard, H. A. (1930). Photoperiodic response of soybeans in relation to temperature and other environmental factors. *Journal of Agricultural Research*, 41, 719-35.

[9] Pascale, A. J. (1969). Tipos agroclimáticos para el cultivo de la soya en la Argentina. *Revista de la Facultad de Agronomía e Veterinaria*, 17, 31-8.

[10] Parker, M. W., & Borthwick, H. A. (1943). Influence of temperature on photoperiodic reactions in leaf blades of Biloxi soybean. *Botanical Gazette*, 104, 612-9.

[11] Chang, J. H. (1974). Climate and agriculture an ecological survey. *Chicago: Transaction Publ.*, 304.

[12] Kendrick, R. E., & Kronenberg, G. H. M. (1994). Photomorphogenesis in Plants. *2th ed. Dordrecht: Academic Publishers*.

[13] Steinberg, R. A., & Gardner, W. W. (1963). Response of certain plants to length of day and temperature under controlled conditions. *Journal of Agricultural Research*, 52, 943-60.

[14] Johnson, H. W., et al. (1960). Effects of photoperiod and time of planting on rates of development of the soybean in various stages of the life cycle. *Botanical Gazette*, 22, 77-95.

[15] Gandolfi, V. H., & Müller, L. (1981). Fisiologia In: Miyasaka S., Medina J.C. (coord.). *A Soja no Brasil. Campinas: ITAL*, 109-174.

[16] Borthwick, H. A., & Parker, M. W. (1938). Photoperodic perception in Biloxi soybeans. *Botanical Gazette*, 100, 374-87.

[17] Parker, M. W., et al. (1945). Action spectrum for the photoperiodic control of floral initiation in Biloxi soybean. *Science*, 2641(102), 152-5.

[18] Howell, R. W. (1963). Physiology of the soybean. In: Norman A.G. (ed.). *The soybean. New York: Academic Press*, 75-124.

[19] Hartwig, E. E. (1970). Growth and reproductive characteristics of soybeans (Glycine max (L.) Merrill) grown under- short day conditions. *Trop. Sci.*, 12, 47-53.

[20] Parker, M. W., & Borthwick, H .A. (1951). Photoperiodic responses of soybean varieties. *Soybean Digest*, 11, 26-30.

[21] Vince-Prue, D. (1975). Photoperiodism in plants. *London: McGraw Hill*, 444.

[22] Major, D. J., et al. (1975). Effects of daylength and temperature on soybean development. *Crop Science*, 15, 174-9.

[23] Jones, P. G., & Laing, D. R. (1978). Simulation of the phenology of soybeans. *Agricultural Systems*, 3, 295-311.

[24] Hodges, T., & French, V. (1985). Soyphen soybean growth stages modeled from temperature, daylength, and water availability. *Agronomy Journal*, 77, 500-5.

[25] Wilkerson, G. G., et al. (1989). Photoperiodically sensitive interval in time to flower of soybean. *Crop Science*, 29, 721-26.

[26] Rodrigues, O., et al. (2001). Resposta quantitativa do florescimento da soja à temperatura e ao fotoperíodo. Pesquisa Agropecuária Brasileira; , 36(3), 431-7.

[27] Rangel, M. A. S. (2012). Minha soja floresceu com porte baixo e agora? *Embrapa artigos.*, http://www.embrapa.gov.br/imprensa/artigos/2006/artigo.2006-01-12.1615609604/, accessed 23 may 2012.

[28] Borthwick, H. A., et al. (1952). The reaction controlling floral initiation. *Proc. Natl. Acad. Science*, 38, 929-34.

[29] Butler, W. L., & Lane, H. C. (1965). Dark transformations of phytochrome in vivo. II. *Plant Physiology*, 40, 13-17.

[30] Sharrock, R. A., & Quail, P. H. (1989). Novel phytochrome sequences in Arabidopsis thaliana: structure, evolution, and differential expression of a plant regulatory photoreceptor family. *Genes & Dev.*, 3, 1745-57.

[31] Pratt, L. H., et al. (1997). The phytochrome gene family in tomato (Lycopersicumsculentum Mill.). *Plant, Cell and Enviroment*, 20, 672-7.

[32] Salisbury, F. B., & Ross, C. W. (1992). Plant Physiology. 4th ed.Belmont: Wadsworth Publishing Company; , 682.

[33] Taiz, L., & Zeiger, E. (2010). Plant Physiology. *Fifth Edition.*, http://www.plantphys.net, accessed 02may 2012.

[34] Davies, P. J. (1995). The plant hormones: their nature, occurrence, and functions. In: Davies P.J. (ed.) Plant hormones: physiology, biochemistry and molecular biology. *2th ed. Amsterdam: Kluwer Academic Publishers*, 1-12.

[35] Zeevaart, J. A. D., et al. (1993). Gibberellin A_1 is required for stem elongation in spinach. Proc. Natl. Acad. Sci. USA; , 90, 7401-5.

[36] Evans, M. M. S., & Poethig, R. S. (1995). Gibberellins promote vegetative phase change and reproductive maturity in maize. *Plant Physiol.*, 108, 475-87.

[37] Zeevaart, J. A. D. (1985). Bryophyllum. In: Halevy A.H. (ed.). *Handbook of Flowering*, 2, Boca Raton-FL: CRC Press, 89-100.

[38] Zeevaart, J. A. D., & Boyer, G. L. (1987). Photoperiodic induction and the floral stimulus in Perilla. In: Atherton J.G. (ed.), Manipulation of Flowering. *London: Butterworths*, 269-277.

[39] Huang, T., et al. (2005). The mRNA of the Arabidopsis gene FT moves from leaf to shoot apex and induces flowering. *Science*, 309, 1694-96.

[40] Miyasaka, S., et al. (1970). Variedades de soja indiferentes ao fotoperíodo e tolerantes a baixas temperaturas. In :SimpósioBrasileiro de Soja, l, Campinas.

[41] Huxley, P. A., et al. (1974). The effect of photoperiod on developement of soybean and cowpea cultivars grown in the U. K. in summer. *Exp. Agric.*, 10(4), 225-39.

[42] Kiihl, R. A. S., & Garcia, A. (1989). The use of the long-juvenile trait in breeding soybean cultivars. In: World Soybean Research Conference, 4., Buenos Aires, Argentina. *Proceedings Buenos Aires: AASOJA.*

[43] Shanmugasundaram, S., & Tsou, S. C. S. (1978). Photoperiodic and critical duration for flower indution in soybean. *Crop Science*, 18, 598-601.

[44] Paludzyszyn, Filho. E., et al. (1993). Desenvolvimento de cultivares de soja na região norte e nordeste do Brasil. In: Arantes N.E., Souza P.I.M. (eds.). *Cultura de Soja nos Cerrados. Piracicaba-SP: POTAFOS*, 255-265.

[45] Martignone, R. A., et al. (2006). Efecto de la fecha de siembra sobre el comportamiento fenológico y agronómico en cultivares de soja de diferentes grupos de maduración. In: CONGRESSO DE SOJA DO MERCOSUL, Rosário. *Libro de Mesas Cientifico-Técnicas Resúmeres Expandidos, Rosário: ACSOJA.*

[46] Câmara, G. M. S., & Heifigg, L. S. (2000). Fisiologia, ambiente e rendimento da cultura da soja. In: Câmara G.M.S. (ed.). *Soja: tecnologia de produção II. Piracicaba: ESALQ/ LPV*, 81-117.

[47] Embrapa- Empresa Brasileira de Pesquisa Agropecuária. (2008). Tecnologias de produção de soja- região central do Brasil 2009 e 2010. *Londrina: Embrapa Soja: Embrapa Cerrados: Embrapa Agropecuária Oeste*, 262, (Embrapa Soja. Sistemas de Produção, 13).

[48] Dutra, J. H. (1986). Comportamento de genótipos de soja (Glycine max (L.) Merrill) em diferentes épocas de plantio em Capinópolis, Minas Gerais. *Dissertação (Mestrado em Agronomia). Viçosa: Universidade Federal.*

[49] Bhering, M. C., et al. (1991). Influência de épocas de plantio sobre algumas características agronômicas da soja (Glycine max (L.) Merrill). *Revista Ceres*, 38(219), 396-407.

[50] Spehar, C. R., Monteiro, P. M. F. O., & Zuffo, N. L. (1993). Melhoramento Genético da Soja na Região Centro-Oeste. In: Arantes N.E., Souza P.I.M. (eds.). *Cultura da Soja nos Cerrados. Piracicaba-SP: POTAFOS*, 229-253.

[51] Weigel, D. (1995). The genetics of flower development: From floral induction to ovule morphogenesis. *Annu. Rev. Genet.*, 29, 19-39.

[52] Reid, J. B., et al. (1996). Physiological genetics of flowering in Pisum. *Sem. Cell Dev. Biol.*, 7, 455-63.

[53] Kiihl, R. A. S. (1976). Inheritance studies of two characteristics in soybeans (Glycine max (L.) Merrill): I. Resistance to soybean mosaic virus; II. Late florewing under short-day conditions. *Doctorate thesis.Mississipi State University Library.*

[54] Hartwig, E. E., & Kiihl, R. A. S. (1979). Identification and utilization of a delayed flowering character in soybeans for short-day conditions. *Field Crops Reseach*, 2, 45-51.

[55] Tisseli, O., Junior. (1981). Inheritance study of the long juvenile characteristic in soybeans under long and short-day conditions. *Ph.D. thesis. Starkville: Mississippi State University*.

[56] Toledo, J. F. F., & Kiih, L. R. A. S. (1982a). Análise do modelo genético envolvido no controle de dias para florescimento em soja. *Pesquisa Agropecuária brasileira*, 17(4), 623-31.

[57] Gilioli, J.L., et al. (1984). Herança do número de dias para floração em quatro mutantes naturais em soja estudada sob condições de dias curtos. In: Anais III Seminário Nacional de Pesquisa de Soja, Campinas. *Londrina: EMBRAPA-CNPSo*.

[58] Hinson, K. (1989). The use of long juvenile trait in cultivar development. In: World Soybean Research Conference, 4., Buenos Aires, Argentina. *Proceedings Buenos Aires: AASOJA*.

[59] Bonato, E. R. (1989). Herança do tempo para florescimento e para maturidade em variantes naturais de soja. Tese de Doutorado. *USP/ESALQ, Piracicaba, SP*.

[60] Bernard, R. L. (1971). Two major genes of flowering and maturity in soybeans. *Crop Science, Madison*, 11, 242-4.

[61] Buzzell, R.I. (1971). Inheritance of soybean flowering response to fluorescent-daylength conditions. Canadian Journal of Genetics and Citology; , 13, 703-7.

[62] Buzzell, R. I., & Voldeng, H. D. (1980). Inheritance of sensibility to long daylength. *Soybean Genetics Newsletter*, 7, 26-9.

[63] Mc Blain, B. A., & Bernard, R. L. (1987). A new gene affecting the time of flowering and maturity in soybeans. *Journal of Heredity*, 78, 160-2.

[64] Ray, D. J., et al. (1995). Genetic control of a long-juvenile trait in soybean. *Crop Science*, 35, 1001-6.

[65] Toledo, J. F. F., & Kiihl, R. A. S. (1982b). Métodos de análise dialélica do modelo genético em controle das características dias para floração e número de folhas trifolioladas em soja. *Pesquisa Agropecuária brasileira*, 17(5), 745-55.

[66] Kiihl, R. A. S., et al. (1985). Strategies for cultivar development in the tropics. In: World Soybean Research Conference III. *Proceedings.Ames, IL, USA*.

[67] Toledo, J. F. F., et al. (1994). Genetics and Breeding. In: Tropical Soybean Improvement and Production. Rome: Plant Production and Protection Series nº 27. *Food and Agriculture Organization of the United Nations*.

[68] Kiihl, R. A. S., et al. (1986). Soybean breeding for the brasilian tropics. In: Soybean in Tropical and Subtropical Cropping Systems. *Proceedings.Tsukuba. Japan. 1983. Revised Edition, AVRDC*.

[69] Kory, Melby. (2012). Brazilian Ag Consulting Services and Investment Tours. *MaTo-PiBa.The New Ag Frontier.*, http://www.brazilintl.com/states/mapitoba/mapitoba.htm, accessed 15july.

[70] Gavioli, E. A., et al. (2006). Analysis of combining ability in soybean cultivars. *Crop Breeding and Applied Biotechnology*, 6, 121-128.

Chapter 17

Climatic Restrictions for Maximizing Soybean Yields

Ana Maria Heuminski de Avila,
José Renato Bouças Farias, Hilton Silveira Pinto and
Felipe Gustavo Pilau

Additional information is available at the end of the chapter

http://dx.doi.org/10.5772/52177

1. Introduction

Soybean (Glycine max L.) is a leguminous species, native to Northeast Asia, and is one of the most important sources of oil and protein for livestock and the human diet. The crop is also being used to make diesel biofuel from the oil extracted from the seed. Today, the world's top producers of soybean are the United States (90 million ton), Brazil (68 million tons), Argentina (52 million tons), China (15 million tons) and India. These countries' production represents more than 90% of global soybean production. In the 2009/2010 harvest the world produced 261.57 million tons of soybean grain on 102.38 million hectares define this by [1]. Also, according to [1], over a period of 50 years, soybean production area increased 330%, rising from 23.81 million in 1961 to the current production area.

Currently with a world population of 7.0 billion people, which is projected to increase to more than 9.0 billion by 2050 and 10 billion by 2100 [2], there is a concern to maintain food. To achieve a gradual increase in world production of soybeans, two alternatives are possible: increased production area and/or increased productivity. Because planted area has increased 37% between 2000 to 2011 [1] and productivity has increased only 17.8%, the second solution looks to be the better option

Among all the factors inherent in agricultural production, the climate are the most difficult to control and it exercises greater limiting action in the maximum yield. It worsens according to the difficult to predict occurrences of adverse weather, the main risk factors in the exploration of major crops. Stress non biotic such as drought, excessive rain, extreme temperatures and low light can significantly reduce yields of crops and restrict the locations, times and soils, where the species most important commercially, can or not be cultivated.

2. Agrometeorological conditioners of soybean productivity

2.1. Solar radiation

In general, soybean production potential and risk is controlled by prevailing climatic conditions and genotypic performance. The basic source for crop production is solar radiation which acts as an energy source for photosynthesis. Light quality also acts to influence plant height and phenological development [3].

The light spectrum duration and quality besides the radiation intensity are determinants of morphological and phenotypic responses striking in soybean, such as plant height, induction of flowering and ontogeny [4]. The final yield of dry matter from the plant depends on the solar radiation absorbed by the leaves and the efficiency with which these convert radiant energy into chemical energy through photosynthesis.

Soybean is a qualitative short-day plant and must receive a certain day length or less so that developmental timing is optimal for the location [5]. This means that adaptability of each cultivar varies with latitude [6]. Movement of soybean to the central and northern regions of the country from the original southern region has faced the challenge of adaptation to shorter day lengths occurring in these regions (from Tropic of Capricorn to Equator Line). To deal with this challenge Brazilian agricultural scientists have sought to introduce the 'long juvenile period'. This characteristic gives soybean a longer developmental period under short days so that it can accrue enough dry matter for optimal yield [7].

Each variety has its critical photoperiod above which flowering is delayed. Flowering occurs anyway, but more rapidly as the days become shorter. It slows progressively, as the photoperiod exceeds the critical period for each genotype.

2.2. Temperature

In addition to photoperiod, temperature also influences growth and developmental timing in soybean. This occurs due to the effect of temperature on the rate of metabolic reactions, the diffusion rate of gases through aqueous media and the solubility of nutrients in the plant. Soybean grows best at temperatures between 20°C and 30°C. Greatest number of pods per plant is obtained under mild temperature conditions having a day/night temperature combination of 26/14°C [8]. Temperatures above 40°C during the vegetative stage (emergence to first flower) reduce growth and hastens flowering. High temperatures during the reproductive phase can cause reductions in seed number and seed weight, thus reducing grain quality and yield. If the high temperature is associated with a drought, the losses on grain production are even higher [7]. On the other hand, cold regions where the temperature is equal or below to 10C is not properly to soybean cultivation, because in these places, the vegetative growth and the development become small or null.

2.3. Water availability

Ninety % of the total soybean fresh weight (biomass) is water and water acts in all physiological and biochemical processes in the plant, working as a solvent, carrying gases, minerals and other plant solutes and acting as a thermal regulator to cool and maintain plant temperature [9].Water availability is important in three periods of soybean development: germination, emergence and flowering-grain filling. During the first period, both: excess or lack of water is detrimental to crop establishment and obtaining a good uniformity of the plant population and the surplus water more limiting than the deficit. Soybean seeds need to absorb at least 50% of its weight in water to ensure a good germination. At this stage, the water content in soil must not exceed 85% of the total maximum available, or must not be less than 50% [10]. The roots can reach over 1.5 m deep, however heavy and compacted soils hinder root penetration, further reducing the effective depth of the root system of soybean plants.

Another serious risk in Brazilian soybean production is drought stress. The problem is more pronounced in tropical and subtropical, semiarid and arid climates where up to 8mm/day may be lost by evapotranspiration.

The necessity of water needs by soybean gradually increase with plant development, peaking at 7 to 8mm/day during flowering through grain filling [11]. The negative effect of water deficiency on yield depends on the phenological stage in which soybean is affected by drought stress. For example, if drought occurs during the germination and emergence stages, plant stand is reduced. In the flowering period it can cause flower abortion and prevents anthesis while in the grain filling stage drought affects seed weight [12]. The lack of water induces these effects through reducing the efficiency of the photosynthesis [13].

An abundance of water is also harmful to yield by causing water logging of the soil. A very wet soil results in low aeration which reduces root growth, can cause nutritional deficiency and promotes the attack of root diseases.

3. General issues

The climate exerts the greatest limiting effect on yield. Thus, proper management of the soybean crop depends on climate information at the site of interest. Agroclimatic zoning is used to provide climate data to the agricultural community [14].

Cardoso et al. [15] verified that in general the precipitation climate forecasts contribution to the improvement of estimated soybean productivity, primarily in periods when the crop is more sensitive to water deficit. The improvement of estimated soybean productivity may give a contribution to agribusiness sector, in order to turn more realistic expectations available and assist on the strategic planning for the decisions maker.

Knowledge of favorable periods for soybean culture, such as timing of the rainy season and duration of intermittent dry wet spells is very important for planning various agronomic operations such as preparing a seedbed, proper maturity, sowing, weeding, harvesting, threshing and drying. This minimizes crop risk and optimizes use of resources such as water, labor, fertilizer, herbicides and insecticides. Knowledge of the critical periods for crop development in conjunction with knowledge of climatic factors is very helpful for adjusting management practices for best crop yield and quality.

As already highlighted climate stresses such as drought, excessive rain, extreme temperatures and low irradiance can significantly reduce yields of crops and restrict areas and timings of production. The most important factors useful to know for soybean production are air temperature, photoperiod and water supply [10].

In Brazil, a system has been developed for identification of suitable areas for planting soybean. The Agricultural Zoning Program has been coordinated by the Ministry of Agriculture and EMBRAPA since 1996. This program has the objective of defining planting seasons guarantee at least an 80% probability of having an adequate water supply for the growing season. The planting periods were defined through the simulation of a climatic water balance that gives an index of water supply called by the Portuguese initials as ISNA (Water Necessity Satisfaction Index) using historical rainfall data, potential evapotranspiration, physiological characteristics of each crop and water retention by the soil [16].

4. Climate change: A challenging

Since 1988, agricultural scientists have been studying climate change mainly according to the rules of IPCC (Intergovernmental Panel on Climate Change). In recent years global studies have shown consistent changes in air temperature and rainfall in many places of the world [17, 18]. The Earth's average temperature can rise between 1.8ºC and 4.0ºC in the next 100 years [19] with more significant increases for minimum than maximum air temperature. However, in the case of South America temperature trends may not be consistent [20]. Contemporary scientific studies have warned of anomalies in temperature and precipitation patterns indicating the occurrence of global change, with direct consequences on human activities, especially those related to agricultural production [18].

Globally, agriculture accounts for 23% of all emissions of greenhouse gases (GHG) that come from human activity. Parts of these (15%) are derived from agricultural practices and the other part (8%) is from changes in land use [21]. However, in Brazil a greater proportion of agriculture's contribution to GHG comes from agricultural practices [22]. Soil is also an important source of carbon emission and sequestration [23]. Soil management practices affect to what degree these processes go on. Conventional tillage operations tend to increase carbon release to the atmosphere, whereas conservation tillage and crop rotation with their increased contribution of crop organic matter to the soil tend to increase carbon sequestration [24]. These practices also improve the physical, chemical, and biological balances in the soil. Current agriculture through its production of fiber, bioenergy and food reduces pollution and mitigates GHG emissions [25].

A good example are the results of a research carried out at Embrapa Soja (Londrina, PR, Brazil) for levels of Carbon (C) and Nitrogen (N) in soil under no-tillage (NT) and conventional sowing

http://dx.doi.org/10.5772/52177

(CT). [26] found that the biggest difference in results between NT and CT systems occurs in the first 30 cm of the soil profile with a 29% increase in total C content of soil in NT vs. CT. Still, disputing claims of U.S. researchers [27], the survey confirmed significant increases in carbon sequestration in the 0-60cm soil layer. They determined that in NT vs. CT there was 18% and 16% increase in the C and N contents, respectively, of the soil organic matter. Within the same 0-60 cm profile, C and N content of the microbial mass increased 35% and 23%, respectively. Over the 20 years of the research, C and N rates of accumulation within the 0-60cm layer of soil were 800 kg C ha^{-1} $year^{-1}$ and 70 kg N ha^{-1} $year^{-1}$, respectively, in NT vs. CT. According to the Agricultural Census [28], in 2006 the total no-tillage area for crops in Brazil was 15.6 million hectares.

Because of the rising temperatures associated with global warming, the adaptability of certain crops to an area may change. In particular, drought problems may become worse and yield potential reduced. Because of this, agricultural scientists have been developing strategies to avoid potential adverse climatic changes, especially those related to drought stress.

Problems with drought and temperature stress have to be resolved with research for adaptation like developing new soybean genotypes and cultivars with heat and water loss tolerance.

In Brazil, the southeast region, which accounts for 40% of soybean production, has suffered severe losses due to the occurrence of droughts. Growing seasons in the years 2003 and 2004 for example, showed yield losses due to drought close to 24% and 44%, respectively. In 2004/05 growing season, Rio Grande do Sul state alone, lost an average 70% of its production. In the last 10 seasons, the direct losses can be estimated by more than $18 billion due to the occurrence of drought [29].

The search for commercial cultivars more drought tolerant through classical breeding are relatively difficult because of the complex mechanisms developed by plants to ameliorate this stress. Biotechnology is also an important ally in the breeding for drought tolerant cultivars. With the sequencing of the soybean genome it becomes possible to understand the function of a specific gene and how it interacts with other genes. This allows for breeding new cultivars with greater resistance to environmental stresses.

5. Final considerations

Currently inside the world agriculture, the increases of yields and the reduces of the costs and risks from failure, have become the basic requirements of competitive. Besides, in terms of economic activity with narrow profit margins, there is no room for risk.

Immense progress has been achieved, adapting the soybeans to obtain high levels of incomes in areas of lower latitudes. The photoperiod limitation was eliminated by the careful selection and also the development of the germoplasma less sensitive to photoperiod. The yields are extremely dependent of the water available and probably are also necessary to increase it to meet the crop transpiration and the productivity enhancement.

Environmental forecasts signaling the increase of the temperatures in the upcoming decades which mean the climate change. The difficulty to feed the world population, (which is in growing fast) using only traditional technologies will be immense shortly. The development of variety tolerant to adverse climatic conditions such as drought and high temperatures, it will be essential, as well as the development of new research tools in the areas of plant biotechnology and ecophysiology. It will enable us to understand the details of the processes involved in physiological and agronomic crops. Only with the development and continued growing in agricultural research, you can ensure food quality and quantity for future generations.

Author details

Ana Maria Heuminski de Avila[1], José Renato Bouças Farias[2], Hilton Silveira Pinto[3] and Felipe Gustavo Pilau[4]

1 CEPAGRI, UNICAMP, Campinas – SP, Brasil

2 CNPSO, EMBRAPA, Londrina – PR, Brasil

3 IB and CEPAGRI, UNICAMP, Campinas – SP, Brasil

4 Departamento de Fitotecnia, CCR, UFSM, Santa Maria – RS, Brasil

References

[1] FAOSTAT (2012). Food and Agriculture Organization of the United Nations. FAOSTAT. www.faostat.org (accessed 27 June 2012).

[2] UN (2012). United Nations Department of Economic and Social Affairs. *Population Division.* http://www.un.org/esa/population/unpop.htm. (accessed 19 June 2012.

[3] Chang, J.H., 1974: *Climate and Agriculture: An Ecological Survey.* Chicago, Aldine Publishing Company.

[4] Thomas, JF (1994). Ontogenetic and morphological plasticity in crop plants. In: BOOTE, KJ et al (comp.) *Physiology and determinations of crop yield.* Madison: ASA/CSSA/SSSA, cap. 7B, p.181-185.

[5] Garner, WW; Allard, H A (1920). Effect of the relative length of day and night and other factors of the environment on growth and reproduction in plants. *Journal of Agricultural Research,* Washington, DC, v. 48, n. 7, p. 553-606.

[6] Mavi HS and Tupper GT (2004). Agrometeorology: principles and applications of climate studies in agriculture, *The Haworth Press, Inc.*, New York, 364 p.

[7] Farias, JRB; Nepomuceno, AL; Neumaier, N (2007). *Ecofisiologia da soja.* Londrina: Embrapa Soja, 2007. 9p. (Embrapa Soja. Circular técnica, 48).

[8] Thomas, JF and CD Raper, Jr. (1981). Day and night temperatures during early reproductive development of soybeans: Influence on carpel initiation and growth. *Bot. Gaz.* 142:183-187

[9] Farias, JRB; Neumaier, N; Nepomuceno, AL (2009). *Soja.* In: Monteiro, J. E. B. A. (Org.). Agrometeorologia dos cultivos: o fator meteorológico na produção agrícola. Brasília: INMET, 2009. p. 263-277.

[10] EMBRAPA (2008). Tecnologias de produção de soja: região central do Brasil, 2008. Londrina, 2008. 280p. (Sistemas de produção/ Embrapa Soja, n. 12).

[11] Berlato, Moacir A (1981). *Bioclimatologia da soja.* In: Miyasaka, S.; Medina, J.C. ed. A soja no Brasil. Campinas:ITAL. p.175_184.

[12] Fageria, NK (1989). *Solos tropicais e aspectos fisiológicos das culturas.* Brasília : Embrapa-DPU, 1989. 425p. (Embrapa-CNPAF. Documentos, 18).

[13] Miyasaka, S, Medina, JC (1981). *A soja no Brasil.* 1ª ed., 479p.

[14] Sentelhas, PC; Monteiro, JEB de A (2009). Agrometeorologia dos Cultivos: Informações para uma Agricultura Sustentável. In: Monteiro, J. E. B. A. (Org.). *Agrometeorologia dos cultivos: o fator meteorológico na produção agrícola.* Brasília: INMET, p. 3-12.

[15] Cardoso, AO; Avila, AMH; Pinto, HS. Assad, E(2011). Use of Climate Forecasts to Soybean Yield Estimates, Soybean Physiology and Biochemistry, Hany A. El-Shemy (Ed.), ISBN: 978-953-307-534-1, InTech, Available from: http://www.intechopen.com/books/soybean-physiology-and-biochemistry/use-of-climate-forecasts-to-soybean-yield-estimates

[16] Farias, JRB; Assad, ED; Almeida, IR; Evangelista, BA; Lazzarotto, C; Neumaier, N; Nepomuceno, A.L (2001). Caracterização de risco de déficit hídrico nas regiões produtoras de soja do Brasil. *Revista Brasileira de Agrometeorologia,* 9, 415-421.

[17] Alexander, LV; Zhang, X; Peterson, TC; Caesar, J; Gleason, B; Tank, AMK.; Haylock, M; Collins, D; Trewin, B; Rahimzadeh, F; Tagipour, A; Kumar, KR; Revadekar, J; Griffiths, G; Vincet, L; Stephenson, DB; Burn, J; Aguilar, E; Brunet, M; Taylor, M; New, M; Zhai, P; Rusticucci, M; Vazquez-Aguirre, JL(2006). Global observed changes in daily climate extremes of temperature and precipitation. *Journal of Geophysical Research*, Washington, v. 111, DO 5109, doi: 10.1029/2005 JD006290.

[18] IPCC Assessment Report, 4., (2007). Valencia, Spain. *Climate change 2007*: the physical science basis: summary for policymakers. [Genebra]: IPCC, 2007b. 18 p. A report of

Working Group I of the Intergovernmental Panel on Climate Change. http://www.ipcc.ch (accessed March 2012).

[19] Solomon, S; Qin, D; Manning, M; Alley, RB; Berntsen, T; BIndoff, NL; Chen, Z; Chidthaisong, A; Gregory, JM; Hegerl, GC; Heimann, M; Hewitson, B; Hoskins, BJ; Joos, F; JouzeL, J; Kattsov, V; Lohmann, U; Matsuno, T; Molina, M; Nicholls, N; Overpeck, J; Raga, G; Ramaswamy, V; Ren, J; Rusticucci, M; Somerville, R; Stocker, TF; Whetton, P; Wood, RA; Wratt, D Technical summary. In: IPCC Assessment Report, 4., 2007, Valencia, Spain. *Climate change 2007*: the physical science basis. [Genebra]: IPCC, 2007. 74 p. A report accepted by Working Group I of the Intergovernmental Panel on Climate Change but not approved in detail. http://www.ipcc.ch/pdf/assessment-report/ar4/wg1/ar4-wg1-ts.pdf (accessed March 2012)

[20] Vincent, LA; Peterson, TC; Barros, VR; Marino, MB; Rusticucci, M; Carrasco, G; Ramirez, E; Alves, LM; Ambrizzi, T; Berlato, MA; Grimm, AM; MArengo, JA; Molion, L Moncunill, DF; Rebello, E; Anunciação, YMT; Quintana, J; Santos, JL; Baez, J; Coronel, G; Garcia, J; Trebejo, I; Bidegain M; Haylock, MR; Karoly, D (2005) Observed trends in indices of daily temperature extreme in South America 1960-2000. *Journal of Climate*, Washington, v. 18, p. 5011-5023, 2005.

[21] Lal, R; Logan, TJ (1995). Agricultural activities and greenhouse gas emissions from soils of the tropics. In: LAL, R.; KIMBLE, J.M.; LEVINE, E.; STEWART, B.A. (Ed.). *Soil management greenhouse effect*. Boca Raton: CRC Press, 1995. p.293-307.

[22] Lima, MA de (2006). Estimativa de emissões de gases de efeito estufa provenientes da queima da cana-de-açúcar no estado de São Paulo no período de 1986 a 2008 / Magda Aparecida Lima, Marcos Antonio Vieira Ligo, Elton César de Carvalho, Maria Conceição Peres Young Pessoa. – Jaguariúna: Embrapa Meio Ambiente, 2010. 23 p. : il. — (Embrapa Meio Ambiente. Documentos; 83).

[23] Bayer, C and Mielniczuk, J (1997). Características químicas do solo afetadas por métodos de preparo e sistemas de cultura. *R. Bras. Ci. Solo*, V. 21, p.105-112.

[24] Ciotta, MN; Bayer, C; Ernani, PR; Fontoura, SMV (2004). Albuquerque, JA; Wobeto, C Acidificação de um Latossolo sob semeadura direta. *Revista Brasileira de Ciência do Solo*, v.26, p.1055-1064.

[25] Costa, FS; Albuquerque, JA; Bayer, C; Fontoura, SMV; Wobeto, C (2003) Propriedades físicas de um Latossolo Bruno afetadas pelos sistemas de semeadura direta e preparo convencional. *Revista Brasileira de Ciência do Solo,* v. 27, p.527-535.

[26] Babujia, LC; Hungria, M; Franchini, JC; Brookes, PC (2010). Microbial biomass and activity at various soil depths in a Brazilian oxisol after two decades of no-tillage and conventional tillage. *Soil Biology & Biochemistry*, v. 42, p. 2174-2181.

[27] Baker, JM; Ochsner, TE; Venterea, RT; Griffs, TJ (2007). Tillage and soil carbon sequestration – What do we really know? *Agriculture, Ecosystems and Environment,* v. 118, p. 1–5. doi: 10.1016/j.agee.2006.05.014

[28] IBGE (2011). Instituto Brasileiro De Geografia e Estatística. Senso 2010. Disponível em: www.ibge.gov.br. Acesso em: agosto de 2011.

[29] Farias, JRB; Nepomuceno, AL; Neumaier, N; Tobita, S; Almeida, IR de. (2006). Restrições de disponibilidade hídrica à obtenção de elevados rendimentos de grãos de soja. In: Congresso Brasileiro de Soja, n.4., 2006, Londrina. *Resumos*. Londrina: Embrapa Soja, p. 32-33.

Chapter 18

Climatic Conditions and Production of Soybean in Northeastern Brazil

Jeandson Silva Viana, Edilma Pereira Gonçalves,
Abraão Cicero Silva and Valderez Pontes Matos

Additional information is available at the end of the chapter

http://dx.doi.org/10.5772/52184

1. Introduction

Brazil has become the second largest soybean producer after the USA. Projections to the year 2020 indicate the country's soybean production will increase to 105 million tons, largely due to Brazil's large expanse of potential arable land. This expansion of soybean production will occur mainly in the North and Northeast parts of the country that includes the states of Maranhão, Piauí, Tocantins and Bahia, a region collectively referred to as the MAPITOBA. Within this region, the Cerrados, an area of stable climate, consistent rainfall, and flat topography, is especially suited for soybean production [17].

Soybean was introduced into Brazil in the late nineteenth century. However,widespreadproduction did not occur until about 60 years ago in the southern part of the country. Cultivation then spread from the southern to the middle and northern sections of the country. In recent years, production has spread into the Cerrado Region, an area in the middle and northern parts of Brazil's interior containing 204 million hectares of land. The northern states of Piaui and Maranhao contain about 21 million hectares of land [6]. The savannas of this region are characterized by having acid soils of low fertility, high average temperatures (25-26 ° C) and average rainfall of 1,200 mm from October to April. However, they are subject to occurrences of dry spells [18].

The expansion of soybean into the low latitudes of northeast Brazil has been facilitated by cultivar development for adaptation to the soils and climate of this region. This process has allowed for a large expansion of soybean production into the Cerrado region of Brazil [1].

Cultivar Tropical was first released in 1980 for production in the low latitude regions of Brazil. This created a continual demand for more adapted cultivars [4]. Other cultivars released for northeastern Brazil between 1982 and 1998 are: Timbira; BR 10 (Teresina) and BR 11 (Carajás); BR-27 (Cariri) and BR-28 (Seridó); BR- 32 (Nova Tropical); BR/ EMGOPA 312 (Potiguar) and BR 35 (Rio Balsas); Embrapa 9 (Bays); Embrapa 30 (Vale do Rio Doce), Embrapa 31 (Mina), Embrapa 33 (Cariri RC) and the Embrapa 34 (Teresina RC); Embrapa 63 (Mirador); MA/BR 64 (Parnaíba) and MA/BR 65 (Sambaíba) and MA/BRS-164 (Patí) and MA/ BRS-165 (Seridó RCH).

Characteristics that have been selected for production in low latitude areas are: suitable maturity, adequate plant height, height to first pod of more than 12 cm ,a non-woody stem, erect plant stature, lodging resistance, pest resistance, good seed quality and high oil content and protein content, high yield, yield stability (robustness), uniform ripening, indehiscent pods, high capacity for N2 fixation, tolerance to aluminum and manganese toxicity, a long juvenile phase and photoperiod insensitivity. Across planting dates and environments, the growing season can vary from 75 to 200 days. Thus a wide range of cultivars have had to be developed for adaptation to many environmental conditions [22, 43]. The average height of the plant can range from 20 to 150 cm or more, according to the cultivar and environment, and the height to first pod may be characteristic of the cultivar. Environmental factors or cultural practices may also affect height to first pod [35], which prevents mechanical harvesting and can cause harvest loss.

2. Production of soybeans in Northeast Brazil

The Northeast region of Brazil is situated between latitudes 1°20'00'' and 19º00'00'' S. Maturity groups adapted to this region range from maturity group 7 to 9.Examples of cultivars developed for this region are BRS 326 (maturity group 8.7) and BRSMA Seridó RCH (9.7). Both cultivars have the long-juvenile trait which reduces the photoperiod sensitivity to the short days of this region; thus allowing soybean to have an adequate developmental period for optimal yield. Much of the success of soybean expansion into Northeast Brazil is due to the development of cultivars having the long-juvenile trait [3]. Development of these cultivars was done through breeding programs using the following cultivars and genotypes: PI 240664, PI 159925, Santa Maria, PR Paranagoiana 77-10001 [24, 20]. Experimental studies in the past century, especially in the United States, led to the classification of soybeans into maturity groups based on their photoperiod response. The maturity groups were: classify groups of soybean according to the response to photoperiod, a total of 13 classes of maturation, which are 000, 00, 0, I, II, III, IV, V, VI, VII, VIII, IX and X [27, 35].

Besides the issue of photoperiodism, soybean expansion into Northeast Brazil is also limited by drought, waterlogging, the harsh environment of the Amazon region, infrastructure, marketing, and the availability of manual labor [11]. Cultivar development was initiated in Piaui state in 1972 and resulted in the release of cultivar Tropical in 1980.This was followed by the development and release of more cultivars for production in Northeast Brazil [3]. Examples are the cultivars DM-Soberana and DM-Nobre released in the central region of Brazil in the states of Bahia and Maranhão. These cultivars were well adapted to the tropical

http://dx.doi.org/10.5772/52184

conditions of this region, demonstrated greater yield than check cultivars, had good yield stability across environments, proper maturity for the growing season, and were adapted for mechanical harvest [39]. In the states of Maranhão, Piauí and Tocantins, a new cultivar, BRSMA-Babacu, was developed for commercial production. Across seven environments in the late 1990's, it demonstrated an average yield of 2,952 kg/ha which was 4% higher than the local cultivar Doko Millionario. It is a late-maturing cultivarwith a maturity of 130 days from emergence to maturity, a plant height of 99 cm, moderately resistant to lodging, purple flower color, brown pod and pod pubescence, yellow seed coat, brown hilum, good resistance theto pod dehiscence, average seed quality and average seed weight of 15.8 g/100 seeds [23]. In these states, the time of planting depends on the rainy season, which occurs in summer (Table 1).

STATE	REGION	PLANTING PERIOD
Maranhão	south (Balsas – Tasso Fragoso city)	Novemberto 15 december
Maranhão	northeast (Chapadinha city)	January
Piauí	south-west (Uruçuí – Bom Jesus city)	November to 15 december

Source: [29].

Table 1. Sowing dates for soybean state and the northeast.

Cultivars	Testes				
	NDF (day)	NDFT (day)	PR (day)	ALTM (cm)	NNP
AVRDC 7	34 c	45 c	43 c	25,0 c	8 c
AVRDC 8	36 c	48 c	42 c	24,4 c	9 c
BRS 155	33 c	44 c	44 c	20,3 c	9 c
JLM 003	41 b	54 b	64 b	41,0 b	10 b
JLM 004	43 b	55 b	61 b	39,4 b	11 ab
Pirarara	57 a	75 a	60 a	62,9 a	12 a
Mean	41	54	52	36	10

Source: [40]
In each column, the same letter not differ at 5% probability by Tukey test.

Table 2. Number of days to flowering (NDF), number of days to fruit set (NDFT), reproductive period (RP), plant height at maturity (ALTM), number of nodes / plant (NPP) Observed in different cultivars of soybean-green. Sand-PB, UFPB, 2004.

Research conducted in the Brazilian states located in Northeast Brazil have shown promising results regarding the adaptation of cultivars to cultivation. Research conducted at the Federal University of Paraiba, Center for Agrarian Sciences, resulted in the development of cultivar Pirarara. This cultivar was adapted to the Entisol soils of the region, as well as the sort days, warm and humid climate, rainfall, and acid soils. It had a plant height of about 63 cm and developed about 12 main stem nodes. Days from emergence to maturity was about 75. Data for the cultivar is shown in Table 2 [40]. Further research with this cultivar demonstrated that when intercropped with corn, it had a vegetative period (days from planting to first flower) of about 55 days. This was a sufficient period to allow for enough dry matter accumulation to optimize yield. Other developed cultivars had shorter vegetative periods, such as Pati (50 days) and JLM 004 (39 days).The significant factor is to allow enough time for node production, since nodes are plant structures from which pods, seeds, and yield are made. This has been demonstrated in research conducted by [19] in which yield was shown to be related to the number of nodes formed. Cultivar Black Alliance had the most at 33, while Kanro had the fewest at 8.

Cultivars	Mean (kg ha^{-1})	li	ij	R^2
BRS TRACAJÁ	1882,87	0,52**	792,69**	99,25
BR JUÇARA	1555,41	1,84**	145934,72**	90,12
BRS MA 165 SERIDÓ	1511,10	1,04ns	21982,14**	95,07
BRR 219 BOA VISTA	1585,36	1,14ns	30844,36**	94,35
BRS SAMBAÍBA	1893,58	0,94ns	145903,06**	70,40
BRS CANDEIA	1552,88	0,91ns	10589,73**	75,27
MA BR 971665	1725,72	0,61**	14432,72**	91,06
CV (%)	22,52			
General mean (kg ha^{-1})	1672,42			

Source: [10]

^{ns} not significant; * Significant at 5% probability level; ** Significant to 1% probability by the F test

Table 3. Average linear regression coefficient (li), misuse of linear regression (ij) and coefficient of determination (R2) According to the methodology of Eberhart and Russell (1996) for grain production of soybean genotypes in Ceará.

Although production of soybean is still restricted in some areas for the northeastern states of Bahia, Maranhão and Piauí, it has shown potential for expansion for other states in the region such as Ceará. Near the city of Pentecost (3° 47′ S Latitude), cultivar trials during 2005 and 2005 have identified genotypes BRS TRACAJÁ, and BRS Sambaíba MA BR 97 1665as having potential to be grown in Ceará (Table 3). Certain soybean cultivars, when intercropped with corn, were capable of producing large biomass relative to other cultivars [42].

Preliminary testing of three soybean cultivars is being conducted by [9] in the coastal plains in the state of Alagoas, near the city of Rio Largo, at a latitude of 9°29'45''. Cultivars used in the study were MG / BR - 46 (Conquista), and BRS Tracajá MA/BRS-165 (Seridó RCH). Agronomic data for plant height is shown in Table 4 and data for yield and seed weight is in Table 5.Although plant height was somewhat lower than that which is usually required for mechanical harvesting, [9] demonstrated that yield production was good for all cultivars ranging from 2,620 to 3,600 kg ha^{-1} (Table 5), being equal to or superior to the traditional regions of production soybean.

Cultivars	Plant height (cm)	
	Plant height stage R2	Plant height stage R9
MG/BR – 46 (conquista)	37.13 a	51.08 a
BRS Tracajá	41.95 a	63.88 a
MA/BRS – 165 (Seriado RCH)	35. 35 a	59.98 a
CV (%)	13.7	12.2

Source: [9]
Means Followed by the same letter not differ to the 5% level of probability by Tukey test.

Table 4. Height of plant growth stages R2 and R9 of three soybean varieties grown in the Coastal Plain region of the state of Alagoas.

Cultivars	Mass of 100 seeds	Grain yield
	--------------(g)-------------	-------------(kg ha^{-1})----------
MG/BR – 46 (conquista)	19.22 a	2,620 a
BRS Tracajá	18.75 a	3,069 a
MA/BRS – 165 (Seridó RCH)	18.89 a	2,829 a
CV (%)	11.2	10.1

Source: [9]
Means Followed by the same letter not differ at the 5% level of probability by the Tukey test.

Table 5. Grain yield and mass of 100 seeds of three varieties grown on Coastal Plain region of the state of Alagoas.

In research conducted by [5], agronomic performance of several soybean cultivars in the rugged areas of the states of Bahia and Sergipe (latitude 10°14'S and 10°55' S). There were two series of screening studies conducted, the first having 14 genotypes and the second having 24. The tests were conducted during the growing seasons of 2006 and 2007 in several cities in remote rural areas of Sergipe and Bahia. Among the many cultivars tested, the ones performing the best were BRS SAMBAÍBA and EMGOPA, followed by BRS TRACAJÁ, BRS CONQUISTA, BRS BELA VISTA, BRS CORISCO and BRS BARREIRAS. All cultivars demonstrated a yield range of 2,285-3,373 kg/ha.

Studies with 18 soybean cultivars, performed by [28] in the state of Piauí, (Lat05°02'40 "S), were planted on 28 February 2010. Based on differences in days to first flower, the cultivars were separated into four groups (Table 6): early (45 to 47.7 days), medium (48-57), late (>57 days), and average.

GENOTYPES*	Cycle (days)		Classification cycle	Reproductive period	
NDF days	NDM days		(PR) days	(% PR)	
BRS SAMBAÍBA	49.00 C	114.00 F	Medium	55.25 C	47.12 B

GENOTYPES*	Cycle (days)		Classification cycle	Reproductive period	
BRS CANDEIA	48.25 C	117.00 E	Medium	66.50 A	55.64 A
BRS 219 BOA VISTA	46.25 D	101.25 I	Early	56.00 C	47.14 B
BRS 271RR	45.75 D	104.50 H	Early	60.00 B	48.62 B
114 BCR336F8	50.50 C	134.00 A	Late	59.25 B	43.40 D
142 SOY94F5G	53.75 B	130.50 B	Late	65.00 A	49.34 B
164 SOY94F5G	51.00 C	114.50 F	Medium	55.50 C	45.63 C
168 BCR1069X7RG	53.25 B	120.75 D	Medium	55.50 C	45.30 C
169 BCR1069X7RG	45.00 D	121.25 D	Medium	55.50 C	45.31 C
170 BCR1069X7RG	48.50 C	123.00 D	Medium	50.00 D	39.52 E
171 BCR1069X7RG	50.25 C	120.25 D	Medium	65.25 A	53.59 A
172 BCR1069X7RG	50.50 C	126.00 C	Late	62.50 B	48.54 B
173 BCR1069X7RG	51.00 C	121.25 D	Medium	56.75 C	46.32 C
174 BCR1069X7RG	47.75 D	122.75 D	Medium	55.00 C	45.17 C
175 BCR1069X7RG	49.00 C	128.00 C	Late	62.00 B	48.06 B
176 BCR1069X7RG	57.25 A	117.50 E	Medium	63.00 B	48.64 B
177 BCR1069X7RG	58.00 A	114.00 F	Medium	56.25 C	45.93 C
179 SOY24F5G	50.25 C	107.50 G	Early	61.25 B	48.62 B
C.V. (%)	4.62	1.55		5.16	4.85
General mean	50.29	118.77		58.91	47.33

Source: [28]
* Means Followed by the same letter not differ by Scott-Knott test ($P \leq 0.05$).

Table 6. Mean values for the number of days to flowering (NDF), number of days to maturity (NDM), reproductive period in days (PR) and percentage (RP%) of different soybean genotypes Evaluated in low latitude, in Teresina - PI.

[28] reported excellent yields with genotypes 171BCR1069X7RG and 174 BCR1069X7RG. They both produced yields of about 4000 kg ha^{-1}. These were much better than other genotypes, the lowest of which yielded below 2000 kg ha^{-1} (Table 7).

GENOTYPES*	PRODUCTIVITY (kg ha^{-1})
171 BCR1069X7RG	4,188.53 A
174 BCR1069X7RG	3,838.66 A
170 BCR1069X7RG	3,464.27 B
177 BCR1069X7RG	3,248.29 B
169 BCR1069X7RG	3,084.99 B
175 BCR1069X7RG	2,802.35 C

GENOTYPES*	PRODUCTIVITY (kg ha^{-1})
176 BCR1069X7RG	2,766.36 C
BRS 219 (BOA VISTA)	2,736.91 C
172 BCR1069X7RG	2,673.71 C
164 SOY94F5G	2,630.52 C
179 SOY24F5G	2,529.99 C
173 BCR1069X7RG	2,446.89 C
142 SOY94F5G	2,403.11 C
168 BCR1069X7RG	2,353.72 C
BRS SAMBAÍBA	2,352.66 C
114 BCR336F8	1,857.68 D
BRS CANDEIA	1,813.55 D
BRS 271RR	1,735.16 D
C.V %	13.21
General mean	2,718.19

Source: [28]
* Means followed by same letter do not differ by Scott-Knott test ($P \leq 0.05$)

Table 7. Mean values of grain yield in different soybean genotypes evaluated in low latitude, Teresina - PI.

Research conducted in western Bahia, near the city of São Desidério (Lat 12°45′S). Growth and yield of soybean at different planting dates are shown in Table 8. Although there were some planting dates at which all cultivars showed good yields, only cultivar M-SOY 8411 yielded well at all four planting dates [9].

Cultivars	Productivity (kg/ha)			
	Ep1	Ep2	Ep3	Ep4
M-SOY 8411	3,924 aA	3,518 aA	2,460 aB	938 aC
BRS Corisco	4,142 aA	2,768 bB	1,745 bC	585 aD
BRS 263	3,956 aA	2,518 bcB	1,163 bcC	659 aC
BRS Barreiras	3,930 aA	1,956 cB	715 cC	642 aC
M-SOY 9350	4,006 aA	2,635 bB	1,029 cC	851 aC
D.M.S = 238,38			C.V. (%)= 14.35	

Source: [9]

Table 8. Average grain yield (kg / ha) from five soybean sown in times Ep1 (29/11/2006), Ep2 (12/14/2006) Ep3 (28/12/2006) and ep4 (01/12/2007), in western Bahia.

Data for Value for Cultivation and Use (VCU) of soybean BRS 278RR for the state of São Paulo indicated that it belongs to a medium maturity group (maturity group 9.4), with the number of days to maturity ranging from 115 to 127 days [26]. According to these authors, the average yield of soybean tested in 24 environments of evaluation was 2,973 kg/ha, and 5.2% more productive than the standard transgenic BRS 271RR, and 5% less productive than the conventional BRS tracajá (Table 9).

Cultivar	Productivity (Kg/ha)				Relative productivity (%)
	2004/05	2005/06	2006/07	Média	
BRS 278RR	2,960	3,059	2,890	**2,973**	**105.2**
BRS Tracajá	3,083	3,298	2,942	**3,115**	**110.2**
BRS 271RR	2,720	3,154	2,568	**2,825**	**100.0**

Source: [26]

Table 9. Average grain yield (kg / ha) and relative yield (%) of BRS 278RR, BRS and BRS 271RR turtle in the agricultural years 2004/05 to 2006/07 of the southern regions of the Maranhão, Piauí southwestern and northern Tocantins.

Because of these pioneering studies to develop cultivars adapted to different regions of northern Brazil, soybean production has expanded rapidly throughout the region. For all the soybean growing regions of Brazil, the greatest expansion has occurred in the North and Northeast regions of the country [8]. Across the region, between the 2008/2009 growing season to the 2009/2010 season, the area of production increased by 8.9% from 2,105,600 to 2,292,600 hectares. A more recent survey by [7] highlights the states of Bahia, Piauí and Maranhão. In the state of Piaui area there was a 16.6% increase in soybean acreage between 2011 and 2012, with the entire state having a production area of 447,300 hectares for soybean. In the state of Bahia, the largest producer of oilseeds in the Northeast, the area planted to soybean increased by 6.6%, from 1,040,000 to 1,110,000 hectares across the same period. The same type of expansion occurred in Maranhao where the growing area increased 12.2%, from 518,200 to 581,400 hectares.

Development of adapted cultivars for Northeast Brazil has greatly aided this expansion into tropical regions. This has made a significant contribution to strengthening the regional agricultural economy. Soybeans provides oil for the food industry and livestock meal for meat production, enhances agricultural development in undeveloped areas, and helps to add a protein-rich source of human food to ameliorate the widespread protein deficiencies common to Northeast Brazil [3].

3. Edaphoclimatic Factors

Expansion of soybean into tropical areas of Brazil has been made possible by development of genotypes having the long juvenile or late flowering trait in short-day photoperiods. This was required to allow soybean to have a sufficiently long developmental period to achieve enough size for optimal yield [37]. Soybean cultivars are affected by photoperiod, humidity, temperature, planting date, altitude, latitude, level of soil fertility and other factors [35]. According to [44], these are the main elements responsible for the variability and difficulties for growing the crop in Brazil. Environmental factors interact with soybean's developmental periods in affecting the growth and yield production of the crop.

Although soybean originated in the temperate world, it has a wide range of adaptability and can be grown well in tropical and subtropical locations. Average temperatures optimal for the best soybean development are between 20 and 35°C. At temperatures outside this range there can be physiological disorders with flowering and rhizobial inoculation, as well as poor growth. [35] claimed that at temperatures below 24°C flowering is retarded by two to three days for each decrease of 0.5°C. Such conditions can happen in the Northeast at altitudes greater than 500 m in the rainy winter period. Such cases occur in Areia-Paraíba, Garanhuns-Pernambuco and Triumfo-Pernambuco, among other cities.

Temperature has a strong influence on the rates of all the metabolic and physiological processes occur during development. This has a direct effect on growth rate and yield. Temperature also has a significant effect on the duration ofthe different developmental periods that make up soybean's crop cycle. In particular, the periods from emergence to flowering and flowering to maturity are affected by it [16]. It is recommended that the planting of soybeans should not be done when the soil temperature is below 20 °C so that germination and plant emergence are not compromised. Once temperatures reach this level, the germination rate increases exponentially with increasing temperature [21]. Thus, the number of days from sowing to the state of emergency (LV) can vary from about 5-15 days, depending on temperature [35]. Also according to the authors, temperature is the main factor influencing plant development, where low temperatures delay and high temperaturesaccelerate seedling emergence and leaf development.

The time course of the vegetative activity of plants is adjusted to local conditions during the growing season. In the dry tropics and subtropics, the growing season is limited by the intensification of water stress when the dry season begins [21]. Under the conditions of northeastern Brazil, the predominant climate is hot and dry. According to some authors [14, 35], for soybean temperatures above 40 °C have an adverse effect on the rate of growth, the rate of formation of the nodes and internode growth and floral initiation. This causes problems with flowering, pod formation and retention, and results in lower yield. These problems are accentuated with the occurrence of water deficits.

Differences in flowering date between yearsby a cultivar sown in the same season are due to temperature variations. Thus, high temperatures can cause soybean to flower too early, which may cause a decrease in plant height,and accelerate the maturity of the crop [14]. The nodulation and nitrogen fixation in soybean are greatly affected by soil temperature and the growth of Bradyrhizobiumjaponicum is limited by temperatures above 33 °C [35]. When facing environmental stress, there is always a genotype x environmental interaction which must be considered during the breeding program for cultivar development [32].Solar radiation is a critical environmental factor because it directly affects the crop growth rate and the ability to obtain enough dry matter for optimal yield potential. Water availability is also very important, because it affects leaf expansion, photosynthetic rate, crop growth rate, and nitrogen fixation [30].

Not only is the level of light important, but length of day, or photoperiod, is also important. Photoperiod affects many developmental processes such as flowering, seed germination, growth of stems and leaves, formation of storage organs and assimilate partitioning [38]. It also affects leaf expansion, photosynthesis, senescence, dormancy of buds, and other processes [16]. Photoperiodism is a term describing all these plant responses to day length or photoperiod. Photoperiod will be determined by the latitude of where the crop is grown and the planting date.For soybean, the effective photoperiod includes the time from sunrise to sunset and some of civil twilight. Three types of photoperiodic response are recognized: short-day response in which flowering is induced and/or accelerated at a certain critical day length or less; long-day response in which flowering is induced and/or accelerated at a certain critical daylength or longer; and day neutral in which flowering time is unaffected by day length. Soybean is a short-day plant.

Since vegetative growth in soybean occurs between emergence and the start of seed filling, it is important that soybean be planted at a latitude and planting date where day lengths will be long enough to allow enough time to the start of seed filling so that dry matter accumulation is large enough for optimal yield potential. The major problem with the expansion of soybean into northern Brazil is in these low latitudes, day length is seldom longer than 12 hours per day. Ideally, it is best to plant in October or November to take advantage of the lengthening days that occur at this time and continue until the summer solstice in late December. Since the critical short-day period to hasten flowering occurs when the photoperiod is shorter than 13.5 to 14.5 hours (the length varies with maturity group and cultivars within maturity groups), the typical soybean planted in this region will have too short a time to flowering and the start of seed filling to achieve enough size for optimal yield. For this reason, much research in Brazil has concentrated on incorporating the "long-juvenile" trait into soybean cultivars. The juvenile period is the initial period in the plant's growth when it's developmental rate does not accelerate in response to short days. Thus, cultivars having this trait will have a longer time to flowering and the start of seed filling compared to those that do not. Incorporation of this trait into Brazilian cultivars has greatly facilitated soybean expansion into northern Brazil. [25, 15, 14] .

The physiology of the mode of action of the long juvenile period is still unknown. The trait is either caused by a delayed ability of the plant to respond to short days or a requirement for more short day cycles to induce and promote flowering relative to other soybeans [14].The problem of short-day induced premature flowering in northern Brazil was approached by selecting genotypes with insensitivity to photoperiod and/or having a long juvenile period [27]. Several breeding programs have contributed to the development of high-yielding cultivars adapted to different agro-climatic conditions of Brazil. For expansion into central and northern Brazil, breeding programs have followed a strategy of developing genotypes adapted to low latitudes, through the incorporation of the long juvenile trait. Research conducted by [33] concluded that the germplasm bank of soybeans genetic variability has remained relatively constant over the last 30 years.

The availability of water is another important environmental stress for soybean culture in northern Brazil. The first critical period is seed germination and seedling development. Soybean seeds need to absorb enough water to achieve 50% moisture content to ensure good germination. Soil water content needs to be between 50-85% of total available water. According to [25], an annual rainfall 700-1200 mm that is well distributed during the soybean cycle (500 to 700 mm) will meet the crops water needs. Soybean's water requirement increases with plant development, reaching a maximum during the flowering and grain filling periods (7-8 mm / day), and decreasing thereafter [12].

The growing season in northern Brazil is largely determined by occurrence of the rainy and dry seasons of the region. The dry season occurs between May to September and any agriculture during this time requires irrigation [16]. Drought is usually the main factor responsible for crop losses. The two most critical periods for drought stress in soybean are during seed emergence to seedling establishment and the grain filling period. During germination, both excess and lack of water are harmful to crop establishment. Soils having a low water storage in general are unfit for soybean production for most cultivars and planting dates [13].

The soybean plant requires more water as growth and development proceed. Peak demand is during flowering and early pod formation and remains high until physiological maturity. The most critical period is during flowering and early pod formation. Drought stress at this time will cause abortion of flowers and pods, resulting in lower seed production and reduced yield [27]. Rainfall data from the state of Bahia indicates that there is a 80% chance for receiving sufficient rainfall to avoid drought stress during the critical periods [36]. The most important variable to be considered when dealing with drought stress is water retention in the soil during the growing season [27]. In summary, obtaining optimal yield in a given environment, depends upon maximizing the genetic potential of a given cultivar [31]. This involves having a long enough time for vegetative growth so that the prevailing photoperiod does not reduce yield, and the avoidance of drought stress [34]. Experimental trials in Brazil have shown there is a genetic yield potential of more than 5,000 kg/ha grain [2]. Genetic potential for most cultivated crops has increased greatly over the last 100 years. However, realizing this potential will depend on coping with environmental stress [27]. Soybean is recognized as having wide genetic diversity and with proper genetic and breeding research it can be adapted to a wide range of environmental conditions. Efforts in Brazil over the last 40 years to deal with problems presented by photoperiodic adaptation and coping with other environmental stresses have shown how this can happen through cooperative research efforts involving breeders, physiologists, geneticists, and agronomists.

4. Conclusion

Brazil has become the second largest producer of soybeans in the world. This has occurred largely because of our success at developing soybean cultivation in the low latitude regions (1°20'00" and 19º00'00" S) in central and northern Brazil. Much of the area opened to soybean production lies in the Cerrado region which is Brazil's agricultural frontier.

Northeast Brazil is also an area for agricultural development. It has a wide variety of environmental zones and is suitable for mechanized agriculture on small- or medium-sized farms. It has potential for production of many crops ranging from pastures for dairy cattle to bio-energy crops like sugar cane (ethanol) and biodiesel (soybean). Our progress in the agricultural development of this region is due to concentrated investment of financial and human resources in genetic improvement of cultivars and their adaptation to different climates in this region. The largest effort has been the development of cultivars with the long-juvenile trait which has overcome the problem of the short photoperiods common to this tropical region. Thus, short-day induced premature flowering is no longer a problem, and soybean can be grown in northern Brazil that achieve a size suitable for optimal yield. The greatest expanse of soybean production in northern Brazil has occurred in the states of "Bahia, Maranhão and Piauí". It is in these states that the public and private research efforts have been focused. Similar efforts have now been started in other northern states such as "Ceará, Sergipe, Paraiba, Pernambuco, Alagoas and Rio Grande do Norte". University and state research efforts are now being conducted to develop adapted soybean cultivars for the small and medium-sized farms of this region. Much effort will also have to be invested into marketing and transport. We firmly believe that with continued research efforts, soybean production in northern Brazil will expand into the future and that Brazil will become the largest soybean producer in the world within the next 20 years.

Author details

Jeandson Silva Viana[1*], Edilma Pereira Gonçalves[1], Abraão Cicero Silva[1] and Valderez Pontes Matos[2]

*Address all correspondence to: jeandson@uag.ufrpe.br edilma@uag.ufrpe.br abraaocicero@yahoo.com.br vpmatos@ig.com.br

1 Unidade Acadêmica de Garanhuns, Universidade Federal Rural de Pernambuco, Brazil

2 Departamento de Fitotecnia, Universidade Federal Rural de Pernambuco, Brazil

References

[1] Almeida, L. A., Kiihl, R. A. S., Miranda, M. A. C., & Campelo, G. J. A. Soybean breeding for low latitude regions. *QUEIRÓZ, M. A. de; GOEDERT, C.O.; RAMOS, S.R.R. (Ed.). Genetic Resources and Plant Breeding for the Brazilian Northeast. (on line). Versão 1.0. Petrolina-PE: Embrapa Semi-Árido/Brasília-DF: Embrapa Recursos Genéticos e Biotecnologia*, Disponível via Word Wide Web, http://www.cpatsa.embrapa.br.

[2] Bonato, E. R. (2000). Stress in soybean. *Passo Fundo: EMBRAPA Trigo*, 254.

[3] Campelo, G. J., Kiihl, R. A. S., & Almeida, L. A. (1998). Soybean: development for low latitude regions. *Teresina: Embrapa Meio-Norte*, 36, (Embrapa Meio-Norte. Documentos, 37).

[4] Campelo, G. J. A., Kiihl, R. A. S., & Almeida, L. A. Morphological and agronomic characteristics of soybean cultivars developed for the low latitude regions. *QUEIRÓZ, M. A. de; GOEDERT, C.O.; RAMOS, S.R.R. (Ed.). Genetic Resources and Plant Breeding for the Brazilian Northeast. (on line). Versão 1.0. Petrolina-PE: Embrapa Semi-Árido/ Brasília-DF: Embrapa Recursos Genéticos e Biotecnologia,* Disponível via Word Wide Web, http://www.cpatsa.embrapa.br.

[5] Carvalho, H. W. L., Oliveira, I. R., Garcia, A., Feitosa, L. F., Rodrigues, C. S., Menezes, A. F., & Melo, K. E. (2009). Evaluation of the agronomic performance of soybean cultivars in the rugged area of the states of Bahia and Sergipe. *Aracajú. Comunicado técnico.*, http://www.cpatc.embrapa.br/publicacoes_2010/cot_91.pdf, accessed: 01 de may de 2012.

[6] Castro, A. A. J. F. (1997). Characteristics of the vegetation of the Mid-North. *CERRADOS SYMPOSIUM ON THE MID-NORTH, 1997. Teresina. Anais... Teresina: Embrapa/ Meio-Norte*, 45-56.

[7] Companhia Nacional de Abastecimento. (2012). CONAB:Raising production and productivity in Brazil. http://www.conab.gov.br/OlalaCMS/uploads/arquivos/ 12_04_11_15_04_18_boletimabril_2012.pdf, accessed: june.

[8] Companhia Nacional de Abastecimento- CONAB. (2010). Monitoring of the Brazilian harvest: grains, the sixth survey. March 2010/ Brasília, 42.

[9] Cruz, T. V., Peixoto, C. P., & Martins, M. C. Growth and yield of soybean at different planting dates in western Bahia. *Scientia Agraria, Curitiba*, 11(1), 033-042.

[10] Dias, F. T. C., Pitombeira, J. B., Teofilo, E. M., & Barbosa, F. S. (2009). Adaptability and phenotypic stability for the character grain yield in soybean cultivars for the State of Ceará. Revista Ciência Agronômica, Fortaleza, jan-mar , 40(1), 129-134.

[11] Dossa, D., Contini, E., & Ávila, A. F. D. (1986). The allocation of resources and profitability of the National Center for Soybean Research. *Londrina, EMBRAPA, CNPS, (EMBRAPA-CNPS. Documentos, 26)*, 35.

[12] EMBRAPA, Empresa Brasileira de Pesquisa Agropecuária. (2010). Soybean production technologies central Brazil in 2011. *Londrina, PR: Embrapa Soja: Embrapa Cerrados: Embrapa Agropecuária Oeste*, 255.

[13] Farias, J. R. B., Assad, E. D., Almeida, I. R., Evangelista, B. A., Lazzarotto, C., Neumaier, N., & Nepomuceno, A. L. (2001). Characterization of risk of drought in the soybean growing regions in Brazil. *Revista Brasileira de Agrometeorologia, Passo Fundo*, 9(3), (N#Special: Agricultural Zoning), 415-421.

[14] Farias, J. R. B., Nepomuceno, A. L., & Neumaier, N. (2007). Ecophysiology soybean. *EMBRAPA/Soja, Londrina*, 9, (Circular Técnica, 48).

[15] Fieltz, M. R., & Rangel, M. A. S. (2004). Effect of water stress and photoperiod on the yield of soybean planted in Dourados, MS. *Dourados: Embrapa Agropecuária Oeste*, 5, (Embrapa Agropecuária Oeste. Comunicado Técnico, 89).

[16] Floss, E. L. (2004). Physiology of crop plants: a study that is behind what you see. *Passo Fundo: UPF*, 536.

[17] Freitas, M. C. M. (2011). The soybean crop in Brazil: the growth of Brazilian production and the emergence of a new agricultural frontier. *EnciclopédiaBiosfera, Goiânia*, 7(12).

[18] Frota, A. B., & Campelo, G. J. A. Evolution and prospects of soybean production in the Middle-North region of Brazil. *QUEIRÓZ, M. A. de; GOEDERT, C.O.; RAMOS, S.R.R. (Ed.). Genetic Resources and Plant Breeding for the Brazilian Northeast. (on line). Versão 1.0. Petrolina-PE: Embrapa Semi-Árido/Brasília-DF: Embrapa Recursos Genéticos e Biotecnologia*, Disponível via Word Wide Web, http://www.cpatsa.embrapa.br.

[19] Guerra, E. P., Destro, D., Miranda, L. A., & Montalvan, R. (1999). Performance of foodtype genotypes and their possibility for adaptation to Brazilian latitudes. Pesquisa Agropecuária Brasileira, Brasília, , 34(4), 575-583.

[20] Hartwig, E. E., & Kliihl, R. A. S. (1979). Identification and utilization of a delayed flowering character in soybean for short-day conditions. *Field Crop Research.*, 2, 145-51.

[21] Larcher, W. (2004). Environmental influences on growth and on development. *LARCHER, W. Plant ecophysiology. São Carlos: RiMa*, 531.

[22] Marchiori, L. F. S., Camara, G. M. S., Peixoto, C. P., & Martins, M. C. (1999). Vegetative performance of soybean cultivars in normal times and off-season. *Scientia Agricola, Piracicaba*, 56(2), 383-390.

[23] Miranda, M. A. C., Kiihl, R. A. S., Almeida, L. A., Campello, G. J. A., & Klepker, D. (1999). Cultivar BRSMA BABAÇU. *SOYBEAN RESEARCH MEETING OF THE CENTRAL REGION OF BRAZIL, 2. Dourados, MS. Abstract... Dourados: Embrapa Agropecuária Oeste, Londrina: Embrapa Soja, 1999*, (Embrapa Agropecuária Oeste. Documentos, 7, Embrapa Soja. Documentos, 134), 202-247.

[24] Miyasaka, S., Guimaraes, G., Kiihl, R. S. A., Lovadini, L. A. C., & Dematte, J. D. (1970). Soybean varieties indifferent to photoperiodism and tolerant to low temperatures. *Bragantia, Campinas*, 29.

[25] Mondine, M. L., Vieira, C. P., & Cambraia, L. A. (2001). Seeding time: an important factor affecting the productivity of soybean. *Dourados: Embrapa Agropecuária Oeste*, 16, (Embrapa Agropecuária Oeste. Documentos, 34).

[26] Moreira, J. U. V., Lambert, E. S., Almeida, L. A., Pereira, M. J. Z., Meyer, M. C., Klepker, D., Montalvan, R., Pipolo, A. E., Toledo, J. F. F., Kaster, M., Arias, C. A. A., Carneiro, G. E. S., Oliveira, M. F., Soares, R. M., Almeida, A. M. R., Dias, W. P., Car-

rao-Panizzi, M. C., Abdelnoor, R. V., Brogin, R. L., Arantes, N. E., Mello, Filho. O. L., El -Husny, J. C., & Gianluppi, V. (2008). Cultivar BRS 278RR: indication for cultivation in the south of Maranhão, Piauí southwestern and northern Tocantins. *XXX Soybean Research Meeting of the Central Region of Brazil.- Londrina: EmbrapaSoja*, 173-174.

[27] Mundstock, C. M. (2005). Soybeans: factors affecting growth and yield. *Department of crop plants, Federal University of Rio Grande doSul: EVANGRAF*, 31.

[28] Neves, J. A. (2011). Agronomic performance of soybean under low latitude in Teresina-PI. *Dissertação. Teresina*, 94.

[29] Pereira, M. J. Z., Klepker, D., & Moreira, J. U. V. (2011). Soybean varieties, northern and northeastern Brazil. *Embrapa soja, Londrina*.

[30] Portugal, A. M., Peixoto, C. P., Costa, J. A., & Cristina, D. (2000). Light intensity and water regime in dry matter accumulation of soybean plants potted [Glycine max (L.) Merril]. *Anais... Brazilian Meeting on management and conservation of soil and water. Ilhéus, CD-Room*.

[31] Potafos, Brazilian Association for Research and Potash Phosphate. (1997). As the soybean plant develops. *Piracicaba*, 20.

[32] Prado, E. E., Hiromoto, D. M., Godinho, V. P. C., Utumi, M. M., & Ramalho, A. R. (2001). Adaptability and stability of soybean cultivars in five planting seasons in Rondônia cerrado. *Pesquisa Agropecuária Brasileira, Brasília*, 36(4), 625-635.

[33] Priolli, R. H. G., Mendes-Junior, C. T., Sousa, S. M. B., Sousa, N. E. A., & Contel, E. P. B. (2004). Genetic diversity of soybean in time and among breeding programs in Brazil. *Pesquisa Agropecuária Brasileira, Brasília*, 39(10), 967-975.

[34] Rodrigues, O., Didonet, A. D., Lhamby, J. C. B., Bertagnolli, P. F., & Luz, J. S. (2001). Quantitative response of soybean flowering to temperature and photoperiod. *Pesquisa Agropecuária Brasileira, Brasília*, 36(3), 431-437.

[35] Sediyama, T. (2009). Technology of production and use of soy. *Londrina, Mecenas*, 314.

[36] Da Silva, S. C., & Assad, E. D. (2001). Climatic risks zoning for upland rice in the states of Goias, Mato Grosso, Mato Grosso do Sul, Minas Gerais, Tocantins and Bahia. *Revista Brasileira de Agrometeorologia, Passo Fundo*, 9(3), (n#Special: Agricultural Zoning), 536-543.

[37] Spehar, C. R. (1999). Breeding soybeans to the latitudes of Brazilian cerrados (savannahs). *Pesquisa Agropecuária Brasileira, Brasília*, 29(8), 1167-1180.

[38] Taiz, L., & Zeiger, E. (2004). plant physiology. *3.ed. Porto Alegre: Artmed*, 613.

[39] Vehrmann, V. K. (1999). Launch of the soybean cultivar'DM-Sovereign "in Bahia. *SOYBEAN RESEARCH MEETING OF THE CENTRAL REGION OF BRAZIL, 2. Dourados, MS. RESUMOS... Dourados: Embrapa Agropecuária Oeste, Londrina: Embrapa Soja, 1999. p247*, (Embrapa Agropecuária Oeste. Documentos, 7, Embrapa Soja. Documentos, 134), 91.

[40] Viana, J. S. (2007). Cultivars and soybean cropping systems in green Areia-PB. *(Doctorate in Agronomy) by the Centre for Agrarian Sciences, Federal University of Paraíba. Areia- PB*, 139.

[41] Viana, J. S., Bruno, R. L. A., Mendonca, J. L., Goncalves, E. P., Alves, E. U., & Braga Jr, J. M. (2009a). Precocity of soybean in cropping systems with maize in sand, State of Paraiba. *Acta Scientiarum Agronomy*, 31(3), 481-488.

[42] Viana, J. S., Bruno, R. L. A., Santos, H. C., Silva, I. F., Oliveira, A. P., & Braga Jr, J. M. (2009b). Dry weight of green soybean cropping systems with corn. *Scientia Agraria, Curitiba*, 10(5), 413-418.

[43] Yokomizo, G. K., Duarte, J. B., & Vello, N. A. (2000). Correlations between grain size and other characters in topcrosses soybean type food with grain type. *Pesquisa Agropecuária Brasileira, Brasília*, 35(11), 2235-2241.

[44] Moraes, AVC, Camargo, MBP, Mascarenhas, HAA, Miranda, MAC & Perera, JCVNA (1998). Testing and analysis of agrometeorological models to estimate yield for the soybean crop in the region of Ribeirão Preto. Bragantia, Campinas, 57 (2), 393-406.

Section 4

Soybean Genetics

Chapter 19

Soybean Proteomics: Applications and Challenges

Alka Dwevedi and Arvind M Kayastha

Additional information is available at the end of the chapter

http://dx.doi.org/10.5772/52408

1. Introduction

Proteomics is one of the most explored areas of research based on global-scale analysis of proteins. It leads to direct understanding of function and regulation of genes. Significant advances in the comprehensive profiling, functional analysis, and regulation of plant proteins have not advanced much as compared to model organisms such as yeast, humans etc. The application of proteomic approaches to plants implicates; comprehensive identification of proteins, their isoforms, as well as their prevalence in each tissue, characterizing the biochemical and cellular functions of each protein and the analysis of protein regulation and its relation to other regulatory networks [1]. Genes of higher eukaryotes (including plants) contain introns which are large and numerous. Therefore, combinational exon usage originating from complex gene structures results in a multitude of splice variants leading to generation of different protein products from a given gene. Thus, the determination of the comprehensive pattern of expression of each protein isoform is a challenging task, most importantly for poorly expressed proteins [2].

The two-dimensional gel electrophoresis (2-DE) is used for profiling protein expression involving separation of complex protein mixtures by molecular charge in the first dimension and by mass in the second dimension. Recent advancement in 2-DE has improved resolution and reproducibility but still automation in high-throughput setting is lagging. The alternative approaches like multi-dimensional protein identification technology involving large-scale proteomics are able to generate a large catalog of proteins present in complex cell extracts. Further, detection of low abundance proteins using sub-cellular fractionation reduces the complexity of protein extracts. These efforts have successfully characterized nuclear, chloroplast, amyloplast, plasma membrane, peroxisome, endoplasmic reticulum, cell wall, and mitochondrial proteomes of a model plant, *Arabidopsis*. Although, high-throughput technologies have helped in characterization of *Arabidopsis* and other organisms' proteomes, characterization of various protein classes including membrane and hydrophobic proteins which are recalcitrant to isolation and analysis is still inaccessible [3].

Food allergy can be a serious nutritional problem in children and adults. Any protein-containing food has the potential to elicit an allergic reaction in the human population. Antibody IgE-mediated reactions are the most prevalent allergic reactions to food. These responses occur after the release of chemical mediators from mast cells and basophils as a result of interactions between food proteins and specific IgE molecules on the surface of these receptor cells. Eight foods or food groups have been identified as the most frequent sources of human food allergens and account for over 90% of the documented food allergies worldwide. These foods are milk, eggs, fish, crustaceans, wheat, peanuts, tree nuts and soy [4]. Despite their well-documented allergenicity, soy derivatives continue to be increasingly used in a variety of food products due to their well-documented health benefits. Soybean has also been one of the selected target crops for genetic modification (GM). For example, the artificial introduction of 5-enolpyruvylshikimate-3-phosphate synthase in soybean crop creates an alternative pathway which is insensitive to glyphosate (most potent herbicide), thus increasing overall crop yield. One of the major concerns regarding the safety of GM foods is the potential allergenicity of the resulting products, namely the possible occurrence of either altered or *de novo* expressed of endogenous allergens after genetic manipulation. This concern justifies careful plant characterization [5]. Proteomics is one of the powerful approaches allowing rapid and reliable protein identification. It can provide information about their post-translational modifications, sub-cellular localization, level of protein expression and protein-protein interactions. Despite the importance of soybean and the availability of powerful tools for the analysis of proteins from sub-cellular organelles, and specifically for the identification of allergens, only a limited number of reports have been published to date.

Soybean is an important source of protein for human and animal nutrition, as well as a major source of vegetable oil. Although soybean is adapted to grow in a range of climatic conditions including adverse environmental and biological factors, still it has been affected with respect to growth, development, and global production For instance, drought reduces the yield of soybean by about 40%, affecting all stages of plant development from germination to flowering thus reducing the quality of the seeds. [6]. Several other abiotic stresses, such as flooding, high temperature, irradiation, or the presence of pollutants in the air and soil have detrimental effects on the growth and productivity of soybean. Along with morphological and physiological studies on the responses of plants to stress conditions, several molecular mechanisms from gene transcription to translation as well as metabolites were investigated. Recent advances in the field of proteomics have created an opportunity for dissecting quantitative traits in a more meaningful way. Proteomics can investigate the molecular mechanisms of plants' responses to stresses and provides a path toward increasing the efficiency of indirect selection for inherited traits. In soybean a comprehensive functional genomics is yet to be performed; therefore, proteomics approaches form a powerful tool for analyzing the functions of complete set of proteins including those involved in stress protection.

http://dx.doi.org/10.5772/52408

2. Proteomics: isolation, identification and classification

In plant proteomics, the type of the plant species, tissues, organs, cell organelles, and the nature of desired proteins affect the techniques that can be used for protein extraction. Furthermore, the extraction process becomes more tedious when the protein is present inside vacuoles, rigid cell walls, or membrane plastids. A perfect protein extraction method involves complete solubilization of total proteins from a given sample and minimizing post-extraction artifact formation, proteolytic degradation as well as removal of non-proteinaceous contaminants. To date, only the proteome of *Arabidopsis* and rice have been studied while less attention has been paid to other plants including soybean. Soybean has high levels of phenolic compounds, proteolytic and oxidative enzymes, terpenes, organic acids, and carbohydrates due to which protein extraction is very tedious. Further it contains contains large quantities of secondary metabolites, *viz.* flavone glycosides (kaempferol and quercetin glycosides), phenolic compounds, lipids and carbohydrates. Thus impedes high-quality protein extraction in turn high-resolution protein separation in 2-DE.

In classical proteome analyses, proteins are initially separated by a 2-DE technique with isoelectric focusing (IEF) as the first dimension and sodium dodecyl sulfate polyacrylamide gel electrophoresis (SDS-PAGE) as the second dimension. A greater resolution in protein separation has been achieved by introducing immobilized pH gradients (IPGs) for the first dimension. Methodological advances in 2-DE have led to the introduction of two-dimensional fluorescence difference gel electrophoresis (2D-DIGE), which has been used for the comparative analysis of the proteome of soybean subjected to abiotic and biotic stresses [7]. The separated proteins can be subsequently identified by sequencing or by mass spectrometry. By introduction of mass spectrometry into protein chemistry, matrix-assisted laser desorption/ionization time-of-flight mass spectrometry (MALDI-TOF MS) and liquid chromatography/tandem mass spectrometry (LC-MS/MS) have become the methods of choice for high-throughput identification of proteins. An alternative technique known variously as 'gel-free proteomics', 'shotgun proteomics', or 'LC-MS/MS-based proteomics' can also be used in high-throughput protein analysis. This approach is based on LC separation of complex peptide mixtures coupled with tandem mass spectrometric analysis. A multidimensional protein identification technology (MudPIT) that usually incorporates separation on a strong cation exchange, reverse-phase column and MS/MS analysis helps the efficient separation of complex peptide mixtures. The gel-free technique have the advantage of being capable of identifying low-abundance proteins, proteins with extreme molecular weights or p*I* values, and hydrophobic proteins that cannot be identified by using gel-based technique. A combination of gel-based and gel-free proteomics has been used for identification of soybean plasma membrane proteins under abiotic stress, *viz.* flooding, osmotic, salinity stress. Methods for protein identification are not usually organism specific, and they can be applied to a wide range of living organisms in addition to soybean. Identification of proteins is normally performed by using a database search engine such as MASCOT or SEQUEST.

Soybean has an estimated genome size of 1115 Mbp, which is significantly larger than those of other crops, such as rice (490 Mbp) or sorghum (818 Mbp). Sequencing of the 1100 Mbp of total soybean genome predicts the presence of 46,430 protein-encoding genes, 70% more than in *Arabidopsis* [8]. The soybean genome database contains 75,778 sequences and 25,431,846 residues have been constructed on the basis of the Soybean Genome Project, DOE Joint Genome Institute; this database is available at http://www.phytozome.net. Although the genome sequence information is almost completed, no high-quality genome assembly is available because the results from the computational gene-modeling algorithm are imperfect. In addition, duplications in the genome of soybean result in nearly 75% of the genes being present as multiple copies, which further complicate the analysis. The soybean proteome database (http://proteome.dc.affrc.go.jp/Soybean) provides valuable information including 2-DE maps and functional analysis of soybean proteins. However, the presence of a considerable number of proteins with unknown functions highlights the limitations of bioinformatics prediction tools and the need for further functional analyses. The cellular proteomics helps in identification of changes in protein expression under different growing condition and treatments. The analytical methodology for the separation and identification of a large numbers of proteins should be authentic and confirmable. The proteome map of mature dry soybean seeds has been prepared by employing robotic automation at subsequent steps of 2-DE. Further, UniGene database was implemented for proteins identifications. Total protein from mature dry soybean (*Glycine max* cv. Jefferson) seed was isolated and 2D-PAGE performed using 13 cm IPG strips and subsequently doing SDS-PAGE. Protein spots were analyzed using Phoretix 2D-Advanced software. Excised protein spots were arrayed into 96-well plates and transferred to a Multiprobe II EX liquid handling station for subsequent destaining, tryptic digestion and peptide extraction. MALDI-TOF MS was operated in the positive ion delayed extraction reflector mode. Peptide spectra were submitted to a MS Fit program of Protein Prospector. Assignments from UniGene contigs were subsequently searched against the NCBI non-redundant database using the BLASTP search algorithm to determine similarity matches [9].

Trichloroacetic acid (TCA)/acetone-based and phenol-based buffers are most frequently used in protein extraction from plants. A comprehensive proteomic study was performed on nine organs from soybean plants in various developmental stages by using three different methods for protein extraction and solubilization. The results showed that the use of an alkaline phosphatase buffer followed by TCA/acetone precipitation caused horizontal streaking in 2-DE while use of a Mg/NP-40 buffer followed by extraction with alkaline phenol and methanol/ammonium acetate produced high-quality proteome maps with well-separated spots, high spot intensities, and high numbers of separate protein spots in 2-DE gels [10, 11]. In the case of organelle proteomics particularly that of membrane proteomics, a different extraction procedure is required that involves modifications to dissolve hydrophobic proteins and additional purification steps. Furthermore, when studying protein–protein interactions, it is necessary to extract protein complexes by using buffers with less or no detergent to get the proteins in their native states. Despite the importance of seed filling in the synthesis of storage reserves for germination, systematic proteomic analysis of this phase in legumes is yet to be carried out.

Total seed proteins of soybean (cv. Maverick) at different stages of flowering (14, 21, 28, 35 and 42 days) were isolated and subsequently 2D-PAGE was done. Initially IPG strips of pH 3 to 10 were taken then narrowed down to pH range to 4 to 7 for high-resolution proteome maps. A total of 488 and 679 proteins were identified from 2D-PAGE gels of pH range 4 to 7 and 3 to 10 gels, respectively. Each of the 679 proteins was excised from reference gels for identification by MALDI-TOF MS and a total of 422 proteins (62%) were identified. One unique protein was often represented by more than one spot on the 2D-PAGE gel, most likely due to post-translational modifications or genetic isoforms. Taking into account this redundancy, 216 unique proteins out of 422 were identified. A total of 82 proteins were associated with metabolism (the largest functional class) and the second largest functional class were comprised of 52 spots assigned to the seed storage proteins β-conglycinin and glycinin. An overall down- and up-regulation was observed for metabolism and storage related proteins, respectively, during seed filling, suggesting metabolic activity curtails as seeds approach maturity. Abundance of proteins related to metabolite transporter, disease and defense, energy production, cell growth and division, signal transduction, protein synthesis and secondary metabolism did not vary significantly. Furthermore, 13 sucrose-binding proteins have been mapped to the same UniGene accession number, suggesting the importance of sucrose as a signaling molecule in seed and embryo development. There were a total of 92 unknown proteins which could not be classified, therefore grouped into five expression profiles [12].

3. Implication of proteomics in understanding soybean stress

Soybean is grown worldwide with an average protein content of 40% (highest protein content with respect to other food crops) and oil content of 20% (which is second only to that of groundnut among the leguminous foods). Furthermore, soybean improves soil fertility by fixing nitrogen from the atmosphere in symbiosis with nitrogen fixing bacteria. It is, however, susceptible to various types of stresses (abiotic and biotic). Tolerance and susceptibility to stresses are complex phenomena because they are quantitatively inherited and can occur during different stages of plant growth and development. Extrinsic stress is regarded as the most important stress agent, which results from changes in abiotic factors such as temperature, climatic factors and chemical components, either naturally occurring or manmade. Further, biotic stresses (occurs as a result of damage done to plants by other living organisms, such as bacteria, viruses, fungi, parasites, beneficial and harmful insects, weeds, bacterial, fungal, algal and viral diseases) can also cause huge deterioration in plant growth and yield. Plants have developed adaptive features against these stresses. The genome remains unchanged to a large extent in any particular cell while proteins change dramatically as genes are turned on or off in response to stress. The proteome determines the cellular phenotype and its plasticity in response to external signals. It is proteins that are directly involved in both normal and stress-associated biochemical processes. Therefore, a more complete understanding of stress in soybean may be gained by looking directly into the proteins within a stressed cell or tissue. Proteomic based techniques that allow large-scale protein profiling are powerful tools for the identification of proteins involved in stress-responses in plants. Extensive studies have evaluated changes in protein levels in plant tissues in response to stresses. Unfortunately, these studies have been mainly focused on non-legume species such as *Arabidopsis* and rice, and only recently have been enlarged to include some legumes. As a result only a handful of studies have been carried out in legumes, although in the next few years there should be a significant increase in the number of legume species and stresses would be analyzed. Recently, proteomic approaches have been applied to various legumes like *M. truncatula*, lentils, lupin, common bean, cowpea and soybean to identify proteins involved in the response to different stresses. Interestingly, many of the induced proteins from these different stresses were common or belonged to overlapping pathways [13].

Considerable amount of research has been carried out during the last decade to find the effect of stress under extreme . These include chloroplast membrane, cell wall and nuclear envelope, while some researchers have focused on individual tissues *viz.* seeds, mitochondria, root tips, vacuoles, chloroplasts and thylakoids. To date, lots of reports have come which emphasize changes in protein expression levels during a particular or integrative stress consequently affecting cellular metabolism. Proteomics provides direct assessment of the biochemical processes of monitoring the actual proteins performing signaling, enzymatic, regulatory and structural functions encoded by the genome and transcriptome.

Following are the different categories of proteins with important properties, which have been shown to play a crucial role against abiotic environmental stress as well as biotic stress. The data so collected from various plants including soybean is based on 2-DE, mass spectrometry and bioinformatics tools.

(a) *Antioxidants Enzymes*

Reactive oxygen species (ROS) in plant cellulars are produced as a consequence of myriad stimuli ranging from abiotic and biotic stress, production of hormonal regulators, as well as cell processes such as polar growth and programmed cell death [14]. These reactive molecules are generated at a number of cellular sites, including mitochondria, chloroplasts, peroxisomes, and at the extracellular side of the plasma membrane. ROS trigger signal transduction events, such as mitogen-activated protein kinase cascades eliciting specific cellular response.s. The influence of these molecules on cellular processes is mediated by both the perpetuation of their production and their amelioration by scavenging enzymes such as superoxide dismutase, ascorbate peroxidase, and catalase. The location, amplitude, and duration of production of these molecules are determined by the specificity of the responses [15]. Accumulation of ROS as a result of various environmental stresses is a major cause of loss of crop productivity worldwide. ROS affect many cellular functions by damaging nucleic acids, oxidizing proteins, and causing lipid peroxidation. It is important to note that whether ROS will act as damaging, protective or signaling factors depends on the delicate equilibrium between ROS production and scavenging at the proper site and time. ROS can damage cells as well as initiate responses such as new gene expression. The cell response evoked is strongly dependent on several factors. The subcellular location for formation of ROS may be especially important for a highly reactive ROS, because it diffuses only a very short distance before reacting with a cellular molecule. Stress-induced ROS accumulation is counteracted by enzymatic antioxidant systems that include a variety of scavengers, such as superoxide dismutase, ascorbate peroxidase, glutathione peroxidase, glutathione S-transferase, catalase and non-enzymatic low molecular metabolites, such as ascorbate, glutathione (red.), α-tocopherol, carotenoids and flavonoids. In addition, proline can now be added to an elite list of non-enzymatic antioxidants that microbes, animals, and plants need to counteract the inhibitory effects of ROS [16]. Plant stress tolerance may therefore be improved by the enhancement of *in vivo* levels of antioxidant enzymes. The antioxidants as described are found in almost all cellular compartments which signify the importance of ROS detoxification for cellular survival. It has also been shown that ROS influence the expression of a number of genes and signal transduction pathways which suggest that cells have evolved strategies to use ROS as biological stimuli and signals that activate and control various genetic stress-response programs. Control of plant pathogens by genetic engineering has targeted ROS for development of pathogen resistant crop varieties [17]. Antisense technology has been used to reduce the capability to scavenge H_2O_2 in case of model plants like *Arabidopsis thaliana* and *Nicotiana tabacum*. In these plants, antioxidant enzymes like catalase and ascorbate peroxidase are under-expressed and it has been found that they were hyper-responsive to pathogen attack. This further confirms that the ability of plant cells to regulate the efficiency in their ROS-removal strategies is a key point in their resistance against pathogens. The technology is yet to be implemented in case of legumes including soybean as intensive research is going on future prospects of the technology.

(b) *Abscissic acid signaling and related protein*

Abscissic acid (ABA) has been implicated in plant response to environmental stress by interfering at different levels with signaling. Its level increases under stress conditions to trigger metabolic and physiological changes [18]. It has become increasingly clear that the isolated abiotic signaling network is controlled by ABA and the biotic network is controlled by salicylic acid, jasmonic acid and ethylene are interconnected at various levels [19]. The concept of marker genes whose expression is believed to be regulated by individual hormones does not do justice to the nature of the network. The apparent cross-talk in stress-hormone signaling makes it difficult to assign a marker gene or a mutant phenotype to a specific hormone-controlled pathway. The signaling network into which the four stress hormones and other signals feed is apparently designed to allow plants to adapt optimally to specific situations by integrating possibly conflicting information from environmental conditions, biotic stress, and developmental as well as nutritional status. Promoter analyses of ABA/stress-responsive genes revealed that a DNA sequence element consisting of ACGTGGC is important for ABA regulation. For the past several years, researchers have been trying to identify transcription factors that regulate the expression of ABA/stress-responsive genes *via* the consensus element, which is generally known as 'Abscisic Acid Response Element' (ABRE). Many basic leucine zipper class DNA-binding proteins that interact with the element have been reported [20]. Researchers have focused on the small subfamily of *Arabidopsis* basic leucine zipper proteins referred to as ABFs (ABRE-binding factors), whose expression is induced by ABA and by various abiotic stresses (i.e., cold, high salt and drought). ABA is involved in responses to environmental stress such as salinity, and is required by the plant for stress tolerance as found recently on soybean studies. The leaf ABA content in salt-tolerant soybean increased significantly under salt stress, while in case of salt sensitive soybean has almost negligible increase in ABA. It is thus possible that ABA enhances salt tolerance in soybean [21].

(c) *GABA-related protein*

γ-Aminobutyric acid (GABA) is a non-protein amino acid that is conserved from bacteria through yeast to vertebrates and was discovered in plants over half-a-century ago. It is mainly metabolized through a short pathway called the GABA shunt, because it bypasses two steps of the tricarboxylic-acid (TCA) cycle. The pathway is composed of three enzymes: the cytosolic and mitochondrial glutamate decarboxylase (GAD), GABA transaminase (GABA-T) and succinic semialdehyde dehydrogenase (SSADH). Although there are differences in the subcellular localization of GABA-shunt enzymes in different organisms have been reported (for e.g. in yeast, SSADH is present inside cytosol) [22]. In an alternative reaction, succinic semialdehyde can be converted to GHB (γ-hydroxybutyric acid) through a GHB dehydrogenase (GHBDH) present in animals and recently identified in plants [23]. Interestingly, research of GABA in vertebrates has focused mainly on its role in the context of plant responses to stress, because of its rapid and dramatic production in response to biotic and abiotic stresses. For example, disruption of the unique SSADH gene in *Arabidopsis* results in plants undergoing necrotic cell death caused by the accumulation of reactive oxygen intermediates (ROIs) when they are exposed to environmental stresses [24]. A recent article reports that a gradient of GABA concentration is essential for the growth and guidance of pollen tubes and suggests that this amino acid plays a role in intercellular signaling in plants, possibly similar to its role in animals. The main question raised by these recent findings is whether GABA itself serves as a signaling molecule in plants. If so, this would imply that GABA is capable of mediating developmental changes and cell guidance by interacting with specialized plant receptors [25].

(d) *Mitogen-activated protein kinase signaling and related proteins*

Like other eukaryotes, plants use mitogen-activated protein kinase (MAPK) cascades to regulate various cellular processes in response to a broad range of biotic and abiotic stress. These cascades promote the transient activation of MAPKs by a dual phosphorylation of Thr and Tyr within the activation loop of the MAPK. Recent studies indicate that MAPKs are not only regulated through phosphorylation by upstream kinases, but also by direct binding of different protein factors [26]. The constitutive activation of MAPKs was found to result in detrimental effects, underlining the importance of a negative regulation of MAPK signaling. MAPK phosphatases (MKPs) are negative regulators of MAPKs. Recent progress in analyzing plant MKP mutants has revealed their important role in fine-tuning MAPK signaling. In particular, the dual-specificity phosphatase MKP1 and the protein tyrosine phosphatase (PTP1) negatively regulate defense responses and resistance to a bacterial pathogen by counter balancing the activation of two MAPKs (MPK3 and MPK6). Interestingly, MKP1 and PTP1 bind CaM, and the phosphatase activity of MKP1 is increased by CaM in a Ca^{2+}-dependent manner. Thus, Ca^{2+} and MAPK signaling pathways appear to be connected through the regulation of plant MAPKs and MKPs by CaM [27].

(e) *Calcium signaling and related proteins*

Plant cells are equipped with highly efficient mechanisms to perceive, transduce and respond to a wide variety of internal and external signals during their growth and development. Perception of signals *via* receptors results in generation or synthesis of non-proteinaceous molecules which are termed as messengers. The messengers include Ca^{2+} ions, small organic molecules such as cyclic nucleotide monophosphates, inositol triphosphates and inorganic molecules such as H_2O_2 and NO. The elements of receptors, messengers, sensors and targets vary depending on the signal received. Identification and functional assignment of these elements in a stimulus-specific signal transduction pathway is a challenging area for plant biologists. With the completion of genome sequences of various organisms, including *Arabidopsis thaliana, Oryza sativa, Medicago truncculata, Glycine max* etc. it has become evident that plants have a large number of motifs containing helix-loop-helix which binds to Ca^{2+} [28]. Further, Ca^{2+} has been implicated in mediating various developmental processes (pollen tube growth, root-hair and lateral root development and nodulation), hormone regulated cellular activities (cell division and elongation, stomatal closure/opening), pathogen- and elicitor-induced defense related processes, and a variety of abiotic stress signal induced gene expression. However, the identity and functions of downstream transducers and mechanisms by which Ca^{2+} mediates a variety of cellular responses are just begin to unravel in plants. In plants, spatially and temporally distinct changes in cellular Ca^{2+} concentrations, designated as "Ca^{2+} signatures" that are evoked in response to different stimuli like drought, salt or osmotic stresses, temperature, light and plant hormones represent a central mechanistic principle to present defined stimulus-specific information [29]. These specific "Ca^{2+} signatures" are formed by the tightly regulated activities of channels and transporters at different membranes and cell organelles. While the identity and function of components of the Ca^{2+} extrusion system are rather well understood in plant cells, the molecular identity of Ca^{2+} specific influx channels has remained unknown. However, non-specific influx of Ca^{2+} mediated by ligand gated cation channels like cyclic nucleotide gated channels and glutamate receptor-like proteins contribute to different Ca^{2+} mediated cellular functions like the response to pathogens, pollen tube growth and abiotic stress. The unique structural composition of Ca^{2+} binding proteins and the complexity of the target proteins regulated by the Ca^{2+} sensors allow the plant to tightly control the appropriate adaptation to its ever changing environment. It is actually still not well understood about interface of information presentation by a specific Ca^{2+} signal and initiation of information decoding by Ca^{2+} sensors that represent a most critical step in specific information processing [30].

4. Significance of proteomics in soybean allergenicity

Soybeans have played a central role in concerns about GM introduced allergens and in using GM to remove intrinsic allergens. Soybean is a rich and inexpensive source of proteins for humans and animals. Soybean milk and dairy product replacement is growing in acceptance, not only by people sensitive to lactose and/or milk proteins, but also for health considerations. Soybean protein is widely used in thousands of processed foods throughout the industrialized world and is a staple crop in Asia. Soybean ranks among the eight most significant food allergens. Soybean sensitivity is estimated to occur in 5-8% of children and

1-2% of adults. The allergic reaction is only rarely life-threatening with the primary adverse reactions to consumption being atopic (skin) reactions and gastric distress. Symptoms of soy allergy usually appear within a few minutes to two hours of eating soy ingredients. People with soy allergies may cross-react with peanuts or other legumes, such as beans or peas. Soy is one of the most common allergens for infants who have not yet begun eating solid foods, because they may be fed soy-based infant formula. It is rare for babies to have a traditional IgE mediated food allergy to soy, but some babies may develop milk-soy protein intolerance [31-34] or food protein induced enterocolitis syndrome [http://foodallergies.about.com/od/soyallergies/a/Soy-Allergy-Overview.htm]. Infants will usually develop these sensitivities within a few months of birth, and most will outgrow them by the age of two. Most people with soy allergies can tolerate the small amount of soy protein that remains in refined soybean oil and soy lecithin. Both of these ingredients may cause allergic reactions in highly sensitized people. There are some data available that describe the natural variation in allergen proteins that occur in soybean. For a better understanding of the variation of allergen proteins that might be expected to occur in GM soybeans, it is important to determine the natural variation of protein composition both in wild and GM soybeans. "Proteomics" approach is the foremost one which allows protein identification and quantification with utmost accuracy.

Biotechnology critics have claimed that an apparent rise in the number of soybean allergic individuals in the UK is correlated with the development of GM soybeans in the American market. GM-soybeans that have been developed in the US include herbicide-resistance (glyphosate) and seeds with higher percentage of essential amino acids, *esp*. methionine. Experiments have directly tested the allergenicity of herbicide-tolerant soybeans using immunological tests with samples from soybean-sensitive people. These assays have shown that herbicide-resistant GM soybeans do not present any measurable differences in allergenicity compared with non-GM soybeans and are, therefore, substantially equivalent by allergenic criteria. Sensitive people remain allergic to GM soybeans, but there is no additional allergenic risk to others. According to some reports protein expressed corresponding to transgene responsible for herbicide-resistance in soybeans has allergenic motifs [35]. On ingestion a portion of the transgene along with the promoter get transferred to human gut bacteria. The transformed bacteria containing transgene continues to produce herbicide-resistance allergenic protein even when the individual is not eating GM soy. Therefore an individual is constantly exposed to potentially allergenic protein, being created within his gut. Further, herbicide-resistant protein is made more allergenic due to its misfolding brought by rearrangement of unstable transgenes. Some reports emphasize the fact that protein allergenicity is due to suppression of pancreatic-enzymes due to which protein remains in the gut for longer duration contributing to allergies. There is insufficient data to support *in vivo* toxicity of herbicide-resistant protein either due to transformation or enzyme suppression [36]. GM-soybeans with enhanced methionine content such as prolamines and 2S albumins were tested for its allergenicity before its commercialization. It was found that allergenicity was much higher with respect to wild soybeans [37]. Consequently the development of GM soybean with enhanced methionine has been abandoned and no product was released, thus nobody was harmed by its adverse reactions. Recently, one of the interesting analyses has been done on GM-soy irrespective of herbicide resistance or enhanced methionine content. It has found that GM transformation process may lead to increment in natural allergens in soybeans. The level of one known allergen is trypsin inhibitor which is 27% higher in raw GM soy varieties with respect to natural varieties [38]. Further, it has also been found that cooked GM soy has sevenfold higher amount of trypsin inhibitor as compared to cooked non-GM soy due to its extreme heat stability. There are several reports including both supportive as well unsupportive towards effects of GM-soy on humankind as well as on other flora and fauna of the environment. It will require intensive research including proteomics before their release into the commercial markets.

Plant biotechnology has not only tried to produce GM-soy which is herbicide resistance or with enhanced methionine content but also aimed to remove naturally occurring allergens in native soy varieties. Presently primary treatment for food allergies is avoidance, but it is unavoidable in case of soybean protein which is present in thousands of products. Therefore, it is very difficult to avoid soybean and its derived products. Research is going on to produce hypoallergenic variants of soybean which has potential to reduce the risk of adverse reactions. Soybeans possess as many as 15 proteins recognized by IgEs from soybean-sensitive people [39]. The immunodominant soybean allergens are the β-subunit of conglycinin and P34 or Gly m Bd 30k (cysteine proteases from papain family). The P34/Gly m Bd 30k protein is a unique member of the papain superfamily lacking the catalytic cysteine residue that is replaced by a glycine which is 70% more allergenic with respect to conglycinin. There are several approaches that have been taken to produce a hypoallergenic soybean. One approach was to search cultivars which lack allergens and then crossing its germplasm to elite germplasm. This approach could not be implemented as there was no soybean cultivar (either domesticated or wild) present which lack P34/Gly m Bd 30k. Immunological assays of P34/Gly m Bd 30k with antibodies from soybean-sensitive people resulted in the identification of 14 contiguous and non-contiguous linear epitopes. The presence of so many distinct linear epitopes means that the probability of a naturally occurring variant with a sufficient number of alterations to disrupt the allergenicity is extremely small. Protein engineering could be performed to alter amino acid sequence by disrupting allergenic sequences. Using linear peptides to test possible modifications, it is straightforward to assay numerous variants and pick one that is not recognized by the IgE population. The epitope modification approach is not feasible to produce an essentially hypoallergenic variant. The problem with this technology is to remove completely the intrinsic allergen and substitute the `hypoallergenic' variant in its place. Further, the modification of the protein to remove the allergenic epitopes may alter the protein's folding, that, in turn, may affect the protein's intracellular targeting, stability and accumulation. All these possibilities will need to be tested for experimentally and, finally; the newly produced hypoallergenic variant will need to be tested to ensure that it too is not a new allergen. For these reasons, substituting a hypoallergenic variant of a plant still has a high technological threshold and has yet to be achieved. The alternative GM approach is to eliminate the allergen by suppression. There have been several attempts to reduce and/or eliminate allergens using gene suppression technology. Gene-silencing techniques involve transgenic soybeans with eliminated immunodominant human allergen P34/Gly m Bd 30k. It involves complete elimination of the P34/Gly m Bd 30k allergen from the initial somatic embryos through the third generation homozygous soybeans. Suppression of the allergen did not introduce any changes in the pattern of growth and development of the plant or seed at both the gross and subcellular level. In order to compare the P34-suppressed soybeans with the wild type, large-scale proteomic analysis was performed. Imaging of the 2D gels identified over 1400 individual elements. Mass spectrometry analysis of about 140 of these spots confirmed that the only overt changes in composition in the transgenic soybeans was the suppression of the P34/Gly m Bd 30k protein with no other proteins induced or suppressed [40]. Further analysis with sera samples from soybean-sensitive people confirmed a loss of the P34 allergen and no induction of any new allergens. The proteome and immunological analysis together confirms that it is feasible to suppress an endogenous allergen without introducing adverse effects on the plant or changing the composition of the soybean seed in any way other than the removal of the targeted protein. This result meets the test of `substantial equivalence' where the GM soybean seed is essentially identical except for the change in the single desired characteristic. Suppressing P34/Gly m Bd 30k in GM soybeans is a first step and a demonstration in addressing the growing concerns about food allergies and its relationship to the development of GM crops. More detailed studies and approaches should provide the tests needed to gain regulatory approval in nations that are currently cautious about this technology. Natarajan *et al.* [41] have compared the profiles of allergen and anti-nutritional proteins both in wild and GM soybean seeds. 2D-PAGE was used for the separation of proteins at two different pH ranges and applied a combined MALDI-TOF-MS and LC-MS analysis for the identification of proteins. Although overall distribution patterns of the allergen and anti-nutritional proteins Gly m Bd 60K (conglycinin), Gly m Bd 30K, Gly m Bd 28K, trypsin inhibitors, and lectin appeared similar, there was remarkable variation in the number and intensity of the protein spots between wild and GM soybean. The wild soybean showed fifteen polypeptides of Gly m Bd 60K and three polypeptides of trypsin inhibitors. GM soybean showed twelve polypeptides of Gly m Bd 60K and two polypeptides of trypsin inhibitors. In contrast, the GM soybean showed two polypeptides of Gly m Bd 30K and three polypeptides of lectin and the wild type showed two and one polypeptides of Gly m Bd 30K and lectin, respectively. The same number of Gly m Bd 28K spots was observed in both wild and GM soybean [41].

The fear of allergic reactions has produced much of the concern about the risks of GM crops. In order to broadly apply genetic modification to crops, there is an urgent need for better biochemical and molecular methods, including animal models, to test for food allergens experimentally so that the supporting data can be provided to evaluate newly proposed and actual GM products. In order to design transgenes, it would be useful to predict allergenicity but, currently, there are no models that would permit accurate assessment of allergenic potential of proteins unrelated to known allergens. Liver represents a suitable model for monitoring the effects of a diet, due to its key role in controlling the whole metabolism. Previous studies on hepatocytes from young female mice fed on GM soybean demonstrated nuclear modifications involving transcription and splicing pathways [42, 43]. The morphofunctional characteristics of the liver of 24-month-old mice, fed from weaning on control or GM soybean, were investigated by combining a proteomic approach with ultrastructural, morphometrical and immunoelectron microscopical analyses. Several proteins belonging to hepatocyte metabolism, stress response, calcium signaling and mitochondria were differentially expressed in GM-fed mice, indicating a more marked expression of senescence markers in comparison to controls. Moreover, hepatocytes of GM-fed mice showed mitochondrial and nuclear modifications indicative of reduced metabolic rate. This study demonstrates that GM soybean intake can influence some liver features, although the mechanisms remain unknown. Therefore, it is required to investigate the long-term consequences of GM-diets, further studies are required for potential synergistic effects with other factors like ageing, stress etc.

5. Challenges and perspectives

Soybean is a species of great agronomic and economic interest. It is one of the most recalcitrant plant species to be used as experimental material in proteomic analysis. Furthermore, there are several difficulties in the study of proteins (irrespective of source) with respect to DNA and RNA. The foremost important thing is the maintenance of secondary and tertiary structure during their analysis. They have problems with easy denaturation on exposure to high temperature, extremes of pH, oxidation, specific chemicals etc. There are some classes of proteins which are difficult to analyze due to their poor solubility. Proteins cannot be amplified like DNA, therefore less abundant species are very difficult to detect. However, many potentially important proteins (in scarce) are lost due to non-specific binding or the co-removal of proteins/peptides intrinsically bound to the high abundant carrier proteins. Following are two methods developed recently to resolve detection of less abundant plant proteins [44]:

- The use of equalizer beads coupled with a combinational library of ligands containing diverse population of beads with equivalent binding capacity to most of the proteins present in a sample.
- The ultra-microarrays have been found to have high specificity and sensitivity with detection levels in the range of attomole (10^{-18} mole).

The current depth of knowledge regarding the soybean proteome is significantly less than that for some other plants. The soybean proteome map which is available in the database (http://proteome.dc.affrc.go.jp/soybean/) corresponds to various types of stresses, allergenicity, and studies on natural product biosynthesis in soybean. The other challenges in plant proteomics including soybean are standardization of methodologies, dissemination of proteomics data into publicly available databases and most importantly its cost expensiveness. Furthermore, most proteomics technologies use complex instrumentation and critical computing power. Currently, there is no expertise available for functional interpretation of data obtained from integration of proteomics with genomics and metabolomics.

The significance of proteomics over genomics and transcriptomics has been debated since the field has emerged. The importance of the proteome cannot be overstated as it is the proteins within the cell that provide structure, produce energy, as well as allow communication, movement, and reproduction. Basically, proteins provide structural and functional framework for cellular life. Genetic information is static while the protein complement of a cell is dynamic. Differential proteomics is a scientific discipline that detects the proteins associated with a diseased state (either due to abiotic or biotic stress, toxicity due to allergenicity, genetic modifications etc.) by means of their altered levels of expression between the control and diseased states. Extensive research towards the development of a soybean proteome map would permit the rapid comparison of soybean cultivars, mutants, and transgenic lines. Moreover, studies of soybean physiology will also benefit from the existence of a detailed and quantitative proteome reference map of the soybean plant. The information obtained from soybean proteomics will be helpful in predicting the function of plant proteins and will aid in molecular cloning of the corresponding genes in the future. The identification of novel genes, the determination of their expression patterns in response to stress, and an understanding of their functions in stress adaptation will provide us with the basis for effective strategies for engineering improved stress tolerance in soybean. With the advancement of new technologies in proteomics combined with advanced bioinformatics, we are currently identifying molecular signatures of diseases based on protein pathways and signaling cascades. Applying these findings will improve our understanding of the roles of individual proteins or the entire cellular pathways in the initiation and development of disease. The abundance of information provided by proteomics research is entirely complementary with the genetic information being generated by genomics research. Proteomics makes a key contribution to the development of functional genomics. The combination of genomics and proteomics will play a major role in understanding molecular mechanisms in plant pathology, and it will have a significant impact on the development of high yield varieties, with better resistance towards adverse environmental factors as well as various pathogenic diseases caused by bacteria, viruses and fungi in the future.

Author details

Alka Dwevedi[1*] and Arvind M Kayastha[2*]

*Address all correspondence to: kayasthabhu@gmail.com

1 Regional Centre for Biotechnology, India

2 School of Biotechnology, Faculty of Science, Banaras Hindu University, India

References

[1] Wu, D. D., Hu, X., Park, E., Wang, X., Feng, J., & Wu, X. (2010). Exploratory analysis of protein translation regulatory networks using hierarchical random graphs. *BMC Bioinformatics*, 11(3), S2.

[2] Ezkurdia, I., del Pozo, A., Frankish, A., Rodriguez, J. M., Harrow, J., Ashman, K., Valencia, A., & Tress, M. L. (2012). Comparative proteomics reveals a significant bias toward alternative protein isoforms with conserved structure and function. *Molecular Biology and Evolution*, 29-2265.

[3] Bertone, P., & Snyder, M. (2005). Prospects and challenges in proteomics. *Plant Physiology*, 138-560.

[4] Hefle, S. L, Nordlee, J. A, & Taylor, S. L. (1996). Allergenic foods. *Critical Reviews in Food Science and Nutrition*, 36, S69-S89.

[5] Metcalfe, D. D. (2005). Genetically modified crops and allergenicity. *Nature Immunology*, 6-857.

[6] Manavalan, L. P, Guttikonda, S. K, Tran, L. S, & Nguyen, H. T. (2009). Physiological and molecular approaches to improve drought resistance in soybean. Plant & Cell Physiology , 50-1260.

[7] Atkinson, N. J, & Urwin, P. E. (2012). The interaction of plant biotic and abiotic stresses: from genes to the field. *Journal of Experimental Botany*, 63-3523.

[8] Nouri-Z, M., Toorchi, M., & Komatsu, S. (2011). Chapter 9: Proteomics approach for identifying abiotic stress responsive proteins in soybean. *In Soybean-molecular aspects of breeding ed. Aleksandra Sudaric. Intech, April 11*, 978-9-53307-240-1.

[9] Mooney, B. P., Krishnan, H. B., & Thelen, J. J. (2004). High-throughput peptide mass fingerprinting of soybean seed proteins: automated workflow and utility of UniGene expressed sequence tag databases for protein identification. *Phytochemistry*, 65-1733.

[10] Natarajan, S., Xu, C., Caperna, T. J., & Garrett, W. M. (2005). Comparison of protein solubilization methods suitable for proteomic analysis of soybean seed proteins. *Analytical Biochemistry*, 342-214.

[11] Hajduch, M., Ganapathy, A., Stein, J. W., & Thelen, J. J. (2005). A systematic proteomic study of seed filling in soybean. Establishment of high-resolution two dimensional reference maps, expression profiles, and an interactive proteome database. Plant Physiology , 137-1397.

[12] Lei, Z, Elmer, A. M., Watson, B. S., Dixon, R. A, Mendes, P. J., & Sumner, L. W. (2005). A two-dimensional electrophoresis proteomic reference map and systematic identification of 1367 protein from a cell suspension culture of the model legume Medicago truncatula. *Molecular and Cellular Proteomics*, 4, 1812-1825.

[13] Nasi, A., Picariello, G., & Ferranti, P. (2009). Proteomic approaches to study structure, functions and toxicity of legume seeds lectins. Perspectives for the assessment of food quality and safety. *Journal of Proteomics*, 72-527.

[14] Bailey-Serres, J., & Mittler, R. (2006). The roles of reactive oxygen species in plant cells. *Plant Physiology*, 141, 311.

[15] Mittler, R. (2002). Oxidative stress, antioxidants and stress tolerance. *Trends in Plant Science*, 7-405.

[16] Daft, J., Vandenabeele, S., Vranová, E., Van Montagu, M., Inzé, D., & Van Breusegem, F. (2000). Dual action of the active oxygen species during plant stress responses. *Cellular and Molecular Life Sciences*, 57-779.

[17] Collinge, D. B., Jørgensen, H. J .L., Lund, O. S., & Lyngkjær, M. F. (2010). Engineering pathogen resistance in crop plants: Current Trends and Future Prospects. *Annual Review of Phytopathology*, 48-269.

[18] Leung, J., & Giraudat, J. (1998). Abscisic acid signal transduction. *Annual Review of Plant Physiology and Plant Molecular Biology*, 49-199.

[19] Mauch-Mani, B., & Mauch, F. (2005). The role of abscisic acid in plant-pathogen interactions. *Current Opinion in Plant Biology*, 8-409.

[20] Choi, H. I., Hong, J. H., Ha, J. O., Kang, J. Y., & Kim, S. Y. (2000). ABFs, a family of ABA-responsive element binding factors. *The Journal of Biological Chemistry*, 275-1723.

[21] Xu , Y. X., Fan, R., Zheng, R., Li, M. C., & Yu, Y. D. (2011). Proteomic analysis of seed germination under salt stress in soybeans. *Journal of Zhejiang University-Science B (Biomedicine & Biotechnology)*, 12-507.

[22] Bouche', N., Lacombe, B., & Fromm, H. (2003). GABA signaling: a conserved and ubiquitous mechanism. *Trends in Cell Biology*, 13-607.

[23] Roberts, M. R. (2007). Does GABA Act as a Signal in Plants? Plant Signaling & Behavior , 2-408.

[24] Bouché, N., Fait, A., Bouchez, D., Møller, S. G., & Fromm, H. (2003). Mitochondrial succinic-semialdehyde dehydrogenase of the γ-aminobutyrate shunt is required to restrict levels of reactive oxygen intermediates in plants. *Proceedings of National Academy of Sciences of United States of America*, 100-6843.

[25] Ludewig, F., Hüser, A., Fromm, H., Beauclair, L., & Bouché, N. (2008). Mutants of GABA transaminase (POP2) suppress the severe phenotype of succinic semialdehyde dehydrogenase (SSADH) mutants in Arabidopsis. Plos One , 3, e3383.

[26] Wrzaczek, M., & Hirt, H. (2001). Plant MAP kinase pathways: how many and what for? *Biology of the Cell*, 93-81.

[27] Zhang, T., Liu, Y., Yang, T., Zhang, L., Xu, L., & An, L. (2006). Diverse signals converge at MAPK cascades in plants. *Plant Physiology and Biochemistry*, 44-274.

[28] Yang, T., & Poovaiah, B. W. (2003). Calcium/calmodulin-mediated signal network in plants. *Trends in Plant Science*, 8-505.

[29] Reddy, V. S., & Reddy, A. S. N. (2004). Proteomics of calcium-signaling components in plants. *Phytochemistry*, 65-1745.

[30] Perochon, A, Aldon, D, Galaud, J. P., & Ranty, B. (2011). Calmodulin and calmodulin-like proteins in plant calcium signaling. *Biochimie*, 93, 2048-2053.

[31] Dwevedi, A., & Kayastha, A. M. (2010). Plant β-galactosidases: Physiological significance and recent advances in technological applications. *Journal of Plant Biochemistry and Biotechnology*, 19-9.

[32] Dwevedi, A., & Kayastha, A. M. (2011). Chapter 11: Soybean, a multifaceted legume with enormous economic capability. *In Soybean: Biochemistry, Chemistry and Physiology, ed. Tzi-Bun Ng. Intech, April 26*, 978-9-53307-219-7.

[33] Dwevedi, A., & Kayastha, A. M. (2009). A β-galactosidase from pea seeds (PsBGAL): purification, stabilization, catalytic energetics, conformational heterogeneity, and its significance. *Journal of Agricultural and Food Chemistry*, 57-7086.

[34] Kishore, D., Talat, M., Srivastava, O. N., & Kayastha, A. M. (2012). Immobilization of β-galactosidase onto functionalized graphene nano-sheets using response surface methodology and its analytical applications towards milk and whey lactose. *Plos One*, 7, e40708.

[35] Burks, A.W., & Fuchs, R. L. (1995). Assessment of the endogenous allergens in glyphosate tolerant and commercial soybean varieties. *Journal of Allergy and Clinical Immunology*, 96-1008.

[36] Mazur, B. J, & Falco, S. C. (1989). The development of herbicide resistant crops. *Annual Review of Plant Physiology and Plant Molecular Biology*, 40-441.

[37] Herman, E. M. (2003). Genetically modified soybeans and food allergies. *Journal of Experimental Botany*, 54-1317.

[38] Domingo, J. L. (2007). Toxicity studies of genetically modified plants. *Critical Reviews in Food Science and Nutrition*, 47721-733.

[39] Burks, A. W., Brooks, J. R, & Sampson, H. A. (1988). Allergenicity of major component proteins of soybean determined by enzyme-linked immunosorbent assay (ELISA) and immunoblotting in children with atopic dermatitis and positive soy challenges. *Journal of Allergy and Clinical Immunology*, 81-1135.

[40] Helm, R. M., Cockrell, G., Herman, E., Burks, A. W., Sampson, H. A., & Bannon, G. A. (1998). Cellular and molecular characterization of a major soybean allergen. *International Archive of Allergy Immunology*, 117-29.

[41] Natarajan, S. S., Xu, C., Bae, H., Caperna, T. J., & Garrett, W. M. (2006). Proteomic analysis of allergen and antinutritional proteins in wild and cultivated soybean. *Journal of Plant Biochemistry and Biotechnology*, 15-103.

[42] Malatesta, M., Caporaloni, C., Gavaudan, S., Rocchi, M. B., Serafini, S., Tiberi, C., & Gazzanelli, G. (2002). Ultrastructural morphometrical and immunocytochemical analyses of hepatocyte nuclei from mice fed on genetically modified soybean. *Cell Structure and Function*, 27-173.

[43] Malatesta, M., Boraldi, F., Annovi, G., Baldelli, B., Battistelli, S., Biggiogera, M., & Quaglino, D. (2008). A long-term study on female mice fed on a genetically modified soybean: effects on liver ageing. *Histochemistry and Cell Biology*, 130-967.

[44] Cho, W. C. S. (2007). Proteomics technologies and challenges. *Genomics, Proteomics & Bioinformatics*, 5-77.

Chapter 20

In vitro Regeneration and Genetic Transformation of Soybean: Current Status and Future Prospects

Thankaraj Salammal Mariashibu,
Vasudevan Ramesh Anbazhagan, Shu-Ye Jiang,
Andy Ganapathi and Srinivasan Ramachandran

Additional information is available at the end of the chapter

http://dx.doi.org/10.5772/54268

1. Introduction

Soybean [*Glycine max* (L.) Merrill], grown for its edible seed protein and oil, is often called the miracle crop because of its many uses. It belongs to the genus Glycine under the family Leguminosae, and is widely cultivated in the tropics, subtropics and temperate zones of the world [1].

Soybean is now an essential and dominant source of protein and oil with numerous uses in feed, food and industrial applications. It is the world's primary source of vegetable oil and protein feed supplement for livestock. The global production of soybeans is 250-260 million tons per year. The US is the largest producer with 90.6 million metric tons. Other major countries such as Brazil, Argentina, China and India contributing 70, 49.5, 15.2 and 9.6 million metric tons, respectively [2]. The US, Brazil and Argentina are the major exporters of beans; while China and Europe are the major importers. The annual world market value is around 2 billion US dollars, which stands second in world food production.

Recent nutritional studies claim that consumption of soybean reduces cancer, blood serum cholesterol, osteoporosis and heart diseases [3]. This has sparked increased demand for the many edible soybean products. The priority for more meat in diets among the world's population has also increased the demand for soybean protein for livestock and poultry feed.

Soybean seeds are comprised of 40% protein, mostly consisting of the globulins β-conglycinin (7S globulin) and glycinin (11S globulin). The oil portion of the seed is composed primarily of five fatty acids. Palmitic and stearic acids are saturated fatty acids and comprise 15% of the oil. Soybean is rich in the unsaturated fatty acids like oleic, linoleic and linolenic, which make up 85% of the oil. Soybeans are a good source of minerals, B vitamins, folic acid and isoflavones, which are credited with slowing cancer development, heart diseases and osteoporosis [4].

The productivity of soybean has been limited due to their susceptibility to pathogens and pests, sensitivity to environmental stresses, poor pollination and low harvest index. Among the abiotic stresses, drought is considered the most devastating, commonly reducing soybean yield by approximately 40% and affecting all stages of plant growth and development; from germination to flowering, and seed filling and development as well as seed quality [5]. It suffers from many kinds of fungal diseases, such as frogeye leaf spot and brown spot [6]. As demand increases for soybean oil and protein, the improvement of soybean quality and production through genetic transformation and functional genomics becomes an important issue throughout the world [7].

The main objectives of soybean improvement include increase in yield, development of resistance to various insects, diseases and nutritional quality. Commercial breeding is still very important for the genetic improvement of the crop. However, breeding is difficult due to the fact that the soybean is a self pollinating crop, and the genetic base of modern soybean cultivars is quite narrow [8]. Most of the current soybean genotypes have been derived from common ancestors; therefore, conventional breeding strategies are limited in capability to expand the soybean genetic base. Recent advances in *in vitro* culture and gene technologies have provided unique opportunities for the improvement of plants, which are otherwise difficult through conventional breeding. The technology of plant transformation is only moderately or marginally successful in many important cultivars of crops, which can be a major limiting factor for the biotechnological exploitation of economically important plant species and the wider application of genomics.

Although numerous methods have been developed for introducing genes into plant genomes, the transformation efficiency for soybean still remains low [9]. Since the first successful transformation of soybean was reported [10], two major methods have been used in soybean transformation: one is particle bombardment of embryogenic tissue and another is *Agrobacterium tumefaciens*-mediated transformation of the cotyledonary node. Both methods have limitations: the former is highly genotype-dependent, requires a prolonged tissue culture period and tends to produce multiple insertion events, while the latter is labour intensive and requires specially trained personnel to undertake the work [9]

For soybean *in vitro* regeneration, two principal methods have been identified: somatic embryogenesis and shoot morphogenesis. Each of these systems presents both advantages and disadvantages for production of transformed plants, and each can be used with both of the predominant transformation systems [11]. A better understanding of physiology and molecular biology of *in vitro* morphogenesis needs focal attention to reveal their recalcitrant nature.

The present review gives an overview on the problems associated with low transformation efficiency, and the research conducted to improve tissue culture and transformation efficiency of soybean during the past (Table 1&2) and also discuss the future prospects, demands of these technologies and upcoming new technologies in soybean improvement.

Year	Explant tissue	Major contribution	Reference
1973	Hypocotyl	Adventitious bud development	Kimball and Bingham, [13]
1980	Cotyledonary node	Shoot morphogenesis	Cheng et al. [14]
1986	Immature embryo	Plant regeneration from callus	Barwale et al. [18]
1986	Cotyledonary node	Multiple shoot formation	Barwale et al. [19]
1986	Cotyledonary node	Multiple shoot formation	Wright et al. [20]
1987	Epicotyl	Callus induction and shoot regeneration	Wright et al. [29]
1988	Cotyledonary node	Transfered *npt* II and *gus* gene by Agrobacterium mediated transformation	Hinchee et al. [10]
1988	Immature seeds	Developed transgenic soybean by Particle bombardment	McCabe et al. [25]
1989	Germinating seeds	Transfered *npt* II gene by Agrobacterium mediated transformation	Chee et al. [45]
1989	Immature seed	Particle bombardment of meristems	Christou et al. [62]
1990	Immature cotyledon	Plant regeneration from protoplast	Luo et al. [127]
1990	Cotyledon, cotyledonary node	Evaluated *Agrobacterium* sensitivity and adventitious shoot formation	Delzer et al. [44]
1990	Immature cotyledon, plumule, cotyledonary node	Analysed plant regeneration efficiency of various explants	Yang et al. [32]
1990	Immature embryo	Organogenesis and plant regeneration	Yeh,[128]
1990	Primary leaf node	Adventitious shoot formation	Kim et al. [27]
1991	Immature cotyledon	Plant regeneration from protoplast	Dhir et al. [129]
1992	Epicotyl and hypocotyl	Investigated the stimulative effect of allantoin and amides on shoot regeneration	Shetty, et al. [21]
1993	Shoot tip	Transfered *gus* gene via particle bombardment	Sato et al. [130]
1994	Primary leaf node	Investigated the synergistic effect of proline and micronutrients on shoot regeneration	Kim et al. [40]
1996	Cotyledonary node	Developed transgenic soybean resistance to bean pod mottle virus (BPMV)	Di et al. [131]
1997	Cotyledonary node and hypocotyl	Multiple shoot induction by TDZ	Kaneda et al. [22]
1998	Cotyledonary node	Evaluation of sonication assisted Agrobacterium mediated transformation (SAAT) for cotyledonary node	Meurer et al. [50]
1998	Hypocotyl	Adventitious shoot regeneration	Dan and Reichert, [33]

Year	Explant tissue	Major contribution	Reference
1999	Cotyledonary node	Assessed the use of glufosinate as a selective agent in *Agrobacterium*-mediated transformation of soybean	Zhang et al. [61]
2000	Cotyledonary node	*Agrobacterium* two T-DNA binary system as a strategy to derive marker free transgenic soybean	Xing et al. [132]
2000	Cotyledonary node	Evaluated the effect of glyphosate as a selective agent for *Agrobacterium* mediated cotyledonary node transformation system	Clemente et al. [60]
2000	Embryonic axes	Used of Imazapyr as selection agent for selection of meristematic soybean cells	Aragao et al. [47]
2001	Cotyledonary node	Investigated the use of thiol compound to increase transformation frequency	Olhoft et al. [56]
2001	Cotyledonary node	Increased *Agrobacterium* infection using L-cystine	Olhoft and Somers, [16]
2001	Cotyledonary node	Developed transgenic soybean plants resistant to soybean mosaic virus (SMV)	Wang et al. [133]
2001	Cotyledonary node	Expressed oxalate oxidase gene for resistant to sclerotinia stem rot caused by *Sclerotinia sclerotiorum*	Donaldson et al. [65]
2003	Hypocotyl	Screened soybean genotype for adventitious organogenic regeneration	Reichert et al. [41]
2003	Cotyledonary node	Assessed the effect of genotype, plant growth regulators and sugars on regeneration from calli	Sairam et al. [1]
2003	Cotyledonary node	Used mixture of thiol compounds and hygromycin based selection for increased transformation efficiency	Olhoft et al. [57]
2004	Cotyledonary node	Assessed glufosinate selection for increased transformation efficiency	Zeng et al. [134]
2004	Cotyledonary node	Investigated the effect of seed vigor of explant source, selection agent and antioxidant on *Agrobacterium* mediated transformation efficiency	Paz et al. [15]
2004	Cotyledonary node	Transferred chitinase gene and the barley ribosome-inactivating protein gene to enhance fungal resistance	Li et al. [6]
2004	Mature and immature cotyledon	Shoot regeneration	Franklin et al. [31]
2004	Embryonic tip	Established regeneration and *Agrobacterium* mediated transformation system	Liu et al. [35]

http://dx.doi.org/10.5772/54268

Year	Explant tissue	Major contribution	Reference
2004	Cotyledonary node	Established liquid medium based system for selection transformed plants	Yun, [58]
2005	Cotyledonary node	Developed repetitive organogenesis system	Shan et al. [23]
2005	Cotyledonary node	Expressed Escherichia coli K99 fimbriae subunit antigen in soybean to use as edible vaccine	Piller et al. [66]
2006	Cotyledonary node	*Agrobacterium* mediated transformation efficiency was improved by using half seed explant from mature seed	Paz et al. [24]
2007	Cotyledonary node	Investigated *Agrobacterium rhizogen* to transform soybean cotyledonary node cells.	Olhoft et al. [59]
2007	Cotyledonary node	Expressed synthetic *Bacillus thuringiensis* cry1A gene that confers a high degree of resistance to Lepidopteran Pests	Miklos et al. [135]
2007	Cotyledonary node and leaf node	Established organogenic callus induction and *Agrobacterium* mediated transformation	Hong et al ., [43]
2007	Half seed	Expressed jasmonic acid carboxyl methyltransferase in soybean to produce methyl jasmonate, which resulted in tolerant to water stress	Xue et al. [67]
2008	Hypocotyl	Used silver nitrate to enhance adventitious shoot regeneration after *Agrobacterium* transformation and developed transgenic soybean producing high oleic acid content by silencing endogenous GmFAD2-1 gene by RNAi	Wang and Xu, [7]
2008	Cotyledonary node	Improved transformation efficiency using surfactant Silwet L-77 during *Agrobacterium* infection and L-cysteine during co-cultivation	Liu et al. [136]
2008	Cotyledonary node	Developed rapid regeneration system using whole cotyledonary node	Ma and Wu, [2008]
2010	Cotyledonary node	Production of isoflavone in callus cell lines by expression of isoflavone synthase gene.	Jiang et al. [69]
2010	Cotyledon and embryo	Developed shoot regeneration from calli of soybean cv.Pyramid	Joyner et al. [39]
2011	Hypocotyl	Transgenic soybean with low phytate content	Yang et al. [70]
2011	Cotyledon	Developed transgenic soybean with increased Vitamin E content by transferring γ-tocopherol methyltransferase (γ-TMT) gene in to seedling cotyledon	Lee et al. [137]

Table 1. Major landmarks in soybean organogenesis and transformation

Explant Tissue	Year	Major Contribution	Reference
Embryonic axes	1983	Embryoids development and plant regeneration *via* suspension culture	Christianson et al.[77]
Immature cotyledon	1984	Somatic embryo Induction	Lippmann & Lippmann, [84]
Immature cotyledon	1985	Plant regeneration *via* somatic embryogenesis	Lazzeri et al. [138]
Immature embryo	1985	Somatic embryogenesis and assessment of genotypic variation	Ranch et al. [139]
Immature embryo, cotyledon and, hypocotyl from germinating seedling	1986	Somatic embryogenesis from callus	Ghazi et al. [140]
Hypocotyl and cotyledon	1986	Embryoids development in suspension culture	Kerns et al. [141]
Immature embryo and cotyledon	1987	Investigated the effect of nutritional, physical, and chemical factors on somatic embryogenesis	Lazzeri et al. [85]
Immature cotyledon	1988	Investigated the effect of auxin and orientation of explant on somatic embryogenesis	Hartweck et al. [142]
Immature cotyledon	1988	Analysed genotype dependency and High concentration of auxin on somatic embryo induction	Komatsuda and Ohyama, [143]
Immature cotyledon	1988	Investigated the interaction between auxin and sucrose during somatic embryogenesis	Lazzeri et al. [86]
Immature cotyledon	1988	Germination frequency of somatic embryo has been improved by reducing the exposure to auxin	Parrott et al. [87]
Immature cotyledon	1988	Developed rapid growing maintainable embryogenic suspension culture	Finer and Nagasawa, [82]
Immature cotyledon	1988	Histological analysis to investigate secondary somatic embryo formation.	Finer, [79]
Immature cotyledon	1989	Demonstrated the effect of genotype on embryogenesis	Parrott et al. [144]
Immature cotyledon	1989	Developed primary transformants expressing zein gene by agrobacterium mediated transformation	Parrott et al. [105]
Immature cotyledon	1989	Assayed somatic embryo maturation for conversion into plantlets	Buchheim et al. [94]
Immature cotyledon	1989	Investigated the developmental aspects of somatic embryogenesis	Christou and Yang, [145]
Immature cotyledon	1990	Screened soybean genotypes for somatic embryo production	Komatsuda et al. [146]
Immature cotyledon	1991	Transformed embryogenic cultures with *gus* and *hpt* gene *via* particle bombardment	Finer and McMullen., [64]

http://dx.doi.org/10.5772/54268

Explant Tissue	Year	Major Contribution	Reference
Immature cotyledon	1991	Analysed the interaction between genotype and sucrose concentration on somatic embryogenesis	Komatsuda et al. [147]
Immature cotyledon	1991	Demonstrated adventitious shoot formation from cotyledonary and torpedo stage embryo	Wright et al. [148]
Immature cotyledon	1992	Somatic embryo proliferation by somatic embryo cycling.	Liu et al. [83]
Immature cotyledon	1993	Improved germination efficiency of somatic embryos of cultivar H7190 by desiccation	Bailey et al. [101]
Immature cotyledon	1993	Demonstrated genotypic effect on induction, proliferation, maturation and germination of somatic embryo	Bailey et al. [96]
Immature cotyledon	1993	Investigated the factors affecting somatic embryogenesis	Lippmann & Lippmann, [149]
Immature cotyledon	1993	Soybean transformation by particle bombardment of embryogenic cultures	Sato et al. [130]
Immature cotyledon	1994	Developed transgenic soybean resistance to insect.	Parrott et al. [150]
Immature embryos	1995	Investigated the effect of glutamine and sucrose on dry matter accumulation and composition of somatic embryo.	Saravitz and Raper, [151]
Immature cotyledon	1996	Demonstrated the significance of embryo cycling for transformation	Liu et al.[152]
Immature cotyledon	1996	Transformed embryogenic cultures with 12 different plansmid *via* particle bombardment	Hadi et al. [115]
Immature cotyledon	1996	Developed transgenic soybean expressing a synthetic Bacillus thuringiensis insecticidal crystal protein gene (BtcryIAc) which is resistance to insects	Stewart et al. [46]
Immature cotyledon	1997	Investigated the effect of ethylene inhibitors on embryo histodifferentiation and maturation	Santos et al. [92]
Epicotyls and primary leaves	1997	Somatic embryogenesis and plant regeneration from cotyledon, epicotyls and primary leaves	Rajasekaran and Pello, [153]
Immature cotyledon	1997	Studied the effect of explant orientiation, pH, solidifying agent and wounding on induction of soybean from immature cotyledons	Santarém et al. [81]
Immature cotyledon	1998	Studied growth characteristics of embryogenic cultures for transformability	Hazal et al. [113]

Explant Tissue	Year	Major Contribution	Reference
Immature cotyledon	1998	Established sonication-assisted *Agrobacterium* mediated transformation of soybean immature cotyledon	Santarem et al.[48]
Immature cotyledon	1998	Established sonication-assisted Agrobacterium mediated transformation of embryogenic suspension culture tissue	Trick and Finer, [108]
Immature cotyledon	1998	Improved proliferation efficiency of embryogenic cultures by modifying sucrose and nitrogen content in medium	Samoylov et al. [89]
Immature cotyledon	1998	Developed liquid medium based system for histodifferentiation of embryogenic cultures	Samoylov et al. [154]
Immature cotyledon	1998	Studied soluble carbohydrate content in soybean somatic and zygotic embryo during development.	Chanprame et al. [155]
Immature cotyledon	1999	Studied the factors influencing transformation of prolific embryogenic cultures using bombardment	Santarem and Finer, [116]
Immature cotyledons	1999	Developed transgenic plants with bovine milk protein, β-casein	Maughan et al. [114]
Immature cotyledons	1999	Transformed GFP into embryogenic suspension culture with the aim to improve transformation and regeneration strategy	Ponappa et al. [156]
Immature cotyledons	2000	Improved somatic embryo development and maturation by application of ABA	Tian and Brown, [157]
Immature cotyledon	2000	Screened genotypes for proliferative embryogenesis	Simmonds and Donaldson, [97]
Immature cotyledons	2000	Studied physical factors influencing somatic embryo development from immature cotyledons.	Bonacin et al. [99]
Immature cotyledon	2000	Investigated the factors affecting *Agrobacterium* mediated transformation soybean	Yan et al. [109]
Immature cotyledon	1989	Investigated maturation of somatic embryo for efficient conversion into plantlets	Buchheim et al. [94]
Immature cotyledon	2000	Developed and evaluated transgenic soybean expressing a synthetic cry1Ac gene from *Bacillus thuringiensis* for resistance to variety of insects	Walker et al. [158]
Immature cotyledon	2001	Effect of polyethylene glycol and sugar alcohols on soybean somatic embryo germination and conversion	Walker and Parrott, [90]

Explant Tissue	Year	Major Contribution	Reference
Immature cotyledon	2000	Developed integrated bombardment and *Agrobacterium* transformation method	Droste et al.[159]
Immature cotyledon	2001	Screened soybean from different location in the US for uniform embryogenic response	Meurer et al. [103]
Immature cotyledon	2001	Studied the effect of osmotica for their influence on embryo maturation and germination	Walker & Parrott, [90]
Immature cotyledon	2001	Developed transgenic plant expressing 15-kD zein protein under β-phaseolin seed specific promoter	Dinkins et al. [125]
Immature cotyledon	2001	Somatic embryogenesis in Brazilian soybean cultivars	Droste et al. [160]
Immature cotyledon	2002	Somatic embryogenesis and particle bombardment for south Brazil cultivars	Droste et al. [100]
Immature cotyledon	2002	Histological analysis of developmental stages of somatic embryogenesis	Fernando et al. [161]
Immature cotyledon	2002	Screened soybean genotypes for somatic embryo induction and maturation capability	Tomlin, [162]
Immature cotyledon	2003	Investigated the effect of proliferation, maturation and desiccation on somatic embryo conversion	Moon and Hildebrand, [88]
Immature cotyledon	2004	Improved transformation efficiently using *Agrobacterium* strain KYRT1 carrying pKYRTI	Ko et al. [111]
Immature cotyledon	2004	Developed transgenic plant containing phytase gene that store (produces) more phosphrous in seed.	Chiera et al. [163]
Immature cotyledon	2004	Developed fertile transplastomic soybean	Dufourmantel et al.[117]
Immature cotyledon	2004	Transferred *chi* and *rip* gene to enhance fungal resistance	Li et al. [6]
Immature cotyledon	2004	Improved transformation efficiency using *Agrobacterium* strain KYRT1	Ko and Korban, [80]
Immature cotyledon	2004	Analysed media components and pH on somatic embryo induction	Hoffmann et al. [80]
Immature cotyledon	2005	Developed transgenic soybean expressing maize γ-zein protein	Li et al. [124]
Immature cotyledon	2005	Modified soybean histodifferentiation and msaturation medium with the aim to improve the protein and lipid composition of somatic embryo	Schmidt et al. [164]
Immature cotyledon	2005	Analysed the effect of carbon source and polyethylene glycol on embryo conversion	Korbes et al. [91]

Explant Tissue	Year	Major Contribution	Reference
Immature cotyledon	2006	Improved fatty acid content	Chen et al. [119]
Immature cotyledon	2006	Investigated the ontogeny of somatic embryogenesis	Santos et al. [165]
Somatic embryo	2006	Developed transgenic soybean resistance to dwarf virus	Tougou et al. [120]
Immature cotyledon	2006	Investigated the influence of antibiotics on embryogenic cultures and *Agrobacterium tumefaciens* suppression in soybean transformation	Wiebke et al. [166]
Immature cotyledon	2006	Developed transgenic soybean for increased production of ononitol and pinitol	Chiera et al. [167]
Immature cotyledon	2007	Developed transgenic soybean resistant to dwarf virus	Tougou et al. [168]
Immature cotyledon	2007	Improved somatic embryogenesis in recalcitrant cultivars by back cross with a highly regenerable cultivar Jack	Kita et al. [104]
Immature cotyledon	2007	Evaluated Japanese soybean genotypes for somatic embryogenesis	Hiraga et al. [102]
Immature cotyledon	2007	Soybean seed over expressing the *Perilla frutescens* γ-tocopherol methyltransferase gene	Tavva et al. [123]
Immature cotyledon	2007	Improved protein quality in transgenic soybean transformed with modified Gy1 proglycinin gene with a synthetic DNA encoding four continuous methionines.	El-Shemy et al. [169]
Immature cotyledon	2007	Analysed the effect of Abscisic acid on somatic embryo maturation and conversion.	Weber et al. [170]
Immature cotyledon	2007	Developed transgenic soybean resistance to soybean mosaic virus	Furutani et al. [121]
Immature cotyledon	2008	Used a new Selectable Marker Gene Conferring resistance to Dinitroanilines	Yemets et al. [171]
Immature cotyledon	2008	Developed strategy for transfer of multiple genes *via* micro projectile-mediated bombardment	Schmidt et al. [172]
Immature cotyledon	2009	Assessed the effect mannitol, abscisic acid and explant age on somatic embryogenesis in Chinese soybean cultivars	Yang et al. [98]
Somatic embryo	2009	Developed transgenic soybean with increased oil content	Rao and Hildebrand, [118]

Explant Tissue	Year	Major Contribution	Reference
Embryonic tip	2010	somatic embryogenesis and plant regeneration from the immature embryonic shoot tip	Loganathan et al. [173]
Immature cotyledon	2010	Developed transgenic soybean with more tryptophan content in seed	Ishimoto et al. [122]
Immature cotyledon	2010	Screening of Brazilian soybean genotypes for embryogenesis	Droste et al. [174]
Immature cotyledon	2011	Demonstrated Metabolic engineering of soybean seed coat for the production of novel biochemicals	Schnell et al. [126]
Immature cotyledon	2011	Investigated developmental profile of storage reserve accumulation in soybean somatic embryos	He et al. [175]
Immature cotyledon	2011	Improved transformation efficiency by Micro wounding with DNA free particle bombardment followed by Agrobacterium mediated transformation.	Wiebke et al. [112]
Immature cotyledon	2012	Developed vacuum infiltration assisted *Agrobacterium* mediated transformation for Indian soybean cultivars.	Mariashibu et al. [176]

Table 2. Major landmarks in soybean somatic embryogenesis and transformation

2. Organogenesis and transformation

Organogenesis is characterized by the production of a unipolar bud primordium with subsequent development of the primordium into a leafy vegetative shoot. A successful plant regeneration protocol requires appropriate choice of explant, definite media formulations, specific growth regulators, genotype, source of carbohydrate, gelling agent, other physical factors including light regime, temperature, humidity and other factors [12]. Plant regeneration by organogenesis in soybean was first reported by Kimball and Bingham, [13] from hypocotyl sections followed by Cheng et al.[14] by culturing seedling cotyledonary node segments. Transfer of T-DNA into cotyledonary node cells by *Agrobacterium* mediated transformation was first reported by Hinchee et al. [10]. Advancement in soybean transformation appears to be slow compared to some of the recent improvement in cereal transformation (Paz et al. 2004). Olhoft et al. [16] stated that the efficiency of soybean transformation has to be improved 5-10 times before one person can produce 300 transgenic lines per year. Soybean transformation efficiency has been improved by optimizing the selection system, enhancing explant-pathogen interaction and improving culture conditions to promote regeneration and recovery of transformed plants.

2.1. Organogenesis

The successful application of biotechnology in crop improvement is based on efficient plant regeneration protocol. Soybean has been considered as recalcitrant to regenerate *in vitro*. Tissue culture responses are greatly influenced by three main factors viz. whole plant physiology of donor, *in vitro* manipulation, and *in vitro* stress physiology [17]. After the first report of adventitious bud regeneration from hypocotyl sections by Kimball and Bingham, [13] researchers have used different parts of the soybean plant as explants for successful shoot morphogenesis in soybean. These include cotyledonary node [10,14,18-24], shoot meristems [25], stem-node [26,27] epicotyls [28], primary leaf [29], cotyledons [30,31], plumules (32), hypocotyls [22,33,34], and embryo axes [25,35]. Plant regeneration *via* organogenesis from cotyledonary node was found to be the most convenient and faster approach in soybean. However, much improvement is needed for the cotyledonary node regeneration system. This limitation is mainly due to low frequency of shoot regeneration, long regeneration period and explant growth difficulties, which prevent the plant from being regeneration-competent[36].

The nutritional requirement for optimal shoot bud induction from different explants has been reported to vary with mode of regeneration. Media compositions have a key role in shoot morphogenesis, the basal medium MS [37] is most commonly used for soybean organogenesis and the medium B5 [38] are useful in some approaches. Benzylaminopurine (BA) has been the most commonly used plant growth regulator either alone or in combination with a low concentration of cytokinins, kinetin or thidiazuron (TDZ) [22, 39]. TDZ was reported to induce multiple bud tissue (MBT) from cotyledonary node axillary meristem which then gives shoots in the presence of BA [23]. The efficiency of shoot bud formation were enhanced by supplementing media with proline, increased level of MS micro nutrients [40], and ureide in the form of allantoin and amides [21].

Adventitious shoot regeneration from cotyledonary node or leaf node is based on proliferation of meristems. Use of pre-existing shoot meristems in transformation procedures can increase the chance of chimerism, so identifying tissues that can produce shoots in the absence of such pre-formed organs would be important [41]. Adventitious soybean shoots have been induced from hypocotyls [13]; cotyledons [18, 20], primary leaves [29] and epicotyls [28]. Hypocotyls of seedlings have been used as explants for adventitious shoot regeneration by Kaneda et al. [22]. Explants cultured on media supplemented with TDZ induced adventitious shoots more efficiently than BA. Histological analysis of adventitious shoot regeneration from the hypocotyl shows shoot primordias, formed from parenchymatous tissues of central pith and plumular trace regions [33]. Hypocotyls of seedlings have seldom been used as explants, even though the shoot regeneration frequency from hypocotyl segments was found to be higher than from cotyledons [22]. Franklin et al. [31] investigated the factors affecting adventitious shoot regeneration from the proximal end of mature and immature cotyledons. The presence of BAP and TDZ in the medium exerted a synergistic effect, in that regeneration efficiency was higher than for either cytokinin alone.

Indirect organogenesis is important as an alternative source of genetic variation in order to recover somaclones with interesting agronomic traits. Callus regeneration is advantageous over direct regeneration for transformation since effective selection of transgenic cells can be achieved [1]. However, the efforts made to regenerate plants from callus have yielded poor results since plants could not be regenerated from any type of soybean callus [42]. Yang et al. [32] compared different explants excised from immature and germinated seeds for callus mediated organogenic regeneration, although induction of organogenic callus was easily achieved by culture of immature cotyledons, development of adventitious buds from these calluses and the subsequent growth of these buds to shoots were inefficient, suggesting that only part of the callus was competent for regeneration. Sairam et al. [1] developed a rapid and efficient protocol for regeneration of genotype-independent cotyledonary nodal callus for cultivars Williams 82, Loda and Newton through manipulation of plant growth regulators and carbohydrates in the medium. Hong et al. [43] reported organogenic callus induction from cotyledonary node and leaf node explants in media supplemented with TDZ and BA, the system has been successfully utilized for *Agrobacterium*-mediated transformation

2.2. Genotype

Among the different factors affecting soybean regeneration, the genotypic dependence is ranked quite high. Since there is strong genotype specificity for regeneration of different soybean genotypes, a major limiting factor, it is pivotal to formulate genotype specific regeneration protocols. Genotype specificity for regeneration in soybean is well documented, although organogenesis is less genotype dependent and has become routine in several laboratories [18,20,28,29&33]. Reichert et al. [41] tested organogenic adventitious regeneration from hypocotyl explants excised from 18 genotypes. Plant formation from hypocotyl explants showed that all genotypes were capable of producing elongated shoots that could be successfully rooted. This study confirmed the genotype independent nature of this organogenic regeneration from the hypocotyl explant. Sairam et al. [1] developed an efficient genotype independent cotyledonary nodal callus mediated regeneration protocol for soybean cultivars Williams 82, Loda and Newton developed through manipulation of plant growth regulators and carbon source. Callus induction and subsequent shoot bud differentiation were achieved from the proximal end of cotyledonary explants on modified MS [37] media containing 2,4-dichlorophenoxyacetic acid (2,4-D) and benzyladenine (BA), respectively. Sorbitol was found to be the best for callus induction and maltose for plant regeneration. The genotypic dependence of regeneration from cotyledon explants could be reduced by the use of combinations of cytokinins (Franklin et al. [31]). Though there was no significant difference in shoot bud formation among different genotypes, but there was significant difference in conversion of the number of regenerated plants in each cultivar (Delzer et al. [44]).

2.3. *Agrobacterium* mediated transformation

Agrobacterium-mediated transformation of soybean was first demonstrated by Hinchee et al. [10] through delivering, T-DNA into cells in the axillary meristems of the cotyledonary-node. After that scientists have attempted to introduce a lot of genes using *Agrobacterium* [25, 45-47]. The cotyledonary-node method is a frequently used soybean transformation system based on *Agrobacterium*-mediated T-DNA delivery into regenerable cells in the axillary meristems of the cotyledonary-node [16]. The efficiency of this transformation system remains low, apparently because of infrequent T-DNA delivery to cells in the cotyledonary-node axillary meristem, inefficient selection of transgenic cells that give rise to shoot meristems, and low rates of transgenic shoot regeneration and plant establishment. The development of an effective Agrobacterium transformation method for soybean depends on several factors including plant genotype, explant vigor, Agrobacterium strain, vector, selection system, and culture conditions [48, 49]. Increased soybean transformation efficiency, may be achieved by further optimizing the selection system, enhancing explant-pathogen interaction and improving culture conditions to promote regeneration and recovery of transformed plants. It has been reported that soybean genotype contributed to variation in susceptibility to *Agrobacterium* and regenerability in tissue culture [50, 51]. In addition, surface sterilization of plant tissue material for *in vitro* tissue culture and transformation is one of the critical steps in carrying out transformation experiments. While a short time of sterilization cannot completely decontaminate explants, prolonged sterilization may cause damage to explants and consequently affect their regenerability [52]. Antioxidant reagents such as cysteine, dithiothreitol, ascorbic acid and polyvinyl pyrrolidone have been used in plant transformation optimization to enhance either tissue culture response or transformation efficiency [53-55]. Recently, high transformation efficiency has also been reported in soybean by adding cysteine and thiol compounds to the cocultivation media [16, 56,57]. Liu et al. [35] established *Agrobacterium* mediated transformation using shoot tip explants of Chinese soybean cultivars. It had the advantage over the cotyledonary node by having no necrosis after infection, and showed more transient *gus* expression as embryonic tips are more sensitive to *Agrobacterium* because they contain promeristems and procambium. Yun, [58] established liquid medium to select transformed plants from the cotyledonary node. Liquid selection has proven to be more efficient than solid selection due to the direct contact of the explants with the medium and the selection agent in the medium. Olhoft et al. [59] transformed soybean cotyledonary nodes using *Agrobacterium rhizogens* strain SHA17 for the first time. The transformation efficiency was as high as 3.5 fold when compared with *Agrobacterium tumefaciens* strain AGL1. Clemente et al. [60] successfully used and evaluated the effect of glyphosate as a selective agent within the *Agrobacterium mediated* cotyledonary transformation system. Imazapyr is a herbicidal molecule that inhibits the enzymatic activity of acetohydroxyacid synthase, which catalyses the initial step in the biosynthesis of isoleucine, leucine and valine. Aragao et al. [47] used Imazapyr as a selection agent for selection of meristematic soybean cells transformed with the *ahas* gene from Arabidopsis. The *bar* gene encodes for phosphinothricin acetyltransferase (PAT) which detoxifies glufosinate, the active ingredient in the herbicide. Zhang et al. [61] successfully used glyphosate to select transformed cells after *Agrobacterium* transformation of cotyledonary node cells.

2.4. Particle bombardment

Even though particle bombardment is a widely used technique for transforming soybean embryogenic cultures, it was rarely explored for shoot morphogenesis. McCabe et al. [25] was the first to report particle bombardment mediated transformation in soybean. Transforming meristems of soybean bu DNA coated gold particles followed by shoot regeneration in the presence of cytokinin, resulting in the development of chimeras. In subsequent studies, non-chimeric plants were obtained through the use of screening methods for the selection of plants that contained transgenic germ-line cells [32,62&63]. Shoot apex transfor-

http://dx.doi.org/10.5772/54268

mation is labour intensive because the meristematic tissue is diffcult to target and, without selection, a large number of plants must be regenerated and analysed [64].

2.5. Genes for trait improvement

Soybean has been improved by *Agrobacterium* mediated transformation followed by shoot regeneration. Wheat germin gene (gf-2.8) encoding an oligomeric protein and oxalate oxidase (*oxo*) genes were introduced into soybean to improve resistance to the oxalate-secreting pathogen *Sclerotina sclerotiorum* [65]. Li et al.[6] successfully utilized *Agrobacterium*-mediated transformation to transfer chitinase gene (*chi*) and the barley ribosome-inactivating protein gene (*rip*) into soybean cotyledonary node cells. Piller et al. [66] investigated the feasibility of expressing the major Enterotoxigenic Escherichia coli K99 fimbrial subunit, FanC, in soybean for use as an edible subunit vaccine. Xue et al. [67] successfully expressed jasmonic acid carboxyl methyltransferase (NTR1) gene from *Brassica campestris* into soybean cv.Jungery that produces methyl jasmonate and showed tolerance to water stress. Soybean oil contains very low level of α-tocopherol which is the most active form of tocopherol. The tocopherols present in the seed are converted into α- and β-tocopherols by overexpressing γ-tocopherol methyltransferase from *Brassica napus* (BnTMT) [68]. Jiang et al. [69] transferred isoflavone synthase (IFS) gene into soybean callus using *Agrobacterium*-mediated transformation and the transgenic plants produced increased levels of the secondary metabolite, isoflavone. Transgenic soybean plant containing PhyA gene of *Aspergillus ficuum* exhibited a lower amount of phytate in different soybean tissues including the leaf, stem and root. This indicated that engineering crop plants with a higher expression level of heterologous phytase could improve the degradation of phytate and potentially in turn mobilize more inorganic phosphate from phytate and thus reduce phosphate load on agricultural ecosystems [70].

3. Somatic embryogenesis and transformation

Somatic embryogenesis is a process by which a plant somatic cell develops into a whole plant without gametic fusion but undergoes developmental changes as that of zygotic embryogenesis [71, 72]. The first demonstration of *in vitro* somatic embryogenesis was reported in *Daucus carota* by Reinert [73]. The concept of embryogenesis has drawn a lot of attention because of its significance in theory and practice. Primarily, somatic embryos can be produced easily and quickly, so that it provides an economical and easy way to study plant development. Secondly, synthetic seeds developed from somatic embryos open the possibility of developing high quality seeds and may allow us to produce seeds from those plants that require a long period for seed production. Somatic embryogenesis is also useful in plant genetic engineering since regeneration *via* somatic embryogenesis is frequently single of cell origin, resulting in a low response of chimeras and high a number of true transgenic regenerants [74, 75].

3.1. Somatic embryogenesis

The first record of soybean somatic embryogenesis was reported by Beversdorf & Bingham [76], followed by Christianson et al. [77] who regenerated plants through the method. The immature cotyledon is the preferred explant for soybean somatic embryogenesis as it has pre-determined embryogenic cells. Somatic embryogenesis is a multi-step regeneration process starting with the formation of proembryogenic cell mass, followed by somatic embryo induction, their maturation, desiccation and finally plant regeneration [78].

Soybean somatic embryos were induced from immature cotyledon explants cultured on medium containing high levels of 2,4-D [79]. Even though NAA induced somatic embryogenesis from immature cotyledons, the mean number of embryos produced on 2,4-D was significantly higher [80]. Explant orientation, pH, solidifying agent, and 2,4-D concentration have a synergic effect on somatic embryo induction [81]. The early-staged somatic embryos can be maintained and proliferated by subculturing the tissue on either semi-solid medium [79] or liquid suspension culture medium [82]. Somatic embryos incubated in a medium containing NAA do not proliferate so well as those produced on a medium containing 2,4-D [83]. Somatic embryos initiated on NAA are more advanced in embryo morphology than those induced on 2,4-D and the efficiency of somatic embryo induction was highest with a medium containing 2-3% sucrose. Cultures initiated on lower sucrose concentrations tended to produce a higher amount of friable embryos, while increased concentrations of this sugar impaired embryo induction [80,84-86]. Histodifferentiation and maturation of somatic embryos doesn't need exogenous auxin or cytokinins [87]. Indeed, poorly developed meristem or swollen hypocotyls may be an undesired outcome of the application of exogenous auxins and cytokinins, respectively. Moon and Hildebrand, [88] investigated the effects of proliferation, maturation, and desiccation methods on conversion of soybean somatic embryos to plants. Somatic embryos proliferated on solid medium showed a higher regeneration rate when compared with the embryos proliferated in liquid medium. The growth period of somatic embryo development can be reduced one month by culturing in a medium devoid of 2,4-D and B_5 vitamins. Carbon source is critical for embryo nutritional health and improves somatic embryo maturation. The effects of carbohydrates on embryo histodifferentiation and maturation on liquid medium were analyzed by Samoylov et al. [89]. FNL medium supplemented with 3% sucrose (FNL0S3) or 3% maltose (FNL0M3) were compared. Data indicated that sucrose promotes embryo growth and significantly increases the number of cotyledon-stage embryos recovered during histodifferentiation and maturation. However, the percentages of plants recovered from embryos differentiated and matured in FNL0S3 was lower than those grown in FNL0M3 (Samoylov et al. 1998b). The quality of somatic embryos can be positively influenced by a low osmotic potential in maturation medium [90, 91]. Carbohydrates can act as an osmotic agent. Polyethylene glycol 4000, mannitol and sorbitol were tested as supplements to a liquid Finer and Nagasawa medium-based histodifferentiation/maturation medium FNL0S3, for soybean (*Glycine max* L. Merrill) somatic embryos of 'Jack' and F138 or 'Fayette'[90]. Overall, 3% sorbitol was found to be the best of the osmotic supplements tested. The ability of histodifferentiation and conversion of somatic embryo have been improved by the use of ethylene inhibitor aminoethoxyvinylglycine [92].

The effects of ethylene on embryo histodifferentiation and conversion were genotype-specific. The germination frequency of soybean embryos is very low [93], and therefore, partial desiccation of somatic embryos was emphasised with a view to improving the germination frequency in soybean [87,94&95]. Desiccation induced a physiological state there by increase the germination ability of somatic embryos [87].

3.2. Genotype

Soybean somatic embryogenesis is highly genotypic when compared to organogenesis. The existence of strong genotype specificity in the regeneration capacity of the different cultivars represents a major limiting factor for the advancement of soybean biotechnology. The embryogenic efficiency of soybean was shown to be different among cultivars at each stage (induction, proliferation, maturation, germination) of somatic embryogenesis [92,96] and it is very challenging to identify genotypes highly responsive to all stages. Simmonds and Donaldson, [97] screened 18 short season soybean genotypes for proliferative embryogenesis. Five genotypes produced embryogenic cultures which were proliferative for at least 6 months. Yang et al. [98] screened 98 Chinese soybean varieties for somatic embryogenesis and selected 12 varieties based on their embryogenic capacity. The greatest average number of plantlets regenerated per explant (1.35) was observed in N25281. Bonacin et al. [99] demonstrated the influence of genotype on somatic embryogenic capability of five Brazilian cultivars. Droste et al. [100] reported somatic embryo induction, proliferation and transformation of commercially grown Brazilian soybean cultivars for the first time. Soybean somatic embryo conversion is genotype dependent; germination frequency of H7190 was approximately three fold lower than that of PI 417138 [101]. Hiraga et al. [102] examined the capacity for plant regeneration through somatic embryogenesis in Japanese soybean cultivars and identified Yuuzuru and Yumeyutaka as having high potential for somatic embryogenesis. Several cultivars were identified as uniformly embryogenic at the primary induction phase at all locations, among which Jack was the best [103]. Kita et al. [104] evaluated somatic embryogenesis, proliferation of embryogenic tissue, and regeneration of plantlets in backcrossed breeding lines derived from cultivar Jack and a breeding line, QF2. The backcrossed breeding lines exhibited an increased capacity for induction and proliferation of somatic embryos and were used successfully to generate transgenic plants.

3.3. *Agrobacterium* mediated transformation

Recovery of the first transgenic plant *via* somatic embryogenesis in soybean was reported by Parrott et al. [105]. Immature cotyledon tissues were inoculated with *Agrobacterium* strain which contained 15 kD zein gene and the neomycin phosphotransferase gene. The explants were placed on medium containing high auxin for somatic embryo induction. Three transgenic plants containing the introduced 15 kD zein gene were regenerated. Unfortunately, these plants were chimeric and the 15 kD zein gene was not transmitted to the progeny. Sonication-assisted *Agrobacterium*-mediated transformation (SAAT) of immature cotyledons tremendously improves the efficiency of *Agrobacterium* infection by introducing large numbers of micro wounds into the target plant tissue [48]. The highest GUS expression was obtained when immature cotyledons were sonicated for 2s in the presence of *Agrobacterium* followed by co-cultivation for 3 days. Trick and Finer, [108] successfully employed Sonication-assisted *Agrobacterium*-mediated transformation of embryogenic suspension culture tissue and when SAAT was not used, no transgenic clones were obtained. Yan et al. [109] demonstrated the feasibility of *Agrobacterium* mediated transformation of cotyledon tissue for the production of fertile transgenic plants by optimising the *Agrobacterium* concentration, using co-cultivation time and selecting proper explant. Ko and Korban, [110] investigated optimal conditions for induction of transgenic embryos followed by *Agrobacterium* mediated transformation. Using cotyledon explants from immature embryos of 5-8mm length, a 1:1 (v/v) concentration of bacterial suspension and 4-day co-cultivation period significantly increased the frequency of transgenic somatic embryos. The *Agrobacterium* tumefaciens strain KYRT1 harboring the virulence helper plasmid pKYRT1 induces transgenic somatic embryos at a high frequency from infected immature soybean cotyledons [111]. Recently, the successful recovery of a high number of soybean transgenic fertile plants was obtained from the combination of DNA- free particle bombardment and *Agrobacterium*-mediated transformation using proliferating soybean somatic embryos as targets [112].

3.4. Particle bombardment

Particle bombardment is a widely used technique for transformation of embryogenic cultures of soybean; the major advantage of this technique over *Agrobacterium* is the removal of biological incompatibilities. Particle bombardment in soybean was first reported by Finer and McMullen [64], in which embryogenic suspension culture tissue of soybean was bombarded with particles coated with plasmid DNAs encoding hygromycin resistance and β-glucuronidase. Analysis of DNA from progeny plants showed genetic linkage for multiple copies of introduced DNA. Using particle bombardment, fertile plants could be routinely produced from the proliferating transgenic embryogenic clones. Hazal et al. [113] studied growth characteristics and transformability of embryogenic cultures and found that cultures bombarded between 2-6 days after transfer to fresh medium showed more transient expression of the reporter gene. Histological analysis showed that the most transformable cultures had cytoplasmic-rich cells in the outermost layers of the tissue. Maughan et al. [114] bombarded embryogenic cultures with plasmid containing 630-bp DNA fragment encoding a bovine milk protein, β-casein. Hadi et al. [115] co-transformed 12 different plasmids into embryogenic suspension culture by particle bombardment. Hybridization analysis of hygromycin resistance clones verified the presence of introduced plasmid DNAs. Santarem and Finer [116] investigated the effect of desiccation of target tissue, period of subculture prior to bombardment and number of bombardments per target tissue for enhancement of transient expression of the reporter gene. Desiccation of proliferating tissue for 10 min, subculture on the same day prior to bombardment and three times bombardment on a single day enhanced the transient expression of β-glucuronidase [116]. Dufourmantel et al. [117] successfully transformed chloroplasts from embryogenic tissue of soybean using DNA carrying spectinomycin resistance gene (*aadA*) by bombardment. All transplastomic T0 plants were fertile and T1 progeny was uniformly spectinomycin resistant, showing the stability of the plastid transgene. *Droste* et al. [100] successfully transformed embryogenic cultures of soybean cultivars recommended for commercial growing in South Brazil by bombardment, and this opened the field for the improvement of this crop in this country by genetic engineering.

3.5. Genes for trait improvement

Li et al. [6] attempted to transform two antifungal protein genes (*chitinase* and ribosome-inactivating protein) by co-transformation. Transgenic soybeans expressing the Yeast SLC1 Gene showed higher oil content [118]. They reported that, compared to controls, the average increase in triglyceride values went up by 1.5% in transgenic somatic embryos and also found that a maximum of 3.2% increase in seed oil content was observed in a T3 line. Transfer of Δ6 desaturase, fatty acid elongase and D5 desaturase into soybean under seed specific expression produced arachidonic acid (ARA) in seeds of soybean [119]. In an attempt to enhance soybean resistance to viral diseases, several groups successfully generated transgenic plants by expressing an inverted repeat of soybean dwarf virus SbDV coat protein (*CP*) genes [120], or soybean mosaic virus (SMV) coat protein gene [121]. The nutritional quality of soybean has been improved for enhanced amino acid, proteins and vitamin production by transgenic technology [114, 122, 123, 124, and 125]. The feasibility of genetically engineering soybean seed coats to divert metabolism towards the production of novel biochemicals was tested by transferring the genes phbA, phbB, phbC from *Ralstonia eutropha*. Each gene was under the control of the seed coat peroxidase gene promoter [126]. The analysis of seed coats demonstrated that polyhydroxybutyrate (PHB) was produced at an averge of 0.12% seed coat dry weight.

4. Conclusion and future prospects

As demands increase for soybean oil and protein, the improvement of soybean quality and production through genetic transformation and functional genomics becomes an important issue throughout the world. Modern genetic analysis and improvement of soybean heavily depend on an efficient regeneration and transformation process, especially commercially important genotypes. The transformation techniques developed until now till date do not allow high-throughput analyses in soybean functional genomics; though significant improvements have been made in the particle bombardment of embryogenic culture and *Agrobacetrium* mediated transformation of the cotyledonary node over the past three decades. However, routine recovery of transgenic soybean plants using either of these two transformation systems has been restricted to a few genotypes with no reports of transformation on other locally available commercial genotypes. Therefore, development of an efficient and consistent transformation protocol for other locally available commercial genotypes, will greatly aid soybean functional genomics and transgenic technology.

Author details

Thankaraj Salammal Mariashibu[1], Vasudevan Ramesh Anbazhagan[1], Shu-Ye Jiang[1], Andy Ganapathi[2] and Srinivasan Ramachandran[1*]

*Address all correspondence to: sri@tll.org.sg

1 Temasek Life Sciences Laboratory, 1 Research Link, the National University of Singapore, Singapore

2 Department of Biotechnology, Bharathidasan University, Tiruchirappalli, Tamilnadu, India

References

[1] Sairam RV, Franklin G, Hassel R, Smith B, Meeker K, Kashikar N, Parani M, Abed Dal, Ismail S, Berry K, Goldman SL. A study on effect of genotypes, plant growth regulators and sugars in promoting plant regeneration via organogenesis from soybean cotyledonary nodal callus. Plant Cell, Tissue and Organ Culture 2003, 75:79-85.

[2] http://www.soystats.com/2011/page_30.htm.

[3] Birt DF, Hendrich S, Anthony M and Alekel DL (2004) Soybeans and the prevention of chronic human disease. Soybeans: Improvement, Production and Uses. J. Specht and R. Boerma, Eds. 3rd ed. pp. 1047–1117 American Society of Agronomy, Madison, WI

[4] Wilson RF. Seed composition. In: H.R. Boerma and J.E. Specht (eds) Soybean: improvement, production and uses. 3rd ed. ASA, CSSA, SSA, Madison, WI 2004. 621-677.

[5] Manavalan LP, Guttikonda SK, Tran LS, Nguyen HT. Physiological and molecular approaches to improve drought resistance in soybean. Plant Cell Physiology 2009, 50:1260-1276.

[6] Li HY, Zhu YM, Chen Q, Conner RL, Ding XD, Li J, Zhang BB. Production of transgenic soybean plants with two anti-fungal protein genes via Agrobacterium and particle bombardment. 2004 48(3):367-374.

[7] Wang G, Xu Y. Hypocotyl-based *Agrobacterium*-mediated transformation of soybean (Glycine max) and application for RNA interference. Plant Cell Reports 2008, 27:1177-1184.

[8] Hiromoto DM, Vello NA. The genetic base of Brazilian soybean (Glycine max (L.) Merrill) cultivars. Brazil Journal of Genetics 1986, 2:295-306.

[9] Cao D, Hou W, Song S, Sun H, Wu C, Gao Y, Han T. Assessment of conditions affecting *Agrobacterium rhizogenes*-mediated transformation of soybean. Plant Cell Tissue and Organ Culture 2009, 96:45-52.

[10] Hinchee MA, Connor-Ward DV, Newell CA, McDonnell RE, Sato SJ, Gasser CS, Fischhoff DA, Re DB, Fraley RT, Horsch RB. Production of transgenic soybean plants using *Agrobacterium*-mediated DNA transfer. Nature Biotechnology 1988, 6:915-922.

[11] Trick HN, Dinkins RD, Santarem, ER, Di R, Samoylov V, Meurer CA, Walker DR, Parrott WA, Finer JJ, Collins GB. Recent advances in soybean transformation. Plant Tissue Culture and Biotechnology 1997, 3(1):9-24.

[12] Kothari SL, Joshi A, Kachhwaha S, Ochoa-Alejo N. Chilli peppers -A review on tissue culture and transgenesis. Biotechnology Advances 2010, 28:35-48.

[13] Kimball SL, Bingham ET. Adventitious bud development of soybean hypocotyl sections in culture. Crop Science 1973, 13:758-760.

[14] Cheng TY, Saka H, Voqui-Dinh TH. Plant regeneration from soybean cotyledonary node segments in culture. Plant Science Letters 1980, 19:91-99.

[15] Paz MM, Shou H, Guo Z, Zhang Z, Banerjee AK, Wang K. Assessment of conditions affecting Agrobacterium-mediated soybean transformation using the cotyledonary node explant. Euphytica 2004, 136:167-179.

[16] Olhoft PM, Somers DA. L.Cysteine increases *Agrobacterium*- mediated T-DNA delivery into soybean cotyledonary-node cells. Plant Cell Reports 2001, 20:706-711.

[17] Benson EE. Special symposium: *in vitro* plant recalcitrance, do free radicals have a role in plant tissue culture recalcitrance? In Vitro Cellular and Developmental Biology Plant 2000, 86:163-70.

[18] Barwale UB, Keans HR, Widholm JM. Plant regeneration from callus cultures of several soybean genotypes via embryogenesis and organogenesis. Planta 1986, 167:473-481.

[19] Barwale UB, Mayer Jr MM, Widholm JM. Screening of Glycine max and Glycine soja genotypes for multiple shoot formation at the cotyledonary node. Theoretical and Applied Genetics 1986, 72:423-428.

[20] Wright MS, Koehler SM, Hinchee MA, Carnes MG. Plant regeneration by organogenesis in Glycine max. Plant Cell Reports 1986, 5:150-154.

[21] Shetty K, Asano Y, Oosawa K. Stimulation of in vitro shoot organogenesis in Glycine max (Merrill.) by allantoin and amides. Plant Science 1992, 81:245-251.

[22] Kaneda Y, Tabei Y, Nishimura S, Harada K, Akihama T, Kitamura K. Combination of thidiazuron and basal media with low salt concentrations increases the frequency of shoot organogenesis in soybeans [Glycine max (L.) Merr.]. Plant Cell Reports 1997, 17 8-12.

[23] Shan Z, Raemakers K, Tzitzikas EN, Ma Z, Visser RGF. Development of a highly efficient, repetitive system of organogenesis in soybean (Glycine max (L.) Merr). Plant Cell Reports 2005, 24:507-512.

[24] Paz MM, Martinez JC, Kalvig AB, Fonger TM, and Wang K. Improved cotyledonary node method using an alternative explant derived from mature seed for efficient Agrobacterium mediated soybean transformation. Plant Cell Reports 2006, 25:206-213.

[25] McCabe DE, Swain WF, Martinell BJ, Christou P. Stable transformation of soybean (Glycine max) by particle acceleration. Biotechnology 1988, 6:923-926.

[26] Saka H, Voqui Dinh TH, Cheng TY. Stimulation of multiple shoot formation on soybean stem nodes in culture. Plant Science Letters 1980, 19 193-201.

[27] Kim JH, Lamotte E Hack E. Plant regeneration in vitro from primary leaf nodes of soybean (Glycine max) seedling. Journal Plant Physiology 1990, 136:664-669.

[28] Wright MS, Williams MH, Pierson PE, Carnes MG. Initiation and propagation of Glycine max L. Merrill; Plants from tissue-cultured epicotyls. Plant Cell Tissue Organ Culture 1987, 8:83-90.

[29] Wright MS, Ward DV, Hinchee MA, Carnes MG, Kaufman RJ. Regeneration of soybean (Glycine max L. Merr.) from cultured primary leaf tissue. Plant Cell Reports 1987, 6:83-89.

[30] Mante S, Scorza R, Cordts J. A simple, rapid protocol for adventitious shoot development from mature cotyledons of Glycine max cv.Bragg. In Vitro Cellular and Developmental Biology Plant 1989, 25(4):385-388.

[31] Franklin G, Carpenter L, Davis E, Reddy CS, Al-Abed D, Alaiwi WA, Parani M, Smith B, Goldman SL, Sairam RV. Factors influencing regeneration of soybean from mature and immature cotyledons. Plant Growth Regulation 2004, 43:73-79.

[32] Yang N and Christou P. Cell type specific expression of a CaMV 35s-GUS gene in transgenic soybean plants. Developmental Genetics 1990, 11:289-293.

[33] Dan, Y, Reichert NA. Organogenic regeneration of soybean from hypocotyl explants. In Vitro Cellular and Developmental Biology Plant 1998, 34 14-21.

[34] Yoshida T. Adventitious shoot formation from hypocotyl sections of mature soybean seeds. Breeding Science 2002, 52:1-8.

[35] Liu HK, Yang C, Wie ZM. Efficient *Agrobacterium tumefaciens* mediated transformation of soybeans using an embryonic Tip regeneration system. Planta 2004, 219:1042-1049.

[36] Ma XH, Wu TL. Rapid and efficient regeneration in soybean [Glycine max (L.) Merrill] from whole cotyledonary node explants. Acta Physiologiae Plantarum 2008, 30:209-216.

[37] Murashige T. and Skoog F. A revised medium for rapid growth and bioassays with tobacco tissue cultures. Physiologia Plantarum 1962, 15:473-479.

[38] Gamborg OL, Miller RA, Ojima K. Nutrient requirements of suspension cultures of soybean root cells. Experimental Cell Research 1968, 50:151-158.

[39] Joyner EY, Boykin LS, Lodhi MA. Callus Induction and Organogenesis in Soybean [Glycine max (L.) Merr.] cv. Pyramid from Mature Cotyledons and Embryos. The Open Plant Science Journal 2010, 4:18-21.

[40] Kim J, Hack E, LaMotte CE. Synergistic effects of proline and inorganic micronutrients and effects of individual micronutrients on soybean (Glycine max) shoot regeneration in vitro. Journal of Plant Physiology 1994, 144:726-734.

[41] Reichert NA, Young MM, Woods AL. Adventitious organogenic regeneration from soybean genotypes representing nine maturity groups. Plant Cell, Tissue and Organ Culture 2003, 75:273-277.

[42] Hu CY, Wang L. In planta soybean trasnformation technologies developed in China: procedure, confirmation and field performance. In Vitro Cellular and Developmental Biology Plant 1999, 35:417-420.

[43] Hong HP, Zhang H, Olhoft P, Hill S, Wiley H, Toren E, Hillebrand H, Jones T, Cheng M. Organogenic callus as the target for plant regeneration and transformation *via Agrobacterium* in soybean (Glycine max (L) Merr). In Vitro Cellular and Developmental Biology Plant, 2007, 43(6):558-568.

[44] Delzer BW, Somers DA, Orf JH. *Agrobacterium tumefaciens* susceptibility and plant regeneration of 10 soybean genotypes in maturity groups 00 to II. Crop Science 1990, 30:320-322.

[45] Chee PP, Fober KA, Slightom JL. Transformation of soybean (Glycine max) by infecting germinating seeds with *Agrobacterium tumefaciens*. Plant Physiology 1989, 91:1212-1218.

[46] Stewart CN Jr, Adang MJ, All JN, Barmy HR, Cardineau G, Tucker D, Parrot WA. Genetic transformation, recovery and characterization of fertile soybean [Glycine max (L.) Merr.] transgenic for a synthetic Bacillus thuringensis cryIAc gene. Plant Physiology 1995, 112:121-129.

[47] Aragao FJL, Sarokin L, Vianna GR, Rech, EL. Selection of transgenic meristematic cells utilizing a herbicidal molecule results in the recovery of fertile transgenic soybean [Glycine max (L) Merrill] plants at a high frequency. Theoretical and Applied Genetics 2000, 101 (1&2):1-6.

[48] Santarem ER Trick HN Essig JS Finer JJ. Sonication-assisted Agrobacterium-mediated transformation of soybean immature cotyledons: optimization of transient expression. Plant Cell Reports 1998, 17:752-759.

[49] Zhang Z, Guo Z, Shou H, Pegg SE, Clemente TE, Staswick PE, Wang K. Assessment of conditions affecting Agrobacterium-mediated soybean transformation and routine

recovery of transgenic soybean. In: A.D. Arencibia (Ed.), Plant Genetic Engineering: Towards the Third Millennium: Proceedings of the International Symposium on Plant Genetic Engineering–10 December 1999, 88-94. Havana, Cuba, Elsevier, Amsterdam / New York 2000.

[50] Meurer CA, Dinkins RD, Collins GB. Factors affecting soybean cotyledonary node transformation. Plant Cell Reports 1998, 18:180-186.

[51] Donaldson PA, Simmonds DH. Susceptibility of Agrobacterium tumefaciens and cotyledonary node transformation in short-season soybean. Plant Cell Reports 2000, 19:478-484.

[52] Maruyama EK, Ishi A, Migita S, Migita K. Screening of suitable sterilization of explants and proper media for tissue culture of eleven tree species of Peru-Amazon forest. Journal of Agricultural Science 1989, 33:252-261.

[53] Perl A, Lotan O, Abu-Abied M, Holland D. Establishment of an Agrobacterium-mediated transformation system for grape (Vitis vinifera L.): the role of antioxidants during grape-Agrobacterium interactions. Nature Biotechnology 1996, 14(5):624-628.

[54] Enriquez-Obregon GA, Prieto-Samsonov DL, de la Riva GA, Perez M, Selman-Housein G, Vasquez-Padron RI. Agrobacterium-mediated Japonica rice transformation: a procedure assisted by antinecrotic treatment. Plant Cell, Tissue and Organ Culture 1999, 59:159-168.

[55] Frame BR, Shou V Chikwamba RK, Zhang Z, Xiang C, Fonger TM, Pegg SEK, Li B, Nettleton DS, Pei D, Wang K. Agrobacterium tumefaciens-mediated transformation of maize embryos using standard binary vector system. Plant Physiology 2002, 129:13-22.

[56] Olhoft PM, Lin K, Galbraith J, Nielsen NC, Somers DA. The role of thiol compounds increasing Agrobacterium-mediated transformation of soybean cotyledonary-node cells. Plant Cell Reports 2001, 20:731-737.

[57] Olhoft PM, Flagel LE, Donovan CM, Somers DA. Efficient soybean transformation using hygromycin B selection in the cotyledonary-node method. Planta 2003, 216:723-735.

[58] Yun CS. High-efficiency Agrobacterium-mediated Transformation of Soybean. Acta Botanica Sinica 2004, 46 (5):610-617.

[59] Olhoft PM, Bernal LM, Grist LB, Hill DS, Mankin SL, Shen YW, Kalogerakis M, Wiley H, Toren E, Song HS, Hillebrand H, Jones T. A novel Agrobacterium rhizogenes-mediated transformation method of soybean [Glycine max (L.) Merrill] using primary-node explants from seedlings. In Vitro Cellular and Developmental Biology Plant 2007, 43:536-549.

[60] Clemente TE, LaVallee BJ, Howe AR, Conner Ward D, Rozman R, Hunter P, Broyles DL, Kasten D, Hinchee MA. Progeny analysis of glyphosate selected transgenic soy-

beans derived from Agrobacterium-mediated transformation. Crop Science 2000, 40:797-803.

[61] Zhang Z, Xing A, Staswick PE, Clemente TE. The use of glufosinate as a selective agent in Agrobacterium-mediated transformation of soybean. Plant Cell Tissue Organ Culture 1999, 56:37-46.

[62] Christou P, Swain WF, Yang NS, McCabe DE. Inheritance and expression of foreign genes in transgenic soybean plants. Proceedings of the National Academy of Sciences of the United States of America 1989, 86:7500-7504.

[63] Christou P, McCabe DE, Martinell BJ, Swain WF. Soybean genetic engineering-commercial products of transgenic plants. Trends in Biotechnology 1990, 8:145-15.

[64] Finer JJ, McMullen MD. Transformation of soybean via particle bombardment of embryogenic suspension culture tissue. In Vitro Cellular and Developmental Biology Plant 1991. 27:175-182.

[65] Donaldson PA, Anderson T, Lane BG, Davidson AL, Simmonds DH. Soybean plants expressing an active oligomeric oxalate oxidase from the wheat gf-2.8 (germin) gene are resistant to the oxalate-secreting pathogen Sclerotina sclerotiorum. Physiol. Mol. Plant Path 2001, 59:297-307.

[66] Piller KJ, Clemente TE, Jun SM, Petty CC, Sato S, Pascual DW, Bost KL. Expression and immunogenicity of an Escherichia coli K99 fimbriae subunit antigen in soybean. Planta 2005, 222:6-18.

[67] Xue RG, Zhang B, Xie HF. Overexpression of a NTR1 in transgenic soybean confers tolerance to water stress. Plant Cell Tissue Organ Culture 2007, 89:177-183.

[68] Chen DF, Zhang M, Wang YQ, Chen XW. Expression of γ-tocopherol methyltransferase gene from Brassica napus increased α-tocopherol content in soybean seed. Biologia Plantarum 2012, 56(1):131-134.

[69] Jiang N, Jeon EH, Pak JH, Ha TJ, Baek IY, Jung WS, Lee JH, Kim DH, Choi HK, Cui Z, Chung YS. Increase of isoflavones in soybean callus by Agrobacterium-mediated transformation. Plant Biotechnology Reporter 2010, 4 253-260.

[70] Yang S, Li G, Li M, Wang J. Transgenic soybean with low phytate content constructed by Agrobacterium transformation and pollen-tube pathway. Euphytica 2011, 177:375-382.

[71] Merkle SA, Parrott WA and Flinn BS. Morphogenic aspects of somatic embryogenesis. In: Thorpe TA (ed.): In vitro Embryogenesis in Plants. Kluwer Academic Publishers 1995, 155-203.

[72] McKersie BD and Brown DCW. Somatic embryogenesis and artificial seeds in Forage legumes. Seed Science Research 1996, 6:109-126.

[73] Reinert J. Morphogenese und ihre kontrolle an gewebekulturen aus carotten. Naturwissenschaften 1958, 45:344-345.

[74] Ammirato PV. The regulation of somatic embryo development in plant cell culture: suspension culture, techniques and hormone requirements. Nature Biotechnology 1983, 1:68-74.

[75] Mathews HR, Litz H, Wilde S, Merkle H and Wetzstein HY. Stable gene expression of β-glcuronidase and npt II genes in mango somatic embryos. In vitro Cellular and Developmental Biology Plant 1992, 28:172-178.

[76] Beversdorf WD, Bingham ET. Degrees of differentiation obtained in tissue cultures of Glycine species. Crop Science 1977, 17(2):307-311.

[77] Christianson ML, Warick DA, Carlson PS. A morphogenetically competent soybean suspension culture. Science 1983, 222:632-634.

[78] Von AS, Sabala I, Bozhkov P, Dyachok J, Filonova L. Developmental pathways of somatic embryogenesis. Plant Cell Tissue Organ Culture 2002, 69:233-249

[79] Finer JJ. Apical proliferation of embryogenic tissue of soybean (Glycine max (L.) Merrill). Plant Cell Reports 1988, 7(4):238-241.

[80] Hoffmann N, Nelson RL, Korban SS. Influence of media components and pH on somatic embryo induction in three genotypes of soybean. Plant Cell, Tissue and Organ Culture 2004, 77(2):157-163.

[81] Santarem ER, Pelissier B, Finer JJ. Effect of explant orientation, pH, solidifying agent and wounding on initiation of soybean somatic embryos. In Vitro Cellular and Developmental Biology-Plant 1997, 33 (1):13-19.

[82] Finer JJ, Nagasawa A. Development of an embryogenic suspension culture of soybean (Glycine max Merrill). Plant Cell, Tissue and Organ Culture 1988, 15:125-136.

[83] Liu W, Moore PJ, Collins GB. Somatic embryogenesis in soybean via somatic embryo cycling. In Vitro Cellular and Developmental Biology Plant 1992, 28:153-160.

[84] Lippmann B, Lippmann G. Induction of somatic embryos in cotyledonary tissue of soybean, Glycine max L. Merr. Plant Cell Reports 1984, 185(3):215-218.

[85] Lazzeri PA, Hildebrand DF, Collins GB. Soybean somatic embryogenesis: effects of hormones and culture manipulations. Plant Cell, Tissue and Organ Culture 1987, 10:197-208.

[86] Lazzeri PA, Hildebrand DF, Sunega J, Williams EG, Collins GB. Soybean somatic embryogenesis: interactions between sucrose and auxin. Plant Cell Reports 1988, 7:517-520.

[87] Parrott WA, Dryden G, Vogt S, Hildebrand DF, Collins GB, Williams EG. Optimization of somatic embryogenesis and embryo germination in soybean. In Vitro Cellular and Development Biology Plant 1988, 24:817-820.

[88] Moon H, Hildebrand DF. Effects of proliferation, maturation, and desiccation methods on conversion of soybean somatic embryos. In Vitro Cellular and Developmental Biology Plant 2003, 39:623-628

[89] Samoylov VM, Tucker DM, Parrott WA. Soybean [Glycine max (L.) Merrill] embryogenic cultures: the role of sucrose and total nitrogen content on proliferation. In Vitro Cellular and Developmental Biology Plant 1998, 34:8-13.

[90] Walker DR, Parrott WA. Effect of polyethylene glycol and sugar alcohols on soybean somatic embryo germination and conversion. Plant Cell, Tissue and Organ Culture 2001, 64(1):55-62.

[91] Korbes AP, Droste A. Carbon sources and polyethylene glycol on soybean somatic embryo conversion. Pesquisa Agropecuária Brasileira 2005, 40(3):211-216.

[92] Santos KGB, Mundstock E, Bodanese Zanettini MH. Genotype-specific normalization of soybean somatic embryogenesis through the use of an ethylene inhibitor. Plant Cell Reports 1997, 16 (12):859-864.

[93] Jang GW, Park RD, Kim KS. Plant regeneration from embryogenic suspension cultures of soybean (Glycine max L. Merrill). Journal of Plant Biotechnology 2001, 3:101-106.

[94] Buchheim JA, Colburn SM, Ranch JP. Maturation of soybean somatic embryos and the transition to plantlet grown. Plant Physiology 1989, 89:768-77.

[95] Durham RE and Parrott WA. Repetitive somatic embryogenesis from peanut cultures in liquid medium. Plant Cell Reports 1992, 11:122-125.

[96] Bailey MA, Boerma HR, Parrott WA. Genotype effects on proliferative embryogenesis and plant regeneration of soybean. In vitro Cellular and Developmental Biology Plant 1993, 29:102-108.

[97] Simmonds DH, Donaldson PA. Genotype screening for proliferative embryogenesis and biolistic transformation of short-season soybean genotypes. Plant Cell Reports 2000, 19:485-490.

[98] Yang C, Zhao T, Yu D, Gai J. Somatic embryogenesis and plant regeneration in Chinese soybean (Glycine max (L.) Merr.) Impacts of mannitol, abscisic acid, and explants age. In Vitro Cellular and Developmental Biology-Plant 2009, 45 (2):180-181.

[99] Bonacin GA, DiMauro AO, Oliveira RC, Perecin D. Induction of somatic embryogenesis in soybean: physicochemical factors influencing the development of somatic embryos. Genetics and Molecular Biology 2000, 23(4):865-868.

[100] Droste A, Pasquali G, Bodanese-Zanettini, MH. Transgenic fertile plants of soybean [Glycine max (L) Merrill] obtained from bombarded embryogenic tissue. Euphytica 2002, 127(3):367-376.

[101] Bailey MA, Boerma HR, Parrott WA. Genotype-specific optimization of plant regeneration from somatic embryos of soybean. Plant Science 1993, 93:117-120.

[102] Hiraga S, Minakawa H, Takahashi K, Takahashi R, Hajika M, Harada K, Ohtsubo N. Evaluation of somatic embryogenesis from immature cotyledons of Japanese soybean cultivars. Plant Biotechnology 2007, 24 (4):435-440.

[103] Meurer CA, Dinkins RD, Redmond CT, Mcallister KP, Tucker DT, Walker DR, Parrott WA, Trick HN Essig JS, Frantz HM, Finer JJ, Collins GB. Embryogenic response of multiple soybean [Glycine max (L.) Merr.] cultivars across three locations. In Vitro Cellular and Developmental Biology Plant 2001, 37(67):62-67.

[104] Kita Y, Nishizawa K, Takahashi M, Kitayama M, Ishimoto M. Genetic improvement of the somatic embryogenesis and regeneration in soybean and transformation of the improved breeding lines. Plant Cell Reports 2007, 26(4):439-447.

[105] Parrott WA, Hoffman LM, Hildebrand DF, Williams EG, Collins GB. Recovery of primary transformants of soybean. Plant Cell Reports 1989, 7:615-61.

[106] Hood EE, Gelvin SB, Melchers LS and Hoekema A. New Agrobacterium helper strains for gene transfer to plants. Transgenic Research 1993, 2:208-218.

[107] Torisky RS, Kovacs L, Avdiushko S, Newman JD, Hunt AG and Collins GB. Development of a binary vector system for plant transformation based on the supervirulent Agrobacterium tumefaciens strain Chry5. Plant Cell Reports 1997, 17:102-108.

[108] Trick HN, Finer JJ. Sonication-assisted Agrobacterium-mediated transformation of soybean (Glycine max) embryogenic suspension culture tissue. Plant Cell Reports 1998, 17:482-488.

[109] Yan B, Reddy MSS, Collins GB, Dinkins RD. *Agrobacterium tumefaciens* mediated transformation of soybean [Glycine max (L) Merrill] using immature zygotic cotyledon explants. Plant Cell Reports 2000, 19(11):1090-1097.

[110] Ko TS, Korban SS. Enhancing the frequency of somatic embryogenesis following Agrobacterium-mediated transformation of immature cotyledons of soybean [Glycine max (L.) Merrill.]. In Vitro Cellular & Developmental BiologyPlant 2004, 40:552-558.

[111] Ko TS, Lee S, Farrand SK, Korban SS. A partially disarmed vir helper plasmid, pKYRT1, in conjunction with 2,4-dichlorophenoxyacetic acid promotes emergence of regenerable transgenic somatic embryos from immature cotyledons of soybean. Planta 2004, 218(4):536-541.

[112] Wiebke Strohm B, Droste A, Pasquali G, Osorio MB, Bucker Neto L, Passaglia LMP, Bencke M, Homrich MS, Margis Pinheiro M, Bodanese Zanettini MH. Transgenic fertile soybean plants derived from somatic embryos transformed via the combined DNA-free particle bombardment and Agrobacterium system. Euphytica 2011, 177(3): 343-354.

[113] Hazel CB, Klein TM, Anis M, Wilde HD, Parrott WA. Growth characteristics and transformability of soybean embryogenic cultures. Plant Cell Reports 1998, 17:765-772.

[114] Maughan PJ, Philip R, Cho MJ, Widholm JM and Vodkin LO. Biolistic transformation, expression, and inheritance of bovine β-casein in soybean (Glycine max). In vitro Cellular and Developmental Biology Plant 1999, 35:334-349.

[115] Hadi MZ, McMullen MD, Finer JJ. Transformation of 12 different plasmids into soybean via particle bombardment. Plant Cell Reports 1996, 15:500-505.

[116] Santarem ER, Finer JJ. Transformation of soybean [Glycine max (L.) Merrill] using proliferative embryogenic tissue maintained on semisolid medium. In Vitro Cellular and Developmental Biology Plant 1999, 35 451-455.

[117] Dufourmantel N, Pelissier B, Garcon F, Peltier G, Jean-Marc F, Tissot G. Generation of fertile transplastomic soybean. Plant Molecular Biology 2004, 55 479-489.

[118] Rao SS, Hildebrand D. Changes in Oil Content of Transgenic Soybeans Expressing the Yeast SLC1 Gene. Lipids 2009, 44:945-951.

[119] Chen R, Matsui K, Ogawa M, Oe M, Ochiai M, Kawashima H, Sakuradani E, Shimizu S, Ishimoto M, Hayashi M, Murooka Y, Tanaka Y. Expression of Delta 6, Delta 5 desaturase and GLELO elongase genes from Mortierella alpina for production of arachidonic acid in soybean [Glycine max (L.) Merrill] seeds. Plant Science 2006, 170:399-406.

[120] Tougou M, Furutani N, Yamagishi N, Shizukawa Y, Takahata Y, Hidaka S. Development of resistant transgenic soybeans with inverted repeat-coat protein genes of soybean dwarf virus. Plant Cell Reports 2006, 25:1213-1218.

[121] Furutani N, Yamagishi N, Hidaka S, Shizukawa Y, Kanematsu S, Kosaka Y. Soybean mosaic virus resistance in transgenic soybean caused by post-transcriptional gene silencing. Breed Science 2007, 57:123-128.

[122] Ishimoto M, Rahman SM, Hanafy MS, Khalafalla MM, El-Shemy HA, Nakamoto Y, Kita Y, Takanashi K, Matsuda F. Murano Y, Funabashi T, Miyagawa H, Wakasa K. Evaluation of amino acid content and nutritional quality of transgenic soybean seeds with high-level tryptophan accumulation. Molecular Breeding 2010, 25:313-326.

[123] Tavva VS, Kim YH, Kagan IA, Dinkins RD, Kim KH and Collins GB. Increased α-tocopherol content in soybean seed over- expressing the Perilla frutescens γ-tocopherol methyltransferase gene. Plant Cell Reports 2007, 26:61-70.

[124] Li Z, Meyer S, Essig JS, Liu Y, Schapaugh MA, Muthukrishnan S, Hainline BE,. Trick HN. High-level expression of maize γ-zein protein in transgenic soybean (Glycine max) Molecular Breeding 2005, 16:11-20.

[125] Dinkins RD, Srinivasa Reddy MS, Meurer CA, Yan B, Trick HN, Finer JJ, Thibaud-Nissen F, Parrott WA and Collins GB. Increased sulfur amino acids in soybean plants overexpressing the maize 15 kDa zein protein. In vitro Cellular and Developmental Biology Plant 2001, 37:742-747.

[126] Schnell JA, Treyvaud-Amiguet V, Arnason JT, Johnson DA. Expression of polyhydroxybutyric acid as a model for metabolic engineering of soybean seed coats. Transgenic Research 2012, 21(4):895-899.

[127] Luo XM, Zhao GL, Jian YY. Plant regeneration from protoplasts of soybean (Glycine max L.). Acta Botanica Sinica 1990, 32:616-621.

[128] Yeh MS. *In vitro* culture of immature soybean embryos II. The abilities of organogenesis and plantlet regeneration from different aged immature embryo in Glycine species. *Journal of the Agricultural Association of China* 1990, 39:73-87.

[129] Dhir SK, Dhir S, Widholm JM. Plantlet regeneration from immature cotyledon protoplasts of soybean (Glycine max L.). Plant Cell Reports 1991, 10:39-43.

[130] Sato S, Newell C, Kolacz K, Tredo L, Finer JJ, Hinchee M. Stable transformation via particle bombardment in two different soybean regeneration systems. Plant Cell Reports 1993, 12(7-8):408- 413.

[131] Di R, Purcell V, Collins GB, Ghabrial SA. Production of transgenic soybean lines expressing the bean pod mottle virus coat protein precursor gene. Plant Cell Reports 1996, 15:746-750.

[132] Xing A, Zhang Z, Sato S, Staswick PE, Clemente TE. The use of the two T-DNA binary system to derive marker-free transgenic soybeans. In Vitro Cellular and Developmental Biology Plant 2000, 36:456-463.

[133] Wang X, Eggenberger AL, Nutter Jr FW, Hill JH. Pathogen-derived transgenic resistance to soybean mosaic virus in soybean. Molecular Breeding 2001, 8:119-127.

[134] Zeng P, Vadnais DA, Zhang Z, Polacco JC. Refined glufosinate selection in Agrobacterium-mediated transformation of soybean [Glycine max (L.) Merrill]. Plant Cell Reports 2004. 22:478-482.

[135] Miklos JA, Alibhai MF, Bledig SA, Connor Ward DC, Gao AG, Holmes BA, Kolacz KH, Kabuye VT, MacRae TC, Paradise MS, Toedebusch AS, Harrison LA. Characterization of soybean exhibiting high expression of a synthetic Bacillus thuringiensis cry1A transgene that confers a high degree of resistance to Lepidopteran pests. Crop Science 2007, 47:148-157.

[136] Liu SJ Wei ZM, Huang JQ. The effect of co-cultivation and selection parameters on Agrobacterium-mediated transformation of Chinese soybean varieties. Plant Cell Reports 2008, 27(3):489-498.

[137] Lee K, Yi BY, Kim KH, Kim JB, Suh SC, Woo HJ, Shin KS, Kweon SJ. Development of Efficient Transformation Protocol for Soybean (Glycine max L.) and Characterization of Transgene Expression after Agrobacterium-mediated Gene Transfer. Journal of the Korean Society for Applied Biological Chemistry 2011, 54:37-45.

[138] Lazzeri PA, Hilderbrand DF, Collins GB. A procedure for plant regeneration from immature cotyledon tissue of soybean. Plant Molecular Biology Reporter 1985, 3(4): 160-167.

[139] Ranch JP, Ogelsby L, Zielinski AC. Plant regeneration from embryo-derived tissue cultures of soybean. In Vitro Cellular and Developmental Biology Plant 1985, 21:653-658.

[140] Ghazi TD, Cheema HV, Nabors MW. Somatic embryogenesis and plant regeneration from embryonic callus of soybean [Glycine max (L.) Merr.]. Plant Cell Reports 1986, 5(6):452-456.

[141] Kerns HR, Barwale VB, Meyer MM. Correlation of cotyledonary node shoot proliferation and somatic embryoid development in suspension cultures of soybean [Glycine max (L.) Merr.]. Plant Cell Reports 1986, 5(2), 140-143.

[142] Hartweck LM, Lazzeri PA, Cui D, Collins GB, Williams EG. Auxin orientation effects on somatic embryogenesis from immature soybean cotyledons. In Vitro Cellular and Developmental Biology Plant 1988, 24(8):821-828.

[143] Komatsuda T, Ohyama K. Genotype of high competence for somatic embryogenesis and plant regeneration in soybean Glycine max. Theoretical and Applied Genetics 1988, 75(5):695-700.

[144] Parrott WA, Williams EG, Hildebrand DF, Collins GB. Effect of genotype on somatic embryogenesis from immature cotyledons of soybean. Plant Cell, Tissue and Organ Culture 1989, 16(1):15-21.

[145] Christou, P, Yang NS. Developmental aspects of soybean (Glycine max) somatic embryogenesis. Annals of Botany 1989, 64(2):225-234.

[146] Komatsuda T, Ko SW. Screening of soybean (Glycine max (L.) Merrill) genotypes for somatic embryo production from immature embryo. Japanese Journal of Breeding 1990, 40:249-251.

[147] Komatsuda T, Kanebo K, Oka S. Genotype × sucrose interactions for somatic embryogenesis in soybean. Crop Science 1991, 31(2):333-337.

[148] Wright MS, Launis KL, Novitzky R, Duesiing JH, Harms CT. A simple method for the recovery of multiple fertile plants from individual somatic embryos of soybean [Glycine max (L.) Merrill]. In Vitro Cellular and Developmental Biology Plant 1991, 27:153-157.

[149] Lippmann B, Lippmann G. Soybean embryo culture: factors influencing plant recovery from isolated embryos. Plant Cell, Tissue and Organ Culture 1993, 32(1):83-90.

[150] Parrott WA, All JN, Adang MJ, Bailey MA, Boerma HR, Stewart CN Jr. Recovery and evaluation of soybean plants transgenic for a Bacillus thuringiensis var.kurstaki insecticidal gene. In Vitro Cellular and Development Biology Plant 1994, 30:144-149.

[151] Saravitz CH, Raper CDJr. Responses to sucrose and glutamine by soybean embryos grown in vitro. Physiologia Plantarum 1995, 93:799-805.

[152] Liu W, Torisky RS, McAllister KP, Avdiushko S, Hildebrand D, Collins GB. Somatic embryo cycling: evaluation of a novel transformation and assay system for seed-specific gene expression in soybean. Plant Cell, Tissue and Organ Culture 1996, 47:33-42.

[153] Rajasekaran K, Pellow JW. Somatic embryogenesis from cultured epicotyls and primary leaves of soybean [Glycine max (L.) Merrill]. In Vitro Cellular and Developmental Biology Plant 1997, 33:88-91.

[154] Samoylov VM, Tucker DM, Parrott WA. A liquid medium-based rapid regeneration from embryogenic soybean cultures. Plant Cell Reports 1998, 18 49-54.

[155] Chanprame S, Kuo TM, Widholm AM. Soluble carbohydrate content of soybean [Glycine max (L.) Merr.] somatic and zygotic embryos during development. In Vitro Cellular and Developmental Biology Plant 1998, 34:64-68.

[156] Ponappa T, Brzozowski AE, Finer JJ. Transient expression and stable transformation of soybean using jellyfish green fluorescent protein. Plant Cell Reports 1999, 19:6-12.

[157] Tian L, Brown DCW. Improvement of soybean somatic embryo development and maturation by abscisic acid treatment. Canadian Journal of Plant Science 2000. 80:721-276.

[158] Walker DR, All JN, McPherson RM, Boerma HR, Parrott WA. Field Evaluation of Soybean Engineered with a Synthetic cry1Ac Transgene for Resistance to Corn Earworm, Soybean Looper, Velvetbean Caterpillar (Lepidoptera: Noctuidae), and Lesser Cornstalk Borer (Lepidoptera: Pyralidae). Journal of Economic Entomology, 2000, 93(3) 613-622.

[159] Droste A, Pasquali G, Bodanese-Zanettini MH. Integrated bombardment and Agrobacterium transformation system: an alternative method for soybean transformation. *Plant Molecular Biology Reporter* 2000, 18:51-59.

[160] Droste A, Leite PCP, Pasquali G, Mundstock EC, Bodanese-Zanettini MH. Regeneration of soybean via embryogenic suspension culture. Scientia Agricola 2001, 58(4): 753-758.

[161] Fernando JA, Vieira MLC, Geraldi IO, Appezzato-da-Gloria B. Anatomical study of somatic embryogenesis in Glycine max (L.) Merrill. Brazilian Archives of Biology and Technology 2002, 45 (3):277-286.

[162] Tomlin ES, Branch SR, Chamberlain D, Gabe H, Wright MS, Stewart CN Jr. Screening of soybean, Glycine max (L.) Merrill, lines for somatic embryo induction and maturation capability from immature cotyledons. In Vitro Cellular and Developmental Biology Plant 2002, 38:543-548.

[163] Chiera JM, Finer JJ, Grabau EA. Ectopic expression of a soybean phytase in developing seeds of Glycine max to improve phosphorus availability. Plant Molecular Biology 2004, 56:895-904.

[164] Schmidt MA, Tucker DM, Cahoon EB, Parrott WA. Towards normalization of soybean somatic embryo maturation. Plant Cell Reports 2005, 24:383- 391.

[165] Santos KGB, Mariath JEA, Moco MCC, Bodanese Zanettini MH. Somatic Embryogenesis from Immature Cotyledons of Soybean (Glycine max (L.) Merr.): Ontogeny of Somatic Embryos. Brazilian Archives of Biology and Technology 2006, 49 (1):49-55.

[166] Wiebke B, Ferreira F, Pasquali G, Bodanese Zanettini MH, Droste A. Influence of antibiotics on embryogenic tissue and Agrobacterium tumefaciens suppression in soybean genetic transformation. Bragantia 2006, 65(4):543-551.

[167] Chiera JM, Streeter JG, Finer JJ. Ononitol and pinitol production in transgenic soybean containing the inositol methyl transferase gene from Mesembryanthemum crystallinum. Plant Science 2006. 171:647-654.

[168] Tougou M, Yamagishi N, Furutani N, Shizukawa Y, Takahata Y, Hidaka S. Soybean dwarf virus-resistant transgenic soybeans with the sense coat protein gene. Plant Cell Reports 2007, 26:1967-1975.

[169] El-Shemy H A, Khalafalla MM, Fujita K, Ishimoto M. Improvement of protein quality in transgenic soybean plants. Biologia Plantarum 2007, 51:277-284.

[170] Weber RLM, Korber AP, Baldasso DA, Callegari Jacques SM, Bodanese Zanettini MH, Droste A. Beneficial effect of abscisic acid on soybean somatic embryo maturation and conversion into plants. Plant Cell Culture and Micropropagation 2007, 3(1): 1-9.

[171] Yemets AI, Radchuk VV, Pakhomov AV, Blume Ya B. Biolistic Transformation of Soybean Using a New Selectable Marker Gene Conferring Resistance to Dinitroanilines. Cytology and Genetics 2008, 42(6):413-419.

[172] Schmidt MA, LaFayette PR, Artelt BA, Parrott WA. A comparison of strategies for transformation with multiple genes via microprojectile-mediated bombardment. In Vitro Cellular and Developmental Biology Plant 2008, 44(3):162-168.

[173] Loganathan M Maruthasalam S, Shiu LY, Lien WC, Hsu WH, Lee PF, Yu CW, Lin CH. Regeneration of soybean (Glycine max L. Merrill) through direct somatic embryogenesis from the immature embryonic shoot tip. In Vitro Cellular and Developmental Biology Plant 2010, 46:265-273.

[174] Droste A, Silva da AM, Souza de IF, Wiebke-Strohm, Bucker Neto L, Bencke M, Sauner MV, Bodanese-Zanettin MH. Screening of Brazilian soybean genotypes with high potential for somatic embryogenesis and plant regeneration. Pesquisa Agropecuaria Brasileira 2010, 45(7):715-720.

[175] He Y, Young TE, Clark KR, Kleppinger-Sparace K F, Bridges WC, Sparace SA. Developmental profile of storage reserve accumulation in soybean somatic embryos. In Vitro Cellular and Developmental Biology Plant 2011, 47:725-733.

[176] Mariashibu TS, Subramanyam K, Arun M, Mayavan S, Rajesh M, Theboral J, Manickavasagam M, Ganapathi A. Vacuum infiltration enhances the Agrobacterium-mediated genetic transformation in Indian soybean cultivars. Acta Physiologiae Plantarum 2012, (published online) DOI 10.1007/s11738-012-1046-3.

Chapter 21

Advancements in Transgenic Soy: From Field to Bedside

Laura C. Hudson, Kevin C. Lambirth,
Kenneth L. Bost and Kenneth J. Piller

Additional information is available at the end of the chapter

http://dx.doi.org/10.5772/52467

1. Introduction

Today biotechnology and the process of genetic modification is emerging and advancing worldwide. At the forefront of this technology is soybean, which has become a popular subject of genetic modification due to its versatility and economical importance as a crop plant. Over 15 years ago the first herbicide resistant soybeans were introduced into the market. By 1997, approximately 8% of all soybeans cultivated for commercial use in the United States were genetically modified. This trend has grown exponentially and by 2011 the percentage of genetically modified soybean rose to 94% in the United States and 81% worldwide. The technology to genetically modify soybean has not only had a huge impact on the commercial agricultural market, but has paved the way to an onset of both traditional and nontraditional uses for soybean as well as opening up many new potential applications for this important crop plant. Soybean has become a popular subject of genetic modification over the past two decades and with the advancement of plant transformation technology, it is now possible to manipulate and or add various traits to soybean.

2. Soybean transformation

There are several protocols used to genetically transform plants with either stable or transient expression. Some of the methods include electroporation, silicon carbide fibers, liposome mediated transformation and *in planta Agrobacterium*-mediated transformation via vacuum infiltration of whole plant. However, many of these methods are not used for soybean transformation because of low transformation efficiencies. Two more commonly used platforms that have been successfully optimized for stable soybean transformation include cotyledonary node–*Agrobacterium*-mediated transformation and somatic embryo–particle-bombardment-mediated transformation.

The *Agrobacterium*-mediated plant transformation method uses a soil dwelling bacteria species called *Agrobacterium tumefaciens* to transfer desirable genes into plants. Using this method, a foreign gene can be placed within the T- DNA boarder regions of the bacterial plasmid which then integrates into a host plant's genome [1]. Wounded plant tissue gives off specific phenolic compounds which induce *Agrobacterium* to express a set of virulence (vir) genes. The expression of the vir genes results in the production of single-stranded DNA that is transferred and integrated into the plant genome.

There are several advantages of *Agrobacterium*-meditated plant transformation including straight forward methodology, minimal equipment cost, and reliable insertion of a single or a low copy transgene number. The first reported transformation of soybean with an *Agrobacterium* strain used co-cultivation followed by organogenesis from cotyledonary nodes [2]. This work was followed by using *Agrobacterium* mediated transformation of immature cotyledons [3], and embryogenic suspension cultures [4]. Since then, several groups have worked to improve these methods, in particular the transformation and regeneration from cotyledonary nodes. Cotyledonary node regions contain axillary meristems at the junction between cotyledon and hypocotyl. Generally, the cotyledonary nodes are pre-wounded and then co-cultivated with *Agrobacterium*. The axillary meristems proliferate and regenerate through the formation of multiple adventitious shoots on culture medium containing a cytokinin. In the United States, public facilities, including the Plant Transformation Facility at Iowa State University and the Plant Transformation Core Facility at the University of Missouri, provide fee for service genetic transformation of soybean for public research, mainly by cotyledonary node *Agrobacterium*-mediated transformation.

The other widely used method of soybean transformation is somatic embryo particle bombardment-mediated transformation also called particle bombardment, or biolistic technology. This method directs small tungsten or gold particles coated with the desired genes toward the target plant cells with enough force to penetrate the cell wall and membrane [5]. Once inside the cell the DNA disassociates from the particle and becomes integrated into the plant genome.

The particle bombardment transformation method was first used in soybean in 1988 by McCabe et al., who successfully transformed immature seed meristem [6] and was followed by the transformation of somatic embryonic tissue [7], and apical meristem [8]. Transformation of somatic embryos has been the most successful method and is induced from immature cotyledons cultured on medium containing moderately high concentrations of an auxin. These cotyledons are used to generate proliferative embryogenic cultures and to recover whole plants. A major advantage of the particle bombardment transformation method relative to *Agrobacterium*-based methods is the removal of biological incompatibilities between tissues of many plant species and the *Agrobacterium* vector. However, it has been shown that certain genotypes are more susceptible to the formation of proliferative embryogenic tissue than others. Limitations of the bombardment process include the requirement of specialized equipment (gene gun), transformation limited to cells at or near the surface, and high copy number events with high levels of recombination which may not be desirable.

The development of soybean transformation methods has paved the way for an extensive amount of research to develop genetically modified soybeans that have been widely adopted for crop improvement purposes. This has been a fast growing field with the addition of many agronomic, nutraceutical, and pharmaceutical traits being developed, the progress of which will be reviewed in this chapter.

3. Agronomical improvements in soybean

3.1. Herbicide tolerance

In 1970, glyphosate, a broad-spectrum foliar herbicide, was discovered [9]. Glyphosate inhibits 5-enolpyruvylshikimate-3-phosphate synthase (EPSPS), a crucial enzyme of the shikimate biosynthetic pathway that is responsible for the production of several essential aromatic amino acids. Glyphosate was commercially introduced in 1974 and by 1995 use had reached 4. 5 million kg in the United States. Due to popularity and broad use of this herbicide by farmers, glyphosate became a candidate for research in creating herbicide resistant soybeans and has led to one of the most well-known examples of herbicide tolerance: The Roundup Ready® soybean developed by Monsanto. Roundup Ready® soybeans were one of the first examples of a commercially viable transgenic plant. These transgenic soybeans express functional EPSPS providing tolerance to the herbicide glyphosate (Roundup™). The popularity of these soybeans grew with farmers since Roundup™ could be applied to a field of Roundup Ready® soybeans to significantly reduce weed populations while leaving the soybean crop unharmed. The development of Roundup Ready® soybeans offered farmers many advantages in a system that was relatively easy to use. The level and consistency of weed control allowed farmers to take advantage of a no-till system, and eliminated the need for cultivation allowing growers to space rows more closely. Narrow row planting results in higher yields due to a more efficient use of space and may result in better weed control, as the canopy closes more quickly providing earlier competition against weeds. In addition, the window of application for Roundup™ is wider than for other post emergence herbicides currently used in soybeans, both in terms of the stage of soybean growth and the ability to achieve effective control of larger weeds. These factors contributed to the popularity of this weed control technology. At present, approximately 90% of the soybeans farmed in the United States utilize this technology.

In response to Monsanto's hugely successful Roundup Ready® crops, Bayer Crop Science released its own herbicide tolerant soybean known as Liberty Link® soybean [10]. Liberty Link® soybeans were developed to express a gene derived from the bacteria *Streptomyces viridochromogenes* called phosphinothricin-N-acetyltransferase (PAT). PAT is a glutamine synthetase inhibitor that binds to glutamate, making plants resistant to the broad-spectrum contact herbicide glufosinate ammonium. This herbicide causes cessation of photosynthesis and plant death by interfering with the biosynthetic pathway of the amino acid glutamine and with ammonia detoxification. Glufosinate ammonium is the active ingredient in phosphinothricin herbicides (Basta®, Ignite®, Rely®, Liberty®, Harvest®, and Finale®) used to control a wide range of weeds after the crop emerges or for total vegetation control on land not used for cultivation. Since glufosinate ammonium-based herbicides function by a different mode of action than glyphosate-based herbicides, the Liberty Link® system provides farmers with an alternative strategy for controlling weeds.

There are several other examples of transgenic soybeans expressing herbicide resistance traits that are in various stages of development. Pioneer has developed a transgenic soybean product that provides tolerance to two different classes of herbicides: glyphosate and acetolactate synthase (ALS)-inhibiting herbicides. These soybean plants express the glyphosate acetyltransferase (GAT4601) and modified version of a soybean acetolactate synthase (GM-HRA) proteins. [11]. The GAT4601 protein confers tolerance to glyphosate-containing herbicides by acetylating glyphosate and thereby rendering it non-phytotoxic. The GM-HRA protein confers tolerance to the ALS-inhibiting class of herbicides. The development of GM soybeans with characteristics controlled by multiple genes leading to the expression of two herbicides is a different approach than previous strategies involving single characteristics controlled by a single gene. These genetically modified soybeans express a combination of herbicides with different modes of action. Inherent crop tolerance will enable more effective management of weed populations.

BASF has used a similar method to introduce a soybean that combines herbicide-tolerant soybean varieties with the broad spectrum imidazolinone class of herbicides. These transgenic soybeans contain the csr1-2 gene derived from *Arabidopsis thaliana* that encodes the imidazolinone-tolerant AHAS-Large subunit (also known as ALS). The AHAS-L subunit interacts with the endogenous soybean small regulatory subunit to form an enzyme complex that catalyzes the first step in the synthesis of branched-chain essential amino acids, valine, leucine, and isoleucine [12]. The AHAS enzymes occur ubiquitously in plants. Imidazolinone herbicides inhibit the native enzymes, resulting in plant death.

To address the potential emergence of other herbicide resistant broadleaf weeds, Monsanto has developed a line of transgenic soybeans that are resistant to treatment with dicamba [13]. Dicamba (3,6-dichloro-2-methoxybenzoic acid) is a low-cost, widely-used, broad leaf herbicide that is environmentally friendly. Soybeans transformed with a genetically engineered bacterial dicambamonooxygenase (DMO) gene were able to inactivate dicamba, making them resistant to this herbicide. Dicamba-resistant soybeans are in the advanced stages of research and development and are predicted to be commercialized soon.

Syngenta and Bayer CropScience are co-developing HPPD-inhibitor tolerant soybeans, a novel herbicide tolerance trait for soy. The event consists of a molecular stack of a gene conferring tolerance to hydroxyphenylpyruvatedioxygenase (HPPD)-inhibiting herbicides as well as a gene for glufosinate tolerance. Inhibition of HPPD stops the catabolic degradation of tyrosine to plastoquinones which is important for photosynthesis, carotenoid biosynthesis, and tocopherol production [14]. This multiple herbicide tolerance stack will enable the use of multiple herbicides and will be an important new tool for soybean growers faced with increasing pressure from resistant weeds. In the future, other innovative molecular strategies can be expected to generate genetically modified (GM) soybeans with novel features to combat weeds and enhance weed resistance. These new GM soybeans will reduce environmental contamination risks and reduce costs for consumers and producers.

3.2. Insect resistance

Insect pest management through the use of chemicals has brought about considerable protection to crop yields over the past several decades. Unfortunately, extensive and indiscriminate usage of chemical pesticides has resulted in environmental degradation, adverse effects on human health and other organisms, eradication of beneficial insects, and development of pest-resistant insects. As farmers move forward with the objective of achieving greater crop productivity it will be imperative to replace chemical inputs with safer alternatives to manage insect pests in agricultural ecosystems. Within agricultural biotechnology, insect resistance is a prime research area that has potential to improve agricultural productivity and provide much needed alternatives to pesticides while being effective against pests, innocuous to non-target organisms, and cost effective. With the advent of biotechnology, the ability to genetically modify plants for insect resistance on a commercial scale is within reach. One of the most extensively studied traits for insect resistance in soybeans involves the Bt gene.

Bacillus thuringiensis (Bt) is a common bacteria found in the environment. It has been used as a biological control agent against lepidopteran insects for more than 50 years. Bt targets a class of compounds responsible for insecticidal activity known as crystalline proteins, or cry proteins (Cry1), that are highly toxic after ingestion. The mode of action for Cry1 toxins is the disruption of midgut cellular membranes leading to cell death. One of the primary advantages of using Bt genes for insect control in transgenic plants is the specific insecticidal action toward insects from the Lepidoptera order leaving beneficial insects, birds, and mammals unharmed. Thus, the insertion of Bt toxins into plants, by genetic modification, is an attractive model for the creation of insect resistant transgenic crops.

To date, many different plant species have been genetically modified to exhibit insect resistance using Bt. While the Bt trait has been commercialized in corn and cotton, it is still in developmental stages in soybean. Transformation of soybean with Bt to induce resistance to lepidopteron species has been performed for over a decade. By 1994 fertile transformed soybeans containing a synthetic Bt (Cry1Ac) were generated [15, 16]. Stewart et al., used detached leaf bioassays to show that transgenic soybean lines were resistant to multiple soybean pests with less than 3% leaf defoliation compared to 20% observed in traditionally bred lepidopteron resistance soybean lines [16].

Other groups have used a similar strategy by evaluating soybeans engineered with Cry1Ac for resistance to lepidopteron species under field conditions. One example compared Bt lines to controls in the field using field cages and artificial infestation with lepidopteron larvae over a three year period [17]. In this case, Bt lines showed up to 9 times less defoliation from pets when compared to control plants. Similarly, Mcpherson and MacRae reported the evaluation of Bt soybean lines for suppression of lepidopteron species in the field over 2 years [18]. In this case, soybean plants expressing Cry1Ac were essentially absent of lepidopteron populations when compared to peak population densities of 20-30 larvae per row in control plots. Furthermore, Bt lines showed <1.5% defoliation when compared to 53% defoliation in control plants.

The utility for Bt soybeans has become evident. This has lead to pyramiding strategies using Cry1Ac with native plant resistance genes to increase plant resistance against insect-pests. Several quantitative trait loci (QTLs) from soybean lines have been described as showing antixenosis and antibiosis resistance towards lepidopteron insects [19, 20]. This work lead to the development of transgenic soybean lines by combining QTLs with synthetic Cry1Ac [21]. In this case, field evaluations and detached leaf bioassays were used to test this multiple resistance gene pyramiding strategy for antibiosis resistance. Based on defoliation in the field, as well as larval weight gain on detached leaves, soybean lines carrying a combination of Cry1Ac and the QTL were significantly more resistant to lepidopteron pests.

While Bt soybean varieties have not been commercialized this body of research has lead Monsanto to the development of soybeans that incorporate the Bt trait stacked with the second generation Roundup Ready germplasm [22]. Bt Roundup Ready 2 Yield seeds are currently in Phase IV trials and are targeted for commercialization in Brazil in 2013. This pyramiding strategy would be the first in-seed insect protection for soybeans and is expected to offer an important technology for farmers who face significant yield loss due to insect damage. Although not universal in its application and total in its protection, Bt will play a central role in protecting the crop from major insect pests.

With the onset and success of Bt crops other avenues have been explored for their possible roles in the development of transgenic insect resistant plants. These approaches include the use of plant defense proteins, lectins, α-amylase inhibitors, insect chitinases, and defensins. The development and implementation of engineered insecticidal soybean varieties is currently in its infancy. The incorporation of a multiple gene stacking strategy will also be important in the future development of insect resistant soybeans. Bt, in combination with other biopesticides, has the potential to drastically reduce the consumption of chemical pesticides, however it will be important to continue research and have a development strategy for a future generation of technology, to ensure that insects do not rapidly develop resistance.

3.3. Disease resistance

The United States, Brazil, and Argentina are the three major soybean-producing countries in the world where more than 50% of all soybeans are harvested. Such a geographic distribution facilitates the spread of insect-pests and diseases. Hence, soybean can be attacked by many different pathogens, including bacteria, viruses, fungi, and nematodes. These pathogens and pests can cause damage in seeds, roots, leaves, stems and pods, and usually are tissue-specific. Therefore, disease resistance is another area of great interest for both researchers and farmers.

Disease control management is currently concentrated on agronomic practices, like planting under tillage, use of lodging resistant varieties, wide row planting, and rotation with non-host crops. Chemical control has poor efficiency because of low penetration and uneven distribution due to an already formed canopy. In addition, chemical application can be

http://dx.doi.org/10.5772/52467

extremely expensive for farmers and unhealthy for the environment. There has been little success with conventional plant breeding for disease resistance in soybean leaving room for other approaches such as the use of biotechnology to produce genetically modified soybeans that have disease resistance.

3.3.1. *Bacterial*

Bacterial infections are widespread diseases that occur mainly in the mid-to-upper and young leaves of the soybean plant. There are several bacteria which cause disease in soybean resulting in large amounts of yield loss and poor seed quality. While there has been promising research in the development of bacterial disease resistance with the use of biotechnology in other crop plants such as rice, tomato, banana, and tobacco, there has been less research on the development of bacterial disease resistance for soybean. This research may lead to new strategies for the development of bacterial disease resistance in soybean.

3.3.2. *Viral*

The development of transgenic soybean that confers viral resistance has been studied a bit more extensively. Viruses in soybean are global pests. Significant resistance to several viruses in a number of plant species have been achieved through pathogen derived resistance by the use of viral coat proteins which, when expressed *in planta,* can interfere with viral assembly. This is the same approach that has been used by several groups to develop transgenic viral resistance in soybean. One of the first groups to investigate this approach was Di et al., who produced a soybean that was resistant to bean pod mottle virus (BPMV) [23]. This was done by introducing a BPMV coat protein into the soybean genome. Transgenic events showed complete resistance to BPMV infection. Another study created soybean lines that were resistant to BPMV by inserting a BPMV capsid polyprotien. Events generated in this case were subjected to infectivity assays and not only exhibited resistance to virus infection, but also exhibited systemic infection, showing little to no visible symptoms [24]. Transgenic lines such as these could lead to future commercial cultivars with resistance to BPMV.

The development of soybean mosaic virus (SMV) resistant soybeans is important since SMV is found in all regions where soybean is grown and infection can cause yield loss up to 90%. Despite progress in other important crop plants, efforts to produce transgenic soybeans resistant to SMV have advanced slowly. In order to produce soybean lines that could confer pathogen derived resistance, plants were produced containing a coat protein gene and the 3' UTR from SMV [25]. Coat protein gene transcripts were detected in transgenic lines and two of the soybean lines were highly resistant to infections with the SMV virus. These results represent the first example of stable genetically engineered SMV resistance in soybean.

The sense coat protein gene of soybean dwarf virus (SbDV) was used to acquire SbDV-resistant soybean plants [26]. These insertions were classified into two types: overexpression of SbDV-CP mRNA, or repression accumulation of SbDV-CP mRNA, and siRNA by RNA analysis prior to SbDV inoculation. In both cases, after infection with SbDV, most plants of these transgenic lines remained symptomless, contained little SbDV-specific RNA and exhibited SbDV-CP-specific siRNA. The possible mechanism of the achieved resistance was thought to be RNA silencing. This same group later used RNA silencing to create resistance for SbDV using inverted repeat-SbDV coat protein (CP) genes spaced by a β-glucuronidase sequence [27]. Upon infection with virus, transgenic plants showed no symptoms of the disease. Transgenic soybeans were shown to contain SbDV-CP-specific siRNA and little to no SbDV-specific RNA, suggesting that resistance to SbDV was achieved by an RNA silencing-mediated process.

3.3.3. *Fungal*

Fungi are the most common soybean pathogens and therefore represent targets for the development of disease resistant transgenic varieties in soybean. One of the more important fungal diseases affecting soybeans grown in the United States and Brazil is *Sclerotinia* stem rot (SSR) caused by the fungus *Sclerotinia sclerotiorum* (white mold). This mold has been associated with the presence of oxalic acid (OA). Treatment of plants with OA induced symptoms whereas metabolism of OA is correlated with fungal tolerance. Cunha et al., generated transgenic soybean lines that overexpressed oxalate decarboxylase (OXDC) [28]. When transgenic soybean lines were infected with white mold the disease progression showed significant reduction of severity that correlated with the level of transgene expression. Transgenic events expressing high levels of OXDC showed complete resistance demonstrating the feasibility of this approach.

Much of the research in the development of fungal disease resistance has focused on overexpression of a single gene to confer protection, though such a method favors co-evolution and pathogenic resistance. An alternative strategy taken by Li et al., was to create multigene resistance by overexpressing multiple anti-fungal genes [29]. Two such genes previously shown to be involved with fungal disease resistance are chitinase (CHI) and the barley ribosome-inactivating protein (RIP). While Li et al., successfully produced transgenic soybean overexpressing both traits, transgenic events were not challenged with fungal infection.

An alternative technology has shown promise with controlling fungal infection through the use of single-chain variable fragment (scFv) antibodies. While plants do not produce endogenous antibodies, they can express and correctly assemble antibody fragments. In fact, antibody production in soybean was first demonstrated in 1998 [30]. A similar antibody approach was recently taken by Brar and Bhattacharyya to control *Fusarium virguliforme* which is responsible for soybean sudden death syndrome (SDS) [31]. Using the pathogenic toxin Tox1 as a target, soybeans were transformed with an antibody gene encoding scFv anti-FvTox1 to create transgenic lines with enhanced foliar SDS resistance compared to control plants. Their results suggest that FvTox1 is a pathogenicity factor for the development of SDS and that expression of a soybean plant scFv antibody can reduce a toxin-induced plant disease. This biotechnology approach may be translational in fighting other plant diseases that are induced by pathogenic toxins.

To date there are no commercially available transgenic soybeans that confer resistance to disease, including fungal pathogens. In 2011 DuPont, Pioneer Hi-Bred, and Evogene announced a collaboration to develop soybean varieties displaying in-plant resistance to soybean rust [32]. This is a major step in the direction of creating the first commercially available transgenic soybean variety that is resistant to a fungal pathogen.

3.3.4. *Nematode*

Plant parasitic nematodes are a significant agricultural problem causing major limitations on crop yield and quality. It is estimated that plant parasitic nematodes cause approximately $157 billion [USD] in damage worldwide. Current approaches used to combat agricultural losses include the use of nematicides, cultivation techniques, and varieties with natural resistance. Nematicides include some of the most hazardous compounds used in agriculture and alternative control is required due to health and environmental concerns over their use. In soybean, the majority of yield loss can be attributed to infection by nematodes of the genus *Meloidogyne* and *Heterodera* commonly referred to as root knot nematodes (RKN) and soybean cyst nematodes (SCN), respectively. RKN and SCN infect plant roots and induce the formation of specialized feeding sites. The establishment and maintenance of feeding sites are crucial to the survival of nematodes making them an obvious target of interest for novel control strategies. One approach that has emerged in recent years is the use of *in planta* RNA interference (RNAi) to target genes of feeding nematodes. Through biotechnology, plants can be engineered to produce dsRNAs that silence essential nematode genes. Ingestion of plant-derived dsRNAs by the feeding nematode would trigger the RNAi process thereby inactivating targeted genes and preventing or limiting nematode infection. There are numerous genes known to be essential for nematode survival, and they have been the subject of past reviews [33, 34]. Many of these appear to be candidates for use in an *in planta* RNAi strategy to control nematode infection.

Steeves et al., was one of the first to demonstrate efficacy of an RNAi-based strategy to control SCN [35]. Transgenic soybeans were generated following transformation with an RNAi expression vector containing inverted repeats of a cDNA clone of the SCN major sperm protein (MSP). RNA silencing was elicited in the cyst nematode following ingestion of dsRNA molecules, and resulted in ~75% suppression of reproductive capabilities. Several years later Li et al., used RNAi to test potential gene targets known to be involved with nematode reproduction and fitness [36]. Soybean roots expressing small interfering RNAs against the SCN genes Cpn-1, Y25, and Prp-17 showed a significant reduction in transcript levels in nematode feeding sites. Furthermore, nematode suppression levels were similar to those observed with conventional resistance. Recently RNAi was used to disrupt genes involved with RKN gall formation [37]. Genes encoding tyrosine phosphatase (TP) and mitochondrial stress-70 protein precursor (MSP) were stably expressed in soybean roots, and following infection with RKN the number of galls was decreased by >90%. Nematode growth within roots was measured and the diameter of nematodes inside transformed soybean roots was reduced 5-fold over that of nematodes inside control roots.

Although there are a few cultivars of soybean that have natural resistance to some species of nematode, there are currently no commercially available soybean varieties that offer genetically modified resistance to nematodes. Over the past 10 years, there have been numerous candidate genes found within the nematode-plant interaction that hold potential for the development of novel genetically modified soybeans using an RNAi-based strategy. Results from the above studies show the potential of RNAi technology for reducing gall formation, limiting nematode reproduction and infection, and ultimately broadening soybean resistance to SCN and RKN. The production and eventual commercialization of nematode resistant soybean will benefit both producers and consumers by decreasing dependence on hazardous nematacides and increasing overall soy grain yield.

4. Soybean trait enhancements

In 2008 Monsanto announced their Sustainable Yield Initiative - a pledge to double the yields of corn, cotton, and soybeans by the year 2030 while simultaneously reducing aggregate key inputs such as water, land, and energy. While this will be an especially difficult task given that the vast majority of high-quality farm land is already in use, several recent reports involving transgenic soybean technologies support the notion that future biotechnological advances will indeed be able to help achieve such goals.

4.1. Crop yield

Crop yield is a highly complex trait, and increases in yield have previously been accomplished through a variety of methods involving traditional breeding and modern biotechnology. The introduction of transgenic crops in 1996 helped improve grain yield by protecting plants from insects and disease pathogens that often result in yield pressure if not treated. While new varieties of soybean combine the latest advances in both modern breeding with genetic modification technologies, there continues to be a search for gene-based approaches with potential to increase soy grain yield. Preuss et al., recently performed a large-scale screening for such yield increasing genes and reported that constitutive expression of an *Arabidopsis thaliana* B-box domain gene (BBX32) resulted in plants with increased plant height, node, flower, pod, and total seed number [38]. More importantly, field grown events showed a 5-8% increase in plant height, 8-10% increase in pod number, and 11-14% increase in total yield relative to control plants. It is believed that overexpression of AtBBX32 modulated circadian clock gene transcripts leading to an increase in the duration of reproductive developmental stages (R3 through R7) of the seed which presumably accounted for the increase in seed yield. Over the next decades, it is likely that seed varieties containing these and other yield traits will be commercialized.

4.2. Drought resistance

Drought is a major abiotic stress factor since it can greatly impact crop productivity and grain yield. Soybeans have developed several adaptive traits to endure periods of dry weather and drought. Inclusion of these traits into quality germplasm continues to be a major goal of traditional and marker-assisted breeding programs. While the genetic basis of drought tolerance is not well understood, researchers have focused on understanding physiological responses associated with drought (i. e. leaf wilting, water use efficiency, nitrogen fixation, and root growth biomass). While overexpression of single downstream gene targets have shown potential for increasing drought tolerance in *Arabidopsis* and tobacco model systems, the majority of these findings have not yet been translated to major crop species. One exception involves the overexpression of an endoplasmic reticulum-resident molecular chaperone binding protein (BiP) which is believed to regulate Ca2+ signaling responses. Valente et al., showed that BiP-overexpressing soybean lines exhibited decreases in leaf wilting, leaf water potential, and stomatal closure under reduced and deprived water conditions [39]. Furthermore, transgenic plants showed decreased rates of photosynthesis and transpiration, steady levels of osmolytes and dry root weight, decreased induction of drought-associated mRNAs, and delayed leaf senescence relative to control plants. While overexpression of BiP shows great potential as a target for increasing drought resistance, it will be important to compare grain yields in field-grown transgenic and control lines.

4.3. Increased oil content

Over the past decade, there has been a growing trend for industrial applications utilizing soybean oil, and these applications compete with those used for edible consumption. One example is the recent spike in soy-based biodiesel production which consumed just over 1 billion gallons of soybean oil in 2011 compared with 5 million gallons in 2001 [40]. The growing demand for soybean oil has sparked an interest in novel technologies that could be used to increase the relative oil content of soybean seeds. The retooling of soybean metabolism to increase oil content is not a simple task given that the absolute levels of seed oil and seed protein seem to be set. Increasing oil content comes at the expense of decreasing protein content, and vice versa. To date, only a few papers have reported successes in this area, and both involved manipulation of enzymes and substrate pools in the Kennedy pathway which is responsible for the production of triacylglycerols (TAGs) - the major component of soybean seed oil. In 2008, Lardizabel et al., overexpressed fungal diacylglycerolactetyltransferase (DGAT2) in soybean seeds [41]. DGAT2 converts diacylglycerols (DAGs) to TAGs. Transgenic soybeans overexpressing DGAT2 were grown at 63 locations within the United States and Argentina over five growing seasons, and showed a 1.5% increase in total seed oil with no reduction of seed protein content or yield. In 2009, Rao and Hildebrand overexpressed the yeast sphingolipid compensation (SLC1) protein in soybean seeds [42]. SLC1 has been shown to have lysophosphatidic acid acyltransferase (LPAT) activity which plays a role in the conversion of lysophosphatidic acid to phosphatidic acid, the precursor to DAG in the Kennedy pathway. Overexpression of yeast SLC1 resulted in soybean somatic embryos with 3. 2% increased oil content and stable transgenic lines with 1.5% increased oil content in seeds. Given current commodity pricing for soybean oil [43], a 1.5% increase in oil adds ~$1.2 billion [USD] in value to the United States soybean crop alone. As soybean oil prices rise it is anticipated that other metabolic engineering strategies will be developed and used to obtain similar increases in seed oil content.

5. Health and nutrition

Soybean is considered a dual use crop since it is a valuable source of seed protein and seed oil. Due to consumer health awareness and an increased demand for vegetable oil, much attention has been drawn to seed oil quality and content. To address these needs, efforts have been made with soybean to increase oxidative stability of soybean oil, enhance fatty acid content of oil, and increase total oil content within the seed. Significant progress has been made by breeders to improve overall yield of soybean, however minimal advancements have been made in the development of high-yield germ lines that have a major shift in carbon flux for increased total protein or oil content in the seed. This may be attributed to the inverse correlation between absolute oil and protein contents within soybean seeds. Biotechnology offers new tools for the development of soybeans that have improved oil quality for use in food, feed, and industrial applications. These nutritional enhancements have been achieved by directed modification of fatty acid biosynthesis to alter amounts of fatty acids that naturally occur in soybean, or to produce novel fatty acids. Two fatty acid profiles that have been targeted through genetic strategies include soybeans with low linolenic acid oil content and high oleic acid oil content.

5.1. Fatty acid content

Linolenic acid (LA) accounts for 10-13% of the total fatty acid content of soybean oil. This fatty acid reduces oxidative stability of oil which results in rancidity and decreased shelf-life. A family of three desaturase genes (*Gm*FAD3) contribute to LA biosynthesis in soybean. Flores et al., employed a targeted gene silencing approach to suppress the *Gm*FAD3 gene family using a single RNAi construct [44]. The down regulation of this gene family resulted in low linolenic soybeans with LA contents below 2%. Oleic acid (OA) is a pre-cursor of LA and is considered a healthy source of fat. Conventional soybean oil contains ~18% OA. While high oleic soybean oil has obvious nutritional value, conventional breeding of high oleic soybean lines have not materialized, in part due to the decrease of yield and environmental instability associated with bred traits. The observed yield drag may be attributed to an alteration in the fatty acid profiles within vegetative tissues of soybean. Progress toward decreasing this yield drag has been made by down regulating Δ^{12} desaturases (FAD2-1A and -1B) which converts OA to LA [45, 46]. Seeds resulting from this genetic approach have an increased OA content to ~80% without alterations of vegetative tissue fatty acid contents. In combination with high levels of OA, elevated steric acid levels in soybean oil is also desirable to meet needs of confectionary applications, and development of such soybeans may be possible in the future using similar approaches.

Nutritional enhancements such as increased ω-3 fatty acid levels are also desired for human food and animal feed consumption. Two fatty acids, γ-linolenic acid (GLA) and stearidonic acid (SDA) are of particular interest since they exhibit pharmacological properties and nutritional value, respectively. Sato et al., were successful at increasing these important fatty acids in soybeans by overexpressing a borage Δ^6desaturase which converts LA and α-linolenic (ALA) to GLA and SDA, respectively [47]. Field studies with transgenic soybean harboring the borage Δ^6desaturase produced GLA to ~27% and SDA to ~3% in seed oil [48]. To increase SDA levels in soy, Eckert et al., pyramided the borage Δ^6desaturase with an *Arabidopsis* Δ^{15}desaturase which converts LA to ALA [49]. It was reasoned that increased pools of ALA would lead to increased levels of SDA via the Δ^6desaturase. This strategy resulted in soybean lines with 21.6% SDA when grown under greenhouse conditions. SDA in turn is a precursor for the long chain polyunsaturated fatty acids eicosapentaenoic acid (EPA) and docosahexaenoic acid (DHA). Diets rich in these fatty acids are associated with cardiovascular fitness. While humans and animals possess the enzymatic machinery to convert SDA to EPA and DHA, diets are often supplemented with EPA and DHA derived from fish oil since vascular plants lack the genes for synthesis of these fatty acids. Attempts have been made to assemble these pathways in soybean by using biosynthetic genes from fungi, algae, and protists along with seed specific co-expression of these genes [50]. In such experiments EPA levels approached 20% and DHA levels represented up to 3% of total seed fatty acid content in soybean. The production of such important fatty acids has significant market potential, particularly with respect to human, poultry, pet, and aquaculture feed applications.

5.2. Tocopherols

Tocopherols are lipid-soluble antioxidants that are extracted during the commercial processing of soybean seeds and add to the stability of the oil. In soybean there are four forms of tocopherols (α, β, γ and δ,) classified by the number of methyl groups present on the molecule. These molecules are collectively referred to as vitamin E, with δ - and γ -tocopherols being the most predominant forms in seeds. Biotechnological enhancements of tocopherols in soybean have mainly focused on increasing the amounts of α-tocopherol since this form has the highest nutritional value. Attempts to increase total levels of tocopherols have only been marginally successful. Examples of studies to increase total levels have focused on upregulating homogentisatephytyltransferase (HPT) activity in seeds. While HPT catalyzes the first step in tocopherol synthesis, overexpressing lines resulted in little increase of tocopherol in transgenic seeds [51, 52]. A modest increase in tocopherol content was achieved by the expression of corresponding HPT genes from *Arabidopsis* and *Synechocystis* with a strong seed specific promoter [52]. This approach resulted in a 1.5-fold increase of total tocopherol content. This same group also expressed a bacterial chorismatemutase-prephenate dehydrogenase (TYRA) gene with several other enzymes under the control of seed specific promoters and observed a >10- fold increase in total vitamin E type molecules.

To generate soybean lines with increased levels of tocopherols a more direct approach has involved overexpression of homogentisategeranylgeranlytransferase (HGGT), an enzyme involved in the biosynthesis of tocopherols in monocots. Recently transgenic expression of rice HGGT was expressed in soybean with a seed-specific and constitutive promoter [53]. Transgenic soybean expressing the HGGT gene had significantly higher levels of antioxidant activities and showed enhanced vitamin E levels associated with the presence of all forms of tocopherols, including tocotrienols (with the exception of the β- form) which are not found naturally in soybean.

Enhanced amounts of α-tocopherol in soybean have proven easier to metabolically engineer. Expression of genes for two enzymes responsible for methylation of tocopherol head groups (VTE3 and VTE4) from *Arabidopsis* were co-expressed within the seed and generated plants with α-tocopherol levels greater than 90% of total tocopherol content [54]. The total levels of tocopherol remained the same in these seeds, showing a shift in tocopherol to mainly the α-form yielding a 5-fold increase in vitamin E activity. This research can lead to soybean oil with enhanced vitamin E and more nutritional value for consumers.

5.3. Dietary amino acids

Soy is also deficient in several essential dietary amino acids, most notably methionine and cysteine due to their high sulfur contents. Albumins from Brazil nuts, sunflowers, and corn have been expressed in soybean and although they resulted in increased methionine and cysteine levels, they are not adequate enough to avoid supplementation of these amino acids in animal feed and human diets. The physical synthesis of cysteine is carried out by the enzyme O-acetylserinesulfhydrylase (OASS). In an attempt to increase sulfur containing amino acids in soybean, Kim et al., overexpressed cytosolic OASS and found that transgenic seeds contained elevated levels of both protein bound cysteine (58-74%) and free cysteine (22-32%) [55]. Another approach used to increase sulfur amino acid content in soybean is the use of a maize zein gene which gives rise to several species of insoluble proteins containing high levels of methionine. Dinkins et al., overexpressed maize zein in soybean seeds and observed a 12-20% increase in methionine and 15-35% increase in cysteine without adverse effects on protein composition [56].

The essential amino acids lysine and threonine have also been explored for their potential to create nutritionally enhanced soybean. An increased level of lysine was observed by genetically engineering the lysine biosynthetic pathway to circumvent the normal feedback regulation of the enzymes aspartokinase and dihydrodipicolinic acid synthase in soybean [57]. In this case, a >100-fold increase of free lysine and 5-fold increase of total seed lysine content was observed. More recently, soybeans showing enhanced threonine levels have been engineered using seed-specific expression of lysine-insensitive variants of aspartate kinases from bacteria [58]. This strategy produced transgenic soybeans with a 100-fold increase in threonine levels and 3.5-fold increase in total free amino acid content without negative impacts on seed morphology or germination. While enhancement of essential amino acids in soybean seeds has clear potential for commercial applications, it will be important to demonstrate that transgenic soybean with increased nutritional enhancement traits also maintain optimal agronomic characteristics when grown in the field under a variety of conditions.

6. Soybeans as bioreactors for pharmaceuticals

Recombinant proteins are widely used in medicine, research laboratories, food and nutrition, and play a key role in important agriculture and biopharmaceutical industries. Since the development of recombinant DNA technology in the early 1970's, the commercial pro-

http://dx.doi.org/10.5772/52467

duction of recombinant proteins has traditionally relied on a variety of protein expression systems, each with intrinsic advantages and disadvantages. Over the years several methods have been used to produce recombinant proteins. Traditionally, prokaryotic systems based on fermentation have been used for the production of biopharmaceuticals and enzymes. The bacteria *Escherichia coli*, being one of the earliest and most widely used host for this method, has been used in the production of human insulin since the 1970's [59]. Other platforms include the use of fungal cells and yeast, which have been used as an expression system that is able to perform many of the post-translational modifications required by recombinant protein production. Recombinant proteins that require more complex modifications can be produced using insect or mammalian cells, or transgenic animals. However, major disadvantages associated with these platforms include the inability to perform complex post-translational modifications, the alteration in glycosylation patterns affecting protein activity, the high overall cost associated with manufacturing, the potential for contamination, and long time commitments associated with production in transgenic animals making these platforms impractical for the production of most proteins.

Over the past two decades, there has been a push for recombinant protein technologies to move towards more effective expression systems. These systems must be safe, cost-effective, and conducive to post-translational modifications and processing methods on a large scale. Transgenic plants represent an economical system for accurate expression of complex recombinant proteins on a large scale. Plant cells combine the potential for full post translational modifications and correct protein folding with simple growth requirements. The use of plants as a platform for recombinant protein production has a low risk of contamination with prions, viruses, and other pathogens that infect mammalian cells, and therefore offer advantages that are not associated with existing expression systems. An important advantage of plants as a bioreactor is that recombinant proteins and biopharmaceuticals may be expressed in multiple plant organs including seeds which naturally accumulate high amounts of stored proteins. In general, crops that have higher protein content are more cost-effective for molecular farming. Among recombinant systems that utilize seeds, soybeans present an exceptionally high endogenous protein content, which can reach up to 40% of the dry seed weight. Soybeans are an ideal source of protein for food and feed thus occupying a unique position as a premier target for genetic engineering, and as a platform for the production of recombinant protein. An important characteristic favoring expression in soybean seeds is that these organs have evolved as specialized compartments to store proteins for embryo nutrition. Based on this, soybean seeds offer an environment with metabolic adaptations that permit the stable and long-term storage of proteins, reducing the requirement for sophisticated and expensive conditions for storage. This makes it possible to stockpile harvested seeds so that the downstream processing can be made available based on the demands of the industry. Similarly, soybean seeds provide a compact compartmentalization biomass, which can considerably reduce overall production costs since purification expenses are typically inversely proportional to the final concentration in the plant biomass.

The concept of a soy-derived pharmaceutical was tested back in 1995 when Cho et al., developed a transformation expression cassette using a soybean seed-specific lectin promoter to test for potential expression of the β-glucuronidase reporter gene [60]. This same expression cassette was used to produce bovine β-casein in soybean which accumulated to 0.1-0.4% of seed total soluble protein (TSP) [61]. A follow-up paper characterized post-translational processing, subcellular localization to the PSV, and purification of transgenic β-casein [62]. These proof-of-concept studies showed that a seed-specific promoter could be used to target stable expression of proteins with commercial value in soybean seed.

6.1. Antibodies

Monoclonal antibodies (mAbs) have played a major role in the advancement of biotechnology and development of mAb-based therapeutics and diagnostics. Plants have great potential to serve as a platform for the production of antibodies for therapeutic use. One of the first reports of a functional plant-based antibody was developed in soybean. In an effort to explore cost effective methods of mucosal immunoprotection against sexually transmitted diseases, Zeitlin et al., expressed a humanized monoclonal anti-herpes simplex virus 2 (HSV-2) antibody in leaf tissue [63]. That study compared purified soy-derived and mammalian cell-derived HSV-2 mAbs and found that both were similar with respect to stability in human semen and cervical mucus over a 24 hour period. Both antibodies were also able to diffuse in human cervical mucus, and were efficacious in preventing vaginal HSV-2 infection in a murine model.

6.2. Vaccines

When plants are mentioned as a platform for the production of pharmaceuticals, the concept of edible vaccines often comes to mind. Edible vaccines are desirable since they would eliminate the use of needles and specialized personnel to administer shots, which may have broad applicability in developing nations. Soybean seeds represent an ideal target for the production of vaccines since soymilk-based formulations are safe and can be easily administered orally. Furthermore, soybean seeds are capable of storing vaccine antigens for many years at ambient temperatures without loss or degradation of the antigen [64-66]. Such features can reduce the need for a cold chain therefore reducing costs.

Vaccines that can be administered at mucosal surfaces offer systemic immunity. Subunit antigens used to vaccinate orally or nasally are often ineffective and require formulation with a mucosal adjuvant for increased efficacy. The heat labile toxin (LT) of *E. coli* is comprised of a single A subunit (LTA) and pentameric B subunit (LTB) and has been shown to act as both a strong mucosal adjuvant as well as an antigen [67]. Moravec et al., targeted LTB expression to the endoplasmic reticulum of seed storage parenchyma cells where it accumulated to levels up to 2.4% of seed TSP [68]. Mice orally immunized with seed extracts containing LTB induced both, systemic IgG and mucosal IgA anti-LTB antibody responses. The soybean derived LTB also increased an antibody response against a co-administered bacterial FimHt antigen by 500-fold demonstrating that soy-derived LTB may function as an oral adjuvant.

Several subunit antigens have been expressed in soybeans that are important to the agricultural industry and could lead to effective vaccines. FanC is a specialized adhesion protein

http://dx.doi.org/10.5772/52467

located on the bacterial surfaces of Enterotoxigenic *E. coli* (ETEC). K99 and other ETEC strains cause acute diarrhea in humans and livestock and can be severe and even cause death if left untreated. ETEC vaccinations are routinely administered parenterally to pregnant farm animals in order to stimulate systemic immunity and offer protection in newborns. An edible form of this vaccine has the potential to increase efficacy by conferring mucosal immunity at sites of pathogen invasion. Piller et al., constitutively overexpressed the bacterial FanC antigen in soybeans and reported stable accumulation to levels representing ~0.4% TSP in both leaves and seeds [69]. Mice immunized with adjuvanted soymilk formulations containing FanC elicited FanC-specific systemic and cellular immune responses demonstrating immunogenicity of the soy-derived antigen.

In another study a soybean-based vaccine was developed against the virus that causes porcine reproductive and respiratory syndrome (PRRS) [70]. PRRS is a serious health problem among breeding swine herds and the current vaccine is not efficacious when applied in the field. Vimolmangkang et al., overexpressed a nucleocapsid protein (PRRSV-ORF7) that accumuolated to 0. 64% of seed TSP. Intragastric immunization of mice with transgenic seed extract, in the absence of adjuvant, induced specific humoral and mucosal immune responses against PRRSV-ORF7 [70].

6.3. Therapeutics

Protein therapeutic use is limited by the shortfalls in manufacturing capacity and the high cost of production. While an aging population is a key driver of the protein therapeutics market, the potential for future growth is dependent largely on the industry overcoming drug delivery challenges and cost issues. Plants are cost-effective systems that excel at producing complex therapeutic proteins and therefore could help address some of these issues. The high protein content of soybean seeds, low costs associated with growth, simplified purification methods, and safety, make soybean a unique platform for the production of protein-based therapeutics.

Russell et al., expressed human growth hormone (hGH) in soybean with transformation cassettes using both the constitutive 35S promoter as well as a soybean seed-specific promoter 7S β-conglycyinin [71]. The resulting expression of hGH, both constitutively and within the seed, was detected at low levels of 0.0008% TSP. More recently hGH was expressed in soybean seeds with a more effective expression cassette utilizing the 7S α' subunit of β-conglycinin promoter and α-coixin signal peptide. In this case, hGH was directed to protein storage vacuoles within the seed and accumulated to 2.9% TSP. Bioassays demonstrated that the soy-derived hGH was fully active [66]. The cost of recombinant *E. coli*-derived hGH is still a very expensive therapy. Having such a high level of bioactive hGH protein expression in soybean seeds demonstrates the potential for high-yield production of recombinant proteins in soybean seeds and could lead to reduce costs for large-scale production of therapeutic molecules.

Human basic fibroblast growth factor (bFGF) is another high value therapeutic that has been expressed in soybean seeds [72]. This therapeutic was expressed under the control of the soybean seed specific G1 promoter and endogenous signal sequence from soybean. The bFGF protein accumulated to levels of ~2.3% of seed TSP and biological activity of the transgenic protein was confirmed by its mitogenic activity in mice.

Recombinant expression of Insulin was first reported using *E. coli* [59] and has since been commercialized. Like many pharmaceuticals derived from other expression systems, the potential for contamination along with high costs associated with production remain considerable for this hormone. To show that a soybean expression system could address some of these issues Cunha et al., used a sorghum γ-kafirin seed storage protein promoter and α-coixin PSV signal peptide to target recombinant proinsulin expression to soybean seeds [65]. Transgenic protein was stably expressed in seeds though accumulation levels were not reported. Transgenic seeds containing proinsulin were stable for up to seven years when stored under ambient storage conditions.

The soybean platform has also been used to produce a therapeutic for reducing systolic blood pressure. Novokinin is a hypotensive peptide that has vasorelaxing activity [73]. Novakinin was expressed in soybean seeds under the control of a modified β-conglycinin promoter and accumulated to 0.5% of seed TSP. A purified soy-derived formulation, as well as a less pure defatted flour formulation, was orally administered to groups of spontaneously hypertensive rats. Both the purified and partially purified formulations successfully reduced systolic blood pressure after a single dose [74].

Haemophilia B is a bleeding disorder that results from a deficiency of human coagulation factor IX (hFIX). The current treatment for this disease is intravenous infusion of plasma-derived or recombinant hFIX protein. While this treatment is effective at preventing and arresting hemorrhage, it is very costly and the protein is difficult to produce in large quantities. Using a biolistic transformation approach, hFIX expression was targeted to soybean seeds using the soy 7S promoter and coixin signal peptide [75]. Recombinant hFIX protein accumulated to 0.23% of seed TSP, and purified protein exhibited blood-clotting activity up to 1. 4% of normal plasma demonstrating functionality and efficacy of the soy-derived protein. The recombinant protein was stable for 6 years when stored at room temperature.

Soybeans are also capable of supporting expression and stable accumulation of large and complex proteins that can be difficult or impossible to express using current expression systems. Human thyroglobulin (hTG) is a 660 kDa homodimeric protein that is used as a protein standard and diagnostic for the detection of thyroid disease. To date, no expression system has been capable of producing a recombinant form of hTG which is likely due to strict requirements for correct post-translational modification and proper folding during protein synthesis. As a result, commercial hTG supplied to manufacturers for their assay kits is derived from cadaver and surgically removed thyroid tissue. The heterogeneity and lack of uniformity of commercially-purified hTG preparations is a major factor of variation between kits of different manufacturers. To explore the potential of soybean as a platform for production of large and complex proteins, Powell et al., used the 7S promoter and endogenous hTG signal peptide to target recombinant expression of hTG to soybean seeds [76]. Transgenic lines showed stable expression of full length hTG dimeric protein over multiple generations, and accumulated the protein to levels approaching 1.5% of seed TSP. Functionality of soy-derived hTG was demonstrated with commercial ELISA kits developed specifically for the detection of hTG in patient sera. The expression of 660 kDa dimerich TG appears to be the largest functional recombinant protein expressed in any plant system to date, and demonstrates the practicality of soy as an alternative system for the expression of proteins that are recalcitrant to expression in traditional systems [76].

Soybeans have a high intrinsic capacity for protein production and storage. Other than variability caused by nutrient modulation or environmental effects, the relative distribution of seed protein is primarily determined by genetics. Several groups have been able to achieve recombinant protein expression in soybean seeds at respectable levels (approaching 3% of seed TSP). However, knowledge of protein distribution in soybeans may help to further maximize expression levels. Schmidt and Herman tested this theory by overexpressing Green Fluorescent Protein (GFP) in soybeans with a β-conglycinin suppression background, to observe whether proteome rebalancing would result in a higher GFP yield [77]. They found that the rebalancing of intrinsic proteins could be exploited to obtain protein yields which increased ~4-fold in suppression backgrounds and approached levels representing >7% seed TSP. Thus, proteome rebalancing may represent a strategy that can be used to develop soybean lines capable of producing high levels of recombinant proteins in the future.

7. Conclusion

Over the past 25 years, soybean production in the United States has grown by nearly 57% while during that same period the number of acres used to grow soybeans has increased by only 24%. It is predicted that the worldwide requirement for grain will rise by 40-70% by the year 2050, driven in large part by the growing world population and the increase in demand for protein-rich diets. Clearly the demand for soybean protein and soybean oil is outpacing grain production, which in turn is outpacing available land for growing soybeans. Over the next 50 years, farmers will need to produce as much food as they did in the previous 10,000 years combined, and with fewer resources. The identification of various agricultural improvements such as herbicide, insect, and disease resistance which will allow farmers to obtain increased yields with reduced environmental inputs will be crucial. Traits that not only increase grain yield, but also improve the absolute levels of soy protein and soy oil within the seed will also be important for producers and consumers worldwide. While soybean is recognized for its high protein content, it is also the most widely grown oil-seed crop in the United States. Enhancing nutritional value of soybean oil will greatly increase the effectiveness and value of soy as a food crop, help meet the needs of a growing population, and improve human health.

Over the past decade, soybean has emerged as an ideal expression platform with potential to address current healthcare needs. These unmet needs include cost-effective alternatives to existing protein-based therapeutics, simplified methods for the administration of therapeutics, and the development of reagents that could lead to better diagnostic assays and novel medical devices. Soybeans are unique with respect to protein expression platforms. They are safe to consume, cost-effective to grow, rich in protein content, and stable for years under ambient storage conditions. They have been engineered to express a variety of potential therapeutics, including mAbs, vaccine antigens and adjuvants, hormones, growth factors, and blood-clotting factors. Seed-based expression of 660 kDa homodimerich TG underscores the potential of the soybean system to produce large and complex proteins that cannot be produced in yeast, insect, and mammalian cell cultures. The efficacy of engineered therapeutics in crude soymilk formulations could lead to oral vaccines and therapies that require little, if any, purification from other seed proteins. Reports demonstrating long-term stability of seed-derived therapeutics in the absence of climate control directly address cold chain issues associated with vaccines and other therapeutics. With seed protein levels of ~40% and transgene expression levels approaching 3% of TSP, a single soybean plant yielding 300 seeds can produce >500 mgs of transgenic protein. To put this in perspective, a single soybean plant can produce 500 doses of a vaccine antigen administered at 1 mg/ml, or alternatively, $50,000 [USD] of a therapeutic valued at $100/mg [USD].

The use of soybean as a platform for the production of therapeutics represents a technology with the potential to revolutionize our current approaches to healthcare. Harnessing the full potential of the soybean platform will depend on further increasing stable transgene expression levels, developing efficient purification methods, obtaining interest from pharmaceutical partners, and overcoming issues associated with commercialization. The production of vaccines, antibodies, and other therapeutic proteins will undoubtedly continue to develop over the next decade. As biotechnology evolves, so does the role of soybean - from the field to the bedside.

Glossary

EPSPS:5-enolpyruvylshikimate-3-phosphate synthase

PAT:Phosphinothricin-N-acetyltransferase

ALS:Acetolactate synthase

DMO: Dicambamonooxygenase

HPPD:Hydroxyphenylpyruvatedioxygenase

GM:Genetically modified

Bt:*Bacillus thuringiensis*

Cry1:Crystalline proteins

QTL:Quantitative trait loci

BPMV:Bean pod mottle virus

SMV:Soybean mosaic virus

SbDV:Soybean dwarf virus

http://dx.doi.org/10.5772/52467

SSR:*Sclerotinia* stem rot

OA:Oxalic acid

OXDC:Oxalate decarboxylase

CHI:Chitinase

RIP:Ribosome-inactivating protein

scFv:Single-chain variable fragment

SDS:Sudden death syndrome

RKN:Root knot nematodes

SCN:Soybean cyst nematodes

RNAi:RNA interference

MSP:Major sperm protein

TP:Tyrosine phosphatase

TAGs:Triacylglycerols

DAGS:Diacylglycerols

SCL1:Yeast sphingolipid compensation protein

LPAT:Lysophosphatidic acid acyltransferase

LA:Linolenic acid

OA:Oleic acid

GLA:γ-linolenic acid

SDA:Stearidonic acid

ALA:α-linolenic

EPA:Eicosapentaenoic acid

DHA:Docosahexaenoic acid

HPT:Homogentisatephytyltransferase

HGGT:Homogentisategeranylgeranlytransferase

OASS:O-acetylserinesulfhydrylase

TSP:Total soluble protein

PSV:Protein storage vacuole

mAbs:Monoclonal antibodies (mAbs)

HSV-2:Herpes simplex virus 2

LT:Heat labile toxin of *E. coli*

ETEC:Enterotoxigenic*E. coli*

PRRS:Porcine reproductive and respiratory syndrome

hGH:Human growth hormone

bFGF:Human basic fibroblast growth factor

hFIX:Human coagulation factor IX

hTG:Human thyroglobulin

GFP:Green Fluorescent Protein

Author details

Laura C. Hudson, Kevin C. Lambirth, Kenneth L. Bost and Kenneth J. Piller

University of North Carolina at Charlotte and SoyMeds, Inc., USA

References

[1] Horsch RB, Fry JE, Hoffmann NL, Eichholtz D, Rogers SG, Fraley RT. A simple and general method for transferring genes into plants. Science. 1985;227:1229–1231.

[2] Hinchee MAW, Conner-Ward DV, Newell CA, McDonnell RE, Sato SJ, Gasser CS, Fischhoff DA, Re DB, Fraley RT, Horsch RB. Production of transgenic soybean plants using *Agrobacterium*-mediated DNA transfer. Nat. Biotechnol. 1988;6:915–922.

[3] Parrott WA, Williams EG, Hildebrand DF, Collins GB. Effect of genotype on somatic embryogenesis from immature cotyledons of soybean. Plant Cell Tissue Organ Cult. 1989;16:15–21.

[4] Trick HN, Finer JJ. Sonication-assisted *Agrobacterium* mediated transformation of soybean [*Glycine max* (L.) Merrill] embryogenic suspension culture tissue. Plant Cell Rep. 1998;17:482–488.

[5] Christou P, McCabe DE, Swain WF. Stable transformation of soybean callus by DNA-coated gold particles. Plant Physiol. 1988;87:671–674.

[6] McCabe DE, Swain WF, Martinell BJ, Christou P. Stable transformation of soybean (*Glycine max*) by particle acceleration. Nat. Biotechnol. 1988;6:923–926.

[7] Finer JJ, McMullen MD. Transformation of soybean via particle bombardment of embryogenic suspension culture tissue. In Vitro Cell. Dev. Biol. 1991;27P:175–182.

[8] Aragão FJL, Sarokin L, Vianna GR, Rech EL. Selection of transgenic meristematic cells utilizing a herbicidal molecule results in the recovery of fertile transgenic soybean [*Glycine max* (L.)Merril] plants at a high frequency. Theor. Appl. Genet. 2000;101:1–6.

[9] Franz JE, Mao MK, Sikorski JA. Glyphosate: A Unique Global Pesticide. Washington, DC: American Chemical Society. 1996.

[10] Bayer Crop Science. http://www.bayercropscience.com.

[11] Mathesius CA, Barnett JF Jr, Cressman RF, Ding J, Carpenter C, Ladics GS, Schmidt J, Layton RJ, Zhang JX, Appenzeller LM, Carlson G, Ballou S, Delaney B. Safety assessment of a modified acetolactate synthase protein (GM-HRA) used as a selectable marker in genetically modified soybeans. RegulToxicolPharmacol. 2009;55(3): 309-320.

[12] Lee YT, Duggleby RG. Mutagenesis studies on the sensitivity of *Escherichia coli*acetohydroxyacid synthase II to herbicides and valine. Biochem. J. 2000;350:69–73.

[13] Behrens MR, Mutlu N, Chakraborty S, Dumitru R, Jiang WZ, LaVallee BJ, Herman PL, Clemente TE, Weeks DP. Dicamba Resistance: Enlarging and preserving biotechnology-based weed management strategies. Science 25 2007;316(5828):1185-1188.

[14] vanAlmsick A. HPPD-Inhibitors - A proven mode of action as a new hope to solve current weed problems. Outlooks on Pest Management. 2009;20(1);27-30.

[15] Parrott WA, All JN, Adang MJ, Bailey MA, Boerma HR, Stewart CN. Recovery and evaluationof soybean plants transgenic for a *Bacillus thuringiensis* var. *kurstaki* insecticide gene. In Vitro Cell. Dev.Biol. 1994;30P:144-149.

[16] Stewart CN, Adang MJ, All JN, Boerma HR, Cardineau G, Tucker D, Parrott WA. Genetic transformation, recovery, and characterization of fertile soybean transgenic for a synthetic *Bacillus thuringiensis* cryIAc gene. Plant Physiol. 1996; 112(1): 121–129. \

[17] Walker DR, All JN, Mcpherson RM, Boerma HR, Parrott WA. Field evaluation of soybean engineered with a synthetic cry1Ac transgene for resistance to corn earworm, soybean looper, velvetbean caterpillar (Lepidoptera: Noctuidae),and lesser cornstalk borer (Lepidoptera: Pyralidae). J Econ Entomol. 2000;93(3);613- 622.

[18] Mcpherson RM, MacRae TC. Evaluation of transgenic soybean exhibiting high expression of a synthetic *Bacillus thuringiensis* cry1A transgene for suppressing lepidopteran population densities and crop injury. J Econ Entomol. 2009;102(4)1640-1648.

[19] Cregan PB, Jarvik T, Bush AL, Shoemaker RC, Lark KG, Kahler AL, Kaya N, vanToai TT, Lohnes DG, Chung J, Specht JE. An integrated genetic linkage map of the soybean genome. Crop Science. 1999;39:1464-1490.

[20] Rector BG, All, JN, Parrott WA, Boerma HR. Quantitative trait loci for antibiosis resistance to corn earworm in soybean. Crop Science. 2000;40:233-238.

[21] Walker DR, Narvel JM, Boerma HR, All JN, Parrott WA. A QTL that enhances and broadens Bt insect resistance in soybean. TheorAppl Genet. 2004;109(5):1051-1057.

[22] Monsanto. http:// www.monsanto.com.

[23] Di R, Purcell V, Collins GB, Ghabiral SA. Production of transgenic soybean lines expressing the bean pod mottle virus coat protein precursor gene. Plant Cell Reports. 1996;15:746-750.

[24] Reddy MS, Ghabrial SA, Redmond CT, Dinkins RD and Collins GB. Resistance to Bean mod mottle virus in transgenic soybean lines expressing the capsid polyprotien. Phytopathology. 2001;91:831-838.

[25] Wang X, Eggenberger AL, Nutter FW Jr., Hill JH. Pathogen-derived transgenic resistance to soybean mosaic virus in soybean. Molecular Breeding. 2001;8(2):119-127.

[26] Tougou M, Furutani N, Yamagishi N, Shizukawa Y, Takahata Y, Hidaka S. Development of resistant transgenic soybeans with inverted repeat-coat protein genes of soybean dwarf virus. Plant Cell Rep. 2006;25(11):1213-1218.

[27] Tougou M, Yamagishi N, Furutani N, Shizukawa Y, Takahata Y, Hidaka S. Soybean dwarf virus-resistant transgenic soybeans with the sense coat protein gene. Plant Cell Rep. 2007;26(11):1967-1975.

[28] Cunha WG, Tinoco MLP, Pancoti HL, Ribeiro RE, Aragão FJL. High resistance to *Sclerotinia sclerotiorum* in transgenic soybean plants transformed to express an oxalate decarboxylase gene. Plant Pathology. 2010;59(4):654-660.

[29] Li HY, Zhu YM, Chen Q, Conner RL, Ding XD, Li J, Zhang BB. Production of transgenic soybean plants with two anti-fungal protein genes via *Agrobacterium* and particle Bombardment. Biologia Plantarum. 2004;48(3):367-374.

[30] Zeitlin L, Olmsted SS, Moench TR, Co MS, Martinell BJ, Paradkar VM, Russell DR, Queen C, Cone RA, Whaley KJ. A humanized monoclonal antibody produced in transgenic plants for immunoprotection of the vagina against genital herpes. Nature Biotechnol. 1998;16:1361–1364.

[31] Brar HK, Bhattacharyya MK. Expression of a single-chain variable-fragment antibody against a *Fusarium virguliforme* toxin peptide enhances tolerance to sudden death syndrome in transgenic soybean plants. MPMI. 2012;25(6);817–824.

[32] Evogen. http://www.evogene.com/News-Events/

[33] Gheysen G, Vanholme B. RNAi from plants to nematodes. Trends Biotechnol. 2007;25(3):89-92.

[34] Lilley CJ, Bakhetia M, Charlton WL, URWIN PE. Recent progress in the development of RNA interference for plant parasitic nematodes. Molecular Plant Pathology. 2007;8:701–711.

[35] Steeves RM, Todd TC, Essig JS, Trick HN. Transgenic soybeans expressing siRNAs specific to a major sperm protein gene suppress *Heterodera glycines* reproduction. Funct Plant Biol. 2006;33:991–999.

[36] Li J, Todd TC, Oakley TR, Lee J, Trick HN. Host-derived suppression of nematode reproductive and fitness genes decreases fecundity of *Heterodera glycines Ichinohe*. Planta. 2010;232(3):775-85.

[37] Ibrahim HM, Alkharouf NW, Meyer SL, Aly MA, Gamal El-Din Ael K, Hussein EH, Matthews BF. Post-transcriptional gene silencing of root-knot nematode in transformed soybean roots. Exp Parasitol. 2011;127(1):90-99.

[38] Preuss SB, Meister R, Xu Q, Urwin CP, Tripodi FA, et al. Expression of the *Arabidopsis thaliana* BBX32 gene in soybean increases grain yield. PLoS ONE. 2012;7(2):e30717.

[39] Valente MA, Faria JA, Soares-Ramos JR, Reis PA, Pinheiro GL, Piovesan ND, Morais AT, Menezes CC, Cano MA, Fietto L.G. et al. The ER luminal binding protein (BiP) mediates an increase in drought tolerance in soybean and delays drought-induced leaf senescence in soybean and tobacco J. Exp. Bot. 2009;60(2):533-546.

[40] Soystats. http://www.soystats.com.

[41] Lardizabal K, Effertz R, Levering C, Mai J, Pedroso MC, Jury T, Aasen E, Gruys K, Bennett K.Expression of *Umbelopsis ramanniana*DGAT2A in seed increases oil in soybean. Plant Physiol. 2008;148:89–96.

[42] Rao, SS, Hildebrand D. Changes in oil content of transgenic soybeans expressing the yeast SLC1 gene. Lipids. 2009;44(10):945-951.

[43] CME Group. http://www.cbot.com.

[44] Flores T, Karpova O, Su X, Zeng P, Bilyeu K, Sleper DA, Nguyen HT, Zhang ZJ. Silencing of GmFAD3 gene by siRNA leads to low alpha-linolenic acids (18:3) of fad3-mutant phenotype in soybean [*Glycine max* (Merr.)]. Transgenic Res. 2008;17(5): 839-850.

[45] Mazur B, Krebbers E, Tingey S. Gene discovery and product development for grain quality traits. Science. 1999;285: 372–375.

[46] Buhr T, Sato S, Ebrahim F, Xing A, Zhou Y, Mathiesen M, Schweiger B, Kinney AJ, Staswick P, Clemente T. Ribozyme termination of RNA transcripts down-regulate seed fatty acid genes in transgenic soybean. Plant J 2002;30:155–163.

[47] Sato S, Xing A, Ye X, Schweiger B, Kinney A, Graef G, Clemente T. Production of g-linolenic acid and stearidonic acid in seeds of marker free transgenic soybean. Crop Sci. 2004;44:646–652.

[48] Clement T, and Calhoon EB. Soybean Oil: Genetic Approaches for Modification of Functionality and Total Content. Plant Physiology. 2009;151(3):1030-1040.

[49] Eckert H, LaVallee BJ, Schweiger BJ, Kinney AJ, Cahoon EB, Clemente T. Co-expression of the borage Δ^6 desaturase and the *Arabidopsis* Δ^{15}desaturase results in high ac-

cumulation of stearidonic acid in the seeds of transgenic soybean. Planta. 2006;224:1050–1057.

[50] Kinney AJ, Cahoon EB, Damude HG, Hitz WD, Kolar CW, Liu ZB. Production of very long chain polyunsaturated fatty acids in oilseed plants. World Patent Application. 2004.

[51] Savidge B, Weiss JD, Wong YH, Lassner MW, Mitsky TA, Shewmaker CK, Post-Beittenmiller D, Valentin HE. Isolation and characterization of homogentisatephytyltransferase genes from *Synechocystis* sp. PCC 6803 and Arabidopsis. Plant Physiol. 2002;129: 321–332.

[52] Karunanandaa B, Qi Q, Hao M, Baszis SR, Jensen PK, Wong YH, Jiang J, Venkatramesh M, Gruys KJ, Moshiri F, Post-Beittenmiller D, Weiss JD, Valentin HE. Metabolically engineered oilseed crops with enhanced seed tocopherol. Metab Eng. 2005;7(5-6):384-400.

[53] Kim YH, Lee YY, Kim YH, Choi MS, Jeong KH, Lee SK, Seo MJ, Yun HT, Lee CK, Kim WH, Lee SC, Park SK, Park HM. Antioxidant activity and inhibition of lipid peroxidation in germinating seeds of transgenic soybean expressing OsHGGT. J Agric Food Chem. 2011;59(2):584-591.

[54] Van Eenennaam AL, Lincoln K, Durrett TP, Valentin HE, Shewmaker CK, Thorne GM, Jiang J, Baszis SR, Levering CK, Aasen ED, Hao M, Stein JC, Norris SR, Last RL. Engineering vitamin E content: from *Arabidopsis* mutant to soy oil. Plant Cell. 2003;15(12):3007-3019.

[55] Kim WS, Chronis D, Juergens M, Schroeder AC, Hyun SW, Jez JM, Krishnan HB. Transgenic soybean plants overexpressing O-acetylserinesulfhydrylase accumulate enhanced levels of cysteine and Bowman-Birk protease inhibitor in seeds. Planta. 2012;235(1):13-23.

[56] Dinkins RD, Reddy MSS, Meurer CA, Yan B, Trick H, Thibaud-Nissen F, Finer JJ, Parrott WA, Collins GB. Increased sulfur amino acids in soybean plants overexpressing the maize 15 kDazein protein. In Vitro Cell. Devel. Biol. Plant. 2001;37(6):742-747.

[57] Falco SC, Guida T, Locke M, Mauvais J, Sanders C, Ward RT, Webber P. Transgenic canola and soybean seeds with increased lysine. Biotechnology. 1995;13:577-582.

[58] Qi Q, Huang J, Crowley J, Ruschke L, Goldman BS, Wen L, Rapp WD. Metabolically engineered soybean seed with enhanced threonine levels: biochemical characterization and seed-specific expression of lysine-insensitive variants of aspartate kinases from the enteric bacterium *Xenorhabdus bovienii*. Plant Biotechnol J. 2011;9(2):193-204.

[59] Goeddel DV, Kleid DG, Bolivar F, Heyneker HL, Yansura DG, Crea R, Hirose T, Kraszewski A, Itakura K, Riggs AD. Expression in *Escherichia coli* of chemically synthesized genes for human insulin. ProcNatlAcadSci U S A. 1979;76(1):106–110.

[60] Cho MJ, Widholm JM, Vodkin LO. Cassettes for seed-specific expression tested in transformed embryogenie cultures of soybean. Plant Mol. Binl. Reporter. 1995;13:225-269.

[61] Maughan PJ, Philip R, Cho MJ, Widholm JM, Vodkin LO. Biolistic transformation, expression, and inheritance of bovine β-casein in soybean (*Glycine max*). In Vitro Cell. Devel. Biol. Plant. 1999;35:344–349.

[62] Philip R, Darnowski DW, Maughan PJ, Vodkin LO. Processing and localization of bovine β-casein expressed intransgenic soybean seeds under control of a soybean lectin expression cassette. Plant Science. 2001;161:323-333.

[63] Zeitlin L, Olmsted SS, Moench TR, Co MS, Martinell BJ, Paradkar VM, Russell DR, Queen C, Cone RA, Whaley KJ. A humanized monoclonal antibody produced in transgenic plants for immunoprotection of the vagina against genital herpes. Nature Biotechnol. 1998;16:1361–1364.

[64] Oakes, JL, Bost, KL, Piller, KJ. Stability of a soybean seed-derived vaccine antigen following long-term storage, processing and transport in the absence of a cold chain. Journal of the Science of Food and Agriculture. 2009;89(13):2191-2199.

[65] Cunha, NB, Araujo, AC, Leite, A, Murad, AM, Vianna, GR, Rech, EL. Correct targeting of proinsulin in protein storage vacuoles of transgenic soybean seeds. Genet Mol Res. 2010;9(2):1163-1170.

[66] Cunha NB, Murad AM, Cipriano TM, Cla´udia A, Araujo G,.Aragaõ FJL, Leite A, Vianna GR, McPhee TR, Souza GHMF, Waters MJ, Elı´bio L. Rech. Expression of functional recombinant human growth hormone in transgenic soybean seeds. Transgenic Res. 2011;20:811–826.

[67] Ryan ET, Crean TI, John M, Butterton JR, Clements JD, Calderwood SB. *In vivo* expression and immunoadjuvancy of a mutant of heat-labile enterotoxin of *Escherichia coli* in vaccine and vector strains of *Vibriocholerae*. Infect Immun. 1999;67(4):1694–1701.

[68] Moravec T, Schmidt MA, Herman EM, Woodford-Thomas T. Production of *Escherichia coli* heat labile toxin (LT) B subunit in soybean seed and analysis of its immunogenicity as an oral vaccine. Vaccine. 2007;19;25(9):1647-1657.

[69] Piller KJ, Clemente TE, Jun SM, Petty CC, Sato S, Pascual DW, Bost KL. Expression and immunogenicity of an Escherichia coli K99 fimbriae subunit antigen in soybean. Planta. 2005;222(1):6-18.

[70] Vimolmangkang S, Gasic K, Soria-Guerra R, Rosales-Mendoza S, Moreno-Fierros L, Korban SS. Expression of the nucleocapsid protein of porcine reproductive and respiratory syndrome virus in soybean seed yields an immunogenic antigenic protein. Planta. 2012;235(3):513-522.

[71] Russell, DA, Spatola, LA, Dian, T, Paradkar, VM, Dufield, DR, Carroll, JA and Schlittler, MR. Host limits to accurate human growth hormone production in multiple plant systems. Biotechnol. Bioeng. 2005;89:775–782.

[72] Ding SH, Huang LY, Wang YD, Sun HC, Xiang ZH. High-level expression of basic fibroblast growth factor in transgenic soybean seeds and characterization of its biological activity. BiotechnolLett. 2006;28(12):869-875.

[73] MatobaN ,Usui H, Fujita H, Yoshikawa M. A novel anti-hypertensive peptide derived from ovalbumin induces nitric oxide-mediated vasorelaxation in an isolated SHR mesenteric artery FEBS Lett. 1999;452:181–184.

[74] Yamada Y, Nishizawa K, Yokoo M, Zhao H, Onishi K, Teraishi M, Utsumi S, Ishimoto M, Yoshikawa M. Anti-hypertensive activity of genetically modified soybean seeds accumulating novokinin. Peptides. 2008;29(3):331-337.

[75] Cunha NB, Murad AM, Ramos GL, Maranhão AQ, Brígido MM, Araújo AC, Lacorte C, Aragão FJ, Covas DT, Fontes AM, Souza GH, Vianna GR, Rech EL. Accumulation of functional recombinant human coagulation factor IX in transgenic soybean seeds. Transgenic Res. 2011;20(4):841-855.

[76] Powell R, Hudson LC, Lambirth KC, Luth D, Wang K, Bost KL, PillerKJ. Recombinant expression of homodimeric 660 kDa human thyroglobulin in soybean seeds: an alternative source of human thyroglobulin. Plant Cell Rep. 2011;30(7):1327-1338.

[77] Schmidt MA, Herman EM. Proteome rebalancing in soybean seeds can be exploited to enhance foreign protein accumulation. Plant Biotechnol J. 2008;6(8):832-842.

Chapter 22

Functional Diversity of Early Responsive to Dehydration (ERD) Genes in Soybean

Murilo Siqueira Alves and Luciano Gomes Fietto

Additional information is available at the end of the chapter

http://dx.doi.org/10.5772/50993

1. Introduction

In many regions of the world, agriculture is the primary consumer of water. As the world population increases, and arid regions become more abundant, water will become an increasingly scarce resource [1]. In 2011, the world's soybean crops produced 263.7 million tons from an area of 103,5 million hectares [2]. This global production required an input of 0,2 to 0,25 inch of water per acre per day during peak demand, which represents a major problem for the producer countries [3]. In Brazil, the second largest soybean producer in the world, there was a 7% reduction in soybean production in 2011/2012 compared to the previous season. This yield loss can be attributed to drought in the soybean-growing regions of the country, which in turn resulted in increased use of irrigation water in an attempt to minimize yield losses [4].

Understanding the molecular consequences of drought on soybean plants can accelerate breeding programs aimed at increasing productivity and decreasing the negative impacts of climate change on this important crop. Several classical physiology reviews from recent decades consolidated knowledge of the relationship between leaf structure and function during drought stress [5,6], the morphology of the root during stress tolerance [5,6] and other aspects of the effects of drought on plant morphology. Understanding the physiological responses of plants undergoing drought stress is essential to understanding their ability to survive the water shortage.

In recent years, due to advancements in plant molecular biology methodologies, molecular aspects of drought tolerance have received special attention from researchers [7]. To date, hundreds of genes that are induced by drought stress have been identified and a range of genetic, biochemical and molecular assays (gene expression profiles, transgenic plants, and various functional assays), are being used to elucidate the roles of these genes in response to drought. However, the complexity of the plant response to drought stress makes it difficult to identify genes that are responsible for drought tolerance [7]. In physiological terms, drought stress is characterized by reduction in plant water content, decrease in water potential, loss of leaf turgor, stomata closure and reduction in cell growth [8]. Conditions of severe and prolonged drought result in cessation of photosynthesis, metabolic disorder, and finally plant death [8].

Many abiotic stresses such as high salt levels and low temperature have similar physiological consequences to drought, and therefore similar signaling pathways are induced [7]. The similarity of the cold and drought stress response is illustrated by the observation that plants subjected to drought stress display an increase in frost tolerance [9]. An increase in osmotic pressure is common to these abiotic stresses [10]. The increased osmolarity induces transcription of genes encoding proteins involved in synthesis of osmo-protective compounds, lipid desaturases and transcription factors [11]. Several of these genes have been frequent targets of genetic engineering in breeding programs aimed at producing cultivars with increased tolerance to these adverse conditions [11]. These genes are also induced by other environmental factors such as high salinity and chemical signals such as abscisic acid (ABA), the main phytohormone related to abiotic stress responses in plants.

ABA serves as an endogenous messenger in response to biotic and abiotic stress in plants. Drought results in production of high levels of ABA, accompanied by a major shift in global gene expression in plant cells and, consequently, an adaptive physiological response to the stress [12]. In addition to stress, ABA also controls other important and finely regulated processes such as growth and development, structure and regulation of stomatal function and seed dormancy [13]. During regulation of plant development, ABA also acts in intricate cross-communication with other important phytohormones, such as gibberellic acid, ethylene, auxin and brassinosteroids [13].

How and what environmental stimuli are perceived and result in changes in physiological levels of ABA is still a difficult issue. Drought stress provides an immediate hydraulic signal to the plant, which activates ABA biosynthesis over a great distance [14]. High humidity activates cytochrome P450 enzymes that catalyze ABA synthesis minutes after perception of the stress [15]. Recent studies have shown the importance of the transport driven by absorption and export of ABA. Upon perception of the stress signal, ABA synthesis is primarily induced in vascular tissues, and ABA is exported from the site of biosynthesis to other cells. The absorption is stimulated by ATP-dependent ABC-family transporters. This mechanism allows rapid distribution of ABA to the surrounding tissues [16,17].

Although expression of many genes is induced by ABA-dependent responses to drought, cold and salinity stresses, upregulated genes can be sub-grouped according to the stress they were found to respond to and also by the timing of induction post stress. Genes included in the RD group (responsive to dehydration) include the drought-induced gene RD26, which encodes a NAC (NAM/ATAF/CUC plant protein domains)-family transcription factor [18], ERD (early responsive to dehydration), which includes a gene that encodes a Clp protease [19]. The COR (cold regulated), LTI (low-temperature induced) groups of genes include LOS2, which encodes a bifunctional enolase [20]. The KIN (cold inducible) group of

http://dx.doi.org/10.5772/50993

genes includes SCOF-1, which encodes a protein with a zinc finger domain [21]. The KIN group also contains groups of genes, which also respond to osmotic stress [22-23, 7]. The products of many of these genes are most likely the main components of the first line of plant defense against potential structural damage, or they may be components of signaling pathways such as transcription factors or protein kinases. An example is induction of the gene COR15a; the Arabidopsis homolog ERD1 prevents the injury to the chloroplast membrane [24]. Another gene, GmERD15, from the ERD15 gene family in soybean, acts as a transcription factor, which regulates gene transcription related to programmed cell death [25].

2. The Early Responsive to Dehydration (ERD) genes and their functional diversity

The ERD genes are defined as those genes that are rapidly activated during drought stress. The encoded proteins show a great structural and functional diversity and constitute the first line of defense against drought stress in plants (Table 1).

To date, a total of 16 complementary DNAs (cDNAs) for ERD genes have been isolated from 1-h-dehydrated Arabidopsis thaliana and only half of these are characterized in soybean. These genes encode proteins that include ClpA/B adenosine triphosphate (ATP)-dependent protease, heat shock protein (HSP) 70-1, S-adenosyl-methionine-dependent methyltransferases, membrane protein, proline dehydrogenase, sugar transporter, senescence-related gene, glutathione-S-transferase, group II LEA (Late Embryogenesis Abundant) protein, chloroplast and jasmonic acid biosynthesis protein, hydrophilic protein, and ubiquitin extension protein.

Gene / GenBank accession number	Function	Reference	Best hit on soybean genome / E value.	Similar genes in soybean
*ERD1/D17582**	ClpA/B ATP-dependent protease	[26]	Glyma04g38050.1/8.9e-54	45
*ERD2**/M23105	Heat shock protein (hsp70-i)	[26]	Glyma12g06910.1/ 1.7e-37	45
ERD3/NP_567575.1*	Methyltransferase PMT21	[27]	Glyma01g35220.4/ 1.7e-99	100
ERD4/NP_564354.1	Integral membrane protein	[28]	Glyma15g09820.1/ 1.8e-61	28
ERD5/D83025	Precursor of proline dehydrogenase	[29]	Glyma18g51400.1/ 2.1e-27	5
ERD6/D89051	Sugar transporter	[30]	Glyma03g40160.1/ 6.15e-2	100

Gene / GenBank accession number	Function	Reference	Best hit on soybean genome / E value.	Similar genes in soybean
ERD7/NP_179374.1	Senescence/ dehydration related protein	[31]	Glyma01g36960.1/ 3.5e-43	9
ERD8/Y11827	Heat shock protein hsp81-2)	[26]	Glyma08g44590.1/0	22
ERD9/NP_172508.4*	Glutathione-S-transferase	[32]	Glyma01g04710.1/ 9.9e-22	87
ERD10/D17714*	Group II LEA protein (lti29/lti45)	[33]	Glyma04g01130.1/ 4e-8	3
ERD11/D17672	Glutathione-S-transferase	[32]	Glyma02g17340.1/ 1.7e-5	52
ERD12/NP_189204.1*	Allene oxide cyclase	[40]	Glyma02g11020.1/ 2.3e-36	6
ERD13/D17673	Glutathione S-transferase	[32]	Glyma08g41960.1/ 3.7e-39	63
ERD14/D17715	Group II LEA protein	[33]	No homologs identified	0
ERD15/D30719*	Hydrophilic protein	[35]	Glyma04g28560.1/ 3.5e-26	4
ERD16/J05507*	Ubiquitin extension protein	[33]	Glyma03g35540.1/ 6.6e-50	100

Table 1. ERD genes and their homologs in soybean. (*) Indicates the characterized genes in soybean.

The ERD gene family has been collectively characterized as genes that are rapidly induced by dehydration [26]. ERD1 encodes a chloroplast ATP-dependent protease [26] and ERD2 encodes a, HSP70 [26], ERD3 encodes a methyltransferase in the pMT21 family [27], ERD4 encodes a membrane protein [28], ERD5 and ERD6 encode a mitochondrial dehydrogenase proline protein and a carbohydrates carrier protein, respectively [29-30]. ERD7 encodes a protein related to senescence and dehydration [31], ERD8 encodes a hsp81-family protein [26], ERD9, 11 and 13 belong to the family of glutathione S-transferase [32], ERD10 and 14 belong to the LEA protein family [33], ERD15 was first classified as a hydrophilic protein [34], which has a PAM2 interaction domain which interacts with poly-A tail binding proteins (PABP) [35].

ERD15 from Arabidopsis has been functionally characterized as a common regulator of the abscisic acid (ABA) response and salicylic acid (SA)-dependent defense pathway [35]. Overexpression of ERD15 reduced ABA sensitivity, as the transgenic plants had reduced drought tolerance and failed to increase their freezing tolerance in response to hormone treatment [35]. In contrast, loss of ERD15 function due to gene silencing caused hypersensitivity to ABA, and the silenced plants displayed enhanced tolerance to both drought and freezing. The antagonistic effect of ERD15 activity on ABA signaling enhanced SA-dependent defense; overexpression of ERD15 was associated with increased resistance to the bacterial necrotroph Erwinia carotovora and enhanced induction of systemic acquired resistance reporter genes [35]. The authors also addressed the antagonistic effect of ABA on SA-mediated defense by demonstrating the enhanced expression of reporter genes for systemic acquired resistance in the plant null mutants abi1-1 and abi2-1, which are defective for ABA metabolism. These results together implicate Arabidopsis ERD15 as a shared component of ABA- and SA-mediated responses. The ERD15 homologs from Solanum licopersicum are 98% identical and belong to the same group as Arabidopsis ERD15, indicating a possible conservation of function [36]. Nevertheless, the tomato protein clearly localizes to the nucleus and confers freezing tolerance when ectopically expressed in transgenic tomato plants. These phenotypes are in marked contrast with the phenotypes displaying by ERD15-overexpressing Arabidopsis lines [35]. These contrasting results in transgene overexpression studies suggest that the Arabidopsis and tomato ERD15 homologs have divergent functions. Finally, a soybean homolog, GmERD15, has been described as an ER stress- and osmotic stress-induced transcription factor that activates the promoter and induces the expression of the NRP-B gene. These results indicate that GmERD15 functions as an upstream component of the NRP-mediated cell death signaling pathway, which is induced by ER and osmotic stress [37].

ERD16 encodes a ubiquitination extension protein [33]. Previous studies also showed that ERD13/AtGSTF10, a plant phi specific class GST (Glutathione S-transferase) is an interaction protein with BAK1 (BRI1 Associated receptor Kinase 1). BAK1 is a co-receptor, which forms a receptor complex with BRI1 (brassinosteroid (BR) receptor) to regulate brassinosteroid signaling in Arabidopsis. Overexpression of AtGSTF10 resulted in plants with increased tolerance to salt stress. In contrast, silencing AtGSTF10 by RNAi caused increased tolerance to abiotic stress and accelerated senescence of the transformants [38]. These findings suggest that modulation of ERD13/AtGSTF10 may regulate plant stress responses by regulating brassinosteroid signaling via interaction of AtGSTF10 with BAK1. ERD10 and 14 have chaperone activity, which aid in protein folding during stress [39]. ERD12 encodes a protein with homology to an allene oxide cyclase [40].

With respect to expression controlled by phytohormones, ERD genes present varied functions and responses in ABA signaling, some being sensitive to ABA during germination and development [41], and /or are involved in stress tolerance [42]. Other genes are induced in response to more than one phytohormone [35]. Early Responsive to Dehydration 15 (ERD15) was characterized as a negative regulator of ABA and is induced by ABA, SA, injury and pathogen infection [35]. ABA application increases the expression of some members of the ERD group including ERD10 and 14 [34] while causing no effect on others, such as ERD2, 8 and 16 [33].

Some contradictory data regarding the induction, as well as the function, of ERD genes are present in the literature [35-37]. Reduced expression of the ERD15 gene in response to wounding was reported [43], while an increased number of ERD15 transcripts were observed by other authors [35]. Furthermore, Arabidopsis plants showed increased tolerance to salt stress through the overexpression of AtSAT32, a key gene in the salinity-tolerance family Arabidopsis. These plants showed an increase in the number of ERD15 transcripts relative to control plants [44]. Transgenic wheat plants over-expressing TaDi19A, a gene responsive to salinity in wheat, exhibited increased expression of ERD15 [45]. In contrast to these findings, Arabidopsis plants over-expressing ERD15 demonstrated susceptibility to drought and freezing [35]. In regard to function, a soybean ERD15 homolog was characterized as a transcription factor [25], a function not previously attributed to this protein family, as reported by Kariola and colleagues [35] and Ziaf and colleagues [36].

3. The ERD genes studied in soybean

In respect to ERD genes described in soybean, the behavior of a group of eight genes (ERD1, ERD2, ERD3, ERD9, ERD10, ERD12, ERD15 gene and ERD16) was studied in response to stress. A soybean cDNA ERD1, homologous to yeast Hsp104, was isolated and characterized [46]. The soybean genes encoding homologs to yeast Hsp104 and Hsp101 have a high level of sequence identity to members of the family Clp [46]. When heterologously expressed in yeast, the soybean Hsp101 gene conferred greater thermotolerance to yeast [46]. Several genes related to Hsp70 (ERD2) have been described in soybean using proteomics studies. The first evidence of an ERD2-like protein in soybean was found during heat shock [47]. The presence of similar proteins was also found in response to osmotic and reticulum stress [48]. In respect to the orthologs of ERD3 in soybean, it was found that the GmIMT gene, which encodes a methyltransferase, acts by methylating the substrate D-ononitol. Its overexpression in Arabidopsis causes an increase in drought and salinity tolerance [49]. When a gene encoding a soybean GST, an ortholog of ERD9, was over-expressed in tobacco plants, it conferred an increase in salinity and drought tolerance [50].

Group 2 LEA (dehydrins or responsive to abscisic acid) proteins, such as ERD10 proteins, are postulated to protect macromolecules from damage by freezing, dehydration, ionic, or osmotic stress. In soybean, proteins of this group were studied for their structural and physio-chemical properties but little is discussed regarding the function of these proteins [51]. Overexpression of a member of the ERD12 family, GmAOC5, significantly increased oxidative stress resistance [52]. Within the ERD15 family, a soybean ortholog, GmERD15 has been functionally characterized as a transcription factor; in response to osmotic stress, GmERD15 acts to control transcription of a gene related to an integrative pathway in soybean [25]. Finally, orthologs of ERD16 studied in soybean genes were identified with differential expression during flood stress and hypoxia [53]. All genes studied related to different levels in response to stress, particularly drought and osmotic stress, demonstrating the conservation of function of this gene family in different plant species.

Some ERD genes not yet studied in soybean deserve special attention because of either the proven involvement of a gene with similar functions in drought response in other organisms, or due to multiple copies of the soybean homolog. ERD5 and ERD7 family members have been characterized by activity in response to drought in other organisms (discussed below). They have not been studied in soybean; however, homology to soybean genes is demonstrated by the phylogenic tree shown in Figure 1.

http://dx.doi.org/10.5772/50993

ERD5 (which encodes a precursor of a proline dehydrogenase), has five orthologous genes in soybean. ERD5 has a proven role in drought response due to its role in accumulation of proline [54], a common occurrence during osmotic stress. All soybean genes are clustered in a group distinct from orthologs in other species (Figure 1), which may reflect a possible functional divergence.

ERD7 (which encodes a protein related to senescence and dehydration) has nine orthologous genes in soybean. It also has a central role in response to drought and osmotic stress and it is related with drought-induced leaf senescence in plants [55]; regulation of this process during drought tolerance has been studied in depth [55]. Phylogenetic analysis of these genes suggests the possibility of functional divergence of these genes within the same organism.

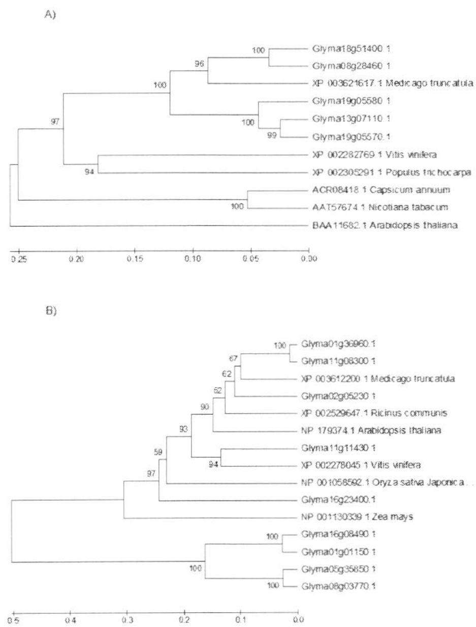

Figure 1. Relatedness of ERD5 (panel A) and ERD7 (panel B) proteins from different plant species. The multiple alignment was made using ClustalW, and the dendrogram was built with the MEGA5 software using the UPGMA method. The numbers at the nodes indicate the bootstrap scores. The proteins accession numbers are indicated.

4. Conclusion

Many studies on the roles and importance of ERD genes in soybean have become necessary due to lack of information about the importance of this group of genes during plant response to drought. The common feature of these genes is that their expression increases rapidly in response to drought stress, suggesting that it is the first line of defense for plants against drought stress. It also suggests these genes may function to regulate expression of effector proteins and signaling pathways in response to stress.

Acknowledgements

This research was supported by the Conselho Nacional de Desenvolvimento Científico e Tecnológico CNPq and Fundação de Amparo a Pesquisa do Estado de Minas Gerais (FAPEMIG). M.S.A. is supported by fellowship from CAPES.

Author details

Murilo Siqueira Alves and Luciano Gomes Fietto*

*Address all correspondence to: lgfietto@ufv.br

Department of Biochemistry and Molecular Biology, Federal University of Viçosa, Viçosa, Minas Gerais, Brazil

References

[1] FAO: Food and agriculture organization of the United Nations. (2012). http://www.fao.org/., (accessed 12 April).

[2] Embrapa Soybean. (2012). http://www.cnpso.embrapa.br, (accessed 12April).

[3] Mississippi Agricultural and Forestry Experiment Station, Mississippi State University Extension Service. (2012). http://msucares.com/pubs/publications/2185 htm., (accessed 12 April).

[4] Brazilian Institute of Geography and Statistics- IBGE. (2012). www.ibge.gov.br/., (accessed 12 April).

[5] Smirnoff, N. (1998). Plant resistance to environmental stress. *Current opinion in Biotechnology*, 9-214.

[6] Gorantla, M., Babu, P. R., Lachagari, V. B. R., Reddy, A. M. M., Wusirika, R., Ennetzen, J. L., & Reddy, A. R. (2007). Identification of stress-responsive genes in anindica rice (Oryza sativa L.) using ESTs generated from drought-stressed seedlings. *Journal of Experimental Botany*, 58(2), 253-265.

[7] Shinozaki, K., & Yamaguchi-Shinozaki, K. (2007). Gene networks involved in drought stress response and tolerance. *Journal of Experimental Botany*, 58-221.

[8] Jaleel, C. A., Manivannan, P., Lakshmanan, G. M. A., Gomathinayagam, M., & Panneerselvam, R. (2008). Alterations in morphological parameters and photosynthetic pigment responses of Catharanthus roseus under soil water deficits. *Colloids and Surfaces B: Biointerfaces*, 61-298.

[9] Guy, C. L., Huber, J. L. A., & Huber, S. C. (1992). Sucrose Phosphate Synthase and Sucrose Accumulation at Low Temperature. *Plant Physiology*, 100-502.

[10] Munns, R., & Tester, M. (2008). Mechanisms of salinity tolerance. *Annual Review in Plant Biology*, 59-651.

[11] Yamagushi-Shinosaki, K., & Shinosaki, K. (2005). Organization of cis-acting regulatory elements in osmotic- and cold-stressresponsive promoters. *TRENDS in Plant Science*, 10(2), 88-94.

[12] Christmann, A., Moes, D., Himmelbach, A., Yang, Y., Tang, Y., & Grill, E. (2006). Integration of abscisic acid signalling into plant responses. *Plant Biology*, 8-314.

[13] Hirayama, T., & Shinozaki, K. (2007). Perception and transduction of abscisic acid signals: keys to the function of the versatile plant hormone ABA. *Trends of Plant Science*, 12-343.

[14] Christmann, A., Weiler, E.W., Steudle, E., & Grill, E. (2007). A hydraulic signal in root-to-shoot signalling of water shortage. *Plant Journal*, 52, 167-174.

[15] Okamoto, M., Tanaka, Y., Abrams, S. R., Kamiya, Y., Seki, M., & Nambara, E. (2009). High humidity induces abscisic acid 80 hydroxylase in stomata and vasculature to regulate local and systemic abscisic acid responses. *Arabidopsis. Plant Physiology.*

[16] Kang, J., Hwang, J.U., Lee, M., Kim, Y.Y., Assmann, S.M., Martinoia, E., & Lee, Y. (2010). PDR-type ABC transporter mediates cellular uptake of the phytohormone abscisic acid. *Proceedings of the National Academy of Sciences USA*, 107, 2355-2360.

[17] Kuromori, T., Miyaji, T., Yabuuchi, H., Shimizu, H., Sugimoto, E., Kamiya, A., Moriyama, Y., & Shinozaki, K. (2010). ABC transporter AtABCG25 is involved in abscisic acid transport and responses. *Proceedings of the National Academy of Sciences USA*, 107-2361.

[18] Fujita, M., Fujita, Y., Maruyama, K., Seki, M., Hiratsu, K., Ohme-Takagi, M., Tran, L. S., Yamaguchi-Shinozaki, K., & Shinozaki, K. (2004). A dehydration-induced NAC protein, RD26, is involved in a novel ABA-dependent stress-signaling pathway. *Plant Journal*, 39(6), 863-876.

[19] Nakashima, K., Kiyosue, T., Yamaguchi-Shinozaki, K., & Shinozaki, K. (1997). A nuclear gene, erd1, encoding a chloroplast-targeted Clp protease regulatory subunit ho-

molog is not only induced by water stress but also developmentally up-regulated during senescence in Arabidopsis thaliana. *Plant Journal*, 12(4), 851-861.

[20] Lee, H., Guo, Y., Ohta, M., Xiong, L., Stevenson, B., & Zhu, J. K. (2002). LOS2, a genetic locus required for cold-responsive gene transcription encodes a bi-functional enolase. *EMBO Journal*, 21(11), 2692-2702.

[21] Kim, J. C., Lee, S. H., Cheong, Y. H., Yoo, C. M., Lee, S. I., Chun, H. J., Yun, D. J., Hong, J. C., Lee, S. Y., Lim, C. O., & Cho, M. J. (2001). A novel cold-inducible zinc finger protein from soybean, SCOF-1, enhances cold tolerance in transgenic plants. *Plant Journal*, 25(3), 247-259.

[22] Ingram, J., & Bartels, D. (1996). The molecular basis of dehydration tolerance in plants. *Annual Reviews in Plant Physiology and Plant Molecular Biology*, 47-377.

[23] Thomashow, M.F. (1999). Plant cold acclimation: freezing tolerance genes and regulatory mechanisms. *Annual Reviews in Plant Physiology and Plant Molecular Biology*, 50-571.

[24] Steponkus, P. L., Uemura, M., Joseph, R. A., Gilmour, S. J., & Thomashow, M. F. (1998). Mode of action of the COR15a gene on the freezing tolerance of Arabidopsis thaliana. *Proceedings of the National Academy of Sciences USA*, 95-14570.

[25] Alves, M. S., Reis, P. A. B., Dadalto, S. P., Faria, J. A. Q. A., Fontes, E. P. B., & Fietto, L. G. (2011). A Novel Transcription Factor, ERD15 (Early Responsive to Dehydration 15), Connects Endoplasmic Reticulum Stress with an Osmotic Stress-induced Cell Death Signal. *Journal of Biological Chemistry*, 286(22), 20020-20030.

[26] Kiyosue, T., Yamaguchi-Shinozaki, K., & Shinozaki, K. (1994). Cloning of cDNAs for genes that are early-responsive to dehydration stress (ERDs) in Arabidopsis thaliana L.: identification of three ERDs as HSP cognate genes. *Plant Molecular Biology*, 25-791.

[27] González-Martínez, S. C., Ersoz, E., Brown, G. R., Wheeler, N. C., & Neale, D. B. (2006). DNA sequence variation and selection of tag single-nucleotide polymorphisms at candidate genes for drought-stress response. *Pinus taeda L. Genetics*, 172(3), 1915-1926.

[28] Froehlich, J. E., Wilkerson, C. G., Ray, W. K., Mc Andrew, R. S., Osteryoung, K. W., Gage, D. A., & Phinney, B. S. (2003). Proteomic study of the Arabidopsis thaliana chloroplastic envelope membrane utilizing alternatives to traditional two-dimensional electrophoresis. *Journal of Proteome Research*, 2(4), 413-425.

[29] Nakashima, K., Satoh, R., Kiyosue, T., Yamaguchi-Shinozaki, K., & Shinozaki, K. (1998). A gene encoding proline dehydrogenase is not only induced by proline and hypoosmolarity, but is also developmentally regulated in the reproductive organs of Arabidopsis. *Plant Physiology*, 118-1233.

[30] Kiyosue, T., Abe, H., Yamaguchi-Shinozaki, K., & Shinozaki, K. (1998). ERD6, a cDNA clone for an early dehydration-induced gene of Arabidopsis, encodes a putative sugar transporter. *Biochimica Biophysica Acta*, 1370-187.

[31] Kimura, M., Yamamoto, Y. Y., Seki, M., Sakurai, T., Sato, M., Abe, T., Yoshida, S., Manabe, K., Shinozaki, K., & Matsui, M. (2003). Identification of Arabidopsis genes regulated by high light-stress using cDNA microarray. *Photochemical and Photobiology*, 77(2), 226-33.

[32] Kiyosue, T., Yamaguchi-Shinozaki, K., & Shinozaki, K. (1993). Characterization of two cDNAs (ERD11 and ERD13) for dehydration inducible genes that encode putative glutathione S-transferases. *Arabidopsis thaliana L. FEBS Letters*, 335-189.

[33] Kiyosue, T., Yamaguchi-Shinozaki, K., & Shinozaki, K. (1994). Characterization of two cDNAs (ERD10 and ERD14) corresponding to genes that respond rapidly to dehydration stress. *Arabidopsis thaliana. Plant Cell Physiology*, 35-225.

[34] Kiyosue, T., Yamaguchi-Shinozaki, K., & Shinozaki, K. (1994). ERD15, a cDNA for a dehydration-inducible gene from Arabidopsis thaliana. *Plant Physiology*, 106-1707.

[35] Kariola, T., Brader, G., Helenius, E., Li, J., Heino, P., & Palva, E. T. (2006). EARLY RESPONSIVE to DEHYDRATION 15, a negative regulator of abscisic acid responses in Arabidopsis. *Plant Physiology*, 142-1559.

[36] Ziaf, K., Loukehaich, R., Gong, P., Liu, H., Han, Q., Wang, T., Li, H., & Ye, Z. (2011). A multiple stress responsive gene ERD15 from Solanum pennelli confers stress tolerance in tobacco. *Plant Cell Physiology*, 52(6), 1055-1067.

[37] Alves, M. S., Fontes, E. P. B., & Fietto, L. G. (2011). EARLY RESPONSIVE to DEHYDRATION 15, a new transcription factor that integrates stress signaling pathways. *Plant signaling & behavior*, 6(12).

[38] Ryu, H. Y., Kim, S. Y., Park, H. M., You, J. Y., Kim, B. H., Lee, J. S., & Nam, K. H. (2009). Modulations of AtGSTF10 expression induce stress tolerance and BAK1 mediated cell death. *Biochemical and Biophysical Research Communications*, 379, 417-422.

[39] Kovacs, D., Kalmar, E., Torok, Z., & Tompa, P. (2008). Chaperon activity of ERD10 and ERD14, two disordered stress-related plant proteins. *Plant Physiology*, 147-381.

[40] Gutiérrez, R. A., Green, P. J., Keegstra, K., & Ohlrogge, J. B. (2004). Phylogenetic profiling of the Arabidopsis thaliana proteome: what proteins distinguish plants from other organisms? *Genome Biology*, 5(8), R53, doi:10.1186/gb-2004-5-8-r53.

[41] Zhang, X. L., Zhang, Z. J., Chen, J., Chen, Q., Wang, X. C., & Huang, R. F. (2005). Expressing TERF1 in tobacco enhances drought tolerance and abscisic acid sensitivity during seedling development. *Planta*, 222-494.

[42] Xiong, L., Lee, B. H., Ishitani, M., Lee, H., Zhang, C., & Zhu, J. K. (2001). FIERY1 encoding an inositol polyphosphate 1 phosphatase is a negative regulator of abscisic acid and stress signaling. *Arabidopsis. Genes Development*, 15, 1971-1984.

[43] Dunaeva, M., & Adamska, I. (2001). Identification of genes expressed in response to light stress in leaves of Arabidopsis thaliana using RNA differential display. *European Journal of Biochemistry*, 268-5521.

[44] Park, M. Y., Chung, M. S., Koh, H. S., Lee, D. J., Ahn, S. J., & Kim, C. S. (2009). Isolation and functional characterization of the Arabidopsis salt-tolerance 32 (AtSAT32) gene associated with salt tolerance and ABA signaling. *Physiologia Plantarum*, 135-426.

[45] Li, S., Xu, C., Yang, Y., & Xia, G. (2010). Functional analysis of TaDi19A, a salt responsive gene in wheat. *Plant Cell Environment*, 33-117.

[46] Lee, Y. R., Nagao, R. T., & Key, J. L. (1994). A soybean 101-kD heat shock protein complements a yeast HSP104 deletion mutant in acquiring thermotolerance. *Plant Cell*, 6(12), 1889-1897.

[47] Roberts, J. K., & Key, J. L. (1991). Isolation and characterization of a soybean Hsp70 gene. *Plant Molecular Biology*, 1(6), 671-683.

[48] Valente, M. A., Faria, J. A., Soares-Ramos, J. R., Reis, P. A., Pinheiro, G. L., Piovesan, N. D., Morais, A. T., Menezes, C. C., Cano, M. A., Fietto, L. G., Loureiro, M. E., Aragão, F. J., & Fontes, E. P. (2009). The ER luminal binding protein (BiP) mediates an increase in drought tolerance in soybean and delays drought-induced leaf senescence in soybean and tobacco. *Journal of Experimental Botany*, 60(2), 533-546.

[49] Ahn, C., Park, U., & Park, P. B. (2011). Increased salt and drought tolerance by D-ononitol production in transgenic Arabidopsis thaliana. *Biochemical and Biophysical Research Communications*, 415(4), 669-674.

[50] Ji, W., Zhu, Y., Li, Y., Yang, L., Zhao, X., Cai, H., & Bai, X. (2010). Over-expression of a glutathione S-transferase gene, GsGST, from wild soybean (Glycine soja) enhances drought and salt tolerance in transgenic tobacco. *Biotechnology Letters*, 32(8), 1173-1179.

[51] Soulages, J. L., Kim, K., Arrese, E. L., Walters, C., & Cushman, J. C. (2003). Conformation of a group 2 late embryogenesis abundant protein from soybean. Evidence of poly(L-proline)-type II structure. *Plant Physiology*, 131(3), 963-975.

[52] Wu, Q., Wu, J., Sun, H., Zhang, D., & Yu, D. (2011). Sequence and expression divergence of the AOC gene family in soybean: insights into functional diversity for stress responses. *Biotechnology Letters*, 33(7), 1351-1359.

[53] Yanagawa, Y., & Komatsu, S. (2012). Ubiquitin/proteasome-mediated proteolysis is involved in the response to flooding stress in soybean roots, independent of oxygen limitation. *Plant Science*.

[54] Verbruggen, N., & Hermans, C. (2008). Proline accumulation. *Plants: a review. Amino Acids*, 35(4), 753-759.

[55] Munné-Bosch, S., & Alegre, L. (2004). Die and let live: leaf senescence contributes to plant survival under drought stress. *Functional Plant Biology*, 31-203.

Chapter 23

An Overview of Genetic Transformation of Soybean

Hyeyoung Lee, So-Yon Park and Zhanyuan J. Zhang

Additional information is available at the end of the chapter

http://dx.doi.org/10.5772/51076

1. Introduction

Soybean (*Glycine max* (L.) Merrill) is a model legume crop, widely grown in the world for human consumption or animal fodder. Moreover, soybeans have gained worldwide research interest in many public laboratories and industrial sectors. Soybean seeds contain protein, oil, carbohydrates, dietary fibers, vitamins, and minerals. For the last few decades the majority of research laboratories have been investigating genetic traits to improve the yield of protein or oil in soybean seeds through genetic engineering, thereby achieving improved quantity and quality of soybean seeds. Until now, most of the transformation experiments have implemented a single functional gene not multiple genes. Those agronomically and economically important traits affect the enhancement of grain quantity and quality [1]. However, the majority of agronomic and genetic traits such as complex metabolic, biological, and pharmaceutical pathways are polygenetic traits and are produced in a complex pathway. Therefore, those traits are encoded and regulated by a number of genes. In an attempt to study and manipulate those pathways, the transfer of multigene or large inserts into plants have been developed by multigene engineering technology and have also been involved in metabolic engineering. Several examples of multigene or large insert transfers have been reported such as the application of carotenogenic genes in rice, canola, and maize [2-4], and of polyunsaturated fatty acid and vitamin E genes in soybean and *Arabidopsis* [5-7]. Therefore, reliable systems for transforming large DNA fragments into plants make it feasible to introduce a natural gene cluster or a series of previously unlinked foreign genes into a single locus.

Over the last two decades, the transfer of DNA into plant cells has been achieved by using several methods. In soybeans, the most frequently employed plant genetic engineering methods are *Agrobacterium*-mediated transformation and particle bombardment. Both systems have successfully been used in genetic transformation of soybean. Since the initial reports of fertile transgenic soybean production [8-9], various efforts have been made to improve the transformation efficiency and to produce transgenic soybean. Particularly, the preferred and reproducible transformation is the use of the cotyledonary node as a plant material, which is based on *Agrobacterium*-mediated gene transfer [10-12]. Nevertheless, new methods have been developed for more efficient soybean transformation. There still remain, however, many challenges for genotype- and tissue- specific independent transformation of soybean. This review provides an overview of historical efforts in developing and advancing soybean regeneration and transformation systems. In addition, recent advances and challenges in soybean transformation are discussed.

2. Different approaches for soybean transformation

In soybean transformation, two major methods are now widely utilized: *Agrobacterium-mediated* transformation of different explant tissues and particle bombardment. The *Agrobacterium*-mediated method, as a simple protocol, does not require any specific or expensive equipment. Moreover, this method usually produces single or low copy numbers of insertions with relatively rare rearrangement [13]. On the other hand, bombardment technique directly introduces desired genes into the target plant cell with small tungsten or gold particles [9]. The success of this approach critically depends upon the ability of the target tissue to proliferate as well as proper pre-cultures to make a target plant.

2.1. Cotyledonary-node-based transformation

The routine regeneration system was first reported by using the mature cotyledonary-node [14]. The multiple adventitious buds and shoots from explant tissues were proliferated and regenerated on culture media containing cytokinin by organogenesis. The transgenic soybean plants have been successfully and reproducibly produced using mature or immature cotyledon explants via *Agrobacterium*-mediated transformation. Hinchee et al. [8] for the first time reported the production of fertile transgenic soybean plants using mature cotyledonary-node by *Agrobacterium*-mediated transformation, but transformation efficiency was very low. The system employed the neomycin phosphotransferase II (*NPT* II) gene as a selectable marker and combined kanamycin as a selective agent. However, this selection was addressed with a problem of regeneration of non-transgenic or chimeric shoots at the shoot formation stage. Moreover, the system was highly genotype-dependent. To overcome the high genotype-dependency and high chimerism problems by the *NPT* II selection and develop a new selection system for soybean transformation, Zhang et al. [10] developed the selection system employing herbicide bialaphos resistance (*bar*) gene as a selectable marker coupled with glufosinate as a selective agent. This system enabled to transform many soybean genotypes with stable transgene inheritance, albeit transformation efficiency remained to be improved. Meanwhile, to solve the escape problem caused by kanamycin selection, Clemete et al. [15] deployed the herbicide glyphosate as a selective agent, leading to high stringent selection and good transgene inheritance. It was discovered later that addition of various thiol compounds in the co-cultivation medium significantly increased the transformation efficiency [11, 16-17]. These thiol compounds, as antioxidants, reduce the oxidative burst that caused tissue browning or necrosis and also promote organogenesis and shoot growth from buds [18].

Recently, an alternative cotyledonary explant derived from mature soybean seed for *Agrobacterium* transformation has been reported by Paz et al. [19]. The term half-seed explants were used as an experiment material and fertile transgenic plants were attained.

In fact, several laboratories have contributed to enhanced soybean transformation using a cotyledonary-node explant. To overcome the low transfer of *Agrobacterium* into plant cell, the infection media were first amended with the phenolic compound, 4'-Hydroxy-3',5'-dimethoxyacetophenone (acetosyringone), to induce expression of the virulence (*Vir*) genes [20-21]. To increase the infection sites, Trick and Finer [22] evaluated cotyledonary node transformation efficiency using a developed sonication assisted *Agrobacterium*-mediated transformation (SAAT) protocol. Although this treatment was not able to obtain fertile transgenic plants, the increase of *Agrobacterium* transfer was shown. Olhoft et al. [16-17] discovered that thiol compounds enhanced *Agrobacterium* infection in soybean. At the same time, however, these compounds caused counter-selection effect when glufosinate was used as a selective agent under previously published selection conditions. To solve this problem, Olhoft et al. [11] developed Hygromycin phosphotransferase (*HPT* II) selection system using hygromycin B as a selective agent. This has led to a substantial increase in transformation frequency. Transformation efficiency with thiol compounds was increased 5-fold by using refined glufosinate selection [12].

Since the transformation process by use of kanamycin or hygromycin B as selection agent has been proven to be genotype-dependent, the most widely used selection system has been the combination of *bar* gene with the herbicide phosphinothricin (glufosinate) [10, 12]. In this selection system, the concentration of agent glufosinate greatly affects the transformation frequency [12], so the appropriate selection schemes can be varied among genotypes, seed vigor and other *in vitro* culture conditions.

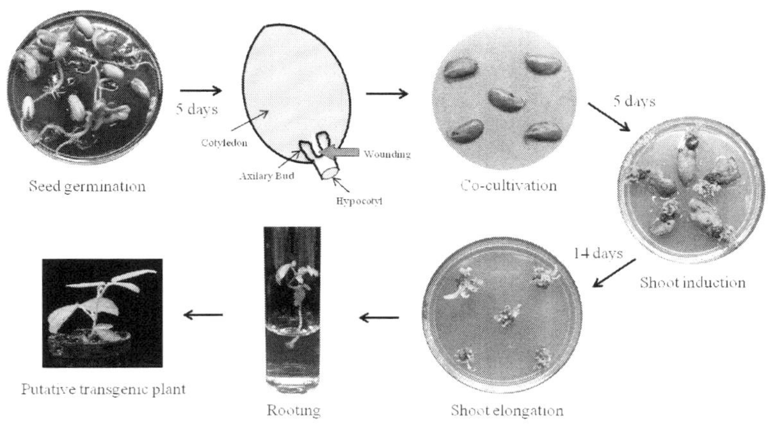

Figure 1. Scheme for genetic transformation of soybean (*Glycine max* (L.) Merrill) cotyledonary nodes.

2.2. Immature embryos-based transformation

The regeneration using immature embryos via somatic embryogenesis was first reported by Christianson et al. [23]. The immature embryos excised from soybean pods were suspended on semi-solid media or liquid media containing high concentration of auxin, 2,4-Dichlorophenoxyacetic acid (2,4-D), and the whole plantlets were recovered [24-25]. After immature embryos were developed as an alternative plant material, transgenic plants were first obtained from this explant tissue via particle bombardment [26]. This system has been exclusively used to produce transgenic soybean such as glyphosate tolerant, hygromycin resistance, and *Bacillus thuringiensis* (BT) transgenic soybean [27-29]. As the formation of proliferative embryogenic tissue depends on genotype, the use of immature embryos for transformation has been limited to few genotypes cultivars including "Jack" and "Williams 82."

The use of particle bombardment with immature embryos tends to be highly variable, and multiple copies of the introduced DNAs are commons. Moreover, this problem has compounded with aged embryogenic suspension cultures from which a high percentage of regenerated plants lost their fertility [29]. In spite of this limitation, the embryogenic cultures have several advantages, one of which is its relatively high transformation efficiency and less chimeric plants recovered.

2.3. Embryogenic shoot tips-based transformation

The embryonic shoot tip explant is another source of explant which has been used for soybean transformation. McCabe et al., [30] first reported the stable transformation using meristemic cell, shoot apex, by particle acceleration. The shoot derived from these meristems via organogenesis has been produced to form multiple shoots prior to mature plants. However, all of the primary transgenic plants were chimeric. Martinell et al., [31] described the successful method using meristemic shoot tip from germinated seedling by *Agrobacterium*-mediated transformation. This system has provided rapid and efficient soybean transformation. Liu et al. [32] also reported the regeneration system using embryonic shoot tips by shoot organogenesis. The explants have been shown the high regeneration and the transformation efficiencies using *Agrobacterium*-mediated with up to 15.8%.

2.4. Immature cotyledonary-nodes

The regeneration capacity of immature cotyledonary-node was found by Parrott et al [33]. Based on this regeneration system, first transgenic soybean plants have been developed by *Agrobacterium tumefaciens* [34]. This system was tested using two different *Agrobacterium* strains, LBA4404 and EHA101 and deploying kanamycin selection. The system utilized auxin 1-Naphthaleneacetic acid (NAA) for plant regeneration. Although these systems allowed development of transgenic plants from the explants, no fertile transgenic plants were recovered. Recently, Ko et al [35] described the efficient transformation system using immature cotyledonary-nodes by *Agrobacterium*-mediated transformation, but transformation efficiency was still very low.

2.5. Hypocotyl based transformation

Another type of explant tissue, hypocotyl, was also investigated with 13 different soybean genotypes. Most of the genotypes initiated shoots from this type of explant [36]. This method was reported to be genotype-independent regeneration protocol via organogenesis and utilized the acropetal end of a hypocotyl section from a 7-day old seedling. Despite inducing adventitious shoots from the explant, most recovered shoot did not matured in the soil. Wang et al [37] reported successful production of fertile transgenic plants using hypocotyl-based *Agrobacterium*-mediated transformation. To improve the transformation system, two different chemicals, cytokinin hormone 6-Benzylaminopurine (BAP) and silver nitrate, were added to the shoot formation media. In spite of the term "hypocotyl" used in the above transformation system, the true tissues responsible for regeneration are actually the preexisting meristem tissues located at the nodal area of the cotyledon, essentially the same source of tissue as cotyledonary-nodes [17] except that cotyledons were removed [45, 46].

2.6. Leaf tissue-based transformation

The reproducible regeneration methods for whole plants from primary leaf tissue or epicotyls were first reported by Wright et al [38]. The multiple shoots from those explants were continually initiated and proliferated with cytokinin BAP hormone. Rajasckaren et al [39] described regeneration of several varieties of soybean by embryogenesis from epicotyls and primary leaf tissues, thereby inducing fertile plants from those explants. Kan et al [40] first tested transformation efficiency using epicotyls and leaf tissues by *Agrobacterium tumefaciens*. To find out proper transformation condition for those explants, they investigated different *Agrobacterium* strains, EHA101 and LBA4404, but also different treatments on inoculation stage, sucrose and mannose.

3. *Agrobacterium*-mediated transformation of soybean

3.1. *Agrobacterium*-mediated transformation mechanism

Agrobacterium is a unique organism to generate transgenic plants and in natural conditions [41]. It allows introduction of a single stranded copy of the bacterial transferred DNA (T-DNA) into a host cell and integration of the genomic DNA of interest, resulting in genetic manipulation of the host. Since the development of disarmed tumour-inducing (Ti) plasmid [42-43], *Agrobacterium* has been used to transform various major crops for genetic modification [44-46].

Agrobacterium recognizes wounded host plant cells which produce penolic compounds such as acetosyringone as inducers of *vir* gene expression [47], and attach to the plant cells to export the T-DNA after virulence (Vir) protein activation. Acetosyringone is now routinely used for improving transformation efficiency. After *vir* gene activation, a single stranded T-DNA copy (T-strand) is transferred into the plant by type IV secretion system (T4SS) which is related to VirB complex [48]. The VirB complex is composed of at least 12 proteins (VirB1-11 and VirD4) which form a multisubunit envelope-spanning structure [49]. Various

Agrobacterium proteins, such as VirD2-T-DNA, VirE2, VirE3, VirF, and VirD5, pass though VirB complex to transfer into plant cells [50-51]. VirE2 and VirD2 interact with cytosolic T-DNA in the plant cells and form a complex which is later imported into the nucleus when it is bound to VIP1 plant protein [52-55]. Recently, Gelvin et al., hypothesized that T-complex (T-DNA, VirE2, VirD2 and VIP1) is imported into the nucleus through actin cytoskeleton and thus myosin may be involved in *Agrobacterium*-mediated transformation [56]. However, the specific mechanism of T-DNA movement through myosin is still unknown.

The T-complex is imported into the nucleus by the phosphorylation of VirE2 Interacting Protein 1 (VIP1), induced by mitogen-activated protein kinase (MAPK), such as MPK3 [55]. After T-complex is imported into the host nucleus, VirE2 and VIP1 need to be degraded before T-DNA integration by a subunit of the SCF (SKP-CUL1-F-box protein) ubiquitin E3 ligase complex. Not only *Agrobacterium* protein VirF but also protein VBF can mark VIP1 protein for the degradation. Furthermore, binding of VIP1-binding F-box (VBF) to T-complex can induce the degradation of VIP1 and VirE2 by the 26S proteosome, and at the end T-strand is integrated into plant genomic DNA and expressed in the host plants [57-58].

3.2. History of *Agrobacterium*-mediated soybean transformation research

Among various transformation technologies, *Agrobacterium*-mediated transformation method has shown to be effective for the production of transgenic soybeans because of straightforward methodology, familiarity to researchers, minimal equipment cost, and reliable insertion of a single transgene or a low copy number [13]. Till now, a number of reports have been published related to the optimum condition to achieve a high yield of soybean transformation; such as *Agrobacterium* inoculation conditions, regeneration media components, etc. For *Agrobacterium*-mediated transformation methods, the susceptibility of soybean to *Agrobacterium* and various *Agrobacterium* strains have been tested to improve the transformation efficiency (Table 1). Also, *Agrobacterium* strains and growth conditions which affect the soybean transformation efficiency have been published [8, 59-62]. After Pederson et al., [46] and Owens et al., [59] showed the susceptibility of certain soybean genotypes against tumor induction, *Agrobacterium* biology study has been advanced to enhance transformation efficiency. In addition to *Agrobacterium* biology study, chemical contents for inoculation have been studied such as varying acetosyringone and syringaldehyde concentrations [63]. For high inoculation efficiency, Mauro et al., [64] tested various *Agrobacterium* biotypes (nopaline, agropine and octopine) to identify the most effective *Agrobacterium* biotype for soybean transformation.

After Hinchee et al., [8] developed *Agrobacterium*-mediated soybean transformation methods, many *Agrobacterium* strains have been tested and employed, such as EHA101, EHA105, LBA4404 and AGL1. Parrott et al., [33] showed that EHA101 was highly potent to transform immature soybean cotyledons, especially PI283332, and had higher recovery of transformed plants over LBA4404. Dang and Wei [65] tested transformation efficiency using embryonic tips instead of cotyledonary explants and somatic embryos, and when embryogenic tips were infected for 20 hours, hypervirulent strain KYRT1 showed increased efficiency over EHA105 and LBA4404.

Recently, *A. tumefaciens* KAT23 (AT96-6) which has an ability to efficiently transfer the T-DNA into soybean, was isolated from peach root. After 20 stains were confirmed by common bean and soybean transformation, Yukawa et al. [66] tested their potential availability as legume super virulent *A. tumefaciens* in various soybean cultivars (Peking, Suzuyutaka, Fayette, Enrei, Mikawashima, WaseMidori, Jack, Leculus, Morocco, Serena, Kentucky Wonder and Minidoka). Without modifying vectors or *vir* function, they showed that KAT23 (AT96-6) has a high potential to function as a common strain to increase soybean transformation efficiency. Therefore, this study identified a novel soybean super virulent *A. tumefaciens* strain which transferred not only the T-DNA of the Ti-plasmid but also introduced T-DNA of the binary vector efficiently. These results indicate that KAT23 (AT96-6) has the ability to transform soybeans at high efficiency.

There has been a significant improvement in soybean transformation over the past two decades. However, the efficiency of soybean transformation is not great enough for practical needs and shows high variation. Thus, considering the potential application of soybean transformation, the importance of *Agrobacterium* can't be over-emphasized.

Strain of *A. tumefacients*	Soybean genotype	Selection		Reference
		Marker	Agent	
A208	Peking, Maple Prest	*npt* II	kanamycin	(8)
AGL1	Bert	*bar*	phosphinothricin	(67)
EHA101	Williams 82	bar	glufosinate	(12)
EHA101	Williams, Williams 79, Peking, Thorne	*bar*	glufosinate or bialaphos	(68)
EHA101	Thorne, Williams, Williams 79, Williams 82	*bar*	glufosinate	(19)
EHA105	AC Colibri	*npt* II	kanamycin	(69)
EHA105	Hefeng 25, Dongnong 42, Heinong 37, Jilin 39, Jiyu 58	*hpt*	hygromycin	(70)
EHA105	A3237	*bar*	glufosinate	(10)
LBA4404	Jungery	*bar*	phosphinothricin	(71)
LBA4404, EHA105	Bert	*hpt*	hygromycin	(11)

Table 1. Summary of cotyledonary-node transformation system.

4. New directions of soybean genetic engineering, skills and vectors

To date, the *Agrobacterium*- and biolistic-mediated transformation methods remain the very successful methods in soybean transformation, whereas other available transformation technologies have not been practical in soybean, which include electroporation-mediated transformation [72], PEG/liposome-mediated transformation [73], silicon carbide-mediated transformation [74], microinjection [75] and chloroplast-mediated transformation [76]. Of these two, *Agrobacterium*-mediated transformation has become more adapted in public laboratories worldwide. On the other hand, there are unintended insertions such as unwanted antibiotic markers and promoters, which can be inserted during transformation. This problem has raised potential biosafety issues related to environmental concerns and human health risks. To overcome these potential risks, methods of developing marker free transgenic plants have been developed, such as cotransformation [77], transposon-mediated transformation [78] and site-specific recombination [79].

Among the various methods, co-transformation system is one of the most commonly used methods to produce marker free transgenic plants. In co-transformation systems, a marker gene and genes of interest are placed on separate DNA molecules and introduced into plant genomes. Then, the non-selectable genes segregate from the marker gene in the progeny generations. Most strains of *A. tumefaciens* have the ability to contain more than one T-DNA, and crown gall tumors were often co-transformed with multiple T-DNAs [42]. As a result, there are two possibilities; Multiple T-DNAs were delivered into plant cells either from a mixture of strains ('mixture methods') or from a single strain ('single-strain methods'). Depicker et al. [80] described that a single strain method was higher in efficiency than a mixture method. For a single-strain method of co-transformation, Kamori et al., [77] tested co-transformation method to develop a suitable superbinary vector system. Using the unique plasmids which carried two T-DNA segments marker free rice and tobacco were produced and evaluated. LBA4404, a derivative of an octopine strain, were used for these co-transformation methods and they hypothesized that LBA4404 may be an important factor contributing to the high frequency of unlinked loci.

To improve plant genetic traits, many soybean research labs have developed tools for soybean functional genomics, such as several libraries containing large inserts of bacterial artificial chromosome (BAC) and plant transformation competent binary plasmids clone (BIBAC) (81). In functional genomic research, bacterial artificial chromosome (BAC) is a single copy artificial chromosome vector and is based on the *E. coli* fertility (F-factor) plasmid. They are not only stable in host cell, but also are used for large scale gene cloning and discovery [82]. However, some BAC libraries that are desirable for functional genomics are often not amenable for transformation directly into plants because of their large subclones. Therefore, binary bacterial artificial chromosome (BIBAC) libraries have been developed for *Agrobacterium*-mediated plant transformation and gene functional complementation. The BIBAC library is based on BAC vector and has both an F-factor plasmid for replication origin of *E.coli* and an Ri plasmid for replication origin of *Agrobacterium rhizogenes*. The vector also has a *sacB* gene as a positive selection for *E. coli* and a selectable marker gene for plant. Since BIBAC vectors were reported, these vectors have been used for plant transformation in some model plant species including tobacco, canola, tomato, and rice [83-86]. Although the transformation efficiency was very low, the BIBAC vectors have been successfully employed to transfer large inserts into those crops as a single locus via *Agrobacterium*-mediated transformation. The introduced T-DNA was stably maintained and inherited through several generations and no gene silencing was observed [85]. However, soybean transformation using a BIBAC vector has not been achieved to date. Moreover, plant transformation with DNA fragments below 20 kb is routine whereas the stable plant transformation with DNA fragments larger than 50 kb is challenging.

5. Zinc finger nucleases (ZFNs) and transcription activator-like effectors (TALEs)

Although many methods have been developed, soybean is considered a recalcitrant plant to transform compared to *Arabidopsis* and rice. Since full genome sequencing data has been rapidly updated in soybean, soybean transformation technology is becoming an essential approach for genomic research. For the phenotypic analysis of genes, knock-out or gene-silencing plants are used to study gene function. However in soybean, making bulk knock-out mutants through conventional mutagenesis approaches is not immediately feasible because of low transformation efficiency. Thus, development of innovative gene targeting methods is necessary to make knock-out plants in soybean.

Zinc finger nucleases (ZFNs) and meganucleases cut specific DNA target sequences *in vivo* and thus are powerful tools for genome modification. In particular zinc finger domain, which predominantly recognize nucleotide triplets, have been widely used in this research. Importantly, ZFNs modification has been reported in soybean (*Glycine max*), maize (*Zea mays*), tobacco (*Nicotiana tabacum*) and *Arabidopsis* [87-90]. Unfortunately, Ramirez et al. [91] found a major disadvantage of ZFNs; they observed that the modular assembly method of engineering zinc-finger arrays has a higher failure rate than previously reported.

To overcome the ZFNs's weakness, in late 2009, a novel DNA binding domain was identified which was a member of the large transcription activator-like (TAL) effector family [92-93]. Transcription activator-like effectors (TALEs) are produced by plant pathogens in the genus *Xanthomonas* as virulence factors and TAL effector–mediated gene induction leads to plant developmental changes [94-95]. The type III secretion system is used by *Xanthomonas* to introduce virulence factors into plant cells [96]. Once inside the plant cell, transcription activator-like (TAL) effectors (TALEs) enter the nucleus, bind effector-specific DNA sequences, and transcriptionally activate gene expression [97-98]. For genomic engineering, two methods of TALEs were developed: TALE nucleases (or TALENs) and TALE transcription factors (or TALE-TFs). Both TALENs and TALE-TFs contain as many as 30 tandem repeats of a 33- to 35-amino-acid-sequence motif (Figure 2). The amino acids in positions 12 and 13 in each 33- to 35-amino-acid-sequence motif have the repeat variable di-residue (RVD). Using this specific ability, two pair (left and right TALENs) of repeats with different RVDs are designed by PCR and bound in the target DNA sequence [92-93]. Fok1 combined with TALE nucleases (or TALENs) make double-strand breaks (DSBs) at specific locations in the genome. These DSBs are repaired by homologous recombination (HR) or non-homologous end-joining (NHEJ) pathways. During DSBs repair, errors in genome via insertion, deletion, or chromosomal rearrangement could be induced by HR and NHEJ (Figure 2). Unlike TALENs, TALE-TFs require only a single TALE construct for activity induction when combined with VP64 activator (derived from the herpex simplex virus activation domain). VP64 binds with RNA polymerase and causes transcriptional activation of the gene of interest [99]. For making transgenic plants, the TALEs technique can be combined with the *Agrobacterium*-mediated transformation method and it is assumed that gene targeting knock-out will be applied in soybean research.

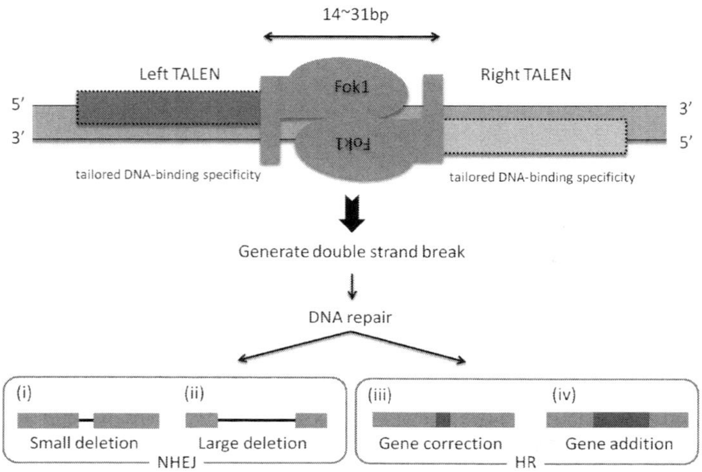

Figure 2. Summary of Transcription activator-like effectors (TALEs) nuclease.

6. Abbreviation

Acetosyringone: 4'-Hydroxy-3',5'-dimethoxyacetophenone

BAC: bacterial artificial chromosome

bar: bialaphos resistance

BAP: 6-Benzylaminopurine

2,4-D: 2,4-Dichlorophenoxyacetic acid

BIBAC: binary bacterial artificial chromosome

BT: *Bacillus thuringiensis*

DSBs: double-strand breaks

HPT II: Hygromycin phosphotransferase

HR: homologous recombination

NAA: Naphthaleneacetic acid

NHEJ: non-homologous end-joining

NPT II: neomycin phosphotransferase II

RVD: repeat variable di-residue

Ti plasmid: tumour-inducing (Ti) plasmid

T4SS: type IV secretion system

TALEs: transcription activator-like effectors

VBF: VIP1-binding F-box

VIP1: VirE2 Interacting Protein 1

Vir protein: virulence protein

ZFNs: Zinc finger nucleases

Author details

Hyeyoung Lee, So-Yon Park and Zhanyuan J. Zhang*

*Address all correspondence to: zhangzh@missouri.edu

Plant Transformation Core Facility, University of Missouri, Columbia, USA

References

[1] Mazur, B., Krebbers, E., & Tingey, S. (1999). Gene Discovery and Product Development for Grain Quality Traits. *Science*, 285(5426), 372-375.

[2] Ye, X., et al. (2000). Engineering the provitamin A (beta-carotene) biosynthetic pathway into (carotenoid-free) rice endosperm. (Translated from eng). *Science*, 287(5451), 303-305, (in eng).

[3] Paine, J. A., et al. (2005). Improving the nutritional value of Golden Rice through increased pro-vitamin A content. (Translated from eng). *Nat Biotechnol*, 23(4), 482-487, (in eng).

[4] Fujisawa, M., et al. (2009). Pathway engineering of Brassica napus seeds using multiple key enzyme genes involved in ketocarotenoid formation. (Translated from eng). *J Exp Bot*, 60(4), 1319-1332, (in eng).

[5] Kinney, A. J., et al. (2004). *COMPANY EIDPDNA WO/2004/071467*.

[6] Qi, B., et al. (2004). Production of very long chain polyunsaturated omega-3 and omega-6 fatty acids in plants. (Translated from eng). *Nat Biotechnol*, 22(6), 739-745, (in eng).

[7] Karunanandaa, B., et al. (2005). Metabolically engineered oilseed crops with enhanced seed tocopherol. (Translated from eng). *Metab Eng*, 7(5-6), 384-400, (in eng).

[8] Hinchee, M. A. W., et al. (1988). Production of Transgenic Soybean Plants Using Agrobacterium-Mediated DNA Transfer. *Nat Biotech*, 6(8), 915-922.

[9] Christou, P., Mc Cabe, D. E., & Swain, W. F. (1988). Stable Transformation of Soybean Callus by DNA-Coated Gold Particles. *Plant Physiology*, 87(3), 671-674.

[10] Zhang, Z., Xing, A., Staswick, P., & Clemente, T. E. (1999). The use of glufosinate as a selective agent in Agrobacterium-mediated transformation of soybean. *Plant Cell, Tissue and Organ Culture* , 56(1), 37-46.

[11] Olhoft, P., Flagel, L., Donovan, C., & Somers, D. (2003). Efficient soybean transformation using hygromycin B selection in the cotyledonary-node method. *Planta*, 216(5), 723-735.

[12] Zeng, P., Vadnais, D. A., Zhang, Z., & Polacco, J. C. (2004). Refined glufosinate selection in Agrobacterium-mediated transformation of soybean [Glycine max (L.) Merrill]. *Plant Cell Reports*, 22(7), 478-482.

[13] Hansen, G., & Wright, M. S. (1999). Recent advances in the transformation of plants. *Trends in Plant Science*, 4(6), 226-231.

[14] Wright, M. S., Koehler, S. M., Hinchee, M. A., & Carnes, M. G. (1986). Plant regeneration by organogenesis in Glycine max. *Plant Cell Reports*, 5(2), 150-154.

[15] Clemente, T. E., et al. (2000). Progeny Analysis of Glyphosate Selected Transgenic Soybeans Derived from-Mediated Transformation. *Crop Sci.*, 40(3), 797-803.

[16] Olhoft, P. O., Lin, K. L., Galbraith, J. G., Nielsen, N. N., & Somers, D. S. (2001). The role of thiol compounds in increasing Agrobacterium-mediated transformation of soybean cotyledonary-node cells. *Plant Cell Reports*, 20(8), 731-737.

[17] Olhoft, P. O., & Somers, D. S. (2001). L-Cysteine increases Agrobacterium-mediated T-DNA delivery into soybean cotyledonary-node cells. *Plant Cell Reports*, 20(8), 706-711.

[18] Dan, Y. (2008). Biological functions of antioxidants in plant transformation. *In Vitro Cellular & Developmental Biology- Plant*, 44(3), 149-161.

[19] Paz, M., Martinez, J., Kalvig, A., Fonger, T., & Wang, K. (2006). Improved cotyledonary node method using an alternative explant derived from mature seed for effi-

cient <i>Agrobacterium</i>-mediated soybean transformation. *Plant Cell Reports*, 25(3), 206-213.

[20] Stachel, S. E., Messens, E., Van Montagu, M., & Zambryski, P. (1985). Identification of the signal molecules produced by wounded plant cells that activate T-DNA transfer in Agrobacterium tumefaciens. *Nature*, 318(6047), 624-629.

[21] Owens, L. D., & Smigocki, A. C. (1988). Transformation of Soybean Cells Using Mixed Strains of Agrobacterium tumefaciens and Phenolic Compounds. (Translated from eng). *Plant Physiol*, 88(3), 570-573, (in eng).

[22] Trick, H. N., & Finer, J. J. (1997). SAAT: sonication-assisted Agrobacterium-mediated transformation. *Transgenic Research*, 6(5), 329-336.

[23] Christianson, M. L., Warnick, D. A., & Carlson, P. S. (1983). A Morphogenetically Competent Soybean Suspension Culture. *Science*, 222(4624), 632-634.

[24] Ranch, J., Oglesby, L., Zielinski, A., & Horsch, R. (1985). Plant regeneration from embryo-derived tissue cultures of soybeans. *In Vitro Cellular & Developmental Biology - Plant*, 21(11), 653-658.

[25] Finer, J. J., & Nagasawa, A. (1988). Development of an embryogenic suspension culture of soybean (Glycine max Merrill.). *Plant Cell, Tissue and Organ Culture* , 15(2), 125-136.

[26] Finer, J., & Mc Mullen, M. (1991). Transformation of soybean via particle bombardment of embryogenic suspension culture tissue. *In Vitro Cellular & Developmental Biology - Plant*, 27(4), 175-182.

[27] Padgette, S. R., et al. (2009). Development, Identification, and Characterization of a Glyphosate-Tolerant Soybean Line. *Crop Sci.*, 35(5), 1451-1461.

[28] Hadi, M. Z., Mc Mullen, M. D., & Finer, J. J. (1996). Transformation of 12 different plasmids into soybean via particle bombardment. *Plant Cell Reports*, 15(7), 500-505.

[29] Stewart, C. N., Jr., et al. (1996). Genetic Transformation, Recovery, and Characterization of Fertile Soybean Transgenic for a Synthetic Bacillus thuringiensis cryIAc Gene. *Plant Physiology*, 112(1), 121-129.

[30] Mc Cabe, D. E., Swain, W. F., Martinell, B. J., & Christou, P. (1988). Stable Transformation of Soybean (Glycine Max) by Particle Acceleration. *Nat Biotech*, 6(8), 923-926.

[31] Martinell, B. J., Horeb, M., Julson, L. S., Mills, L., Emler, C. A., Huang, Y., Mc Cabe, D. E., & Williams, E. J. (2002). *Patent No. US 6384301.*

[32] Liu, H.-K., Yang, C., & Wei, Z.-M. (2004). Efficient Agrobacterium tumefaciens-mediated transformation of soybeans using an embryonic tip regeneration system. *Planta*, 219(6), 1042-1049.

[33] Parrott, W., Williams, E., Hildebrand, D., & Collins, G. (1989). Effect of genotype on somatic embryogenesis from immature cotyledons of soybean. *Plant Cell, Tissue and Organ Culture*, 16(1), 15-21.

[34] Parrott, W. A., Hoffman, L. M., Hildebrand, D. F., Williams, E. G., & Collins, G. B. (1989). Recovery of primary transformants of soybean. *Plant Cell Reports*, 7(8), 615-617.

[35] Ko-S, T., Lee, S., Krasnyanski, S., & Korban, S. S. (2003). Two critical factors are required for efficient transformation of multiple soybean cultivars: <i>Agrobacterium</i> strain and orientation of immature cotyledonary explant. *TAG Theoretical and Applied Genetics* , 107(3), 439-447.

[36] Dan, Y., & Reichert, N. (1998). Organogenic regeneration of soybean from hypocotyl explants. *In Vitro Cellular & Developmental Biology- Plant*, 34(1), 14-21.

[37] Wang, G., & Xu, Y. (2008). Hypocotyl-based Agrobacterium-mediated transformation of soybean (Glycine max) and application for RNA interference. *Plant Cell Reports*, 27(7), 1177-1184.

[38] Wright, M. S., Williams, M. H., Pierson, P. E., & Carnes, M. G. (1987). Initiation and propagation of Glycine max L. Merr.: Plants from tissue-cultured epicotyls. *Plant Cell, Tissue and Organ Culture*, 8(1), 83-90.

[39] Rajasekaran, K., & Pellow, J. (1997). Somatic embryogenesis from cultured epicotyls and primary leaves of soybean [<i>Glycine max</i> (L.) Merrill]. *In Vitro Cellular & Developmental Biology- Plant*, 33(2), 88-91.

[40] Khan, R. (2006). AG SP.

[41] De Cleene, M., & De Ley, J. (1976). The host range of crown gall. *The Botanical Review*, 42(4), 389-466.

[42] Hooykaas, P. J. J., & Schilperoort, R. A. (1992). Agrobacterium and plant genetic engineering. *Plant Molecular Biology*, 19(1), 15-38.

[43] Newell, C. (2000). Plant transformation technology. *Molecular Biotechnology*, 16(1), 53-65.

[44] Ishida, Y., et al. (1996). High efficiency transformation of maize (Zea mays L.) mediated by Agrobacterium tumefaciens. *Nat Biotech*, 14(6), 745-750.

[45] Hiei, Y., Komari, T., & Kubo, T. (1997). Transformation of rice mediated by Agrobacterium tumefaciens. *Plant Molecular Biology*, 35(1), 205-218.

[46] Pedersen, H. C., Christiansen, J., & Wyndaele, R. (1983). Induction and in vitro culture of soybean crown gall tumors. *Plant Cell Reports*, 2(4), 201-204.

[47] Stachel, S. E., & Nester, E. W. (1986). The genetic and transcriptional organization of the vir region of the A6 Ti plasmid of Agrobacterium tumefaciens. (Translated from eng). *EMBO J*, 5(7), 1445-1454, (in eng).

[48] Cascales, E., & Christie, P. J. (2003). The versatile bacterial type IV secretion systems. (Translated from eng). *Nat Rev Microbiol*, 1(2), 137-149, (in eng).

[49] Christie, P. J., Atmakuri, K., Krishnamoorthy, V., Jakubowski, S., & Cascales, E. (2005). Biogenesis, architecture, and function of bacterial type IV secretion systems. (Translated from eng). *Annu Rev Microbiol*, 59, 451-485, (in eng).

[50] Vergunst, A. C., et al. (2000). VirB/D4 -dependent protein translocation from Agrobacterium into plant cells. (Translated from eng). *Science*, 290(5493), 979-982, (in eng).

[51] Vergunst, A. C., et al. (2005). Positive charge is an important feature of the C-terminal transport signal of the VirB/D4 -translocated proteins of Agrobacterium. (Translated from eng). *Proc Natl Acad Sci U S A*, 102(3), 832-837, (in eng).

[52] Gelvin, S. B. (2003). Agrobacterium-mediated plant transformation: the biology behind the "gene-jockeying" tool. (Translated from eng). *Microbiol Mol Biol Rev*, 67(1), 16-37, table of contents (in eng).

[53] Christie, P. J., Ward, J. E., Winans, S. C., & Nester, E. W. (1988). The Agrobacterium tumefaciens virE2 gene product is a single-stranded-DNA-binding protein that associates with T-DNA. (Translated from eng). *J Bacteriol*, 170(6), 2659-2667, (in eng).

[54] Citovsky, V., Zupan, J., Warnick, D., & Zambryski, P. (1992). Nuclear localization of Agrobacterium VirE2 protein in plant cells. (Translated from eng). *Science*, 256(5065), 1802-1805, (in eng).

[55] Djamei, A., Pitzschke, A., Nakagami, H., Rajh, I., & Hirt, H. (2007). Trojan horse strategy in Agrobacterium transformation: abusing MAPK defense signaling. (Translated from eng) . *Science*, 318(5849), 453-456, (in eng).

[56] Gelvin, S. B. (2012). Traversing the cell: Agrobacterium T-DNAs journey to the host genome. (Translated from English). *Frontiers in Plant Science*, 3, (in English).

[57] Zaltsman, A., Krichevsky, A., Loyter, A., & Citovsky, V. (2010). Agrobacterium induces expression of a host F-box protein required for tumorigenicity. (Translated from eng). *Cell Host Microbe*, 7(3), 197-209, (in eng).

[58] Tzfira, T., Vaidya, M., & Citovsky, V. (2004). Involvement of targeted proteolysis in plant genetic transformation by Agrobacterium. (Translated from eng). *Nature*, 431(7004), 87-92, (in eng).

[59] Owens, L. D., & Cress, D. E. (1985). Genotypic variability of soybean response to agrobacterium strains harboring the ti or ri plasmids. (Translated from eng). *Plant Physiol*, 77(1), 87-94, (in eng).

[60] Delzer, B. W., Somers, D. A., & Orf, J. H. . Agrobacterium tumefaciens Susceptibility and Plant Regeneration of 10 Soybean Genotypes in Maturity Groups 00 to II. *Crop Sci.*, 30(2), 320-322.

[61] Olhoft, P. M., & Somers, D. A. (2007). *Soybean Transgenic Crops VI. Biotechnology in Agriculture and Forestry, eds Pua E-C & Davey MR*, Springer, Berlin, Heidelberg, 61, 3-27.

[62] Somers, D. A., Samac, D. A., & Olhoft, P. M. (2003). Recent Advances in Legume Transformation. *Plant Physiology*, 131(3), 892-899.

[63] Smigocki, A. C., & Owens, L. D. (1988). Cytokinin gene fused with a strong promoter enhances shoot organogenesis and zeatin levels in transformed plant cells. (Translated from eng). *Proc Natl Acad Sci U S A*, 85(14), 5131-5135, (in eng).

[64] Mauro, A. O., Pfeiffer, T. W., & Collins, G. B. (2009). Inheritance of Soybean Susceptibility to Agrobacterium tumefaciens and Its Relationship to Transformation. *Crop Sci.*, 35(4), 1152-1156.

[65] Dang, W., & Wei, Z-m. (2007). An optimized Agrobacterium-mediated transformation for soybean for expression of binary insect resistance genes. *Plant Science*, 173(4), 381-389.

[66] Yukawa, K., Kaku, H., Tanaka, H., Koga-Ban, Y., & Fukuda, M. (2007). Characterization and host range determination of soybean super virulent Agrobacterium tumefaciens KAT23. *Bioscience, biotechnology, and biochemistry* , 71(7), 1676-1682.

[67] Olhoft, P. O., & Somers, D. S. (2001). L-Cysteine increasesAgrobacterium-mediated T-DNA delivery into soybean cotyledonary-node cells. *Plant Cell Reports*, 20(8), 706-711.

[68] Paz, M., et al. (2004). Assessment of conditions affecting Agrobacterium-mediated soybean transformation using the cotyledonary node explant. *Euphytica*, 136(2), 167-179.

[69] Donaldson, P. A., & Simmonds, D. H. (2000). Susceptibility to Agrobacterium tumefaciens and cotyledonary node transformation in short-season soybean. *Plant Cell Reports*, 19(5), 478-484.

[70] Liu, S.-J., Wei, Z.-M., & Huang, J.-Q. (2008). The effect of co-cultivation and selection parameters on Agrobacterium-mediated transformation of Chinese soybean varieties. *Plant Cell Reports*, 27(3), 489-498.

[71] Xue, R.-G., Xie, H.-F., & Zhang, B. (2006). A multi-needle-assisted transformation of soybean cotyledonary node cells. *Biotechnology Letters*, 28(19), 1551-1557.

[72] Salmenkallio-Marttila, M., et al. (1995). Transgenic barley (Hordeum vulgare L.) by electroporation of protoplasts. *Plant Cell Reports*, 15(3), 301-304.

[73] Davey, M. R., Anthony, P., Power, J. B., & Lowe, K. C. (2005). Plant protoplasts: status and biotechnological perspectives. *Biotechnology Advances*, 23(2), 131-171.

[74] Kaeppler, H. F., Somers, D. A., Rines, H. W., & Cockburn, A. F. (1992). Silicon carbide fiber-mediated stable transformation of plant cells. *TAG Theoretical and Applied Genetics*, 84(5), 560-566.

[75] Crossway, A., et al. (1986). Integration of foreign DNA following microinjection of tobacco mesophyll protoplasts. *Molecular and General Genetics MGG*, 202(2), 179-185.

[76] Boynton, J., et al. (1988). Chloroplast transformation in Chlamydomonas with high velocity microprojectiles. *Science*, 240(4858), 1534-1538.

[77] Komari, T., Hiei, Y., Saito, Y., Murai, N., & Kumashiro, T. (1996). Vectors carrying two separate T-DNAs for co-transformation of higher plants mediated by Agrobacterium tumefaciens and segregation of transformants free from selection markers. *The Plant Journal*, 10(1), 165-174.

[78] Yan, H., & Rommens, C. M. (2007). Transposition-Based Plant Transformation. *Plant Physiology*, 143(2), 570-578.

[79] Bai, X., Wang, Q., & Chu, C. (2008). Excision of a selective marker in transgenic rice using a novel Cre/ <i>loxP</i> system controlled by a floral specific promoter. *Transgenic Research*, 17(6), 1035-1043.

[80] Depicker, A., Herman, L., Jacobs, A., Schell, J., & Montagu, M. (1985). Frequencies of simultaneous transformation with different T-DNAs and their relevance to the <i>Agrobacterium</i>/plant cell interaction. *Molecular and General Genetics MGG*, 201(3), 477-484.

[81] Wu, C., et al. (2004). A BAC- and BIBAC-based physical map of the soybean genome. (Translated from eng). *Genome Res*, 14(2), 319-326, (in eng).

[82] Zhang-B, H., & Wu, C. (2001). BAC as tools for genome sequencing. *Plant Physiology and Biochemistry*, 39(3-4), 195-209.

[83] Hamilton, C. M., Frary, A., Lewis, C., & Tanksley, S. D. (1996). Stable transfer of intact high molecular weight DNA into plant chromosomes. (Translated from eng). *Proc Natl Acad Sci U S A*, 93(18), 9975-9979, (in eng).

[84] Cui, Y., Bi, Y. M., Brugiere, N., Arnoldo, M., & Rothstein, S. J. (2000). The S locus glycoprotein and the S receptor kinase are sufficient for self-pollen rejection in Brassica. (Translated from eng). *Proc Natl Acad Sci U S A*, 97(7), 3713-3717, (in eng).

[85] Frary, A., & Hamilton, C. M. (2001). Efficiency and stability of high molecular weight DNA transformation: an analysis in tomato. (Translated from eng). *Transgenic Res*, 10(2), 121-132, (in eng).

[86] He, R. F., et al. (2006). Development of transformation system of rice based on binary bacterial artificial chromosome (BIBAC) vector. (Translated from eng). *Yi Chuan Xue Bao*, 33(3), 269-276, (in eng).

[87] Shukla, V. K., et al. (2009). Precise genome modification in the crop species Zea mays using zinc-finger nucleases. *Nature*, 459(7245), 437-441.

[88] Townsend, J. A., et al. (2009). High-frequency modification of plant genes using engineered zinc-finger nucleases. *Nature*, 459(7245), 442-445.

[89] Curtin, S. J., et al. (2011). Targeted Mutagenesis of Duplicated Genes in Soybean with Zinc-Finger Nucleases. *Plant Physiology*, 156(2), 466-473.

[90] Zhang, F., et al. (2010). High frequency targeted mutagenesis in Arabidopsis thaliana using zinc finger nucleases. *Proceedings of the National Academy of Sciences*, 107(26), 12028-12033.

[91] Ramirez, C. L., et al. (2008). Unexpected failure rates for modular assembly of engineered zinc fingers. *Nat Meth*, 5(5), 374-375.

[92] Boch, J., et al. (2009). Breaking the Code of DNA Binding Specificity of TAL-Type III Effectors. *Science*, 326(5959), 1509-1512.

[93] Moscou, M. J., & Bogdanove, A. J. (2009). A Simple Cipher Governs DNA Recognition by TAL Effectors. *Science*, 326(5959), 1501.

[94] Bogdanove, A. J., Schornack, S., & Lahaye, T. (2010). TAL effectors: finding plant genes for disease and defense. *Current Opinion in Plant Biology*, 13(4), 394-401.

[95] Kay, S., & Bonas, U. (2009). How Xanthomonas type III effectors manipulate the host plant. *Current Opinion in Microbiology*, 12(1), 37-43.

[96] Boch, J., & Bonas, U. (2010). Xanthomonas AvrBs3 Family-Type III Effectors: Discovery and Function. *Annual Review of Phytopathology*, 48(1), 419-436.

[97] Gürlebeck, D., Szurek, B., & Bonas, U. (2005). Dimerization of the bacterial effector protein AvrBs3 in the plant cell cytoplasm prior to nuclear import. *The Plant Journal*, 42(2), 175-187.

[98] Römer, P., Recht, S., & Lahaye, T. (2009). A single plant resistance gene promoter engineered to recognize multiple TAL effectors from disparate pathogens. *Proceedings of the National Academy of Sciences*, 106(48), 20526-20531.

[99] Sanjana, N. E., et al. (2012). A transcription activator-like effector toolbox for genome engineering. *Nat. Protocols*, 7(1), 171-192.

Chapter 24

Gene Duplication and RNA Silencing in Soybean

Megumi Kasai, Mayumi Tsuchiya and
Akira Kanazawa

Additional information is available at the end of the chapter

http://dx.doi.org/10.5772/51053

1. Introduction

Soybean, *Glycine max* (L.) Merr., is considered to be a typical paleopolyploid species with a complex genome [1-3]. Approximately 70 to 80% of angiosperm species have undergone polyploidization at some point in their evolutionary history, which is a well-known mechanism of gene duplication in plants [4]. The soybean genome actually possesses a high level of duplicate sequences, and furthermore, possesses homoeologous duplicated regions, which are scattered across different linkage groups [5-8]. Based on the genetic distances estimated by synonymous substitution measurements for the pairs of duplicated transcripts from expressed sequence tag (EST) collections of soybean and *Medicago truncatula*, Schlueter et al. estimated that soybean probably underwent two major genome duplication events: one that took place 15 million years ago (MYA) and another 44 MYA [9].

Gene duplication is a major source of evolutionary novelties and can occur through duplication of individual genes, chromosomal segments, or entire genomes (polyploidization). Under the classic model of duplicate gene evolution, one of the duplicated genes is free to accumulate mutations, which results in either the inactivation of transcription and/or a function (pseudogenization or nonfunctionalization) or the gain of a new function (neofunctionalization) as long as another copy retains the requisite physiological functions [10; and references therein]. However, empirical data suggest that a much greater proportion of gene duplicates is preserved than predicted by the classic model [11].

Recent advances in genome study have led to the formulation of several evolutionary models: a model proposed by Hughes [12] suggests that gene sharing, whereby a single gene encodes a protein with two distinct functions, precedes the evolution of two functionally distinct proteins; the duplication–degeneration–complementation model suggests that duplicate genes acquire debilitating yet complementary mutations that alter one or more sub-functions of the single gene progenitor, an evolutionary consequence for duplicated loci referred to as subfunctionalization [4, 11, 13]. In addition to this notion, models involving epigenetic silencing of duplicate genes [14] or purifying selection for gene balance [15, 16] have also been proposed. In soybean, differential patterns of expression have often been detected between homoeologous genes [17, 18], which indicates that subfunctionalization has occurred in these genes.

When the extent of subfunctionalization is limited, mutations in only one of multiple cognate gene copies do not often result in phenotypic changes. Therefore, methods that allow suppression of all copies of the duplicated gene are required for analyzing gene function or engineering novel traits. RNA silencing refers collectively to diverse RNA-mediated pathways of nucleotide-sequence-specific inhibition of gene expression, either at the posttranscriptional or transcriptional level, which provides a powerful tool to downregulate a gene or a gene family [19, 20]. Suppression of gene expression through RNA silencing is particularly useful for analyzing the function(s) of duplicated genes or engineering novel traits because it allows silencing of multiple cognate genes having nucleotide sequence identity. In fact, to produce soybean lines that have a novel trait, researchers have frequently used RNA silencing induced by a transgene.

In this review, we describe application of RNA silencing to understand the roles of genes or engineering novel traits in soybean. We describe methods to induce simultaneous silencing of duplicated genes and selective silencing of each copy of duplicated genes through RNA silencing. In addition to intentionally induced RNA silencing, we also refer to naturally occurring RNA silencing. Based on our knowledge of RNA silencing in soybean, we propose a hypothesis that plants may have used subfunctionalization of duplicated genes as a means to avoid the occurrence of simultaneous silencing of duplicated genes, which could be deleterious to the organism.

2. Mechanisms and diverse pathways of RNA silencing

Gene silencing is one of the regulatory mechanisms of gene expression in eukaryotes, which refers to diverse RNA-guided sequence-specific inhibition of gene expression, either at the posttranscriptional or transcriptional level [19, 20]. Post-transcriptional gene silencing (PTGS) was first discovered in transgenic petunia plants whose flower color pattern was changed as a consequence of overexpression of a gene that encodes the key enzyme for anthocyanin biosynthesis in 1990 [21, 22]. Similar phenomena have also been reported for plants transformed with various genes, which include virus resistance of plants that have gene or gene segments derived from the viral genome [23, 24]. Because of these findings, gene silencing is thought to have developed to defend against viruses. Several lines of research in plants indicated that double-stranded RNA (dsRNA) is crucial for RNA degradation [25, 26]. The potency of dsRNA to induce gene silencing was demonstrated in *Caenorhabditis elegans* by injecting dsRNA into cells in 1998 [27], and the phenomenon was termed RNA interference (RNAi).

http://dx.doi.org/10.5772/51053

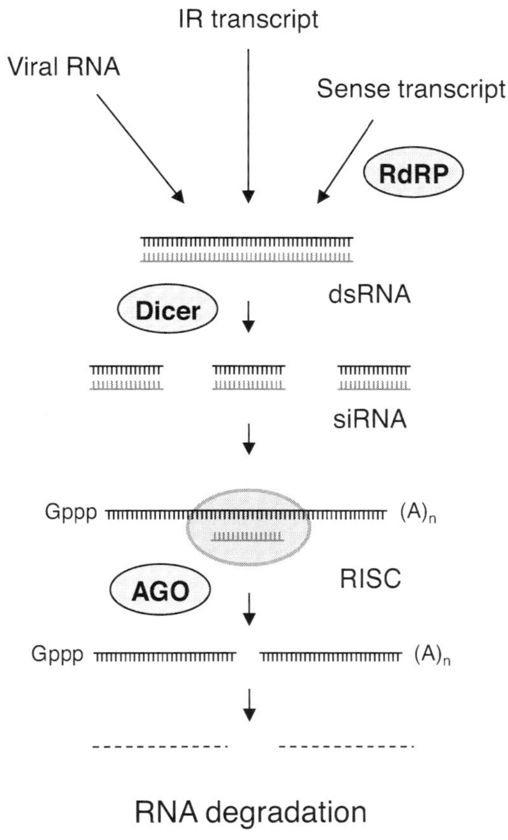

Figure 1. Pathways of RNA silencing used to downregulate a target gene through RNA degradation. Posttranscriptional gene silencing is triggered by dsRNA. Transcripts from transgenes that have an IR sequence can form dsRNA. Sense transcripts can produce dsRNA through the synthesis of complementary strand by RdRP. The replication intermediate or duplex structures formed within single-stranded RNA of the viral genome can also provide dsRNA. These dsRNAs are processed into siRNAs by the endonuclease Dicer. The siRNA is loaded into the RISC complex that contains AGO and guides the RISC complex to the mRNA by base-pairing. The RISC complex cuts the mRNA, which is subsequently degraded. Abbreviations: IR, inverted repeat; RdRP, RNA-dependent RNA polymerase; dsRNA, double-stranded RNA; siRNA, short interfering RNA; RISC, RNA-induced silencing complex; AGO, Argonaute.

Subsequent genetic and biochemical analyses in several organisms revealed that PTGS and RNAi share the same pathway and consist of two main processes: (i) processing of dsRNA into 20–26-nt small RNA molecules (short interfering RNA; siRNA) by an enzyme called Dicer that has RNaseIII-like endonuclease activity; (ii) cleavage of RNA guided by siRNA at a complementary nucleotide sequence in the RNA-induced silencing complex (RISC) containing the Argonaute (AGO) protein (Figure 1) [28]. The formation of dsRNA from single-stranded sense RNA was explained by the synthesis of its complementary strand by RNA-dependent RNA polymerase (RdRP). This process provides templates for Dicer cleavage that produces siRNAs and consequently allows amplification of silencing [29]. siRNA is responsible for not only induction of sequence-specific RNA degradation but also epigenetic changes involving DNA methylation and histone modification in the nucleus, which leads to transcriptional gene silencing (TGS) [30]. It has become evident that siRNA plays a role in systemic silencing as a mobile signal [31, 32]. In addition to siRNA, small RNA molecules called micro RNAs (miRNAs) are also involved in negative regulation of gene expression [33]. These gene silencing phenomena that are induced by sequence-specific RNA interaction are collectively called RNA silencing [34, 35].

RNA silencing plays an important role in many biological processes including development, stability of the genome, and defense against invading nucleic acids such as transgenes and viruses [20, 29, 30]. It can also be used as a tool for analyzing specific gene functions and producing new features in organisms including plants [36-38].

3. Methods of the induction of RNA silencing in soybean

3.1. Transgene-induced RNA silencing

Engineering novel traits through RNA silencing in soybean has been done using transgenes or virus vectors (Figure 1). RNA silencing in some transgenic soybean lines was induced by introducing a transgene that transcribes sense RNA homologous to a gene present in the plant genome, a phenomenon termed co-suppression [21]. This type of silencing was first discovered in transgenic petunia plants that had silencing of *CHS-A* for chalcone synthase [21, 22], in which mRNA transcribed from both *CHS-A* transgene and endogenous *CHS-A* gene was degraded. When sense transcripts from a transgene trigger RNA degradation, the pathway is also referred to as sense (S)-PTGS [19]. To obtain plants that have RNA silencing of a particular gene target, it is possible to generate co-suppressed plant lines as a byproduct of a transformation to overexpress the gene under the control of a strong promoter. However, a more promising method to induce RNA degradation is to transform plants with a construct comprising an inverted repeat (IR) sequence of the target gene, which forms dsRNA upon transcription (IR-PTGS) [39, 40]. This idea was based on the understanding of general mechanisms of RNA silencing in which dsRNA triggers the reaction of RNA degradation. The majority of transgene-induced RNA silencing in soybean have actually been done using such an IR construct. IR-PTGS can also be induced when multiple transgenes are integrated in the same site in the genome in an inverted orientation and fortuitous read-through transcription over the transgenes produces dsRNA.

An interesting finding reported in soybean is that RNA silencing is induced by a transgene that transcribes inverted repeats of a fatty acid desaturase *FAD2-1A* intron [41]. This result is contrary to the earlier belief that RNA silencing is a cytoplasmic event and intron does not trigger RNA degradation, which has been shown, for example, by using viral vector in plants [42] or by dsRNA injection to *C. elegans* cells [27], although irregular nuclear processing of primary transcripts associated with PTGS/RNAi has been reported previously [43]. The *FAD2-A1* intron-induced RNA silencing led to the understanding that RNA degradation can take place in the nucleus [44]. Although whether RNA degradation in the nucleus is inducible for other genes or in other plants has not been known, this phenomenon is intriguing because the involvement of nuclear events has been assumed for amplification of RNA silencing via transitivity [45] or intron-mediated suppression of RNA silencing [46].

Transcribing a transgene with a strong promoter tends to induce RNA silencing more frequently than that with a weak promoter [47]. For obtaining a higher level of transcription in soybean plants, the *Cauliflower mosaic virus* (CaMV) promoter has been used as in other plant species. Seed-specific promoters, such as those derived from the genes encoding subunits of β-conglycinin, glycinin, or Kunitz trypsin inhibitor, have also been used in soybean to induce seed-specific silencing, one feature that is exploited for metabolic engineering in soybean.

A gene construct that induces RNA silencing has been introduced to the soybean genome using either *Agrobacterium tumefaciens* infection or particle bombardment, which can produce stable transgenic soybean lines that have altered traits. In addition, RNA silencing can be induced in soybean roots using *A. rhizogenes*-mediated transformation, which has been used for gene functional analysis. Methods for soybean transformation have been reviewed elsewhere [48].

3.2. Virus-induced gene silencing (VIGS)

RNA silencing has also been induced using a virus vector in soybean. Plants intrinsically have the ability to cope with viruses through the mechanisms of RNA silencing. When plants are infected with an RNA virus, dsRNA of the viral genome is degraded by the infected plants [49, 50]. The dsRNA in the virus-infected cells is thought to be the replication intermediate of the viral RNA [51] or a duplex structure formed within single-stranded viral RNA [52]. The viral genomic RNA can be processed into siRNAs, then targeted by the siRNA/RNase complex. In this scenario, if a nonviral segment is inserted in the viral genome, siRNAs would also be produced from the segment. Therefore, if the insert corresponds to a sequence of the gene encoded in the host plant, infection by the virus results in the production of siRNAs corresponding to the plant gene and subsequently induces loss of function of the gene product (Figure 2). This fact led to the use of a virus vector as a source to induce silencing of a specific gene in the plant genome, which is referred to as virus-induced gene silencing (VIGS) [42, 53, 54]. So far, at least 11 RNA viruses and five DNA viruses were developed as a plant virus vector for gene silencing, as listed previously [37]. Three vectors are now available in soybean: those based on *Bean pod mottle virus* (BPMV) [55], *Cucumber mosaic virus* (CMV) [56], and *Apple latent spherical virus* (ALSV) [57].

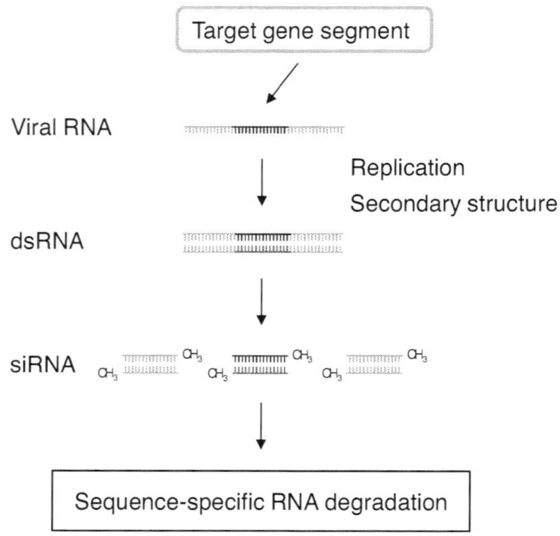

Figure 2. Virus-induced silencing of plant endogenous gene. When plants are infected with an RNA virus, dsRNA of the viral genome is degraded by the infected plants. The dsRNA in the virus-infected cells is thought to be the replication intermediate or secondary-structured viral RNA. The viral genomic RNA can be processed into siRNAs. If a plant gene segment is inserted in the viral genome, siRNAs corresponding to the plant gene are produced and subsequently induce sequence-specific RNA degradation of the plant gene.

4. Examples of RNA silencing reported in soybean

4.1. Metabolic engineering by transgene-induced RNA silencing

To the authors' knowledge, 28 scientific papers that describe metabolic engineering by transgene-induced RNA silencing in soybean have been published up to 2011 [58]. Because soybean seeds are valued economically for food and oil production, most modifications to transgenic soybean plants using RNA silencing are focused on seed components. Metabolic pathways in developing seeds have been targeted in terms of altering nutritional value for human or animals, e.g., changing seed storage protein composition [59, 60], reducing phytic acids [61, 62], saponin [63] or allergens [64], and increasing isoflavone [65]. Metabolic engineering has also targeted oil production [66-72]. These modifications were done by inhibiting a step in a metabolic pathway to decrease a product or by blocking a competing branch pathway to increase a product.

RNA silencing can be induced efficiently in soybean roots using *A. rhizogenes*-mediated root transformation. This method has been used for analyzing roles of gene products in nodule development and/or function, which occurs as a consequence of interaction between legume plants and the nitrogen-fixing symbiotic bacterium *Bradyrhizobium japonicum* [73-78]. The

http://dx.doi.org/10.5772/51053

hairy root system was also used for analyzing roles of a MYB transcription factor in isoflavonoid biosynthesis [79].

Transgene-induced RNA silencing has also been induced in leaf tissues for the β-glucuronidase gene [80] or the senescence-associated receptor-like kinase gene [81] and in calli for the amino aldehyde dehydrogenase gene to induce the biosynthesis of 2-acetyl-1-pyrroline [82].

4.2. Disease resistance acquired by transgene-induced RNA silencing

Another focus of modifying soybean plants through RNA silencing is resistance against diseases, particularly to those caused by viruses. Resistance to viruses was achieved by transforming plants with genes or segments of genes derived from viruses and was referred to as pathogen-derived resistance [23, 24, 83, 84]. The resistance did not need protein translated from the transgene [85-87], which led to the understanding that RNA is the factor that conferred resistance to the plants and that the enhanced resistance is acquired via a mechanism analogous to that involved in co-suppression. Using this strategy, soybean plants resistant to *Soybean mosaic virus* (SMV) [88-90], or *Soybean dwarf virus* [91, 92] have been produced.

In addition to resistance against a virus, transgenic soybean plants resistant to cyst nematode (*Heterodera glycines*) have also been produced using RNA silencing [93], in which an inverted repeat of the major sperm protein gene from cyst nematode was transcribed from the transgene. RNA silencing was elicited in cyst nematode after nematode ingestion of dsRNA molecules produced in the soybean plants; as a consequence, reproductive capabilities of the cyst nematode were suppressed. The effects of RNA silencing on controlling *H. glycines* [94] or root-knot nematode (*Meloidogyne incognita*) [95] infection have been assayed in soybean roots using *A. rhizogenes*-mediated transformation. On the other hand, this root transformation method has also been used for analyzing a role of host genes in resistance against diseases caused by *Phytophthora sojae* [96, 97], *Fusarium solani* [98] or cyst nematode [99].

4.3. Gene functional analysis by VIGS

An advantage of VIGS is its ease for making a gene construct and introducing nucleic acids to cells. In addition, the effect of silencing can be monitored within a short time after inoculating plants with the virus. Because of these features, VIGS is suitable for gene function analysis [51, 100, 101] and has been used for gene identification via downregulating a candidate gene(s) responsible for a specific phenomenon in soybean. VIGS was used to demonstrate that genes present in the genetically identified loci actually encode the genes responsible for the phenotype: VIGS of the putative *flavonoid 3'-hydroxylase* (*F3'H*) gene resulted in a decrease in the content of quercetin relative to kampferol, which indicated that the putative gene actually encodes the F3'H protein [56]; VIGS of the *GmTFL1b* gene, a soybean orthologue of *Arabidopsis TERMINAL FLOWER1* (*TFL1*) and a candidate gene for the genetically identified locus *Dt1*, induced an early transition from vegetative to reproductive phases, which indicated the identity between *Dt1* and *GmTFL1b* [102]. VIGS has also been used to identify genes involved in resistance of soybean plants against pathogens such as SMV, BPMV, *Pseudomonas syringae* or *Phakopsora pachyrhizi* [103-107].

4.4. Naturally occurring RNA silencing

In addition to artificially induced RNA silencing, naturally occurring RNA silencing has also been known in soybean. Naturally occurring RNA silencing, involving mRNA degradation induced as a consequence of certain genetic changes, has been detected based on phenotypic changes. Most commercial varieties of soybean produce yellow seeds due to loss of pigmentation in seed coats, and this phenotype has been shown to be due to PTGS of the *CHS* genes [108, 109]. In cultivated soybean, there are varieties producing seeds with yellow seed coats and those producing seeds with brown or black seed coats in which anthocyanin and proanthocyanidin accumulate. In contrast, wild soybeans (*Glycine soja*), an ancestor of the cultivated soybean, have exclusively produced seeds with pigmented seed coats in thousands of accessions from natural populations in East Asia that we have screened (unpublished data). Thus, the nonpigmented seed coat phenotype was probably generated after domestication of soybean, and humans have maintained the plant lines that have *CHS* RNA silencing. The genetic change that induced *CHS* RNA silencing has been attributed to a structural change in the *CHS* gene cluster, which allows production of inverted repeat *CHS* RNA [110].

The occurrence of RNA silencing that leads to changes in pigmentation of plant tissues has also been reported for the *CHS* genes in maize [111] and petunia [112]. In petunia, a variety 'Red Star' produces flowers having a star-type red and white bicolor pattern, which resembles the flower-color patterns observed in transgenic petunias with co-suppression of the *CHS* genes [113], and in fact, the phenotype was demonstrated to be due to RNA silencing of the *CHS* genes in the white sectors [112]. Breeding of petunia was launched in the 1830s by crossing among wild species. The generation of the star-type petunia plants as a consequence of hybridizations between plant lines suggests that RNA silencing ability can be conferred via shuffling of genomes that are slightly different from each other. These phenomena also resemble the RNA silencing in a seed storage protein gene in rice, which is associated with a structural change in the gene region induced by mutagenesis [114], a case of RNA silencing in nontransgenic plants.

5. Diagnosis of an RNA silencing-induced phenotype using viral infection

In the course of the analysis of *CHS* RNA silencing, the function of a virus-encoded protein called suppressor protein of RNA silencing was used to visually demonstrate the occurrence of RNA silencing [108, 111, 112, 115]. These suppressor proteins affect viral accumulation in plants. The ability of the suppressor protein to allow viral accumulation is due to its inhibition of RNA silencing by preventing the incorporation of siRNAs into RISCs or by interfering with RISCs [116]. Because of these features, RNA silencing can be suppressed in plants infected with a virus that carries the suppressor protein. When a soybean plant that has a yellow seed coat is infected with CMV, the seed coat restores pigmentation [108]. This phenomenon is due to the activity of gene silencing suppressor protein called 2b encoded by the CMV. This example typically indicates that, using the function of viral suppressor protein, we can "diagnose" whether an observed phenotypic change in a plant is caused by RNA silencing. A similar phenomenon has also been detected in maize [111] and petunia [112] lines, both of which have phenotypic changes through naturally occurring RNA silencing of an endogenous *CHS* gene, or a transgenic petunia line that has *CHS* co-suppression [115].

6. What do phenotypic changes induced by RNA silencing in soybean indicate?

Soybean is thought to be derived from an ancestral plant(s) with a tetraploid genome, and as a consequence, large portions of the soybean genome are duplicated [7], with nearly 75% of the genes present in multiple copies [117]. In addition, genes in the soybean genome are sometimes duplicated in tandem [118-121]. Our recent studies have indeed shown functional redundancy of duplicated genes in soybean [122, 123]. Such gene duplication can be an obstacle to producing mutants by conventional methods of mutagenesis. In this regard, the gene silencing technique is particularly useful because it allows silencing of multiple cognate genes having nucleotide sequence identity.

Changes in phenotypes as a consequence of inducing RNA silencing have been successful for many genes in soybean as mentioned above. Considering that many genes are duplicated in soybean genome, this fact indicates either that RNA silencing worked on all duplicated genes that have the same function or that the genes were subfunctionalized after duplication, so that RNA silencing of even a single gene of the duplicated genes resulted in the phenotypic changes.

It is of interest to understand whether duplicated genes have identical or diversified functions, which may depend on the time after duplication event and/or the selection pressure on the genes. To analyze the functions of each copy of the duplicated genes, we need to silence a specific copy of the duplicated genes. If the duplicated genes are expressed in different tissues, RNA silencing of both genes can lead to understanding the function of each gene. PTGS by transcribing inverted repeat with a constitutive promoter or VIGS will be suitable for this analysis. An example of such an approach is the VIGS of duplicated *TFL1* orthologues, which are expressed in different tissues. A specific role of one of the *TFL1* orthologues has been identified by VIGS as mentioned earlier [102].

7. Methods to induce selective RNA silencing of duplicated genes

When duplicated genes are subfunctionalized with only limited nucleotide changes and are expressed in overlapping tissues, specific silencing of each gene will be necessary for understanding their function(s). Silencing a specific copy of duplicated genes can be achieved by targeting a gene portion whose nucleotide sequence is differentiated between the duplicated genes. A condition that allows this type of silencing involves a lack of silencing of the other copy of duplicated genes even when they have the same sequence in the other portions.

In plants, miRNAs or siRNAs promote production of secondary siRNAs from the 5′ upstream region and/or the 3′ downstream region of the initially targeted region via production of dsRNA by RdRP. These secondary siRNAs can lead to silencing of a secondary target that is not directly targeted by the primary silencing trigger [124]. Studies so far have indicated that such a spread of RNA silencing, called transitive RNA silencing, does not occur with the majority of endogenous genes, although it can happen to a transgene [45; and references therein]. Assuming the lack of transitive RNA silencing, it is possible to induce silencing of a specific copy of a duplicated gene. Targeting a region specific for each copy, e.g., the 3′ untanslated region (UTR), can induce silencing of the gene copy only, whereas targeting a region conserved in duplicated gene copies can induce silencing of the multiple gene copies simultaneously (Figure 3). Such selective RNA silencing was successful in a gene family of rice [125] and this strategy may work for analyzing functional diversification of duplicated genes in any plant species.

An alternative approach to suppress gene expression in plants is the use of artificial miRNAs (Figure 4) (amiRNAs; also called synthetic miRNAs) [38, 126]. This approach involves modification of plant miRNA sequence to target specific transcripts, originally not under miRNA control, and downregulation of gene expression via specific cleavage of the target RNA. Melito *et al.* have used amiRNA to downregulate the leucine-rich repeat transmembrane receptor-kinase gene in soybean [99]. miRNA has been extensively studied in soybean [127-130], information of which may be useful for designing amiRNAs. Because of its specificity, this method will be useful for silencing a limited copy of duplicated genes in soybean.

Induction of TGS by targeting dsRNA to a gene promoter can also be the method of choice. Gene silencing through transcriptional repression can be induced by dsRNA targeted to a gene promoter (Figure 4). However, until recently, no plant has been produced that harbors an endogenous gene that remains silenced in the absence of promoter-targeting dsRNA. We have reported for the first time that TGS can be induced by targeting dsRNA to the endogenous gene promoters in petunia and tomato plants, using a *Cucumber mosaic virus* (CMV)-based vector and that the induced gene silencing is heritable. Efficient silencing depended on the function of the 2b protein encoded in the vector, which facilitates epigenetic modifications through the transport of siRNA to the nucleus [131, 132]. The progeny plants do not have any transgene because the virus is eliminated during meiosis. Therefore, plants that are produced by this system have altered traits but do not carry a transgene, thus constituting a novel class of modified plants [131, 132]. We have also developed *in planta* assay systems to detect inhibition of cytosine methylation using plants that contain a transgene transcriptionally silenced by an epigenetic mechanism [133]. Using these systems, we found that genistein, a major isoflavonoid compound rich in soybean seeds, inhibits cytosine methylation and restores the transcription of epigenetically silenced genes [133]. Whether developing soybean seeds are resistant (or susceptible) to epigenetic modifications is an interesting issue in terms of both developmental control of gene expression and intentionally inducing TGS through epigenetic changes.

http://dx.doi.org/10.5772/51053

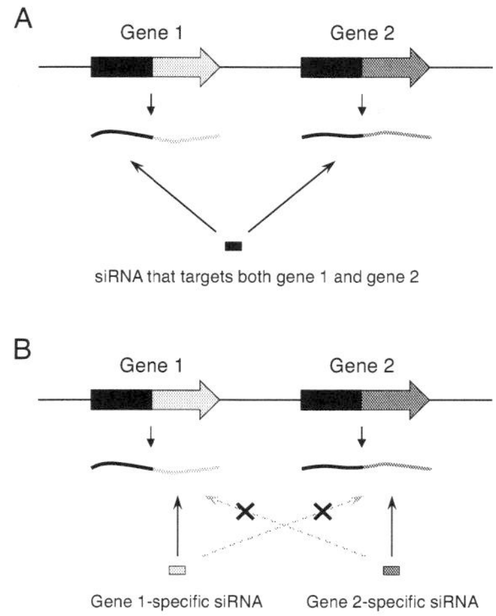

Figure 3. Selective RNA silencing of duplicated genes. The gene 1 and gene 2 are produced as a consequence of gene duplication. They share a highly conserved nucleotide sequence in the 5′ region, while they have a different sequence in the 3′ region. When siRNAs corresponding to the conserved region are produced, they can induce RNA degradation of the transcripts from both genes (A). On the other hand, siRNAs corresponding to the 3′ region can induce gene 1-specific or gene 2-specific RNA degradation (B). A combination of these different approaches enables functional analysis of duplicated genes.

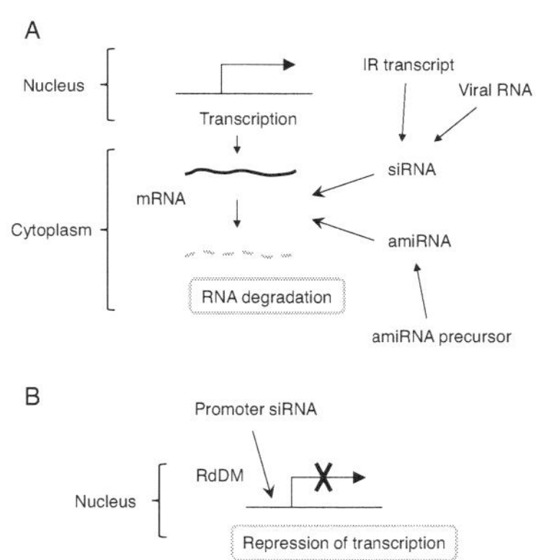

Figure 4. Various pathways of RNA silencing that can be intentionally induced to suppress gene expression in plants. Transcripts from transgenes that have an IR sequence of a plant gene segment or viral genomic RNA that carries the segment can form dsRNA. These dsRNAs are subsequently processed into siRNAs in the cytoplasm. Similarly, amiRNA precursors transcribed from the amiRNA gene are processed into amiRNAs. These small RNAs can cause degradation of target gene transcripts, a process termed PTGS (A). When siRNAs corresponding to a gene promoter are produced, they can induce RdDM in the nucleus, thereby TGS of the target gene can be induced (B). Abbreviations: amiRNA, artificial microRNA; PTGS, posttranscriptional gene silencing; RdDM, RNA-directed DNA methylation; TGS, transcriptional gene silencing.

8. Differentiation of duplicated genes and induction of RNA silencing

How much sequence difference will be necessary to induce selective RNA silencing? A factor that affects induction of RNA silencing is the extent of sequence identity between the dsRNA that triggers RNA silencing and its target gene. IR-PTGS could be induced by IR-transcripts that can form 98-nt or longer dsRNAs [39]. In VIGS, the lower size limit of the inserted fragments required for inducing PTGS is 23-nt, a size almost corresponding to that of siRNAs [134], and that for inducing TGS is 81-91 nt [135]. Silencing a gene probably requires sequence identity longer than the size of siRNAs between dsRNA and its target, although the efficiency of silencing may depend on the system of silencing induction.

We previously induced *CHS* VIGS in soybean [56]. In soybean seed coats, the *CHS7/CHS8* genes, which share 98% nucleotide sequence identity in the coding region, are predominantly expressed among the eight members of the *CHS* gene family [136, 137]. We have induced the silencing using a virus vector that carried a 244-nt fragment of the *CHS7* gene [56]. The *CHS* mRNA levels in the seed coats and leaf tissues of plants infected with the virus were reduced to 12.4% and 47.0% of the control plants, respectively. One plausible explanation for the differential effects of VIGS on these tissues may be that the limited sequence homology (79%-80%) between the *CHS7* and the *CHS1-CHS3* genes, the transcripts of which make up approximately 40% of the total *CHS* transcript content of leaf tissues [137], results in the degradation of the *CHS1-CHS3* transcripts at a lower efficiency than the degradation of *CHS7/CHS8* transcripts. Consistent with these results, naturally occurring *CHS* RNA silencing, in which *CHS7/CHS8* genes are silenced in seed coat tissues, is thought to be induced by inverted repeat transcripts of a *CHS3* gene segment [110]. In terms of the practical use of transgene-induced RNA silencing, these results suggest that a portion of genes whose sequence identity between duplicated genes is lower than 79%-80% should be chosen as a target for inducing selective RNA silencing.

The naturally occurring RNA silencing of the *CHS* genes in soybean may indicate relationships between diversification of duplicated genes and RNA silencing. Gene duplication can be a cause of RNA silencing because it may sometimes result in the production of dsRNA, which triggers RNA silencing through read-through transcription [114, 115]. In the *CHS* silencing in soybean, the extent of mRNA decrease differs between different copies of the gene family. These observations may indicate that plants use subfunctionalization of duplicated genes as a means to avoid the occurrence of simultaneous silencing of duplicated genes, which may have a deleterious effect on the organism.

9. Conclusion and perspectives

RNA silencing has been used as a powerful tool to engineer novel traits or analyze gene function in soybean. Soybean plants that have engineered a metabolic pathway or acquired resistance to diseases have been produced by transgene-induced gene silencing. VIGS has been used as a tool to analyze gene function in soybean. In addition to RNA silencing, site-directed mutagenesis using zinc-finger nucleases has been applied to mutagenizing duplicated genes in soybean [138]. Such reverse genetic approaches may be supplemented by forward genetic approaches such as high linear energy transfer radiation-based mutagenesis, e.g., irradiation of ion beam [139] and fast neutron [140]. Similarly, gene tagging systems using maize Ds transposon [141] and rice *mPing* transposon [142] have also been developed in soybean. Aside from using RNA silencing as a tool to engineer novel traits, analysis of mutants in combination with reverse genetic approaches may facilitate the identification of causative gene(s) of the mutation. An interesting feature of RNA silencing is its inducible nature, which allows downregulation of a gene in a tissue-specific manner. This strategy is particularly advantageous for analyzing the function of genes whose mutation or ubiquitous downregulation is lethal. Another feature of RNA silencing is that it allows analysis of biological phenomena that involve the effect of a difference in the mRNA level of the gene. The dependence of pigmentation in soybean pubescence on the mRNA level of the *F3′H* gene has actually been shown by utilizing VIGS [143]. In this regard, selective RNA silencing of duplicated genes may reveal the presence of additive effects of the expression levels of duplicated genes in soybean.

Abbreviations

AGO, Argonaute; ALSV, *Apple latent spherical virus*; amiRNA, artificial miRNA; BPMV, *Bean pod mottle virus*; CaMV, *Cauliflower mosaic virus*; CHS, chalcone synthase; CMV, *Cucumber mosaic virus*; dsRNA, double-stranded RNA; EST, expressed sequence tag; F3′H, flavonoid 3′-hydroxylase; IR, inverted repeat; miRNA, micro RNA; MYA, million years ago; PTGS, post-transcriptional gene silencing; RdRP, RNA-dependent RNA polymerase; RISC, RNA-induced silencing complex; RNAi, RNA interference; siRNA, short interfering RNA; SMV, *Soybean mosaic virus*; TFL1, TERMINAL FLOWER1; TGS, transcriptional gene silencing; UTR, untanslated region; VIGS, virus-induced gene silencing

Acknowledgements

Our work is supported by Grants-in-Aid for Scientific Research from the Ministry of Education, Culture, Sports, Science and Technology of Japan.

Author details

Megumi Kasai, Mayumi Tsuchiya and Akira Kanazawa*

*Address all correspondence to: kanazawa@res.agr.hokudai.ac.jp

Research Faculty of Agriculture, Hokkaido University, Japan

References

[1] Lackey, J. (1980). Chromosome-numbers in the Phaseoleae (Fabaceae, Faboideae) and their relation to taxonomy. *Am J Bot*, 67, 595-602.

[2] Hymowitz, T. (2004). Speciation and cytogenetics. In: Boerma HR, Specht JE. (eds) Soybeans: improvement, production, and uses, Ed 3, Agronomy Monograph No. 16. American Society of Agronomy, Inc., Crop Science Society of America, Inc., Soil Science Society of America, Inc. Madison, Wisconsin, USA; 2004. p97-136.

[3] Shoemaker, R. C., Schlueter, J., & Doyle, J. J. (2006). Paleopolyploidy and gene duplication in soybean and other legumes. *Curr Opin Plant Biol*, 9, 104-9.

[4] Moore, R. C., & Purugganan, M. D. (2005). The evolutionary dynamics of plant duplicate genes. *Curr Opin Plant Biol*, 8, 122-8.

[5] Lohnes, D., Specht, J., & Cregan, P. (1997). Evidence for homoeologous linkage groups in the soybean. *Crop Sci*, 37, 254-7.

[6] Zhu, T., Schupp, J. M., Oliphant, A., & Keim, P. (1994). Hypomethylated sequences: characterization of the duplicate soybean genome. *Mol Gen Genet*, 244, 638-45.

[7] Shoemaker, R. C., Polzin, K., Labate, J., Specht, J., Brummer, E. C., Olson, T., et al. (1996). Genome duplication in soybean (*Glycine* subgenus *soja*). *Genetics*, 144, 329-38.

[8] Lee, J., Bush, A., Specht, J., & Shoemaker, R. (1999). Mapping of duplicate genes in soybean. *Genome*, 42, 829-36.

[9] Schlueter, J. A., Dixon, P., Granger, C., Grant, D., Clark, L., Doyle, J. J., et al. (2004). Mining EST databases to resolve evolutionary events in major crop species. *Genome*, 47, 868-76.

[10] Lynch, M., & Conery, J. S. (2000). The evolutionary fate and consequences of duplicate genes. *Science*, 290, 1151-5.

[11] Force, A., Lynch, M., Pickett, F. B., Amores, A., Yan, Y. L., & Postlethwait, J. (1999). Preservation of duplicate genes by complementary, degenerative mutations. *Genetics*, 151, 1531-45.

[12] Hughes, A. L. (1994). The evolution of functionally novel proteins after gene duplication. *Proc Biol Sci*, 256, 119-24.

[13] Lynch, M., & Force, A. (2000). The probability of duplicate gene preservation by subfunctionalization. *Genetics*, 154, 459-73.

[14] Rodin, S. N., & Riggs, A. D. (2003). Epigenetic silencing may aid evolution by gene duplication. *J Mol Evol*, 56, 718-29.

[15] Freeling, M., & Thomas, B. C. (2006). Gene-balanced duplications, like tetraploidy, provide predictable drive to increase morphological complexity. *Genome Res*, 16, 805-14.

[16] Birchler, J. A., & Veitia, R. A. (2007). The gene balance hypothesis: from classical genetics to modern genomics. *Plant Cell*, 19, 395-402.

[17] Schlueter, J. A., Scheffler, B. E., Schlueter, S. D., & Shoemaker, R. C. (2006). Sequence conservation of homeologous bacterial artificial chromosomes and transcription of homeologous genes in soybean (*Glycine max* L. Merr.). *Genetics*, 174, 1017-28.

[18] Schlueter, J., Vaslenko-Sanders, I., Deshpande, S., Yi, J., Siegfried, M., Roe, B., et al. (2007). The FAD2 gene family of soybean: insights into the structural and functional divergence of a paleopolyploid genome. *Crop Sci*, 47, S14-S26.

[19] Brodersen, P., & Voinnet, O. (2006). The diversity of RNA silencing pathways in plants. *Trends Genet*, 22, 268-80.

[20] Vaucheret, H. (2006). Post-transcriptional small RNA pathways in plants: mechanisms and regulations. *Genes Dev*, 20, 759-71.

[21] Napoli, C., Lemieux, C., & Jorgensen, R. (1990). Introduction of a chimeric chalcone synthase gene into petunia results in reversible co-suppression of homologous genes *in trans*. *Plant Cell*, 2, 279-89.

[22] van der Krol, A. R., Mur, L. A., Beld, M., Mol, J. N., & Stuitje, A. R. (1990). Flavonoid genes in petunia: addition of a limited number of gene copies may lead to a suppression of gene expression. *Plant Cell*, 2, 291-9.

[23] Wilson, T. M. (1993). Strategies to protect crop plants against viruses: pathogen-derived resistance blossoms. *Proc Natl Acad Sci U S A*, 90, 3134-41.

[24] Baulcombe, D. C. (1996). Mechanisms of pathogen-derived resistance to viruses in transgenic plants. *Plant Cell*, 8, 1833-44.

[25] Metzlaff, M., O'Dell, M., Cluster, P. D., & Flavell, R. B. (1997). RNA-mediated RNA degradation and chalcone synthase A silencing in petunia. *Cell*, 88, 845-54.

[26] Waterhouse, P., Graham, H., & Wang, M. (1998). Virus resistance and gene silencing in plants can be induced by simultaneous expression of sense and antisense RNA. *Proc Natl Acad Sci U S A*, 95, 13959-64.

[27] Fire, A., Xu, S., Montgomery, M. K., Kostas, S. A., Driver, S. E., & Mello, C. C. (1998). Potent and specific genetic interference by double-stranded RNA in *Caenorhabditis elegans*. *Nature*, 391, 806-11.

[28] Matzke, M., Matzke, A. J., & Kooter, J. M. (2001). RNA: guiding gene silencing. *Science*, 293, 1080-3.

[29] Baulcombe, D. (2004). RNA silencing in plants. *Nature*, 431, 356-63.

[30] Matzke, M., Kanno, T., Daxinger, L., Huettel, B., & Matzke, A. J. (2009). RNA-mediated chromatin-based silencing in plants. *Curr Opin Cell Biol*, 21, 367-76.

[31] Dunoyer, P., Schott, G., Himber, C., Meyer, D., Takeda, A., Carrington, J. C., et al. (2010). Small RNA duplexes function as mobile silencing signals between plant cells. *Science*, 328, 912-6.

[32] Molnár, A., Melnyk, C. W., Bassett, A., Hardcastle, T. J., Dunn, R., & Baulcombe, D. C. (2010). Small silencing RNAs in plants are mobile and direct epigenetic modification in recipient cells. *Science*, 328, 872-5.

[33] Mallory, A. C., & Vaucheret, H. (2006). Functions of microRNAs and related small RNAs in plants. *Nat Genet*, 38(S31-6).

[34] Voinnet, O. (2002). RNA silencing: small RNAs as ubiquitous regulators of gene expression. *Curr Opin Plant Biol*, 5, 444-51.

[35] Matzke, M., Aufsatz, W., Kanno, T., Daxinger, L., Papp, I., Mette, M. F., et al. (2004). Genetic analysis of RNA-mediated transcriptional gene silencing. *Biochim Biophys Acta*, 1677, 129-41.

[36] Mansoor, S., Amin, I., Hussain, M., Zafar, Y., & Briddon, R. W. (2006). Engineering novel traits in plants through RNA interference. *Trends Plant Sci*, 11, 559-65.

[37] Kanazawa, A. (2008). RNA silencing manifested as visibly altered phenotypes in plants. *Plant Biotechnol*, 25, 423-35.

[38] Frizzi, A., & Huang, S. (2010). Tapping RNA silencing pathways for plant biotechnology. *Plant Biotechnol J*, 8, 655-77.

[39] Wesley, S. V., Helliwell, C. A., Smith, N. A., Wang, M. B., Rouse, D. T., Liu, Q., et al. (2001). Construct design for efficient, effective and high-throughput gene silencing in plants. *Plant J*, 27, 581-90.

[40] Helliwell, C. A., & Waterhouse, P. M. (2005). Constructs and methods for hairpin RNA-mediated gene silencing in plants. *Methods Enzymol*, 392, 24-35.

[41] Wagner, N., Mroczka, A., Roberts, P. D., Schreckengost, W., & Voelker, T. (2011). RNAi trigger fragment truncation attenuates soybean *FAD2-1* transcript suppression and yields intermediate oil phenotypes. *Plant Biotechnol J*, 9, 723-8.

[42] Ruiz, M. T., Voinnet, O., & Baulcombe, D. C. (1998). Initiation and maintenance of virus-induced gene silencing. *Plant Cell*, 10, 937-46.

[43] Metzlaff, M., O'Dell, M., Hellens, R., & Flavell, R. B. (2000). Developmentally and transgene regulated nuclear processing of primary transcripts of *chalcone synthase A* in petunia. *Plant J*, 23, 63-72.

[44] Hoffer, P., Ivashuta, S., Pontes, O., Vitins, A., Pikaard, C., Mroczka, A., et al. (2011). Posttranscriptional gene silencing in nuclei. *Proc Natl Acad Sci U S A*, 108, 409-14.

[45] Vermeersch, L., De Winne, N., & Depicker, A. (2010). Introns reduce transitivity proportionally to their length, suggesting that silencing spreads along the pre-mRNA. *Plant J*, 64, 392-401.

[46] Christie, M., Croft, L. J., & Carroll, B. J. (2011). Intron splicing suppresses RNA silencing in *Arabidopsis*. *Plant J*, 68, 159-67.

[47] Que, Q., Wang, H. Y., English, J. J., & Jorgensen, R. A. (1997). The frequency and degree of cosuppression by sense chalcone synthase transgenes are dependent on transgene promoter strength and are reduced by premature nonsense codons in the transgene coding sequence. *Plant Cell*, 9, 1357-68.

[48] Yamada, T., Takagi, K., & Ishimoto, M. (2012). Recent advances in soybean transformation and their application to molecular breeding and genomic analysis. *Breeding Sci*, 61, 480-94.

[49] Covey, S., AlKaff, N., Langara, A., & Turner, D. (1997). Plants combat infection by gene silencing. *Nature*, 385, 781-2.

[50] Al-Kaff, N. S., Covey, S. N., Kreike, M. M., Page, A. M., Pinder, R., & Dale, P. J. (1998). Transcriptional and posttranscriptional plant gene silencing in response to a pathogen. *Science*, 279, 2113-5.

[51] Lu, R., Martin-Hernandez, A. M., Peart, J. R., Malcuit, I., & Baulcombe, D. C. (2003). Virus-induced gene silencing in plants. *Methods*, 30, 296-303.

[52] Molnár, A., Csorba, T., Lakatos, L., Várallyay, E., Lacomme, C., & Burgyán, J. (2005). Plant virus-derived small interfering RNAs originate predominantly from highly structured single-stranded viral RNAs. *J Virol*, 79, 7812-8.

[53] Kumagai, M. H., Donson, J., della -Cioppa, G., Harvey, D., Hanley, K., & Grill, L. K. (1995). Cytoplasmic inhibition of carotenoid biosynthesis with virus-derived RNA. *Proc Natl Acad Sci U S A*, 92, 1679-83.

[54] Purkayastha, A., & Dasgupta, I. (2009). Virus-induced gene silencing: a versatile tool for discovery of gene functions in plants. *Plant Physiol Biochem*, 47, 967-76.

[55] Zhang, C., & Ghabrial, S. A. (2006). Development of *Bean pod mottle virus*-based vectors for stable protein expression and sequence-specific virus-induced gene silencing in soybean. *Virology*, 344, 401-11.

[56] Nagamatsu, A., Masuta, C., Senda, M., Matsuura, H., Kasai, A., Hong, J. S., et al. (2007). Functional analysis of soybean genes involved in flavonoid biosynthesis by virus-induced gene silencing. *Plant Biotechnol J*, 5, 778-90.

[57] Yamagishi, N., & Yoshikawa, N. (2009). Virus-induced gene silencing in soybean seeds and the emergence stage of soybean plants with *Apple latent spherical virus* vectors. *Plant Mol Biol*, 71, 15-24.

[58] Kasai, M., & Kanazawa, A. (2012). RNA silencing as a tool to uncover gene function and engineer novel traits in soybean. *Breeding Sci*, 61, 468-79.

[59] Kinney, A. J., Jung, R., & Herman, E. M. (2001). Cosuppression of the α subunits of β-conglycinin in transgenic soybean seeds induces the formation of endoplasmic reticulum-derived protein bodies. *Plant Cell*, 13, 1165-78.

[60] Schmidt, M. A., Barbazuk, W. B., Sandford, M., May, G., Song, Z., Zhou, W., et al. (2011). Silencing of soybean seed storage proteins results in a rebalanced protein composition preserving seed protein content without major collateral changes in the metabolome and transcriptome. *Plant Physiol*, 156, 330-45.

[61] Nunes, A. C., Vianna, G. R., Cuneo, F., Amaya-Farfán, J., de Capdeville, G., Rech, E. L., et al. (2006). RNAi-mediated silencing of the *myo*-inositol-1-phosphate synthase gene (*GmMIPS1*) in transgenic soybean inhibited seed development and reduced phytate content. *Planta*, 224, 125-32.

[62] Shi, J., Wang, H., Schellin, K., Li, B., Faller, M., Stoop, J. M., et al. (2007). Embryo-specific silencing of a transporter reduces phytic acid content of maize and soybean seeds. *Nat Biotechnol*, 25, 930-7.

[63] Takagi, K., Nishizawa, K., Hirose, A., Kita, A., & Ishimoto, M. (2011). Manipulation of saponin biosynthesis by RNA interference-mediated silencing of β-amyrin synthase gene expression in soybean. *Plant Cell Rep*, 30, 1835-46.

[64] Herman, E. M., Helm, R. M., Jung, R., & Kinney, A. J. (2003). Genetic modification removes an immunodominant allergen from soybean. *Plant Physiol*, 132, 36-43.

[65] Yu, O., Shi, J., Hession, A. O., Maxwell, C. A., McGonigle, B., & Odell, J. T. (2003). Metabolic engineering to increase isoflavone biosynthesis in soybean seed. *Phytochemistry*, 63, 753-63.

[66] Kinney, A. (1996). Development of genetically engineered soybean oils for food applications. *J Food Lipids*, 3, 273-92.

[67] Chen, R., Matsui, K., Ogawa, M., Oe, M., Ochiai, M., Kawashima, H., et al. (2006). Expression of Δ6, Δ5 desaturase and GLELO elongase genes from *Mortierella alpina* for production of arachidonic acid in soybean [*Glycine max* (L.) Merrill] seeds. *Plant Sci*, 170, 399-406.

[68] Flores, T., Karpova, O., Su, X., Zeng, P., Bilyeu, K., Sleper, D. A., et al. (2008). Silencing of *GmFAD3* gene by siRNA leads to low alpha-linolenic acids (18:3) of fad3-mutant phenotype in soybean [*Glycine max* (Merr.)]. *Transgenic Res*, 17, 839-50.

[69] Schmidt, M. A., & Herman, E. M. (2008). Suppression of soybean oleosin produces micro-oil bodies that aggregate into oil body/ER complexes. *Mol Plant*, 1, 910-24.

[70] Wang, G., & Xu, Y. (2008). Hypocotyl-based Agrobacterium-mediated transformation of soybean (*Glycine max*) and application for RNA interference. *Plant Cell Rep*, 27, 1177-84.

[71] Lee, J., Welti, R., Schapaugh, W. T., & Trick, H. N. (2011). Phospholipid and triacylglycerol profiles modified by PLD suppression in soybean seed. *Plant Biotechnol J*, 9, 359-72.

[72] Wagner, N., Mroczka, A., Roberts, P. D., Schreckengost, W., & Voelker, T. (2011). RNAi trigger fragment truncation attenuates soybean *FAD2-1* transcript suppression and yields intermediate oil phenotypes. *Plant Biotechnol J*, 9, 723-8.

[73] Lee, M. Y., Shin, K. H., Kim, Y. K., Suh, J. Y., Gu, Y. Y., Kim, M. R., et al. (2005). Induction of thioredoxin is required for nodule development to reduce reactive oxygen species levels in soybean roots. *Plant Physiol*, 139, 1881-9.

[74] Subramanian, S., Stacey, G., & Yu, O. (2006). Endogenous isoflavones are essential for the establishment of symbiosis between soybean and *Bradyrhizobium japonicum*. *Plant J*, 48, 261-73.

[75] Hayashi, S., Gresshoff, P. M., & Kinkema, M. (2008). Molecular analysis of lipoxygenases associated with nodule development in soybean. *Mol Plant Microbe Interact*, 21, 843-53.

[76] Dalton, D. A., Boniface, C., Turner, Z., Lindahl, A., Kim, H. J., Jelinek, L., et al. (2009). Physiological roles of glutathione S-transferases in soybean root nodules. *Plant Physiol*, 150, 521-30.

[77] Govindarajulu, M., Kim, S. Y., Libault, M., Berg, R. H., Tanaka, K., Stacey, G., et al. (2009). GS52 ecto-apyrase plays a critical role during soybean nodulation. *Plant Physiol*, 149, 994-1004.

[78] Libault, M., Zhang, X. C., Govindarajulu, M., Qiu, J., Ong, Y. T., Brechenmacher, L., et al. (2010). A member of the highly conserved *FWL* (tomato *FW2.2-like*) gene family is essential for soybean nodule organogenesis. *Plant J*, 62, 852-64.

[79] Yi, J., Derynck, M. R., Li, X., Telmer, P., Marsolais, F., & Dhaubhadel, S. (2010). A single-repeat MYB transcription factor, GmMYB176, regulates *CHS8* gene expression and affects isoflavonoid biosynthesis in soybean. *Plant J*, 62, 1019-34.

[80] Reddy, M. S., Dinkins, R. D., & Collins, G. B. (2003). Gene silencing in transgenic soybean plants transformed via particle bombardment. *Plant Cell Rep*, 21, 676-83.

[81] Li, X. P., Gan, R., Li, P. L., Ma, Y. Y., Zhang, L. W., Zhang, R., et al. (2006). Identification and functional characterization of a leucine-rich repeat receptor-like kinase gene that is involved in regulation of soybean leaf senescence. *Plant Mol Biol*, 61, 829-44.

[82] Arikit, S., Yoshihashi, T., Wanchana, S., Uyen, T. T., Huong, N. T., Wongpornchai, S., et al. (2011). Deficiency in the amino aldehyde dehydrogenase encoded by *GmAMADH2*, the homologue of rice *Os2AP*, enhances 2-acetyl-1-pyrroline biosynthesis in soybeans (*Glycine max* L.). *Plant Biotechnol J*, 9, 75-87.

[83] Prins, M., & Goldbach, R. (1996). RNA-mediated virus resistance in transgenic plants. *Arch Virol*, 141, 2259-76.

[84] Goldbach, R., Bucher, E., & Prins, M. (2003). Resistance mechanisms to plant viruses: an overview. *Virus Res*, 92, 207-12.

[85] Smith, H. A., Swaney, S. L., Parks, T. D., Wernsman, E. A., & Dougherty, W. G. (1994). Transgenic plant virus resistance mediated by untranslatable sense RNAs: expression, regulation, and fate of nonessential RNAs. *Plant Cell*, 6, 1441-53.

[86] Mueller, E., Gilbert, J., Davenport, G., Brigneti, G., & Baulcombe, D. (1995). Homology-dependent resistance: transgenic virus resistance in plants related to homology-dependent gene silencing. *Plant J*, 7, 1001-13.

[87] Sijen, T., Wellink, J., Hiriart, J. B., & Van Kammen, A. (1996). RNA-mediated virus resistance: role of repeated transgenes and delineation of targeted regions. *Plant Cell*, 8, 2277-94.

[88] Wang, X., Eggenberger, A., Nutter, F., & Hill, J. (2001). Pathogen-derived transgenic resistance to soybean mosaic virus in soybean. *Mol Breed*, 8, 119-27.

[89] Furutani, N., Hidaka, S., Kosaka, Y., Shizukawa, Y., & Kanematsu, S. (2006). Coat protein gene-mediated resistance to soybean mosaic virus in transgenic soybean. *Breeding Sci*, 56, 119-24.

[90] Furutani, N., Yamagishi, N., Hidaka, S., Shizukawa, Y., Kanematsu, S., & Kosaka, Y. (2007). Soybean mosaic virus resistance in transgenic soybean caused by posttranscriptional gene silencing. *Breeding Sci*, 57, 123-8.

[91] Tougou, M., Furutani, N., Yamagishi, N., Shizukawa, Y., Takahata, Y., & Hidaka, S. (2006). Development of resistant transgenic soybeans with inverted repeat-coat protein genes of soybean dwarf virus. *Plant Cell Rep*, 25, 1213-8.

[92] Tougou, M., Yamagishi, N., Furutani, N., Shizukawa, Y., Takahata, Y., & Hidaka, S. (2007). Soybean dwarf virus-resistant transgenic soybeans with the sense coat protein gene. *Plant Cell Rep*, 26, 1967-75.

[93] Steeves, R., Todd, T., Essig, J., & Trick, H. (2006). Transgenic soybeans expressing siRNAs specific to a major sperm protein gene suppress *Heterodera glycines* reproduction. *Funct Plant Biol*, 33, 991-9.

[94] Li, J., Todd, T. C., Oakley, T. R., Lee, J., & Trick, H. N. (2010). Host-derived suppression of nematode reproductive and fitness genes decreases fecundity of *Heterodera glycines* Ichinohe. *Planta*, 232, 775-85.

[95] Ibrahim, H. M., Alkharouf, N. W., Meyer, S. L., Aly, M. A., Gamal El-Din, AlK., Hussein, E. H., et al. (2011). Post-transcriptional gene silencing of root-knot nematode in transformed soybean roots. *Exp Parasitol*, 127, 90-9.

[96] Subramanian, S., Graham, M. Y., Yu, O., & Graham, T. L. (2005). RNA interference of soybean isoflavone synthase genes leads to silencing in tissues distal to the transformation site and to enhanced susceptibility to *Phytophthora sojae*. *Plant Physiol*, 137, 1345-53.

[97] Graham, T. L., Graham, M. Y., Subramanian, S., & Yu, O. (2007). RNAi silencing of genes for elicitation or biosynthesis of 5-deoxyisoflavonoids suppresses race-specific resistance and hypersensitive cell death in *Phytophthora sojae* infected tissues. *Plant Physiol*, 144, 728-40.

[98] Lozovaya, V. V., Lygin, A. V., Zernova, O. V., Ulanov, A. V., Li, S., Hartman, G. L., et al. (2007). Modification of phenolic metabolism in soybean hairy roots through down regulation of chalcone synthase or isoflavone synthase. *Planta*, 225, 665-79.

[99] Melito, S., Heuberger, A. L., Cook, D., Diers, B. W., MacGuidwin, A. E., & Bent, A. F. (2010). A nematode demographics assay in transgenic roots reveals no significant impacts of the *Rhg1* locus LRR-Kinase on soybean cyst nematode resistance. *BMC Plant Biol*, 10, 104.

[100] Metzlaff, M. (2002). RNA-mediated RNA degradation in transgene- and virus-induced gene silencing. *Biol Chem*, 383, 1483-9.

[101] Burch-Smith, T. M., Anderson, J. C., Martin, G. B., & Dinesh-Kumar, S. P. (2004). Applications and advantages of virus-induced gene silencing for gene function studies in plants. *Plant J*, 39, 734-46.

[102] Liu, B., Watanabe, S., Uchiyama, T., Kong, F., Kanazawa, A., Xia, Z., et al. (2010). The soybean stem growth habit gene *Dt1* is an ortholog of Arabidopsis *TERMINAL FLOWER1*. *Plant Physiol*, 153, 198-210.

[103] Kachroo, A., Fu, D. Q., Havens, W., Navarre, D., Kachroo, P., & Ghabrial, S. A. (2008). An oleic acid-mediated pathway induces constitutive defense signaling and enhanced resistance to multiple pathogens in soybean. *Mol Plant Microbe Interact*, 21, 564-75.

[104] Fu, D. Q., Ghabrial, S., & Kachroo, A. (2009). *GmRAR1* and *GmSGT1* are required for basal, R gene-mediated and systemic acquired resistance in soybean. *Mol Plant Microbe Interact*, 22, 86-95.

[105] Meyer, J. D., Silva, D. C., Yang, C., Pedley, K. F., Zhang, C., van de Mortel, M., et al. (2009). Identification and analyses of candidate genes for *Rpp4*-mediated resistance to Asian soybean rust in soybean. *Plant Physiol*, 150, 295-307.

[106] Pandey, A. K., Yang, C., Zhang, C., Graham, M. A., Horstman, H. D., Lee, Y., et al. (2011). Functional analysis of the Asian soybean rust resistance pathway mediated by *Rpp2*. *Mol Plant Microbe Interact*, 24, 194-206.

[107] Singh, A. K., Fu, D. Q., El-Habbak, M., Navarre, D., Ghabrial, S., & Kachroo, A. (2011). Silencing genes encoding omega-3 fatty acid desaturase alters seed size and accumulation of *Bean pod mottle virus* in soybean. *Mol Plant Microbe Interact*, 24, 506-15.

[108] Senda, M., Masuta, C., Ohnishi, S., Goto, K., Kasai, A., Sano, T., et al. (2004). Patterning of virus-infected *Glycine max* seed coat is associated with suppression of endogenous silencing of chalcone synthase genes. *Plant Cell*, 16, 807-18.

[109] Senda, M., Kurauchi, T., Kasai, A., & Ohnishi, S. (2012). Suppressive mechanism of seed coat pigmentation in yellow soybean. *Breeding Sci*, 61, 523-30.

[110] Kasai, A., Kasai, K., Yumoto, S., & Senda, M. (2007). Structural features of GmIRCHS, candidate of the I gene inhibiting seed coat pigmentation in soybean: implications for inducing endogenous RNA silencing of chalcone synthase genes. *Plant Mol Biol*, 64, 467-79.

[111] Della Vedova, C. B., Lorbiecke, R., Kirsch, H., Schulte, M. B., Scheets, K., Borchert, L. M., et al. (2005). The dominant inhibitory chalcone synthase allele *C2-Idf* (*Inhibitor diffuse*) from *Zea mays* (L.) acts via an endogenous RNA silencing mechanism. *Genetics*, 170, 1989-2002.

[112] Koseki, M., Goto, K., Masuta, C., & Kanazawa, A. (2005). The star-type color pattern in *Petunia hybrida* 'Red Star' flowers is induced by sequence-specific degradation of chalcone synthase RNA. *Plant Cell Physiol*, 46, 1879-83.

[113] Jorgensen, R. A. (1995). Cosuppression, flower color patterns, and metastable gene expression States. *Science*, 268, 686-91.

[114] Kusaba, M., Miyahara, K., Iida, S., Fukuoka, H., Takano, T., Sassa, H., et al. (2003). *Low glutelin content1*: a dominant mutation that suppresses the *glutelin* multigene family via RNA silencing in rice. *Plant Cell*, 15, 1455-67.

[115] Kasai, M., Koseki, M., Goto, K., Masuta, C., Ishii, S., Hellens, R. P., et al. (2012). Coincident sequence-specific RNA degradation of linked transgenes in the plant genome. *Plant Mol Biol*, 78, 259-73.

[116] Silhavy, D., & Burgyán, J. (2004). Effects and side-effects of viral RNA silencing suppressors on short RNAs. *Trends Plant Sci*, 9, 76-83.

[117] Schmutz, J., Cannon, S. B., Schlueter, J., Ma, J., Mitros, T., Nelson, W., et al. (2010). Genome sequence of the palaeopolyploid soybean. *Nature*, 463, 178-83.

[118] Yoshino, M., Kanazawa, A., Tsutsumi, K., Nakamura, I., Takahashi, K., & Shimamoto, Y. (2002). Structural variation around the gene encoding the α subunit of soybean β-conglycinin and correlation with the expression of the α subunit. *Breeding Sci*, 52, 285-92.

[119] Matsumura, H., Watanabe, S., Harada, K., Senda, M., Akada, S., Kawasaki, S., et al. (2005). Molecular linkage mapping and phylogeny of the chalcone synthase multigene family in soybean. *Theor Appl Genet*, 110, 1203-9.

[120] Schlueter, J. A., Scheffler, B. E., Jackson, S., & Shoemaker, R. C. (2008). Fractionation of synteny in a genomic region containing tandemly duplicated genes across *Glycine max*, *Medicago truncatula*, and *Arabidopsis thaliana*. *J Hered*, 99, 390-5.

[121] Kong, F., Liu, B., Xia, Z., Sato, S., Kim, B. M., Watanabe, S., et al. (2010). Two coordinately regulated homologs of *FLOWERING LOCUS T* are involved in the control of photoperiodic flowering in soybean. *Plant Physiol*, 154, 1220-31.

[122] Liu, B., Kanazawa, A., Matsumura, H., Takahashi, R., Harada, K., & Abe, J. (2008). Genetic redundancy in soybean photoresponses associated with duplication of the phytochrome A gene. *Genetics*, 180, 995-1007.

[123] Kanazawa, A., Liu, B., Kong, F., Arase, S., & Abe, J. (2009). Adaptive evolution involving gene duplication and insertion of a novel *Ty1/copia*-like retrotransposon in soybean. *J Mol Evol*, 69, 164-75.

[124] Voinnet, O. (2008). Use, tolerance and avoidance of amplified RNA silencing by plants. *Trends Plant Sci*, 13, 317-28.

[125] Miki, D., Itoh, R., & Shimamoto, K. (2005). RNA silencing of single and multiple members in a gene family of rice. *Plant Physiol*, 138, 1903-13.

[126] Ossowski, S., Schwab, R., & Weigel, D. (2008). Gene silencing in plants using artificial microRNAs and other small RNAs. *Plant J*, 53, 674-90.

[127] Zhang, B., Pan, X., & Stellwag, E. J. (2008). Identification of soybean microRNAs and their targets. *Planta*, 229, 161-82.

[128] Chen, R., Hu, Z., & Zhang, H. (2009). Identification of microRNAs in wild soybean (*Glycine soja*). *J Integr Plant Biol*, 51, 1071-9.

[129] Song, Q. X., Liu, Y. F., Hu, X. Y., Zhang, W. K., Ma, B., Chen, S. Y., et al. (2011). Identification of miRNAs and their target genes in developing soybean seeds by deep sequencing. *BMC Plant Biol*, 11, 5.

[130] Turner, M., Yu, O., & Subramanian, S. (2012). Genome organization and characteristics of soybean microRNAs. *BMC Genomics*, 13, 169.

[131] Kanazawa, A., Inaba, J., Shimura, H., Otagaki, S., Tsukahara, S., Matsuzawa, A., et al. (2011). Virus-mediated efficient induction of epigenetic modifications of endogenous genes with phenotypic changes in plants. *Plant J*, 65, 156-68.

[132] Kanazawa, A., Inaba, J., Kasai, M., Shimura, H., & Masuta, C. R. N. (2011). RNA-mediated epigenetic modifications of an endogenous gene targeted by a viral vector: A potent gene silencing system to produce a plant that does not carry a transgene but has altered traits. *Plant Signal Behav*, 6, 1090-3.

[133] Arase, S., Kasai, M., & Kanazawa, A. (2012). *In planta* assays involving epigenetically silenced genes reveal inhibition of cytosine methylation by genistein. *Plant Methods*, 8, 10.

[134] Thomas, C. L., Jones, L., Baulcombe, D. C., & Maule, A. J. (2001). Size constraints for targeting post-transcriptional gene silencing and for RNA-directed methylation in *Nicotiana benthamiana* using a potato virus X vector. *Plant J*, 25, 417-25.

[135] Otagaki, S., Kawai, M., Masuta, C., & Kanazawa, A. (2011). Size and positional effects of promoter RNA segments on virus-induced RNA-directed DNA methylation and transcriptional gene silencing. *Epigenetics*, 6, 681-91.

[136] Kasai, A., Watarai, M., Yumoto, S., Akada, S., Ishikawa, R., Harada, T., et al. (2004). Influence of PTGS on chalcone synthase gene family in yellow soybean seed coat. *Breeding Sci*, 54, 355-60.

[137] Tuteja, J. H., Clough, S. J., Chan, W. C., & Vodkin, L. O. (2004). Tissue-specific gene silencing mediated by a naturally occurring chalcone synthase gene cluster in *Glycine max*. *Plant Cell*, 16, 819-35.

[138] Curtin, S. J., Zhang, F., Sander, J. D., Haun, W. J., Starker, C., Baltes, N. J., et al. (2011). Targeted mutagenesis of duplicated genes in soybean with zinc-finger nucleases. *Plant Physiol*, 156, 466-73.

[139] Arase, S., Hase, Y., Abe, J., Kasai, M., Yamada, T., Kitamura, K., et al. (2011). Optimization of ion-beam irradiation for mutagenesis in soybean: effects on plant growth and production of visibly altered mutants. *Plant Biotechnol*, 28, 323-9.

[140] Bolon, Y. T., Haun, W. J., Xu, W. W., Grant, D., Stacey, M. G., Nelson, R. T., et al. (2011). Phenotypic and genomic analyses of a fast neutron mutant population resource in soybean. *Plant Physiol*, 156, 240-53.

[141] Mathieu, M., Winters, E. K., Kong, F., Wan, J., Wang, S., Eckert, H., et al. (2009). Establishment of a soybean (*Glycine max* Merr. L) transposon-based mutagenesis repository. *Planta*, 229, 279-89.

[142] Hancock, C. N., Zhang, F., Floyd, K., Richardson, A. O., Lafayette, P., Tucker, D., et al. (2011). The rice miniature inverted repeat transposable element *mPing* is an effective insertional mutagen in soybean. *Plant Physiol*, 157, 552-62.

[143] Nagamatsu, A., Masuta, C., Matsuura, H., Kitamura, K., Abe, J., & Kanazawa, A. (2009). Down-regulation of flavonoid 3'-hydroxylase gene expression by virus-induced gene silencing in soybean reveals the presence of a threshold mRNA level associated with pigmentation in pubescence. *J Plant Physiol*, 166, 32-9.

Chapter 25

Proteomics and Its Use in Obtaining Superior Soybean Genotypes

Cristiane Fortes Gris and Alexana Baldoni

Additional information is available at the end of the chapter

http://dx.doi.org/10.5772/51353

1. Introduction

Soybean (*Glycine max* L. Merrill) is one of the most important and most cultivated crops in the world, with significant quantities of proteins being found in their yield composition, around 40% of their yield dry matter. This expressive quantity of proteins, and also a considerable percentage of oil, around 21% of their dry matter, has turned this grain into a product of great importance for the industrial sector, whether it be for food, cosmetics or, more recently, biofuels. Thus, soybean breeding programs directed toward these areas become ever more important, together with agronomic characteristics that allow greater productivity in sustainability with the environment in which they are produced.

The achievement of soybean genome sequencing [1], facilitated by identification of the genetic base, lead to advances in obtaining improved cultivars through knowledge of the complete sequence of expressed genes. Nevertheless, this information is not sufficient to identify which proteins are really being expressed in the cell at a given moment and under a certain condition since, through the phenomenon of splicing, different proteins may be produced by alteration of the command of a single gene. Thus, the complementary DNA (cDNA) and the messenger RNA (mRNA) have come to be the main focus of study for obtaining information regarding genetic expression or transcriptome. Nevertheless, due to post-translational regulation mechanisms, the quantity of expressed protein is not necessarily proportional to the quantity of its corresponding mRNA, which often raises questions regarding the role of this gene in cellular metabolism.

The reason for this is that control of gene expression occurs from mRNA transcription up to post-translational modifications like glycosylation and phosphorylation, among other processes, which alter protein activity (Figure 1).

In recent years, for the purpose of complementing the information obtained by means of genome sequencing and transcriptome, proteomics, one of the dimensions of the post-genome era [2], arises with a set of highly powerful techniques for separation and identification of proteins in biological samples, allowing better understanding of the networks of cellular operation and regulation upon representing the link between the genotype and the phenotype of an organism.

For the aforementioned reasons, proteomic analysis is now one of the most efficient means for functional study of the genes and genomes of complex organisms [3]. This has generated new data, as well as validated, complemented and even corrected information obtained through other approaches, thus contributing to better understanding of plant biology.

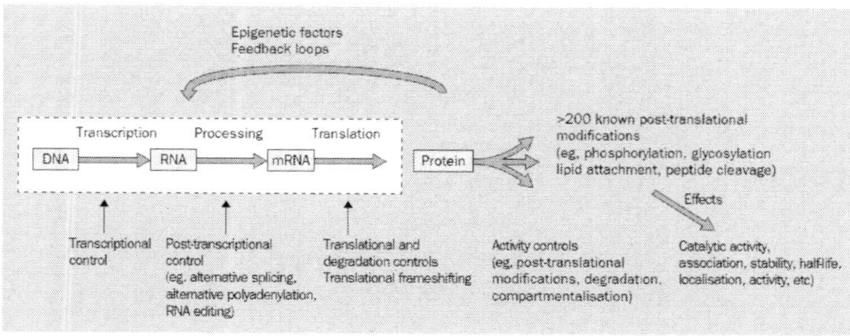

Figure 1. Pathways in which gene and protein expression may be regulated or modified in transcription or in post-translation [13].

Its study involves the entire set of proteins expressed by the genome of a cell, or only those that are expressed differentially under specific conditions. Also it is directed to the set of protein isoforms and post-translational modifications, to the interactions among them, as well as to the structural description of molecules and their complexes.

Bidimensional electrophoresis and mass spectrometry are the core technologies of proteomics, although new methodologies are being applied to plants for specific studies [4,5,6]. Among the most recent proteomic techniques are Difference Gel Electrophoresis (DIGE) and Multi-dimensional Protein Identification Tecnology (MudPIT), used in separation of proteins from a complex mixture. Other methods involved are Stable Isotopic Labeling using Amino Acids in Cell Culture (SILAC), Isotope Coded Affinity Tag (ICAT) and Isobaric Tag for Relative and Absolute Quantitation (iTRAQ) are based on labeling with isotopes for quantification of molecules by mass spectrometry.

In spite of the recent nature of research in this area, diverse studies with soybeans using proteomic tools are being performed throughout the world, showing this to be a promising area for selection of genotypes for genetic breeding programs [7,8]. Moreover, the study of plant responses to infections from pathogens has supplied significant data for understanding the signaling process that triggers the defense response in plants [9]. Additionally, there are studies characterizing the proteome of plants in response to different stress conditions arising from both abiotic factors [10] and biotic factors [11]. These comparative studies of contrasting genotypes for a determined type of stress allow identification of the proteins that respond to stress by means of changes in their levels of expression. Identifying these molecules and their respective functions, the work of breeding is directed and should have continuity only with those molecules that perform roles related to the characteristic of stress tolerance. For that reason, it is essential to cross the proteomic data with information also obtained by genomics, transcriptomics and metabolomics, so as to verify the correlation of the candidate proteins with the desired characteristic.

In relation to products derived from genetically modified foods, proteomic techniques have been applied to allow a broad approach and the analysis of many variables simultaneously in a single sample. There are also other studies relating the proteome expressed during development of the plants, as well as research in which soybeans have been the target of investigations regarding nutritional, toxicological and allergenic aspects, above all on genetically modified varieties [12].This makes for increased use of this technique in biosecurity studies. In this context, the objective of this chapter is to present the main technologies used in proteomic studies in diverse areas of activity, as well as the main scientific results obtained in the search for superior soybean genotypes.

2. Technologies used in proteomic studies.

Execution of a proteomic study involves the integration of many technologies which permeate the fields of molecular biology, biochemistry, physiology, statistics and bioinformatics, among other areas. The key steps in this type of study are separation of complex mixtures of proteins and their identification.

Separation is performed through the use of electrophoresis a term created by Michaelis in 1909. The first electrophoresis of proteins (Figure 2) was performed in 1937. Alfenas (1998) [14] explains that electrophoresis aims at separation of molecules in terms of their electrical charges, their molecular weights and their conformations, in porous supports and appropriate buffers, under the influence of a continuous electrical field. Molecules with a preponderance of negative charges migrate in the electrical field to the positive pole (anode), and molecules with excess of positive charges migrate to the negative pole (cathode). The preponderant charge of a proteic molecule is in accordance with its amino acids.

Many of the technologies currently used in proteomics were developed much before the beginning of proteomics, as is the case of electrophoresis. Nevertheless, it was the advance in protein sequencing technology by means of mass spectrometry that allowed its emergence and development [15].

The study of proteomics may be performed by means of techniques like two-dimensional electrophoresis in polyacrylamide gel (2D PAGE) followed by mass spectrometry (MS) (Figure 3), or furthermore, more recently, by the association of ionization and chromatographic methods, among others, which increase detection sensitivity even more. Nevertheless, the point of departure has still been the exposure of a large number of proteins from a cell line or organism in two-dimensional polyacrylamide gels [16,17,18].

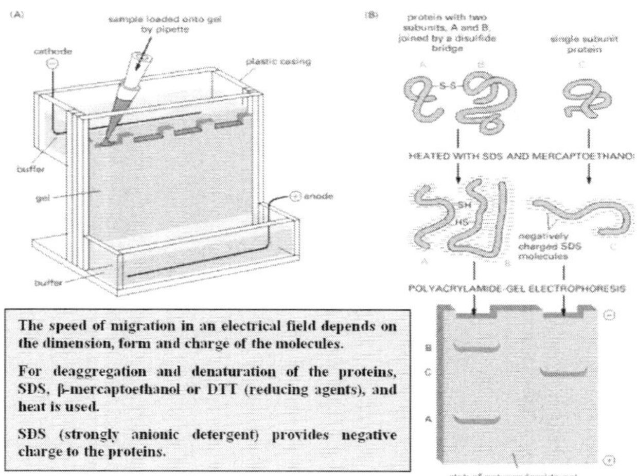

Figure 2. Polyacrylamide-gel electrophoresis (SDS-PAGE) used in proteome analysis [19].

2.1. Two-dimensional polyacrylamide gel electrophoresis (2D PAGE).

Two-dimensional polyacrylamide gel electrophoresis constitutes an analytical method capable of separating hundreds of proteins in a single analytical run. In this case, the gel, with the sample already applied, is submitted to an electrical field for two-dimensional separation. In the first dimension, separation occurs through isoelectric focalization, in which physical separation of the proteins occurs in terms of their respective isoelectric points on a strip of polyacrylamide with continuous gradation and known pH (IPG - immobilized pH gradient) submitted to increasing voltage. In the second dimension, the proteins under focus are submitted to polyacrylamide gel electrophoresis in the presence of SDS (SDS-PAGE) for separation according to their specific molecular masses (Figure 4). Thus, this is a technique that separates the proteins through different charges and masses.

The result of two-dimensional electrophoresis is a profile of spot distribution formed by single proteins or simple mixtures of proteins [21]. Each spot visualized in the gel may be considered as an orthogonal coordinate of a protein that migrated specifically in accordance with its isoelectric point (x axis) and its molecular mass (y axis), as shown in Figure 4.

The next step consists of staining the gel with silver, Coomassie blue, fluorescence, radioactive labeling or specific markers for phosphoproteins and glycoproteins, among others. This allows visualization of the protein expression pattern and photodocumentation of the gel

(Figure 5). After that, sectioning and digestion of selected spots of the gel are carried out and, finally, proteins of interest are identified by mass spectrometry integrated with a bioinformatics tool.

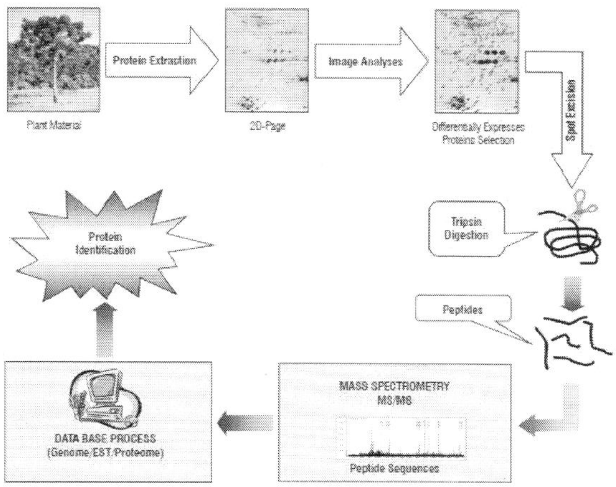

Figure 3. Stages of plant proteomics, using interface two-dimensional electrophoresis (2D-PAGE) and mass spectrometry [20].

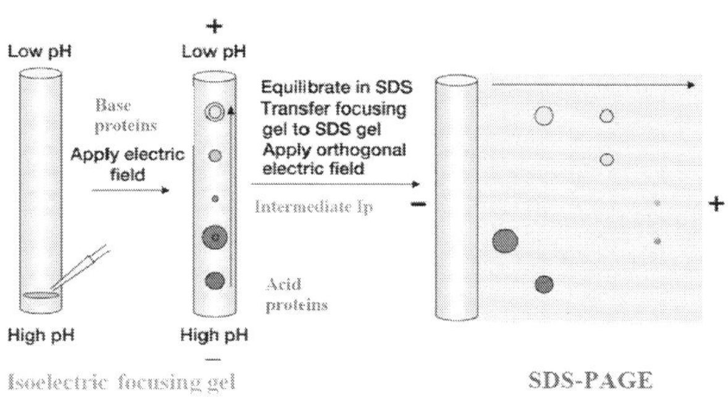

Figure 4. Two-dimensional electrophoresis 2D-PAGE used in analysis of proteomes [19].

Two-dimensional electrophoresis gels reflect the protein expression pattern of the biological sample analyzed and allow detection of variation of even a single amino acid between two isoforms or covalent modifications in the same protein thanks to change in the position of the spot.

It is important to highlight that each sample, depending on its nature, requires a specific type of processing for extraction and focalization. Therefore, it is expected that the user checks beforehand in related publications as to the protocols and methodologies that best suit the experimental needs.

Some limitations are associated with two-dimensional electrophoresis, such as low reproducibility and little power of automation. Nevertheless, reproducibility may be increased by defining optimal conditions for the electrophoresis, while automation of the process is only possible in relation to analysis of gels. Gel analysis software determines the spots and identifies those expressed differentially and their volumes, inferring a relative quantification of expression of that protein in comparison to the same spot of another gel [22]. Thus, by a process of subtraction, the differences among the different samples are revealed, as, for example, the presence, absence or intensity of the proteins. Thus, the proteins of interest may then be identified based on knowledge of the isoelectric point and of apparent molecular weight, determined by the two-dimensional gels [23].

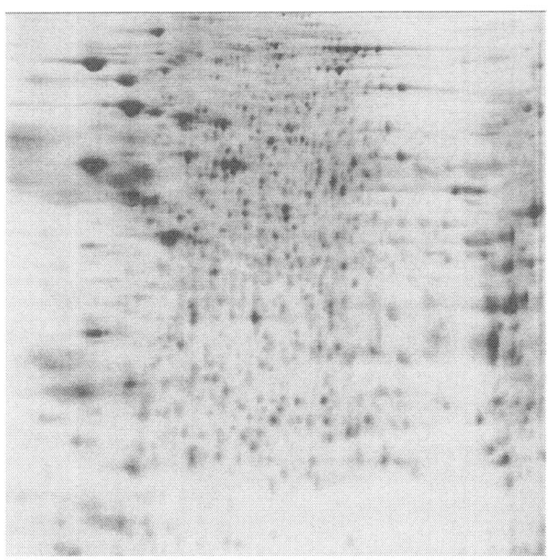

Figure 5. Proteins extracted and separated by two-dimensional (2D) gel electrophoresis and stained with Coomassie blue [24].

2.2. Differential in gel electrophoresis (DIGE).

An efficient procedure in the attempt to eliminate variation from gel to gel is use of the technique of differential in gel electrophoresis or DIGE (Figure 6), which allows analysis of up to three proteomes in a single gel. These results in one internal pattern common to all the gels and two different samples labeled with distinct fluorophores (CyDye) [25]. That way, only the proteins labeled with their own fluorophore are visualized. In addition, this technique uses labeling of proteins with a broad dynamic range of detection and has sensitivity greater than staining of the gels by silver methods, allowing proteomic studies of a quantitative nature to be performed with greater precision, accuracy and sensitivity [26].

2.3. Liquid chromatography

Another form used for separation of proteins is by means of liquid chromatography. The sample that is, for example, a mixture of peptides generated by proteolytic digestion from a protein extract passes through a first separation, by means of liquid chromatography, where the enriched peptide fractions are collected and applied in the spectrometer. As complete automation is the main target of the methods for large scale analyses, methods of separation were developed free of gel by reverse phase liquid chromatography connected with tandem mass spectrometry (LC/MS/MS). In Figure 7 the operational and equipment sequence involved in a typical analysis via LC/MS/MS is shown.

Greater automation is possible with multidimensional liquid chromatography, which uses different characteristics of the proteins in columns of distinct properties or in a single two-phase column [29]. The fraction eluted in the first column is directly introduced in the second column, which may be directly connected to the mass spectrometer. This technique, called MudPIT, is inserted in the context of the shotgun proteomic, in which greater resolution of the proteomes is possible, facilitating identification of the less abundant proteins frequently lost when gels are used [30].

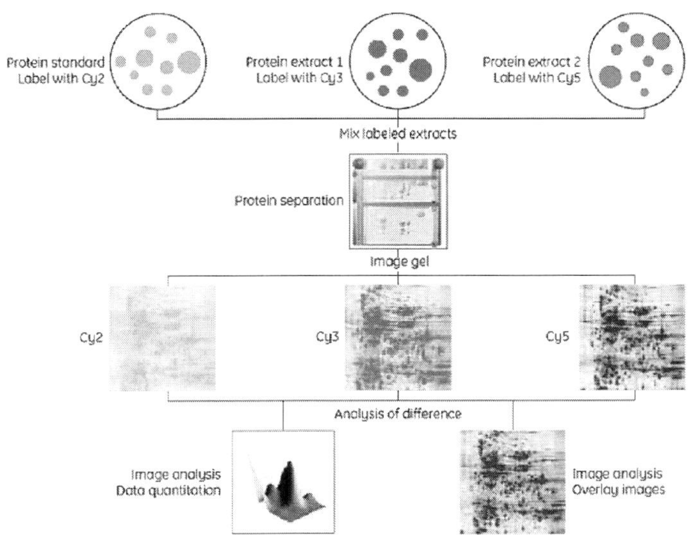

Figure 6. Differential in gel electrophoresis technique or DIGE [27].

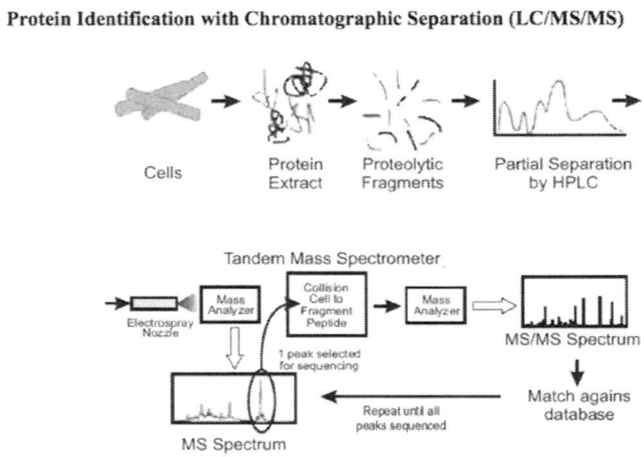

Figure 7. Protein identification with chromatographic separation (LC/MS/MS) [28].

2.4. Protein identification methods.

After separation of proteins, the next stage consists of their characterization and identification using mass spectrometry, which is a technique where the ratio between the mass and the charge (m/z) of ionized molecules in the gas phase is measured. In general, a mass spectrometer consists of an ionization source, a mass analyzer, a detector and a data acquisition system.

The great variety of spectrometers found on the market is the result of different combinations of types of sources of ionization and mass analyzers, which provide certain levels of sensitivity and accuracy in the results. At the ionization source, the molecules are ionized and transferred to the gas phase. In the mass analyzer, the ions formed are separated in accordance with their m/z ratios and later detected, usually by electron multiplier [31].

With the development of ever more specialized equipment for proteins, mass spectrometry has become a revolutionary tool in modern protein chemistry. This technology has allowed identification of proteins by a methodology called peptide mass fingerprinting. Rocha et al. (2003) [3], state that this methodology is based on protein digestion to be identified by a proteolytic enzyme, for example trypsin, producing fragments called peptides. The masses of these peptides obtained form a kind of fingerprinting of the protein, which are then determined with great acuity (0.1 to 0.5 Da) by mass spectrometry.

Special software allows comparing the peptide mass fingerprinting of the protein one wishes to identify with those theoretically generated for all the protein sequences present in the databases. If the protein sequence problem is in the database, it will immediately be identified [32].

2.5. Relative protein quantification

Large scale protein quantification methods make an estimate of relative expression possible by means of labeling with radioactive isotopes, fluorescents and light/heavy, allowing the same protein to be quantified in a relative way among differently labeled samples. Some of the most used radioactive isotopes are the iCAT (Isotopic coded affinity tag), iTRAQ (isobaric tags) and H_2O^{18}.

The iCAT consists of addition of a label that has affinity for cysteine residues and which has a bonded molecule of eight atoms of hydrogen or eight atoms of deuterium. One sample is labeled with the tag containing hydrogen and the other sample with the tag containing deuterium. After digestion of the proteins, the resulting peptides are identified by mass spectrometry. Equal peptides labeled in the two samples are identified by overlap of the peaks that show distinct m/z due to the type of bonded isotope, with the ratio between the area of the two peaks being a relative measure of the expression of that protein. According to Yi & Goodlett (2003) [33], the main problems associated with this technique are the need for the presence of cysteine residues, the high cost of the reagents and the greater time necessary for sequencing.

In the iTRAQ technique, labeling of proteins with tags and identification by mass spectrometry is also used. The tags bond to all the free amino groups at the N terminal of all the peptides and on the internal side chains with lysine residues and vary according to the reporter group they carry, and they may have 114, 115, 116 or 117Da, thus allowing for the quantification of proteins in up to four types of samples at the same time. The relative quantification is carried out in the same way as in the iCAT, but high cost has restricted its use [34].

The aforementioned techniques require the consumption of specific and expensive reagents. Nevertheless, the same goal may be achieved with a simpler labeling method in which the proteins are labeled with one or two atoms of O_2. These are incorporated in the carboxyl terminal by simply supplying a solution with H_2O for one sample and a solution with H_2O^{18} for the other sample. Thus, the relative abundance of the peptides that will differ by 2Da is estimated [35].

Another quantification technique is Stable isotope labeling by amino acids in cell culture, (SILAC) which, together with mass spectrometry and bioinformatics resources, has proven to be quite adequate in proteomic studies. It is a technique that detects differences in the abundance of proteins among cell cultures by means of isotopic labeling of proteins. Labeling with stable isotopes is obtained by supplying isotopically enriched amino acids to a cell culture and natural amino acids to the culture to be compared (Figure 8).

2.6. Analysis of post-translational modifications (PTM's).

Another area of great interest in plant proteomics is in regard to characterization of post-translational modifications or PTM's, essential for proteins to play their roles in the varied cell events, producing different proteins from the same gene.

These modifications occur at specific sites in the proteins [37] changing their physical, chemical and biological properties [38]. They may occur by means of cleavages or by the addition of a chemical group to one or more amino acids [39]. The main goals of PTM studies in proteomics are identifying the proteins that have them, mapping the sites where these modifications occur, quantifying their occurrence at the different sites and characterizing cooperative PTM's [40].

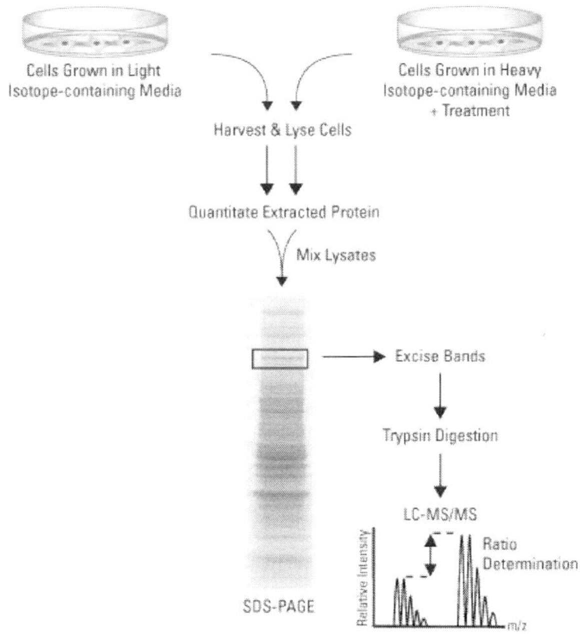

Figure 8. General outline of the SILAC technique [36].

The fact that covalent modifications result in changes in the protein molecular masses makes it possible for these modifications and the amino acids that carry them to be identified by mass spectrometry, allowing more than 300 different types of PTM's to be identified until now with the aid of this technique. Nevertheless, according to Mann and Jensen (2003) [41], mass spectrometry has reduced power of resolution of PTM's because they occur at low stoichiometric levels. This problem may be resolved by adopting fractioning methods prior to sequencing that allow enrichment of the sample for the proteins that have a certain type of PTM. Large scale modified protein enrichment systems are generally carried out by means of affinity chromatography.

One example is the IMAC system – a column of immobilization through affinity to a metal for isolation of phosphorylated proteins in which metal ions of Fe(III) are joined to the matrix to promote the isolation of proteins that have phosphorylate residues since the Fe(III) ion is capable of interacting in a reversible manner with the phosphate group of the modified peptide keeping it attached to the column [41].

Contrary to that which occurs with the reversible yet permanent PTM's, like glycosylation, low stoichiometry does not occur, but the addition of carbohydrates hinders the proteolytic digestion necessary for identification by mass spectrometry [21]. In addition, when the modified peptide is fragmented for sequencing, it loses sugar residues, impeding the identification of the modified amino acids. To resolve this problem, digestion of the proteins is performed so as to remove the sugar residues and produce a modification in the modified site that makes it identifiable [42].

Electrophoresis gels may also be used in enrichment of samples for PTM's as performed for detection of phosphorylations and glycosylations with commercially available kits. The modified proteins, specifically labeled in the gel, are visualized and excised for identification by mass spectrometry. One important aspect of the use of gels for identification of PTM is the possibility of visualizing the spots differentially expressed among samples that have the PTM.

3. Research dealing with proteomics in soybeans.

3.1. Food safety

In the case of food, proteins are especially important for evaluation of food safety because they may place consumer health at risk. That is because proteins may be involved in synthesis of toxins and antinutrients, as well as being a toxin, an antinutrient or even an allergenic [43].

Soybeans are an important source of food throughout the world, being consumed in daily meals of all types. It has also been widely used as a food substitute by people that have intolerance to lactose or other milk proteins [44]. Nevertheless, in this species are also found proteins considered allergenic. Thus, knowledge regarding the proteins with toxic/antinutritional potential present in this grain becomes fundamental for development of biotechnological strategies that would have the target of elimination or inactivation in the genome of these species of genes that codify for these proteins.

Therefore, application of proteomic analysis in this type of study has been widely discussed. In relation to products derived from genetically modified (GM) foods, proteomic techniques have been applied because they allow a wide-ranging approach and analysis of many variables simultaneously in the same sample [45]. Ocana et al. (2007) [46], studying GM proteins present in soybean and maize samples using proteomic analysis, identified the protein CP4 EPSPS, which confers tolerance to glyphosate herbicide. These samples were submitted to specific separation techniques followed by two-dimensional electrophoresis and mass spectrometry for detection and characterization of the proteome.

Related to allergies, various allergens belonging to the superfamily of cupins and prolamins have been identified in soybeans [47]. Research has suggested that a heterogeneous group of soybean proteins bond to the IgE antibody and are potential allergens as, for example, Gly m Bd 30k, β-conglycinin, Gly m Bd 28k, glycinin, Kunitz type protease inhibitor, some proteins present in the hull (Gly m 1.0101, Gly m 1.0102 e Gly m 2), profilin (Gly m 3), SAM 22 (Gly m 4), and other allergens like lectin and lipoxygenase [47,48]. According to Wilson et al. (2005) [49], in spite of the allergens identified in soybeans, the challenge of food researchers is developing a process for eradicating the immunodominant allergens, maintaining the functionality, nutritional value and effectiveness in the subsequent products derived from soybeans. For that reason, research has been developed using genetic engineering for silencing the soybean gene responsible for synthesis of the protein Gly m Bd 30K, one of the main soybean proteins that develop allergic reactions with serums of sensitive patients [44].

3.2. Biotic and abiotic factors

In a similar manner, various studies have shown that the proteomic approach is highly useful for investigation of crop response to environmental stresses because it compares the way the proteome is affected by different physiological conditions.

Saline stress is one of the many types of abiotic stresses that affect plants and compromise their yield. Salinity is a common agricultural problem in arid and semiarid regions and creates large unproductive areas. There has been an ever greater search for cultivars adaptable to this condition. Sobhanian et al. (2010) [10], used proteomic techniques to evaluate the metabolism of proteins in leaves, hypocotyls and roots submitted to different NaCl concentrations (Figure 9), thus leading to saline stress.

Results in soybeans suggest that, in adaptation to saline conditions, proteins perform different roles in each organ, and the proteins most affected by saline stress are those related to photosynthesis. Therefore, there is less energy production, and, consequently, reduction in plant growth. The conclusion suggests that the gene Glyceraldehyde-3-phosphate dehydrogenase may be, in the future, one of the target genes to improve tolerance to saline stress in this species.

Another type of abiotic stress studied in soybeans in which a proteomic approach is used is flooding stress [50,51]. Growing this species in areas subject to flooding makes the root environment anoxic, affecting nodulation or root growth. That way, plants respond with greater or less efficiency, allowing the distinction between cultivars which are tolerant and intolerant to this stress.

Proteomic analyses of soybean seedlings in response to flooding were undertaken by Shi et al. (2008) [52] to identify the key proteins involved in this process. To identify the first proteins produced in response to flooding, the roots of the seedlings were used for extraction of the proteins. The two-dimensional gel results suggest that cytosolic ascorbate peroxidase 2 (cAPX 2) is involved in response to flooding stress in young soybean seedlings.

In the case of drought stress, up-regulation of reactive oxygen species (ROS) scavengers such as superoxide dismutase (SOD) was reported in soybean seedlings [53]. The proteome analysis of two-day-old soybean seedlings subjected to drought stress by withholding of water for two days revealed a variety of responsive proteins involved in metabolism, disease/defense and energy including protease inhibitors [53]. The major reason for loss of crop yields under drought stress is a decrease in carbon gain through photosynthesis. Proteome analysis of soybean root under drought condition showed that two key enzymes involved in carbohydrate metabolism, UDP- glucose pyrophosphorylase and 2,3-bisphosphoglycerate independent phosphoglycerate mutase, were down-regulated upon exposure to drought [54]. The identification of proteins such as UDP-glucose pyrophosphorylase and 2,3- bisphosphoglycerate has provided new insights that may lead to a better understanding of the molecular basis of responses to drought stress in soybean

Figure 9. Soybean seedlings submitted to different concentrations of NaCl [10].

Stress by toxicity caused by the presence of high quantities of aluminum in the soil has also been investigated in soybeans from the perspective of proteomics [55,56]. Duressa et al. (2011) [56], studying cultivars tolerant and susceptible to high doses of aluminum, made proteomic analyses of roots, arriving at the conclusion that the greatest expression of enzymes involved with citrate synthesis would be a good strategy in the search for cultivars tolerant to this mineral (Figure 10).

Another focus of the study within the context of selection of superior soybean genotypes using the proteomic approach is exposure to ultraviolet radiation, which has gained importance with the prominent worldwide concern for global warming and the consequent degradation of the ozone layer. Xu et al. (2007) [57], studied the proteome of soybean leaves to investigate the protective role of flavonoids against the incidence of UV-B radiation. The authors suggest that high levels of flavonoid reduce the sensitivity of the plant to this radiation.

In relation to biotic stresses caused by pathogens like fungi, bacteria, nematodes and viruses, proteomic tools are also greatly used because they allow understanding of the plant-pathogen relationship [11,58,59,60] and also how the nodulation process occurs by means of symbiosis between the soybean roots and rhizobia [61]. In these cases, proteomic analysis provides the information that will be used by genetic breeding in the search for cultivars resistant to various diseases.

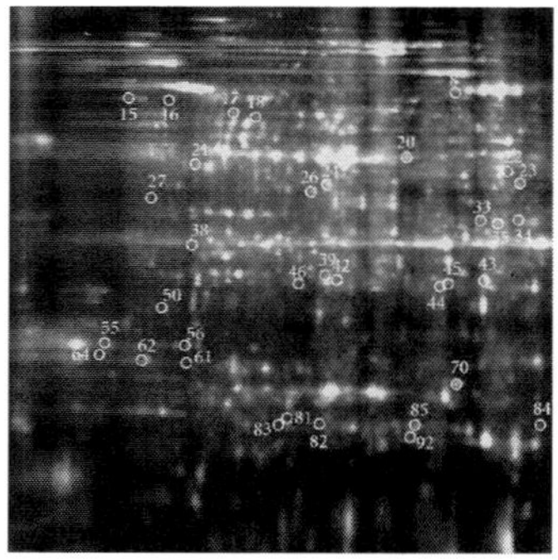

Figure 10. profile of aluminum regulated-proteins in PI 416937 72 h posttreatment [56].

Zhang et al. (2011) [58] evaluated the responses of cultivars tolerant and susceptible to the fungus *Phytophthora sojae* by means of two-dimensional electrophoresis. The authors observed 46 proteins being expressed (Figure 11), among which only 11% were related to plant defense.

In addition, proteomic studies that deal with seed development also play an essential role [62]. The data obtained may help to interpret the function of genes that determine protein concentration, considered as a key characteristic for genetic breeding of soybeans. Moreover, differential proteomic analyses designed to describe the changes that occur from maturation to senescence in organs and organelles have been reported. There is also already a soybean proteome database, providing information on the proteins involved in the soybean response to stress caused by drought, salinity and, principally, flooding [63].

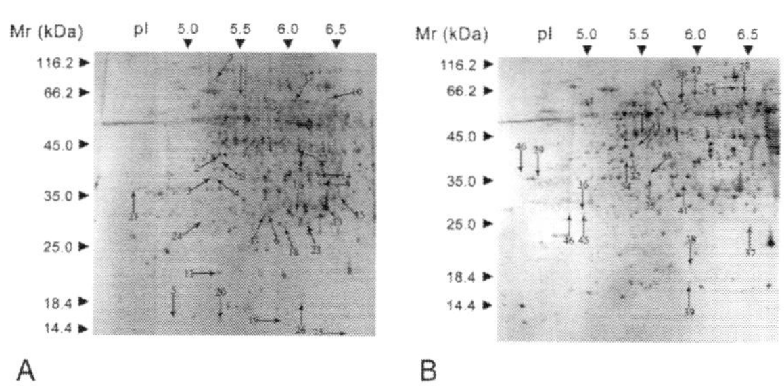

Figure 11. Identification of 26 and 20 protein spots from Yudou25 (A) and NG6255 (B), respectively. The numbers with arrows indicate the differentially expressed protein spots. Ip and Mr are shown on the gels [58].

4. Final considerations

In light of the above, proteomics in soybean studies contributes to diverse biotechnological applications, with its approach proving to be fundamental. Its use in the search for superior soybean materials has the purpose of comparing and contrasting genotypes for a determined type of stress and identifying the proteins that respond to the stress by means of changes in their levels of expression. The identification of these molecules and their respective functions will allow direction of breeding work, which should continue only with those that perform roles related to the characteristic of stress tolerance.

For that reason, it is essential to cross proteomic data with information also gathered from genomics, transcriptomics and metabolomics so as to check the correlation of the candidate proteins with the desired characteristic. The following stage aims to evaluate these proteins (genes) in regard to their segregation for the characteristic of interest or quantitative trait locus (QTL), that is, determine how much each one of them contributes to the characteristic of tolerance. Finally, the selected genes may be integrated in marker assisted selection (MAS) or in genetic transformation programs.

Author details

Cristiane Fortes Gris[1] and Alexana Baldoni[2]

1 Federal Institute of Southern Mines

2 Federal University of Lavras, Brazil

References

[1] Schmutz, J., Cannon, S. B., Schlueter, J., , J., Mitros, T., Nelson, W., Hyten, D. L., Song, Q., Thelen, J. J., Cheng, J., et al. (2010). Genome sequence of the paleopolyploid soybean. *Nature*, 463, 178-183, http://www.nature.com/nature/journal/463n7278/full/nature08670.html, (accessing 20 february 2012).

[2] Pandey, A., & Mann, M. (2000). Proteomics to study genes and genome. Nature .http://65.199.186.23/nature/journal/ n6788/abs/405837a0.html (accessing 20 february 2012) , 405(6788), 837-846.

[3] Rocha, T. L., Costa, P. H. A., Magalhaes, J. C. C., Evaristo, R. G. S., Vasconcelos, E. A. R., Coutinho, M. V., Paes, N. S., Silva, M. C. M., & Rossi-de-Sa, M. F. (2005). Eletroforese bidimensional e análise de proteomas. *Comunicado Técnico. Embrapa Recursos Genéticos e Biotecnologia*, 136, 1-12.

[4] Mann, M. (2006). Functional and quantitative proteomics using SILAC. *Nature Review Molecular Cell Biology*, 7(12), 952-958, http://patf.lf1.cuni.cz/proteomika/silac_2006.pdf, (accessed 07 march 2012).

[5] Mcdonald, T., Sheng, S., Stanley, B., Cheng, D., Ko, Y., Cole, R. N., Pedersen, P., & Van Eyk, J. E. (2006). Expanding the subproteome of the inner mitochondria using protein preparation technologies. One- and two-dimensional liquid chromatography and two-dimensional gel electrophoresis. *Molecular & Cellular Proteomics*, 5(12), 2392-2411, http://www.mcponline.org/content/5/12/2392.full, (accessed 05 march 2012).

[6] Maor, R., Jones, A., Nuhse, T. S., Studholme, D. J., Peck, S. C., & Shirasu, K. (2007). Multidimensional protein identification technology (MudPIT) analysis of ubiquitinated proteins in plants. *Molecular & Cellular Proteomics*, 6, 601-610, http://www.mcponline.org/content/6/4/601.full, (accessed 2 march 2012).

[7] Natarajan, S., Xu, C., Caperna, T. J., & Garrett, W. M. (2005). Comparison of protein solubilization methods suitable for proteomic analysis of soybean seed proteins. *Analytical Biochemistry*, 342(2), 214-220, http://www.sciencedirect.com/science/article/pii/S0003269705003593, (accessed 2 march 2012).

[8] Krishnan, H. B., & Nelson, R. L. (2011). Proteomic analysis of high protein soybean (Glycine max) accessions demonstrates the contribution of novel glycinin subunits. *Journal of Agricultural and Food Chemistry*, 59, 2432-2439, http://pubs.acs.org/doi/abs/10.1021/jf104330n, (accessed 2 march 2012).

[9] Qureshi, M. I., Qadir, S., & Zolla, L. (2007). Proteomics-based dissection of stress response pathways in plants. *Journal of plant physiology*, 01, 1-13, http://www.sciencedirect.com/science/article/pii/S0176161707000995, (accessed 3 march 2012).

[10] Sobhanian, H., Razavizadeh, R., Nanjo, Y., Ehsanpour, A. A., Jazii, F. R., Motamed, N., & Komatsu, S. (2010). Proteome analysis of soybean leaves, hypocotyls and roots under salt stress. *Proteome Science*, 8, 19-33, http://www.biomedcentral.com/1477-5956/8/19, (accessed 26 february 2012).

[11] Liu, D., Chen, L., & Duan, Y. (2011). Differential proteomic analysis of the resistant soybean infected by soybean cyst nematode. *Heterodera glycines race 3. Journal of Agricultural Science*, 3(4), 160-167, http://www.ccsenet.org/journal/index.php/jas/article/view/9658, (accessed 26 february 2012).

[12] Castro, V. A. O. T., & Finardi, F. F. (2008). Análise Proteômica de amostras de soja GM e parental. I XXI Congresso Brasileiro de Ciência e Tecnologia de Alimentos e XV Seminário Latino Americano e do Caribe de Ciência e Tecnologia de Alimentos, Belo Horizonte, Brasil XXI SBCTA, 2008.

[13] Banks, R. E., Dunn, M. J., Hochstrasser, D. F., Sanchez, J., Blackstock, W., Pappin, D. J., & Selby, P. J. (2000). Proteomics: new perspectives, new biomedical opportunities. *The Lancet*, 356, 1749-1756, http://cmbi.bjmu.cn/cmbidata/proteome/reviews/02.pdf, (accessed 26 february 2012).

[14] Alfenas, A.C. (1998). Eletroforese de isoenzimas e proteínas afins: fundamentos e aplicações em plantas e microorganismos. *Viçosa, MG: UFV.*, 574.

[15] Tyers, M., & Mann, M. (2003). From genomics to proteomics. *Nature*, 422(6928), 193-197, http://www.ncbi.nlm.nih.gov/pubmed/12634792, (accessed 2 march 2012).

[16] Anderson, N. G., & Anderson, N. L. (1996). Twenty years of Two-dimensional electrophoresis: past, present and future. *Electrophoresis*, 17, 443-453, http://siba.unipv.it/fisica/articoli/E/Electrophoresis1996_17_443.pdf, (accessed 26 february 2012).

[17] Wilkins, M. R., Sanchez, J. C., Williams, K. L., & Hochstrasser, D. F. (1996). Current challenges and future applications for protein maps and posttranslational vector maps in proteome projects. *Electrophoresis*, 17(5), 830-838, http://onlinelibrary.wiley.com/doi/10.1002/elps.1150170504/abstract , (accessed 12 mai 2012).

[18] Wilkins, M. R., Willians, K. L., Appel, R. D., & Hochstrasser, D. (1997). *Proteome research: new frontiers in functional genomics.*, Germany, Springer-Verlag, 243.

[19] Stracham, T., & Read, A. P. (2004). *Human Molecular Genetics 3.*, New York, Garland Science.

[20] Balbuena, T. S., Dias, L. L. C., Martins, M. L. B., Chiquieri, T. B., Santa-Catarina, C., Floh, E. I. S., & Silveira, V. (2011). Challenges in proteome analyses of tropical plants. *Brazilian Journal of Plant Physiology*, 23(2), http://www.scielo.br/scielo.php?pid=S167704202011000200001 &script=sci_arttext, (accessed 09 march 2012).

[21] Pennington, S. R., & Dunn, M. J. (2001). *Proteomics: from protein sequence to function*, New York, Springer-Verlag e BIOS scientific Plubishers, 1v.

[22] López, J. L., Marina, A., Vásquez, J., & Alvarez, G. (2002). A proteomic approach to the study of the marine mussels. *Mytilus edulis and M. galloprovincialis. Marine Biology*, 141, 217-223, http://www.whoi.edu/science/B/people/mhahn/Lopez.pdf, (accessed 13 april 2012).

[23] James, P. (1997). Protein identification in the post-genome era: the rapid rise of proteomics. *Quaterly reviews of biophysics*, 30, 279-331.

[24] Applied Biomics: 2D Gel Staining. (2012). http://www.appliedbiomics.com/proteomics_differentstaining.html, (accessed 10may).

[25] Unlu, M., Morgan, M. E., & Minden, J. S. (1997). Difference gel electrophoresis: A single gel method for detecting changes in protein extracts. *Electrophoresis*, 18(11), 2071-2077, http://onlinelibrary.wiley.com/doi/10.1002/elps.1150181133/abstract, (accessed 15 april 2012).

[26] Tong, R., Shaw, J., Middleton, B., Rowlinson, R., Rayner, S., Young, J., Pognan, F., Hawkins, E., Currie, I., & Davison, M. (2001). Validation and development of fluorescent two-dimensional gel electrophoresis proteomics technology. *Proteomics*, 1, 377-396, http://www.ncbi.nlm.nih.gov/pubmed/11680884, (accessed 12 april 2012).

[27] Healthcare, G.E. (2012). *Life Sciences 2D Electrophoresis Principles and Methods*, https://www.gelifesciences.com/gehcls_images/GELS/Related%20Content/Files/1335426794335/litdoc80642960_20120426103127.pdf, (accessed 10 april 2012).

[28] Iowa State University. (2012). Q-Star Tandem Mass Spectrometry. http://www.protein.iastate.edu/q-star.html, (accessed 15 april).

[29] Motoyama, A., & Yates, J. R. (2012). Multidimensional L.C. separations in shotgun proteomics. *Analytical Chemistry*, 80(19), 7187-7193, http://pubs.acs.org/doi/full/10.1021/ac8013669, (accessed 12 mai 2012).

[30] Washburn, M. P., Wolters, D., & Yates, J. R. (2008). Large-scale analysis of the yeast proteome by multidimensional protein identification technology. *Nature Biotechnology*, 19, 242-247, http://proteome.gs.washington.edu/classes/Genome490/papers/Washburn_et_al_Nat_Biotech_2001.pdf, (accessed 28 april 2012).

[31] Fenn, J. B., Mann, M., Meng, C. K., Wong, S. F., & Whitehouse, C. M. (1989). Electrospray ionization for mass spectrometry of large biomolecules. *Science*, 246, 64-71, http://www.sciencemag.org/content/246/4926/64.abstract, (accessed 25 february).

[32] De Sousa, M. V., Fontes, W., & Ricart, C. A. O. (2003). Análise de Proteomas: O despertar da era pós-genômica. *Revista on line-Biotecnologia Ciência e Desenvolvimento*, 7, 12-14, http://www.biotecnologia.com.br/revista/bio07/analise.pdf, (accessed 10 april 2012).

[33] Yi, E. C., & Goodlett, D. R. Quantitative protein profile comparisons using the isotope-coded affinity tag method. *Current Protocols in Protein Science*, 23, 2.1-23.2.11, http://onlinelibrary.wiley.com/doi/10.1002/0471140864.ps2302s34/abstract?systemMessage=Wiley+Online+Library+will+be+disrupted+on+9+June+from +10%3A00-12%3A00+BST+%2805%3A00-07%3A00+EDT%29+for+essential+maintenance, (accessed 28 april 2012).

[34] Schmidt, C., & Urlaub, H. (2009). Itraq-labeling of in-gel digested proteins for relative quantification. *Methods in Molecular Biology*, 564(4), 207-226, http://www.springerlink.com/content/t864281256610gr4/#section=78222&page=1&locus=0, (accessed 28 april 2012).

[35] Ye, X., Luke, B., Andresson, T., & Blonder, J. (2009). O stable isotope labeling in MS-based proteomics. *Briefings in Functional Genomics & Proteomics*, 8(2), 136-144, http://bfg.oxfordjournals.org/content/8/2/136.full, (accessed 15 april 2012).

[36] Thermo Scientific: Pierce Protein Biology Products. (2012). SILAC Protein Quantitation Kits. http://www.piercenet.com/browse.cfm?fldID=01040418, (accessed 10 may).

[37] Blom, N. (2004). Prediction of post-translational glycosylation and phosphorylation of proteins from the amino acid sequence. *Proteomics*, 4(6), 1633-1649, http://onlinelibrary.wiley.com/doi/10.1002/pmic.200300771/abstract , (accessed 15 april 2012).

[38] Nesaty, V. J., & Suter, M. J. F. (2008). Analysis of environmental stress response on the proteome level. *Mass Spectrometry Reviews*, 27(6), 556-574, http://onlineli-

brary.wiley.com/doi/10.1002/mas.20177/abstract;jsessionid=C9FCCA3D6997D990B13A7CAE38CB4F3E.d03t02 , (accessed 05 mai 2012).

[39] Mann, M., & Jensen, O. N. (2003). *Nature Biotechnology*, 21, 255-261, http://genomics.wsu.edu/pages/teaching/MPS570%20PMB/ArticlesPDFs/MannandJensen.pdf, (accessed 02 april 2012).

[40] Seo, J., & Lee, K. J. (2004). Post-translational modifications and their biological functions: proteomic analysis and systematic approaches. *Journal of Biochemistry and Molecular Biology*, 37(1), 35-44, http://bmbreports.org/jbmb/jbmb_files/%5B37-1%5D0401271834_035-044.pdf, (accessed 12 april 2012).

[41] Vener, A. V., Harms, A., Sussman, M. R., & Vierstra, R. D. (2001). Mass spectrometric resolution of reversible protein phosphorylation in photosynthetic membranes of Arabidopsis thaliana. *Journal of Biological Chemistry*, 276(10), 6959-6966, http://www.ufv.br/dbv/pgfvg/BVE684/htms/pdfs_revisao/genomica/protome_articles/Mass%20Spectrometric%20Resolution%20of%20Reversible%20Protein%20Phosphorylation.pdf, (accessed 10 april 2012).

[42] Cantin, G. T., & Yates, J. R. (2004). Strategies for shotgun identification of post-translational modifications by mass spectrometry. *Journal of Chromatography A*, 1053(1-2), 7-14, http://www.sciencedirect.com/science/article/pii/S0021967304009719, (accessed 25 march 2012).

[43] Ruebelt, M. C., Leimgruber, N. K., Lipp, M., Reynolds, T. L., Nemeth, M. A., Astwood, J. D., Engel, K. H., & Jany, K. D. (2006). Application of two-dimensional gel electrophoresis to interrogate alterations in the proteome of genetically modified crops. 1. Assessing analytical validation. *Journal of Agricultural Food Chem*, 54(6), 2154-2161, http://pubs.acs.org/doi/abs/10.1021/jf0523566 , (accessed 12 march 2012).

[44] Herman, E. (2003). Genetically modified soybeans and food allergies. *Journal of Experimental Botany*, 54(386), 1317-1319, http://www.ask-force.org/web/Soya/Herman-GM-Soy-Allergies-2003.pdf, (acessed 12 march 2012).

[45] EFSA. GMO (2008). Panel Working Group on Animal Feeding Trials. Safety and nutritional assessment of GM plants and derived food and feed: The role of animal feeding trials. Food and Chemical Toxicology (suppl. , 46(1), s2-s70.

[46] Ocana, M. F., Fraser, P. D., Patel, R. K., Halket, J. M., & Bramley, P. M. (2007). Mass spectrometric detection of CP4EPSPS in genetically modified soya and maize. Rapid Communication. *Mass Spectrometry*, 21(3), 319-328, http://onlinelibrary.wiley.com/doi/10.1002/rcm.2819/abstract , (accessed 13 march 2012).

[47] Boxtel, E.L.V. (2007). Protein quaternary structure and aggregation in relation to allergenicity. *Dissertation, Graduate School VLAG, chapter 1.*

[48] L'Hocine, L., & Boye, J. I. (2007). Allergenicity of soybean: New developments in identification of allergenic proteins, cross-reactivities and hypoallergenization technologies. *Critical Reviews in Food Science and Nutrition*, 47(2), 127-143, http://

www.mendeley.com/research/allergenicity-soybean-new-developments-identification-allergenic-proteins-crossreactivities-hypoallergenization-technologies/#, (accessed 13 march 2012).

[49] Wilson, S., Blaschek, K., & De Mejia, E. G. (2005). Allergenic Proteins in Soybean: Processing and Reduction of 34 Allergenicity. (accessed 10 march 2012). *Nutrition Reviews,* 63(2), 47-58, http://onlinelibrary.wiley.com/doi/10.1111/j.1753-4887.2005.tb00121.x/abstract.

[50] Komatsu, S., Kobayashi, Y., Nishizawa, K., Nanjo, Y., & Furukawa, K. (2010). Comparative proteomics analysis of differentially expressed proteins in soybean cell wall during flooding stress. *Amino Acids,* 39(5), 1435-1449, http://www.springerlink.com/content/e017718020408605/, (accessed 12 march 2012).

[51] Nanjo, Y., Maruyama, K., Yasue, H., Yamaguchi-Shinozaki, K., Shinozaki, K., & Komatsu, S. (2011). Transcriptional responses to flooding stress in roots including hypocotyl of soybean seedlings. Plant Molecular Biology http://www.springerlink.com/content/j1783j7116046564/ (accessed 11 march 2012) , 77(1-2), 129-144.

[52] Shi, F., Yamamoto, R., Shimamura, S., Hiraga, S., Nakayama, N., Nakamura, T., Yukawa, K., Hachinohe, M., Matsumoto, H., & Komatsu, S. (2008). Cytosolic ascorbate peroxidase 2 (cAPX 2) is involved in the soybean response to flooding. *Phytochemistry,* 69, 1295-1303, http://www.sciencedirect.com/science/article/pii/S0031942208000356, (accessed 10 mai 2012).

[53] Toorchi, M., Yukawa, K., Nouri, M.Z., & Komatsu, S. (2009). Proteomics approach for identifying osmotic-stress-related proteins in soybean roots. *Peptides,* 30, 2108-2117, http://www.sciencedirect.com/science/article/pii/S0196978109003647, (accessed 09 mai 2012).

[54] Alam, I., Sharmin, S. A., Kim, K. H., Yang, J. K., Choi, M. S., & Lee, B. H. (2010). Proteome analysis of soybean roots subjected to short-term drought stress. *Plant and Soil,* 333(1-2), 491-505, http://www.springerlink.com/content/g9813jp3t5687604/, (accessed 12 april 2012).

[55] Zhen, Y., Qi, J. L., Wang, S. S., Su, J., Xu, G. H., Zhang, M. S., Miao, L., Peng, X. X., Tian, D., & Yang, Y. H. Comparative proteome analysis of differentially expressed proteins induced by Al toxicity in soybean. *Physiologia Plantarum,* 131(4), 542-554, http://onlinelibrary.wiley.com/doi/10.1111/j.13993054.2007.00979.x/abstract?systemMessage=Wiley+Online+Library+will+be+disrupted+on+9+June+from +10%3A00-12%3A00+BST+%2805%3A00-07%3A00+EDT%29+for+essential+maintenance&userIsAuthenticated=false&deniedAccessCustomisedMessage=, (accessed 10 march 2012).

[56] Duressa, D., Soliman, K., Taylor, R., & Senwo, Z. (2011). Proteomic Analysis of Soybean Roots under Aluminum Stress. *International Journal of Plant Genomics,* doi: 10.1155/2011/282531., http://www.hindawi.com/journals/ijpg/2011/282531/cta/, (accessed 10 march 2012).

[57] Xu, C., Sullivan, J. H., Garrett, W. M., Caperna, T. J., & Natarajan, S. (2008). Impact of solar Ultraviolet-B on the proteome in soybean lines differing in flavonoid contents. *Phytochemistry*, 69(1), 38-48, http://www.sciencedirect.com/science/article/pii/S0031942207004025, (accessed 10 march 2012).

[58] Zhang, Y., Zhao, J., Xiang, Y., Bian, X., Zuo, Q., Shen, Q., Gai, J., & Xing, H. (2011). Proteomics study of changes in soybean lines resistant and sensitive to Phytophthora sojae. *Proteome Science*, 9(52), http://www.proteomesci.com/content/9/1/52, (accessed 26 february 2012).

[59] Mithofer, A., Muller, B., Wanner, G., & Eichacker, L. A. (2002). Identification of defence-related cell wall proteins in Phytophthora sojae-infected soybean roots by ESI-MS/MS. *Molecular Plant Pathology*, 3(3), 163-166, http://onlinelibrary.wiley.com/doi/10.1046/j.13643703.2002.00109.x/abstract;jsessionid=436F73B1434802655BD95CC22314118C.d01t04?deniedAccessCustomisedMessage=&userIsAuthenticated=false , (accessed 26 february 2012).

[60] Afzal, A. J., Natarajan, A., Saini, N., Iqbal, M. J., Geisler, M., El Shemy, H. A., Mungur, R., & Willmitzer, L. (2009). Lightfoot DA: The nematode resistance allele at the rhg1 locus alters the proteome and primary metabolism of soybean roots. *Plant Physiology*, 151(3), 1264-1280, http://www.ncbi.nlm.nih.gov/pmc/articles/PMC2773059/, (accessed 10 march 2012).

[61] Wan, J., Torres, M., Ganapathy, A., Thelen, J., Da, Gue. B., Mooney, B., Xu, D., & Stacey, G. (2005). Proteomic analysis of soybean root hairs after infection by Bradyrhizobium japonicum. *Molecular Plant-Microbe Interactions*, 18(5), 458-467, https://mospace.umsystem.edu/xmlui/bitstream/handle/10355/3478/ProteomicAnalysisSoybeanRootHairs.pdf?sequence=1, (accessed 27 april 2012).

[62] Pandurangan, S., Pajak, A., Molnar, S. J., Cober, E. R., Dhaubhadel, S., Hernandez-Sebastia, C., Kaiser, W. M., Nelson, R. L., Huber, S. C., & Marsolais, F. (2012). Relationship between asparagine metabolism and protein concentration in soybean seed. *Journal of Experimental Botany*, 63(8), 3173-3184, http://jxb.oxfordjournals.org/content/63/8/3173, (accessed 11 march 2012).

[63] Sakata, K., Ohyanagi, H., Nobori, H., Nakamura, T., Hashiguchi, A., Nanjo, Y., Mikami, Y., Yunokawa, H., & Komatsu, S. (2009). Soybean Proteome Database: A data resource for plant differential omics. *Journal of Proteome Research*, 8(7), 3539-3548, http://pubs.acs.org/doi/abs/10.1021/pr900229k , (accessed 11 march 2012).

Chapter 26

Use of Organelle Markers to Study Genetic Diversity in Soybean

Lidia Skuza, Ewa Filip and Izabela Szućko

Additional information is available at the end of the chapter

http://dx.doi.org/10.5772/52028

1. Introduction

Soybean is the most important crop provider of proteins and oil used in animal nutrition and for human consumption. Plant breeders continue to release improved cultivars with enhanced yield, disease resistance, and quality traits. It is also the most planted genetically modified crop. The narrow genetic base of current soybean cultivars may lack sufficient allelic diversity to counteract vulnerability to shifts in environmental variables. An investigation of genetic relatedness at a broad level may provide important information about the historical relationship among different genotypes. Such types of study are possible thanks to different markers application, based on variation of organelle DNA (mtDNA or cpDNA).

2. Mitochondrial genome

2.1. Genomes as markers

Typically, all sufficiently variable DNA regions can be used in genetic studies of populations and in interspecific studies. Because of in seed plants chloroplasts and mitochondria are mainly inherited uniparentally, organelle genomes are often used because they carry more information than nuclear markers, which are inherited biparentally. The main benefit is that there is only one allele per cell and per organism, and, consequently, no recombination between two alleles can occur. With different dispersal distances, genomes inherited biparentally, maternally and paternally, also reveal significant differences in their genetic variability among populations. In particular, maternally inherited markers show diversity within a population much better [1].

In gymnosperms the situation is somewhat different. Here, chloroplasts are inherited mainly paternally and are therefore transmitted through pollen and seeds, whereas mitochondria are largely inherited maternally and are therefore transmitted only by seeds [2]. Since pollen is distributed at far greater distances than seeds [3], mitochondrial markers show a greater population diversity than chloroplast markers and therefore serve as important tools in conducting genetic studies of gymnosperms [4]. Mitochondrial markers are also sometimes used in conjunction with cpDNA markers [5].

Mitochondrial regions used in interspecific studies of plants, mainly gymnosperms, include, for example, introns of the NADH dehydrogenase gene *nad1* [4, 5, 6], the *nad7* intron 1 [7], the *nad5* intron 4 [3] and an internally transcribed spacer (ITS) of mitochondrial ribosomal DNA [8, 9].

In addition to the aforementioned organelle markers, microsatellite markers [10, 11] and simple sequence repeats (SSR) are often used in population biology, and sometimes also in phylogeographic studies. Microsatellites are much less common in plants than in animals [12]. However, they are present in both the nuclear genome and the organelle genome. Microsatellites may reveal a high variability, which may be useful in genetic studies of populations, whereas other sequences or methods such as fingerprinting do not detect mutations sufficiently [9,10,13]. Inherited only uniparentally, organelle markers have a certain quality in phylogeographic analyses. Since they are haploid, the effective population size should be reduced after the analysis using these markers as compared to those in which nuclear markers are used [1, 14]. Smaller effective populations sizes should bring about faster turnover rates for newly evolving genotypes, resulting in a clearer picture of past migration history than those obtained using nuclear markers [15-17].

Initially, it was mainly in phylogeographic studies of animal species that mitochondrial markers were used [18]. These studies have provided some interesting data on the beginnings and the evolutionary history of human population [19]. In contrast to studies of animals, using mitochondrial markers in studies of plants, especially angiosperms, is limited [20]. Presently, cpDNA markers are most commonly used in phylogeographic studies of angiosperms, whereas mitochondrial markers are prevalent in studies of gymnosperms.

2.2. Plant mitochondrial DNA

Mitochondrial genomes of higher plants (208-2000 kbp) are much larger than those of vertebrates (16-17 kbp) or fungi (25-80 kbp) [21, 22]. In addition, there are clear differences in size and organization of mitochondrial genomes between different species of plants. Intramolecular recombination in mitochondria leads to complex reorganizations of genomes, and, in consequence, to alternating arrangement of genes, even in individual plants, and the occurrence of duplications and deletions are common [23]. In addition, the nucleotide substitution rate in plant mitochondria is rather low [24], causing only minor differences within certain loci between individuals or even species. Extensively characterized circular animal mitochondrial genomes are highly conservative within a given species; they do not contain introns and have a very limited number of intergenic sequences [25]. Plant mitochondrial DNA (mtDNA) contains introns in multiple genes and several additional genes undergoing expression when compared to animal mitochondria, but most of the additional sequences in plants are not expressed and they do not seem to be esssentials [26]. The completely sequenced mitochondrial genomes are available for several higher plants, including *Arabidopsis thaliana* [27] or *Marchantia polymorpha* [28].

Restriction maps of nearly all plant mitochondrial genomes provide for the occurrence of the master circle with circular subgenomic molecules that arise after recombination among large direct repeats (> 1 kbp) [21, 29-36], which are present in most mitochondrial genomes of higher plants. However, such molecules, whose sizes can be predicted, are very rare or very difficult to observe. It can be explained by the fact that plant mitochondrial genomes are circularly permuted as in the phage T4 [37, 38]. Oldenburg and Bendich reported that mostly linear molecules in *Marchantia* mtDNA are circularly permuted with random ends [39]. It shows that plant mtDNA replication occurs similarly to the mechanism of recombination in the T4 [38].

Many reports that have appeared in recent years indicate that mitochondrial genome of yeasts and of higher plants exist mainly as linear and branched DNA molecules with variable size which is much smaller than the predicted size of the genomes [39-44]. Using pulsed field gel electrophoresis (PFGE) of in-gel lysed mitochondria from different species revealed that only about 6-12% of the molecules are circular [41, 44]. The observed branched molecules are very similar to the molecules seen in yeast in the intermediate stages of recombination of mtDNA [45] or the phage T4 DNA replication [37, 38].

In all but one known case (*Brassica hirta*) [46], plant mitochondrial genomes contain repeat recombinations. These sequences, ranging in length from several hundred to several thousand nucleotides (nt) exist at two different loci in the master circle, yet in four mtDNA sequence configurations [47]. These four configurations correspond to the reciprocal exchange of sequences 5' and 3' surrounding the repeat in the master circle, which suggests that the repeat mediates homologous recombination. Depending on the number and orientation of repeats, the master circle is a more or less complex set of subgenomic molecules [48].

Maternally inherited mutations, which are associated with mitochondria in higher plants, most often occur as a result of intra- and intergenic recombination. This happens in most cases of cytoplasmic male sterility (cms) [41, 49-51], in *chm*-induced mutation in *Arabidopsis* [52] and in non-chromosomal stripe mutations in maize [53]. In this way, it is assumed that the recombination activity explains the complexity of the variations detected in the mitochondrial genomes of higher plants.

2.3. Mitochondrial genome of soybean

The size of soybean mtDNA has been estimated to be approximately 400 kb [54-56]. Spherical molecules have also been observed by electron microscopy [55, 57].

Repeated sequences 9, 23 and 299 bp have been characterized in soybean mitochondria [58, 59]. Also, numerous reorganizations of genome sequences have been characterized among different cultivars of soybean. It has been demonstrated that they occur through homologous recombination produced by these repeat sequences [58, 60, 61], or through short elements that are part of 4.9kb PstI fragment of soybean mtDNA [62]. The 299 bp repeat sequence has been found in several copies of mtDNA of soybean and in several other higher plants, suggesting that this repeated sequence may represent a hot spot for recombination of mtDNA in many plant species [59, 62]. Previous results suggested that active homologous recombinations of mtDNA are present in at least some species of plants. Recently (2007) amitochondrial-targeted homolog of the *Escherichia coli recA* gene in *A. thaliana* has been identified [63]. However, the data on recombnation activity in plant mitochondria is still missing. The first data on such an activity in soybean was obtained in 2006 [64]. This discovery is supported by an analysis of mtDNA of soybean using electron microscopy and 2D-electrophoresis. The results suggest that only a small portion of mtDNA molecules undergoes recombination at any given time. Therefore the question is whether this recombination is essential to the functioning of mitochondria and to plant growth.

The repeated sequences of the *atp6, atp9* and *coxII* genes have been also characterized, but their recombination activity has not been analysed [65].

The first data for the restriction map of soybean mtDNA were obtained from the analysis of loci of the *atp4* gene [48]. In the vicinity of this gene two repeated sequences that show characteristics of recombination repeats have been found [47, 48]. Active recombination repeats were also identified in circular molecules smaller than 400 kb [55, 66]. These observations suggest that soybean mtDNA has multipartite structure that is similar to other plant mitochondrial genomes containing recombination repeats.

In the mitochondrial genome of cultivar Williams 82, recombinantly active repeats 1 kb and 2 kb have been described [48]. In a different repeat of 10 kb, surrounding both 1 kb and 2 kb repeats, two breakpoints have been identified. This recombination of smaller and larger repeats probably leads to the complex structure of genomes.

The analysis of restriction fragment length polymorphism (RFLP) of mtDNA seems to be a useful method in studying phylogenetic relationships within species.

Grabau et al. (1992) analyzed the genomes of 138 soybean cultivars [60]. Using 2.3 kb HindIII mtDNA probe from Williams 82 soybean cultivar revealed restriction fragment length polymorphisms (RFLPs), which allowed for the division of many soybean cultivars into four cytoplasmic groups: Bedford, Arksoy, Lincoln and soja-forage.

Subsequent analyses showed variations within, and adjacent to, the 4.8 kb repeats. Bedford cytoplasm turned out to be the only one that contains copies of the repeat in four different genomic environments, which indicates its recombination activity [61]. Lincoln and Arksoy cytoplasms contain two copies of the repeat and a unique fragment that appear to result from rare recombination events outside, but near, the repeat. In contrast, forage-soja cytoplasm contains no complete repeat, but it contains a unique truncated version of the repeat [61]. Sequence analysis revealed that truncating is caused by the recombination with a repeat of 9 bp CCCCTCCCC. The structural reorganization that occurred in the region around 4.8 kb repeat may provide a way to analyze the relationships between species and evolution within the soybean subgenus.

In order to determine the sources of cytoplasmic variability, Hanlon and Grabau (1995) studied the old cultivars of soybeans with the same 2.3-kb *Hind*III fragment and with a mtDNA fragment containing the *atp6* gene [62]. They showed that mtDNA RFLP analysis with these probes is useful for the classification of mitochondrial genomes of soybean. Grabau and Davies (1992) made a general classification of wild soybean using the 2.3-kb *Hind*III as a probe [68].

Mt type	Probe	*coxI*			*coxII*			*atp6*		Reference
	Enzyme	*Hind*III	*Bam*HI	*Eco*RI	*Hind*III	*Bam*HI	*Eco*RI	*Bam*HI	*Eco*RI	
Ic					1,6	5,8		5,0		[69]
Id					1,6	5,8		5,0;6,0 ;12,0		[69]
Ie					1,6	5,8		5,0; 12,0		[69]
Ik					1,6	5,8		5,0; 5,4; 5,8		[69]
IIg					1,3	7,0		1,0; 2,6		[69]
IIIa					1,2	8,5		2,4; 5,0		[69]
IIIb					1,2	8,5		2,9; 5,0		[69]
IIId					1,2	8,5		5,0;6,0; 12,0		[69]
IVa					3,5	8,1		2,4; 5,0		[69]
IVb					3,5	8,1		2,9; 5,0		[69]
IVc					3,5	8,1		5,0		[69]
IVf					3,5	8,1		2,4; 3,5; 5,0		[69]
IVh					3,5	8,1		2,6; 2,9		[69]
IVi					3,5	8,1		5,2; 12,0		[69]
Va					5,8	8,1		2,4; 5,0		[69]
V'j					5,8	15,0		5,0; 6,0		[69]
VIg					1,7	5,8		1,0; 2,6		[69]
VIIg					8,5	15,0		1,0; 2,6		[69]
mtI					1,6	5,8				[69]
mtII					1,3	7,0				[69]
mtIII					1,2	8,1				[69]
mtIV					3,5					[69]
mtV					5,8					[69]
mt-a								2,4; 5,0		[87]
mt-b								2,9; 5,0		[87]
mt-c								5,0		[87]
mt-d								5,0; 6,0; 12,0		[87]
mt-e								5,0; 12,0		[87]

Mt type	Probe	*coxI*			*coxII*			*atp6*		Reference
	Enzyme	*Hind*III	*Bam*HI	*Eco*RI	*Hind*III	*Bam*HI	*Eco*RI	*Bam*HI	*Eco*RI	
mt-f								2,4; 3,5; 5,0		[87]
mt-g								1,0; 2,6		[87]
mt-h								2,6; 2,9		[87]
mt-m								2,9		[87]
mt-n								12,0		[87]
Ic		5,6	0,8; 2,5; 5,0	10,5	1,6	5,8	1,9	5,0	8,2; 12,0	[58]
Id		5,6	0,8; 2,5; 5,0	10,5	1,6	5,8	1,9	5,0; 6,0; 12,0	2,8; 6,0; 12,0	[58]
Ie		5,6	0,8; 2,5; 5,0	10,5	1,6	5,8	1,9	5,0; 12,0	2,8; 6,0; 12,0	[58]
Ik		5,6	0,8; 2,5; 5,0	10,5	1,6	5,8	1,9	5,0; 5,4; 5,8	2,8; 6,0; 12,0	[58]
IIg		8,5	0,8; 2,5; 5,0	9,0	1,3	7,0	4,8	1,0; 2,6	2,8; 3,0; 9,5	[58]
IIIb		5,6	0,8; 2,5; 5,0	10,5	1,2	8,5	6,2; 6,5	2,9; 5,0	6,0; 8,2; 12,0	[58]
IIId		5,6	0,8; 2,5; 5,0	10,5	1,2	8,5	6,2; 6,5	5,0; 6,0; 12,0	3,2; 6,2; 12,0	[58]
Iva		5,6	0,8; 2,5; 5,0	10,5	3,5	8,1	5,0	2,4; 5,0	3,0; 6,0; 12,0	[58]
IVb		5,6	0,8; 2,5; 5,0	10,5	3,5	5,8	5,0	2,9; 5,0	6,0; 8,2; 12,0	[58]
IVc		5,6	0,8; 2,5; 5,0	10,5	3,5	5,8	5,0	5,0	8,2; 12,0	[58]
IVf		5,6	0,8; 2,5; 5,0	10,5	3,5	5,8	5,0	2,4; 3,5; 5,0	3,2; 6,2; 12,0	[58]
IVh		5,6	0,8; 2,5; 5,0	10,5	3,5	5,8	5,0	2,6; 2,9	3,2; 6,2; 12,0	[58]
IVi		5,6	0,8; 2,5; 5,0	10,5	3,5	5,8	5,0	5,2; 12,0	3,2; 6,2; 12,0	[58]
Va		5,6	0,8; 2,5; 5,0	10,5	5,8	5,8	12,0	2,4; 5,0	3,0; 6,0; 12,0	[58]
Vb		5,6	0,8; 2,5; 5,0	10,5	5,8	5,8	12,0	2,9; 5,0	6,0; 8,2; 12,0	[58]
Vc		5,6	0,8; 2,5; 5,0	10,5	5,8	5,8	12,0	5,0	8,2; 12,0	[58]
V'j		5,6	0,8; 2,5; 5,0	10,5	5,8	15,0	1,6	5,0; 6,0	2,8; 6,0; 12,0	[58]

http://dx.doi.org/10.5772/52028

Mt type	Probe	*coxI*			*coxII*			*atp6*		Reference
	Enzyme	***Hind*III**	***Bam*HI**	***Eco*RI**	***Hind*III**	***Bam*HI**	***Eco*RI**	***Bam*HI**	***Eco*RI**	
VIg		5,6; 8,5	0,8; 2,5; 5,0; 5,2	9,0; 10,5	1,7	5,8	4,5	1,0; 2,6	2,8; 3,0; 4,3; 9,5; 12,0	[58]
VIIg		8,5	0,8; 5,0; 5,2	9,0	8,5	15,0	1,6	1,0; 2,6	2,8; 3,0; 9,5	[58]
VIIIc		5,6	0,8; 2,5; 5,0	10,5	8,5; 10,0	11,0; 15,0	1,6	5,0	8,2; 12,0	[58]
					Combined chloroplast and mitochondrial genome type					
cpI +mtIIIb					1,2	8,5		2,9; 5,0		[89]
cpI +mtIVb					3,5	8,1		2,9; 5,0		[89]
cpI+mtIVc					3,5	8,1		5,0		[89]
cpII +mtIVb					3,5	8,1		2,9; 5,0		[89]
cpII +mtIVc					3,5	8,1		5,0		[89]
cpIII+mtIe					1,6	5,8		5,0; 12,0		[89]
cpIII +mtIVa					3,5	8,1		2,4; 5,0		[89]
cpIII +mtVIIIc					8,5; 10,0	11,0; 15,0		5,0		[89]

Table 1. Classification of mitochondrial genome types based on RFLPs using coxI, *coxII* and *atp6* as probes. Sizes of hybridization signals (kb) are shown.

In their research Tozuka et al. (1998) used two fragments of mtDNA as probes: the 0.7-kb *Hind*lll-*Nco*I fragment containing the *coxII* (the gene encoding the mitochondrial cytochrome oxidase subunit II) of wild soybean and the 0.66-kb *Sty*I fragment containing the *atp6* (the gene encoding the mitochondrial ATPase subunit 6) from *Oenothera* [69, 70] (Table 1).

Based on the RFLPs detected in gel-blot analysis with the *coxII* and *atp6* probes, the harvested plants were divided into 18 groups. Five mtDNA types were described in 94% of the surveyed plants. The geographical distribution of mtDNA types revealed that in many regions soybean growing wild in Japan consisted of a mixture of plants with different types of mtDNA, sometimes even within a single location. Some of these mtDNA types have shown marked geographic clines among the regions. In addition, some wild soybeans had mtDNA types that were identical to those described in cultivated soybeans. These results suggest that mtDNA analysis could resolve maternal origin among of the genus *Glycine* subgenus *Soja* [69].

Kanazawa et al. (1998) gathered 1097 *G. soja* plants from all over Japan and analyzed their RFLP of mitochondrial DNA (mtDNA) using five probes (*coxI, coxII, atp6, atp9, atp1=atpA*) [58] (Table 1). 20 different types of mitochondrial genomes labeled as combinations of types I to VII and types from a to k were identified and characterized in this study. Nearly all the mtDNA types described for soybean cultivars also occurred in wild soybean.

The mitochondrial *atpA* gene was also analysed [48]. It was shown that in soybean this gene has a sequence in 90-97% identical with mitochondrial genes of other plants [71-81]. Sequence similarity is limited to the *atpA* coding region. An intriguing feature of the *atpA* open reading frame of soybean is an 642 nt overlap in the putative translation termination site onto an unidentified open reading frame of the *orf214*. The ends of the open reading frame contain four tandems of UGA codon that covers four tandems of AUG codon that initiates an unidentified *orf214* frame. The *atpA-orf 214* region was found in soybean mtDNA in multiple sequence contexts. This can be attributed to the presence of two recombination repeats.

The open reading frame shares 79% of nucleotide identity with the *orf214* and is located in the same *atpA* locus position as in common bean *orf209* [82]. Since such organization is a repeat of overlapping the *atpB* and *atpE* reading frames in several chloroplast genes [83, 84], the probability that the *orf214* codes a different ATPase subunit cannot be evaluated because small ATPase subunits are poorly conserved [85].

So for a total of 26 mtDNA haplotypes of wild soybeans have been identified based on RFLP with probes from two mitochondrial genes: *cox2* and *atp6* [69, 86] (Table 1). The three most common haplotypes (Id, IVa and Va) are present in 43 populations. The distribution of mtDNA haplotypes varies among opulations [87]. Recently Shimamoto (2001) analyzed the genetic polymorphisms of mitochondrial genes subgenus *Soja* originating from China and Japan [88] (Table 1). As a result of these studies, 6 types of mitochondrial genomes were distinguished.

3. Chloroplast genome

As the result of the extensive research conducted in the past two decades, cpDNA analysis brought about fundamental changes to the systematics of plants. The chloroplast genome is ideal for phylogenetic analyses of plants for several reasons. First, it occurs abundantly in plant cells and is taxonomically ubiquitous. And since it is well researched, it can be easily tested in the laboratory conditions and analyzed in comparative programs. Moreover, it often contains marker structural features cladistically useful, and, above all, it exhibits moderate or low rate of nucleotide substitution [89]. In regard to the mitochondrial genome, and also to cpDNA, researchers use in their studies two distinct phylogenetic approaches [90], namely taxonomic checking of specific traits features of molecular cpDNA and sequencing of specific genes or regions.

3.1. Chloroplast genome of soybean

In estimating the phylogeny of plants belonging to *Glycine*, particular attention was paid to unusual and specific features of cpDNA. In the course of many studies on the variability of chloroplast genome, a breakthrough came in 1993, with a study on assessing phylogeny of seed plants. The study used a huge database of the nucleotide sequences of the *rbcL* gene [91], encoding the ribulose-1,5-bisphosphate carboxylase, large subunit. The accumulation of a number of comparative data on this chloroplast gene made it a frequent object of research. This is due to the fact that this gene's locus is large (> 1400 bp), and provides many phylogenetically informative traits. The rate of the *rbcL* evolution proved to be appropriate for assessing issues related to phylogeny of plants, especially on the medium and high taxonomic levels. Over the years other sequences from other species as well as many other genes with another chloroplast *atpB* gene coding H+ -ATPase subunits [92-95]. The *atpB-rbcL* sequence reaches different lengths in *Glycine* as well as in other seed plants. The study by Chiang (1998) shows that the size of the *atpB-rbcL* space in the studied species ranges from 524 bp to 1000 bp [5], where in the non-coding region the occurrence of deletions and insertions, as well as a number of nucleotide substitutions is a common phenomenon, which can also be observed in *Glycine*. In *Glycine max*, its chloroplast genome differs from the core set chloroplast DNA genes because of the presence of a single, large inversion of approximately 51 kb, in the area between the *rbcL* gene and the *rps16* intron [96]. This inversion is also present in other legumes: the mutation was reported in *Lotus* and *Medicago* [96]. In addition, the non-coding *atpB-rbcL* region is rich in AT, due to which most non-coding regions rich in these base pairs show a small number of functions [97, 98]. Therefore, this predisposes them for faster evolution, and hence for use in molecular systematics.

The summary phylogeny was based on sequence of several cpDNA genes from hundreds of spermatophytes including *Glycine* (Table 2). These genes can be divided into three classes. The genes encoding the photosynthetic apparatus structure form the first class. The second class includes the rRNA genes and genes encoding the chloroplast genetic apparatus. The last class consists of an average of about 30 tRNA encoding genes [99], although their number can vary from 20 to 40 [100, 101].

Genes	Products
	Genes for the photosynthesis system
rbcL	Ribulose -1,5- bisphosphate carboxylase, large subunit
psaA, B	Photosystem I, P700 apoproteins A1, A2
psaC	9kDa protein
psaA	Photosystem II, D1 protein
psaB	47kDa chlorophyll a-binding protein
psaC	43 kDa chlorophyll a-binding protein
psaD	D2 protein
psaE	Cytochrome b559 (8kDa protein)
psaF	Cytochrome b559 (4kDa protein)
psaH	10 kDa phosphoprotein

Genes	Products
	Genes for the photosynthesis system
psaI, J, K, L, M, N	–J, -K, -L, -M, -N-proteins
atpA, B, E	H^+ -ATPase, CF_1 subunits α, β, ε
atp F, H, I	CF_0 subunits I, III, IV
petB, D	Cytochrome b_6 /*f* complex, subunit b_6, IV
nadA- K	NADH Dehydrogenase, subunits ND 1, NDI 1
	Genes for the genetic system
16S rRNA	16S rRNA
23S rRNA	23S rRNA
trnA -UGC	Alanine tRNA (UGC)
trnG- UCC	Gliycine tRNA (UCC)
rnH- GUG	Histidine tRNA (GUG)
trnI- GAU	Isoleucine tRNA (GAU)
trnK- UUU	Lysine tRNA (UUU)
trnL- UAA	Leucine tRNA (UAA)
rps2, 7, 12, 16	30S: ribosomal proteins CS2, CS7, CS12, CS16
rp12, 20, 32	50S: ribosomal proteins CL2, CL 20, CL32
rpoA, B, C1, C2	RNA polymerase, subunits α, β, β', β''
matK	Maturase –like protein
sprA	Small plastid RNA
	Others
clpP	ATP-dependent protease, proteolytic subunit
irf168 (ycf3)	Intron- containing Reading frame (168 codons)

Table 2. Chloroplast genes for the photosynthesis system, for the genetic and others.

The complete size of the *Glycine max* chloroplast genome is 152,218 bp. It contains 25,574 bp of inverted repeats (IRa and IRb), which are separated by a unique small single copy (SSC) region (17,895 bp) [98]. In addition, this genome consists of a large single region (LSC) of unique sequences with 83,175 bp. The IR extends from the *rps19* gene up to the *ycf1*. The *Glycine* chloroplast genome contains 111 unique genes and 19 duplicate copies in the IR, amounting to a total of 130 genes. The cpDNA analysis has showed the presence of 30 different tRNAs in it and 7 of them are repeated within the IR regions. The genes are composed in 60% of encoding regions (52% are protein coding genes and 8% are RNA genes), and in 40% of non-coding regions, including both intergenic spacers and introns. The total content of GC and AT pairs in the *Glycine* chloroplast genes is 34% and 66% respectively. Distinctly higher percentage of AT pairs (70%) was observed in non-coding regions than in coding regions (62% AT) [98].

In comparison with other eukaryotic genomes, cpDNA is highly concentrated, for example, only 32% of the rice genome is non-coding. In *Glycine max* it is slightly more – 40%. Most of the non-coding DNA is found in very short fragments that separate functional genes. Some studies have shown complex patterns of mutational changes in the non-coding regions. Some of the best known regions in the chloroplast genome is the farther region of the *rbcL* gene in many legumes. This non-coding sequence is flanked by the *rbcL* and *psaI* (the gene encoding the polypeptide I of photosystem I).

3.2. Extent of IR in Glycine

Analysis of the IR (inverted repeats) regions in *Glycine max* has shown that they are separated by a large region and a small region of a unique sequence. In cpDNA repeated sequences are usually located asymmetrically, which results in the formation of long and short regions of a unique sequence [102]. The IR in *Glycine* is a region with 25,574 bp containing 19 genes. At the IR/LSC junction, at the ends of the 5' IR, there is the repeated *rps19* gene (68 bp), and at the junction of the IR/SSC and 5' ends the duplicated *ycf1* gene (478 bp) is located. In the course of study it was shown that comparing cpDNA IR region in *Medicago, Lotus, Glycine* and *Arabidopsis* indicates that there are changes within the IR in the two legumes. *Glycine* and *Lotus* have 478 bp and 514 bp of the *ycf1* duplicated, whereas *Arabidopsis* has 1,027 bp duplicated in the IR. This contraction of the IR in these legumes accounts for the smaller size of their IR and larger size of the SSC. In addition, contraction of the IR boundary in legumes, IRa has been lost in *Medicago*. This loss has resulted in *ndhF* (usually located in the SSC) being adjacent to *trnH* (usually the first gene in the LSC at the LSC/IRa junction). Loss of one copy of the IR in some legumes provides support for monophyly of six tribes [103-106]. Wolfe (1988) identified duplicated sequences of portions of two genes, 40 bp of *psbA* and 64 bp of *rbcL*, in the region of the IR deletion between *trnH* and *ndhF* in *Pisum sativum* and these duplications were later identified in broad bean (*Vicia faba*) [104,107]. According to many researchers, the IR region is considered the most conserved part of the chloroplast genome, and thus, it is responsible for stabilizing the plastid DNA molecules [108, 109]. Thus the loss of IR can be phylogenetically informative at the local level, as well as misleading at the global phylogeny level, because the IR loss likely occurred independently in more than one group of plants. Coniferous and some legumes (*Pisum sativum, Vicia faba, Medicago sativa*), for example, contain only one IR. Perhaps the lack of repeat sequences in these plants is associated with an increased incidence of rearrangement of chloroplast genomes [109].

Introns or intergenic sequences in legume chloroplast DNA have become extremely important tools in phylogenetic analyses aimed at systematizing of this species [110, 111]. Moreover, their microstructural changes occur with great frequency in the regions of cpDNA. The body of existing research suggests that mutations in the non-coding regions and relatively fast evolution of the organelle genome encoding regions can serve as valuable markers for the separation species in their evolutionary origin [110, 111]. The systematics of plants generally considers chloroplast indeles to be phylogenetic markers, because of their low prevalence in comparison with nucleotide substitutions [5].

3.3. CpDNA markers

There are many methods of generating molecular markers that rely on site-specific amplification of a selected DNA fragment using polymerase chain reaction (PCR) and its further processing (restriction analysis, sequencing). Initially the research on the plant genome (mostly phylogenetic studies) used non-coding and coding sequences of chloroplast DNA. With time, the genes or DNA segments located in the nuclear DNA, mitochondrial (mtDNA) and chloroplast (cpDNA) found a prominent place among plant DNA markers. Fully automated DNA sequencing made it possible to subject ever-newer regions of plant DNA to comparative sequencing.

One of the most frequently sequenced cpDNA fragments in plant phylogeny of spermatophytes is the *rbcL* gene encoding a large ribulose bisphosphate carboxylase subunit (RUBISCO), whose length in most plants is 1,428, 1,431 and 1,434 bp, and insertions and deletions within it are extremely rare [94]. For many years this gene has been the subject of many comprehensive phylogenetic analyses of subgenus *Glycine* [112-114]. The *rbcL* is most commonly used in the analyses at the family and genus levels, but there also exists research at the lower levels, cultivars and wild soybean [98, 115, 116]. A marker with very similar characteristics to those of the *rbcL* (the rate of evolution, the length of 1497 bp) is a gene encoding the ATP synthase β subunit – the *atpB* [94].The *matK* gene sequence, encoding maturase involved in splicing of the type II introns, and whose length is 1,550 bp is characterized by a rapid rate of evolution that allows to use it in research at the species and genus levels [117, 118]. Frequent mutations in this gene make it unsuitable for studies at higher taxonomic levels. Other popular cpDNA sequences used in phylogenetic studies of legumes include the *ndhF* (the gene encoding the NADH protein, which is a dehyd98rogenase subunit), 16S rDNA, the non-coding *atpB-rbcL* region [94], or the *trnL* (UAA) intron and mediator between the *trnL* (UAA) exon and the *trnF* (GAA) gene [96, 117- 119].

It should be noted that the rate of evolution for a specific DNA region to be used as a marker can vary significantly not only among systematic groups, but also within these groups [98]. Moreover, each DNA fragment within the same group has a different rate of evolution, such as the *ndhF* cpDNA sequence in the *Solanaceae* family, which provides about 1.5 times more information in terms of parsimony than the *rbcL* [90]. Therefore each gene or any other DNA fragment used as a genetic marker has a typical range of "taxonomic" or phylogenetic applications, which can vary significantly within a taxon. For this reason, the *rbcL* sequence has been widely used in *Gycine* for many years at the species and genus levels [104, 117, 118].

3.4. The genetic diversity of soybeans

The importance of genetic variations in facilitating plant breeding and/or conservation strategies has long been recognized [121]. Molecular markers are useful tools for assaying genetic variation and provide an efficient means to link phenotypic and genotypic variation [122].

In recent years, the progress made in the development of DNA based marker systems has advanced our understanding of genetic resources. These molecular markers are classified as: (i) hybridization based markers i.e. restriction fragment length polymorphisms (RFLPs), (ii) PCR-based markers i.e. random amplification of polymorphic DNAs (RAPDs), amplified fragmentlength polymorphisms (AFLPs), inter simple sequence repeats (ISSRs) and microsatellites or simple sequence repeats (SSRs), and (iii) sequence based markers i.e. single nucleotide polymorphisms (SNPs) [121, 123]. Majority of these molecular markers have been developed either from genomic DNA library (e.g. RFLPs or SSRs) or from random PCR amplification of genomic DNA (e.g. RAPDs) or both (e.g. AFLPs) [123]. Availability of an array of molecular marker techniques and their modifications led to comparative studies among them in many crops including soybean, wheat and barley [124-126]. Among all these, SSR markers have gained considerable importance in plant genetics and breeding owing to many desirable attributes including hypervariability, multiallelic nature, codominant inheritance, reproducibility, relative abundance, extensive genome coverage (including organellar genomes), chromosome specific location, amenability to automation and high throughput genotyping [127]. In contrast, RAPD assays are not sufficiently reproducible whereas RFLPs are not readily adaptable to high throughput sampling. AFLP is complicated as individual bands are often composed of multiple fragments mainly in large genome templates [123]. The general features of DNA markers are presented in Table 3.

	Molecular markers			
	EST–SSRs	**SSRs**	**RFLPs**	**RAPDs/AFLPs/ISSRs**
Need for sequence data	Essential	Essential	Not required	Not required
Level of polymorphism	Low	High	Low	Low-moderate
Dominance	Co-dominant	Co-dominant	Co-dominant	Co-dominant
Interspecific transferability	High	Low-moderate	Moderate-high	Low-moderate
Utility in Marker assisted selection	High	High	Moderate	Low-moderate
Cost and labour involved in generation	Low	High	High	Low-moderate

Table 3. Important features of different types of molecular markers.

The genetic diversity of wild and cultivated soybeans has been studied by various techniques including isozymes [128], RFLP [87], SSR markers [124], and cytoplasmic DNA markers [87, 128, 129]. Based on haplotype analysis of chloroplast DNA, cultivated soybean appears to have multiple origins from different wild soybean populations [129, 130].

Using PCR-RFLP method soybean chloroplast DNAs were classified into three main haplotype groups (I, II and III) [113, 130, 131]. Type I is mainly found in the species of cultivated soybean (*Glycine max*), while types II and III are often found in both the cultivated and wild forms of soybean (*Glycine soja*). Type III is by far the most dominant in the wild soybean species [113]. In *Glycine*, these types are widely used in evaluating cpDNA variability and in determining phylogenetic relationships between different types of cpDNA using different marker systems. According to Chen and Hebert (1999) [133] analysis of cpDNA sequence is not sufficient for when the analysis of population genetics, and so cpDNA polymorphism assessment methods must be constantly complemented with methods such as single-strand conformation polymorphism (SSCP) [134], or dideoxy fingerprinting (ddF) [135], and directed termination and polymerase chain reaction (DT-PCR). However, some researchers point out that there are many disadvantages of these methods, mainly because of their high cost and large amount of work necessary for obtaining the results. In their view, a single change in the regions of *Glycine* chloroplast DNA at the species and genus levels should be located on a local-specific markers, for example, non-coding regions, using PCR and sequencing.

Analyses of non-coding regions of cpDNA have been employed to elucidate phylogenetic relationship of different taxa [90]. Compared with coding regions, non-coding regions may provide more informative characters in phylogenetic studies at the species level because of their high variability due to the lack of functional constraints. Non-coding regions of cpDNA have been assayed either by direct sequencing [136-141], or by restriction-site analysis of PCR products (PCR-RFLP) [142-146]. In Small's opinion (1998) non-coding regions, which include introns and intergenic sequences, often show greater variability at nucleotides than at the encoding regions, which makes the non-coding regions good phylogenetic markers [139]. Mutations in the form of insertions and deletions are accumulated in noncoding regions at the same rate as nucleotide substitutions, and such kinds of mutations significantly accelerate changes in these regions. In many cases, insertions or deletions are related to short repeat sequences. Therefore, many researchers continually focus on the analysis of non-coding regions. Using RFLP method, Close et al. (1989) found six cpDNA haplotypes and described them in types, ranging from group I to VI, including cultivated and wild soybeans [147]. In the course of their research they found that groups I and II diverge from groups III to VI, thus dividing subgenus *Soja* into two main groups. They presented a hypothesis that group II can be distinguished form group III by two independent mutations. Similar groups of haplotypes in legumes were also obtained by Shimamoto et al, (1992) [128] and Kanazawa et al, (1998) [148], using a combination of *EcoR*I and *Cla*I RFLPs. In their classification, Kazanawa et al. (1998) relied on sequential analysis and found that differences in the three types described by Shimamoto et al. (1992) resulted from two single-base substitutions: one in the non-coding region, between the *rps11 and rpl36,* and the other in the 3' part of the coding region of the *rps3*. Based on the existing reports, Xu et al. (2000) sequenced nine non-coding regions of cpDNA for seven cultivars and 12 wild forms of soybean (*Glycine max, Glycine soja, Glycine tabacina, Glycine tomentella, Glycine microphylla, Glycine clandestina*) in order to verify earlier classification of *Glycine* [113]. In the course of their studies, they located eleven single-base changes (substitutions and deletions) in the collected 3849 database. They located five mutations in the distinguished haplotypes I and II, and seven mutations in type III. In addition, haplotypes I and II were identical and clearly different from the taxons in type III. This research has not yielded significant results, because different types of cpDNA could not originate monophyletically, but it contributed to finding a common ancestor in the course of evolution of *Glycine*. A neighbor joining tree resulting from the sequence data revealed that the subgenus *Soja* connected with *Glycine microphylla,* which formed a distinct clad from *Clycine clandestine* and the tetraploid cytotypes of *Glycine tabacina* and *Glycine tomentella*. Several informative length mutations of 54 to 202 bases, due to insertions or deletions, were also detected among the species of the genus *Glycine*.

3.5. Non-coding regions of the chloroplast genome as site-specific markers in Glycine

In the chloroplast genomes of legumes, including soybean, there are many non-coding regions, which are characterized by a faster rate of evolution when compared to the coding regions. As mentioned earlier some of the chloroplast genes have introns, yet their structure differs from those occuring in the nuclear genes, since in the case of cpDNA introns have a tendency to adopt secondary structure, which affects the model in which cpDNA introns evolve and it is enforeced by the secondary structure. This restriction in changes caused by mutations affects the functional requirements related to the formation of introns [98, 108]. As there are no adequate studies on the evolution of introns, it can be assumed that their evolution is similar to that of the protein-encoding genes. The loss of introns in the course of the evolution of chloroplast DNA is an interesting process. It has been discovered that *O. sativa* has 3 introns less in cpDNA than *M. polymorpha* and *N. tabacum*. The loss of an intron in the *rpl2* gene was researched in 340 species representing 109 families of angiosperms including *Glycine* [149]. When trying to determine the taxonomic position, the absence of this intron in a given gene shows that it was lost at least six times in the evolution of angiosperms. In *Glycine* 23 introns have been identified while in *Arabidopsis thaliana* there are 26 introns, mostly located in the same genes and in the same locations within those genes [98, 102].

Non-coding regions in chloroplast DNA have become a major source for phylogenetic studies within the species *Glycine* and in many other seed plants. Earlier, the most popular phylogeny sequences included encoding regions, such as the *rbcL* gene sequences that were designed to determine the phylogenetic relationships between species in major taxonomic groups [113, 136-141]. According to Taberlet et al. (1991) [119] the potential ability of non-coding regions of cpDNA was reserved for species located in the lower taxonomic levels while the non-coding regions, which include introns and intergenic sequences, often show greater variability at nucleotides than is evident in the coding regions, which predisposes them to be used in population studies involving *Glycine,* and others [139,142].

As the result, many studies on phylogenetic utility of non-coding regions have been published [110]. For example [150]: *trnH-psbA, trnS-tang;* [148]: *rps11-rpl36, rpl16-rps3,* [113]: *trnT-trnL.*

In cpDNA analysis of many plants, very conservative regions flanking areas with high variability are used. The more conservative regions, the higher the chance for the primers designed in the PCR reaction, which will be able to join the broader taxonomic group [96, 113]. The region occurring between the *trnT* (UGU) and the *trnF* (GAA) genes is a large single copy wich is suitable because of the conservativeness of the *trn* genes and several hundred base pairs of noncoding regions. The intergenic space between the *trnT* (UGU) and the *trnL* (UAA) 5' exon ranged from 298 bp to about 700 bp in the species studied by [119]. In the plant genomes completely sequenced by Sugiute, the length of this region is different and amounts to 770 bp in rice and 710 bp in tobacco. In *Marchantia polymorpha* it is 188 bp [151]. This region is located between the tRNA genes, just as the non-coding sequence located between the *trnL* (UAA) 3' exon and the *trnF* (GAA). Due to its catalytic properties and its secondary structure, the *trnL* (UAA) intron, which belongs to type I introns, is less variable and therefore of better utility for evolutionary studies at higher taxonomic levels [113]. Moreover, depending on the species, they show high frequency of insertions or deletions, which makes them potentially useful as genetic markers.

Region	Primer sequence (5 - 3)*	Annealing temperature	Reference
trnH–psbA	f:TGATCCACTTGGCTACATCCGCC	60°C	[99] (tobacco)
	r: GCTAACCTTGGTATGGAAGT		[150] (soybean)
trnS–trnG	f: GATTAGCAATCCGCCGCTTT	60°C	[99] (tobacco)
	r: TTACCACTAAACTATACCCGC		[99] (tobacco)
trnT–trnL	f: GGATTCGAACCGATGACCAT	60°C	[113] (soybean)
	r: TTAAGTCCGTAGCGTCTACC		[99] (tobacco)
trnL–trnF	f: TCGTGAGGGTTCAAGTCC	56°C	[99] (tobacco)
	r: AGATTTGAACTGGTGACACG		[99] (tobacco)
atpB–rbcL	f: GAAGTAGTAGGATTGATTCTC	58°C	[99] (tobacco)
	r: CAACACTTGCTTTAGTCTCTG		[99] (tobacco)
psbB–psbH	f: AGATGTTTTTGCTGGTATTGA	56°C	[99] (tobacco)
	r: TTCAACAGTTTGTGTAGCCA		[99] (tobacco)
rps11–rpl36	f:GTATGGATATATCCATTTCGTG	50°C	[148] (soybean)
	r: TGAATAACTTACCCATGAATC		[148] (soybean)
rpl16–rps3	f: ACTGAACAGGCGGGTACA	50°C	[148] (soybean)
	r: ATCCGAAGCGATGCGTTG		[148] (soybean)
ndhD–ndhE	f: GAAAATTAAGGAACCCGCAA	56°C	[99] (tobacco)
	r: TCAACTCGTATCAACCAATC		[99] (tobacco)

*f, forward primer; r, reverse primer

Table 4. Primers used for amplification of nine non-coding regions of soybean cpDNA.

In most studied species, the *trnL* (UAA) intron ranges in size from 254 - 767 bp. Its smaller fragment – the P6 loop – reaches a length of 10 - 143 bp. It is commonly applied in DNA barcoding. Its main limitation lies in its low homologousness with the species from the Gene Bank, which amounts to 67.3%, while the homologousness of the P6 loop is 19.5%. However, it also has some advantages: conservative primers projected form and trouble-free amplification process. Amplification of the P6 loop can be performed even in a very degraded DNA. The intron is well known and its sequences are used to determine phylogenetic relationships between closely related species or to identify a plant species [152]. The first universal primers for this region were designed more than 20 years ago [119]. However, it does not belong to the most variable non-coding regions in chloroplast DNA [108]. The *trnL* (UAA) intron is the only one belonging to group I introns in chloroplast DNA, which means that its secondary structure is highly conservative, with a possibility of changes in its conservative [113] and variability in regions [99, 153]. Consequently, comparing the diversity of the *trnL* intron sequences allows to obtain new primers that contain conservative regions and amplify short sections contained between them [152].

Thus, in angiosperms, using non-coding regions in research at lower levels of the genome is a routine practice [108]. A large number of non-coding regions of cpDNA has been located in angiosperms, some of which are highly variable, whereas others show relatively small variability [108]. In studying the chloroplast genome, many researchers looked for universal primers that would allow amplification of many non-coding regions of cpDNA (Table 4) [111, 113, 148, 150].

4. Conclusion

In phylogenetic and population studies of *Glycine*, genetic information contained not only in cpDNA but also in mtDNA are often analysed. Organelle DNA can be used to find species-specific molecular markers. Molecular markers are an important tool to systematize the species because their use allows for detecting the differences in the genes directly. The selection of appropriate sequences, which depends on the taxonomic level at which reconstruction of the origin is carried out is very important. The initial selection concerns non-conservative sequences, which are subject to fast evolution, because the more related the specimen are, the more changeable the region should be. The relatively slow rate of evolution of certain sequences may exclude statistically significant analyses within families or species, while the study of relationship between species, which phylogenetically are very distant, using more slowly evolving sequences can be very useful. Non-coding sequences show a faster rate of evolution than the coding sequences. These regions accumulate a greater number of insertion/deletion or substitution than the non-coding regions, and therefore may be more suitable for research at inter-or intra-genus levels.

Author details

Lidia Skuza*, Ewa Filip and Izabela Szućko

*Address all correspondence to: skuza@univ.szczecin.pl

Cell Biology Department, Faculty of Biology, University of Szczecin, Poland

References

[1] Petit, R. J., Duminil, J., Fineschi, S., Hampe, A., Salvini, D., & Vendramin, G. G. (2005). Comparative organization of chloroplast, mitochondrial and nuclear diversity in plant populations. *Molecular Ecology*, 14, 689-701.

[2] Wagner, D. B. (1992). Nuclear, chloroplast, and mitochondrial DNA polymorphisms as biochemical markers in population genetic analyses of forest trees. *New Forests*, 6, 373-390.

[3] Liepelt, S., Bialozyt, R., & Ziegenhagen, B. (2002). Wind-dispersed pollen mediates postglacial gene flow among refugia. USA. *Proceedings of the Natlional Academy of Sciences*, 99, 14590-14594.

[4] Johansen, A. D., & Latta, R. G. (2003). Mitochondrial haplotype distribution, seed dispersal and patterns of postglacial expansion of ponderosa pine. *Molecular Ecology*, 12, 293-298.

[5] Chiang, Y. C., Hung, K. H., Schaal, B. A., Ge, X. J., Hsu, T. W., & Chiang, T. Y. (2006). Contrasting phylogeographical patterns between mainland and island taxa of the Pinus luchuensis complex. *Molecular Ecology*, 15, 765-779.

[6] Jaramillo-Correa, J. P., Beaulieu, J., & Bousquet, J. (2004). Variation in mitochondrial DNA reveals multiple distant glacial refugia in black spruce (Picea mariana), a transcontinental North American conifer. *Molecular Ecology*, 13, 2735-2747.

[7] Godbout, J., Jaramillo, J. P., Beaulieu, J., & Bousquet, J. (2005). A mitochondrial DNA minisatellite reveals the postglacial history of jack pine (Pinus banksiana), a broad-range North American conifer. *Molecual Ecology*, 14, 3497-3512.

[8] Huang, S., Chiang, Y. C., Schaal, B. A., Chou, C. H., & Chiang, T. Y. (2001). Organelle DNA phylogeography of Cycas taitungensis, a relict species in Taiwan. *Molecular Ecology*, 10, 2669-2681.

[9] Pleines, T., Jakob, S. S., & Blattner, D. B. (2009). Application of non-coding DNA regions in intraspecific analyses. *Plant Systematic and Evolution*, 282, 281-294.

[10] Tautz, D. (1989). Hypervariability of simple sequences as a general source for polymorphic DNA markers. *Nucleic Acids Research*, 17, 6463-6471.

[11] Tautz, D., Trick, M., & Dover, G. A. (1986). Cryptic simplicity in DNA is a major source of genetic variation. *Nature*, 322, 652-656.

[12] Lagercrantz, U., Ellegren, H., & Andersson, L. (1993). The abundance of various polymorphic microsatellite motifs differs between plants and vertebrates. *Nucleic Acids Research*, 21, 1111-1115.

[13] Powell, W., Morgante, M., Mc Devitt, R., Vendramin, G. G., & Rafalski, J. A. (1995). Polymorphic simple sequence repeat regions in chloroplast genomes: applications to

the population genetics of pines. USA. *Proceedings of the Natlional Academy of Sciences*, 92, 7759-77763.

[14] Birky, C. W., Fuerst, P., & Maruyama, T. (1989). Organelle gene diversity under migration, mutation, and drift: equilibrium expectations, approach to equilibrium, effects of heteroplasmic cells, and comparison to nuclear genes. *Genetics*, 121, 613-627.

[15] Rendell, S., & Ennos, R. A. (2002). Chloroplast DNA diversity in Calluna vulgaris (heather) populations in Europe. *Molecular Ecology*, 11, 69-78.

[16] Hudson, R. R., & Coyne, J. A. (2002). Mathematical consequences of the genealogical species concept. *Evolution*, 56, 1557-1565.

[17] Kadereit, J. W., Arafeh, R., Somogyi, G., & Westberg, E. (2005). Terrestrial growth and marine dispersal? Comparative phylogeography of five coastal plant species at a European scale. *Taxon*, 54, 861-876.

[18] Avise, J. C. (2000). *Phylogeography: the history and formation of species*, Cambridge, Harvard University Press.

[19] Richards, M. B., Macauly, V. A., Bandelt, H. J., & Sykes, B. C. (1998). Phylogeography of mitochondrial DNA in western Europe. *Annals of Human Genetics*, 62, 241-260.

[20] Tomaru, N., Takahashi, M., Tsumura, Y., Takahashi, M., & Ohba, K. (1998). Intraspecific variation and phylogeographic patterns of Fagus crenata (Fagaceae) mitochondrial DNA. *American Journal of Botany*, 85, 629-636.

[21] Mackenzie, S., & Mc Intosh, L. (1999). Higher plant mitochondria. *Plant Cell*, 11, 571-585.

[22] Schuster, W., & Brennicke, A. (1994). The plant mitochondrial genome: physical structure, information content, RNA editing, and gene migration to the nucleus. *Annual Review of Plant Physiology and Plant Molecular Biology*, 45, 61-78.

[23] Palmer, J. D. (1992). Mitochondrial DNA in plant systematics: applicationsand limitations. *In: Soltis PS, Soltis DE, Doyle JJ. (ed.) Molecular systematics of plants*, New York, Chapman and Hall, 26-49.

[24] Wolfe, K. H., Li, W. H., & Sharp, P. M. (1987). Rates of nucleotide substitution vary greatly among plant mitochondrial, chloroplast, and nuclear DNAs. USA. *Proceedings of the Natlional Academy of Sciences*, 84, 9054-9058.

[25] Larsson, N. G., & Clayton, D. A. (1995). Molecular genetic aspects of human mitochondrial disease. *Annual Review of Genetics*, 29, 151-178.

[26] Binder, S., Marchfelder, A., & Brennicke, A. (1996). Regulation of gene expression in plant mitochondria. *Plant Molecular Biology*, 32, 303-314.

[27] Unseld, M., Marienfeld, J., Brandt, P., & Brennicke, A. (1997). The mitochondrial genome of Arabidopsis thaliana contains 57 genes in 366,924 nucleotides. *Nature Genetics*, 15, 57-61.

[28] Oda, K., Yamato, K., Ohata, E., Nakamura, Y., Takemura, M., Nozato, N., Akashi, K., Kanrgae, T., Ogura, Y., Kohchi, T., & Ohyama, K. (1992). Gene organization from the complete sequence of liverwort, Marchantia polymorpha, mitochondrial DNA: a primitive form of plant mitochondrial genome. *Journal of Molecular Biology*, 223, 1-7.

[29] Backert, S., Nielsen, B. L., & Borner, T. (1997). The mystery of the rings: structure and replication of mitochondrial genomes from higher plants. *Trends in Plant Science*, 2, 477-483.

[30] Andre, C. P., & Walbot, V. (1995). Pulsed-field gel mapping of maize mitochondrial chromosomes. *Molecular and General Genetics*, 247, 255-263.

[31] Bendich, A. J. (1993). Reaching for the ring: the study of mitochondrial genome structure. *Current Genetetics*, 24, 279-290.

[32] Fauron, C., Casper, M., Gao, Y., & Moore, B. (1995). The maize mitochondrial genome: dynamic, yet functional. *Trends in Genetics*, 11, 228-235.

[33] Fauron, C. M. R., Casper, M., Gesteland, R., & Albertsen, M. (1992). A multi-recombination model for the mtDNA rearrangements seen in maize cmsT regenerated plants. *The Plant Journal*, 2, 949-958.

[34] Palmer, J. D., & Herbon, L. A. (1988). Plant mitochondrial DNA evolves rapidly in structure, but slowly in sequence. *Journal of Molecular Evolution*, 28, 87-97.

[35] Palmer, J. D., & Shields, C. R. (1984). Tripartite structure of the Brassica campestris mitochondrial genome. *Nature*, 307, 437-440.

[36] Palmer, J. D., Makaroff, C. A., Apel, I. J., & Shirzadegan, M. (1990). Fluid structure of plant mitochondrial genomes: evolutionary and functional implications. *In: Clegg MT, O'Brien SJ (ed.) Molecular Evolution*, New York, Alan R. Liss, 85-96.

[37] Mosig, G. (1987). The essential role of recombination in phage T4 growth. *Annual Review of Genetics*, 21, 347-371.

[38] Mosig, G. (1998). Recombination and recombinationdependent DNA replication in bacteriophage T4. *Annual Review of Genetics*, 32, 379-413.

[39] Oldenburg, D. J., & Bendich, A. J. (2001). Mitochondrial DNA from the liverwort Marchantia polymorpha: circularly permuted linear molecules, head-to-tail concatemers, and a 50 protein. *Journal of Molecular Biology*, 310, 549-562.

[40] Backert, S., Dorfel, P., & Borner, T. (1995). Investigation of plant organellar DNA by pulsed-field gel electrophoresis. *Current Genetics*, 28, 390-399.

[41] Backert, S., Lurz, R., Oyarzabal, O. A., & Borner, T. (1997). High content, size and distribution of singlestranded DNA in the mitochondria of Chenopodium album (L.). *Plant Molecular Biology*, 33, 1037-1050.

[42] Bendich, A. J., & Smith, S. B. (1990). Moving pictures and pulsed-field gel electrophoresis show linear DNA molecules from chloroplasts and mitochondria. *Current Genetics*, 17, 421-425.

[43] Bendich, A.J. (1996). Structural analysis of mitochondrial DNA molecules from fungi and plants using moving pictures and pulsed-field gel electrophoresis. *Journal of Molecular Biology*, 225, 564-588.

[44] Oldenburg, D. J., & Bendich, A. J. (1996). Size and structure of replicating mitochondrial DNA in cultured tobacco cells. *Plant Cell*, 8, 447-461.

[45] Sena, E. P., Revet, B., & Moustacchi, E. (1986). In vivo homologous recombination intermediates of yeast mitochondrial DNA analyzed by electron microscopy. *Molecular and General Genetics*, 202, 421-428.

[46] Palmer, J. D., & Herbon, L. A. (1987). Unicircular structure of the Brassica hirta mitochondrial genome. *Current Genetics*, 11, 565-570.

[47] Stern, D. B., & Palmer, J. D. (1984). Recombination sequences in plant mitochondrial genomes: Diversity and homologies to known mitochondrial genes. *Nucleic Acid Research*, 12, 6141-6157.

[48] Chanut, F. A., Grabau, E. A., & Gesteland, R. F. (1993). Complex organization of the soybean mitochondrial genome: recombination repeats and multiple transcripts at the atpA loci. *Current Genetics*, 23, 234-247.

[49] Dewey, R. E., Levings, C. S., III., & Timothy, D. H. (1986). Novel recom- binations in the maize mitochondrial genome produce a unique transcriptional unit in the Texas male-sterile cytoplasm. *Cell*, 44, 439.

[50] Laver, H. K., Reynolds, S. J., Moneger, F., & Leaver, C. J. (1991). Mitochondrial genome organization and expression associated with cytoplasmic male sterility in sunflower (Helianthus annuus). *Plant Journal*, 1, 185-193.

[51] Johns, C., Lu, M., Lyznik, A., & Mackenzie, S. (1994). A mitochondrial DNA sequence is associated with abnormal pollen development in cytoplasmic male sterile bean plants. *Plant Cell*, 4, 435-449.

[52] Martinez-Zapater, J. M., Gil, P., Capel, J., & Somerville, C. R. (1992). Mutations at the Arabidopsis CHM locus promote rearrangements of the mitochondrial genome. *Plant Cell*, 4, 889-899.

[53] Newton, K., Knudsen, C., Gabay-Laughnan, S., & Laughnan, J. (1990). An abnormal growth mutant in maize has a defective mitochondrial cytochrome oxidase gene. *Plant Cell*, 2, 107-113.

[54] Levings, C. S., III., & Pring, D. R. (1979). Mitochondrial DNA of higher plants and genetic engineering. In: Setlow JK, Hollaender A (ed.). *Genetic engineering*, 1, New York, Plenum Press, 205-222.

[55] Bailey-Serres, J., Leroy, P., Jones, S. S., Wahleithner, J. A., & Wolstenholme, D. R. (1987). Size distribution of circular molecules in plant mitochondrial DNAs. *Current Genetics*, 12, 49-53.

[56] Grabau, E., Davis, W. H., & Gengenbach, B. G. (1989). Restriction fragment length polymorphism in a subclass of the mandarin soybean. *Crop Science*, 29, 1554-1559.

[57] Synenki, R. M., Levings, C. S., III., & Shah, D. M. (1978). Physicochemical characterization of mitochondrial DNA from soybean. *Plant Physiology*, 61, 460-464.

[58] Kanazawa, A., Tozuka, A., Kato, S., Mikami, T., Abe, J., & Shimamoto, Y. (1998). Small interspersed sequences that serve as recombination sites at the cox2 and atp6 loci in the mitochondrial genome of soybean are widely distributed in higher plants. *Current Genetics*, 33, 188-198.

[59] Kato, S., Kanazawa, A., Mikami, T., & Shimamoto, Y. (1998). Evolutionary changes in the structures of the cox2 and atp6 loci in the mitochondrial genome of soybean involving recombination across small interspersed sequences. *Current Genetics*, 34, 303-312.

[60] Grabau, E. A., Davis, W. H., Phelps, N. D., & Gengenbach, B. G. (1992). Classification of soybean cultivars based on mitochondrial DNA restriction fragment length polymorphisms. *Crop Science*, 32, 271-274.

[61] Moeykens, C. A., Mackenzie, S. A., & Shoemaker, R. C. (1995). Mitochondrial genome diversity in soybean: repeats and rearrangements. *Plant Molecular Biology*, 29, 245-254.

[62] Hanlon, R. W., & Grabau, E. A. (1997). Comparison of mitochondrial organization of four soybean cytoplasmic types by restriction mapping. *Soybean Genetics Newsletter*, 24, 208-210.

[63] Khazi, F. R., Edmondson, A. C., & Nielsen, B. L. (2003). An Arabidopsis homologue of bacterial RecA that complements an E. coli recA deletion is targeted to plant mitochondria. *Molecular and General Genetics*, 269, 454-463.

[64] Manchekar, M., Scissum-Gunn, K., Song, D., Khazi, F., Mc Lean, S. L., & Nielsen, B. L. (2006). DNA Recombination Activity in Soybean Mitochondria. *Journal of Molecular Bioliology*, 356, 288-299.

[65] Grabau, E. A., Havlik, M., & Gesteland, R. F. (1988). Chimeric Organization of Two Genes for the Soybean Mitochondrial Atpase Subunit 6. *Current Genetics*, 13, 83-89.

[66] Wissinger, B., Brennicke, A., & Schuster, W. (1992). Regenerating good sense: RNA editing and trans-splicing in plant mitochondria. *Trends in Genetics*, 8, 322-328.

[67] Hanlon, R., & Grabau, E. A. (1995). Cytoplasmic diversity in old domestic varieties of soybean using two mitochondrial markers. *Crop Science*, 35, 1148-1151.

[68] Grabau, E. A., & Davis, W. H. (1992). Cytoplasmic diversity in Glycine soja. *Soybean Genetics Newslett*, 19, 140-144.

[69] Tozuka, A., Fukushi, H., Hirata, T., Ohara, M., Kanazawa, A., Mikami, T., Abe, J., & Shimamoto, Y. (1998). Composite and clinal distribution of Glycine soja in Japan revealed by RFLP analysis of mitochondrial. *DNA Theoretical and Applied Genetics*, 96, 170-176.

[70] Schuster, W., & Brennicke, A. (1987). Nucleotide sequence of the Oenothera ATPase subunit 6 gene. Nucleic Acids Research ., 15, 9092.

[71] Braun, C. J., & Levings, C. S., III. (1985). Nucleotide Sequence of the F(1)-ATPase alpha Subunit Gene from Maize Mitochondria. *Plant Physiology*, 79(2), 571-577.

[72] Isaac, G. I., Brennicke, A., Dunbar, M., & Leaver, C. J. (1985). The mitochondrial genome of fertile maize (Zea mays L.) contains two copies of the gene encoding the alpha-subunit of the F1-ATPase. *Current Genetics*, 10(4), 321-328.

[73] Schuster, W., & Brennicke, A. (1986). Pseudocopies of the ATPase a-subunit gene in Oenothera mitochondria are present on different circular molecules. *Molecular and General Genetics*, 204, 29-35.

[74] Morikami, A., & Nakamura, K. (1987). *Structure and expression of pea mitochondrial F1ATPase alpha-subunit gene and its pseudogene involved in homologous recombination, J Biochem*, 101, 967-976.

[75] Chaumont, F., Boutry, M., Briquet, M., & Vassarotti, A. (1988). Sequence of the gene encoding the mitochondrial F1ATPase alpha subunit from Nicotiana plumbaginifolia. *Nucleic Acids Research*, 16(13), 6247.

[76] Schulte, E., Staubach, S., Laser, B., & Kück, U. (1989). Wheat mitochondrial DNA: organization and sequences of the atpA and atp9 genes. *Nucleic Acids Research*, 17(18), 7531.

[77] Kadowaki, K., Kazama, S., & Suzuki, S. (1990). Nucleotide sequence of the F,-ATPase a subunit genes from rice mitochondria. *Nucleic Acids Research*, 18, 1302.

[78] Köhler, R. H., Lössl, A., & Zetsche, K. (1990). Nucleotide sequence of the F1ATPase alpha subunit gene of sunflower mitochondria. *Nucleic Acids Research*, 18, 4588.

[79] Makaroff, C. A., Apel, I. J., & Palmer, J. D. (1990). Characterization of radish mitochondrial atpA: influence of nuclear background on transcription of atpA-associated sequences and relationship with male sterility. *Plant Molecular Biology*, 15(5), 735-746.

[80] Siculella, L., D'Ambrosio, L., De Tuglie, A. D., & gauerani, R. (1990). Minor differences in the primary structures of atpa genes coded on the mt DNA of fertile and male sterile sunflower lines. *Nucleic Acids Research*, 18(15), 4599.

[81] Handa, H., & Nakajima, K. (1991). Nucleotide sequence and transcription analyses of the rapeseed (Brassica napus L.) mitochondrial F1ATPase alpha-subunit gene. *Plant Molecular Biology*, 16, 361-364.

[82] Chase, C. D., & Ortega, V. M. (1992). Organization of ATPA coding and 3' flanking sequences associated with cytoplasmic male sterility in Phaseolus vulgaris. *L. Current Genetics*, 22, 147-153.

[83] Zurawski, G., Bottomley, W., & Whitfeld, P. R. (1982). Structures of the genes for the beta and c subunits of spinach chloroplast ATPase indicates a dicistronic mRNA and an overlapping translation stop/start signal. USA. *Proceedings of the Natlional Academy of Sciences*, 79, 6260-6264.

[84] Krebbers, E. T., Larrinua, I. M., Mc Intosh, L., & Bogorad, L. (1982). The maize chloroplast genes for the beta and epsilon subunits of the photosynthetic coupling factor CF1 are fused. *Nucleic Acids Research*, 10(16), 4985-5002.

[85] Walker, J. E., Fearnley, I. M., Gay, N. J., Gibson, B. W., Northrop, F. D., Powell, S. J., Runswick, M. J., Saraste, M., & Tybulewicz, V. L. J. (1985). Primary structure and subunit stoichiometry of F1-ATPase from bovine mitochondria. *Journal of Molecular Biology*, 184(4), 677-701.

[86] Shimamoto, Y., Fukushi, H., Abe, J., Kanazawa, A., Gai, J., Gao, Z., & Xu, D. (1998). RFLPs of chloroplast and mitochondrial DNA in wild soybean, Glycine soja, growing in China. *Genetic Resources and Crop Evolution*, 45, 433-439.

[87] Abe, J. (1999, 13th-15th October). The genetic structure of natural populations of wild soybeans revealed by isozymes and RFLP's of mitochondrial DNS's: possible influence of seed dispersal, cross pollination and demography. Japan. *In: The Seventh Ministry of Agriculture, Forestry and Fisheries (MAFF), Japan International Workshop on Genetic Resources Part I Wild Legumes*, National Institute of Agrobiological Resources Tsukuba, Ibaraki, 143-159.

[88] Shimamoto, Y. (2001). Polymorphism and phylogeny of soybean based on chloroplast and mitochondrial DNA analysis. *Japan Agricultural Research Quarterly (JARQ)*, 35(2), 79-84.

[89] Cleeg, M. T., & Zurawski, G. (1992). Chloroplast DNA and the study of plant phylogeny. New York, *In: Soltis PS, Soltis DE, Doyle JJ (eds.) Present status and future procpects in Molecular Systematics of Plants*, Chapman &Hall, 1-13.

[90] Olmstead, R. G., & Palmer, J. D. (1994). Chloroplast DNA systematics: a review of method and data analysis. *American Journal of Botany*, 81, 1205-1224.

[91] Chase, M., & 41others, . (1993). Phylogenetics of seed plants: An analysis of nucleotide sequences from the plastid gene rbcL. *Annals Missouri Botany Gardens*, 80, 528-580.

[92] Parkinson, C. L., Adams, L., & Palmer, D. (1999). Multigene analyses identify the three earliest lineages of extant flowering plants. *Current Biology*, 9, 1485-1488.

[93] Qui, Y. L. (1999). The earliest angiosperms: Evidence from mitochondrial, plastid, and nuclear genomes. *Nature*, 402, 404-407.

[94] Soltis, D. E., Soltis, P. S., & Doyle, J. J. (1998). Molecular systematic of Plants II. *DNA sequencing*, Boston MA, Kluwer.

[95] Soltis, P. S., Soltis, D. E., & Chase, M. W. (1999). Angiosperm phylogeny inferred from multiple genes as a tool for comparative biology. *Nature*, 402, 402-404.

[96] Kato, T., Kaneko, T., Sato, S., Nakamura, Y., & Tabata, S. (2000). Complete structure of the chloroplast genome of a legume, Lotus japonicus. *DNA Research*, 7, 323-330.

[97] Li, W. H. (1997). *Molecular Evolution*, Sunderland, Sinauer Associates, 487.

[98] Saski, C.h., Lee, S. B., Daniell, H., Wood, T. C., Tomkins, J., Kim, H. G., & Jansen, R. K. (2005). Complete chloroplast genome sequence of Glycine max and comparative analyses with other legume genomes. *Plant Molecular Biology*, 59, 309-322.

[99] Shinozaki, K., Ohme, M., Tanaka, M., Wakasugi, T., Hayashida, N., Matsubayashi, T., Zaita, N., Chunwongse, J., Obokata, J., Yamaguchi-Shinozaki, K., Ohto, C., Torazawa, K., Yamada, K., Kusuda, J., Takaiwa, F., Kato, A., Tohdoh, N., Shimada, H., & Sugiura, M. (1986). The complete nucleotide sequence of tobacco chloroplast genome: its gene organization and expression. *EMBO Journal*, 5, 2043-2049.

[100] Sugiura, M., & Shimada, H. (1991). Fine- structural features of the chloroplast genome- comparison of the sequenced chloroplast genomes. *Nucleic Acids Research*, 19, 983-995.

[101] Sugiura, M., & Sugita, M. (1992). The chloroplast genome. *Plant Molecular Biology*, 19, 149-168.

[102] Woźny, A. (1990). *Wykłady i ćwiczenia z biologii komórki*, Warszawa, Wydawnictwo Naukowe PWN.

[103] Palmer, J. D. (1985). Evolution of chloroplast and mitochondrial DNA in plants and algae. *In: MacIntyre RJ. (ed.) Monographs in Evolutionary Biology*, New York:, Molecular Evolutionary Genetics Plenum Press, 131-240.

[104] Wolfe, K.H. (1988). The site of deletion of the inverted repeat is pea chloroplast DNA contains duplicated gene fragments. *Current Genetic*, 13, 97-99.

[105] Palmer, J. D., Osorio, B., Aldrich, J., & Thompson, W. F. (1987). Chloroplast DNA evolution among legumes: loss of a large inverted repeat occurred prior to other sequence rearrangements. *Current Genetic*, 11, 275-286.

[106] Lavin, M., Doyle, J. J., & Palmer, J. D. (1990). Evolutionary significance of the loss of the chloroplast-DNA inverted repeat in the Leguminosae subfamily Papilionoideae. *Evolution*, 44, 390-402.

[107] Herdenberger, F., Pillay, D. T. N., & Steinmetz, A. (1990). Sequence of the trnH gene and the inverted repeat structure deletion of the broad bean chloroplast genome. *Nucleic Acids Research*, 18, 1297.

[108] Shaw, J., Lickey, E. B., Beck, J. T., Farmer, S. B., Liu, W., Miller, J., Siripun, K. C., Winder, C. T., Schilling, E. E., & Small, R. L. (2005). The tortoise and the here II: rela-

tive utility of 21 noncoding chloroplast DNA sequences for phylogenetic analysis. *American Journal of Botany*, 92, 142-166.

[109] Ravi, V., Khurana, J. P., Tyagi, A. K., & Khurana, P. (2008). An update on chloroplast genomes. *Plant Systematics and Evolution*, 271101-122.

[110] Kelchner, S. A. (2000). The evolution of non-coding chloroplast DNA and its application in plant systematics. *Annals of the Missouri Botanical Garden*, 87, 482-498.

[111] Perry, A. S., Brennan, S., Murphy, D. J., & Wolfe, K. H. (2006). Evolutionary re-organisation of a large operon in Adzuki bean chloroplast DNA caused by inverted repeat movement. *DNA Research*, 9, 157-162.

[112] Close, P. S., Shoemaker, R. C., & Keim, P. (1989). Distribution of restriction site polymorphism within the chloroplast genome of the genus Glycine, subgenus Soja. *Theoretical and Applied Genetics*, 77, 768-776.

[113] Xu, D. H., Abe, J., Kanazawa, A., Sakai, M., & Shimamoto, Y. (2000). Sequence variation of non-coding regions of chloroplast DNA of soybean and related wild species and its implications for the evolution of different chloroplast haplotypes. *TAG Theoretical and Applied Genetics*, 101, 5-6, 724-732.

[114] Kajita, T., Ohashi, H., Tateishi, Y., Bailey, C. D., & Doyle, J. J. (2001). rbcL and legume phylogeny, with particular reference to Phaseoleae, Millettieae, and allies. *Systematic Botany*, 26, 515-536.

[115] Nickret, D. L., & Patrick, J. A. (1998). The nuclear ribosomal DNA intergenic spacer of wild and cultivated soybean have low variation and cryptic subrepeats. *NRC Canada. Genome*, 41183-192.

[116] Palmer, J. D., Adams, K. L., Cho, Y., Parkinson, C. L., Qiu, Y. L., & Song, K. (2000). Dynamic evolution of plant mitochondrial genomes: mobile genes and introns and highly variable mutation rates. U.S.A., *Proceedings of the National Academy of Sciences*, 97, 6960-6966.

[117] Pennington, R. T., Klitgaard, B. B., Ireland, H., & Lavin, M. (2000). New insights into floral evolution of basal Papilionoideae from molecular phylogenies. *In: PS Herendeen A Bruneau (ed.) Advances in Legume Systematics Kew*, UK, 9, 233-248.

[118] Wojciechowski, M. F., Lavin, M., & Sanderson, M. J. (2004). A phylogeny of legumes (Leguminosae) based on analysis of the plastid matK gene resolves many well-supported subclades within the family. *American Journal of Botany*, 91, 1846-1862.

[119] Taberlet, P., Gielly, L., Pautou, G., & Bouvet, J. (1991). Universal primers for amplification of three non-coding regions of chloroplast DNA. *Plant Molecular Biology*, 17, 1105-1109.

[120] Wolfe, A. D., Elisens, W. J., Watson, L. E., & Depamphilis, C. W. (1997). Using restriction-site variation of PCR-amplified cpDNA genes for phylogenetic analysis of Tribe Cheloneae (Scrophulariaceae). *American Journal of Botany*, 84, 555-564.

[121] Sehgal, D., & Raina, S. N. (2008). DNA markers and germplasm resource diagnostics: new perspectives in crop improvement and conservation strategies. *In: Arya ID, Arya S. (ed.) Utilization of biotechnology in plant sciences*, Microsoft Printech (I) Pvt. Ltd., Dehradun, 39-54.

[122] Varshney, R. K., Graner, A., & Sorrells, M. E. (2005). Genic microsatellite markers in plants: features and applications. *Trends Biotechnology*, 23, 48-55.

[123] Varshney, R. K., Thudi, M., Aggarwal, R., & Börner, A. (2007). Genic molecular markers in plants: development and applications. *In: Varshney RK, Tuberosa R (ed.) Genomics assisted crop improvement: genomics approaches and platforms*, Springer, Dordrecht, 1, 13-29.

[124] Powell, W., Machray, G. C., & Provan, J. (1996). Polymorphism revealed by simple sequence repeats. *Trends Plant Sciences*, 1, 215-222.

[125] Russell, J. R., Fuller, J. D., Macaulay, M., Hatz, B. G., Jahoor, A., Powell, W., & Waugh, R. (1997). Direct comparison of levels of genetic variation among barley accessions detected by RFLPs, AFLPs, SSRs and RAPDs. *Theoretical and Applied Genetics*, 95, 714-722.

[126] Bohn, M., Utz, H. F., & Melchinger, A. E. (1999). Genetic similarities among winter wheat cultivars determined on the basis of RFLPs, AFLPs, and SSRs and their use for predicting progeny variance. *Crop Sciences*, 39, 228-237.

[127] Parida, S. K., Kalia, S. K., Sunita, K., Dalal, V., Hemaprabha, G., Selvi, A., Pandit, A., Singh, A., Gaikwad, K., Sharma, T. R., Srivastava, P. S., Singh, N. K., & Mohapatra, T. (2009). Informative genomic microsatellite markers for efficient genotyping applications in sugarcane. *Theoretical and Applied Genetics*, 118, 327-338.

[128] Shimamoto, Y., Hasegawa, A., Abe, J., Ohara, M., & Mikami, T. (1992). Glycine soja germplasm in Japan: isozymes and chloroplast DNA variation. *Soybean Genet Newsletter*, 19, 73-77.

[129] Xu, D. H., Abe, J., Gai, J. Y., & Shimamoto, Y. (2002). Diversity of chloroplast DNA SSRs in wild and cultivated soybeans: evidence for multiple origins of cultivated soybean. *Theoretical and Applied Genetics*, 105, 645-653.

[130] Xu, D. H., Abe, J., Kanazawa, A., Gai, Y., & Shimamoto, Y. (2001). Identification of sequence variations by PCR-RFLP and its application to the evaluation of cpDNA diversity in wild and cultivated soybeans. *Theoretical and Applied Genetics*, 102, 683-688.

[131] Hirata, T., Abe, J., & Shimamoto, Y. (1996). RFLPs of chloroplast and motochondrial genomes in summer and autumn maturing cultivar groups of soybean in Kyushu district of Japan. *Soybean Genetic Newsletter*, 23, 107-111.

[132] Shimamoto, A., Fukushi, H., Abe, J., Kanazawa, A., Gai, J., Gao, Z., & Xu, D. (1998). RFLPs of chloroplast and mitochondrial DNA in wild soybean, Glycine soja, growing in China. *Genetic Resources and Crop Evolution*, 45, 433-439.

[133] Chen, J., & Hebert, P. D. N. (1999). Directed termination of the polymerase chain reaction: kinetics and application in mutation detection. *Genome*, 42, 72-79.

[134] Orita, M., Suzuki, T. S., & Hayashi, K. (1989). Rapid and sensitive detection of point mutation and DNA polymorphisms using the polymerase chain reaction. *Genomics*, 5, 874-879.

[135] Sarkar, G., Yoon, H., & Sommer, S. S. (1995). Dideoxy fingerprinting (ddF): a rapid and efficient screen for the presence of mutations. *Genomics*, 13, 441-443.

[136] Manen, J. F., & Natall, A. (1995). Comparison of the evolution of ribulose- 1, 5-bisphosphate carboxylase (rbcL) and atpB-rbcL non-coding spacer sequences in a recent plant group, the tribe Rubieae (Rubiaceae). *Journal of Molecular Evolution*, 41, 920-927.

[137] Jordan, W. C., Courtney, M. W., & Neigel, J. E. (1996). Low levels of intraspecific genetic variation at a rapidly evolving chloroplast DNA locus in North American duckweeds (Lemnaceae). *American Journal of Botany*, 83, 430-439.

[138] Sang, T., Crawford, D. J., & Stuessy, T. F. (1997). Chloroplast DNA phylogeny, reticulate evolution, and biogeography of Paeonia (Paeoniaceae). *American Journal of Botany*, 84, 1120-1136.

[139] Small, R. L., Ryburn, J. A., Cronn, R. C., Seelanan, T., & Wendel, J. F. (1998). The tortoise and the hare: choosing between noncoding plastome and nuclear Adh sequences for phylogeny reconstruction in a recently diverged plant group. *American Journal of Botany*, 85, 1301-1315.

[140] Mc Dade, L. A., & Moody, M. L. (1999). Phylogenetic relationship among Acanthaceae: evidence from noncoding trnL-trnFF chloroplast DNA sequences. *American Journal of Botany*, 86, 70-80.

[141] Molvray, M., Kores, P. J., & Chase, M. W. (1999). Phylogenetic relationships within Korthalsella (Viscaceaek) based on nuclear ITS and plastid trnL-F sequence data. *American Journal of Botany*, 86, 249-260.

[142] Demesure, B., Comps, B., & Petit, R. J. (1996). Chloroplast DNA phylogeography of the common beach (Fagus sylvatica L.) in Europe. *Evolution*, 50, 2515-2520.

[143] Wolfe, A. D., Elisens, W. J., Watson, L. E., & Depamphilis, C. W. (1997). Using restriction-site variation of PCR-amplified cpDNA genes for phylogenetic analysis of Tribe Cheloneae (Scrophulariaceae). *American Journal of Botany*, 84, 555-564.

[144] Asmussen, C. B., & Liston, A. (1998). Chloroplast DNA characters, phylogeny, and classification of Lathyrus (Fabaceae). *American Journal of Botany*, 85, 387-40.

[145] Cipriani, G., Testolin, R., & Gardner, R. (1998). Restriction-site variation of PCR-amplified chloroplast DNA regions and its implication for the evolution of Actinidia. *Theoretical and Applied Genetics*, 96, 389-396.

[146] Friesen, N., Pollner, S., Bachmann, K., & Blattner, F. R. (1999). RAPDs and noncoding chloroplast DNA reveal a single origin of the cultivated Allium fistulosum from A. altaicum (Alliaceae). *American Journal of Botany*, 86, 554-562.

[147] Close, P. S., Shoemaker, R. C., & Keim, P. (1989). Distribution of restriction site polymorphism within the chloroplast genome of the genus Glycine, subgenus Soja. *Theoretical and Applied Genetics*, 77, 768-776.

[148] Kanazawa, A., Tozuka, A., & Shimamoto, Y. (1998). Sequence variation of chloroplast DNA that involves EcoRI and ClaI restriction site polymorphisms in soybean. *Genes Genetic Systems*, 73, 111-119.

[149] Downie, S. R., Olmstead, R. G., Zurawski, G., Soltis, D. E., Soltis, P. S., Watson, C., & Palmer, J. D. (1991). Six independents losses of the chloroplast DNA rpl2 intron in dicotyledons: Molecular and phylogenetic implications. *Evolutions*, 45, 1245-1259.

[150] Spielmann, A., & Stutz, E. (1983). Nucleotide sequence of soybean chloroplast DNA regions which contain the psbA and trnH genes and cover the ends of the large single-copy region and one end of the inverted repeats. *Nucleic Acids Research*, 11, 7157-7167.

[151] Sugiura, M., & Shimada, H. (1991). Fine- structural features of the chloroplast genome- comparison of the sequenced chloroplast genomes. *Nucleic Acids Research*, 19, 983-995.

[152] Taberlet, P., Coissac, E., Popanon, F., Gielly, L., Miquel, C., Valentini, A., Vermat, T., Corthier, G., Brochmann, C., & Willerslev, E. (2006). Power and limitations of the chloroplast trnL (UAA) intron for plant DNA barcoding. *Nucleic Acids Research*, 35(3).

[153] Palmer, J. D. (1991). Plastid chromosomes: Structure and Evolution. *In: Bogorag L., Vasil IK (ed.) Cell culture and somatic cell genetic of plants*, New York, Academic Press, 7, 5-53.

Chapter 27

Comparative Studies Involving Transgenic and Non-Transgenic Soybean: What is Going On?

Marco Aurélio Zezzi Arruda,
Ricardo Antunes Azevedo,
Herbert de Sousa Barbosa,
Lidiane Raquel Verola Mataveli,
Silvana Ruella Oliveira,
Sandra Cristina Capaldi Arruda and
Priscila Lupino Gratão

Additional information is available at the end of the chapter

http://dx.doi.org/10.5772/52212

1. Introduction

There is no doubt that soybean [*Glycine max* (L.) Merrill] has a global importance with widespread applicability (food, biodiesel, secondary metabolites, among others) and economic value of its products in the global market [1] (*ca.* US$ 38.9 billion is the estimated crop value in 2010 [2]). In terms of crop production, USA and Brazil occupy the first and second position in the world, with *ca.* 83 and 70 million ton, respectively [3]. These facts may explain the significant interest by biotechnological industries and research institutes in enhancing some characteristics of this crop such as nutrient quality, resistance to pests, and other subjects.

One of the most effective processes for attaining this task is the genetic modification, and clearly the one conferring resistance to glyphosate (the number one selling herbicide worldwide since the 80's [4]) is considered the main soybean genetic modification. This process involves the insertion of a gene (cp4 EPSPS: 5-enolpyruvylshikimate-3 phosphate synthase), via biobalistic (acceleration of metallic particles recovered with genetic material), which is responsible to the production of cp4-EPSPS enzyme from *Agrobacterium sp.* Such enzyme confers tolerance to glyphosate (N-fosfometil glycine), once that this substance inhibits the action of 5-enolpyruvylshiquimate synthase – EPSPS enzyme, which is involved in the biosynthesis of aromatic amino acids. Its inhibition provokes delay in the development of plants, amino acids unbalance, and death of plants [5]. As a consequence it is easy to rationalize that plants genetically modified can normally develop in the presence of this herbicide, being the excellent result obtained in terms of soybean crop production also attributed to the transgenic cultures used. Exemplifying, *ca.* 49 million hectares (60% of the cultivated world area) are occupied nowadays by transgenic soybean culture [6].

In this way, our hypothesis is that the genetic modification itself is contributing for changing a variety of characteristics of this organism, producing alterations, in a cascade manner, to the metabolism. As a result of these modifications, the genetically modified soybean is apparently searching a new equilibrium as a living organism in nature.

In order to evaluate this hypothesis, our research group has been, for the last eight years, carrying out some comparative studies taking into account alterations in proteins, metalloproteins, metals and enzymes. Since the content of proteins in soybean seeds is high (*ca.* 40%) [7], hundreds of proteins are expected to be found after a separation process, making our proposal a hard task. Then, the utilization of the most up-to-date analytical techniques presenting high resolution in terms of enzymes, proteomics and metallomics approaches is almost imperative.

In this chapter, the concept of plant stress is mainly one related to oxidative stress, followed by a variety of examples regarding soybean. Additionally, basic concepts of proteomics and metallomics will be described, followed by a compilation of the results from all strategies and techniques that we have been adopting along the period devoted to the study of transgenic soybean, which were utilized for corroborating our previous hypothesis. Other examples in the literature are also presented in order to support our data. Then, techniques based on bidimensional chromatographic and non-chromatographic protein separations (*i.e.* 2D-HPLC, 2D PAGE), image analysis for protein expression evaluations (*i.e.* 2D DIGE), inorganic mass spectrometry for identification/quantification of metals (*i.e.* HR-SF-ICP-MS, ICP-MS, LA-ICP-MS), organic mass spectrometry for characterization of proteins (*i.e.* MALDI-QTOF-MS, ESI-LC-MS-MS), and hyphenated techniques for improving the quality on protein information (*i.e.* 2D-HPLC-ICP-MS) will be also emphasized. In the end of this chapter, a section of future Trends is provided, putting in evidence, in our point of view, some other strategies to be adopted for an in-depth investigation of this transgenic crop.

Finally, it is important to stress that the main goal of this chapter, and also of our studies, is only to present those results found within a series of projects developed by our research group concerning transgenic soybean. Despite the awareness of a public disagreement about the cultivation and commercialization of transgenic soybean, this chapter does not have the intention neither to defend genetic modification nor to make any criticisms to it.

2. Plant stress

In this section we have decided to focus our attention on the antioxidant responses trigged by some key biotic and abiotic stresses that have more significant information available, based on recent publications.

http://dx.doi.org/10.5772/52212

When oxidative stress is taken into account, it is interesting to mention firstly, the role of molecular oxygen (O_2) in our environment. Due to the presence of oxygen and its reactions, both positive and negative aspects inherent to the process can occur, which is called oxidative stress.

Among all planets in our solar system, Earth is the only one that contains O_2, and the only one able to support aerobic life as the way that we understand its meaning. According to [8], the concentration of 21% (v/v) O_2 on Earth's atmosphere is derived from the photosynthetic activities of cyanobacterias and plants. The reference [9] commented that by an estimate, the total amount of O_2 in the Earth is about 410 x 10^3 Erda moles and from this value, 38.4 x 10^3 Erda moles is in the water form. When the aerobic life is concerned, these authors commented that this specific style of life is responsible for the major portion of O_2 turnover: photosynthesis is the main input of O_2, and respiration the main output.

Oxygen is relatively non reactive, but in some situations (as normal metabolic activity or when under environmental disturbance), it is able to switch to an excited state, producing free radicals and similar forms [9-10]. Then, it is clear in this scenario that adaptation processes to environmental changes are crucial for plant growth and survival. In view of its importance, it is interesting to remember the processes which lead to the reduction of molecular O_2. According to [11] such processes occur following four steps and generate several O_2 species. The first one requires an extra energy but the subsequent steps are exothermic, occurring spontaneously. The reaction products (H_2O_2; $O_2^{\bullet}$; HO_2^{-} ; $OH^{\cdot}$) can act in different ways in the cellular environment.

Hydrogen peroxide is a relatively long-lived molecule and can diffuse from its site of production [12]. Beside this, its toxicity has long been known. The $O_2^{\bullet}$ radical half-life is short (2-4 μs), but it is highly reactive and can form hydroperoxides and can oxidize histidine, methionine and tryptophan. When this radical is in the cellular environment, it causes lipid peroxidation as a consequence of oxidative deterioration of membrane polyunsaturated lipids. So, the hydrogen peroxide is not only toxic to cells, but in an extracellular medium it may react with transition metals, such as iron and copper, generating hydroxyls, which can cause cell damage. Beside this, when the levels of lipid peroxidation are higher (normally lipid peroxidation values are estimated by the concentration of malondialdehyde in samples) it suggests indirectly the establishment of a condition of oxidative stress. The hydroxyl radical ($OH^{\cdot}$) has a very strong potential and half-life of less than 1μs, and as a consequence, it has very high affinity for biological molecules [13]. What is particularly interesting about these species is that all of them can be generated by molecular oxygen reduction and they may play roles as toxic molecules or they can be excellent candidates for events/studies involving plant cell signaling [8, 13-16].

In these terms, the production of reactive oxygen species (ROS) is generally described as harmful due to their potential to cause irreversible damage to photosynthetic components in plants. However, despite this potential in causing harmful oxidation, modulation of ROS-antioxidant interaction plays a role in many stresses, as well as other responses to the environment. Additionally, this system can be considered as a powerful signaling process to molecules involved in the control of plant growth and development as well as priming acclimatory responses to stress stimuli [17-18]. In these terms, oxidative stress can be described as a central factor in abiotic and biotic stress that occurs due to imbalances in any cell compartment between the production of ROS and antioxidant defense [16, 19].

As indicated in [18] it is possible to verify that the pathways of ROS signaling are made by homeostatic regulation which can be achieved by the antioxidant redox buffering, making possible the determination of lifetime and the specificity of the ROS signal. It is interesting to emphasize that plants which demonstrate low activities for catalase (CAT) and cytosolic ascorbate peroxidase (APX), two key enzymes involved in the breakdown of H_2O_2, show less severe stress symptoms when compared to the ones where one of these enzymes is missing [20].

Talking about antioxidant defense systems, it can be attested that, in plants, the first line of defense against oxidative stress is the avoidance of ROS production [17] and once formed, ROS must be detoxified in order to either avoid or minimize eventual damages. In this way, the detoxification mechanisms can be considered as a second line of defense against the detrimental effects of ROS [21]. Beside this, some antioxidant enzymes can be considered as a second defense line against oxidative stress, since they act either as a catalyzer in ROS reaction or are involved in directing ROS processing [22]. The repair of oxidatively damaged proteins can be considered as the third line of defense against ROS [23].

According to [24] ROS species are commonly generated under stress conditions and due to its strong oxidative capacity, it acts on all types of biomolecules. In terms of the interactive effects of these species, it is possible to say that it can react with each other and with other molecules. For example, $O_2^{\bullet-}$ may react with lipids peroxides or nitric oxide, leading to the formation of peroxynitrite, which is less reactive than peroxides. In the same context, [13] pointed out that plants may favor the formation of one or other reactive species by preferentially scavenging peroxide (H_2O_2) with antioxidants or, in contrast, accumulating peroxide by the activation of superoxide dismutase (SOD).

The oxidative response in plants can be exacerbated by stressful conditions [16]. At the molecular level, the extent and nature of this response can differ among species and even among those closely related varieties of the same species. For example, 24 differentially expressed genes in soybean leaves were observed after glyphosate treatment when comparing tolerant and non tolerant soybean lines [25]. Therefore, oxidative responses are not only linked to the genetic expression. The reference [26] shows that some biochemical parameters (such as total soluble amino acid content and CAT activity in soybean roots) were also altered as a response to differential glyphosate application. The increase in the enzyme activities indicates ROS generation and a subsequent antioxidant response. Alterations in the antioxidative system of suspension-cultured soybean cells were observed [27], which were induced by oxidative stress using a peroxidizing herbicide (oxyfluorfen). Ascorbate and glutathione (non-enzymatic cellular antioxidants) showed different responses and the activities of some enzymes involved in cellular defense were also altered. For instance, peroxidase and catalase increased by 40 – 70% while glutathione S-transferase (GST) exhibited a 6-fold increase under oxyfluorfen stress.

Stress-induced ROS accumulation is counteracted either by enzymatic oxidant systems that include a variety of scavengers, such as superoxide dismutase (SOD), ascorbate peroxidase (APX), glutathione peroxidase (GPX), glutathione S-transferase (GST), and catalase (CAT), or by non enzymatic low molecular weight metabolites, such as carotenoids and flavonoids [16, 28-29]. As an example related to the influence of the enzymatic machinery under a stress situation, the reference [30] pointed out that as a response to stress, plants may increase the activities of some enzymes such as glutathione S-transferase (GST), involved in the detoxification of xenobiotics. These authors also investigated in detail the mechanisms of interaction between the GST enzyme and its substrates, indicating that the information might help in the engineering of new GSTs with improved detoxification efficiency [30].

In the context so far referred, and particularly with soybean, which is the main focus of this chapter, molecular and biochemical studies have explored several aspects related to the manipulation of metabolic pathways towards adaptation responses which can help to mitigate oxidative stress.

If ROS scavenging pathways in plants are the main focus, the involvement of at least 3 cycles have to be considered: a) the water-water cycle in chloroplasts, including SOD; b) the ascorbate-glutathione cycle in chloroplasts, cytosol, mitochondria, appoplast and peroxisomes; and c) GPX and CAT in the peroxisomes [31]. The equilibrium between the production and the scavenging of ROS may be altered by biotic and abiotic stress factors such as UV radiation, temperature, air pollution, pathogen attack, heavy metals, nutrient deficiency, and herbicides, among others [32].

The clear understanding of the mechanisms by which some endogenous or exogenous agents can lead to plant toxicity and how plants answer to this specific situation, is essential. Besides this, understanding how toxicity occurs, what kind of alterations occur in plant structure and metabolism among other situations are important steps for genetic breeding programs, when searching for new varieties susceptible or tolerant to stress factors and even for bioremediation/phytoremediation programs [32].

Although there has been a rapid progress in recent years in the field of plant stress studies, there is a consensus among researches that there are still many uncertainties in understanding how effectively ROS affects the stress response of plants [32-33].

A short list of examples that will emphasize the detoxification mechanism involved in plant stress defense and a diversity of enzymes that can be involved in the dismutation of ROS is then presented. For example, soybean has been shown to be highly sensitive to ozone (O_3) and the oxidation of some proteins may cause alterations in the activities of enzymes across nitrogen and sulfur nutrient assimilation pathways linked to stress responses [34]. The chronic exposure to high O_3 may lead to increased expression or oxidation of proteins, including APX, GSTs [34] and decrease the activities of monodehydroascorbate reductase (MDHAR) and glutathione reductase (GR) [35], indicating a fundamental role of these enzymes in stress response when soybean is subjected to O_3. Furthermore, soybean submitted to chronic high O_3 concentration and then exposed to an acute O_3 stress provided evidence that there was an immediate transcriptional reprogramming that allowed for maintained or increased ascorbate (AA) content in plants grown at high O_3 [36]. In another study using two tropical soybean varieties (PK 472 and Bragg) exhibiting differential sensitivity to O_3, reference [37] showed that the CAT activity decreased whereas peroxidase increased in both varieties upon exposure to O_3, but reflecting the greatest sensitivity of PK 472 in relation to the high magnitude of the reductions in the levels of antioxidants, metabolites and nutrients. Besides this, the damage O_3 effects produced were found to be more prominent during the reproductive than the vegetative growth stage.

Although soybean plants have shown a positive and significant correlation between activity of antioxidant enzymes and the osmolyte proline (Pro) content to water deficit stress [38], the metabolic reasons associated with the differential sensitivity of soybean cultivars to water deficit stress are not well understood [39]. According to the authors, water deficit stress increased antioxidant enzyme activities of SOD, CAT and GPX more at mild than at high water deficit stress [39]. In addition, soybean plants have shown protective mechanisms associated to proline concentration and GR, APX, and CAT activities under salt stress [40].

The increase in soybean productivity has been also accounted to the development and widespread use of improved cultivars with increased resistance to stressful conditions. A promising technique for agricultural improvement in arid and semiarid areas is the use of a pretreatment of soybean dry seeds with a low dose of gamma rays (20 Gy) before planting, enhancing drought tolerance and minimizing the yield losses caused by a water deficit condition [41].Overall, application of a low dose of gamma irradiation (20 Gy) increased the activities of phosphoenol pyruvate carboxylase and ribulose-1,5-bisphosphate carboxylase/oxygenase (RUBISCO) under drought stress, avoiding the destructive effects of water deficits on chloroplasts [41]. Furthermore, the manipulation of Pro can affect the (h)GSH, amino acids concentrations and APX activity, contributing to the detoxification of ROS in soybean subjected to simultaneous drought and heat stresses [42].

The regulation of thiol metabolism has become important for optimizing crop yield and quality of soybean [43]. The sulfur assimilatory pathway in soybean metabolism can be metabolized into molecules that protect plants against oxidative stress. The genetic manipulation of the cytosolic isoform of O-acetylserine sulfhydrylase (OASS), an enzyme involved in the sulfur assimilatory pathway, resulted in high levels of thiols and increased tolerance of plants to metal toxicity [44].

It is also important that information concerning changes in antioxidant capacity in immature seeds harvested at different reproductive stages [45], exhibited decreases in free radical scavenging activity and total antioxidant capacity with the advancement of maturity. This occurred concomitant with increased concentration of tocopherol and isoflavone isomers. Therefore, it is important to take into consideration that not only organs or tissues may present distinct responses to stress, but the plant stage of development is also important.

Reference [46] reported that different metals may act and induce different levels of copper-zinc superoxide dismutase (Cu-Zn/SOD) expression in soybean plants exposed to Cd and Pb. Also, Cd caused the induction of Cu-Zn/SOD mRNA accumulation for all Cd concentrations and Pb-treated roots showed induction of these isoenzymes only at medium metal concentrations.

It is also important to bring into the scene of stressful condition for plants, the soil type used. In a recent work soybean exposed to Cd and Ba [47] showed that the activity of antioxidant enzymes can change depending on the soil type, time-length of exposure and metal concentration [47]. For instance, GR and SOD activities in the leaves of soybean plants grown in an Oxisol soil contaminated with Cd decreased over time, whilst remaining high in an Entisol soil. The changes of enzyme activities were mainly dependent on buffering capacity of the soils with the Entisol exhibiting a lower capacity, with the plants suffering higher oxidative stress than those plants grown in a clay soil such as presented by an Oxisol soil [47].

Moreover, it is also important to investigate the effect of stressful conditions in soybean productivity taking into consideration more than one environmental contaminant or stress factor in the same agricultural region. In this context, the combination of Cd and acid rain pollution damaged the cell membrane, decreased the activities of POX and CAT, showing a higher potential threat to soybean seed germination than the single separate effect of each contaminant [48]. In another study, a correlation between the rate of ROS generation and antioxidant enzyme activities was established under hypoxia and high CO_2 concentration [49]. The CAT activity in soybean plants increased during the first hours of hypoxia whereas peroxidase activity started to play a more key role in cell defense only after a longer exposure to hypoxia. In this study, the processes of ROS accumulation and antioxidant enzymes were induced by the higher CO_2 content, indicating that CO_2 can switch on plant adaptation to hypoxic stress [49].

Another interesting study involving the combination of distinct stressor agents was carried out by combining Al and Cd with both leading to synergistic effects on plant growth and antioxidant responses in two soybean cultivars with different Al tolerance levels [50]. According to the authors, the Al treatments and low pH value (4.0) caused reduction in chlorophyll content and net photosynthetic rate, leading to growth reduction. The increased SOD and peroxidase activities were detected in the plants submitted to both metals, especially in the Al-sensitive cv. Zhechun 2, which also exhibited significantly higher Al and Cd contents than the Al tolerant cv. Liao-1. Moreover, Cd supplementation increased Al content in the plants exposed to Al+Cd stress [50]. Such an observation confirms another key aspect that should receive attention which is how the elements of the soil interact and can define an uptake profile by the plant root system possibly resulting in an induction of stress condition. Such studies are also of the upmost importance when considering the use of phytoremediation as a technique to recover a contaminated soil.

Similarly, studies about the interactions between plant roots and beneficial metal-tolerant microorganisms are gaining importance and may be an important approach to be considered in studies about plant adaptation and alleviation to a variety of environmental stresses [51]. For example, soybean plants inoculated with arbuscular mycorrhizal (AM) fungi showed reduced MDA content and increased APX activity to the oxidative stress generated by paraquat (PQ) [52]. In another study, activities of SOD and peroxidase were increased in the shoots of soybean plants with mycorrhizal (M) fungi grown under

NaCl salinity [53]. Once again, a more integrated view is needed and deserves attention. These two studies commented above indicate the importance of mycorrhizal fungi regulation as a general strategy to protect plants from stress. If soil type is added to this equation, a much more complex situation is created and such an integrated study reflects the reality of many agronomic situations. This also raises the question over the use of hydroponic systems to study oxidative stress in plants, particularly when induced by non-essential elements, since it is not necessarily the real field condition. Yet, is not our intention to say that such studies under hydroponic conditions are not important. On the contrary, they also have advantages. However, a more dynamic or integrated type of study should be considered in our point of view.

Curiously, grafting, which is a well-known agronomic technique largely used in agriculture, has not been used much in studies of stress in plants. The grafting technique has a tremendous potential to add further important understanding about stress signaling, assimilation, transport and accumulation of metals, opening a new perspective to study these grafted plants at the biochemical and molecular levels. Unfortunately, very few examples are available in the literature focusing on the investigation of plant stress responses. An example of such a study is the one carried out by [54] who showed that Cd seed concentration can be influenced by the difference in translocation of Cd from soil to the seed and in Cd accumulation capacity of roots among soybean cultivars by the use of grafting.

Nowadays, the development of plant manipulation techniques, for example the production and use of transgenic plants, has contributed to studies involving plant antioxidant responses induced either by exogenous or endogenous factors (such as herbicide, metals, pollution). Studies involving the mechanisms leading to stress-tolerant plants are important and needed, since they can aid understanding and create new possibilities for the use of these kinds of plants. The knowledge provided by the "omics studies" such as proteomics, metabolomics, metallomics and genomics, added to enzymatic evaluation, can provide information that can decisively help in answering many questions related to oxidative stress and ROS control [32].

Taking into account the importance of "omics" platforms, as well as their use for corroborating our initial hypothesis, the following sections will focus on these important strategies. They will be divided into proteomics and metallomics, with brief descriptions of each one, as well as some discussions and examples regarding transgenic cultures, but always concentrating the focus on soybean.

3. Proteomics

Proteomics can be defined as the large-scale study of proteins, including not only their identifications and quantifications, but also the determination of their localizations, modifications, interactions, activities, and functions [55]. This information is extremely important to evaluate interactions between different proteins, or between proteins and other molecules, and may reveal the functional role of proteins [56]. In this sense, proteomics is an important part of plant science, providing essential tools for understanding the functions of many plant-specific biological processes at the molecular level [57]. Currently, plant proteomic studies are focused on understanding the impact of different conditions of plant physiology such as the characterization of plant defense under biotic and abiotic stress [58,59], the characterization of subcellular, cellular or plant organ proteomes [60-61], the characterization of genetic modifications [62-63], as well as others.

The insertion of exogen DNA fragments into the DNA of the target organism, to confer some enhanced characteristics to the latter, describes the process termed genetic modification [64]. Focusing on plants, improved productiveness, enhanced tolerance to herbicides, synthesis of new substances and others can have consequences related to genetic modification [64]. The natural responses to this process are known to change the protein map of an organism [65]. In this sense, comparative proteomics become the strategy of choice, being useful for establishing qualitative and quantitative differences between genetically and non-genetically modified organisms [66]. In this way, studies of protein changes are frequently carried out through polyacrylamide gels by evaluation of their images, providing relevant information for comparative proteomic studies [67,68] as well as using appropriate mass spectrometric techniques for evaluating the identity of the studied proteins [69-70].

For proteomic studies, gel electrophoresis separations are the most used platform, due to their high resolution, allowing either high efficiency protein separation or the identification of potential protein spots with differences in concentration or expression in the gels evaluated [67,68]. The gel electrophoresis technique can be applied in proteomic studies as follow: (1) one-dimensional gel electrophoresis (SDS-PAGE) [71], (2) two-dimensional gel electrophoresis (2-D PAGE) followed by manual image analysis [66] and (3) two-dimensional difference gel electrophoresis (2-D DIGE) followed by automatic image analysis [72]. The application of these techniques in comparative studies involving transgenic soybeans has been little explored, where variation of different proteomic profiles in soybean genotypes [73], abiotic environmental stress [74], osmotic stress [75] and improvement of protein quality in transgenic soybean plants [76] are examples found. In this way, the use of these separation techniques in combination with mass spectrometry were applied in our research group to comparative proteomic studies in transgenic and non-transgenic soybean seeds, and it will be discussed below.

4. One-dimensional gel electrophoresis (SDS-PAGE)

The separation of the proteins using SDS-PAGE technique is based on their molecular mass, covering a broad separation range [67]. In our research group, this technique was used with mass spectrometry (ESI-QTOF MS/MS) for the identification of the enzyme cp4 EPSPS, in order to prove that the soybean in question was genetically modified. For this task, the protein band corresponding to a mass of 47 kDa was cut, the proteins reduced, alkylated, and subjected to two enzymatic digestion protocols: trypsin and chymotrypsin.

As a result, the enzyme cp4 EPSPS was identified using the SDS-PAGE technique and using either trypsin or chymotrypsin as a cleavage enzyme. However, trypsin showed the best results in terms of score and coverage (as a percentage). Moreover, the enzyme was identified in the database containing sequences from the soil bacterium *Agrobacterium* sp, the origin of the gene used in genetic modification. This method proved to be simple and very efficient, without needing sample prefractionation using chromatographic columns [77].

5. Two-dimensional gel electrophoresis: 2-D PAGE

In 2-D PAGE technique, the proteins are first separated on the basis of their net charges by isoelectric focusing (IEF) and then separated on the basis of their molecular mass by polyacrylamide gel electrophoresis in the presence of sodium dodecyl sulfate (SDS) [67].

The combination of 2-D PAGE and mass spectrometry is one excellent strategy to obtain proteomic maps [78]. Furthermore, taking into account that we were finding some differential proteins when investigating transgenic and non-transgenic soybean seeds, this technique showed itself to be excellent for this task. In this sense, our experience in terms of this kind of study will be commented.

Reference [79] used the combination of 2-D PAGE and mass spectrometry for obtaining a proteomic map for transgenic soybean seeds. The literature reports [12] that the number of protein spots present in the linear pH range from 4 to 7 was higher than the number of spots present in the linear range from 3 to 10 for transgenic soybean seeds. Therefore, for the range from 4 to 7, a higher number of spots were selected when compared with the 3-10 range. As result, a total of 192 proteins from transgenic soybean seeds were identified, 179 of them identified within the pH range from 4 to 7, and 13 of them identified within the 3-10 pH range. Regarding the pH range from 4 to 7, 49% of the spots present in the gel were identified in the database showing good efficiency with a similar study involving soybean published in the literature [80]. Regarding their biological functions, 50% were related to storage function, 18% related in growth/cell division process, 9% involved in metabolic/energy process, 6% related to protein transport, 4% corresponding to proteins involved in the disease/defense category and 21% in the category of non-classified proteins.

The application of the 2-D PAGE technique in comparative proteomic studies can lead to some problems due to the intrinsic characteristics of the electrophoretic systems such as sample preparation strategies, the natural variations when considering biological systems, gel-to-gel variance, labor intensiveness and possible identification of several proteins from one spot [67,81]. In this sense, 2-DE technologies need to be evaluated critically.

In a pioneering work, reference [66] evaluates some parameters that influence the comparisons of the protein map after different gel runs, establishing comparative image analysis after 2-D PAGE of transgenic and non-transgenic soybean seeds for identifying possible differences in protein expression. In that work, two pH ranges were used: 3-10 and 4-7. For improving accuracy, image treatments were made by the same analyst and concomitantly carried out for each pair of gels in the same electrophoretic run (4 pairs of gels with the optimized loaded mass) for avoiding possible variations between evaluations of the gel images.

In relation to detection and selection of the protein spots, the choice of the parameters of image analysis is extremely important. Differences between manual and automatic detection of the spots were obtained, showing the importance in editing the images to avoid erroneous interpretations not only in terms of the quantities of the detected spots, but also in terms of the intensities and/or volume of each protein detected.

The matching study is of utmost importance for those ones where the target is to find possible changes in protein expression as well as to establish the similarities in protein distribution in sets of gels. For gels obtained in the same run and within 3 to 10 and 4 to 7 pH ranges, 163±37 (79±4% match) and 287±28 spots (77±2% match) were respectively obtained from 4 pairs of gels (transgenic *x* non-transgenic). However, when gels were obtained from different runs, even considering the same sample (transgenic seeds), high variation was detected in terms of matches (39±6% and 58±13% for 3–10 and 4–7 pH ranges gels, respectively). Similar results for non-transgenic seeds were obtained (40±10% and 62±18% for 3–10 and 4–7 pH gel ranges, respectively). In this way, it is preferable to acquire the gels in the same run in order to produce high matches. The use of these procedures points out that elimination of gel-to-gel variation is mandatory in image analysis.

Proteins were considered as up or down regulated when the ratio between spot volume and/or intensity for non-transgenic and transgenic soybean seeds changed from < 0.55 to >1.8 (*ca.* 90% variation). Thus, 3 and 7 spots from 3 to 10 and 4 to 7 pH ranges were respectively highlighted and characterized by MALDI-QTOF MS. From this total, 8 proteins were identified as: glycinin G2/A2B1 precursor, glycinin G1 precursor, α-subunit of β-conglycinin (03 spots), allergen Gly mBd 28 K (fragment), actin (fragment) and sucrose binding protein.

Then, it is easy to observe that well optimized conditions for acquiring images from 2D gels are an important tool in the identification of possible biomarkers for genetically modified organisms.

6. Two-dimensional difference gel electrophoresis: 2-D DIGE

A promising alternative for circumventing possible variations in the technique already described (2-D PAGE) is the two-dimensional difference gel electrophoresis (2-D DIGE). This technique, which is based on fluorescent cyanine dyes, allows comparisons between two exact quantitative proteomic samples, which are resolved on the same gel, minimizing the problems previously mentioned [82]. Moreover, there is the advantage of the high sensitivity of these dyes (*ca.* 1 fmol of protein), which enables the detection of low abundance proteins when compared to other dyes used in the detection of protein spots, such as Coomassie Brilliant Blue (CBB) and silver staining [82]. Frequently, three samples are labeled in 2-D DIGE: two of them are experimental samples whereas the third one is composed of a mixture of equal amounts of all experimental samples (*i.e.*, a pooled internal standard). This creates a standard for each protein during analysis. After 2D separation, different protein samples labeled can be visualized separately by exciting the different dyes at their specific excitation wavelengths. Therefore, from the images generated for each dye, the signals from labeled protein spots are determined and the normalized intensities or spot volumes for each spot from different dyes (Cy2, Cy3, Cy5) are compared in order to identify differentially expressed proteins between the samples [82-83].

Once this technique is finely developed for finding possible biomarkers, reference [79] applied the 2-D DIGE technique and mass spectrometry (ESI-QTOF MS/MS) to assess differences in proteomic profiles of transgenic and non-transgenic soybean seeds. Three biological replicates were analyzed. A regulation factor of 1.5 (50% variation) was chosen as determined by the image analysis program and statistically significant differences in expression were determined ($p \leq 0.05$, according to the Student *t* test). The program of image analysis uses the automatic detection of the spots, and does not require any manual editing, either in adding or in altering the area defining the spots, in contrast to other programs for 2-D PAGE image.

As a result, a total of four proteins were differentially expressed between transgenic and non-transgenic soybean seeds, where two were overexpressed, being more highly expressed in transgenic soybean, and two were underexpressed, being less expressed in transgenic soybean. Thus, these four spots were selected for identification by mass spectrometry. As results, the spots were identified as: Actin (fragment) (*Glycine max*), involved in various types of cell motility, widely expressed in all eukaryotic cells and binds to ATP and other proteins [84]; cytosolic glutamine synthetase (*Cucumis melo*) (Figure 1a), considered as a ligand enzyme, being highly expressed in many types of roots, binds ATP molecules and is responsible for the primary assimilation of ammonia in all living organisms, participates in nitrogen fixation [84]; Glycinin subunit G1 (*Glycine max*), responsible for the nutritional, physicochemical, and physiological characteristics of soybean seeds [85] and Glycine-rich RNA-binding protein (*Glycine max*) (Figure 1b), involved in cellular response to environmental and developmental conditions [84]. It is noteworthy that the actin protein was also detected by Brandão et al. [66] working with 2-D PAGE and image analysis, and with the same sample.

The results obtained in reference [79], comments about some differential proteins found, establishing a relationship between oxidative stress (ROS production) and genetic modification. In this way, spectrophotometric enzymatic assays demonstrate that soybean transgenic seeds (for glyphosate resistance) exhibited higher activities for APX, CAT and GR enzymes compared to non-transgenic. Considering these results, the authors concluded that the oxidative stressful condition in transgenic seeds resulted in an increase of H_2O_2, which is probably controlled by the action of APX and CAT and even GR. Related to SOD, reference [79] showed the results for SOD activity in nondenaturing polyacrylamide gel electrophoresis, and it was possible to observe eight SOD isoenzymes detected in both transgenic and non-transgenic soybean seeds, one as Mn-SOD, two as Fe-SOD and five as Cu/Zn-SOD. The authors commented that the reduction in SOD activity in transgenic seeds was much more a result of a reduction in the Fe-SOD isoenzymes activities. Finally, and as a conclusion, the genetic modification itself might have induced extra ROS generation.

Proteins involved in the RNA processing and alternative splicing, RNA transport, messenger RNA (mRNA) translation, mRNA stability, and mRNA silencing mechanisms have been shown to be required for normal plant development and the responses of plants to altered environments [86-87]. In our case, just the glycine-rich RNA-binding protein was differentially found after DIGE analysis, and this protein correlated to ROS production according to different articles [88, 89]. As already mentioned, the cytosolic glutamine synthetase is involved in nitrogen fixation. Oxidative stress can also control the expression of nitrogen-metabolism genes as recently demonstrated [90], demonstrating that cytosolic glutamine synthetase can be altered because of the oxidative stress observed in the transgenic soybean line [90].

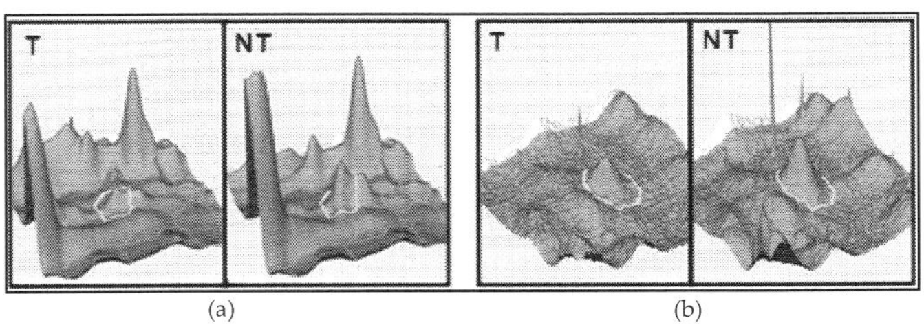

Figure 1. Examples of spots with expression variation among samples of transgenic (T) and non-transgenic (NT) soybean seeds: a - cytosolic glutamine synthetase (*Cucumis melo*) and b - Glycine-rich RNA-binding protein (*Glycine max*) [modified from reference 79].

7. Metallomics studies involving HPLC coupled to ICP-MS

In the last years, many soybean varieties have been genetically modified for adaptation to different geographical regions, to increase quality and productivity. Due to these genetic modifications, the proteins composition and profile can be affected, causing changes in the species proteome [91-92]. As previously described, the knowledge of the soybean genotype alone does not show enough information about the protein modifications due to environmental interactions. For better understanding of the consequences of a genetic manipulation, the elucidation of the protein map composition is necessary because it is directly related to the phenotype [93]. Since the proteome can be affected, it is assumed that the metallome can also be affected somehow by the genetic modification [94].

The metallome is defined as the entirety of metals and metalloid species, present in a cell or tissue type [95]. Deciphering the metallome provides information such as: (i) how an element is distributed among the cellular compartments; (ii) its coordination environment, in which the biomolecule is incorporated or by which bioligand it is complexed, and (iii) the concentration of the individual metal species present [96]. The majority of metals present in biological fluids and organs are linked to proteins, called metalloproteins. It is believed that every third protein require a metal cofactor, such as Cu, Fe and Zn, to develop their functions correctly [97]. The determination of an organism metallome involves separation techniques associated to microanalytic processes, such as mass spectrometry. These are the two key steps for general proteomics: separation and posterior identification of the proteins [80].

Metallomics studies were already performed in our group, being that one involving comparative metallomics of transgenic and non-transgenic soybean seeds [94], the first published in the literature. Soybean proteins were separated using two-dimensional polyacrylamide gel electrophoresis (2-D PAGE), tryptically digested, characterized using matrix assisted laser desorption ionization - quadrupole time of flight – mass spectrometry (MALDI-Q-TOF-MS) and mapped using synchrotron X-ray fluorescence radiation (SR-XRF). The following metallic ions were found: Ca(II), Cu(II), Fe(II), Mn(II), Ni(II) and Zn(II), and the quantitative profile was acquired using atomic absorption spectrometry, showing changes in metal contents of transgenic and non-transgenic soybean seeds. Although promising results could be found in this study, the canonical analytical approaches for proteomics (such as 2-D PAGE) and metabolomics studies usually do not consider the existence of metal complexes with proteins and metabolites.

In this way, the use of high performance liquid chromatography (HPLC), an analytical technique used to separate a mixture in solution in its individual components, should be considered. Distinctly compared from 2-D PAGE, HPLC is based on different protein-surface interactions [98]. The principal HPLC separation mechanisms used for bioinorganic studies include size exclusion chromatography (SEC), reversed-phase (RP) and ion-exchange (IEX) chromatography, and because of the complex nature of the metal-molecule interaction, a combination of these mechanisms is often necessary to identify the elemental species correctly.

Together with HPLC, an element-specific detector must be used, and since the 80's, inductively coupled plasma mass spectrometry (ICP-MS) is being widely applied for studying elements at low concentrations. It is robust for multielementar determinations, allowing to reach extremely low detection limits and giving isotopic information for identification and quantification of the species, besides being easily coupled to classic separation techniques, such as HPLC, readily realized since the chromatographic flow (0.5-1.0 mL min^{-1}) is compatible to common ICP-MS nebulizers [97,99].

The excitation source of this technique is argon inductively coupled plasma, which is used to form ions which are transferred to a high vacuum region through an interface containing small orifices. Ions are focalized using ionic lenses and directed to the mass spectrometer in order to be separated by *m/z* ratio. The *m/z* ratio analyzer generally used in ICP-MS is a quadrupole, ideal for quantitative analysis [100].

The elements detected using ICP-MS include metal coordination complexes with larger proteins and metallothioneins, as well as selenoproteins and metal/semi-metals linked to carbohydrates. A great amount of examples are found in the literature where ICP-MS is used to detect and quantify metallic ions bounded to biomolecules.

The attempts to avoid metal-ligand denaturation make the SEC mechanism the most used for metallomics studies [97]. It separates molecules according to their hydrodynamic volume, determined by their Stokes ratio. It results in the partial exclusion of analytes that pass through defined size pores due to their molecular sizes [101]. When the mobile phase passes through the column, those particles with small hydrodynamic volumes are transported through a larger path because they equilibrate in the pores more frequently than the ones with higher hydrodynamic volumes, resulting in separation. Elution volume is determined by the molecule size, directly related to their molecular weight, so this volume can be used to determine the molecular weight of an unknown compound. For carrying out that task, the relation between molecular weight and elution volume, obtained empirically by injecting standards with known molecular weight and measuring their elution volumes, must be known [102-103].

SEC is especially suitable for separation of element species presenting limited stability frequently found in protein-rich matrices. The main advantages of SEC are simplicity of application, tolerance to biological matrices, compatibility of mobile phases with specific demands of certain biological samples and the possibility of estimation of molecular weights of the compounds. It is widely used for protein separation, including soybean proteins and, although considered a low resolution method [104], it is often applied as the first separation method of fractions containing metallo-biomolecules of interest followed by another separation step with element-specific detection or MS identification. Its uses alone is very helpful to study the distribution of elements in different molecular weight fractions, and the coupling SEC-ICP-MS is being accepted as a hyphenated technique for speciation studies to evaluate the association of elements to compounds present in the sample. These studies are considered the initial point for a deeper evaluation of the nature of the species found.

In a work developed in our group [105], a comparison between elution profiles from transgenic and non-transgenic soybean seeds was carried out, using SEC coupled to high resolution ICP-MS. The elution profiles were similar between the samples, and the conclusions are in agreement to the ones discussed by reference [106]. It was found that areas of the most abundant peaks for Cu and Fe in transgenic soybean seeds were 3- and 2-fold higher, respectively, than those found in non-transgenic samples. This, summed up with total element analysis results in the same article, where the concentrations of Cu and Fe had statistically significant differences between transgenic and non-transgenic soybean seeds, could lead to the conclusion that Cu and Fe are associated with compounds more expressed in transgenic soybean seeds.

Reference [107] used different parts of the soybean plants to analyze Se elution profile using SEC-ICP-MS. The authors concluded that the bean had the most interesting profile, since it absorbed most of the Se from the shoots and presented a very intense peak for this element at higher molecular weight fraction. These data showed that the soybean plants convert selenite (used to enrich the plant) to high molecular weight species, which, according to the authors, can add nutritional value to the plant. Another work from the same group [108] used reversed phase coupled to ICP-MS to study the Se-Hg antagonism, and they found that in plants enriched with Hg, more Se was assimilated, indicating a possible protective response mechanism to the Hg.

As already commented, it is known that purity of peaks in SEC is poor, and even if a single species of a given element is present, matrix components may co-elute. They are invisible to the element specific detector, but if the goal is the identification of the organic specie linked to the element, they will be detected by the organic MS instrument. Also, matching the elution volume with a standard in this case is not definitive, due to the small number of theoretical plates found in SEC. For these reasons, SEC is usually followed by a second chromatographic separation (2nd dimension) using an orthogonal separation mechanism, such as ion-exchange, reversed-phase or hydrophilic interaction chromatography, before the identification of the components.

Multidimensional liquid chromatography is an efficient tool and an alternative procedure for the classic methods based on unidimensional HPLC. The multidimensional chromatography can be carried out *online* or *off-line*. In the *off-line* mode, fractions eluted from the 1st dimension are collected manually or using a fraction collector, and then are re-injected in the second chromatographic column. *Online* techniques are automated using a selector valve, which can enhance reliability and sample processing. The limitation here is that the mobile phases used for both dimensions must be compatible [109].

The selectivity in a multidimensional system can be enhanced only if the chromatographic dimensions are based in different separation mechanisms. The second dimension must not decrease the resolution obtained using the previous one. For the separation mechanisms to be different, the columns must have different stationary phases, allowing the less efficient separation attained in the first dimension to be improved in the second [110].

In the case of multidimensional liquid chromatography [103,111] coupled to ICP-MS, fractions isolated mainly using SEC, can be fractionated again using an independent separation mechanism to provide more detailed results, and also to attain metal species pure enough to be characterized using molecular mass spectrometry.

Many stationary phases can be used for a second chromatographic dimension. Among then, reversed phase (RP) [111], the most popular liquid chromatography separation mechanism, should be highlighted. It has great efficiency and is able to separate a great range of compounds with different polarities. The separation is obtained through partition of the analyte between a non-polar stationary phase and a polar mobile phase.

Ion exchange chromatography [112] (IEX) can also be used to separate biomolecules based in charge differences. It can be considered a highly selective technique, able to separate, for example, proteins differing in only one charged group. It is a widely used technique in bioseparations, since peptides, proteins, nucleic acids and related biopolymers have ionizable chemical domains, making them susceptible to enhancement or diminishment of their charges as a function of pH and ionic strength changes. It can be used to separate large biomolecules, with more than 60 kDa.

Finally, polar compounds can be efficiently separated using polar/hydrophilic stationary phases using normal phase aqueous chromatography (aqNPC), also called hydrophilic interaction chromatography (HILIC) [113]. Here, retention times tend to be longer as high as is the hydrophobicity of the solutes, indicating potential for small metallic complexes separation.

Concerning multidimensional chromatographic separations and soybeans, a recent work from reference [107] used IEX as second dimension for the separation of proteins from selenium-enriched soybean. Here, the target was only Se, and the 26 fractions collected from the second dimension were pure enough to allow the identification of a considerable number of proteins in the soybean databank.

In our group [114], IEX was also used as second dimension, generating a number of different fractions for both transgenic and non-transgenic soybean seeds used in that research. Taking as an example cobalt, the SEC separation (Figure 2a) provided 3 peaks, divided into F1, F2, and F3. When F3 is separated again using IEX (Figure 2b), the wide peak found using SEC was separated into two narrow peaks, showing that the separation resolution increased.

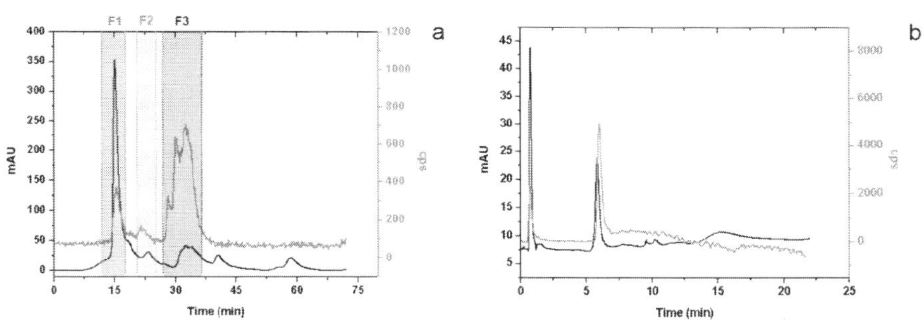

Figure 2. Chromatograms for UV absorption at 280 nm (—) and Co signal in the ICP-MS (—) for (a) SEC, separated in three fractions: F1, F2 and F3; and (b) IEX separation of F3 using transgenic soybean seeds [modified from supplementary material of reference 114].

Nowadays, mass spectrometry associated to bioinformatics has become essential in studies involving proteins, not only due to their sensitivity, but also to the total of information that can be obtained [69]. Electrospray ionization (ESI) is the most used technique for protein identification, allowing the formation of ions in the gas phase using a soft ionization process, making possible the analysis of non-volatile and thermolabile compounds [115]. As a consequence, ESI facilitated the analysis of large biomolecules, as well as drugs and their metabolites.

To improve metallomics information concerning transgenic and non-transgenic soybean seeds, our group [114] used the multidimensional chromatography strategy, as pointed out previously in this text. Total amounts of Fe and Cu were already found to be higher in transgenic soybeans, and in an attempt to link these metals to proteins, an ESI-MS/MS analysis was carried out. As results, more than 20 proteins were identified, encompassing 4 different functional categories. Among them, β-conglycinin, a protein previously associated to metals, was identified in three fractions, and one metalloprotein that binds Fe, lypoxigenase 1, was found in a high molecular weight fraction, the only fraction where an Fe peak was separated.

8. Future trends

Currently, the comparative studies concerning alterations in proteins, metalloproteins, metals and enzymes have demonstrated significant differences among transgenic and non-transgenic soybean. These differences have indicated that these genetic modifications provide not only tolerance to herbicide but also cause many changes in the whole metabolism of the transgenic plants.

Carefully taking into account all the results presented, it is possible to raise the following question: what are the future trends in comparative studies involving transgenic and non-transgenic soybean? Since the whole metabolism of transgenic soybean plants seems to be different to the non-transgenic one, this is a promising research area, and too much work is still needed. Much more information is still ahead of us for a better comprehension of the specific aspects of the transgenic soybean plant metabolism. In this way, investigations into techniques and novel approaches, quantitative proteomics, imaging and mapping of elemental distribution, tracer experiments employing stable isotopes and also in natural variation in the isotopic composition of the elements may possibly be the future trends in this topic. This will contribute to elucidation and expansion of our knowledge about transgenic soybean.

Since the proper functioning of life depends on the elements in a variety of processes, the understanding of molecular mechanisms of the elements and information on its chemical forms present in a living organism are very important. In this context, studies about identification and/or quantification of one or more chemical species of elements in transgenic and non-transgenic soybean samples are able to generate valuable information about their metabolisms. Therefore, it would be useful if more efforts were devoted to this topic. A novel technique that has an unexplored potential for speciation analysis is travelling wave ion mobility spectrometry coupled to mass spectrometry (TWIMS-MS) [116]. The use of this technique in speciation analysis of metals associated with biomolecules should increase due to its capability of differentiation of ions by shape and size, besides mass and charge. Until now, the studies employing ion mobility are concentrated in proof-of-the-concept using isolated species commercially available and its application to complex matrices certainly will be a big challenge, but very helpful to elucidate many questions.

The main objective of quantitative proteomics is to quantify protein expression alterations in response to a variety of changes, and, nowadays, one of the most challenging and emerging area of proteomics involves the developments of accurate quantitative methods for proteins. The quantitative proteomics is divided in absolute and relative subjects. In the absolute quantification, changes in protein expression are determined in exact amount or concentration of each protein present. The relative one determines the up- or down-regulation of a protein relative to the control sample, and the results are presented as 'fold' increases or decreases. The 2-D DIGE is an example of relative quantification technique that is applied to intact proteins and the differential expression determination is based on fluorescence as commented earlier in this chapter. Taking into account the soybean comparative studies the application of quantitative proteomics by 2-D DIGE or by other technique could continue establishing the differences in protein expressions accurately [117].

According to the results presented earlier, some elements are present at higher concentrations in transgenic soybean seeds than in non-transgenic ones [66,94,105]. The transgenic seed seems to have ability to take up higher amounts of some metals from the soil and this is a sign that the processes involved in intake, transport and storage of essential and toxic metals and metalloids probably are suffering changes due to genetic modification. Various new queries take place with this information, such as: The other transgenic plant parts (roots, stems and leaves), are also taking up, transporting and storing higher amounts of these metals? Other plant parts try to eliminate some excess of these metals? Are these higher amounts really an excess for a transgenic plant or not? Are there differences in the distributions of these metals among transgenic and non-transgenic soybean? A potential tool for obtaining a better insight in these processes can be to use tracer experiments employing stable isotopes. In the last few years the use of stable isotopes and their isotope ratio measurements have gained importance for tracer experiments in biological and medical research [118]. In these studies stable isotopic tracers with an isotopic composition sufficiently different from the corresponding natural one is added to the studied system and changes in the selected isotope ratio monitored. The absorption or bioavailability of an element can be determined with this approach as well as information about element redistribution over various compartments of an organism [118-119]. According to our knowledge, no tracer study for essential or toxic metals evaluating transgenic and non-transgenic soybean is found in the literature and therefore there is a great amount of work to perform in this challenging area.

LA-ICP-MS offers *in situ* analysis of solid samples with respect to metals and nonmetals at trace concentration level mostly without sample preparation and without charging effects during the measurements. This technique can also be applied to the imaging of soft tissues such as plant leaves with relatively high spatial resolution and good sensitivity [120] and therefore, some investigations involving metals distribution by LA-ICP-MS in transgenic and non-transgenic soybean would also be a future trend.

Another challenging issue that can provide evidences supporting the hypothesis that genetic modification is affecting the metabolism of soybean plants involves the investigation of natural variation in the isotopic composition of the elements. Even though isotopic abundances are assumed to be almost constant in nature, small isotopic or mass fractionation effects occur in both natural and industrial processes [118]. Since the isotopes present the same number of electrons, they show basically the same chemical behavior. However, there is a small discrepancy in their physicochemical behavior due to the mass difference, which may leave isotopes of the same element to take part with slightly different efficiencies in physical processes or (bio)chemical reactions, and, consequently, to result in variations in the isotopic composition [118-119]. These differences in efficiency are associated to a minor distinction in equilibrium for each different isotopic molecule - thermodynamic effect or in the rate with which the isotopes participate in a process or reaction - kinetic effect. Lighter elements, such as H, C, N, O and S suffer more pronounced isotopic variations because of the high relative mass difference between their isotopes. Nevertheless, heavier elements are subject to isotope fractionation, even though the change is minor [121].

As relative abundances cannot be measured directly, these studies are based on measuring the isotope ratio of an element because it is experimental accessible. The isotope ratio measured in a particular sample (R_x) is compared to the corresponding one in another sample, frequently a reference sample (R_{RF}) [118]. The differences found are frequently very small and thus high reproducibility/repeatability is required. Thus, the ICP-MS technique is becoming the more advantageous choice for most applications employing isotope ratios, mainly considering the recent instrumental developments. As the elements are subject to isotope fractionation in nature, the genetic modification could also provoke or intensify this effect.

In view of that comment here, it is easy to rationalize that many aspects can be explored when focusing on studies related to transgenic soybean.

9. Conclusion

The initial hypothesis formulated that the genetic modification itself is stressing the soybean, is apparently right, once the plant is searching a new equilibrium as living organism. The results presented in this chapter demonstrate that not only is the proteomic map changed with some proteins increasing and others decreasing, but also chromatographic separations are altered when transgenic and non-transgenic soybeans are compared. Examples are activities of some enzymes (as CAT, SOD, GPx, among others) involved in neutralization of ROS, as well as the possible capacity in taking metals from the soil (mainly for Fe and Cu). Because of these modifications that occur when both transgenic and non-transgenic organisms are compared, the theme of genetic modification could be even better explained with some alternative strategies, such as quantitative proteomics, image analysis, tracer experiments with stable isotopes, and other possibilities.

Finally, in our point of view, one of the key points for the success of studies involving transgenic organisms is not only to involve good technology, but also a transdisciplinary view, involving different areas of expertise. With this strategy, it will be easier to understand this area of investigation, making possible the demystification of the genetic modification that have occurred, and allowing answers for some questions that still remain unknown.

10. Nomenclature and acronyms

2-D-HPLC Two-Dimensional High Performance Liquid Chromatography

2-D PAGE Two-Dimensional Gel Electrophoresis

http://dx.doi.org/10.5772/52212

2-D DIGE Two-Dimensional Difference Gel Electrophoresis

ICP-MS Inductively Coupled Plasma Mass Spectrometry

HR-SF-ICP-MSHighResolutionSectorFieldInductivelyCoupledPlasmaMassSpectrometry

LA-ICP-MS Laser Ablation Inductively Coupled Plasma Mass Spectrometry

MALDI-QTOF-MS Matrix-Assisted Laser Desorption Ionisation Quadrupole-Time-of-Flight Mass Spectrometry

ESI-LC-MS-MS ElectroSpray Ionization Liquid Chromatography Mass Spectrometry-Mass Spectrometry

2-D-HPLC-ICP-MS Two-Dimensional High Performance Liquid Chromatography Inductively Coupled Plasma Mass Spectrometry

SDS-PAGE Sodium Dodecyl Sulfate Polyacrylamide Gel Electrophoresis

ESI-QTOFMS-MSElectroSprayIonization-Time-of-FlightMassSpectrometry-MassSpectrometry

SEC-ICP-MSSizeExclusionChromatography-InductivelyCoupledPlasmaMassSpectrometry

Acknowledgements

The authors are grateful to the Fundação de Amparo a Pesquisa do Estado de São Paulo (FAPESP, São Paulo, Brazil), Conselho Nacional de Desenvolvimento Científico e Tecnológico (CNPq, Brasília, Brazil), and Coordenação de Aperfeiçoamento de Pessoal de Nível Superior (CAPES, Brasília, Brazil), for financial support.

Author details

Marco Aurélio Zezzi Arruda[1], Ricardo Antunes Azevedo[2], Herbert de Sousa Barbosa[1], Lidiane Raquel Verola Mataveli[1], Silvana Ruella Oliveira[1], Sandra Cristina Capaldi Arruda[2] and Priscila Lupino Gratão[3]

1 Institute of Chemistry, National Institute of Science and Technology for Bioanalytics and Department of Analytical Chemistry, University of Campinas, Campinas, Brazil

2 Department of Genetics, Laboratory of Genetics Biochemistry of Plants, University of São Paulo, ESALQ, Piracicaba, Brazil

3 Department of Applied Biology to Agricultural, Universidade Estadual Paulista, FCAV, Jaboticabal, Brazil

References

[1] Sharma S, Kaur M, Goysil R, Gil BS. Physical Characteristics and Nutritional Composition of Some New Soybean (Glycine max (L.) Merrill) genotypes. Journal of Food Science and Technology 2011; DOI:10.1007/s13197-011-0517-7.

[2] The American Soybean Association. SoyStats. http://www.soystats.com (accessed 3 May 2012).

[3] Rural Centro. http://www.ruralcentro.com.br/noticias/53562/usda-reduz-producao-de-soja-da-safra-201112-para-2515-milhoes (accessed 21 March 2012).

[4] Cavusoglu K, Yapar K, Oruc E, Yalcin E. Protective Effect of Ginkgo Biloba L. Leaf Extract Against Glyphosate Toxicity in Swiss Albino Mice. Journal of Medicinal Food 2011;14(10) 1263-1272.

[5] Windels P., Tavernies I., Depicker A., Van Bockstaele E., De Loose M. Characterisation of the Roundup Ready Soybean Insert. European Food Research and Technology 2001;213(2) 107-112.

[6] Monsanto do Brasil. http://www.monsanto.com.br (acessed May 15 2012).

[7] Saz JM, Marina ML. High Performance Liquid Chromatography and Capillary Electrophoresis in the Analysis of Soybean Proteins and Peptides in Foodstuffs. Journal of Separation Science 2007;30(4) 431-451.

[8] Scandalios JG. Oxidative Stress: Molecular Perception and Transduction of Signals Triggering Antioxidant Gene defenses. Brazilian Journal of Medical and Biological Research 2005;38(7) 995-1014.

[9] Scandalios J.G., Guan L., Polidoros A.N. Catalases in Plants:Gene Structure, Properties, Regulation, and Expression. In: Scandalios J.G. (ed.) Oxidative Stress and the Molecular Biology of Antioxidant Defenses. New York: Cold Spring Harbor Laboratory Press Plainview; 1997. p343-406.

[10] Fridovich I. Superoxide Radical and Superoxide Dismutases. Annual Review of Biochemistry 1995;64 87-112.

[11] Hippeli S, Heiser I, Elstner EF. Activated Oxygen and Free Oxygen Radicals in Pathology:New Insights and Analogies Between Animals and Plants. Plant Physiology Biochemistry 1999;37(3) 167-178.

[12] Levine A, Tenhaken R, Dixon R, Lamb C. H2O2 from the Oxidative Burst Orchestrates the Plant Hypersensitive Disease Resistance Response. Cell 1994;79(4) 583-593.

[13] Dat J, Vandenabeele S, Vranová E, Van Montagu M, Inzé D, Van Breusegem F. Dual Action of the Active Oxygen Species During Plant Stress Responses. Cellular and Molecular Life Sciences 2000;57(5) 779-795.

[14] Foyer, CH, Shigeru S. Understanding Oxidative Stress and Antioxidant Functions to Enhance Photosynthesis. Plant Physiology 2011;155(1) 93-100.

[15] Foyer, CH, Noctor G. Redox Homeostasis and Antioxidant Signaling: A Metabolic Interface between Stress Perception and Physiological Responses. The Plant Cell 2005;17(7) 1866-1875

[16] Gratão PL, Polle A, Lea PJ, Azevedo RA. Making the Life of Heavy Metal-Stressed Plants a Little Easier. Functional Plant Biology 2005;32(6) 481-494.

[17] Jones DP. Redefining Oxidative Stress. Antioxidant Redox Signaling 2006;8(9-10) 1865-1879.

[18] Foyer CH, Noctor G. Oxidant and Antioxidant Signaling in Plants: a Re-Evaluation of the Concept of Oxidative Stress in a Physiological Context. Plant Cell Environment 2005;28(8) 1056-1071.

[19] Foyer CH, Noctor G. Oxygen Processing in Photosynthesis: Regulation and Signaling. New phytologist 2000;146(3) 359-388.

[20] Rizhsky L, Hallak-Herr E, Van Breusegem F, Rachmilevitch S, Barr JE, Rodermel S, Inzé D, Mitler R. Double Antisense Plants Lacking Ascorbate Peroxidase and Catalase are Less Sensitive to Oxidative Stress than Single Antisense Plants Lacking Ascorbate Peroxidase or Catalase. Plant Journal 2002;32(3) 329-342.

[21] Moller IM. Plant Mitochondria and Oxidative Stress: Electron Transport, NADPH Turnover, and Metabolism of Reactive Oxygen Species. Annual Review of Plant Physiology and Plant Molecular Biology 2001;52 561-591.

[22] Medici LO, Azevedo RA, Smith RJ, Lea PJ. The Influence of Nitrogen Supply on Antioxidant Enzymes in Plant Roots. Functional Plant Biology 2004;31(1) 1-9.

[23] Bechtold U, Murphy DJ, Mullineaux PM. Arabidopsis Peptide Methionine Sulfoxide Reductase 2 Prevents Cellular Oxidative Damage in Long Nights. The Plant Cell 200;16(4) 908-919.

[24] Wojtaszek P. Oxidative Burst: an Early Plant Response to Pathogen Infection. Biochemical Journal 1997;322(3) 681-692.

[25] Yu W, Zhang R, Runzhi L, Sandui G. Isolation and Characterization of Glyphosate-Regulated Genes in Soybean Seedlings. Plant Science 2007;172(3) 497-504.

[26] Moldes CA, Médici LO, Abrahão OS, Tsai SM, Azevedo RA. Biochemical Responses of Glyphosate Resistant and Susceptible Soybean Plants Exposed to Glyphosate. Acta Physiologiae Plantarum 2008;30(4) 469-479.

[27] Knörzer O, Burner J, Boger P. Alterations in the Antioxidative System of Suspension-Cultured Soybean Cells (Glycine Max) Induced by Oxidative Stress. Physiologia Plantarum 2008;97(2) 388-396.

[28] Gill SS, Nafees AK, Narendra T. Differential Cadmium Stress Tolerance in Five Indian Mustard (Brassica Juncea L.) Cultivars. An evaluation of the role of antioxidant machinery. Plant Signaling and Behavior 2011;6(2) 293-300.

[29] Mittler R, Blumwald E. Genetic Engineering for Modern Agriculture: Challenges and Perspectives. Annual Review in Plant Biology. 2010;61 443-462

[30] Ghelfi A, Gaziola SA, Cia MC, Chabregas SM, Falco MC, Kuser-Falcão PR, Azevedo RA. Cloning, Expression, Molecular Modeling and Docking Analysis of Glutathione Transferase from Saccharum Officinarum. Annals of Applied Biology 2011;159(2) 267-280.

[31] Delaunay A, Pflieger D, Barrault MB, Vinh J, Toledano MB. A Thiol Peroxidase is an H2O2 Receptor and Redox-Transducer in Gene Activation. Cell 2002;111(4) 471-481.

[32] Arruda MAZ., Azevedo RA. Metallomics and Chemical Speciation: Towards a Better Understanding of Metal-Induced Stress in Plants. Annals of Applied Biology 2009;155(3) 301-307.

[33] Apel K, Hirt H. Reactive Oxygen Species: Metabolism, Oxidative Stress and Signal Transduction. Annual Review on Plant Biology 2004;55 373-399.

[34] Galant A, Koester RP, Ainsworth EA, Hicks LM. From climate change to molecular response: redox proteomics of ozone-induced responses in soybean. New Phytologist 2012;194(1) 220-229.

[35] Zhao T, Wang JL, Wang Y, Cao Y. Effects Of Antioxidant Enzymes of Ascorbate-Glutathione Cycle in Soybean (Glycine Max) Leaves Exposed to Ozone. Advanced Research on Industry, Information Systems and Material Engineering, pts 1-7 Advanced Materials Research 2011;204-210(1-7) 672-677.

[36] Gillespie KM, Rogers A, Ainsworth EA. Growth at Elevated Ozone or Elevated Carbon Dioxide Concentration Alters Antioxidant Capacity and Response to Acute Oxidative Stress in Soybean (Glycine Max). Journal of Experimental Botany 2011;62(8) 2667-2678.

[37] Singh E, Tiwari S, Agrawal M. Variability in Antioxidant and Metabolite Levels, Growth and Yield of Two Soybean Varieties: an Assessment of Anticipated Yield Losses under Projected Elevation of Ozone. Agriculturall Ecosystems & Environment 2010;135(3) 168-177.

[38] Masoumi H, Darvish F, Daneshian J, Normohammadi G, Habibi D. Effects of Water Deficit Stress on Seed Yield and Antioxidants Content in Soybean (Glycine Max L.) Cultivars. African Journal of Agricultural Research 2011;6(5) 1209-1218.

[39] Masoumi H, Darvish F, Daneshian J, Nourmohammadi G, Habibi D. Chemical and Biochemical Responses of Soybean (Glycine Max L.) Cultivars to Water Deficit Stress. Australian Journal of Crop Science 2011;5(5) 544-553.

[40] Dogan M. Antioxidative and Proline Potentials as a Protective Mechanism in Soybean Plants under Salinity Stress. African Journal of Biotechnology. 2011;10(32) 5972-5977.

[41] Moussa HR. Low Dose of Gamma Irradiation Enhanced Drought Tolerance in Soybean. Bulgarian Journal of Agricultural Science 2011;17(1) 63-72.

[42] Kocsy G, Laurie R, Szalai G, Szilagyi V, Simon-Sarkadi L, Galiba G, de Ronde JA. Genetic Manipulation of Proline Levels Affects Antioxidants in Soybean Subjected to Simultaneous Drought and Heat Stresses. Physiologia Plantarum 2005;124(2) 227-235.

[43] Yi H, Ravilious GE, Galant A, Krishnan HB, Jez JM. From Sulfur to Homoglutathione: Thiol Metabolism in Soybean. Amino Acids 2010;39(4) 963-978.

[44] Kim WS, Chronis D, Juergens M, Schroeder AC, Hyun SW, Jez JM, Krishnan HB. Transgenic Soybean Plants Overexpressing O-Acetylserine Sulfhydrylase Accumulate Enhanced Levels of Cysteine and Bowman-Birk Protease Inhibitor in Seeds. Planta 2012;235(1) 13-23.

[45] Kumar V, Rani A, Dixit AK, Bhatnagar D, Chauhan GS. Relative Changes in Tocopherols, Isoflavones, Total Phenolic Content, and Antioxidative Activity in Soybean Seeds at Different Reproductive Stages. Journal of Agricultural and Food Chemistry 2009;57(7) 2705-2710.

[46] Pawlak S, Firych A, Rymer K, Deckert J. Cu,Zn-Superoxide Dismutase is Differently Regulated by Cadmium and Lead in Roots of Soybean Seedlings. Acta Physiologiae Plantarum 2009;31(4) 741-747.

[47] Melo LCA, Alleoni LRF, Carvalho G, Azevedo RA. Cadmium- and Barium-Toxicity Effects on Growth and Antioxidant Capacity of Soybean (Glycine Max L.) Plants, Grown in Two Soil Types with Different Physicochemical Properties. Journal of Plant Nutrition and Soil Science 2011;174(5) 847-859.

[48] Liu TT, Wu P, Wang LH, Zhou Q. Response of Soybean Seed Germination to Cadmium and Acid Rain. Biological Trace Element Research 2011;144(1-3) 1186-1196.

[49] Ershova AN, Popova NV, Berdnikova OS. Production of Reactive Oxygen Species and Antioxidant Enzymes of Pea and Soybean Plants under Hypoxia and High CO2 Concentration in Medium. Russian Journal of Plant Physiology 2011;58(6) 982-990.

[50] Shamsi IH, Wei K, Zhang GP, Jilani GH, Hassan MJ. Interactive Effects of Cadmium and Aluminum on Growth and Antioxidative Enzymes in Soybean. Biologia Plantarum 2008;52(1) 165-169.

[51] Sousa NR, Ramos MA, Marques APGC, Castro PML. The Effect of Ectomycorrhizal Fungi Forming Symbiosis with Pinus Pinaster Seedlings Exposed to Cadmium. Science of the Total Environment 2012;414 63-67.

[52] Bressano M, Curetti M, Giachero L, Gil SV,Cabello M, March G, Ducasse DA, Luna CM. Mycorrhizal Fungi Symbiosis as a Strategy against Oxidative Stress in Soybean Plants. Journal of Plant Physiology 2010;167(18) 1622-1626.

[53] Ghorbanli M, Ebrahimzadeh H, Sharifi M. Effects of Nacl and Mycorrhizal Fungi on Antioxidative Enzymes in Soybean. Biologia Plantarum 2004;48(4) 575-581.

[54] Sugiyama M, Ae N, Arao T. Role of Roots in Differences in Seed Cadmium Concentration among Soybean Cultivars - Proof by Grafting Experiment. Plant Soil 2007;295(1-2) 1–11.

[55] Fields S. Proteomics in Genomeland. Science, 2001;291(5507) 1221-1224.

[56] Kersten B, Bürkle L, Kuhn EJ, Giavalisco P, Konthur Z, Lucking A, Walter G, Eickhoff H, Schneider U. Large-Scale Plant Proteomics. Plant Molecular Biology 2002;48 133-141.

[57] Job D, Haynes PA, Zivy M. Plant Proteomics. Proteomics 2011;11(9) 1557-1558.

[58] Wang Y, Kim SG, Kim ST, Agrawal GK, Rakwal R, Kang KY. Biotic Stress-Responsive Rice Proteome: an Overview. Journal of Plant Biology 2011;54(4) 219-226.

[59] Hossain Z, Nouri MZ, Komatsu S. Plant Cell Organelle Proteomics in Response to Abiotic Stress. Journal of Proteome Research 2012;11(1) 37-48.

[60] Komatsu S, Konishi H, Hashimoto M. The Proteomics of Plant Cell Membranes. Journal of Experimental Botany 2007;58(1) 103-112.

[61] Ito J, Batth TS, Petzold CJ, Redding-Johanson AM, Mukhopadhyay A, Verboom R, Meyer EH, Millar AH, Heazlewood JL. Analysis of the Arabidopsis Cytosolic Proteome Highlights Subcellular Partitioning of Central Plant Metabolism. Journal of Proteome Research 2011;10(4) 1571-1582.

[62] Cortleven A, Noben JP, Valcke R. Analysis of the Photosynthetic Apparatus in Transgenic Tobacco Plants with Altered Endogenous Cytokinin Content: A Proteomic Study. Proteome Science 2011;9(33) 1-14.

[63] Satoh R, Nakamura R, Komatsu A, Oshima M, Teshima R. Proteomic Analysis of Known and Candidate Rice Allergens between Non-Transgenic and Transgenic Plants. Regulatory Toxicology and Pharmacology 2011;59(3) 437-444.

[64] Uzogara SG. The Impact of Genetic Modification of Human Foods in the 21st Century: a Review. Biotechnology Advances 2000;18(3) 179-206.

[65] Kim Y, Choi SJ, Lee H, Moon TW. Quantitation of CP4 5-Enolpyruvylshikimate-3-Phosphate Synthase in Soybean by Two-Dimensional Gel Electrophoresis. Journal of Microbiology and Biotechnology 2006;16(1) 25-31.

[66] Brandão AR, Barbosa HS, Arruda MAZ. Image Analysis of Two-Dimensional Gel Electrophoresis for Comparative Proteomics of Transgenic and Non-Transgenic Soybean Seeds. Journal of Proteomics 2010;73(8) 1433-1440.

[67] Rabilloud T, Lelong C. Two-Dimensional Gel Electrophoresis in Proteomics: a Tutorial. Journal of Proteomics 2011;74(10) 1829-1841.

[68] Garcia JS, Souza GHMF, Eberlin MN, Arruda MAZ. Evaluation of Metal-Ion Stress in Sunflower (Helianthus Annus L.) Leaves through Proteomic Changes. Metallomics 2009;1(1) 107-113.

[69] Domon B., Aebersold R. Mass Spectrometry and Protein Analysis. Science 2006;312(5771) 212-217.

[70] Aebersold R, Mann M. Mass Spectrometry-Based Proteomics. Nature 2003;422(6928) 198-207.

[71] Sussulini A, Garcia JS, Mesko MF, Moraes DP, Flores EMM, Perez CA, Arruda MAZ. Evaluation of Soybean Seed Protein Extraction Focusing on Metalloprotein Analysis. Microchimica Acta 2007;158(1-2) 173-180.

[72] Arruda SCC, Barbosa HS, Azevedo RA, Arruda MAZ. Two-Dimensional Difference Gel Electrophoresis Applied for Analytical Proteomics: Fundamentals and Applications to the Study of Plant Proteomics. Analyst 2011;136(20) 4119-4126.

[73] Natarajan SS. Natural variability in abundance of prevalent soybean proteins. Regulatory Toxicology and Pharmacology 2010;58(3) 26-29.

[74] Cheng L, Gao X, Li S, Shi M, Javeed H, Jing X, Yang G, He G. Proteomic Analysis of Soybean [Glycine Max (L.) Meer.] Seeds During Inhibition at Chilling Temperature. Molecular Breeding 2010;26(1) 1-17.

[75] Toorchi M, Yukawa K, Nouri MZ, Komatsu S. Proteomics Approach for Identifying Osmotic-Stress-Related Proteins in Soybean Roots. Peptides 2009;30(12) 2108-2117.

[76] El-shemy HA, Khalafalla MM, Fujita K, Ishimoto M. Improvement of Protein Quality in Transgenic Soybean Plants. Biologia Platarum 2007;51(2) 277-284.

[77] Ocana MF, Fraser PD, Patel RKP, Halket JM, Bramley PM. Mass Spectrometric Detection of CP4 EPSPS in Genetically Modified Soya and Maize. Rapid Communication in Mass Spectrometry 2007;21(3) 319-328.

[78] Wittmann-Liebold B, Graack HR, Pohl T. Two-Dimensional Gel Electrophoresis as Tool for Proteomics Studies in Combination with Protein Identification by Mass Spectrometry. Proteomics 2006;6(17) 4688-4703.

[79] Barbosa HS, Arruda SCC, Azevedo RA, Arruda MAZ. New Insights on Proteomics of Transgenic Soybean Seeds: Evaluation of Differential Expressions of Enzymes and Proteins. Analytical and Bioanalytical Chemistry 2012;402(1) 299-314.

[80] Xu C, Garrett WM, Sullivan J, Caperna T, Natarajan SS. Separation and Identification of Soybean Leaf Proteins by Two-Dimensional Gel Electrophoresis and Mass Spectrometry. Phytochemistry 2006;67(22) 2431-2440.

[81] Rabilloud T, Chevallet M, Luche S, Lelong C. Two-Dimensional Gel Electrophoresis in Proteomics: Past, Present and Future. Journal of Proteomics 2010;73(11) 2064-2077.

[82] Timms JF, Cramer R. Difference Gel Electrophoresis. Proteomics, 2008;8(23-24) 4886-4897.

[83] Marouga R, David S, Hawkins E. The Development of the DIGE System: 2D Fluorescence Difference Gel Analysis Technology. Analytical and Bioanalytical Chemistry 2005;382(3) 669-678.

[84] Wang W, Vinocur B, Altman A. Plant Responses to Drought, Salinity and Extreme Temperatures: Towards Genetic Engineering for Stress Tolerance. Planta 2003;218(1) 1-14.

[85] Thanh VH, Shibasaki K. Major Proteins of Soybean Seeds. Subunit Structure of B-Conglycinin. Journal of Agricutural and Food Chemistry 1978;26(3) 692-695.

[86] Cheng Y, Chen X. Posttranscriptional Control of Plant Development. Current Opinion in Plant Biology 2004;7(1) 20-25.

[87] Wang QL, Li ZH. The Functions of Micrornas in Plants. Frontiers in Bioscience 2007;12 3975-3982.

[88] Schmidt F, Marnef A, Cheung MK, Wilson I, Hancock J, Staiger D, Ladomery M. A Proteomic Analysis of Oligo(Dt)-Bound Mrnp Containing Oxidative Stress-Induced Arabidopsis Thaliana RNA-Binding Proteins ATGRP7 and ATGRP8. Molecular Biology Reports 2010;37(2) 839-845.

[89] Kim YO, Pan S, Jung CH, Kang H. A Zinc Finger-Containing Glycine-Rich RNA-Binding Protein, Atrz-1a, has a Negative Impact on Seed Germination and Seedling Growth of Arabidopsis Thaliana under Salt or Drought Stress Conditions. Plant and Cell Physiology 2007;48(8)1170-1181.

[90] Pageau K, Reisdorf-Cren M, Morot-Gaudry JF, Masclaux-Daubresse C. The Two Senescence-Related Markers, GS1 (Cytosolic Glutamine Synthetase) and GDH (Glutamate Dehydrogenase), Involved in Nitrogen Mobilization, are Differentially Regulated during Pathogen Attack and by Stress Hormones and Reactive Oxygen Species in Nicotiana Tabacum L. Leaves. Journal of Experimental Biology 2006;57(3) 547-557.

[91] Funke T, Han H, Fried MLH, Fischer M, Schonbrünn E. Molecular Basis for the Herbicide Resistance of Roundup Ready Crops. Proceedings of the National Academy of Sciences of the United States of America 2006;103(35) 13010-13015.

[92] Natarajan S, Xu C, Caperna TJ, Garrett WM. Comparison of Protein Solubilization Methods Suitable for Proteomic Analysis of Soybean Seed Proteins. Analytical Biochemistry 2005;342 214-220.

[93] Natarajan S, Xu C, Bae H, Bailey BA, Cregan P, Caperna TJ, Garrett WM, Luthria D. Proteomic Analysis of Glycin Subunits of Sixteen Soybean Genotypes. Plant Physiology and Biochemistry 2007;45(6-7) 436-444.

[94] Sussulini A, Souza GHMF, Eberlin MN, Arruda MAZ. Comparative Metallomics for Transgenic and Non-Transgenic Soybeans. Journal of Analytical Atomic Spectrometry 2007;22(12) 1501-1506.

[95] Mounicou S, Szpunar J, Lobinski R. Metallomics: The Concept and Methodology. Chemical Society Reviews 2009;38(4) 1119-1138.

[96] Szpunar J. Advances in Analytical Methodology for Bioinorganic Speciation Analysis: Metallomics, Metalloproteomics and Heteroatom-Tagged Proteomics and Metabolomics. Analyst 2005;130(4) 442-465.

[97] Mounicou S, Lobinski R. Challenges to Metallomics and Analytical Chemistry Solutions. Pure and Applied Chemistry 2008;80(12) 2565-2575.

[98] Garcia JS, Magalhães CS, Arruda MAZ. Trends in Metal-Binding and Metalloproteins Analysis. Talanta 2006;69(1) 1-15.

[99] Sanz-Mendel A, Montes-Bayón M, Sánchez MLF. Trace Elements Speciation by ICP-MS in Large Biomolecules and its Potential for Proteomics. Analytical and Bioanalytical Chemistry 2003;377(2) 236-247.

[100] Thomas RA. A Beginners Guide to ICP-MS. Spectroscopy 2002;17(1) 36-41.

[101] Wagner K, Racaityte K, Unger KK, Miliotis T, Edholm LE, Bischoff R, Marko-Varga G. Protein Mapping by Two-Dimensional High Performance Liquid Chromatography. Journal of chromatography A 2000;893(2) 293-305.

[102] Neue UD. HPLC Columns: Theory, Technology and Practice. New York: Wiley-VCH Inc.; 1997.

[103] Silva MAO, Mataveli LRV, Arruda MAZ. Liquid Chromatography for Biosseparations: Fundamentals, Developments and Applications. Brazilian Journal of Analytical Chemistry 2011;1(5) 234-245.

[104] Wuilloud RG, Kannankumarath SS, Caruso JA. Speciation of Nickel, Copper, Zinc and Manganese in Different Edible Nuts: a Comparative Study of Molecular Size Distribution by SEC-UV-ICP-MS. Analytical and Bioanalytical Chemistry 2004;379(3) 495-503.

[105] Mataveli LRV, Pohl P, Mounicou S, Szpunar J. A Comparative Study of Elements Concentration and Binding in Transgenic and Non-Transgenic Soybeans Seeds. Metallomics 2010;2(12) 800-805.

[106] Koplik R, Pavelková H, Cincibuchová J, Mestek O, Kvasnicka F, Suchánek M. Fractionation of Phosphorus and Trace Elements Species in Soybean Flour and Common White Beans by Size Exclusion Chromatography-Inductively Coupled Plasma Mass Spectrometry. Journal of Chromatography B 2002;770(1-2) 261-273.

[107] Chan Q, Caruso JA. A Metallomics Approach Discovers Selenium-Containing Proteins in Selenium-Enriched Soybeans. Analytical and Bioanalytical Chemistry 2012;403(5) 1311-1321.

[108] Chan Q, Afton SE, Caruso JA. Selenium Speciation Profiles in Selenite-Enriched Soybean (Glycine Max) by HPLC-ICPMS and ESI-ITMS. Metallomics 2010;2(2) 147-153.

[109] dos Santos Neto AJ. Multidimensional Liquid Chromatography and Tandem Mass Spectrometry for the Direct Analysis of Drugs in Biofluids: from the Conventional to the Miniaturized Scale. PhD thesis. University of São Paulo; 2007.

[110] Stroink T, Ortiz MC, Bult A, Lingeman H, de Jong JG, Underberg WJM. On-Line Multidimensional Liquid Chromatography and Capillary Electrophoresis Systems for Peptides and Proteins. Journal of Chromatography B 2010;817(1) 449-466.

[111] Lobinski R, Schaumlöffel D, Szpunar J. Mass Spectrometry in Bioinorganic Analytical Chemistry. Mass Spectrometry Reviews 2006;25(2) 255-289.

[112] Ouerdane L, Mari S, Czernic P, Lebrun M, Lobinski R. Speciation of Non-Covalent Nickel Species in Plant Tissue Extracts by Electrospray Q-TOFMS/MS after their Isolation by 2D Size Exclusion-Hydrophilic Interaction LC (SEC-HILIC) Monitored by ICP-MS. Journal of Analytical Atomic Spectrometry 2006;21(7) 676-683.

[113] Mounicou S, Szpunar J, Lobinski R, Andrey D, Blake CJ. Bioavailability of Cadmium and Lead in Cocoa: Comparison of Extraction Procedures Prior to Size-Exclusion Fast-Flow Liquid Chromatography with Inductively Coupled Plasma Mass Spectrometric Detection (SEC-ICP-MS). Journal of Analytical Atomic Spectrometry 2002;17(8) 880-886.

[114] Mataveli LRV, Fioramonte M, Gozzo FC, Arruda MAZ. Improving Metallomics Information Related to Transgenic and Non-Transgenic Soybean Seeds Using 2D-HPLC-ICP-MS and ESI-MS/MS. Metallomics 2012;4(4) 373-378.

[115] Cozzolino R, de Giulio B. Application of ESI and MALDI-TOF-MS for Triacylglycerols Analysis in Edible Oils. European Journal of Lipid Science and Technology 2011;113(2) 160-167.

[116] Pessôa GS, Pilau EJ, Gozzo FC, Arruda MAZ. Ion Mobility Mass Spectrometry: an Elegant Alternative Focusing on Speciation Studies. Journal of Analytical and Atomic Spectrometry 2011;26(1) 201-206.

[117] Elliot MH, Smith DS, Parkera CE, Borchers, C. Current Trends in Quantitative Proteomics. Journal of Mass Spectrometry 2009;44(12), 1637-1660.

[118] Rodríguez-Cea A, de la Campa MRF, Alonso, IG, Sanz-Medel A. The Use of Enriched 111Cd as Tracer to Study De Novo Cadmium Accumulation and Quantitative Speciation in Anguilla Anguilla Tissues. Journal of Analytical and Atomic Spectrometry 2006;21(3) 270-278.

[119] Vanhaecke F., Balcaen L., Taylor P. Use of ICP-MS for Isotope Ratio Measurements. In: Hill S.J. (ed.) Inductively Coupled Plasma Spectrometry and its Applications. Oxford: Blackwell Publishing; 2007. p160-225.

[120] Vanhaecke F, Balcaen L, Malinovsky, D. Use of Single-Collector and Multi-Collector ICP-Mass Spectrometry for Isotopic Analysis. Journal of Analytical and Atomic Spectrometry 2009;24(7) 863-886.

[121] Wu B, Zoriy M, Chen Y, Becker JS. Imaging of Nutrient Elements in the Leaves of Elsholtzia Splendens by Laser Ablation Inductively Coupled Plasma Mass Spectrometry (LA-ICP-MS). Talanta 2009;78(1) 132-137.

[122] Vanhaecke F., Kyser K. The Isotopic Composition of the Elements. In: Vanhaecke F., Degryse, P. (ed.) Isotopic Analysis: Fundamentals and Application Using ICP-MS. Weinheim: Wiley VCH; 2012. p1-29.